Composite Structures
Design, Mechanics, Analysis, Manufacturing, and Testing

Composite Structures
Design, Mechanics, Analysis, Manufacturing, and Testing

Manoj Kumar Buragohain

CRC Press
Taylor & Francis Group
Boca Raton London New York

CRC Press is an imprint of the
Taylor & Francis Group, an **informa** business

MATLAB® is a trademark of The MathWorks, Inc. and is used with permission. The MathWorks does not warrant the accuracy of the text or exercises in this book. This book's use or discussion of MATLAB® software or related products does not constitute endorsement or sponsorship by The MathWorks of a particular pedagogical approach or particular use of the MATLAB® software.

CRC Press
Taylor & Francis Group
6000 Broken Sound Parkway NW, Suite 300
Boca Raton, FL 33487-2742

© 2017 by Taylor & Francis Group, LLC
CRC Press is an imprint of Taylor & Francis Group, an Informa business

No claim to original U.S. Government works

Printed on acid-free paper

International Standard Book Number-13: 978-1-138-03540-9 (Hardback)
978-1-138-74667-1 (Paperback)

This book contains information obtained from authentic and highly regarded sources. Reasonable efforts have been made to publish reliable data and information, but the author and publisher cannot assume responsibility for the validity of all materials or the consequences of their use. The authors and publishers have attempted to trace the copyright holders of all material reproduced in this publication and apologize to copyright holders if permission to publish in this form has not been obtained. If any copyright material has not been acknowledged please write and let us know so we may rectify in any future reprint.

Except as permitted under U.S. Copyright Law, no part of this book may be reprinted, reproduced, transmitted, or utilized in any form by any electronic, mechanical, or other means, now known or hereafter invented, including photocopying, microfilming, and recording, or in any information storage or retrieval system, without written permission from the publishers.

For permission to photocopy or use material electronically from this work, please access www.copyright.com (http://www.copyright.com/) or contact the Copyright Clearance Center, Inc. (CCC), 222 Rosewood Drive, Danvers, MA 01923, 978-750-8400. CCC is a not-for-profit organization that provides licenses and registration for a variety of users. For organizations that have been granted a photocopy license by the CCC, a separate system of payment has been arranged.

Trademark Notice: Product or corporate names may be trademarks or registered trademarks, and are used only for identification and explanation without intent to infringe.

Visit the Taylor & Francis Web site at
http://www.taylorandfrancis.com

and the CRC Press Web site at
http://www.crcpress.com

To my parents…

Author's parents
Late Bijoy Krishna Buragohain and Smt Tilottama Buragohain

Contents

Preface xxi
Author xxiii
Book Road Map xxv

PART I Introduction, Mechanics, and Analysis

Chapter 1 Introduction to Composites 3

    1.1   Chapter Road Map..3
    1.2   Introduction 3
    1.3   History of Composites..3
    1.4   Characteristics of Composite Materials..................................4
         1.4.1   Definition 4
         1.4.2   Classification 5
               1.4.2.1   Polymer Matrix Composites.....................5
               1.4.2.2   Metal Matrix Composites.........................5
               1.4.2.3   Ceramic Matrix Composites 6
               1.4.2.4   Carbon/Carbon Composites 6
               1.4.2.5   Particulate Composites.............................7
               1.4.2.6   Short Fiber Composites............................7
               1.4.2.7   Flake Composites.....................................8
               1.4.2.8   Unidirectional Composites 8
               1.4.2.9   3D Composites 8
               1.4.2.10 Laminated Composites............................8
               1.4.2.11 Sandwich Composites.............................8
         1.4.3   Characteristics and Functions of Reinforcements and Matrix....8
         1.4.4   Composites Terminologies 9
    1.5   Advantages and Disadvantages of Composites 10
         1.5.1   Advantages 10
         1.5.2   Disadvantages..13
    1.6   Applications of Composites.. 14
    1.7   Summary 18
    Exercise Problems 19
    References and Suggested Reading... 19

Chapter 2 Basic Solid Mechanics..21

    2.1   Chapter Road Map..21
    2.2   Principal Nomenclature 21
    2.3   Introductory Concepts 23
         2.3.1   Solid Mechanics and Continuum............................23
         2.3.2   Spatial Point, Material Point, and Configuration...23
         2.3.3   Fundamental Principles and Governing Equations 23
    2.4   Kinematics..24
         2.4.1   Normal Strain and Shear Strain 25
         2.4.2   Types of Strain Measures: 1D Approach..........................26
               2.4.2.1   Engineering Strain 26

		2.4.2.2	True Strain	26
		2.4.2.3	Green Strain	27
		2.4.2.4	Almansi Strain	27
	2.4.3	Displacement at a Point		28
	2.4.4	Deformation Gradient and Displacement Gradient		30
	2.4.5	Infinitesimal Strain and Finite Strain Theories		32
	2.4.6	Infinitesimal Strain at a Point		33
	2.4.7	Finite Strain at a Point		36
		2.4.7.1	Finite Strain Tensor	36
		2.4.7.2	Physical Meaning of Finite Strain Tensor Components	38
	2.4.8	Strain–Displacement Relations in Cylindrical Coordinates		40
	2.4.9	Transformation of Strain Tensor		41
	2.4.10	Compatibility Conditions		42
2.5	Kinetics			43
	2.5.1	Forces on a Body		43
	2.5.2	Cauchy's Stress Principle and Stress Vector		43
	2.5.3	State of Stress at a Point and Stress Tensor		44
	2.5.4	Transformation of Stress Tensor		46
	2.5.5	Stress Tensor–Stress Vector Relationship		49
	2.5.6	Principal Stresses		50
	2.5.7	Equilibrium Equations		53
2.6	Thermodynamics			54
2.7	Constitutive Modeling			56
	2.7.1	Idealization of Materials		56
	2.7.2	Elastic Materials		57
	2.7.3	Generalized Hooke's Law		58
		2.7.3.1	Symmetry of Stress and Strain Tensors	59
		2.7.3.2	Symmetry of Elastic Stiffness Matrix	61
		2.7.3.3	Anisotropic Materials	61
		2.7.3.4	Monoclinic Materials	61
		2.7.3.5	Orthotropic Materials	63
		2.7.3.6	Transversely Isotropic Materials	67
		2.7.3.7	Cubic Symmetry	68
		2.7.3.8	Isotropic Materials	68
2.8	Plane Elasticity Problems			69
	2.8.1	Plane Stress		69
		2.8.1.1	Plane Stress Problem in Orthotropic Materials	70
		2.8.1.2	Plane Stress Problem in Isotropic Materials	71
	2.8.2	Plane Strain		72
		2.8.2.1	Plane Strain Problem in Orthotropic Materials	73
		2.8.2.2	Plane Strain Problem in Isotropic Materials	73
2.9	Summary			74
Exercise Problems				74
References and Suggested Reading				78

Chapter 3 Micromechanics of a Lamina 79

3.1	Chapter Road Map	79
3.2	Principal Nomenclature	79
3.3	Introduction	82

	3.4	Basic Micromechanics...84		
		3.4.1	Assumptions and Restrictions 84	
		3.4.2	Micromechanics Variables 84	
			3.4.2.1	Elastic Moduli and Strengths of Fibers and Matrix 84
			3.4.2.2	Volume Fractions..84
			3.4.2.3	Mass Fractions..86
		3.4.3	Representative Volume Element..87	
	3.5	Mechanics of Materials-Based Models 88		
		3.5.1	Evaluation of Elastic Moduli 88	
			3.5.1.1	Longitudinal Modulus (E_{1c})...............................88
			3.5.1.2	Transverse Modulus (E_{2c})..................................90
			3.5.1.3	Major Poisson's Ratio (ν_{12c}) 93
			3.5.1.4	In-Plane Shear Modulus (G_{12c}) 95
		3.5.2	Evaluation of Strengths..99	
			3.5.2.1	Longitudinal Tensile Strength $(\sigma_{1c}^T)_{ult}$............. 100
			3.5.2.2	Longitudinal Compressive Strength $(\sigma_{1c}^C)_{ult}$...... 107
			3.5.2.3	Transverse Tensile Strength $(\sigma_{2c}^T)_{ult}$ 110
			3.5.2.4	Transverse Compressive Strength $(\sigma_{2c}^C)_{ult}$ 113
			3.5.2.5	In-Plane Shear Strength $(\tau_{12c})_{ult}$.....................114
		3.5.3	Evaluation of Thermal Coefficients................................116	
		3.5.4	Evaluation of Moisture Coefficients 119	
	3.6	Elasticity-Based Models 123		
	3.7	Semiempirical Models... 124		
		3.7.1	General Form of Halpin–Tsai Equations 125	
		3.7.2	Halpin–Tsai Equations for Elastic Moduli..................... 125	
			3.7.2.1	Longitudinal Modulus 125
			3.7.2.2	Transverse Modulus 126
			3.7.2.3	Major Poisson's Ratio 126
			3.7.2.4	In-Plane Shear Modulus 127
	3.8	Summary 128		
	Exercise Problems 129			
	References and Suggested Reading.. 131			
Chapter 4	Macromechanics of a Lamina.. 133			
	4.1	Chapter Road Map... 133		
	4.2	Principal Nomenclature 133		
	4.3	Introduction to Lamina.. 134		
	4.4	Constitutive Equations of a Lamina 135		
		4.4.1	Specially Orthotropic Lamina 136	
			4.4.1.1	Constitutive Relation 136
			4.4.1.2	Restrictions on Elastic Constants 143
		4.4.2	Generally Orthotropic Lamina 144	
	4.5	Engineering Constants of a Generally Orthotropic Lamina........ 152		
	4.6	Strength 164		
		4.6.1	Strength of an Orthotropic Lamina 165	
		4.6.2	Failure Criteria 167	
			4.6.2.1	Maximum Stress Failure Criterion 167
			4.6.2.2	Maximum Strain Failure Criterion.................. 170
			4.6.2.3	Tsai–Hill Failure Criterion 172

                4.6.2.4 Tsai–Wu Failure Criterion 175
                4.6.2.5 Discussion on Failure Criteria 179
        4.7 Hygrothermal Effects 187
                4.7.1 Hygrothermal Effects in Specially Orthotropic Lamina 188
                4.7.2 Hygrothermal Effects in Generally Orthotropic Lamina ... 189
        4.8 Summary 193
        Exercise Problems 194
        References and Suggested Reading 196

Chapter 5 Macromechanics of a Laminate 197
        5.1 Chapter Road Map 197
        5.2 Principal Nomenclature 197
        5.3 Laminate Codes 198
        5.4 Classification of Laminate Analysis Theories 200
        5.5 Classical Laminated Plate Theory 201
                5.5.1 Basic Assumptions 201
                5.5.2 Kinematics of CLPT: Strain–Displacement Relations 202
                5.5.3 Kinetics of CLPT: Force and Moment Resultants 205
                5.5.4 Constitutive Relations in CLPT 210
        5.6 Classical Laminated Shell Theory 223
                5.6.1 Geometry of the Middle Surface 224
                5.6.2 Kinematics of CLST: Strain–Displacement Relations 226
                5.6.3 Kinetics of CLST: Force and Moment Resultants 228
                5.6.4 Constitutive Relations in CLST 230
        5.7 Hygrothermal Effects in a Laminate 231
                5.7.1 Hygrothermal Constitutive Relations 231
                5.7.2 Coefficients of Thermal Expansion and Coefficients
                        of Moisture Expansion of a Laminate 236
        5.8 Special Cases of Laminates 240
                5.8.1 Significance of Stiffness Matrix Terms 240
                5.8.2 Single-Ply Laminate 242
                        5.8.2.1 Single Isotropic Ply 242
                        5.8.2.2 Single Specially Orthotropic Ply 243
                        5.8.2.3 Single Generally Orthotropic Ply 244
                5.8.3 Symmetric Laminate 244
                5.8.4 Antisymmetric Laminate 246
                5.8.5 Balanced Laminate 247
                5.8.6 Cross-Ply Laminate 248
                5.8.7 Angle-Ply Laminate 248
                5.8.8 Quasi-Isotropic Laminate 248
        5.9 Failure Analysis of a Laminate 252
                5.9.1 First Ply Failure and Last Ply Failure 252
                5.9.2 Progressive Failure Analysis 252
        5.10 Other Topics in a Laminate Analysis 260
                5.10.1 Interlaminar Stress 260
                5.10.2 Shear Deformation Theories 261
                5.10.3 Layerwise Theories 263
        5.11 Summary 264
        Exercise Problems 264
        References and Suggested Reading 267

Chapter 6 Analysis of Laminated Beams, Columns, and Rods 269

- 6.1 Chapter Road Map 269
- 6.2 Principal Nomenclature 269
- 6.3 Introduction 270
- 6.4 Bending of a Laminated Beam (Solid Rectangular Cross Section: Plies Normal to Loading Direction) 272
  - 6.4.1 Basic Assumptions and Restrictions 272
  - 6.4.2 Governing Equations 273
  - 6.4.3 In-Plane Stresses 276
  - 6.4.4 Interlaminar Stresses 277
  - 6.4.5 Specific Cases of Beam Bending 279
    - 6.4.5.1 Simply Supported Beam under Point Load 279
    - 6.4.5.2 Simply Supported Beam under Uniformly Distributed Load 285
    - 6.4.5.3 Fixed Beam under Point Load 286
    - 6.4.5.4 Fixed Beam under Uniformly Distributed Load 289
    - 6.4.5.5 Cantilever Beam under Point Load 291
    - 6.4.5.6 Cantilever Beam under Uniformly Distributed Load 293
- 6.5 Bending of a Laminated Beam (Solid Rectangular Cross Section: Plies Parallel to Loading Direction) 294
- 6.6 Bending of a Laminated Composite Beam (Thin-Walled Cross Section) 299
  - 6.6.1 T-Section 299
  - 6.6.2 I-Section 307
  - 6.6.3 Box-Section 309
- 6.7 Buckling of a Column 311
  - 6.7.1 Concept of Buckling 311
  - 6.7.2 Governing Equations 312
  - 6.7.3 Specific Cases of Column Buckling 314
    - 6.7.3.1 Simply Supported Column 314
    - 6.7.3.2 Fixed-Free Column 316
    - 6.7.3.3 Fixed-Fixed Column 317
- 6.8 Vibration of a Beam 319
  - 6.8.1 Concept of Vibration 319
  - 6.8.2 Governing Equations 320
  - 6.8.3 Specific Cases 322
    - 6.8.3.1 Simply Supported Beam 322
    - 6.8.3.2 Fixed-Free Beam 323
    - 6.8.3.3 Fixed-Fixed Beam 324
- 6.9 Summary 325
- Exercise Problems 326
- References and Suggested Reading 329

Chapter 7 Analytical Solutions for Laminated Plates 331

- 7.1 Chapter Road Map 331
- 7.2 Principal Nomenclature 331
- 7.3 Introduction 332
  - 7.3.1 Rectangular Laminated Plate under General Loading 333

    7.3.2   Governing Equations for Bending, Buckling, and
            Vibration of Laminated Plates 334
            7.3.2.1   Equilibrium Equations for Laminated Plate
                      Bending 334
            7.3.2.2   Buckling Equations for Laminated Plates 339
            7.3.2.3   Vibration Equations for Laminated Plates 343
    7.3.3   Boundary Conditions in a Laminated Plate 344
            7.3.3.1   Simply Supported Boundary Condition 345
            7.3.3.2   Clamped Boundary Condition 346
    7.3.4   Solution Methods 346
            7.3.4.1   Navier Method 347
            7.3.4.2   Levy Method 349
            7.3.4.3   Ritz Method 350
7.4 Solutions for Bending of Laminated Plates 355
    7.4.1   Specially Orthotropic Plate with All Edges Simply
            Supported: Navier Method for Bending 355
            7.4.1.1   Deflection of Middle Surface 355
            7.4.1.2   In-Plane Stresses 358
            7.4.1.3   Interlaminar Stresses 359
    7.4.2   Specially Orthotropic Plate with Two Opposite Edges
            Simply Supported: Levy Method for Bending 360
            7.4.2.1   Deflection of Middle Surface 360
            7.4.2.2   In-Plane and Interlaminar Stresses 363
    7.4.3   Specially Orthotropic Plate with All Edges Simply
            Supported: Ritz Method for Bending 364
            7.4.3.1   Deflection of Middle Surface 364
            7.4.3.2   In-Plane and Interlaminar Stresses 366
            7.4.3.3   Approximation Functions for General
                      Boundary Conditions 367
    7.4.4   Symmetric Angle-Ply Laminated Plate with
            All Edges Simply Supported: Ritz Method for
            Bending 368
    7.4.5   Antisymmetric Cross-Ply Laminated Plate with
            All Edges Simply Supported: Navier Method for
            Bending 369
    7.4.6   Antisymmetric Angle-Ply Laminated Plate with
            All Edges Simply Supported: Navier Method for
            Bending 372
7.5 Solutions for Buckling of Laminated Plates 374
    7.5.1   Specially Orthotropic Simply Supported Plate
            under In-Plane Uniaxial Compressive Loads: Navier
            Method for Buckling 374
    7.5.2   Specially Orthotropic Simply Supported Plate under
            In-Plane Uniaxial Compressive Loads: Ritz Method
            for Buckling 376
    7.5.3   Symmetric Angle-Ply Laminated Simply Supported
            Plate under In-Plane Uniaxial Compressive Loads:
            Ritz Method for Buckling 377
    7.5.4   Antisymmetric Cross-Ply Laminated Simply
            Supported Plate under In-Plane Uniaxial
            Compressive Loads: Navier Method for Buckling 379

		7.5.5	Antisymmetric Angle-Ply Laminated Simply Supported Plate under In-Plane Uniaxial Compressive Load: Navier Method for Buckling 382

      7.6 Solutions for Vibration of Laminated Plates 384
           7.6.1 Specially Orthotropic Simply Supported Plate: Navier Method for Free Vibration 384
           7.6.2 Specially Orthotropic Simply Supported Plate: Ritz Solution for Free Vibration 386
           7.6.3 Symmetric Angle-Ply Laminated Plate with All Four Edges Simply Supported: Ritz Method for Free Vibration 387
           7.6.4 Antisymmetric Cross-Ply Laminated Simply Supported Plate: Navier Method for Free Vibration 389
           7.6.5 Antisymmetric Angle-Ply Laminated Simply Supported Plate: Navier Method for Free Vibration 392
      7.7 Summary 395
      Exercise Problems 395
      References and Suggested Reading 398

Chapter 8 Finite Element Method 399

      8.1 Chapter Road Map 399
      8.2 Principal Nomenclature 399
      8.3 Introduction 400
      8.4 Basic Concepts in Finite Element Method 402
           8.4.1 Elements and Nodes 402
           8.4.2 Discretization 403
           8.4.3 Approximating Function and Shape Function 404
           8.4.4 Element Characteristic Matrices and Vectors 407
           8.4.5 Derivation of Element Characteristic Matrices 408
           8.4.6 Finite Element Equations by the Variational Approach 409
           8.4.7 Coordinate Transformation 416
           8.4.8 Assembly 417
           8.4.9 Solution Methods 421
      8.5 Basic Finite Element Procedure 424
      8.6 Development of Elements 425
           8.6.1 One-Dimensional Elements 425
                  8.6.1.1 Bar Element 425
                  8.6.1.2 Torsion Element 430
                  8.6.1.3 Planar Beam Element 431
                  8.6.1.4 General Beam Element 434
           8.6.2 Two-Dimensional Elements 436
                  8.6.2.1 Rectangular Membrane Element 436
                  8.6.2.2 Rectangular Bending Plate Element 439
                  8.6.2.3 Rectangular General Plate Element 443
                  8.6.2.4 Rectangular General Plate Element with Laminated Composites 447
      8.7 Summary 450
      Exercise Problems 451
      References and Suggested Reading 453

PART II Materials, Manufacturing, Testing, and Design

Chapter 9 Reinforcements and Matrices for Polymer Matrix Composites 457
- 9.1 Chapter Road Map 457
- 9.2 Polymers 457
  - 9.2.1 Thermosets 458
  - 9.2.2 Thermoplastics 458
  - 9.2.3 Rubber 458
- 9.3 Common Thermosets for PMCs 459
  - 9.3.1 Epoxy Resins 459
    - 9.3.1.1 Base Epoxy Resin 459
    - 9.3.1.2 Applications 462
  - 9.3.2 Polyester Resins 462
    - 9.3.2.1 Polyester Oligomer 462
    - 9.3.2.2 Applications 464
  - 9.3.3 Vinyl Ester Resins 464
  - 9.3.4 Phenolic Resins 465
- 9.4 Reinforcements 466
- 9.5 Common Reinforcements for PMCs 467
  - 9.5.1 Glass Fibers 467
    - 9.5.1.1 Types of Glass Fibers 467
    - 9.5.1.2 Production of Glass Fiber 468
    - 9.5.1.3 Forms of Glass Fiber Reinforcements 469
    - 9.5.1.4 Properties of Glass Fibers 469
    - 9.5.1.5 Applications of Glass Fibers 469
  - 9.5.2 Carbon Fibers 470
    - 9.5.2.1 Types of Carbon Fiber 470
    - 9.5.2.2 Production of Carbon Fiber 471
    - 9.5.2.3 Forms of Carbon Fiber Reinforcements 473
    - 9.5.2.4 Properties of Carbon Fibers 473
    - 9.5.2.5 Applications of Carbon Fibers 474
  - 9.5.3 Aramid Fibers 476
    - 9.5.3.1 Types of Aramid Fibers 476
    - 9.5.3.2 Production of Aramid Fibers 476
    - 9.5.3.3 Forms of Aramid Fibers 477
    - 9.5.3.4 Properties of Aramid Fibers 477
    - 9.5.3.5 Applications of Aramid Fibers 478
  - 9.5.4 Boron Fibers 478
  - 9.5.5 Extended Chain Polyethylene Fibers 479
  - 9.5.6 Ceramic Fibers and Whiskers 479
  - 9.5.7 Natural Fibers 479
- 9.6 Physical Forms of Reinforcements 480
  - 9.6.1 Continuous and Short Fibers 480
  - 9.6.2 Fabrics and Mats 480
  - 9.6.3 Preforms 481
  - 9.6.4 Molding Compounds 482
    - 9.6.4.1 Sheet Molding Compounds 482
    - 9.6.4.2 Bulk Molding Compounds 482
    - 9.6.4.3 Injection Molding Compounds 483
  - 9.6.5 Prepregs 483
- 9.7 Summary 484

Contents

    Exercise Problems 485
    References and Suggested Reading 485

Chapter 10 Manufacturing Methods for Polymer Matrix Composites 489

   10.1 Chapter Road Map 489
   10.2 Introduction 489
   10.3 Basic Processing Steps 490
     10.3.1 Impregnation 490
     10.3.2 Lay-Up 490
     10.3.3 Consolidation 490
     10.3.4 Solidification 491
   10.4 Composites Manufacturing Processes 491
     10.4.1 Open Mold Processes 491
       10.4.1.1 Wet Lay-Up 492
       10.4.1.2 Prepreg Lay-Up 494
       10.4.1.3 Spray-Up 495
       10.4.1.4 Rosette Lay-Up 496
     10.4.2 Closed Mold Processes 497
       10.4.2.1 Compression Molding Process 497
       10.4.2.2 Resin Transfer Molding Process 499
     10.4.3 Continuous Molding Processes 501
       10.4.3.1 Pultrusion 501
       10.4.3.2 Tape Winding 503
       10.4.3.3 Fiber Placement 504
   10.5 Filament Winding 506
     10.5.1 Filament Winding Fundamentals and the Basic
       Process 506
       10.5.1.1 Impregnation 506
       10.5.1.2 Lay-Up 507
       10.5.1.3 Consolidation 507
       10.5.1.4 Solidification 507
       10.5.1.5 Basic Processing Steps 508
     10.5.2 Computational Aspects of Filament Winding 509
       10.5.2.1 Geodesic and Nongeodesic Windings ... 509
       10.5.2.2 Helical, Hoop, and Polar Windings 510
       10.5.2.3 Programming Basics 511
     10.5.3 Basic Raw Materials 514
     10.5.4 Tooling and Capital Equipment 515
     10.5.5 Advantages and Disadvantages 517
   10.6 Curing 517
     10.6.1 Tools and Equipment 517
     10.6.2 Vacuum Bagging 518
     10.6.3 Curing of Epoxy Composites 519
     10.6.4 Curing of Phenolic Composites 520
   10.7 Manufacturing Process Selection 520
     10.7.1 Configuration of the Product 522
     10.7.2 Size of the Product 522
     10.7.3 Structural Property Requirement 523
     10.7.4 Surface Finish 523
     10.7.5 Reliability and Repeatability 524
     10.7.6 Production Requirement 524

    10.7.7 Tooling Requirements.. 525
    10.7.8 Automation and Skilled Manpower Needs................... 525
    10.7.9 Cycle Time.. 526
    10.7.10 Cost... 526
  10.8 Other Topics in Composites Manufacturing............................. 527
    10.8.1 Process Modeling 527
    10.8.2 Machining of Composites... 529
      10.8.2.1 Requirements of Composites Machining 529
      10.8.2.2 Critical Aspects of Composites Machining...... 530
  10.9 Summary 531
 Exercise Problems 532
 References and Suggested Reading 534

Chapter 11 Testing of Composites and Their Constituents................................ 537
  11.1 Chapter Road Map.. 537
  11.2 Introduction 537
    11.2.1 Test Objectives... 538
    11.2.2 Building Block Approach 539
    11.2.3 Test Standards.. 540
  11.3 Tests on Reinforcement 541
    11.3.1 Nonmechanical Tests on Reinforcement 541
      11.3.1.1 Density of Fiber 542
      11.3.1.2 Moisture Content 543
      11.3.1.3 Filament Diameter 543
      11.3.1.4 Tex 543
      11.3.1.5 Fabric Construction 544
      11.3.1.6 Areal Density of Fabric 544
    11.3.2 Mechanical Tests on Reinforcement............................ 544
      11.3.2.1 Tensile Properties by Single-Filament
           Tensile Testing.. 544
      11.3.2.2 Tensile Properties by Tow Tensile Testing 546
      11.3.2.3 Breaking Strength of Fabric 547
  11.4 Tests on Matrix 547
    11.4.1 Nonmechanical Tests on Matrix................................... 547
      11.4.1.1 Density 547
      11.4.1.2 Viscosity 548
      11.4.1.3 Glass Transition Temperature........................... 549
    11.4.2 Mechanical Tests on Matrix 550
      11.4.2.1 Tensile Properties 550
      11.4.2.2 Compressive Properties 550
      11.4.2.3 Shear Properties 551
  11.5 Tests for Lamina/Laminate Properties 551
    11.5.1 Nonmechanical Tests on Laminae................................ 552
      11.5.1.1 Density of Composites....................................... 552
      11.5.1.2 Constituent Content 552
      11.5.1.3 Void Content 554
      11.5.1.4 Glass Transition Temperature........................... 555
    11.5.2 Tests for Mechanical Properties of a Lamina................ 555
      11.5.2.1 Tension Testing.. 556
      11.5.2.2 Compression Testing 560
      11.5.2.3 Shear Testing.. 565

            11.5.2.4 Flexural Testing 572
            11.5.2.5 Fracture Toughness Test 575
            11.5.2.6 Fatigue Testing 580
        11.5.3 Note on Tests for Laminate Properties 581
    11.6 Tests for Element-Level Properties 582
        11.6.1 Open-Hole Tests 582
        11.6.2 Bolted Joint 585
            11.6.2.1 Bearing Strength 585
            11.6.2.2 Bearing/By-Pass Strength 586
            11.6.2.3 Shear-Out Strength 586
            11.6.2.4 Fastener Pull-Through Strength 586
        11.6.3 Bonded Joint 586
            11.6.3.1 Adhesive Characterization 587
            11.6.3.2 Bonded Joint Characterization 587
    11.7 Tests at Component Level 588
        11.7.1 Subscale Component Testing 588
        11.7.2 Full-Scale Component Testing 588
    11.8 Summary 589
    Exercise Problems 589
    References and Suggested Reading 591

Chapter 12 Nondestructive Testing of Polymer Matrix Composites 595

    12.1 Chapter Road Map 595
    12.2 Introduction 595
        12.2.1 Defects in Polymer Matrix Composites 596
        12.2.2 NDT Techniques 597
    12.3 Ultrasonic Testing 598
        12.3.1 Basic Concept of Ultrasonic Testing 598
        12.3.2 Test Equipment 599
        12.3.3 Through-Transmission Technique 600
        12.3.4 Pulse-Echo Technique 601
        12.3.5 Data Representation: A-Scan, B-Scan, and C-Scan 601
            12.3.5.1 A-Scan 601
            12.3.5.2 B-Scan 601
            12.3.5.3 C-Scan 602
        12.3.6 Advantages and Disadvantages 603
    12.4 Radiographic Testing 604
        12.4.1 Basic Concept of Radiographic Testing 604
        12.4.2 Radiographic Test Setup 605
            12.4.2.1 X-Ray Radiography 605
            12.4.2.2 Gamma Ray Radiography 606
        12.4.3 Real-Time Radiography 606
        12.4.4 Computed Tomography 607
        12.4.5 Advantages and Disadvantages 607
    12.5 Acoustic Emission 608
        12.5.1 Basic Concept of Acoustic Emission 608
            12.5.1.1 Acoustic Emission 608
            12.5.1.2 AE Sources 608
            12.5.1.3 Kaiser Effect and Felicity Effect 609
        12.5.2 AE Test Setup 609
        12.5.3 Data Acquisition 609

        12.5.4 Data Analysis...610
        12.5.5 Advantages and Disadvantages 611
   12.6 Infrared Thermography...611
        12.6.1 Basic Concept of IR Thermography 611
        12.6.2 Types of Active Thermographic Methods 612
               12.6.2.1 Pulse Thermography 612
               12.6.2.2 Lock-In Thermography 612
               12.6.2.3 Vibrothermography 613
        12.6.3 Advantages and Disadvantages 613
   12.7 Eddy Current Testing...613
        12.7.1 Basic Concept of Eddy Current Testing 613
        12.7.2 Advantages and Disadvantages 614
   12.8 Shearography...615
        12.8.1 Basic Concept of Shearography....................................615
        12.8.2 Advantages and Disadvantages 615
   12.9 Summary 615
   Exercise Problems 616
   References and Suggested Reading..616

Chapter 13 Metal Matrix, Ceramic Matrix, and Carbon/Carbon Composites 619

   13.1 Chapter Road Map...619
   13.2 Introduction 619
   13.3 Metal Matrix Composites 620
        13.3.1 Characteristics of MMCs...620
        13.3.2 Matrix Materials for MMCs 622
        13.3.3 Reinforcing Materials for MMCs 622
        13.3.4 Manufacturing Methods for MMCs 623
               13.3.4.1 Powder Metallurgy Methods 625
               13.3.4.2 Consolidation Diffusion Bonding....................626
               13.3.4.3 Liquid Metal Infiltration Process 627
               13.3.4.4 Stir Casting Method 628
               13.3.4.5 Spray Casting 629
               13.3.4.6 Deposition Methods..630
               13.3.4.7 *In Situ* Methods 631
        13.3.5 Applications of MMCs 632
   13.4 Ceramic Matrix Composites..634
        13.4.1 Characteristics of CMCs..634
        13.4.2 Matrix Materials for CMCs...635
        13.4.3 Reinforcing Materials for CMCs...................................636
        13.4.4 Manufacturing Methods for CMCs 636
               13.4.4.1 Powder Consolidation Methods.......................637
               13.4.4.2 Slurry Infiltration..637
               13.4.4.3 Liquid Infiltration 638
               13.4.4.4 Sol–Gel Technique 639
               13.4.4.5 CVI and CVD...640
               13.4.4.6 Polymer Infiltration and Pyrolysis...................641
               13.4.4.7 Reaction Bonding Processes 642
        13.4.5 Applications of CMCs 642
   13.5 Carbon/Carbon Composites 642
        13.5.1 Characteristics of C/C Composites................................642
        13.5.2 Manufacturing Methods for C/C Composites................644

	13.6 Summary	645
	Exercise Problems	646
	References and Suggested Reading	646

Chapter 14 Design of Composite Structures 649

- 14.1 Chapter Road Map 649
- 14.2 Introduction 649
- 14.3 Basic Features of Structural Design 651
  - 14.3.1 Requirements 651
    - 14.3.1.1 Strength 651
    - 14.3.1.2 Stiffness 651
    - 14.3.1.3 Other Design Requirements 652
  - 14.3.2 Resources 652
    - 14.3.2.1 Material 652
    - 14.3.2.2 Manufacturing Technology 652
    - 14.3.2.3 Computing Technology 653
    - 14.3.2.4 Human Resources 653
  - 14.3.3 Constraints 653
    - 14.3.3.1 Weight 653
    - 14.3.3.2 Cost 654
    - 14.3.3.3 Assembly Requirements 654
    - 14.3.3.4 Manufacturing Feasibility 654
- 14.4 Design versus Analysis 654
- 14.5 Composites Structural Design 656
  - 14.5.1 Generation of Specifications 657
  - 14.5.2 Materials Selection 658
    - 14.5.2.1 Selection of the Composite Material 658
    - 14.5.2.2 Selection of the Reinforcements 659
    - 14.5.2.3 Selection of the Matrix 659
  - 14.5.3 Configuration Design 660
  - 14.5.4 Analysis Options 661
  - 14.5.5 Manufacturing Process Selection 661
  - 14.5.6 Testing and NDE Options 662
  - 14.5.7 Design of Laminate and Joints 662
- 14.6 Laminate Design 662
  - 14.6.1 Scope of Laminate Design 662
  - 14.6.2 Laminate Design Concepts 663
    - 14.6.2.1 Load Definitions 663
    - 14.6.2.2 Design Allowables 664
    - 14.6.2.3 Factor of Safety, Margin of Safety, Buckling Factor, and Knockdown Factor 665
  - 14.6.3 Laminate Design Process 666
    - 14.6.3.1 Laminate Selection 667
    - 14.6.3.2 Laminate Analysis and Measurement 672
    - 14.6.3.3 Laminate Design Criteria 672
- 14.7 Joint Design 673
  - 14.7.1 Introduction 673
  - 14.7.2 Types of Joints 673
  - 14.7.3 Bonded Joints 673
    - 14.7.3.1 Introduction to Bonded Joints 673
    - 14.7.3.2 Failure Modes in Bonded Joints 675

            14.7.3.3 Advantages and Disadvantages of Bonded
                     Joints... 676
            14.7.3.4 General Design Considerations 677
      14.7.4 Mechanical Joints 677
            14.7.4.1 Introduction to Mechanical Joints 677
            14.7.4.2 Failure Modes in Mechanical Joints................ 679
            14.7.4.3 Advantages and Disadvantages of
                     Mechanical Joints.. 679
            14.7.4.4 General Design Considerations 680
      14.7.5 Other Joints... 681
14.8 Stiffened Structures... 682
      14.8.1 Introduction 682
      14.8.2 Failure Modes in a Stiffened Structure 682
      14.8.3 Design of Stiffeners 683
14.9 Optimization... 684
14.10 Design Examples 685
      14.10.1 Design of a Tension Member 685
            14.10.1.1 Micromechanics-Based approach................. 685
            14.10.1.2 Macromechanics-Based approach 686
      14.10.2 Design of a Compression Member 695
      14.10.3 Design of a Torsion Member 698
      14.10.4 Design of a Beam 698
      14.10.5 Design of a Flat Panel under In-Plane Loads 702
      14.10.6 Design of a Pressure Vessel under Internal Pressure...... 708
            14.10.6.1 Introduction 708
            14.10.6.2 Advantages of Composite Pressure Vessels... 709
            14.10.6.3 Configuration of a Pressure Vessel............... 709
            14.10.6.4 End Domes 711
            14.10.6.5 Metallic End Fittings 711
            14.10.6.6 Ply Design 712
14.11 Summary 715
Exercise Problems 716
References and Suggested Reading 718

Index.. 721

Preface

The composites industry has grown multifold in recent times; it continues to grow and further growth is expected in the future as well. Composite materials and products are now regularly used in a wide range of applications across various industrial sectors. Naturally, there has been an increased demand for trained personnel in the field of composites. The subject of composites, which used to be taught only at a few select universities and institutes a couple of decades ago, is offered today by many other universities and institutes both at the undergraduate as well as the postgraduate levels. It is also offered as short-term courses as a part of continuing education program by certain institutes. Also, there are a large number of practicing professionals who do self-study.

One of the primary objectives of composites education is to equip the student with adequate know-how in the area of development of composite products. Composites, in general, and composite product development, in particular, are interdisciplinary subjects that draw resources from a number of subfields, namely, material science, mechanics, analysis, design, tooling, manufacturing, and testing. These topics have been extensively covered in a number of excellent books. Depending on the content, the books on composites can be broadly placed in three categories. The first category includes several excellent texts on mechanics of composites. In the second category, composites are treated as a part of material science. The third category includes the literature on manufacturing methods and shop floor and lab activities in composites.

The topics in composites mentioned above, however, cannot be considered in isolation and an integrated approach is essential for successful execution of a composite product development program. This book is a humble effort to present the concepts in composites in an integrated manner.

The contents of this book are organized in two parts. Part I is devoted to the topics related to mechanics, analytical methods in composites, and basic finite element procedure. An introductory discussion on the characteristic features of composites is given first. Basic concepts of solid mechanics are reviewed and it is followed up by discussions on the concepts of micromechanics and macromechanics. Analytical methods are excellent tools in understanding the behavior of composite structural elements. Some of these methods in the simple cases of beams and plates are presented next. Finite element method is the most popular tool for analysis; understanding of the underlying concepts and the basic procedure is essential for effective use of this method and a brief presentation on the same is given to complete the discussions in Part I.

Part II of this book is devoted to the topics on materials, manufacturing processes, testing, and design. These are the aspects in composites that the shop floor man is directly concerned with. The author is of the firm belief that composites design is not a closed door activity and a general understanding of the concepts of mechanics, analysis tools, available materials, manufacturing processes, tooling, and destructive and nondestructive test methods is essential for doing an efficient design. With this in mind, a discussion on composites design is given in the end.

The primary objective of this book is to expose the reader to the complete cycle of development of a composite product. I sincerely hope that this book will be an excellent guide to a student who wants to make a career in composites. I also expect that it will be an excellent companion to a practicing professional in the field of composites.

Finally, I take this opportunity to place on record my sincerest gratitude to all my teachers who molded me—right from my early school days to my doctoral study at IIT Madras; all that I present in this book belongs to them.

This book would not have seen the light of day without very professional guidance and support from CRC Press; my sincere thanks to Dr. Gagandeep Singh (Commissioning Editor), Mouli Sharma, Hector Mojena, Renee Nakash, Rachael Panthier, and Shashikumar Veeran, all of whom have been directly associated with the editing and production of this book.

I would like to place on record my sincere gratitude to Dr. Tessy Thomas, Outstanding Scientist and Director, Advanced Systems Laboratory, Hyderabad, for her encouragement and support in publishing this book. My sincerest thanks are also due to my colleagues with whom I have had long hours of invaluable interactions developing composite products.

I take this opportunity to express my gratitude to my parents, who brought me up in a small sleepy town yet taught me to be ambitious. I thank all my family members and near and dear ones for their support in this humble endeavor. Life is a long journey and the past six to seven years, were special and tough too; I spent long hours working on the manuscript of this book; my wife Ainu managed family affairs and son Beli grew up silently. I would like to express my love and gratitude to my loving family for their sacrifice and support and for standing patiently by my side in the hours of need. I indeed remain indebted to them.

Manoj Kumar Buragohain
Hyderabad

MATLAB® is a registered trademark of The MathWorks, Inc. For product information, please contact:

The MathWorks, Inc.
3 Apple Hill Drive
Natick, MA 01760-2098 USA
Tel: 508 647 7000
Fax: 508-647-7001
E-mail: info@mathworks.com
Web: www.mathworks.com

Author

Dr. Manoj Kumar Buragohain is a scientist at the Advanced Systems Laboratory, Hyderabad of Defense Research and Development Organization, India. He earned his BSc Engg from the Regional Engineering College Rourkela; MTech and PhD from the Indian Institute of Technology Madras; and PGDFA from The Institute of Chartered Financial Analysts of India, Hyderabad.

Dr. Buragohain has well over two decades of hands-on experience in the design and development of composite products. His primary research interests are in the fields of geodesic and nongeodesic filament winding, tape winding, rosette lay-up, and contact lay-up. Some of his major contributions have been in large size composite pressure vessels, grid-stiffened composite structures, tubular structures, ablative liners, and composite rotor blade. He has several journal and conference publications to his credit and is a life member of the Indian Society for Advancement of Materials and Process Engineering (ISAMPE) and Aeronautical Society of India (AeSI). His contributions to the field of composites have been well recognized and he was awarded the Laboratory Scientist of the Year Award, National Science Day Commendation Certificate & Silicon Medal, DRDO Award for Performance Excellence as a team member, and Agni Award for Excellence in Self-reliance as a team leader.

Book Road Map

The topics for this book have been chosen keeping in mind the primary objective of the book, that is, to expose the reader to the complete cycle of development of a composite product—from design to manufacturing to testing and evaluation. The chapters, as depicted pictorially in the figure given below, are organized into two parts and placed by and large in a chronological order of reading.

PART I

Part I is devoted to the introductory concepts and the topics on mechanics, analytical methods, and analysis; these topics are primarily computational in nature.

The objective of Chapter 1 is to introduce the subject of composite materials and structures. Toward this, we shall begin our journey with a discussion on the characteristic features that define a composite material, their advantages and disadvantages, and their typical applications.

The mechanics of composite materials is an important subject; a good understanding of the concepts of mechanics is essential for understanding the analytical methods and analysis tools, which in turn are essential for the efficient design of a composite product. Chapters 2 through 5 present discussions on basic solid mechanics and mechanics of composites. Composites are anisotropic in nature and, as a result, the

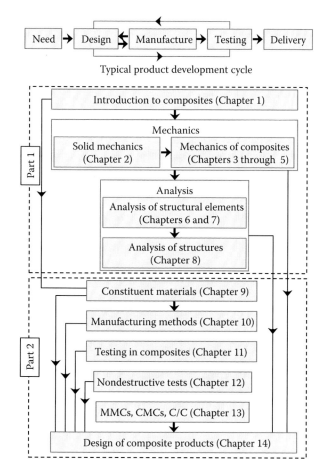

Structure of this book

mechanics of composite material is more involved than that for conventional metallic materials. The concepts of solid mechanics provide the foundation on which the subject of mechanics of composite materials is built. A detailed review of the basic solid mechanics concepts is presented in Chapter 2. A composite structure is built with composite laminates. A laminate is made by combining several laminae and a lamina consists of reinforcements and matrix. The laminae are the building blocks and we shall address them in Chapters 3 and 4. Chapter 3 presents the micromechanics of a lamina; the interaction of the individual constituents and their effect on the behavior of the lamina are discussed in this chapter. The macromechanics of a lamina, that is, the study of the gross behavior of a lamina without making a distinction between the constituents, is presented in Chapter 4. The macromechanics of a laminate is discussed next in Chapter 5.

Analytical methods and analysis tools play an important role in the design process by providing estimates of the response of a structure to applied loads; these topics are presented in Chapters 6 through 8. Analytical methods are available for the solution of simple structural elements under simple loading; we shall discuss such analytical tools for composite beams and plates in Chapters 6 and 7, respectively. These methods, however, are not suitable for most real-life situations, where a structure as well as the applied loads are rather complex. In such cases, numerical methods such as the finite element method are invariably used. The finite element method is the most popular tool used for the analysis of structures. Several general-purpose finite element software are commercially available. A basic understanding of the method is essential for the proper use of these software. We shall wind up Part I of this book with a brief discussion on the basic concepts and general procedure in the finite element method in Chapter 8.

PART II

There are several aspects in the overall cycle of a composite product development, where the engineer is primarily involved with shop-floor-related activities. Part II of this book is devoted to these topics. These topics are materials, manufacturing methods, testing of composites and their constituents, and nondestructive evaluation. In addition to these, other major classes of composites, viz. metal matrix composites (MMCs), ceramic matrix composites (CMCs), and carbon/carbon composites (C/C composites) are included in this part. Also, a discussion on the design of composite products is given in the end.

The major raw materials used in the polymer matrix composites industry are presented in Chapter 9. Raw materials play a key role in any product development exercise. Two primary categories of raw materials needed to make a composite product are the reinforcements and the matrix. The general characteristics and the mechanical and physical properties of common fibers and resins are presented. We shall also briefly present the principles of manufacturing methods for these materials. It is expected that this chapter will be able to guide the designer in selecting the appropriate reinforcement and matrix materials for specific applications.

Composites technology is process-intensive and a good knowledge of manufacturing processes is essential for anyone in this field. Similar to materials selection, the manufacturing process selection is a critical decision to be made in the design of a composite product. With a view to getting solutions to such issues, we shall address manufacturing methods in polymer composites in Chapter 10. Several manufacturing processes are regularly employed in the composites industry; they can be categorized into open mold, closed mold, and continuous molding processes. The basic processing steps, some of the popular manufacturing processes, and the manufacturing process selection are presented in this chapter.

Another major aspect of composites technology is testing, various aspects of which are addressed in Chapter 11. We will see there that testing is an inseparable part in any composite product development program. It is done with either one or more of the following as objectives—design data generation, quality control, and development of new materials. Testing in composites is unique and typically a building-block approach is adopted. Tests are done at various levels—constituent raw materials testing to full-scale component testing. These tests are destructive in nature and the specimen gets consumed/damaged during testing. In contrast to destructive testing, nondestructive testing neither destroys nor causes any damage to the part, and the utility of the part remains intact. We shall briefly review some of the common nondestructive evaluation techniques in Chapter 12.

MMCs, CMCs, and C/C composites complement polymer matrix composites in the overall composites industry. The scope of this book is limited to mainly polymer matrix composites. However, familiarity with these sister composite materials helps a polymer matrix composite professional immensely in the design and development of a product. The introductory concepts covering general characteristics, raw materials, and manufacturing methods with regard to MMCs, CMCs, and C/C composites are presented in Chapter 13.

Finally, we shall acquaint ourselves with various aspects of design in Chapter 14. Design is a common term, yet very often misunderstood. It is an art, yet certain set patterns and key features can be associated with it. The concept of design as a solution to meet certain requirements using available resources within certain constrains is introduced in this chapter. The fundamental features of composites structural design process, laminate design, joint design, and some important design issues are presented. Design examples are provided to help in the assimilation of the concepts. It is a phase that comes fairly early in the overall product development program. However, it is a subject that demands a reasonable level of insight into various other aspects of composites technology; inputs from mechanics, analysis estimates, materials data, manufacturing, testing, and evaluation are required in the design process. Accordingly, we shall deliberate on it in the end.

SUGGESTED PLAN FOR READING

There are 14 chapters in this book and they can be read in a sequential manner. However, it will be difficult to cover the entire book in the time frame of a single semester. From the points of view of (i) organizing the contents in one-semester courses and (ii) effective self-study, the following study plans are suggested:

First, a basic course on the mechanics of composites can be planned based on the sequence: Chapter 1 → Chapter 2 → Chapter 3 → Chapter 4 → Chapter 5. Some selected sections from Chapters 6 through 8 can be added.

Second, an advanced course on the mechanics and analysis of composites can be planned based on the sequence: Chapter 1 → Chapter 4 → Chapter 5 → Chapter 6 → Chapter 7. Some selected sections from Chapters 2, 3, and 8 can be added.

Third, a course on manufacturing and testing of composites can be planned based on the sequence: Chapter 1 → Chapter 9 → Chapter 10 → Chapter 12 → Chapter 13. Some selected sections from Chapters 3 through 5 and 11 can be added.

Fourth, a generalized course on the design of composite products can be planned based on the sequence: Chapter 1 → Chapter 4 → Chapter 5 → Chapter 9 → Chapter 10 → Chapter 14. Some selected sections from Chapters 2, 3, 6 through 8, and 11 through 13 can be added.

Part I

Introduction, Mechanics, and Analysis

Introduction to Composites

1.1 CHAPTER ROAD MAP

The objective of this chapter is to give the reader an overview of composite materials. Advanced composites are a relatively new class of materials, but the concept of composites is rather old. A brief historical note on composites is provided. Composites are unique materials; the characteristic features that differentiate them from conventional metallic materials are presented next. A classification of these materials is provided so as to establish a link within the overall system of materials. A brief description of different types of composites is given for a proper understanding of classification of composites. Composites are multiphase material systems and their behavior is dependent on the constituents; a note on the general functions and characteristics of the constituents is given. For successful use of composites in product design, it is important to know their advantages as well as disadvantages; the general advantages and disadvantages associated with composites are presented, followed by a discussion on their applications in various industrial sectors.

1.2 INTRODUCTION

Materials have always played a major role in the development and growth of human civilization. Composite materials are no exception. The advent of advanced composites has influenced almost every aspect of modern life and today, major impacts are felt in aerospace and aviation sector, automobile industry, sports goods industry, naval applications, civil engineering, etc.

Composites have their own unique features. There are advantages as well as disadvantages. While exceptionally high mechanical and thermal properties can be achieved in a composite material, translation of such high levels of material properties to composite structures is equally important and highly challenging. Basic material science, process engineering, and design and analysis of composite materials and structures are inherently related.

1.3 HISTORY OF COMPOSITES

The use of composite materials can be traced back to 2000 BC or even earlier. Straw-reinforced mud bricks were used in Egypt and Mesopotamia. Straws were also used for making reinforced pottery. Composite bows were used in ancient Mongolia and other places across Asia. Evidence exists on the use of composites in ancient Japan, where laminated metals were used by the Samurais to make swords.

The development of modern advanced composites has been greatly influenced by the developments in the fields of raw materials, viz. reinforcements and resins, and composites manufacturing processes [1–4].

Glass fiber was first commercially produced in the 1930s. Around the same time, unsaturated polyester resins were also developed and commercialized. The first glass fiber-reinforced plastic (GFRP) boats and radomes were built in the early 1940s. Rapid progress took place in the 1950s with increasing use of GFRP in boat hulls, car bodies, electrical components, etc. GFRP products were the first advanced composite products, and they still constitute a very large proportion of today's composites' market.

The next phase of development of composites was marked by the development of high-performance composites using carbon, boron, and aramid fibers. High-performance carbon fibers and boron fibers were introduced around the same time in the late 1950s and early 1960s. They were followed by the development of aramid fibers in the early 1970s. Epoxy resins have been available since the 1930s, and with the advent of these high-performance fibers, composites industry received a major boost. Development in the composites industry was also pushed hard by demands from aerospace and defense sectors for lighter and more efficient structures. Further, technological developments in respect of processing equipment and machinery such as filament winding machine, computer numerical controls, autoclave, etc. have been some of the major features of the growth of composites. Another noteworthy point is the development in the field of analytical tools for composites product design and analysis.

1.4 CHARACTERISTICS OF COMPOSITE MATERIALS

1.4.1 Definition

Broadly, four types of materials are used for making a structural element. These are metals, polymers, ceramics, and composites. In a general sense, a composite material is one that has two or more constituent materials in it. The constituent materials in a composite material are metals, polymers, ceramics, or a combination of these three. A definition of composites can be found by identifying the characteristics of the constituents and the process of combining them [5–7]. We check them as follows:

- The constituent materials differ in composition and form. Their combination results in two phases in a composite material: reinforcement—a discontinuous phase, which is usually hard and strong, and matrix—a continuous phase, which binds the reinforcements together.
- The reinforcing material is embedded in the matrix material at a macroscopic level. Thus, the constituent materials do not dissolve or merge together and they retain their individual properties.
- The matrix binds the reinforcements in such a way as to form a distinct interphase between them.
- The reinforcements and the matrix, as individual materials, may not be of any engineering use; it is the process of combining them that transforms them into a new material, which is a useful and efficient one. The interphase helps the reinforcement and the matrix act in unison and the resultant composite material often exhibits better properties than the constituent materials.

Thus, keeping the above points in mind, we can arrive at a definition as the following: A composite material is a useful and efficient material system that is made by macroscopically combining two constituents—a reinforcement and a matrix, in such a way that the constituents do not dissolve or merge together and retain their individual properties, yet they act in unison to exhibit better engineering properties.

1.4.2 Classification

Composites can be classified primarily in two ways. As mentioned earlier, there are four broad types of structural materials, of which any one of the first three, namely, metals, polymers, and ceramics, can be used as the matrix for making a composite material. Thus, the first way to classify a composite material is based on the type of matrix material. From this angle, composite materials are classified into:

- Polymer matrix composites (PMCs)
- Metal matrix composites (MMCs)
- Ceramic matrix composites (CMCs)
- Carbon/carbon composites (C/C composites)

1.4.2.1 Polymer Matrix Composites

PMCs have been a subject of great interest for basic as well as applied research. They possess several advantages over monolithic metals and today, products of wide variety, in terms of shape and size, are efficiently designed, fabricated, and used. In a PMC material, a polymer such as epoxy is used as the matrix material that is reinforced with very fine diameter fibers such as carbon, glass, etc. The reinforcing fibers can be either continuous or discontinuous. Continuous fibers can be used in forms such as strands, roving, fabric, etc. Discontinuous fibers can be particulate, whiskers, or flakes.

Mechanical properties such as strength and stiffness of PMCs are directly dependent on the reinforcement properties. Matrix, on the other hand, binds the reinforcements together and helps in load transfer. Thus, in general, the principal philosophy in designing a composite part is to orient the reinforcements in the direction of load so that the composite properties are efficiently exploited. Of course, manufacturing constraints need to be given due consideration.

The processing techniques in composites are a critical part in the study of PMCs. Several processing techniques are available for the manufacture of structural elements using PMCs. A common objective in all these processing techniques is to place the reinforcements as per design requirement and wet them properly with the matrix. An essential step is to cure the composite, during which the matrix solidifies through a process of cross-linking and it binds the reinforcements.

PMCs play a dominant role in the overall market for composites, including MMCs, CMCs, and C/C composites. MMCs, CMCs, and C/C composites are introduced in this chapter. We shall discuss them in some detail in Chapter 13. However, our emphasis in this book would be on PMCs. Thus, unless otherwise specifically stated, composites in the remainder of this book would mean PMCs.

1.4.2.2 Metal Matrix Composites

In an MMC material, a metal or an alloy is the continuous phase in which the reinforcements are embedded [8]. Addition of reinforcements in a monolithic metal greatly improves its mechanical and other properties to suit specific design requirements. MMCs have certain advantages over PMCs as well as monolithic metals. These advantages include high transverse strength and modulus, high shear strength and modulus, high service temperature, low thermal expansion, very low moisture absorption, dimensional stability, high electrical and thermal conductivities, better fatigue and damage resistance, ease of joining and resistance to most radiations including ultraviolet (UV) radiation, etc. These benefits have been made use of and MMCs have found many applications in several sectors [9–12].

1.4.2.3 Ceramic Matrix Composites

CMCs are a class of structural materials in which either continuous or discontinuous reinforcements are embedded in a monolithic ceramic material [13]. Ceramics, as a class of materials by themselves, characteristically have very high temperature resistance but low fracture toughness. Low fracture toughness makes ceramics susceptible to catastrophic failure under tensile or impact loads. In CMCs, fracture properties are improved. Other advantages of CMCs include low density, chemical inertness, hardness, and high strength. Thus, these materials are suitable for applications where high mechanical properties are desired at high service temperatures [14].

1.4.2.4 Carbon/Carbon Composites

In C/C composites, carbon fiber reinforcements are embedded in a carbon matrix [15]. Carbon by itself is brittle and sensitive to material defects. By reinforcing carbon with carbon fibers, the properties are greatly improved. C/C composites are typically very highly temperature resistant. They retain high tensile and compressive strengths and high fatigue strength at high temperatures, and they are used in several aerospace and other high-end applications such as brake disks for aircrafts, nose cone of reentry vehicles, nozzle throat, etc. [16]. C/C composites, however, involve long and complex processing cycle.

We had mentioned earlier that the reinforcements form a discontinuous phase. Different shapes and sizes of reinforcements can be used and the final composite structural forms can be different. Thus, based on the geometry and shape of the reinforcements and the structural form of the composites, the following classification of composites is possible (Figure 1.1):

- Phased composites
  - Particulate composites
  - Short fiber composites
  - Flake composites
  - Unidirectional composites
- Layered composites
  - Laminated composites
  - Sandwich composites

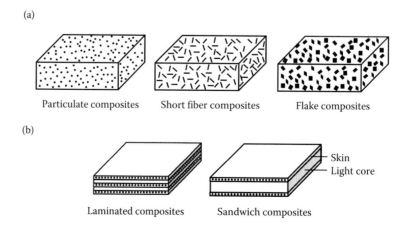

FIGURE 1.1 Types of composites based on shape of reinforcements and form of composites. (a) Phased composites. (b) Layered composites.

TABLE 1.1

Common Particulate Reinforcements

Metal	Aluminum, tungsten
Nonmetal	
Ceramic	Alumina (Al_2O_3), silicon carbide (SiC), silicon nitride (Si_3N_4), titanium carbide (TiC), titanium diboride (TiB_2), boron carbide (B_4C)
Others	Sand, rock particles

1.4.2.5 Particulate Composites

In a particulate composite material, reinforcing particles are added in a continuous matrix. A particle is an object such that no dimension is more than about five times the other two dimensions. The particles are added at random and due to the random orientation of the particles, particulate composites are isotropic in nature. Both metallic and nonmetallic particles are used as reinforcements in particulate composites. Table 1.1 lists common particulate reinforcements. These particles can be combined with either metallic matrix or nonmetallic matrix materials and thus we have the following four possible combinations in a particulate composite material:

- Metal particles in metal matrix
- Metal particles in nonmetal matrix
- Nonmetal particles in metal matrix
- Nonmetal particles in nonmetal matrix

Examples of particulate composites are given in Table 1.2.

Particulate composites possess several advantages such as improved strength, stiffness, and toughness compared to the unreinforced metal or ceramic material. They also possess typically higher operating temperatures and in certain cases, specific beneficial properties are infused into the matrix by adding particulate reinforcements. Further, their processing is cheap and simple. However, they exhibit generally inferior mechanical properties as compared to the other types of composites.

1.4.2.6 Short Fiber Composites

In short fiber composites, either short fibers or whiskers are added in a continuous matrix. The reinforcements are discontinuous and they are mixed in the matrix at random. The fibers have highly direction-dependent properties and thus these composites are anisotropic. However, owing to the random orientation of the fibers/whiskers, these composites depict isotropic behavior at a product level.

TABLE 1.2

Examples of Particulate Composites

Metal particles in metal matrix	Tungsten/Al
Metal particles in nonmetal matrix	Solid propellant (Al/rubber)
Nonmetal particles in metal matrix	Graphite/Al, SiC/Al
Nonmetal particles in nonmetal matrix	Concrete (sand/cement)

Note: The convention followed in this table and the remainder of this book to represent a composite material is to put the name of the reinforcement first followed by a front slash and the name of the matrix. Thus, SiC/Al is silicon carbide-reinforced aluminum matrix composite, and so on.

1.4.2.7 Flake Composites

Unlike fibers, flakes have a two-dimensional (2D) structure—strength and stiffness properties are high in two directions. Two types of flake composites are used. In the first type of flake composites, nonmetallic flakes such as mica or glass are embedded, usually parallel to one another, in a matrix material. The resulting composite material exhibits highly direction-dependent behavior. In the second type of flake composites, preimpregnated fabric cut pieces are randomly mixed. Although the fabric cut pieces are 2D with direction-dependent properties, due to the random orientation, the resulting composite part is isotropic.

1.4.2.8 Unidirectional Composites

These are composites with long continuous reinforcements. Strength and stiffness properties of these composites are very high in the direction of the reinforcements but poor in the other two directions.

1.4.2.9 3D Composites

Three-dimensional (3D) composites are a special variety of composites that are reinforced with long continuous reinforcements oriented in all the three dimensions. Most frequently, these composites are made from preforms of oriented fibers into which the resin matrix is injected and cured.

1.4.2.10 Laminated Composites

Laminated composites are made up of several thin plies (layers) stacked and bonded together. Different types of laminated composites such as fiber-reinforced plastic, bimetals, laminated wood, etc. are used. In bimetallic laminated composites, layers of two different metals of usually significantly different thermal coefficients are bonded. In fiber-reinforced laminated composites, each ply is a plastic layer reinforced with usually continuous fibers. The reinforcements are unidirectional, bidirectional, or even multidirectional. In certain cases, short fiber-reinforced plies are also used, in which case the fibers are normally randomly oriented. Fiber-reinforced laminated composites are widely used, and in the remainder of this book, the term "laminated composites" would be used to mean fiber-reinforced laminated composites.

1.4.2.11 Sandwich Composites

These are basically panels of lightweight core sandwiched between two relatively thin but hard and strong skins. The core material may be low-density foam or honeycomb.

1.4.3 Characteristics and Functions of Reinforcements and Matrix

We have learnt that the reinforcements and matrix do not react with each other, retain their individual characteristics, act in unison, and offer better resultant composite properties. Now, we need to understand that the reinforcements and matrix have their own characteristics and specific functions.

Fibers used as reinforcements for advanced composites are typically very fine in diameter, and their volume per unit length is low. Thus, in the fiber form, flaws are far less than in the bulk form. Also, many fiber manufacturing processes involve spinning and stretching operations, during which a high degree of microstructural orientation takes place. As a result, mechanical properties such as strength and modulus of fibrous reinforcements are very high. Thus, structural functions are performed primarily by the reinforcements. In addition to possessing very high mechanical properties, some fibers also possess some specific characteristics. In such cases, fibers impart specific characteristics to the composite material. For example, carbon fibers have negative

Introduction to Composites

longitudinal thermal coefficients of expansion, and carbon fiber composites can be designed to make dimensionally stable part across a wide temperature range. Similarly, silica fibers can be used for making thermal insulator.

In general, we can list a few key functions of fiber reinforcements as follows:

- Reinforcements are the primary load-bearing element in a composite material.
- Reinforcements provide stiffness to the composite material.
- Reinforcements provide thermal stability.
- Reinforcements provide electrical and thermal conductivity (or insulation).

The reinforcements alone, without the matrix, are meaningless as a structural part. Matrix is inferior to fibers in terms of mechanical properties; however, it influences a number of composite mechanical properties such as transverse modulus and strength, shear modulus and strength, compressive strength, fatigue characteristics, interlaminar shear strength, and coefficient of thermal expansion (CTE). The matrix, in a composite material, has several critical functions, of which the following may be noted:

- The matrix acts as glue and holds the reinforcing fibers together, and gives shape and rigidity to the composite material as a structural part.
- The matrix transfers load between the reinforcing fibers.
- The matrix provides good protection to the reinforcing fibers against chemical attack and mechanical wear and tear.
- The matrix provides good surface finish to the part.
- Transverse mechanical properties of composite materials are greatly influenced by the matrix.

1.4.4 Composites Terminologies

1. *Isotropic material*: An isotropic material is one that has equal or same material properties in all directions at a point. In other words, material properties are not dependent on directions in an isotropic material. Conventional metallic materials such as steel, aluminum, etc. are isotropic.
2. *Anisotropic material*: An anisotropic material is one that has unequal or dissimilar material properties in different directions at a point. In other words, material properties are dependent on directions in an anisotropic material.
3. *Orthotropic material*: An orthotropic material is one with material properties that are different in three mutually perpendicular directions at a point.

    We will learn about planes of material property symmetry in Chapter 2. We will learn that an isotropic material has infinite numbers of planes of material property symmetry, an orthotropic material has three, and anisotropic material has none.

    Composite materials are generally anisotropic. The degree of anisotropy in a composite material is highly dependent upon the reinforcement. The matrix is typically isotropic in nature but the reinforcements may exhibit highly directional properties. For example, carbon fibers are very strong and stiff in the longitudinal direction. Laminated composite materials are reinforced with such reinforcements and these composites are exceptionally strong and stiff in the direction of fibers as compared to the transverse directions. Particulate composites, with uniform dispersion of the reinforcing particles, are isotropic. Flakes and short fibers, as reinforcements, are highly anisotropic with high strength and stiffness in the direction of the fibers. However, these reinforcements are generally randomly oriented in the matrix and thus, the resulting composites are isotropic at a macro level.
4. *Homogeneous material*: A homogeneous material is one that has equal or same material properties in a specified direction at all points. In other

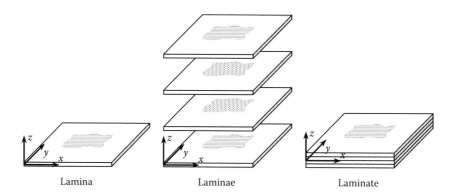

FIGURE 1.2 Schematic representations of lamina and laminate.

words, material properties are not dependent on location in a homogeneous material.
 5. *Nonhomogeneous material*: A nonhomogeneous material is one that has unequal or dissimilar material properties in a specified direction at different points. In other words, the material properties are dependent on location in a nonhomogeneous material.

    Composite materials, in a strict sense, are almost always nonhomogeneous. These are multiphase materials and the properties at a location in the matrix are mostly different from those at a location in the reinforcements. However, at a macro level, a composite material is mostly homogeneous.
 6. *Lamina*: A lamina (laminae in plural) is a single layer or ply in a laminated composite material. It is the building block of a laminated composite structure. It can be either flat or curved and is made up of unidirectional, bidirectional, multidirectional, or randomly oriented fibers in the matrix material.
 7. *Laminate*: A laminate is a laminated composite structural element that is made by a number of laminae. As shown in Figure 1.2, typically, the reinforcements in the laminae are oriented w.r.t. the coordinate system of the structural element and the laminae are stacked as per certain ply sequence.
 8. *Micromechanics*: It is the study of a composite material wherein the constituents of the composite material are considered as distinct phases and their interaction with each other is analyzed to determine the gross properties of the composite material. Thus, in micromechanics, we study the behavior of the composite material based on micro-level properties or the properties of the constituents.
 9. *Macromechanics*: It is the study of a composite material wherein the constituents of the composite material are not considered as distinct phases; rather, the gross or apparent properties of the laminae and the interaction between laminae are analyzed to determine the laminate behavior. Thus, in macromechanics, we study the behavior of composite material based on macro-level properties or the apparent properties of the laminae.

1.5 ADVANTAGES AND DISADVANTAGES OF COMPOSITES

1.5.1 Advantages

There are several advantages associated with composite materials that make them more attractive than other traditional materials in many applications where high performance and light weight are essential requirements. These are briefly discussed below:

1. High tensile strength and stiffness
2. High specific strength and specific stiffness
3. High fatigue strength
4. Inherent material damping and good impact properties
5. Tailorable properties
6. Design flexibility
7. Less corrosion
8. Simple manufacturing techniques
9. Near net shape part and lower part count
10. Cost-effective product development

1. *High tensile strength and stiffness*: Strength and stiffness properties of monolithic polymers are generally low compared to metals and ceramics. These properties can be greatly improved by reinforcing with suitable reinforcements such as glass or carbon. Thus, PMCs exhibit comparable or better tensile strength and stiffness in the direction of reinforcements than conventional metals or ceramics. Mechanical properties of MMCs and CMCs are also higher as compared to those of their monolithic counterparts. Table 1.3 gives a comparison of typical mechanical properties of some of the common metals, polymers, ceramics, and composites.
2. *High specific strength and specific modulus*: Specific strength and specific modulus are the ratios of strength to density and modulus to density, respectively. Composites, on account of their low densities and high strength and modulus, possess very high specific strengths and specific moduli. As a result, structural parts made by using composites are typically lighter than those made from metals. In aerospace vehicles, light weight is a key requirement that is associated with longer range, higher payload, and fuel saving. There are specific applications in other sectors as well where weight saving is beneficial to the overall performance. In all these applications, composites are suitable due to their high specific strength and stiffness.
3. *High fatigue strength*: Fatigue strength or endurance limit refers to the failure stress under cyclic loads. Under cyclic loads, most materials fail at lower levels of stress than under static loads. Composites exhibit higher fatigue strength than conventional metals such as steel and aluminum. Fatigue strengths of metals are far lower compared to the respective static strengths—as low as 35% for aluminum and 50% for steel and titanium. On the other hand, unidirectional composites exhibit high fatigue strengths of about 90% of static strengths.
4. *Inherent material damping and good impact properties*: Composite materials, due to the presence of fiber–matrix interface, exhibit better damping characteristics than conventional metals. A composite part, under the action of an impact load, develops numerous microcracks. During the process of formation of the microcracks, the energy of impact is absorbed, and catastrophic failure is avoided.
5. *Tailorable properties*: In a composite material, reinforcements can be aligned in the direction of principal direction of load; plies can be stacked in a desired sequence such that material properties are utilized in the most efficient manner. This characteristic of composites is very unique, one that allows the engineer to design the material system itself or virtually tailor the material properties as per product requirement.
6. *Design flexibility*: Design flexibility is a key advantage of composite materials. Metallic parts are designed using raw materials that are readily available

TABLE 1.3

Comparison of Typical Mechanical Properties of Common Engineering Materials

Material	Specific Gravity	Tensile Modulus (GPa)	Tensile Strength (MPa)	Specific Tensile Modulus (GPa/g/cc)	Specific Tensile Strength (MPa/g/cc)
Metals					
Steel	7.9	205	275–1880	26	35–238
Aluminum	2.7	70	60–700	26	22–259
Polymers					
Epoxy	1.2	2.5–4.5	50–130	2.1–3.8	42–108
Polyester	1.2	2.5–4.0	20–80	2.1–3.3	17–67
Phenolic	1.3	3.0–4.0	35–70	2.3–3.1	27–54
Polyimide	1.4	3.0–4.0	70–80	2.1–2.9	50–57
Ceramics					
Alumina	3.9	380	330	97	85
Magnesia	3.6	205	230	57	64
Unidirectional PMCs ($V_f = 0.5$)					
Glass/epoxy	1.8–1.9	30–45	550–1350	16–25	289–750
Carbon/epoxy	1.4–1.7	105–460	875–2760	62–329	515–1971
Kevlar/epoxy	1.3–1.4	70–76	1065–1380	50–58	761–1062
MMCs					
SiC/Al	2.7–2.9	82–228	210–700	28–84	72–259
CMCs					
SiC/SiC	2.3–2.4	190–210	280–340	79–91	117–148
C/C composites					
UD C/C composites	1.7	125–220	570–600	74–129	335–353

Note: The material properties given in the table are only representative. Certain materials such as carbon fibers, depending upon their subtypes, have wide variation in their strength and stiffness properties. As a consequence, composites made by using such a reinforcing fiber exhibit wide variation in their mechanical properties. Similarly, the tensile strength of monolithic materials such as steel and aluminum vary widely depending upon their composition and processing parameters. Specific properties from the manufacturer's data sheet should be used in an actual design and analysis exercise.

in the market. These raw materials such as bar stocks, sheets, sections, etc. are available with standard specifications; thus, in general, the designer's final choice in respect of the structural elements is influenced by the available raw materials. On the contrary, composite parts are designed along with the design of the material system itself. Several classes of reinforcements with wide range of properties are commercially available in various physical forms. Similarly, many matrix materials with their own characteristics are available. Also, there has been extensive technological development in the field of composite processing and many efficient processing techniques are available today. Thus, the designer has a wide range of choices of combinations of raw materials, stacking sequence, processing techniques, etc. that enables him to meet the end requirement in the most efficient way.

7. *Less corrosion*: Composite materials, in general, offer better corrosion resistance than metallic materials. As a result, composite structures have longer storage life.

8. *Simple manufacturing techniques*: Manufacturing techniques available in the broad field of composites vary widely in terms of their complexity, equipment and machinery required, cycle time, and cost. In general, manufacturing techniques for PMCs are much simpler compared to those for MMCs, CMCs, and

C/C composites. There are many areas where the manufacture of a PMC part is simpler than a similar metallic part. Today, a wide variety of manufacturing techniques are available for PMCs; some of these techniques do not need application of high temperature or pressure. Similarly, equipment and machinery required for PMC manufacture need not be very expensive and complex. As a result, PMCs have grown very fast in terms of volume as well as variety of applications.

9. *Near net shape part and lower part count*: In the case of composite materials, parts are made by adding materials, and parts with near net shape can be manufactured. Parts with complex shapes and large sizes can be realized, as a result of which, the number of parts in the overall assembly comes down drastically. This feature also enables the engineer to eliminate/reduce machining operations and reduce manufacturing cycle time.

10. *Cost-effective product development*: Composites are preferred due to their low cost as well. Different elements of cost in the development and commercialization of a composite product include raw material cost, equipment and machinery cost, processing cost, design and analysis cost, marketing cost, etc. Raw material cost depends on the type of reinforcements and matrix materials—while raw materials such as carbon fibers and high-end epoxy resins are expensive, many other reinforcements and resins such as E-glass fiber and polyester resins for commercial applications are economically priced. Similarly, simpler manufacturing methods lead to low costs of equipment and machinery and processing. The overall cost is also reduced due to reduced processing cost (near net shape part), lower assembly cost (lower part count), and longer storage life (less corrosion).

1.5.2 Disadvantages

While there is a long list of advantages associated with composites, there are certain disadvantages as well. Special care has to be taken to overcome these limitations. They are enumerated below:

1. Low service temperature
2. Sensitivity to radiation and moisture
3. Low elastic properties in the transverse direction
4. Complex design and analysis
5. Complex mechanical characterization
6. High cost of raw materials and fabrication
7. Difficulty in jointing

1. *Low service temperature*: PMCs, in general, degrade at relatively low temperatures above room temperature, and have low service temperature. PMCs are of two broad types—structural composites and thermal/ablative composites. Polymer matrix structural composites tend to lose strength and stiffness properties as the ambient temperature increases. For high-temperature applications, these composites need protection in the form of insulating or ablative lining. Ablative composites can withstand high temperatures, but they are nonstructural in nature.

2. *Sensitivity to radiation and moisture*: Many polymers when exposed to radiations such as UV radiation, etc. degrade. Moisture absorption is also a problem for many PMCs. Protective coatings are applied to increase the service life of these composite parts. In certain other cases, additives such as UV-resistant

fillers, etc. are added to the matrix to reduce degradation of the composites when exposed to harmful radiations.
3. *Low elastic properties in the transverse direction*: Strength and stiffness of unidirectional composites in transverse direction, that is, normal to the direction of the reinforcements, are controlled by the matrix. Polymer matrix materials have low elastic properties. Thus, unidirectional composites are associated with rather low strength and stiffness properties in the transverse direction. However, this drawback of unidirectional composites is successfully overcome by tailoring of the plies and efficient fiber directional properties in the design of the laminates.
4. *Complex design and analysis*: There are only two independent elastic constants in an isotropic material. On the other hand, more numbers of elastic constants are required for describing an anisotropic material. (We will discuss about elastic constants of anisotropic materials in Chapter 2.) Similarly, strength parameters are also more in anisotropic materials. Composite materials are anisotropic in nature, and thus, the number of parameters to be considered in the design and analysis of a part using composite material is more. Fiber-reinforced composites are typically layered. While we can consider different combinations of ply sequences, tailor the material properties and exploit the material system, this flexibility also increases the complexity of the design and analysis procedure.
5. *Complex mechanical characterization*: Owing to the presence of more numbers of elastic constants and strength properties, mechanical characterization procedure, which involves laminate making, coupon preparation, and testing, is complex and time consuming. Further, individual raw materials, viz. resin, curing agents, and fiber, also have to be evaluated for mechanical and physical properties.
6. *High cost of raw materials and fabrication*: While some raw materials such as E-glass fiber, polyester resin, etc. are cheap, several others such as carbon fibers, high-performance epoxy resin, etc. are expensive. Similarly, some of the composite manufacturing processes such as autoclave molding, filament winding, etc. are rather expensive, and these processes are suitable for high-end applications where cost is not a primary criterion.
7. *Difficulty in jointing*: Joints in composite parts are a major area of concern. Conventional jointing methods using nut and bolt, rivets, threaded holes, etc. are not directly applicable in composites. Utmost care and caution and innovative thinking are required for designing efficient and reliable composite joints.

1.6 APPLICATIONS OF COMPOSITES

The benefits of composites are well recognized today, and the use of composite materials in different industrial sectors is steadily growing. Industrial sectors that use composites can be broadly listed as aerospace, automotive, building and construction, chemical, consumer goods, electrical and electronics, marine, and others. It is important to note that each sector has its own characteristics in respect of functional requirement, demand for goods, and many other parameters. Depending upon the particular needs of a sector, composite materials, their design, and manufacturing processes are exploited suitably. Thus, characteristic features of composite structures vary from one industrial sector to another. Some of the common applications of composite materials are listed in Table 1.4 [5,17–24].

Composites are used in both commercial and military aircrafts. Typical benefits include (i) weight reduction leading to higher speeds, increased payloads, longer range, and fuel economy, (ii) reduced part count leading to simpler assembly and reduced

TABLE 1.4
Applications of PMCs

Sector	Applications	Typical Materials, Processes, and Benefits
Aerospace		

Aircraft	• Primary structures
	  – Fuselage, forward fuselage, mid-fuselage, rear fuselage
	  – Wing box
	  – Empennage box
	• Control components
	  – Flaps
	  – Ailerons
	  – Spoilers
	  – Slats
	  – Horizontal stabilizer
	  – Vertical stabilizer
	  – Elevator
	  – Rudder
	• Exterior parts
	  – Radome
	  – Landing gear hatches
	  – Karmans
	  – Storage room doors
	  – Fairings
	  – Propeller blades
	• Interior parts
	  – Floors
	  – Doors
	  – Partitions
	  – Bulkheads
	  – Brake disks
		Materials:
		• CFRP, AFRP, and GFRP with epoxy
		• UD and BD prepregs
		• GFRP mainly in light aircrafts
		• CFRP in modern aircrafts
		Processes:
		• Automated prepreg lay-up
		• Vacuum bagging, autoclave curing
		• Filament winding and pultrusion
		• Honeycomb sandwich, stiffened structures
		• Adhesive bonding of skins to core
		Benefits:
		• Weight reduction leading to higher speeds, increased payloads, longer range, and fuel economy.
		• Reduced part count leading to simpler assembly and reduced overall cost
		• Reduced radar reflection and heat radiation leading to stealth capability in military aircrafts
		• Higher fatigue resistance
		• Higher corrosion resistance

Helicopter	• Rotor blades
	  – Spar
	  – Skin
	  – Core
	• Rotor hub
		Materials:
		• CFRP and GFRP with epoxy, polyimide, and phenolics
		Processes:
		• Filament winding and molding processes
		Benefits:
		• Reduced weight
		• Enhanced dynamic characteristics
		• Manufacturing ease

Launch vehicles and missiles	• Solid rocket motor
	  – Rocket motor case
	  – Insulating and ablative nozzle liners
	• Airframe structures
	  – Interstage section
	  – Payload adapters
	  – Fairings
	• Control surfaces
	  – Fins
	• Reentry vehicle components
	• Launch canisters
		Material:
		• Carbon/epoxy, Kevlar/epoxy for rocket motor casing
		• Carbon/phenolic and glass/phenolic for nozzle liners
		• Carbon/phenolic for reentry vehicle liners
		• Carbon/epoxy for satellite applications
		Process:
		• Filament winding
		• Tape winding, compression molding for nozzle liners
		• Advanced grid-stiffened shells and panels for airframe structures, payload adapters and fairings
		Benefits:
		• Weight reduction
		• Reduced part count
		• Reduced cycle time
		• Manufacturing flexibility

Satellite	• Tubings • Brackets and fittings • Shear panels • Bus panel • Flywheels	

(Continued)

TABLE 1.4 (Continued)
Applications of PMCs

Sector	Applications	Typical Materials, Processes, and Benefits
Automotive		

Car, bus, and truck	• Structural components
	  – Chassis parts
	  – Leaf springs
	  – Floor elements
	• Body components
	  – Roof
	  – Doors
	  – Hood cover
	  – Bumper
	• Interior components
	  – Seat frames
	  – Side panel and central console
	  – Dash board
	• Components under the hood
	  – Motor and gear box parts
	  – Battery support
	  – Head light support
	  – Transmission shafts
		Materials:
		• E-glass/polyester and E-glass/vinyl ester SMCs for most body parts
		• E-glass/epoxy for leaf springs
		Processes:
		• Compression molding
		• Structural reaction injection molding
		Benefits:
		• Reduced weight leading to fuel efficiency
		• Reduced tooling cost
		• Corrosion resistance
		• Lower part count

Chemical industry	• Corrosion-resistant tanks • Pipes, industrial vessels, sewer lines • Waste water treatment equipment • Pollution control equipment	*Materials*: • GFRP with vinyl ester *Processes*: • Contact lay-up, filament winding *Benefits*: • Corrosion resistance
Civil engineering structures		
Buildings and houses	• Modular house • Doors • Bathtubs • Bathroom fixtures	*Materials*: • GFRP with polyester, vinyl ester, and epoxy *Processes*: • Contact lay-up—manual and automated
Infrastructures	• Bridges	• Sandwich construction • Pultrusion for sections • Adhesive bonding for repair of old and damaged concrete bridges *Benefits*: • Corrosion resistance • Weight reduction leading to ease of transportation and installation, longer span, etc.
Marine		

Small crafts	• Hulls of
	  – Lifeboats
	  – Pleasure boats
	  – Fishing boats
	  – Speed boats
		Materials:
		• GFRP with polyester and vinyl ester
		• AFRP
		• CFRP in high-performance applications
		Processes:

Large crafts	• Hulls of
	  – Military and commercial hovercrafts
	  – Mine countermeasure ships
	  – Yachts
	• Sonar domes
	• Fairings
	• Superstructures of ships
	• Radomes
	• Rudders
	• Masts
		• Contact molding
		• Honeycomb sandwich construction
		Benefits:
		• Weight reduction leading to greater speeds, better maneuverability and fuel efficiency

(Continued)

TABLE 1.4 (*Continued*)
Applications of PMCs

Sector	Applications	Typical Materials, Processes, and Benefits
Offshore oil exploration	• Oil platforms	
Piping system	• Pipes • Pumps • Valves • Heat exchangers	
Others		
Wind turbines	• Rotor blades	*Materials*:
Sporting goods	• Golf shafts, tennis rackets, snow skis, fishing rods, sports bike, pole vault, etc.	• GFRP with polyester for rotor blades • CFRP with epoxy for tennis rackets, golf club shafts, fishing rods, bicycle frames, etc.
Consumer goods	• Chairs, tables, desert air cooler body, computer, printer, washing machine, etc.	• GFRP with epoxy for pole vault *Processes*: • Contact molding
Electrical and electronics	• Circuit boards, insulators, switch gears, appliance covers	• Pultrusion *Benefits*: • Weight reduction • Better strength/stiffness • Better damping characteristics energy absorption

Note: (i) CFRP = carbon fiber-reinforced plastic, GFRP = glass fiber-reinforced plastic, AFRP = aramid fiber-reinforced plastic. UD = unidirectional, BD = bidirectional. (ii) Materials and process options indicated in the table are indicative and not exhaustive.

overall cost, (iii) reduced radar reflection and heat radiation leading to stealth capability in military aircrafts, (iv) higher fatigue resistance, and (v) higher corrosion resistance, etc. Initial applications include boron/epoxy composites in skins of the horizontal stabilizers of F-14 in the 1960s. Airbus A310 was the first commercial aircraft to have extensive composites (about 10% of the total weight). Since then, the use of composites has steadily increased and today they are used in significant proportions in many aircrafts. For example, Airbus A380 and Boeing 787 Dreamliner use about 25% and 50% of composites, respectively. Carbon fiber-reinforced plastic (CFRP), glass fiber-reinforced plastic (GFRP), and aramid fiber-reinforced plastic (AFRP) with epoxy resin are all used, of which CFRP is the dominant composite material system. Both unidirectional and bidirectional carbon/epoxy prepregs are used employing automated tape laying, vacuum bagging, autoclave curing, filament winding, pultrusion, and adhesive bonding as common manufacturing processes. Structural concepts such as honeycomb sandwich and conventional stiffened panel and grid-stiffened panel are employed.

CFRP and GFRP are used in helicopter rotor blades and rotor hub. Rotor blades are typically made by filament winding and molding processes. The principal advantages of composites in rotor blade are (i) reduced weight, (ii) enhanced dynamic characteristics, and (iii) manufacturing ease.

Composites are used in many space vehicle applications. While the primary objective of using composites in space applications is weight reduction, several other benefits such as reduced part count, reduced cycle time, manufacturing flexibility, etc. can also be exploited. Carbon/epoxy and Kevlar/epoxy filament-wound rocket motor cases and carbon/phenolic and glass/phenolic nozzle liners are common in many solid propulsion systems in rockets and missiles. Advanced grid-stiffened shells and panels have been adopted in airframe structures, payload adapters, and fairings. Filament-wound and tape-wound carbon/phenolic liners are used for thermal protection in reentry vehicles.

Carbon/epoxy composites are used in satellite applications like tubings, brackets and fittings, bus panel, etc. CFRP composites can be designed to yield near-zero CTE, which helps achieve dimensional stability across a wide range of temperature variations.

Composites have unique applications in the automotive industry. E-glass fiber-reinforced epoxy leaf springs are the first major structural applications of composites in automobiles. Other structural applications include chassis components, drive shafts, etc. However, these applications have somewhat limited acceptability. On the other hand, the major automotive applications of composites are in respect of the body components such as roof, doors, hood cover, etc. These are made by compression molding of discontinuous E-glass fiber-reinforced sheet molding compounds (SMCs), in which the resin is either polyester or vinyl ester. In addition to compression molding, structural reaction injection molding (SRIM), a variant of resin transfer molding (RTM) is also employed in the manufacture of the automobile body parts. The major advantages of using composites in automobiles are (i) reduced weight leading to fuel efficiency, (ii) reduced tooling cost, (iii) corrosion resistance, and (iv) lower part count.

Civil engineering applications of composites are broadly of two types—housing sector and infrastructure. GFRP prefabricated modular house, bunk house, cabin, mobile toilet cabin, etc. are some of the commercially available products today. In the infrastructure sector, construction of new bridges and repair of old bridges have been the major applications of composites. In this regard, corrosion resistance of composites is the main attraction. Weight saving is not the main objective; however, it has some indirect advantages like ease of transportation and installation, longer span, etc. E-glass fiber-reinforced polyester composite laminates are used as facing sheets in sandwich construction to make bridge decks. The core is typically glass/polyester tubes. Pultruded sections, resin transfer molded panels, etc. are other forms of composites in bridge construction.

Glass/polyester and glass/vinyl ester composites are routinely used in the production of different types of small and large yachts. In some cases, aramid fibers are also used these days. The primary attraction of composites is weight reduction, which leads to greater speeds, better maneuverability, and fuel efficiency. Hulls of these boats are made typically by contact molding. In some high-performance applications such as racing boats, high specific strength and stiffness are essential. In such cases, hulls, decks, masts, etc. are made using carbon/epoxy laminates and honeycomb sandwich construction with carbon/epoxy skins. There are other marine applications of composites that include submarines, offshore oil exploration, etc.

Composites are also extensively used in other sectors, including energy sector, sporting goods, consumer goods, chemical industry, etc. Rotor blades of wind turbines are made by using glass fiber composites. Carbon/epoxy composites are used in tennis rackets, golf club shafts, fishing rods, bicycle frames, etc. Weight reduction coupled with better strength/stiffness and damping characteristics are the primary attraction of composites in these sporting goods. Glass/epoxy composites are also used, for example, in pole vaults for better energy absorption. Glass/polyester composites are predominant players in the consumer goods sector, where chairs, tables, desert air cooler, etc. are made typically by using chopped strand mat (CSM).

Note: Materials and manufacturing processes referred to in this section shall be discussed subsequently in Chapters 9 and 10.

1.7 SUMMARY

An introduction to composite materials is given in this chapter. We have seen that there are several unique features that differentiate composite materials from conventional materials. They are a class of useful materials made by macroscopic combination of reinforcements and matrix. The reinforcements and the matrix retain their individual characteristics; they have their own individual functions, and as a whole, the resultant composite material exhibits better properties that the individual reinforcements and matrix do not possess.

Composites are classified based on the type of matrix used. They are also classified based on the physical form of the reinforcements.

Composites are associated with many advantages that include high mechanical properties, low densities, tailorable properties, design and manufacturing flexibility, less corrosion, and cost-effective product development. There are certain limitations as well, which need to be addressed in the design and manufacture of composite structures.

Applications of composites are no longer limited to high-end aerospace and defense sectors. Today, PMCs are regularly used in many industrial sectors, including aerospace and defense, automotive, chemical engineering, civil engineering, marine, and others. These applications in each industrial sector, in general, have their own characteristics w.r.t. materials and manufacturing processes.

EXERCISE PROBLEMS

1.1 Define composite materials. What are the characteristic features and functions of the reinforcements and matrix in a composite material?
1.2 List the various classifications of composite materials.
1.3 Write a short note giving details of the advantages and disadvantages associated with polymer matrix composites.
1.4 Write a note on the applications of polymer matrix composites.

REFERENCES AND SUGGESTED READING

1. D. J. Vaughan, Fiberglass reinforcement, *Handbook of Composites* (S. T. Peters, ed.), second edition, Chapman & Hall, London, 1998, pp. 131–155.
2. C. D. Dudgeon, Polyester resins, *Engineering Materials Handbook, Vol. 1, Composites* (T. J. Reinhart, Technical Chairman), ASM International, Ohio, 1987, pp. 90–96.
3. P. J. Walsh, Carbon fibers, *ASM Handbook, Vol. 21, Composites* (D. B. Miracle and S. L. Donaldson, Vol. Chairs), ASM International, Ohio, 2001, pp. 35–40.
4. M. A. Boyle, C. J. Martin, and J. D. Neuner, Epoxy resins, *ASM Handbook, Vol. 21, Composites* (D. B. Miracle and S. L. Donaldson, Vol. Chairs), ASM International, Ohio, 2001, pp. 78–89.
5. R. M. Jones, *Mechanics of Composite Materials*, second edition, Taylor & Francis, New York, 1999.
6. P. K. Mallick, *Fiber-Reinforced Composites: Materials, Manufacturing and Design*, third edition, CRC Press, Boca Raton, FL, 2013.
7. F. L. Matthews and R. D. Rawlings, *Composite Materials: Engineering and Science*, CRC Press, Boca Raton, FL, 1999.
8. N. Chawla and K. K. Chawla, *Metal Matrix Composites*, second edition, Springer, New York, 2013.
9. A. Evans, C. San Marchi, and A. Mortensen, *Metal Matrix Composites in Industry—An Introduction and a Survey*, Kluwer Academic Publishers, Dordrecht, The Netherlands, 2003.
10. M. J. Koczak, S. C. Khatri, J. E. Allison, and M. G. Bader, Metal-matrix composites for ground vehicle, aerospace and industrial applications, *Fundamentals of Metal-Matrix Composites* (S. Suresh, A. Mortensen, and A. Needleman, eds.), Butterworth-Heinemann, Boston, 1993, pp. 297–326.
11. D. B. Miracle, Aeronautical applications of metal-matrix composites, *ASM Handbook, Vol. 21, Composites* (D. B. Miracle and S. L. Donaldson, Vol. Chairs), ASM International, Ohio, 2001, pp. 1043–1048.
12. W. H. Hunt, Jr. and D. B. Miracle, Automotive applications of metal-matrix composites, *ASM Handbook, Vol. 21, Composites* (D. B. Miracle and S. L. Donaldson, Vol. Chairs), ASM International, Ohio, 2001, pp. 1029–1032.
13. K. K. Chawla, *Ceramics Matrix Composites,* second edition, Kluwer Academic Publishers, Dordrecht, The Netherlands, 2002.
14. J. R. Davis, Applications of ceramic matrix composites, *ASM Handbook, Vol 21, Composites* (D. B. Miracle and S. L. Donaldson, Vol. Chairs), ASM International, Ohio, 2001, pp. 1101–1109.
15. P. Morgan, *Carbon Fibers and Their Composites*, Taylor & Francis, Boca Raton, FL, 2005.

16. K. M. Kearns, Applications of carbon–carbon composites, *ASM Handbook, Vol. 21, Composites* (D. B. Miracle and S. L. Donaldson, Vol. Chairs), ASM International, Ohio, 2001, pp. 1067–1070.
17. D. Gay, S. V. Hoa, and S. W. Tsai, *Composite Materials—Design and Applications*, CRC Press, Boca Raton, FL, 2003.
18. G. C. Krumweide and Eddy A. Derby, Aerospace equipment and instrument structure, *Handbook of Composites* (S. T. Peters, ed.), second edition, Chapman & Hall, London, 1998, pp. 1004–1021.
19. S. W. Shalaby and R. A. Latour, Composite biomaterials, *Handbook of Composites* (S. T. Peters, ed.), second edition, Chapman & Hall, London, 1998, pp. 957–966.
20. V. P. McConnell, Scientific applications of composites, *Handbook of Composites* (S. T. Peters, ed.), second edition, Chapman & Hall, London, 1998, pp. 967–981.
21. W. C. Tucker and T. Juska, Marine applications, *Handbook of Composites* (S. T. Peters, ed.), second edition, Chapman & Hall, London, 1998, pp. 916–930.
22. S. N. Loud, Commercial and industrial applications of composites, *Handbook of Composites* (S. T. Peters, ed.), second edition, Chapman & Hall, London, 1998, pp. 931–956.
23. D. L. Denton, Land transportation applications, *Handbook of Composites* (S. T. Peters, ed.), second edition, Chapman & Hall, London, 1998, pp. 905–915.
24. G. Eckold, *Design and Manufacture of Composite Structures*, Woodhead Publishing, Cambridge, 1994.

2

Basic Solid Mechanics

2.1 CHAPTER ROAD MAP

For the design and development of a composite product, it is imperative, on the part of the designer, to have a good grasp of the available materials and manufacturing processes. In addition to these, the composites' engineer must have a thorough knowledge of the behavior of a composite structure under loads. A composite product, like any other structural element, is subjected to different types of loads. Study of the response of a composite structure to different types of loads is the focus area in the field of mechanics of composite materials. One of the primary objectives of composite mechanics is to develop appropriate tools for analysis of composite structures. Basic knowledge of solid mechanics is essential for understanding the topics on analysis of composite lamina and laminate. In this chapter, we review the basic solid mechanics concepts. Subsequently, we will have detailed discussions on mechanics of composite materials in Chapters 3 through 5.

The basic concepts of solid mechanics are well developed and we present a brief discussion on these concepts. Next, the governing equations that are required for the development of analytical tools are discussed; the various concepts under kinematics, kinetics, and constitutive modeling are also addressed. Generalized Hooke's law is reduced to various specialized cases such as orthotropic and isotropic materials. Plane elasticity idealizations are made use of in composite lamina and laminate analysis and these topics are introduced toward the end of this chapter.

Solid mechanics concepts are fundamental requirements in the fields of composites mechanics and analysis; this chapter will be a prerequisite to subsequent Chapters 3 through 8 and 14.

2.2 PRINCIPAL NOMENCLATURE

$\mathcal{B}$	Body force
b	Position vector of deformed coordinate system w.r.t. undeformed coordinate system
C, C_{ijkl}	Generalized fourth-order tensor of elastic constants
$[D], \mathbf{D}, D_{ij}$	Displacement gradient in the component form, vector notation, and indicial notation, respectively
$d\varepsilon$	Incremental strain
E_x, E_y, E_z	Young's moduli in the x-, y-, and z-directions, respectively
e_x, e_y, e_z	Unit vectors along x-, y-, and z-directions, respectively
$[F], \mathbf{F}, F_{ij}$	Deformation gradient in the component form, vector notation, and indicial notation, respectively
G_{xy}, G_{yz}, G_{zx}	In-plane shear moduli in the xy-, yz-, and zx-planes, respectively
$\mathbf{I}, I_{ij}$	Unit tensor in vector form and indicial notation, respectively

K	Change in kinetic energy
L, l	Undeformed and deformed lengths, respectively
$\boldsymbol{n}$	Unit normal vector
n_x, n_y, n_z	Components of the unit normal vector
Q	Heat input to the body during loading
r, θ, z	Cylindrical coordinate axes
S	Material compliance matrix
$\boldsymbol{T}_n, \boldsymbol{T}_x, \boldsymbol{T}_y, \boldsymbol{T}_z$	Stress vectors or traction vectors normal to planes represented by unit vectors $\boldsymbol{n}, \boldsymbol{e}_x, \boldsymbol{e}_y, \boldsymbol{e}_z$, respectively
$T_{nx}, T_{ny}, T_{nz}, T_{xx}, T_{xy}, T_{xz}, T_{yx},$ $T_{yy}, T_{yz}, T_{zx}, T_{zy}, T_{zz}$	Components of the stress vectors $\boldsymbol{T}_n, \boldsymbol{T}_x, \boldsymbol{T}_y, \boldsymbol{T}_z$ in the x-, y-, and z-directions
t	Time
U	Change in internal energy
U_0	Strain energy density function
$\boldsymbol{u}, u_i$	Displacement vector of a point (x, y, z) in the vector form and indicial notation, respectively
u_r, u_θ, u_z	Displacement vector components in the r-, θ-, and z-directions, respectively
u_X, u_Y, u_Z	Displacement vector components in the X-, Y-, and Z-directions, respectively
u_x, u_y, u_z	Displacement vector components in the x-, y-, and z-directions, respectively
W	Total work done by surface traction and body forces
$\boldsymbol{X}, X_i$	Position vector of a point (X, Y, Z) in the undeformed configuration in vector and indicial notation, respectively
X, Y, Z	Cartesian coordinate axes in the initial undeformed configuration
$\boldsymbol{x}, x_i$	Position vector of a point (x, y, z) in the deformed configuration in vector and indicial notation, respectively
x, y, z	Cartesian coordinate axes in the deformed configuration
$[\alpha]$	Transformation matrix
$a_{x'x}, a_{x'y}, a_{x'z}, a_{y'x},$ $a_{y'y}, a_{y'z}, a_{z'x}, a_{z'y}, a_{z'z}$	Direction cosines (elements of transformation matrix)
ΔA	Infinitesimal area
$\Delta \boldsymbol{F}$	Force on an infinitesimal area
ΔL	Change in length
$\Delta \boldsymbol{M}$	Moment on an infinitesimal area
∇	Differential operator
$[\mathcal{E}], \boldsymbol{\mathcal{E}}, \mathcal{E}_{ij}$	Finite strain tensor in the component form, vector notation, and indicial notation, respectively
$\mathcal{E}_{rr}, \mathcal{E}_{r\theta}, \mathcal{E}_{rz}, \mathcal{E}_{\theta\theta}, \mathcal{E}_{\theta z}$	Finite strains in the cylindrical coordinate system
$\varepsilon_{rr}, \varepsilon_{r\theta}, \varepsilon_{rz}, \varepsilon_{\theta\theta}, \varepsilon_{\theta z}$	Infinitesimal strains in the cylindrical coordinate system
$\mathcal{E}_{XX}, \mathcal{E}_{YY}, \mathcal{E}_{ZZ}$	Finite normal strains in the Cartesian coordinate system
$\mathcal{E}_{XY}, \mathcal{E}_{YZ}, \mathcal{E}_{ZX}$	Finite shear strains in the Cartesian coordinate system
$\varepsilon_A, \varepsilon_E, \varepsilon_G, \varepsilon_L$	Almansi strain, engineering strain, Green strain, and logarithmic strain, respectively
$\varepsilon_{xx}, \varepsilon_{yy}, \varepsilon_{zz}$	Infinitesimal normal strains in the Cartesian coordinate system
$\varepsilon_{xy}, \varepsilon_{yz}, \varepsilon_{zx}$	Infinitesimal tensorial shear strains in the Cartesian coordinate system

$\gamma_{xy}, \gamma_{yz}, \gamma_{zx}$	Infinitesimal engineering shear strains in the Cartesian coordinate system
ν_{xy}, ν_{yz}	Major Poisson's ratio in the xy- and yz-planes, respectively
ϕ_x, ϕ_y	Change in angle
σ	Stress tensor
$\sigma_1, \sigma_2, \sigma_3, \sigma_4, \sigma_5, \sigma_6$	Components of the stress vector in the contracted notation
$\sigma_{xx}, \sigma_{yy}, \sigma_{zz}$	Normal stress components of the stress tensor
$\tau_{xy}, \tau_{yz}, \tau_{zx}$	Shear stress components of the stress tensor

2.3 INTRODUCTORY CONCEPTS

2.3.1 Solid Mechanics and Continuum

Mechanics is the study of the response (motion and deformation) of a body to applied forces. This is based on two approaches:

- Physical approach
- Phenomenological approach

In the physical approach, adopted in solid state physics, the structure of a body is studied at the atomic and molecular levels.

On the other hand, the phenomenological approach is adopted in solid mechanics. It is based on the fundamental concept of continuum. Continuum is a state, in which it is assumed that the material is continuously distributed, without any crack or flaw, in the body. Thus, properties such as mass and displacement associated with the body can be defined as continuous functions or piecewise continuous functions inside the body. Governing equations are developed by considering the behavior of the solid body at a macroscopic level.

2.3.2 Spatial Point, Material Point, and Configuration

A spatial point or simply a point is a point fixed in space. A particle is a very small volumetric element with mass concentrated in it. It is a material point and it should be clearly differentiated from a spatial point. A particle or a material point occupies a certain spatial point at a certain instant in time.

A body, on the other hand, is a collection of particles that are constrained and bounded within certain volume. It has a definite mass and volume. Solid mechanics is concerned with solid bodies. A solid body is made up of particles that are geometrically bounded within a certain boundary. It, at a particular instant in time, occupies a certain region in physical space. This region with certain geometrical shape is the configuration of the body at that instant in time. Under the application of forces, the body undergoes motion and deformation and the particles move from one spatial point to another. Solid mechanics is concerned about this movement of material points.

2.3.3 Fundamental Principles and Governing Equations

In solid mechanics, we are concerned with the determination of the response of a body to the applied loads. While the loads are mechanical, thermal, or both, generally, the response of a body is expressed in terms of stress, strain, displacement, and temperature distribution. Often, we make mathematical models to represent the physical problem in terms of differential equations and solve the same to obtain the response. Mathematical models are based on fundamental principles of physics and assumptions supported by experimental observations.

Fundamental principles of physics are the result of centuries of research. A detailed discussion on these principles is beyond the scope of this chapter. Instead, we name here four key fundamental laws of physics that are often employed in solid mechanics:

- Principle of conservation of mass
- Principle of conservation of linear momentum
- Principle of conservation angular momentum
- Principle of conservation of energy

Now, we turn our attention to the governing equations in solid mechanics. These equations can be broadly categorized into four classes:

- Kinematics
- Kinetics
- Constitutive relations
- Thermodynamics

Kinematics is the study of geometric changes or deformation in a body. The factors that cause such deformations are not considered and attention is paid only to the initial and final configurations of the body. The basis for kinematic study is geometrical considerations and no fundamental principles of mechanics are involved. The variables involved in kinematics are the displacements and strains; strain–displacement relations are the primary output of kinematics.

Kinetics is the study of forces and moments acting on a body in static or dynamic equilibrium. It is based on the principles of conservation of linear and angular momenta. Conservation of linear momentum results in the equilibrium equations or the equations of motion. However, conservation of angular momentum leads to symmetry of stress tensor. (We shall discuss about stress tensor in the section on kinetics.)

In thermodynamics, we study the relations between thermodynamic state variables such as strain tensor, temperature, etc. (We shall discuss about strain tensor in the section on kinematics.) Thermodynamic state variables are governed by the first law (conservation of energy) and the second law of thermodynamics.

Constitutive relations are based on experimental observations on material behavior. They relate the dependent variables of kinematics to those of kinetics. These relations are not independent and they are governed by thermodynamic principles.

Table 2.1 presents the governing equations in solid mechanics. As we can see (and it will be clear by the end of the section on constitutive modeling) that we have 15 governing equations from kinematics, kinetics, and constitutive modeling in a 3D structure. We will see that there are 15 unknowns (six stress components, six strain components, and three displacement components) corresponding to these 15 equations. Thus, we get a complete solution on the deformation and force distribution in a solid. The numbers of equations and unknowns reduce in the case of 2D and one-dimensional (1D) problems.

Keeping in view the overall objective of this book, solid mechanics topics are discussed here only in an introductory manner; for more details, interested reader may refer to References 1–4, for instance.

2.4 KINEMATICS

A solid body has a certain configuration at a particular instant in time. Under the action of forces, the body undergoes rigid body motion, deformation, or a combination of both. A rigid body is one that does not change its configuration under loads, and the relative distance between any two material points in it remains unchanged. Under the action of forces, a rigid

TABLE 2.1
Governing Equations in Solid Mechanics

Subject	Basis	Output Equations	Key Parameters	Number of Solid Mechanics Equations		
				3D	2D	1D
Kinematics	Geometrical considerations	Strain–displacement relations	Strains, displacements	6	3	1
Kinetics	Conservation of linear momentum	Equations of motion	Stresses	3	2	1
	Conservation of angular momentum	Symmetry of stress tensor	Stresses	–	–	–
Thermodynamics	First law of thermodynamics	Energy equation	Stresses, temperature, heat flux, velocities	–	–	–
Constitutive modeling	Experimental observations	Stress–strain relations (Hooke's law)	Stresses, strains, temperature, heat flux	6	3	1

Source: Adapted with permission from J. N. Reddy, *An Introduction to Continuum Mechanics—With Applications*, Cambridge University Press, Cambridge, 2010.

body undergoes rigid body translation and rigid body rotation. Rigid body, however, is a mathematical concept and in reality, all bodies are deformable. A deformable body, under the action of forces, changes its configuration; the material points undergo displacements and the relative distance between two arbitrary material points in the body changes. In simple term, deformation of a body produces change in shape and size of the body.

Strain, on the other hand, is a quantitative measure of relative deformation of a body w.r.t. its undeformed or initial configuration.

Study of deformation and strain is necessary for three reasons. First, governing equations obtained from considerations of stress and force alone are insufficient to obtain a solution of a solid body and complete stress picture cannot be obtained. Deformation and strains have to be considered for evolving additional equations. Second, from functional angle, in many applications deformations are required to be known. Third, stress is an abstract quantity and it cannot be seen. Strain can be evaluated experimentally; stress can be related to strain and based on strain data, stress distribution in a body can be indirectly obtained.

2.4.1 Normal Strain and Shear Strain

Strain is a measure of relative deformation of a body. Two modes of deformation can be identified—first, change in size and second, change in shape. Strains that cause only change in size but not shape are normal strains or direct strains. Let us consider an elemental cuboid as shown in Figure 2.1. For simplicity, let us assume that the cuboid deforms only in the x- and y-directions. (It is the case of plane strain as discussed in Section 2.8.2.) Under normal strains, the cuboid changes its size but not shape and the rectangular faces remain rectangular after deformation. Also, normal strain in a particular direction would cause change in length of a line segment in that direction. Thus, line segments such as *OA* and *OB* change in length to *OA'* and *OB'*, respectively, but the angle between *OA* and *OB* does not change.

On the other hand, shear strains cause change in shape. Such a change in shape can be expected under the action of shear forces in the x- and y-directions. Note that the angle between *OA* and *OB* changes.

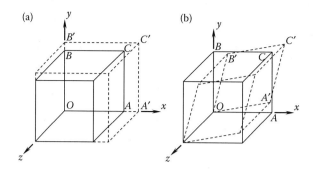

FIGURE 2.1 Normal and shear strains. (a) Deformation under normal strains in the x- and y-directions. (b) Deformation under shear strain in the xy-plane.

2.4.2 Types of Strain Measures: 1D Approach

Strain measurement schemes are somewhat arbitrary and several types are in vogue. We shall define some of the common strain measures in this section. For the sake of simplicity, we shall adopt a 1D approach first, which can be extended to two and three dimensions. Let us consider a bar as shown in Figure 2.2. The undeformed length of the bar is L, which changes by ΔL to l after deformation.

2.4.2.1 Engineering Strain

Engineering strain is the most common measure of strain used in structural engineering. It is defined as the change in length of the bar per unit undeformed length. Thus,

$$\varepsilon_E = \frac{l - L}{L} = \frac{\Delta L}{L} \tag{2.1}$$

Here, we have taken a bar for easy visualization. We can also consider an elemental material line segment and define engineering normal strain as the ratio of the change in the length to the original length of the line segment.

2.4.2.2 True Strain

Engineering strain takes only the initial undeformed and final deformed configurations into account. True strain, also known as logarithmic strain or natural strain, takes into

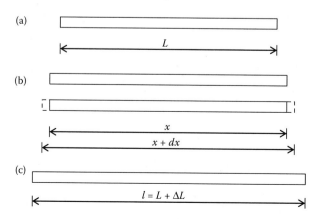

FIGURE 2.2 1D strain in a bar. (a) Undeformed configuration. (b) Intermediate configuration. (c) Final deformed configuration. (Adapted from A. K. Singh, *Mechanics of Solids*, PHI Learning, New Delhi, 2011.)

account the intermediate configurations as well. Now, w.r.t. the bar in Figure 2.2, the incremental strain at an intermediate configuration is defined as

$$d\varepsilon = \frac{dx}{x} \quad (2.2)$$

where x is the intermediate length of the bar. Then, the true strain at the final deformed configuration is given by

$$\varepsilon_L = \int_L^l \frac{dx}{x} = \ln\left(\frac{l}{L}\right) \quad (2.3)$$

2.4.2.3 Green Strain

Green strain represents change in square of the length w.r.t. the undeformed length. Thus, Green strain in one dimension is given by

$$\varepsilon_G = \frac{l^2 - L^2}{2L^2} \quad (2.4)$$

2.4.2.4 Almansi Strain

Almansi strain is similar to the Green strain; however, it is defined w.r.t. the deformed configuration. Thus, Almansi strain in one dimension is given by

$$\varepsilon_A = \frac{l^2 - L^2}{2l^2} \quad (2.5)$$

We have adopted a 1D approach for defining normal strain in different strain measures. Shear strain is a measure of change in angle and thus we need to adopt a 2D approach. Let us consider two initially mutually orthogonal line segments as in Figure 2.3. (Note that the coordinate axes in the deformed and undeformed configurations are superimposed.) Under shearing action, the line segments change their orientations. Engineering shear strain is defined as the total change in angle. On the other hand, true shear strain (tensorial shear strain) is defined as half of absolute change in angle in radian. Thus,

Engineering shear strain,

$$\gamma_{xy} = \phi_x + \phi_y \quad (2.6)$$

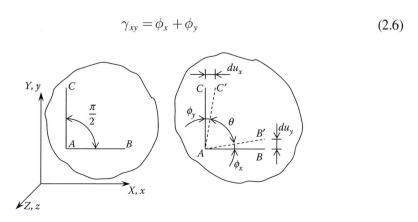

FIGURE 2.3 Definition of shear strain. (Adapted with permission from J. N. Reddy, *An Introduction to Continuum Mechanics—With Applications*, Cambridge University Press, Cambridge, 2010.)

True shear strain,

$$\varepsilon_{xy} = \frac{1}{2}(\phi_x + \phi_y) \qquad (2.7)$$

2.4.3 Displacement at a Point

Let us consider a body as shown in Figure 2.4. The undeformed or initial configuration $\mathcal{B}_0$ at time $t = t_0$ changes to the deformed or current or final configuration $\mathcal{B}$ at time $t = t$, and in this process of deformation, a particle at P_0 in the undeformed configuration moves to P in the deformed configuration.

Let us consider two Cartesian coordinate systems: O-XYZ with unit vectors $\boldsymbol{e}_X$, $\boldsymbol{e}_Y$, and $\boldsymbol{e}_Z$ for the undeformed configuration and o-xyz with unit vectors $\boldsymbol{e}_x$, $\boldsymbol{e}_y$, and $\boldsymbol{e}_z$ for the deformed configuration. The origins of the two coordinate systems are connected by the vector $\boldsymbol{b}$.

Position vectors of the points $P_0(X, Y, Z)$ and $P(x, y, z)$ are given, respectively, by

$$\boldsymbol{X} = X\boldsymbol{e}_X + Y\boldsymbol{e}_Y + Z\boldsymbol{e}_Z = \begin{Bmatrix} X \\ Y \\ Z \end{Bmatrix}^T \begin{Bmatrix} \boldsymbol{e}_X \\ \boldsymbol{e}_Y \\ \boldsymbol{e}_Z \end{Bmatrix} \qquad (2.8)$$

and

$$\boldsymbol{x} = x\boldsymbol{e}_x + y\boldsymbol{e}_y + z\boldsymbol{e}_z = \begin{Bmatrix} x \\ y \\ z \end{Bmatrix}^T \begin{Bmatrix} \boldsymbol{e}_x \\ \boldsymbol{e}_y \\ \boldsymbol{e}_z \end{Bmatrix} \qquad (2.9)$$

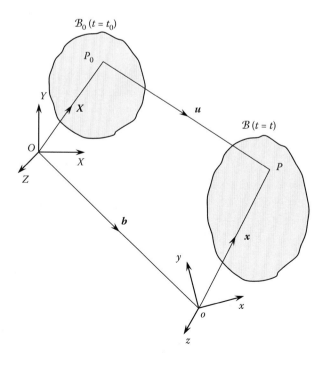

FIGURE 2.4 Deformation of a solid body: a particle in the undeformed and deformed configurations. (Adapted from G. E. Mase, *Theory and Problems of Continuum Mechanics*, McGraw-Hill, New York, 1970.)

Basic Solid Mechanics

The vector $\boldsymbol{u}$ joining the points P_0 and P is the displacement vector, and in terms of its Cartesian components, it is given by

$$\boldsymbol{u} = u_X \boldsymbol{e}_X + u_Y \boldsymbol{e}_Y + u_Z \boldsymbol{e}_Z = \begin{Bmatrix} u_X \\ u_Y \\ u_Z \end{Bmatrix}^T \begin{Bmatrix} \boldsymbol{e}_X \\ \boldsymbol{e}_Y \\ \boldsymbol{e}_Z \end{Bmatrix} \qquad (2.10)$$

or

$$\boldsymbol{u} = u_x \boldsymbol{e}_x + u_y \boldsymbol{e}_y + u_z \boldsymbol{e}_z = \begin{Bmatrix} u_x \\ u_y \\ u_z \end{Bmatrix}^T \begin{Bmatrix} \boldsymbol{e}_x \\ \boldsymbol{e}_y \\ \boldsymbol{e}_z \end{Bmatrix} \qquad (2.11)$$

The displacement vector is related to the position vectors as follows:

$$\boldsymbol{u} = \boldsymbol{b} + \boldsymbol{x} - \boldsymbol{X} \qquad (2.12)$$

If the initial and final position vectors of the particle are known for the chosen coordinate systems, the displacement vector can be determined from Equation 2.12. The motion of a point and deformation of a continuum can be studied in two ways—*Lagrangian* or *material* description of motion and *Eularian* or *spatial* description of motion.

In the Lagrangian description, motion of a body is referred to a reference configuration. The initial configuration is usually chosen as the reference configuration. Thus, current coordinates of a particle are expressed as functions of the coordinates the particle occupied at time $t = t_0$. In other words, Lagrangian description is deformation mapping of the initial configuration onto the final or current configuration. Mathematically, in the Lagrangian description,

$$\boldsymbol{x} = \boldsymbol{x}(\boldsymbol{X}, t) \qquad (2.13)$$

On the other hand, in the Eularian description, the undeformed configuration is expressed in terms of the deformed configuration. Thus, initial coordinates of a particle at time $t = t_0$ are expressed as functions of the coordinates the particle occupies at time $t = t$. Thus, Eularian description is a mapping of the final configuration onto the initial configuration such that the original position of a particle can be traced from the current position. Mathematically, in the Eularian description,

$$\boldsymbol{X} = \boldsymbol{X}(\boldsymbol{x}, t) \qquad (2.14)$$

Solid mechanics generally uses the Lagrangian description, whereas, in fluid mechanics, the Eularian description is used. Further, in solid mechanics, the two coordinate systems are often superimposed. We shall use such superimposed coordinate systems for which, it may be noted, $\boldsymbol{b} = 0$. Then, from Equation 2.12, we write the relation between position vectors and displacement vector as follows:

In the component form,

$$\begin{Bmatrix} x \\ y \\ z \end{Bmatrix} = \begin{Bmatrix} X + u_x \\ Y + u_y \\ Z + u_z \end{Bmatrix} \qquad (2.15)$$

In the vector form,

$$x = X + u \tag{2.16}$$

In the indicial notation,

$$x_i = X_i + u_i \tag{2.17}$$

Note: Indicial notation is very helpful in concise and clear representation of solid mechanics expressions involving vectors and matrices; the reader is urged to get acquainted with it (see, for instance, References 3–5).

2.4.4 Deformation Gradient and Displacement Gradient

Let us consider the undeformed and deformed configurations of a body as shown in Figure 2.5. The material points A and B in the undeformed or initial configuration of the body get displaced to the new positions A' and B', respectively, and the infinitesimal line segment, represented by the vector dX deforms to the line segment dx after deformation.

Deformation gradient and displacement gradient are two important quantities in the analysis of deformation and strain. Deformation gradient connects the deformed configuration of a body to its undeformed configuration.

The two coordinate systems O-XYZ and O-xyz share the same origin and are aligned. For convenience, we shall use only the lower case letters for coordinate axes. Thus, the undeformed and deformed line segments are given by

$$d\mathbf{X} = dX\mathbf{e}_x + dY\mathbf{e}_y + dZ\mathbf{e}_z = \begin{Bmatrix} dX \\ dY \\ dZ \end{Bmatrix}^T \begin{Bmatrix} \mathbf{e}_x \\ \mathbf{e}_y \\ \mathbf{e}_z \end{Bmatrix} \tag{2.18}$$

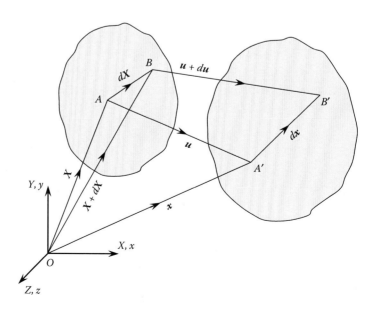

FIGURE 2.5 Deformation of a solid body: a line segment in the undeformed and deformed configurations. (Adapted from G. E. Mase, *Theory and Problems of Continuum Mechanics*, McGraw-Hill, New York, 1970.)

Basic Solid Mechanics

$$dx = dxe_x + dye_y + dze_z = \begin{Bmatrix} dx \\ dy \\ dz \end{Bmatrix}^T \begin{Bmatrix} e_x \\ e_y \\ e_z \end{Bmatrix} \quad (2.19)$$

Using chain rule of differentiation, we can express the components of the deformed line segment as

$$\begin{aligned} dx &= \frac{\partial x}{\partial X}dX + \frac{\partial x}{\partial Y}dY + \frac{\partial x}{\partial Z}dZ \\ dy &= \frac{\partial y}{\partial X}dX + \frac{\partial y}{\partial Y}dY + \frac{\partial y}{\partial Z}dZ \\ dz &= \frac{\partial z}{\partial X}dX + \frac{\partial z}{\partial Y}dY + \frac{\partial z}{\partial Z}dZ \end{aligned} \quad (2.20)$$

Equation 2.20 can be written as follows:
In the component form,

$$\begin{Bmatrix} dx \\ dy \\ dz \end{Bmatrix} = \begin{bmatrix} \frac{\partial x}{\partial X} & \frac{\partial x}{\partial Y} & \frac{\partial x}{\partial Z} \\ \frac{\partial y}{\partial X} & \frac{\partial y}{\partial Y} & \frac{\partial y}{\partial Z} \\ \frac{\partial z}{\partial X} & \frac{\partial z}{\partial Y} & \frac{\partial z}{\partial Z} \end{bmatrix} \begin{Bmatrix} dX \\ dY \\ dZ \end{Bmatrix} \quad (2.21)$$

In the vector form,

$$d\boldsymbol{x} = \boldsymbol{F}.d\boldsymbol{X} \quad (2.22)$$

In the indicial notation,

$$dx_i = \frac{\partial x_i}{\partial X_j}dX_j \quad (2.23)$$

Now, w.r.t. Equations 2.21 through 2.23, deformation gradient is defined as follows:
In the component form,

$$[\boldsymbol{F}] = \begin{bmatrix} \frac{\partial x}{\partial X} & \frac{\partial x}{\partial Y} & \frac{\partial x}{\partial Z} \\ \frac{\partial y}{\partial X} & \frac{\partial y}{\partial Y} & \frac{\partial y}{\partial Z} \\ \frac{\partial z}{\partial X} & \frac{\partial z}{\partial Y} & \frac{\partial z}{\partial Z} \end{bmatrix} \quad (2.24)$$

In the vector form,

$$\mathbf{F} = \left[\frac{\partial \mathbf{x}}{\partial \mathbf{X}}\right]^T \quad (2.25)$$

In the indicial notation,

$$F_{ij} = \frac{\partial x_i}{\partial X_j} \quad (2.26)$$

On the other hand, the displacement gradient is defined as follows:
In the component form,

$$[D] = \begin{bmatrix} \frac{\partial u_x}{\partial X} & \frac{\partial u_x}{\partial Y} & \frac{\partial u_x}{\partial Z} \\ \frac{\partial u_y}{\partial X} & \frac{\partial u_y}{\partial Y} & \frac{\partial u_y}{\partial Z} \\ \frac{\partial u_z}{\partial X} & \frac{\partial u_z}{\partial Y} & \frac{\partial u_z}{\partial Z} \end{bmatrix} \quad (2.27)$$

In the vector form,

$$D - \left[\frac{\partial u}{\partial X}\right]^T \quad (2.28)$$

In the indicial notation,

$$D_{ij} = \frac{\partial u_i}{\partial X_j} \quad (2.29)$$

Now, from Equations 2.15, 2.16, and 2.17, we note that for the superimposed coordinates the position vector of a material point in its final configuration is related to its position vector in the initial configuration as $x = X + u$. Both sides of these equations are operated by the differential operator ∇, given by

$$\nabla \equiv \begin{bmatrix} \frac{\partial}{\partial X} & \frac{\partial}{\partial Y} & \frac{\partial}{\partial Z} \end{bmatrix} \quad (2.30)$$

and the relation between deformation gradient and the displacement gradient is obtained as follows:
In the component form,

$$[F] = [I] + [D] \quad (2.31)$$

In the vector form,

$$F = I + D \quad (2.32)$$

In the indicial notation,

$$F_{ij} = I_{ij} + D_{ij} \quad (2.33)$$

2.4.5 Infinitesimal Strain and Finite Strain Theories

Deformation of a body can be classified into small deformation and large deformation.
Small deformation is the one in which the deformed and undeformed configurations of the body are nearly identical. In this case, displacement gradient terms are far

smaller than unity, that is, $D_{ij} \ll 1$. This class of deformation is governed by the infinitesimal strain theory (also known as small strain theory and small deformation theory), wherein the strain–displacement relations are linear. Deformation characteristics of many engineering materials exhibiting elastic behavior, such as metals and composites, belong to this category. Engineering strains, defined in the previous section are used in the analysis of strains as per the small strain theory.

On the other hand, many materials such as elastomers, fluids, etc., which exhibit plastic deformation, undergo large deformations under loads. In such a case, the deformed and undeformed configurations are grossly different. Finite strain theory (also known as large strain theory or large deformation theory) is used in the strain analysis of such materials. Strain–displacement relations are nonlinear and the displacement gradient terms are not small such that squares of these terms are not negligible. Engineering strains are not applicable in this class of deformations and other more complex definitions such as logarithmic strain, Green strain, and Almansi strain are used.

2.4.6 Infinitesimal Strain at a Point

State of strain at a material point is given by changes in lengths per unit length of all the possible infinitesimal line segments and changes in angle between all the possible pairs of orthogonal line segments at that material point. Fortunately, however, we do not need to consider all these possible line segments or pairs of line segments. We rather consider three mutually perpendicular axes and three mutually perpendicular planes formed by these three axes passing through the point, and express the state of strain at that point by means of three unique normal strains and three unique shear strains. For strains in any other directions or plane, we need to resort to strain transformation.

We shall first discuss the case of normal strains. Let us go back to Figure 2.5 and consider the infinitesimal line segment AB. The line segment moves to $A'B'$ after deformation. Normal strain in the direction of the line segment would only cause change in its length. Then, by following the definition of engineering strain, normal strain in the direction of the line segment is given by

$$\varepsilon = \frac{A'B' - AB}{AB} = \frac{|d\mathbf{x}| - |d\mathbf{X}|}{|d\mathbf{X}|} \tag{2.34}$$

The length of the undeformed line segment can be expressed as

$$|d\mathbf{X}| = \sqrt{(dX)^2 + (dY)^2 + (dZ)^2} \tag{2.35}$$

We shall first find an expression for the infinitesimal normal strain in the x-direction. Toward this, let us align the line segment in the x-direction such that $dY = dZ = 0$. Thus,

$$|d\mathbf{X}| = dX \tag{2.36}$$

The length of the deformed line segment is

$$|d\mathbf{x}| = \sqrt{(dx)^2 + (dy)^2 + (dz)^2} \tag{2.37}$$

Now, we see from Equations 2.22 and 2.32 that $d\mathbf{x} = \mathbf{F}.d\mathbf{X} = (\mathbf{I} + \mathbf{D}).d\mathbf{X}$. Thus,

$$\begin{Bmatrix} dx \\ dy \\ dz \end{Bmatrix} = \left(\begin{bmatrix} 1 & 0 & 0 \\ 0 & 1 & 0 \\ 0 & 0 & 1 \end{bmatrix} + \begin{bmatrix} \dfrac{\partial u_x}{\partial X} & \dfrac{\partial u_x}{\partial Y} & \dfrac{\partial u_x}{\partial Z} \\ \dfrac{\partial u_y}{\partial X} & \dfrac{\partial u_y}{\partial Y} & \dfrac{\partial u_y}{\partial Z} \\ \dfrac{\partial u_z}{\partial X} & \dfrac{\partial u_z}{\partial Y} & \dfrac{\partial u_z}{\partial Z} \end{bmatrix} \right) \begin{Bmatrix} dX \\ 0 \\ 0 \end{Bmatrix} \quad (2.38)$$

or

$$dx = \left(1 + \dfrac{\partial u_x}{\partial X}\right) dX$$
$$dy = \dfrac{\partial u_y}{\partial X} dX \quad (2.39)$$
$$dz = \dfrac{\partial u_z}{\partial X} dX$$

So,

$$|d\mathbf{x}| = dX \sqrt{\left(1 + \dfrac{\partial u_x}{\partial X}\right)^2 + \left(\dfrac{\partial u_y}{\partial X}\right)^2 + \left(\dfrac{\partial u_z}{\partial X}\right)^2} \quad (2.40)$$

Substituting Equations 2.36 and 2.40 in Equation 2.34, we get the following for infinitesimal normal strain in the *x*-direction:

$$\varepsilon_{xx} = \sqrt{\left(1 + \dfrac{\partial u_x}{\partial X}\right)^2 + \left(\dfrac{\partial u_y}{\partial X}\right)^2 + \left(\dfrac{\partial u_z}{\partial X}\right)^2} - 1 \quad (2.41)$$

The terms under the square root can be expanded as a binomial series. Now, for infinitesimal strains, displacement gradients are so small that higher order terms of displacement gradients can be ignored when compared to unity. Thus, ignoring the second and third terms inside the square root in Equation 2.41, we get

$$\varepsilon_{xx} = \dfrac{\partial u_x}{\partial X} \quad (2.42)$$

Next, we align the line segment in the *y*- and *z*-directions, respectively, and we can obtain the expressions for respective infinitesimal normal strains. Further, for infinitesimal strains, the partial derivatives of a displacement component w.r.t. x and X are nearly equal to each other, that is,

$$\dfrac{\partial(\ )}{\partial X} \approx \dfrac{\partial(\ )}{\partial x}, \quad \dfrac{\partial(\ )}{\partial Y} \approx \dfrac{\partial(\ )}{\partial y}, \quad \text{and} \quad \dfrac{\partial(\ )}{\partial Z} \approx \dfrac{\partial(\ )}{\partial z} \quad (2.43)$$

The space inside the brackets in the above expressions can be filled with any displacement component. Thus, we can write the expressions for infinitesimal normal strains as

Basic Solid Mechanics

$$\varepsilon_{xx} = \frac{\partial u_x}{\partial x} \tag{2.44}$$

$$\varepsilon_{yy} = \frac{\partial u_y}{\partial y} \tag{2.45}$$

$$\varepsilon_{zz} = \frac{\partial u_z}{\partial z} \tag{2.46}$$

Next, we shift our attention to the infinitesimal shear strains and go back to Figure 2.3. Let us consider two mutually orthogonal line segments AB and AC. Let AB and AC be aligned in the x- and y-directions, respectively. Shear forces causing shear strains in the xy-plane would change the angle BAC to $B'AC'$. Engineering shear strain is defined as the change in the angle. Thus,

$$\gamma_{xy} = \phi_x + \phi_y \tag{2.47}$$

For small angles, the angles are equal to the tangents of the respective angles, that is, $\phi_x \approx \tan\phi_x$ and $\phi_y \approx \tan\phi_y$.

We note that

$$\tan\phi_x = \frac{du_y}{dX}, \quad \tan\phi_y = \frac{du_x}{dY} \tag{2.48}$$

Now, in the line segment AB, $dY = dZ = 0$ and

$$du_y = \frac{\partial u_y}{\partial X} dX \tag{2.49}$$

Similarly, in the line segment AC, $dX = dZ = 0$ and

$$du_x = \frac{\partial u_x}{\partial Y} dY \tag{2.50}$$

Thus, from Equation 2.47, together with Equations 2.48 through 2.50,

$$\gamma_{xy} = \frac{\partial u_y}{\partial X} + \frac{\partial u_x}{\partial Y} \tag{2.51}$$

Then, considering the orthogonal line segments in the yz- and zx-planes, respectively, we can arrive at the expressions for infinitesimal shear strains in the other two planes. Further, like in the case of infinitesimal normal strains, partial derivatives of the displacement components w.r.t. x are nearly equal to those w.r.t. X. Thus, we can write the expressions for infinitesimal shear strains as

$$\gamma_{xy} = \frac{\partial u_y}{\partial x} + \frac{\partial u_x}{\partial y} \tag{2.52}$$

$$\gamma_{yz} = \frac{\partial u_z}{\partial y} + \frac{\partial u_y}{\partial z} \qquad (2.53)$$

$$\gamma_{zx} = \frac{\partial u_x}{\partial z} + \frac{\partial u_z}{\partial x} \qquad (2.54)$$

2.4.7 Finite Strain at a Point

2.4.7.1 Finite Strain Tensor

Let us once again go back to Figure 2.5. Let us consider the deformation of the solid body and the arbitrarily chosen infinitesimal line segment at material point A. The change in the square of the length of the infinitesimal line segment from the undeformed configuration to the deformed configuration is the quantity used for analysis of deformation in the finite strain theory. Let us note carefully that the coordinates of the points A, B, A', and B' as indicated below:

$$A \to (X, Y, Z)$$
$$B \to (X + dX, Y + dY, Z + dZ)$$
$$A' \to (X + u_x, Y + u_y, Z + u_z)$$
$$B' \to (x + dx, y + dy, z + dz)$$

Or,

$$(X + dX + u_x + du_x, Y + dY + u_y + du_y, Z + dZ + u_z + du_z)$$

The length of the line segment in its undeformed configuration is given by

$$AB = |d\boldsymbol{X}| = \sqrt{(dX)^2 + (dY)^2 + (dZ)^2} \qquad (2.55)$$

Similarly, length of the deformed line segment is given by

$$A'B' = |d\boldsymbol{x}| = \sqrt{(dX + du_x)^2 + (dY + du_y)^2 + (dZ + du_z)^2} \qquad (2.56)$$

Using Equations 2.55 and 2.56, we can express the quantity $(|d\boldsymbol{x}|)^2 - (|d\boldsymbol{X}|)^2$ as

$$(|d\boldsymbol{x}|)^2 - (|d\boldsymbol{X}|)^2 = 2(dX du_x + dY du_y + dZ du_z) + (du_x)^2 + (du_y)^2 + (du_z)^2 \qquad (2.57)$$

The displacement differentials can be expressed as

$$\begin{Bmatrix} du_x \\ du_y \\ du_z \end{Bmatrix} = \begin{bmatrix} \dfrac{\partial u_x}{\partial X} & \dfrac{\partial u_x}{\partial Y} & \dfrac{\partial u_x}{\partial Z} \\ \dfrac{\partial u_y}{\partial X} & \dfrac{\partial u_y}{\partial Y} & \dfrac{\partial u_y}{\partial Z} \\ \dfrac{\partial u_z}{\partial X} & \dfrac{\partial u_z}{\partial Y} & \dfrac{\partial u_z}{\partial Z} \end{bmatrix} \begin{Bmatrix} dX \\ dY \\ dZ \end{Bmatrix} \qquad (2.58)$$

Basic Solid Mechanics

Substituting Equation 2.58 in Equation 2.57 and by rearranging the terms, we get the following:

$$(|d\boldsymbol{x}|)^2 - (|d\boldsymbol{X}|)^2$$

$$= 2\left[\frac{\partial u_x}{\partial X} + \frac{1}{2}\left\{\left(\frac{\partial u_x}{\partial X}\right)^2 + \left(\frac{\partial u_y}{\partial X}\right)^2 + \left(\frac{\partial u_z}{\partial X}\right)^2\right\}\right]dX^2$$

$$+ 2\left[\left(\frac{\partial u_y}{\partial Y}\right) + \frac{1}{2}\left\{\left(\frac{\partial u_x}{\partial Y}\right)^2 + \left(\frac{\partial u_y}{\partial Y}\right)^2 + \left(\frac{\partial u_z}{\partial Y}\right)^2\right\}\right]dY^2$$

$$+ 2\left[\left(\frac{\partial u_z}{\partial Z}\right) + \frac{1}{2}\left\{\left(\frac{\partial u_x}{\partial Z}\right)^2 + \left(\frac{\partial u_y}{\partial Z}\right)^2 + \left(\frac{\partial u_z}{\partial Z}\right)^2\right\}\right]dZ^2$$

$$+ 2\left[\frac{\partial u_y}{\partial X} + \frac{\partial u_x}{\partial Y} + \frac{\partial u_x}{\partial X}\frac{\partial u_x}{\partial Y} + \frac{\partial u_y}{\partial X}\frac{\partial u_y}{\partial Y} + \frac{\partial u_z}{\partial X}\frac{\partial u_z}{\partial Y}\right]dX\,dY$$

$$+ 2\left[\frac{\partial u_z}{\partial Y} + \frac{\partial u_y}{\partial Z} + \frac{\partial u_x}{\partial Y}\frac{\partial u_x}{\partial Z} + \frac{\partial u_y}{\partial Y}\frac{\partial u_y}{\partial Z} + \frac{\partial u_z}{\partial Y}\frac{\partial u_z}{\partial Z}\right]dY\,dZ$$

$$+ 2\left[\frac{\partial u_x}{\partial Z} + \frac{\partial u_z}{\partial X} + \frac{\partial u_x}{\partial Z}\frac{\partial u_x}{\partial X} + \frac{\partial u_y}{\partial Z}\frac{\partial u_y}{\partial X} + \frac{\partial u_z}{\partial Z}\frac{\partial u_z}{\partial X}\right]dZ\,dX \quad (2.59)$$

The coefficients of the terms dX^2, dY^2, dZ^2, $dXdY$, $dYdZ$, and $dZdX$ in Equation 2.59 are of special significance and we rewrite Equation 2.59 as follows:

$$(|d\boldsymbol{x}|)^2 - (|d\boldsymbol{X}|)^2 = 2\mathcal{E}_{XX}dX^2 + 2\mathcal{E}_{YY}dY^2 + 2\mathcal{E}_{ZZ}dZ^2$$
$$+ 4\mathcal{E}_{XY}dXdY + 4\mathcal{E}_{YZ}dYdZ + 4\mathcal{E}_{ZX}dZdX \quad (2.60)$$

where the coefficients are given by

$$\mathcal{E}_{XX} = \frac{\partial u_x}{\partial X} + \frac{1}{2}\left[\left(\frac{\partial u_x}{\partial X}\right)^2 + \left(\frac{\partial u_y}{\partial X}\right)^2 + \left(\frac{\partial u_z}{\partial X}\right)^2\right]$$

$$\mathcal{E}_{YY} = \frac{\partial u_y}{\partial Y} + \frac{1}{2}\left[\left(\frac{\partial u_x}{\partial Y}\right)^2 + \left(\frac{\partial u_y}{\partial Y}\right)^2 + \left(\frac{\partial u_z}{\partial Y}\right)^2\right]$$

$$\mathcal{E}_{ZZ} = \frac{\partial u_z}{\partial Z} + \frac{1}{2}\left[\left(\frac{\partial u_x}{\partial Z}\right)^2 + \left(\frac{\partial u_y}{\partial Z}\right)^2 + \left(\frac{\partial u_z}{\partial Z}\right)^2\right]$$

$$\mathcal{E}_{XY} = \frac{1}{2}\left[\frac{\partial u_y}{\partial X} + \frac{\partial u_x}{\partial Y} + \frac{\partial u_x}{\partial X}\frac{\partial u_x}{\partial Y} + \frac{\partial u_y}{\partial X}\frac{\partial u_y}{\partial Y} + \frac{\partial u_z}{\partial X}\frac{\partial u_z}{\partial Y}\right]$$

$$\mathcal{E}_{YZ} = \frac{1}{2}\left[\frac{\partial u_z}{\partial Y} + \frac{\partial u_y}{\partial Z} + \frac{\partial u_x}{\partial Y}\frac{\partial u_x}{\partial Z} + \frac{\partial u_y}{\partial Y}\frac{\partial u_y}{\partial Z} + \frac{\partial u_z}{\partial Y}\frac{\partial u_z}{\partial Z}\right]$$

$$\mathcal{E}_{ZX} = \frac{1}{2}\left[\frac{\partial u_x}{\partial Z} + \frac{\partial u_z}{\partial X} + \frac{\partial u_x}{\partial Z}\frac{\partial u_x}{\partial X} + \frac{\partial u_y}{\partial Z}\frac{\partial u_y}{\partial X} + \frac{\partial u_z}{\partial Z}\frac{\partial u_z}{\partial X}\right]$$

(2.61)

Thus, Equation 2.60, with the help of Equations 2.61, can be rearranged further as follows:

In the component form,

$$|d\boldsymbol{x}|^2 - |d\boldsymbol{X}|^2 = 2\begin{Bmatrix} dX \\ dY \\ dZ \end{Bmatrix}^T [\boldsymbol{\mathcal{E}}] \begin{Bmatrix} dX \\ dY \\ dZ \end{Bmatrix} \tag{2.62}$$

In the vector form,

$$|d\boldsymbol{x}|^2 - |d\boldsymbol{X}|^2 = 2 d\boldsymbol{X} \cdot \boldsymbol{\mathcal{E}} \cdot d\boldsymbol{X} \tag{2.63}$$

In the indicial notation,

$$|d\boldsymbol{x}|^2 - |d\boldsymbol{X}|^2 = 2\mathcal{E}_{ij} dX_i dX_j \tag{2.64}$$

where

$$[\boldsymbol{\mathcal{E}}] = \begin{bmatrix} \mathcal{E}_{XX} & \mathcal{E}_{XY} & \mathcal{E}_{ZX} \\ \mathcal{E}_{XY} & \mathcal{E}_{YY} & \mathcal{E}_{YZ} \\ \mathcal{E}_{ZX} & \mathcal{E}_{YZ} & \mathcal{E}_{ZZ} \end{bmatrix} \tag{2.65}$$

Here, the second-order tensor $[\boldsymbol{\mathcal{E}}]$ or $\boldsymbol{\mathcal{E}}$ or $\mathcal{E}_{ij}$ is known as the *Green's* (or *Lagrangian*) *finite strain tensor*.

The components of the finite strain tensor are given in the explicit component forms by Equations 2.61. A very convenient way to express the finite strain tensor components is in the indicial notation as

$$\varepsilon_{ij} = \frac{1}{2}\left[\frac{\partial u_i}{\partial X_j} + \frac{\partial u_j}{\partial X_i} + \frac{\partial u_k}{\partial X_i}\frac{\partial u_k}{\partial X_j}\right] \tag{2.66}$$

2.4.7.2 Physical Meaning of Finite Strain Tensor Components

We consider three mutually orthogonal infinitesimal line segments PA, PB, and PC, aligned in the x-, y-, and z-directions, respectively, in the undeformed configuration as shown in Figure 2.6. After deformation, the line segments move to $P'A'$, $P'B'$, and $P'C'$.

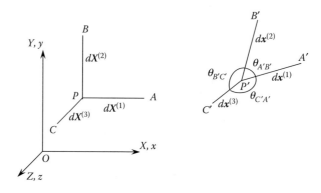

FIGURE 2.6 Physical meaning of finite strain tensor components.

Basic Solid Mechanics

Let us consider the line segment *PA*; we note that for *PA*,

$$|d\boldsymbol{X}^{(1)}| = dX^{(1)} \neq 0 \quad \text{and} \quad dY^{(1)} = dZ^{(1)} = 0 \tag{2.67}$$

Now, we substitute Equation 2.67 in Equation 2.59, and with the help of Equation 2.61, obtain the following:

$$\frac{|d\boldsymbol{x}^{(1)}|^2 - |d\boldsymbol{X}^{(1)}|^2}{2|d\boldsymbol{X}^{(1)}|^2} = \frac{\partial u_x}{\partial X} + \frac{1}{2}\left[\left(\frac{\partial u_x}{\partial X}\right)^2 + \left(\frac{\partial u_y}{\partial X}\right)^2 + \left(\frac{\partial u_z}{\partial X}\right)^2\right] = \mathcal{E}_{XX} \tag{2.68}$$

By definition, the expression on the left-hand side in Equation 2.68 is the Green's normal strain in the *X*-direction. In a similar way, we can consider the line segments *PB* and *PC* and conclude that the three diagonal elements in the finite strain tensor in Equation 2.65 are the Green's normal strains in the *X*-, *Y*- and *Z*-directions.

To check the physical meaning of the off-diagonal elements in the strain tensor, let us consider the infinitesimal line segments *PA* and *PB* in the *XY*-plane in the undeformed configuration. In the undeformed configuration, the line segments are perpendicular to each other, whereas in the deformed configuration, the included angle changes to $\theta_{A'B'}$. From basic coordinate geometry, we know that the included angle $\theta_{A'B'}$ is given by

$$\cos(\theta_{A'B'}) = \frac{dx^{(1)}dx^{(2)} + dy^{(1)}dy^{(2)} + dz^{(1)}dz^{(2)}}{|d\boldsymbol{x}^{(1)}||d\boldsymbol{x}^{(2)}|} \tag{2.69}$$

Noting that $d\boldsymbol{x}^{(1)} = d\boldsymbol{X}^{(1)} + d\boldsymbol{u}$ and $d\boldsymbol{x}^{(2)} = d\boldsymbol{X}^{(2)} + d\boldsymbol{u}$, we can arrive at the following:

$$\begin{aligned}
dx^{(1)} &= \left(1 + \frac{\partial u_x}{\partial X}\right)dX^{(1)} & dx^{(2)} &= \frac{\partial u_x}{\partial Y}dY^{(2)} \\
dy^{(1)} &= \frac{\partial u_y}{\partial X}dX^{(1)} & dy^{(2)} &= \left(1 + \frac{\partial u_y}{\partial Y}\right)dY^{(2)} \\
dz^{(1)} &= \frac{\partial u_z}{\partial X}dX^{(1)} & dz^{(2)} &= \frac{\partial u_z}{\partial Y}dY^{(2)} \\
|d\boldsymbol{x}^{(1)}| &= \sqrt{(dx^{(1)})^2 + (dy^{(1)})^2 + (dz^{(1)})^2} \\
|d\boldsymbol{x}^{(2)}| &= \sqrt{(dx^{(2)})^2 + (dy^{(2)})^2 + (dz^{(2)})^2}
\end{aligned} \tag{2.70}$$

Utilizing Equations 2.70 and 2.61, we get from Equation 2.69

$$\cos(\theta_{A'B'}) = \frac{2\mathcal{E}_{XY}}{\sqrt{1 + 2\mathcal{E}_{XX}}\sqrt{1 + 2\mathcal{E}_{YY}}} \tag{2.71}$$

Now, let us denote the change in angle between the line segments in the *PA* and *PB* by α_{XY}, etc. Then,

$$\sin(\alpha_{XY}) = \sin\left(\frac{\pi}{2} - \theta_{A'B'}\right) = \cos(\theta_{A'B'}) \tag{2.72}$$

For small angle, $\sin(\alpha_{XY}) \approx \alpha_{XY}$. Thus,

$$\alpha_{XY} = \cos(\theta_{A'B'}) \tag{2.73}$$

Then, from Equation 2.71 and by considering the other two possible combinations of pairs of line segments, it can be shown that

$$\begin{aligned} 2\mathcal{E}_{XY} &= \alpha_{XY}\sqrt{1+2\mathcal{E}_{XX}}\sqrt{1+2\mathcal{E}_{YY}} \\ 2\mathcal{E}_{YZ} &= \alpha_{YZ}\sqrt{1+2\mathcal{E}_{YY}}\sqrt{1+2\mathcal{E}_{ZZ}} \\ 2\mathcal{E}_{ZX} &= \alpha_{ZX}\sqrt{1+2\mathcal{E}_{ZZ}}\sqrt{1+2\mathcal{E}_{XX}} \end{aligned} \tag{2.74}$$

Equation 2.74 shows that the off-diagonal elements, that is, shear strain components in the finite strain tensor depend on change in the angle between the corresponding line segments as well as normal strains in the line segments.

We have discussed both the infinitesimal as well as finite strains and derived the strain–displacement relations. At this juncture following points may be noted:

- We have not made any assumption of smallness in any quantity in the finite strain theory.
- Engineering shear strains are twice tensorial shear strains.
- In the case of infinitesimal strains, the displacement gradient terms are so small compared to unity that second or higher order terms of displacement gradient can be neglected compared to unity. Further, we replace X with x in the partial derivatives of displacements. Thus, by ignoring higher order terms of displacement gradients in the expressions for finite strain tensor components, we can obtain the expressions for the components of the infinitesimal strain tensor. In the indicial form, we can write the expression for infinitesimal strain tensor as

$$\varepsilon_{ij} = \frac{1}{2}\left[\frac{\partial u_i}{\partial x_j} + \frac{\partial u_j}{\partial x_i}\right] \tag{2.75}$$

2.4.8 Strain–Displacement Relations in Cylindrical Coordinates

Strain–displacement relations are a set of very useful equations that are used frequently in solid mechanics. We have discussed them in detail in the previous sections and arrived at the expressions in the Cartesian coordinate system. In this section, these relations in the cylindrical coordinate system (Figure 2.7) are presented [1,6]. Finite strain–displacement relations in cylindrical coordinates are as follows:

$$\mathcal{E}_{rr} = \frac{\partial u_r}{\partial r} + \frac{1}{2}\left[\left(\frac{\partial u_r}{\partial r}\right)^2 + \left(\frac{\partial u_\theta}{\partial r}\right)^2 + \left(\frac{\partial u_z}{\partial r}\right)^2\right] \tag{2.76}$$

$$\mathcal{E}_{r\theta} = \frac{1}{2}\left[\frac{\partial u_\theta}{\partial r} + \frac{1}{r}\left(\frac{\partial u_r}{\partial \theta} - u_\theta + \frac{\partial u_r}{\partial r}\frac{\partial u_r}{\partial \theta} + \frac{\partial u_\theta}{\partial r}\frac{\partial u_\theta}{\partial \theta} + \frac{\partial u_z}{\partial r}\frac{\partial u_z}{\partial \theta} + \frac{u_r \partial u_\theta}{\partial r} - \frac{u_\theta \partial u_r}{\partial r}\right)\right] \tag{2.77}$$

$$\mathcal{E}_{rz} = \frac{1}{2}\left[\frac{\partial u_r}{\partial z} + \frac{\partial u_z}{\partial r} + \frac{\partial u_r}{\partial r}\frac{\partial u_r}{\partial z} + \frac{\partial u_\theta}{\partial r}\frac{\partial u_\theta}{\partial z} + \frac{\partial u_z}{\partial r}\frac{\partial u_z}{\partial z}\right] \tag{2.78}$$

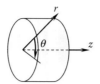

FIGURE 2.7 Cylindrical coordinate system.

$$\mathcal{E}_{\theta\theta} = \frac{1}{r}\left(u_r + \frac{\partial u_\theta}{\partial \theta}\right) + \frac{1}{2r^2}\left[\left(\frac{\partial u_r}{\partial \theta}\right)^2 + \left(\frac{\partial u_\theta}{\partial \theta}\right)^2 + \left(\frac{\partial u_z}{\partial \theta}\right)^2 \right.$$
$$\left. + (u_r)^2 + (u_\theta)^2 + 2\left(u_r \frac{\partial u_\theta}{\partial \theta} - u_\theta \frac{\partial u_r}{\partial \theta}\right)\right] \quad (2.79)$$

$$\mathcal{E}_{\theta z} = \frac{\gamma_{\theta z}}{2} = \frac{1}{2}\left[\frac{\partial u_\theta}{\partial z} + \frac{1}{r}\left(\frac{\partial u_z}{\partial \theta} + \frac{\partial u_r}{\partial \theta}\frac{\partial u_r}{\partial z} + \frac{\partial u_\theta}{\partial \theta}\frac{\partial u_\theta}{\partial z} + \frac{\partial u_z}{\partial \theta}\frac{\partial u_z}{\partial z}\right) - u_\theta \frac{\partial u_r}{\partial z} + u_r \frac{\partial u_\theta}{\partial z}\right] \quad (2.80)$$

$$\mathcal{E}_{zz} = \frac{\gamma_{zz}}{2} = \frac{\partial u_z}{\partial z} + \frac{1}{2}\left[\left(\frac{\partial u_r}{\partial z}\right)^2 + \left(\frac{\partial u_\theta}{\partial z}\right)^2 + \left(\frac{\partial u_z}{\partial z}\right)^2\right] \quad (2.81)$$

Infinitesimal strain–displacement relations in cylindrical coordinates are as follows:

$$\varepsilon_{rr} = \frac{\partial u_r}{\partial r} \quad (2.82)$$

$$\varepsilon_{r\theta} = \frac{1}{2}\left[\frac{\partial u_\theta}{\partial r} + \frac{1}{r}\left(\frac{\partial u_r}{\partial \theta} - u_\theta + \frac{u_r \partial u_\theta}{\partial r} - \frac{u_\theta \partial u_r}{\partial r}\right)\right] \quad (2.83)$$

$$\varepsilon_{rz} = \frac{1}{2}\left[\frac{\partial u_r}{\partial z} + \frac{\partial u_z}{\partial r}\right] \quad (2.84)$$

$$\varepsilon_{\theta\theta} = \frac{1}{r}\left(u_r + \frac{\partial u_\theta}{\partial \theta}\right) + \frac{1}{2r^2}\left[(u_r)^2 + (u_\theta)^2 + 2\left(u_r \frac{\partial u_\theta}{\partial \theta} - u_\theta \frac{\partial u_r}{\partial \theta}\right)\right] \quad (2.85)$$

$$\varepsilon_{\theta z} = \frac{\gamma_{\theta z}}{2} = \frac{1}{2}\left[\frac{\partial u_\theta}{\partial z} + \frac{1}{r}\left(\frac{\partial u_z}{\partial \theta} - u_\theta \frac{\partial u_r}{\partial z} + u_r \frac{\partial u_\theta}{\partial z}\right)\right] \quad (2.86)$$

$$\varepsilon_{zz} = \frac{\gamma_{zz}}{2} = \frac{\partial u_z}{\partial z} \quad (2.87)$$

2.4.9 Transformation of Strain Tensor

Transformation of strain tensor is similar to that of stress tensor. We shall discuss transformation of stress tensor in Section 2.5.4. Here, we merely present the strain transformation equations. Then, w.r.t. the Cartesian coordinate systems as shown in Figure 2.11, the small strain tensor transformation is given by

$$[\varepsilon]_{(x',y',z')} = [\alpha][\varepsilon]_{(x,y,z)}[\alpha]^T \quad (2.88)$$

In the explicit component form, the strain transformation is given by

$$\begin{bmatrix} \varepsilon_{x'x'} & \varepsilon_{x'y'} & \varepsilon_{z'x'} \\ \varepsilon_{x'y'} & \varepsilon_{y'y'} & \varepsilon_{y'z'} \\ \varepsilon_{z'x'} & \varepsilon_{y'z'} & \varepsilon_{z'z'} \end{bmatrix} = \begin{bmatrix} a_{x'x} & a_{x'y} & a_{x'z} \\ a_{y'x} & a_{y'y} & a_{y'z} \\ a_{z'x} & a_{z'y} & a_{z'z} \end{bmatrix} \begin{bmatrix} \varepsilon_{xx} & \varepsilon_{xy} & \varepsilon_{zx} \\ \varepsilon_{xy} & \varepsilon_{yy} & \varepsilon_{yz} \\ \varepsilon_{zx} & \varepsilon_{yz} & \varepsilon_{zz} \end{bmatrix} \begin{bmatrix} a_{x'x} & a_{y'x} & a_{z'x} \\ a_{x'y} & a_{y'y} & a_{z'y} \\ a_{x'z} & a_{y'z} & a_{z'z} \end{bmatrix} \quad (2.89)$$

Note: The finite strain transformation is the same as small strain transformation.

2.4.10 Compatibility Conditions

As we had mentioned earlier, in kinematics, we are concerned about the initial and the final configurations of a body. We had related an arbitrary line segment in the initial configuration to the line segment in the final configuration and arrived at strain–displacement equations. No constraint was put regarding the configuration that the body can assume. Physically, in solid mechanics, deformation of a body does not produce any void or gap. Also, deformation cannot result in a configuration, in which a single spatial point is occupied by more than one material points, that is, one portion of the body cannot penetrate into another. Compatibility conditions are the equations, which ensure that these physical requirements are met.

We know that there are six strain–displacement equations and only three displacement components. Thus, given the components of the strain tensor, if we have to find the displacement components, we face a problem which is overdeterminate. Compatibility equations, also known as St. Venant's compatibility equations, ensure that a unique displacement field is obtained from a given strain field. The compatibility equations can be derived by differentiating the strain–displacement equations [7–9]. There are six compatibility equations. However, it can be proved that they are not independent, and they can be reduced to only three. Here, we present the compatibility equations for small strains. In the explicit component form, these equations are as follows:

$$\frac{\partial^2 \varepsilon_{xx}}{\partial y^2} + \frac{\partial^2 \varepsilon_{yy}}{\partial x^2} = \frac{\partial^2 \gamma_{xy}}{\partial x \partial y} \quad (2.90)$$

$$\frac{\partial^2 \varepsilon_{yy}}{\partial z^2} + \frac{\partial^2 \varepsilon_{zz}}{\partial y^2} = \frac{\partial^2 \gamma_{yz}}{\partial y \partial z} \quad (2.91)$$

$$\frac{\partial^2 \varepsilon_{zz}}{\partial x^2} + \frac{\partial^2 \varepsilon_{xx}}{\partial z^2} = \frac{\partial^2 \gamma_{zx}}{\partial z \partial x} \quad (2.92)$$

$$2\frac{\partial^2 \varepsilon_{xx}}{\partial y \partial z} + \frac{\partial^2 \gamma_{yz}}{\partial x^2} = \frac{\partial^2 \gamma_{zx}}{\partial x \partial y} + \frac{\partial^2 \gamma_{xy}}{\partial z \partial x} \quad (2.93)$$

$$2\frac{\partial^2 \varepsilon_{yy}}{\partial z \partial x} + \frac{\partial^2 \gamma_{zx}}{\partial y^2} = \frac{\partial^2 \gamma_{xy}}{\partial y \partial z} + \frac{\partial^2 \gamma_{yz}}{\partial x \partial y} \quad (2.94)$$

$$2\frac{\partial^2 \varepsilon_{zz}}{\partial x \partial y} + \frac{\partial^2 \gamma_{xy}}{\partial z^2} = \frac{\partial^2 \gamma_{yz}}{\partial z \partial x} + \frac{\partial^2 \gamma_{zx}}{\partial y \partial z} \quad (2.95)$$

Basic Solid Mechanics

2.5 KINETICS

2.5.1 Forces on a Body

Forces can be broadly divided into two types—body forces and surface forces.

Body forces are the results of characteristic properties of a body, and they act on all the points of the volume of the body. Examples of body forces include gravitational force, inertia, magnetic force, centrifugal force, etc. These forces are expressed as force per unit volume. They can also be expressed in terms of force per unit mass.

Surface forces are results of interaction between two bodies. These forces act on a portion or whole of the bounding surface of the volume of a structural element. From the point of view of mathematical convenience, surface forces can be considered to be acting on a surface, a line, or a point. Thus, these forces are expressed in units such as N, N/mm, N/mm^2, etc. Examples of surface forces are plenty; they include contact forces between bodies and nearly all our day-to-day experiences such as carrying a bag of grocery items, opening a door, pushing a car, and so on.

2.5.2 Cauchy's Stress Principle and Stress Vector

In a very simple way, stress is known as force per unit area. It has a magnitude and orientation. Thus, it is a vector. Let us consider a body under applied surface forces, F, and body forces, $\mathcal{B}$ (Figure 2.8). These forces are transmitted from one point in the body to another, and reacted at the restrained boundary by reaction forces, Ω, and the body is in static equilibrium. The transmission of forces within the body results in internal forces. Let us consider an arbitrary plane that divides the body into two halves. Each half is kept in equilibrium by the internal forces acting on the dividing plane and the surface forces and body forces acting on that portion of the body.

We intend to find the stress vector at a point O in the dividing plane. Let us consider a small area, ΔA, in the arbitrary dividing plane around the point, O. The internal forces acting on the area, ΔA, can be expressed as a resultant force, ΔF, and resultant moment, ΔM, acting at the point, O. As per Cauchy's principle, as $\Delta A \to 0$, the

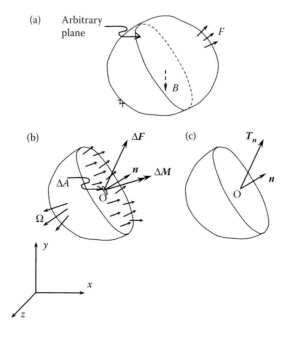

FIGURE 2.8 Cauchy's stress principle. (a) A solid body under applied surface forces and body forces. (b) Internal forces and resultant force and moment on a small area. (c) Stress vector at a point.

limiting value of the resultant moment per unit area vanishes, whereas, the limiting value of the resultant force per unit area has a finite value, and it is called the stress vector at that point. Mathematically,

$$\lim_{\Delta A \to 0} \frac{\Delta \boldsymbol{M}}{\Delta A} = 0 \qquad (2.96)$$

$$\lim_{\Delta A \to 0} \frac{\Delta \boldsymbol{F}}{\Delta A} = \boldsymbol{T}_n \qquad (2.97)$$

$\boldsymbol{T}_n$ in Equation 2.97 is called the stress vector or traction vector. The subscript n indicates that the stress vector $\boldsymbol{T}_n$ is associated with a plane whose unit outward normal vector at the point O is $\boldsymbol{n}$. The stress vector can be resolved into two components—one normal to the plane, called normal stress and the other along the plane, called shear stress. The shear stress can be further resolved into two components. Thus, the stress vector has one normal stress component and two shear stress components.

2.5.3 State of Stress at a Point and Stress Tensor

We, now, focus our attention to the state of stress at a point. Cauchy's stress principle gives us the stress vector at a point on a surface element represented by its unit normal vector. The state of stress at the point is given by all the possible combinations of stress vectors and associated unit outward normal vector. However, we do not need to consider all these pairs of stress vectors and the associated unit normal vectors. We, rather, consider three mutually orthogonal planes at the point and determine the stress vectors on these three planes. Stress transformation equations can then be applied to determine stress vectors on any other plane.

Let us consider a Cartesian coordinate system as shown in Figure 2.9. $\boldsymbol{e}_x$, $\boldsymbol{e}_y$, and $\boldsymbol{e}_z$ are the unit vectors along the respective axes. The stress vector at a point O on a plane normal to axis x is $\boldsymbol{T}_x$ and the plane is represented by its unit normal vector $\boldsymbol{e}_x$. We can consider two more planes through the point O with unit normal vectors $\boldsymbol{e}_y$ and $\boldsymbol{e}_z$ and associated stress vectors $\boldsymbol{T}_y$ and $\boldsymbol{T}_z$, respectively. These three stress vectors associated with the three unit vectors are sufficient to express the total state of stress at the point O. Mathematically,

$$\boldsymbol{T}_x = T_{xx}\boldsymbol{e}_x + T_{xy}\boldsymbol{e}_y + T_{xz}\boldsymbol{e}_z = \begin{Bmatrix} T_{xx} \\ T_{xy} \\ T_{xz} \end{Bmatrix}^T \begin{Bmatrix} \boldsymbol{e}_x \\ \boldsymbol{e}_y \\ \boldsymbol{e}_z \end{Bmatrix} \qquad (2.98)$$

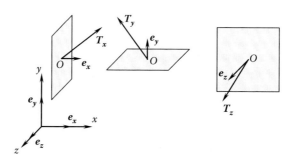

FIGURE 2.9 Stress vectors at a point on three mutually orthogonal planes.

Basic Solid Mechanics

$$\boldsymbol{T}_y = T_{yx}\boldsymbol{e}_x + T_{yy}\boldsymbol{e}_y + T_{yz}\boldsymbol{e}_z = \begin{Bmatrix} T_{yx} \\ T_{yy} \\ T_{yz} \end{Bmatrix}^T \begin{Bmatrix} \boldsymbol{e}_x \\ \boldsymbol{e}_y \\ \boldsymbol{e}_z \end{Bmatrix} \qquad (2.99)$$

$$\boldsymbol{T}_z = T_{zx}\boldsymbol{e}_x + T_{zy}\boldsymbol{e}_y + T_{zz}\boldsymbol{e}_z = \begin{Bmatrix} T_{zx} \\ T_{zy} \\ T_{zz} \end{Bmatrix}^T \begin{Bmatrix} \boldsymbol{e}_x \\ \boldsymbol{e}_y \\ \boldsymbol{e}_z \end{Bmatrix} \qquad (2.100)$$

Combining the three stress vectors,

$$\begin{Bmatrix} \boldsymbol{T}_x \\ \boldsymbol{T}_y \\ \boldsymbol{T}_z \end{Bmatrix} = \begin{bmatrix} T_{xx} & T_{xy} & T_{xz} \\ T_{yx} & T_{yy} & T_{yz} \\ T_{zx} & T_{zy} & T_{zz} \end{bmatrix} \begin{Bmatrix} \boldsymbol{e}_x \\ \boldsymbol{e}_y \\ \boldsymbol{e}_z \end{Bmatrix} \qquad (2.101)$$

The nine stress vector components in Equation 2.101 constitute a second-order Cartesian tensor, called stress tensor. These components are commonly expressed in the following way: σ_{xx} for T_{xx}, σ_{yy} for T_{yy}, σ_{xy} for T_{xy}, σ_{zy} for T_{zy}, and so on. Thus, the stress tensor is expressed as

$$[\boldsymbol{\sigma}] = \begin{bmatrix} \sigma_{xx} & \sigma_{xy} & \sigma_{xz} \\ \sigma_{yx} & \sigma_{yy} & \sigma_{yz} \\ \sigma_{zx} & \sigma_{zy} & \sigma_{zz} \end{bmatrix} \qquad (2.102)$$

The first letter in the subscript indicates the axis to which the concerned plane is normal, and the second letter indicates the direction of the stress component. σ_{xx}, σ_{yy}, and σ_{zz} are the normal stresses. Remaining six stress components are the shear stresses. For shear stresses, it is common to use the symbol τ. Also, it can be shown that the stress tensor is symmetric, that is, $\tau_{xy} = \tau_{yx}$, $\tau_{xz} = \tau_{zx}$, and $\tau_{yz} = \tau_{zy}$. Thus, the stress tensor becomes

$$[\boldsymbol{\sigma}] = \begin{bmatrix} \sigma_{xx} & \tau_{xy} & \tau_{zx} \\ \tau_{xy} & \sigma_{yy} & \tau_{yz} \\ \tau_{zx} & \tau_{yz} & \sigma_{zz} \end{bmatrix} \qquad (2.103)$$

The stress tensor components are conveniently expressed pictorially by considering an infinitesimal cube as shown in Figure 2.10. We intend to find the stress tensor at a point O and the cube is constructed such that the point is at its centroid and the sides of the cube are parallel to the axes of the Cartesian coordinates. On each of the six sides, one normal stress and two shear stress components act. The convention for denoting the stress components is as follows:

- σ's are normal stress and τ's are shear stress components.
- The first letter in the suffix stands for the plane and the second letter for the direction of the stress component.
- Normal stress is positive if it is in the outward direction (producing tension in the cube).
- Shear stress is positive is if it has the same sense as the corresponding normal stress. Thus, on a plane where the normal stress is positive and in the direction of the coordinate axis, positive shear stresses are also in the direction of the

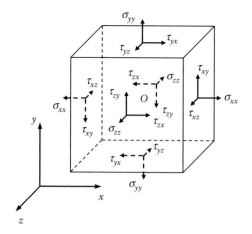

FIGURE 2.10 State of stress at a point.

corresponding coordinate axes. On the other hand, if a positive normal stress is in the opposite direction of the axis, positive shear stresses are also in the opposite directions to the coordinate axes.

Note: The stress behavior we have studied so far is on the deformed configuration of the body under loads. Stress tensor on the deformed configuration is called the Cauchy stress tensor.

2.5.4 Transformation of Stress Tensor

Let us consider two Cartesian coordinate systems O-xyz and O-$x'y'z'$ as shown in Figure 2.11. Our aim is to express the stress tensor in the O-$x'y'z'$ system in terms of the stress tensor in the O-xyz system. Direction cosines are used for stress transformation and these are: direction cosine of x' w.r.t. x is $a_{x'x} = \cos \alpha$, direction cosine of x' w.r.t. y is $a_{x'y} = \cos \beta$, direction cosine of x' w.r.t. z is $a_{x'z} = \cos \gamma$, and so on. Direction cosines are given in a tabular form in Table 2.2.

Let us consider a tetrahedron with three mutually orthogonal planes and one inclined plane as shown in Figure 2.12. The inclined plane is chosen in such a way that the axis x' is along its normal. The areas of the orthogonal triangles can be related to that of the inclined triangle in the following way:

$$\frac{\triangle OBC}{\triangle ABC} = a_{x'x} \qquad (2.104)$$

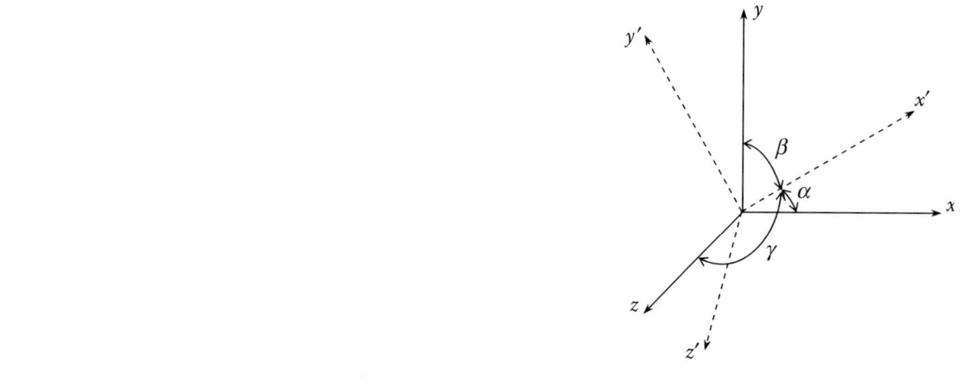

FIGURE 2.11 Cartesian coordinate systems for stress/strain transformation.

Basic Solid Mechanics

TABLE 2.2
Direction Cosines

Axes	x	y	z
x'	$a_{x'x}$	$a_{x'y}$	$a_{x'z}$
y'	$a_{y'x}$	$a_{y'y}$	$a_{y'z}$
z'	$a_{z'x}$	$a_{z'y}$	$a_{z'z}$

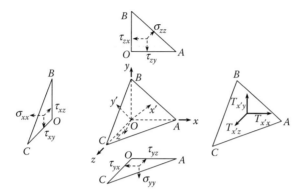

FIGURE 2.12 Stress components on the planes of an octahedron (exploded view).

$$\frac{\Delta OAC}{\Delta ABC} = a_{x'y} \qquad (2.105)$$

$$\frac{\Delta OAB}{\Delta ABC} = a_{x'z} \qquad (2.106)$$

Components of the stress tensor in the O-xyz system are shown on the three orthogonal planes. The resultant stress vector on the inclined plane is resolved into three components parallel to the x-, y-, and z-axes. Considering static equilibrium of forces acting on the tetrahedron, we get

$$T_{x'x}(\Delta ABC) = \sigma_{xx}(\Delta OBC) + \tau_{yx}(\Delta OAC) + \tau_{zx}(\Delta OAB) \qquad (2.107)$$

$$T_{x'y}(\Delta ABC) = \tau_{xy}(\Delta OBC) + \sigma_{yy}(\Delta OAC) + \tau_{zy}(\Delta OAB) \qquad (2.108)$$

$$T_{x'z}(\Delta ABC) = \tau_{xz}(\Delta OBC) + \tau_{yz}(\Delta OAC) + \sigma_{zz}(\Delta OAB) \qquad (2.109)$$

Dividing both the sides with the area of the inclined triangle and using Equations 2.104 through 2.106 and noting that $\tau_{xy} = \tau_{yx}$, $\tau_{yz} = \tau_{zy}$, and $\tau_{zx} = \tau_{xz}$, we get

$$T_{x'x} = \sigma_{xx}a_{x'x} + \tau_{xy}a_{x'y} + \tau_{zx}a_{x'z} \qquad (2.110)$$

$$T_{x'y} = \tau_{xy}a_{x'x} + \sigma_{yy}a_{x'y} + \tau_{yz}a_{x'z} \qquad (2.111)$$

$$T_{x'z} = \tau_{zx}a_{x'x} + \tau_{yz}a_{x'y} + \sigma_{zz}a_{x'z} \qquad (2.112)$$

In the matrix form, we can write

$$\begin{Bmatrix} T_{x'x} \\ T_{x'y} \\ T_{x'z} \end{Bmatrix} = \begin{bmatrix} \sigma_{xx} & \tau_{xy} & \tau_{zx} \\ \tau_{xy} & \sigma_{yy} & \tau_{yz} \\ \tau_{zx} & \tau_{yz} & \sigma_{zz} \end{bmatrix} \begin{Bmatrix} a_{x'x} \\ a_{x'y} \\ a_{x'z} \end{Bmatrix} \qquad (2.113)$$

The stress resultant on the inclined plane can also be resolved into a normal stress component $\sigma_{x'x'}$ and two shear stress components $\tau_{x'y'}$ and $\tau_{z'x'}$. We can obtain the components of $T_{x'x}$, $T_{x'y}$, and $T_{x'z}$ in the x'-, y'-, and z'-directions by multiplying them with the respective direction cosines. Thus,

$$\sigma_{x'x'} = T_{x'x} a_{x'x} + T_{x'y} a_{x'y} + T_{x'z} a_{x'z} \qquad (2.114)$$

$$\tau_{x'y'} = T_{x'x} a_{y'x} + T_{x'y} a_{y'y} + T_{x'z} a_{y'z} \qquad (2.115)$$

$$\tau_{z'x'} = T_{x'x} a_{z'x} + T_{x'y} a_{z'y} + T_{x'z} a_{z'z} \qquad (2.116)$$

In the matrix form,

$$\begin{Bmatrix} \sigma_{x'x'} \\ \tau_{x'y'} \\ \tau_{z'x'} \end{Bmatrix} = \begin{bmatrix} a_{x'x} & a_{x'y} & a_{x'z} \\ a_{y'x} & a_{y'y} & a_{y'z} \\ a_{z'x} & a_{z'y} & a_{z'z} \end{bmatrix} \begin{Bmatrix} T_{x'x} \\ T_{x'y} \\ T_{x'z} \end{Bmatrix} \qquad (2.117)$$

Combining Equations 2.113 and 2.117, we get the following:

$$\begin{Bmatrix} \sigma_{x'x'} \\ \tau_{x'y'} \\ \tau_{z'x'} \end{Bmatrix} = \begin{bmatrix} a_{x'x} & a_{x'y} & a_{x'z} \\ a_{y'x} & a_{y'y} & a_{y'z} \\ a_{z'x} & a_{z'y} & a_{z'z} \end{bmatrix} \begin{bmatrix} \sigma_{xx} & \tau_{xy} & \tau_{zx} \\ \tau_{xy} & \sigma_{yy} & \tau_{yz} \\ \tau_{zx} & \tau_{yz} & \sigma_{zz} \end{bmatrix} \begin{Bmatrix} a_{x'x} \\ a_{x'y} \\ a_{x'z} \end{Bmatrix} \qquad (2.118)$$

We have considered a tetrahedron with an inclined plane whose normal is along x'. Now, we consider two more tetrahedrons with normals along y' and z'. Following a similar procedure, we get

$$\begin{Bmatrix} \tau_{x'y'} \\ \sigma_{y'y'} \\ \tau_{y'z'} \end{Bmatrix} = \begin{bmatrix} a_{x'x} & a_{x'y} & a_{x'z} \\ a_{y'x} & a_{y'y} & a_{y'z} \\ a_{z'x} & a_{z'y} & a_{z'z} \end{bmatrix} \begin{bmatrix} \sigma_{xx} & \tau_{xy} & \tau_{zx} \\ \tau_{xy} & \sigma_{yy} & \tau_{yz} \\ \tau_{zx} & \tau_{yz} & \sigma_{zz} \end{bmatrix} \begin{Bmatrix} a_{y'x} \\ a_{y'y} \\ a_{y'z} \end{Bmatrix} \qquad (2.119)$$

$$\begin{Bmatrix} \tau_{z'x'} \\ \tau_{y'z'} \\ \sigma_{z'z'} \end{Bmatrix} = \begin{bmatrix} a_{x'x} & a_{x'y} & a_{x'z} \\ a_{y'x} & a_{y'y} & a_{y'z} \\ a_{z'x} & a_{z'y} & a_{z'z} \end{bmatrix} \begin{bmatrix} \sigma_{xx} & \tau_{xy} & \tau_{zx} \\ \tau_{xy} & \sigma_{yy} & \tau_{yz} \\ \tau_{zx} & \tau_{yz} & \sigma_{zz} \end{bmatrix} \begin{Bmatrix} a_{z'x} \\ a_{z'y} \\ a_{z'z} \end{Bmatrix} \qquad (2.120)$$

Basic Solid Mechanics

Now, combining the three Equations 2.118 through 2.120, we get the following:

$$\begin{bmatrix} \sigma_{x'x'} & \tau_{x'y'} & \tau_{z'x'} \\ \tau_{x'y'} & \sigma_{y'y'} & \tau_{y'z'} \\ \tau_{z'x'} & \sigma_{y'z'} & \sigma_{z'z'} \end{bmatrix}$$

$$= \begin{bmatrix} a_{x'x} & a_{x'y} & a_{x'z} \\ a_{y'x} & a_{y'y} & a_{y'z} \\ a_{z'x} & a_{z'y} & a_{z'z} \end{bmatrix} \begin{bmatrix} \sigma_{xx} & \tau_{xy} & \tau_{zx} \\ \tau_{xy} & \sigma_{yy} & \tau_{yz} \\ \tau_{zx} & \tau_{yz} & \sigma_{zz} \end{bmatrix} \begin{bmatrix} a_{x'x} & a_{y'x} & a_{z'x} \\ a_{x'y} & a_{y'y} & a_{z'y} \\ a_{x'z} & a_{y'z} & a_{z'z} \end{bmatrix} \quad (2.121)$$

Equation 2.121 can be written as

$$[\sigma]_{(x',y',z')} = [\alpha][\sigma]_{(x,y,z)}[\alpha]^T \quad (2.122)$$

where

$$[\sigma]_{(x',y',z')} \equiv \begin{bmatrix} \sigma_{x'x'} & \tau_{x'y'} & \tau_{z'x'} \\ \tau_{x'y'} & \sigma_{y'y'} & \tau_{y'z'} \\ \tau_{z'x'} & \tau_{y'z'} & \sigma_{z'z'} \end{bmatrix}$$

is the stress tensor in the $O\text{-}x'y'z'$ coordinate system

$$[\sigma]_{(x,y,z)} \equiv \begin{bmatrix} \sigma_{xx} & \tau_{xy} & \tau_{zx} \\ \tau_{xy} & \sigma_{yy} & \tau_{yz} \\ \tau_{zx} & \tau_{yz} & \sigma_{zz} \end{bmatrix}$$

is the stress tensor in the $O\text{-}xyz$ coordinate system

$$[\alpha] \equiv \begin{bmatrix} a_{x'x} & a_{x'y} & a_{x'z} \\ a_{y'x} & a_{y'y} & a_{y'z} \\ a_{z'x} & a_{z'y} & a_{z'z} \end{bmatrix}$$

is the transformation matrix of direction cosines.

2.5.5 Stress Tensor–Stress Vector Relationship

In Equation 2.113, the vector of direction cosines is also the vector of the unit normal to the inclined plane. Thus, the stress vector at a point on a surface is related to the stress tensor at that point as follows:

In the component form,

$$\{T\} = [\sigma]\{n\} \quad (2.123)$$

In the vector form,

$$T_n = \sigma.n \quad (2.124)$$

In the indicial notation,

$$T_i = \sigma_{ji} n_j \qquad (2.125)$$

where

$$\{T\} = \begin{Bmatrix} T_{nx} \\ T_{ny} \\ T_{nz} \end{Bmatrix}$$

is the stress vector at a point on a plane whose unit normal is $\boldsymbol{n}$

$$[\sigma] = \begin{bmatrix} \sigma_{xx} & \tau_{xy} & \tau_{zx} \\ \tau_{xy} & \sigma_{yy} & \tau_{yz} \\ \tau_{zx} & \tau_{yz} & \sigma_{zz} \end{bmatrix}$$

is the stress tensor at the point

$$\{n\} = \begin{Bmatrix} n_x \\ n_y \\ n_z \end{Bmatrix}$$

is the unit normal vector with components (n_x, n_y, n_z) such that $n_x^2 + n_y^2 + n_z^2 = 1$
Note that $[\sigma]$ is symmetric; thus, $[\sigma] = [\sigma]^T$.

2.5.6 Principal Stresses

The stress tensor at a point gives the state of stress at that point w.r.t. a set of three mutually orthogonal planes. Each of these planes is associated with a stress vector that has one normal stress and two shear stress components. Theoretically, innumerable planes and the corresponding stress vectors can be thought of at a point; however, from design and analysis point, we are more concerned about the maximum normal and shear stresses and the associated planes at that point.

The normal stress is the maximum when the stress vector $\boldsymbol{T}_n$ is parallel to the unit normal vector $\boldsymbol{n}$. Let λ be the magnitude of the stress vector. Then, for a stress vector, which is parallel to the unit normal vector, we can write

$$\boldsymbol{T}_n = \lambda \boldsymbol{n} = \lambda \boldsymbol{I}.\boldsymbol{n} \qquad (2.126)$$

$\boldsymbol{I}$ being an unit tensor
We know from Equation 2.124, that the stress vector is related to the stress tensor as

$$\boldsymbol{T}_n = \boldsymbol{\sigma}.\boldsymbol{n} \qquad (2.127)$$

Thus, from the above two equations, we get the following:
In the component form,

$$([\sigma] - \lambda[\boldsymbol{I}])\{n\} = 0 \qquad (2.128)$$

Basic Solid Mechanics

or

$$\left(\begin{bmatrix} \sigma_{xx} & \tau_{xy} & \tau_{zx} \\ \tau_{xy} & \sigma_{yy} & \tau_{yz} \\ \tau_{zx} & \tau_{yz} & \sigma_{zz} \end{bmatrix} - \lambda \begin{bmatrix} 1 & 0 & 0 \\ 0 & 1 & 0 \\ 0 & 0 & 1 \end{bmatrix}\right) \begin{Bmatrix} n_x \\ n_y \\ n_z \end{Bmatrix} = 0 \qquad (2.129)$$

or

$$\begin{bmatrix} \sigma_{xx} - \lambda & \tau_{xy} & \tau_{zx} \\ \tau_{xy} & \sigma_{yy} - \lambda & \tau_{yz} \\ \tau_{zx} & \tau_{yz} & \sigma_{zz} - \lambda \end{bmatrix} \begin{Bmatrix} n_x \\ n_y \\ n_z \end{Bmatrix} = 0 \qquad (2.130)$$

In the vector form,

$$(\sigma - \lambda I).n = 0 \qquad (2.131)$$

and, in the indicial notation,

$$(\sigma_{ij} - \lambda \delta_{ij})n_i = 0 \qquad (2.132)$$

Equations 2.130 through 2.132 are an eigenvalue problem. The solution of this problem is obtained by equating the determinant of the square matrix to zero, that is,

$$\begin{vmatrix} \sigma_{xx} - \lambda & \tau_{xy} & \tau_{zx} \\ \tau_{xy} & \sigma_{yy} - \lambda & \tau_{yz} \\ \tau_{zx} & \tau_{yz} & \sigma_{zz} - \lambda \end{vmatrix} = 0 \qquad (2.133)$$

Equation 2.133 is a cubic equation for λ (called the characteristic equation), solving which we get three eigenvalues λ. These eigenvalues are the principal stresses and the associated eigenvector, that is, the unit normal vector associated with each principal stress represents the corresponding principal plane.

EXAMPLE 2.1

Let the stress tensor at a point be given by

$$[\sigma] = \begin{bmatrix} 10 & 4 & 0 \\ 4 & 4 & 0 \\ 0 & 0 & 4 \end{bmatrix} \text{MPa}$$

Find the principal stresses and the principal planes.

Solution

Corresponding to the principal stresses and principal planes,

$$|\sigma - \lambda I| = 0$$

or

$$\begin{vmatrix} 10-\lambda & 4 & 0 \\ 4 & 4-\lambda & 0 \\ 0 & 0 & 4-\lambda \end{vmatrix} = 0$$

On solving, we get

$$\lambda = 2, 4, 12$$

Thus, the principal stresses are

$$\sigma_1 = 2 \text{ MPa}$$

$$\sigma_2 = 4 \text{ MPa}$$

$$\sigma_3 = 12 \text{ MPa}$$

Now, let us find the principal planes. Note that each eigenvalue has got an associated eigenvector.

First, for $\sigma_1 = 2$,

$$\begin{bmatrix} 10-2 & 4 & 0 \\ 4 & 4-2 & 0 \\ 0 & 0 & 4-2 \end{bmatrix} \begin{Bmatrix} n_x \\ n_y \\ n_z \end{Bmatrix} = 0$$

which gives us

$$2n_x + n_y = 0$$

$$n_z = 0$$

Also,

$$n_x^2 + n_y^2 + n_z^2 = 1$$

Thus, on solving, we get

$$n_x = \frac{1}{\sqrt{5}}$$

$$n_y = -\frac{2}{\sqrt{5}}$$

$$n_z = 0$$

which means

$$\boldsymbol{n}_1 = \frac{1}{\sqrt{5}}(e_x - 2e_y)$$

Second, for $\sigma_2 = 4$,
Following a similar procedure, we get

$$n_x = 0$$

$$n_y = 0$$

$$n_z = 1$$

Basic Solid Mechanics

that is,

$$n_2 = e_z$$

Third, for $\sigma_3 = 12$,
Here, we get

$$n_x = \frac{2}{\sqrt{5}}$$
$$n_y = \frac{1}{\sqrt{5}}$$
$$n_z = 0$$

that is,

$$n_3 = \frac{1}{\sqrt{5}}(2e_x + e_y)$$

Note: Let us consider the following cross-product $n_1 \times (n_2 \times n_3)$, which is equal to zero. We can consider any other similar possible combination. It shows that the three unit vectors are mutually orthogonal.

2.5.7 Equilibrium Equations

Let us consider an infinitesimal cuboid with point O at its centroid as shown in Figure 2.13. By Newton's second law of motion, the sum of all forces in any direction on a body in dynamic equilibrium is equal to the mass of the body multiplied by its acceleration in the same direction. Now, the sum of all forces in the x-direction results in

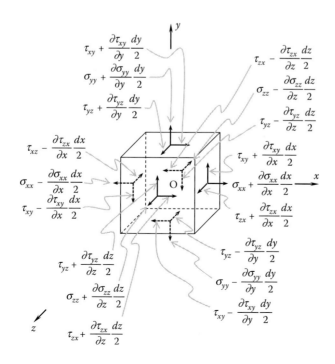

FIGURE 2.13 Infinitesimal cuboid in equilibrium.

$$\left(\sigma_{xx} + \frac{\partial \sigma_{xx}}{\partial x}\frac{dx}{2}\right)dy\,dz - \left(\sigma_{xx} - \frac{\partial \sigma_{xx}}{\partial x}\frac{dx}{2}\right)dy\,dz$$
$$+ \left(\tau_{xy} + \frac{\partial \tau_{xy}}{\partial y}\frac{dy}{2}\right)dz\,dx - \left(\tau_{xy} - \frac{\partial \tau_{xy}}{\partial y}\frac{dy}{2}\right)dz\,dx$$
$$+ \left(\tau_{zx} + \frac{\partial \tau_{zx}}{\partial z}\frac{dz}{2}\right)dx\,dy - \left(\tau_{zx} - \frac{\partial \tau_{zx}}{\partial z}\frac{dz}{2}\right)dx\,dy$$
$$+ \mathcal{B}_x\,dx\,dy\,dz = \rho a_x\,dx\,dy\,dz \tag{2.134}$$

Upon simplifying and generalizing for all the three directions, we get the following:
In the component form,

$$\frac{\partial \sigma_{xx}}{\partial x} + \frac{\partial \tau_{xy}}{\partial y} + \frac{\partial \tau_{zx}}{\partial z} + \mathcal{B}_x = \rho a_x$$
$$\frac{\partial \sigma_{yy}}{\partial y} + \frac{\partial \tau_{yz}}{\partial z} + \frac{\partial \tau_{xy}}{\partial x} + \mathcal{B}_y = \rho a_y \tag{2.135}$$
$$\frac{\partial \sigma_{zz}}{\partial z} + \frac{\partial \tau_{zx}}{\partial x} + \frac{\partial \tau_{yz}}{\partial y} + \mathcal{B}_z = \rho a_z$$

In the vector form,

$$\nabla \cdot \boldsymbol{\sigma} + \boldsymbol{\mathcal{B}} = \rho \boldsymbol{a} \tag{2.136}$$

In the indicial notation,

$$\frac{\partial \sigma_{ji}}{\partial x_j} + \mathcal{B}_i = \rho a_i \tag{2.137}$$

Equations 2.135 through 2.137 are the equations of motion. Note that for static equilibrium, $\boldsymbol{a} = 0$.

2.6 THERMODYNAMICS

Strain tensor and stress tensor in kinematics and kinetics, respectively, have been discussed. In the next section, we shall relate these variables by means of constitutive relations. Constitutive relations, however, are not independent, and thermodynamic principles put constraints on them. In this section, without going into the details of mathematical derivation, we shall briefly state these thermodynamic principles.

Let us consider a body as shown in Figure 2.14. Let us apply surface traction and heat on the body. Then, as per the *first law of thermodynamics*, also known as the *principle of conservation of energy*, which states that the sum of the work done by external forces and the heat input to a body per unit time is equal to the change in stored energy, which we can write as

$$W + Q = K + U \tag{2.138}$$

where
  W: Total work done by the surface traction and body forces
  Q: Heat input to the body during loading

Basic Solid Mechanics

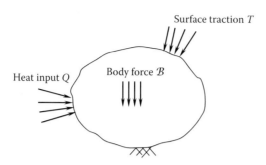

FIGURE 2.14 Body under thermo-mechanical loads.

K: Change in kinetic energy during the same period
U: Change in internal energy during the same period

If the forces are applied in a quasi-static manner, $K = 0$. Also, for an adiabatic process, no heat transfer takes place between the body and the surrounding, and $Q = 0$. Thus, for quasi-static loading under adiabatic conditions,

$$W = U \tag{2.139}$$

The sum of work done by surface traction (W_t) and work done by body forces (W_b) during the process of deformation is given by

$$W = W_t + W_b = \int\int_S \int_u T_i du_i \, dS + \int\int_V \int_u \mathcal{B}_i du_i \, dV \tag{2.140}$$

By Cauchy's stress formula, $T_i = \sigma_{ji} n_j$. Substituting in the above equation, we get

$$W = \int\int_S \int_u \sigma_{ji} n_j du_i \, dS + \int\int_V \int_u \mathcal{B}_i du_i \, dV \tag{2.141}$$

The first term in the above equation is a surface integral and, by employing Gauss divergence theorem, it can be converted into a volume integral. Then,

$$W = \int\int_V \int_u \frac{\partial(\sigma_{ji} du_i)}{\partial x_j} dV + \int\int_V \int_u \mathcal{B}_i du_i \, dV \tag{2.142}$$

Expanding and rearranging the terms, we get

$$W = \int\int_V \int_u \left(\frac{\partial \sigma_{ji}}{\partial x_j} + \mathcal{B}_i\right) du_i \, dV + \int\int_V \int_u \sigma_{ji} \frac{\partial du_i}{\partial x_j} dV \tag{2.143}$$

From equation of motion (Equation 2.137), the first term is zero for a quasi-static loading ($a_i = 0$). Thus,

$$W = \int\int_V \int_u \sigma_{ji} \frac{\partial du_i}{\partial x_j} dV \tag{2.144}$$

Stress tensor is symmetric, that is, $\sigma_{ij} = \sigma_{ji}$. Then, using strain–displacement relation, it can be shown that

$$\sigma_{ji}\frac{\partial du_i}{\partial x_j} = \frac{1}{2}\sigma_{ij}\left(\frac{\partial du_i}{\partial x_j} + \frac{\partial du_j}{\partial x_i}\right) = \sigma_{ij}d\varepsilon_{ij} \qquad (2.145)$$

Then, substituting Equation 2.145 in Equation 2.144 and combining with Equation 2.139, we obtain

$$U = \int_V \int_\varepsilon \sigma_{ij}d\varepsilon_{ij}\, dV \qquad (2.146)$$

Equation 2.146 gives the expression for the total internal energy of a body for adiabatic process under quasi-static loading. The internal energy is the strain energy and we can define a scalar function called *strain energy density function* U_0, such that

$$dU_0 = \sigma_{ij}d\varepsilon_{ij} \qquad (2.147)$$

Thus, we can express the stress tensor as

$$\sigma_{ij} = \frac{\partial U_0}{\partial \varepsilon_{ij}} \qquad (2.148)$$

We have arrived at the concept of strain energy density function from thermodynamic principles and its existence implies that the energy stored is recoverable, the deformation is reversible, and the body is elastic. So far, we have considered the first law of thermodynamics. Now, we can apply the second law of thermodynamics and, from considerations of entropy, it can finally be shown that the strain energy density function is positive. It puts restriction on constitutive modeling.

2.7 CONSTITUTIVE MODELING

Let us consider a body under the action of surface traction and body force. Let displacements be given on part of the boundary. A typical solid mechanics problem is to find the following unknowns: displacement vector, strain tensor, and the stress tensor. Thus, in three dimensions, we have a total of 15 unknowns (three displacement components, six strain components, and six stress components). We have nine equations from kinematics and kinetics—six strain–displacement relations and three equilibrium equations. Thus, we need six more equations for a complete solution of our problem. An insight into the problem tells us that these equations have to come from the relations between stress and strain. Our objective in this section is to find these six equations of stress–strain relations.

Constitutive modeling is mathematical modeling on the response of material to external loads. Constitutive equations relate primary field variables with secondary field variables. In the present context, stress–strain relations (or force–displacement relations) are derived from constitutive modeling. Mathematical modeling is based on assumptions regarding different aspects of the subject. Constitutive modeling is about material and the assumptions made are based on experimental observations on material behavior.

2.7.1 Idealization of Materials

There are many parameters, which are studied by experiments on materials. In this section, we are concerned with stress–strain behavior of materials, and the common tests conducted are the uniaxial tension test, torsion test, and triaxial test. The results of these tests are typically expressed as stress–strain curves where stress and strain are plotted along *y*-axis (ordinate) and *x*-axis (abscissa), respectively. These curves vary widely for different materials. Even for the same material, under different loading environments such as different rates of loading, temperature, etc. these curves vary. Thus, for simplicity and as an aid to design of structures, idealized materials are constructed and used in mathematical modeling. Some of the idealized materials are as follows:

- Linear elastic material
- Nonlinear elastic material
- Linear elastic perfectly plastic material
- Rigid material
- Rigid perfectly plastic material, etc.

We shall restrict our discussions to elastic materials.

2.7.2 Elastic Materials

Elastic materials are those that regain their original shape and size once the applied loads are removed. For these materials, the constitutive behavior depends only on the current state of deformation. Many materials such as metals exhibit linear relationship between load and deformation below the yield point. Stress–strain relation of these materials below the yield point can be idealized as linear elastic (Figure 2.15a). On the other hand, materials such as rubber exhibit nonlinear behavior between load and deformation and their stress–strain relationship can be idealized as nonlinear elastic (Figure 2.15b). In both these cases, the material regains its original shape and size once the applied loads, which caused deformations, are removed. Following points should be noted:

- Elastic deformation is instantaneous and time independent. Upon loading, an elastic body deforms instantaneously without any time gap between loading and deformation. Thus, elastic deformation is time-independent and time is not a parameter in the constitutive modeling of elastic material.
- Elastic deformation does not involve loss of energy. Upon loading, an elastic body deforms and the work done is stored in the body as strain energy. Once the load is removed, the strain energy is fully recovered.
- Elastic deformation is reversible. Upon loading, an elastic body deforms. Once the load is removed, the body regains its original configuration. There is no permanent deformation.

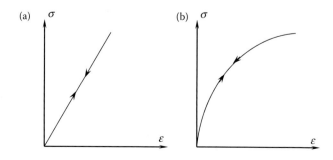

FIGURE 2.15 Idealized stress–strain curves for (a) linear elastic and (b) nonlinear elastic materials.

- Elastic deformation is such that there is one-to-one relation between state of stress and state of strain in the current configuration.

2.7.3 Generalized Hooke's Law

We stated above that for an elastic solid, there is one-to-one relation between state of stress and state of strain in the current configuration. It is possible to relate the stress tensor σ_{ij} to the strain tensor ε_{kl} by a one-to-one function f_{ij} as follows:

$$\sigma_{ij} = f_{ij}(\varepsilon_{kl}) \tag{2.149}$$

Such materials are called *Cauchy elastic material*. These materials are not based on thermodynamic principles, and it can be shown that reversibility of energy is not satisfied by these materials.

Reversibility of energy is ensured by assuming the existence of the strain energy density function U_0. Such materials are called *hyperelastic materials* or *Green elastic materials*. U_0 can be expanded in the Taylor's series about $\varepsilon = 0$. For linear elastic materials, cubic and higher order terms in the Taylor's series expansion are neglected and a quadratic form of U_0 is obtained. By partial differentiation of this quadratic form w.r.t. ε_{ij}, for a linear elastic body with zero stress prior to load application, we can relate stress to strain as

$$\sigma_{ij} = C_{ijkl}\varepsilon_{kl} \tag{2.150}$$

Equation 2.150 gives the most generalized relation between stress tensor and strain tensor for a linear elastic material and it is known as the *generalized Hooke's law*. C_{ijkl} are the components of a fourth-order tensor $\boldsymbol{C}$ with $81(=3^4)$ elastic constants.

σ_{ij} and ε_{ij} are second-order tensors; however, it is convenient to adopt an alternate notation, in which, we write them as 9×1 vectors as follows:

$$\sigma_{ij} = \begin{Bmatrix} \sigma_{xx} \\ \sigma_{yy} \\ \sigma_{zz} \\ \tau_{yz} \\ \tau_{zx} \\ \tau_{xy} \\ \tau_{zy} \\ \tau_{xz} \\ \tau_{yx} \end{Bmatrix} \text{ and } \varepsilon_{ij} = \begin{Bmatrix} \varepsilon_{xx} \\ \varepsilon_{yy} \\ \varepsilon_{zz} \\ \varepsilon_{yz} \\ \varepsilon_{zx} \\ \varepsilon_{xy} \\ \varepsilon_{zy} \\ \varepsilon_{xz} \\ \varepsilon_{yx} \end{Bmatrix} = \begin{Bmatrix} \varepsilon_{xx} \\ \varepsilon_{yy} \\ \varepsilon_{zz} \\ \gamma_{yz}/2 \\ \gamma_{zx}/2 \\ \gamma_{xy}/2 \\ \gamma_{zy}/2 \\ \gamma_{xz}/2 \\ \gamma_{yx}/2 \end{Bmatrix} \tag{2.151}$$

On the other hand, C_{ijkl} is a fourth-order tensor and, as per the alternate notation, we write it as a 9×9 matrix. In this way, the generalized Hooke's law can be written as

Basic Solid Mechanics

$$\begin{Bmatrix} \sigma_{xx} \\ \sigma_{yy} \\ \sigma_{zz} \\ \tau_{yz} \\ \tau_{zx} \\ \tau_{xy} \\ \tau_{zy} \\ \tau_{xz} \\ \tau_{yx} \end{Bmatrix} = \begin{bmatrix} C_{xxxx} & C_{xxyy} & C_{xxzz} & C_{xxyz} & C_{xxzx} & C_{xxxy} & C_{xxzy} & C_{xxxz} & C_{xxyx} \\ C_{yyxx} & C_{yyyy} & C_{yyzz} & C_{yyyz} & C_{yyzx} & C_{yyxy} & C_{yyzy} & C_{yyxz} & C_{yyyx} \\ C_{zzxx} & C_{zzyy} & C_{zzzz} & C_{zzyz} & C_{zzzx} & C_{zzxy} & C_{zzzy} & C_{zzxz} & C_{zzyx} \\ C_{yzxx} & C_{yzyy} & C_{yzzz} & C_{yzyz} & C_{yzzx} & C_{yzxy} & C_{yzzy} & C_{yzxz} & C_{yzyx} \\ C_{zxxx} & C_{zxyy} & C_{zxzz} & C_{zxyz} & C_{zxzx} & C_{zxxy} & C_{zxzy} & C_{zxxz} & C_{zxyx} \\ C_{xyxx} & C_{xyyy} & C_{xyzz} & C_{xyyz} & C_{xyzx} & C_{xyxy} & C_{xyzy} & C_{xyxz} & C_{xyyx} \\ C_{zyxx} & C_{zyyy} & C_{zyzz} & C_{zyyz} & C_{zyzx} & C_{zyxy} & C_{zyzy} & C_{zyxz} & C_{zyyx} \\ C_{xzxx} & C_{xzyy} & C_{xzzz} & C_{xzyz} & C_{xzzx} & C_{xzxy} & C_{xzzy} & C_{xzxz} & C_{xzyx} \\ C_{yxxx} & C_{yxyy} & C_{yxzz} & C_{yxyz} & C_{yxzx} & C_{yxxy} & C_{yxzy} & C_{yxxz} & C_{yxyx} \end{bmatrix} \times \begin{Bmatrix} \varepsilon_{xx} \\ \varepsilon_{yy} \\ \varepsilon_{zz} \\ \varepsilon_{yz} \\ \varepsilon_{zx} \\ \varepsilon_{xy} \\ \varepsilon_{zy} \\ \varepsilon_{xz} \\ \varepsilon_{yx} \end{Bmatrix} \quad (2.152)$$

The number of elastic constants in Equation 2.152 is 81 and it can be drastically reduced under different criteria. In the following sections, we explore some of these cases.

2.7.3.1 Symmetry of Stress and Strain Tensors

σ_{ij} and ε_{ij} are symmetric tensors. In their vector forms, the seventh, eighth, and ninth rows can be deleted and the (9 × 1) vectors are replaced with (6 × 1) vectors. Similarly, seventh, eighth, and ninth rows and columns in C_{ijkl} are deleted. Thus, the number of elastic constants reduces from 81 to 36. Now, we can rewrite the generalized Hooke's law as

$$\begin{Bmatrix} \sigma_{xx} \\ \sigma_{yy} \\ \sigma_{zz} \\ \tau_{yz} \\ \tau_{zx} \\ \tau_{xy} \end{Bmatrix} = \begin{bmatrix} C_{xxxx} & C_{xxyy} & C_{xxzz} & C_{xxyz} & C_{xxzx} & C_{xxxy} \\ C_{yyxx} & C_{yyyy} & C_{yyzz} & C_{yyyz} & C_{yyzx} & C_{yyxy} \\ C_{zzxx} & C_{zzyy} & C_{zzzz} & C_{zzyz} & C_{zzzx} & C_{zzxy} \\ C_{yzxx} & C_{yzyy} & C_{yzzz} & C_{yzyz} & C_{yzzx} & C_{yzxy} \\ C_{zxxx} & C_{zxyy} & C_{zxzz} & C_{zxyz} & C_{zxzx} & C_{zxxy} \\ C_{xyxx} & C_{xyyy} & C_{xyzz} & C_{xyyz} & C_{xyzx} & C_{xyxy} \end{bmatrix} \begin{Bmatrix} \varepsilon_{xx} \\ \varepsilon_{yy} \\ \varepsilon_{zz} \\ \gamma_{yz} \\ \gamma_{zx} \\ \gamma_{xy} \end{Bmatrix} \quad (2.153)$$

Note that ε_{yz}, ε_{zx}, and ε_{xy} have been replaced by $2\varepsilon_{yz}(=\gamma_{yz})$, $2\varepsilon_{zx}(=\gamma_{zx})$, and $2\varepsilon_{xy}(=\gamma_{xy})$, respectively.

We make further changes and adopt an alternate (usually referred to as the contracted) notation. As per the contracted notation,

$$xx \to 1 \quad yy \to 2 \quad zz \to 3 \quad yz \to 4 \quad zx \to 5 \quad xy \to 6 \quad (2.154)$$

$$\sigma_{xx} = \sigma_1 \quad \sigma_{yy} = \sigma_2 \quad \sigma_{zz} = \sigma_3 \quad \tau_{yz} = \sigma_4 \quad \tau_{zx} = \sigma_5 \quad \tau_{xy} = \sigma_6 \quad (2.155)$$

$$\varepsilon_{xx} = \varepsilon_1 \quad \varepsilon_{yy} = \varepsilon_2 \quad \varepsilon_{zz} = \varepsilon_3 \quad \gamma_{yz} = \varepsilon_4 \quad \gamma_{zx} = \varepsilon_5 \quad \gamma_{xy} = \varepsilon_6 \quad (2.156)$$

Then, we can write Hooke's law as follows:
In the component form,

$$\{\sigma\} = [C]\{\varepsilon\} \quad (2.157)$$

In the vector form,

$$\boldsymbol{\sigma} = \boldsymbol{C} : \boldsymbol{\varepsilon} \tag{2.158}$$

In the indicial notation (Summation on repeated indices is implied from 1 to 6.),

$$\sigma_i = C_{ij} \varepsilon_j \tag{2.159}$$

where

$$\{\boldsymbol{\sigma}\} = \begin{Bmatrix} \sigma_1 \\ \sigma_2 \\ \sigma_3 \\ \sigma_4 \\ \sigma_5 \\ \sigma_6 \end{Bmatrix}$$

is the vector of stress components. (Note that this vector is different from the stress vector or traction vector defined in the section on kinetics.)

$$[\boldsymbol{C}] = \begin{bmatrix} C_{11} & C_{12} & C_{13} & C_{14} & C_{15} & C_{16} \\ C_{21} & C_{22} & C_{23} & C_{24} & C_{25} & C_{26} \\ C_{31} & C_{32} & C_{33} & C_{34} & C_{35} & C_{36} \\ C_{41} & C_{42} & C_{43} & C_{44} & C_{45} & C_{46} \\ C_{51} & C_{52} & C_{53} & C_{54} & C_{55} & C_{56} \\ C_{61} & C_{62} & C_{63} & C_{64} & C_{65} & C_{66} \end{bmatrix}$$

is called the elastic stiffness matrix.

$$\{\boldsymbol{\varepsilon}\} = \begin{Bmatrix} \varepsilon_1 \\ \varepsilon_2 \\ \varepsilon_3 \\ \varepsilon_4 \\ \varepsilon_5 \\ \varepsilon_6 \end{Bmatrix}$$

is the vector of strain components.

Now, let us go back to Table 2.1 in Section 2.3.3. We mentioned there that we need 15 independent equations for determining 15 unknowns in a 3D problem. We obtained nine equations from kinematics and kinetics. Here, we see that the remaining six equations are the stress–strain relations obtained from constitutive modeling.

Equation 2.157 is invertible and we can express strains in terms of the stresses. Thus,

$$\{\boldsymbol{\varepsilon}\} = [\boldsymbol{S}]\{\boldsymbol{\sigma}\} \tag{2.160}$$

Basic Solid Mechanics

where $[S] = [C]^{-1}$ is called the material compliance matrix. And, we can write

$$\begin{Bmatrix} \varepsilon_1 \\ \varepsilon_2 \\ \varepsilon_3 \\ \varepsilon_4 \\ \varepsilon_5 \\ \varepsilon_6 \end{Bmatrix} = \begin{bmatrix} S_{11} & S_{12} & S_{13} & S_{14} & S_{15} & S_{16} \\ S_{21} & S_{22} & S_{23} & S_{24} & S_{25} & S_{26} \\ S_{31} & S_{32} & S_{33} & S_{34} & S_{35} & S_{36} \\ S_{41} & S_{42} & S_{43} & S_{44} & S_{45} & S_{46} \\ S_{51} & S_{52} & S_{53} & S_{54} & S_{55} & S_{56} \\ S_{61} & S_{62} & S_{63} & S_{64} & S_{65} & S_{66} \end{bmatrix} \begin{Bmatrix} \sigma_1 \\ \sigma_2 \\ \sigma_3 \\ \sigma_4 \\ \sigma_5 \\ \sigma_6 \end{Bmatrix} \quad (2.161)$$

2.7.3.2 Symmetry of Elastic Stiffness Matrix

For elastic materials, strain energy density function takes a quadratic form and it can be shown that

$$C_{klij} = \frac{\partial^2 U_0}{\partial \varepsilon_{ij} \partial \varepsilon_{kl}} = \frac{\partial^2 U_0}{\partial \varepsilon_{kl} \partial \varepsilon_{ij}} = C_{ijkl} \quad (2.162)$$

In other words, the tensor C_{ijkl} is symmetric in ij and kl. Thus, the elastic stiffness matrix C_{ij} is symmetric in i and j. Thus, the total number of independent elastic constants for a general anisotropic elastic material reduces to 21 and we can write the generalized Hooke's law as

$$\begin{Bmatrix} \sigma_1 \\ \sigma_2 \\ \sigma_3 \\ \sigma_4 \\ \sigma_5 \\ \sigma_6 \end{Bmatrix} = \begin{bmatrix} C_{11} & C_{12} & C_{13} & C_{14} & C_{15} & C_{16} \\ C_{12} & C_{22} & C_{23} & C_{24} & C_{25} & C_{26} \\ C_{13} & C_{23} & C_{33} & C_{34} & C_{35} & C_{36} \\ C_{14} & C_{24} & C_{34} & C_{44} & C_{45} & C_{46} \\ C_{15} & C_{25} & C_{35} & C_{45} & C_{55} & C_{56} \\ C_{16} & C_{26} & C_{36} & C_{46} & C_{56} & C_{66} \end{bmatrix} \begin{Bmatrix} \varepsilon_1 \\ \varepsilon_2 \\ \varepsilon_3 \\ \varepsilon_4 \\ \varepsilon_5 \\ \varepsilon_6 \end{Bmatrix} \quad (2.163)$$

The number of independent elastic constants can be further reduced by consideration of material planes of symmetry. We shall see in the following sections simpler forms of Hooke's law corresponding to different classes of materials.

2.7.3.3 Anisotropic Materials

Equation 2.163 gives the Hooke's law for general anisotropic elastic material. As mentioned before, it has 21 independent elastic constants.

Some materials exhibit directional symmetry in properties w.r.t. certain planes. In the most general case, in which there is no material plane of symmetry, the material is called anisotropic. The presence of material plane(s) of symmetry reduces the number of elastic constants as discussed below.

2.7.3.4 Monoclinic Materials

A monoclinic material has one material plane of symmetry. Figure 2.16 shows a monoclinic material in which the material plane of symmetry is normal to the z-direction. For the axes shown in the figure, the transformation matrix (refer to Equation 2.122) is given by

$$[\alpha] = \begin{bmatrix} 1 & 0 & 0 \\ 0 & 1 & 0 \\ 0 & 0 & -1 \end{bmatrix} \quad (2.164)$$

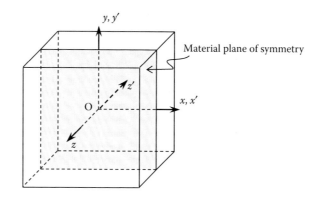

FIGURE 2.16 Monoclinic material.

For this transformation, we can find from Equations 2.88 and 2.122 that

$$\begin{Bmatrix} \sigma'_1 \\ \sigma'_2 \\ \sigma'_3 \\ \sigma'_4 \\ \sigma'_5 \\ \sigma'_6 \end{Bmatrix} = \begin{Bmatrix} \sigma_1 \\ \sigma_2 \\ \sigma_3 \\ -\sigma_4 \\ -\sigma_5 \\ \sigma_6 \end{Bmatrix} \text{ and } \begin{Bmatrix} \varepsilon'_1 \\ \varepsilon'_2 \\ \varepsilon'_3 \\ \varepsilon'_4 \\ \varepsilon'_5 \\ \varepsilon'_6 \end{Bmatrix} = \begin{Bmatrix} \varepsilon_1 \\ \varepsilon_2 \\ \varepsilon_3 \\ -\varepsilon_4 \\ -\varepsilon_5 \\ \varepsilon_6 \end{Bmatrix} \quad (2.165)$$

Now,

$$\sigma_1 = C_{11}\varepsilon_1 + C_{12}\varepsilon_2 + C_{13}\varepsilon_3 + C_{14}\varepsilon_4 + C_{15}\varepsilon_5 + C_{16}\varepsilon_6 \quad (2.166)$$

$$\sigma'_1 = C'_{11}\varepsilon'_1 + C'_{12}\varepsilon'_2 + C'_{13}\varepsilon'_3 + C'_{14}\varepsilon'_4 + C'_{15}\varepsilon'_5 + C'_{16}\varepsilon'_6 \quad (2.167)$$

Employing Equation 2.165 in Equation 2.167,

$$\sigma_1 = C'_{11}\varepsilon_1 + C'_{12}\varepsilon_2 + C'_{13}\varepsilon_3 - C'_{14}\varepsilon_4 - C'_{15}\varepsilon_5 + C'_{16}\varepsilon_6 \quad (2.168)$$

Comparing Equation 2.166 with Equation 2.168, we get

$$\varepsilon_1(C_{11} - C'_{11}) + \varepsilon_2(C_{12} - C'_{12}) + \varepsilon_3(C_{13} - C'_{13}) \\ + \varepsilon_4(C_{14} + C'_{14}) + \varepsilon_5(C_{15} + C'_{15}) + \varepsilon_6(C_{16} - C'_{16}) = 0 \quad (2.169)$$

Equation 2.169 is valid for all values of strain components. It is possible when,

$$C_{14} = C_{15} = 0$$

Extending the process to other stress components, it can be shown that

$$C_{24} = C_{25} = 0 \quad \text{and} \quad C_{34} = C_{35} = 0 \quad \text{and} \quad C_{46} = C_{56} = 0$$

Thus, the number of independent elastic constants reduces to 13 and Hooke's law for monoclinic material can be written as given below

$$\begin{Bmatrix} \sigma_1 \\ \sigma_2 \\ \sigma_3 \\ \sigma_4 \\ \sigma_5 \\ \sigma_6 \end{Bmatrix} = \begin{bmatrix} C_{11} & C_{12} & C_{13} & 0 & 0 & C_{16} \\ C_{12} & C_{22} & C_{23} & 0 & 0 & C_{26} \\ C_{13} & C_{23} & C_{33} & 0 & 0 & C_{36} \\ 0 & 0 & 0 & C_{44} & C_{45} & 0 \\ 0 & 0 & 0 & C_{45} & C_{55} & 0 \\ C_{16} & C_{26} & C_{36} & 0 & 0 & C_{66} \end{bmatrix} \begin{Bmatrix} \varepsilon_1 \\ \varepsilon_2 \\ \varepsilon_3 \\ \varepsilon_4 \\ \varepsilon_5 \\ \varepsilon_6 \end{Bmatrix} \quad (2.170)$$

2.7.3.5 Orthotropic Materials

An orthotropic material has three mutually orthogonal material planes of symmetry. Arguments similar to those for monoclinic materials can be extended and it can be shown that the number of elastic constants is reduced to nine in the elastic stiffness matrix. Thus, Hooke's law for orthotropic materials takes the form as below

$$\begin{Bmatrix} \sigma_1 \\ \sigma_2 \\ \sigma_3 \\ \sigma_4 \\ \sigma_5 \\ \sigma_6 \end{Bmatrix} = \begin{bmatrix} C_{11} & C_{12} & C_{13} & 0 & 0 & 0 \\ C_{12} & C_{22} & C_{23} & 0 & 0 & 0 \\ C_{13} & C_{23} & C_{33} & 0 & 0 & 0 \\ 0 & 0 & 0 & C_{44} & 0 & 0 \\ 0 & 0 & 0 & 0 & C_{55} & 0 \\ 0 & 0 & 0 & 0 & 0 & C_{66} \end{bmatrix} \begin{Bmatrix} \varepsilon_1 \\ \varepsilon_2 \\ \varepsilon_3 \\ \varepsilon_4 \\ \varepsilon_5 \\ \varepsilon_6 \end{Bmatrix} \quad (2.171)$$

Similarly, in terms of the compliance matrix, for orthotropic materials, we can write

$$\begin{Bmatrix} \varepsilon_1 \\ \varepsilon_2 \\ \varepsilon_3 \\ \varepsilon_4 \\ \varepsilon_5 \\ \varepsilon_6 \end{Bmatrix} = \begin{bmatrix} S_{11} & S_{12} & S_{13} & 0 & 0 & 0 \\ S_{12} & S_{22} & S_{23} & 0 & 0 & 0 \\ S_{13} & S_{23} & S_{33} & 0 & 0 & 0 \\ 0 & 0 & 0 & S_{44} & 0 & 0 \\ 0 & 0 & 0 & 0 & S_{55} & 0 \\ 0 & 0 & 0 & 0 & 0 & S_{66} \end{bmatrix} \begin{Bmatrix} \sigma_1 \\ \sigma_2 \\ \sigma_3 \\ \sigma_4 \\ \sigma_5 \\ \sigma_6 \end{Bmatrix} \quad (2.172)$$

Orthotropic materials are an important class of materials, especially in the field of composites. We have noted that there are nine independent constants in the elastic matrix. We shall now see that there are nine engineering elastic constants that describe an orthotropic material.

Engineering constants are experimentally determined by tests such as tension test and torsion test. These tests can be mathematically represented by means of application of stress in the respective direction. Figure 2.17 shows a cuboid under normal stress and shear stress states.

Let us first consider a stress state in Figure 2.17a, in which, a normal stress is applied in the x-direction.

$$\sigma_1 \neq 0, \quad \sigma_2 = \sigma_3 = \sigma_4 = \sigma_5 = \sigma_6 = 0 \quad (2.173)$$

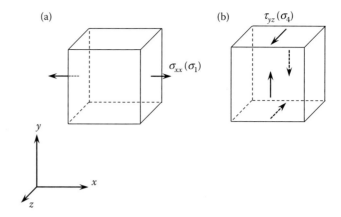

FIGURE 2.17 A cuboid under (a) normal stress and (b) shear stress (stresses within brackets are as per our alternate notation).

Then, from Equation 2.172, the strain components can be obtained as

$$\varepsilon_1 = S_{11}\sigma_1 \quad \text{or} \quad \varepsilon_{xx} = S_{11}\sigma_{xx} \tag{2.174}$$

$$\varepsilon_2 = S_{12}\sigma_1 \quad \text{or} \quad \varepsilon_{yy} = S_{12}\sigma_{xx} \tag{2.175}$$

$$\varepsilon_3 = S_{13}\sigma_1 \quad \text{or} \quad \varepsilon_{zz} = S_{13}\sigma_{xx} \tag{2.176}$$

$$\varepsilon_4 = \varepsilon_5 = \varepsilon_6 = 0 \quad \text{or} \quad \gamma_{yz} = \gamma_{zx} = \gamma_{xy} = 0 \tag{2.177}$$

Thus, we see that there are normal strains in all the three orthogonal directions and no shear strain. Under this applied stress, the normal strain in the x-direction is due to the direct stress and the normal strains in the y- and z-directions are due to Poisson's effect.

Now, in a general case, Young's modulus E_i is defined as the ratio of the direct stress in the i-direction to the normal strain in the same direction. Similarly, Poisson's ratio ν_{ij}, $i \neq j$ is defined as the ratio of the transverse strain in the j-direction to the normal strain in the i-direction when the applied stress is in the i-direction. (Note: i and j take the values x, y, z. No indicial notation is implied here.)

Thus, Young's modulus in the x-direction and Poisson's ratios in the xy- and xz-planes are given by

$$E_x = \frac{\sigma_{xx}}{\varepsilon_{xx}} \quad \text{and} \quad \nu_{xy} = -\frac{\varepsilon_{yy}}{\varepsilon_{xx}} \quad \text{and} \quad \nu_{xz} = -\frac{\varepsilon_{zz}}{\varepsilon_{xx}} \tag{2.178}$$

In a similar way, by applying normal stress in the y- and z-directions, respectively, we can obtain the following:

$$E_y = \frac{\sigma_{yy}}{\varepsilon_{yy}} \quad \text{and} \quad \nu_{yz} = -\frac{\varepsilon_{zz}}{\varepsilon_{yy}} \quad \text{and} \quad \nu_{yx} = -\frac{\varepsilon_{xx}}{\varepsilon_{yy}} \tag{2.179}$$

$$E_z = \frac{\sigma_{zz}}{\varepsilon_{zz}} \quad \text{and} \quad \nu_{zx} = -\frac{\varepsilon_{xx}}{\varepsilon_{zz}} \quad \text{and} \quad \nu_{zy} = -\frac{\varepsilon_{yy}}{\varepsilon_{zz}} \tag{2.180}$$

Basic Solid Mechanics

Using Equations 2.174 through 2.176 in Equations 2.178 through 2.180 and extending the procedure to the y- and z-directions, we get the following:

$$E_x = \frac{1}{S_{11}} \quad \text{and} \quad \nu_{xy} = -\frac{S_{12}}{S_{11}} \quad \text{and} \quad \nu_{xz} = -\frac{S_{13}}{S_{11}} \qquad (2.181)$$

$$E_y = \frac{1}{S_{22}} \quad \text{and} \quad \nu_{yz} = -\frac{S_{23}}{S_{22}} \quad \text{and} \quad \nu_{yx} = -\frac{S_{12}}{S_{22}} \qquad (2.182)$$

$$E_z = \frac{1}{S_{33}} \quad \text{and} \quad \nu_{zx} = -\frac{S_{13}}{S_{33}} \quad \text{and} \quad \nu_{zy} = -\frac{S_{23}}{S_{33}} \qquad (2.183)$$

Next, as shown in Figure 2.17b, let us consider a stress state in which a shear stress is applied.

$$\sigma_1 = \sigma_2 = \sigma_3 = \sigma_5 = \sigma_6 = 0, \quad \sigma_4 \neq 0 \qquad (2.184)$$

Then, from Equation 2.172, the strain components can be obtained as

$$\varepsilon_1 = \varepsilon_2 = \varepsilon_3 = 0 \quad \text{or} \quad \varepsilon_{xx} = \varepsilon_{yy} = \varepsilon_{zz} = 0 \qquad (2.185)$$

$$\varepsilon_4 = S_{44}\sigma_4 \quad \text{or} \quad \gamma_{yz} = S_{44}\tau_{yz} \qquad (2.186)$$

$$\varepsilon_5 = \varepsilon_6 = 0 \quad \text{or} \quad \gamma_{zx} = \gamma_{xy} = 0 \qquad (2.187)$$

Thus, we see that there is only one shear strain and all other strains are zero.

Now, in a general case, shear modulus G_{ij}, $i \neq j$ is defined as the ratio of the shear stress in the ij-plane to the shear strain in the same plane. (Note: i and j take the values x, y, z. No indicial notation is implied here.)

Thus, shear modulus in the yz-plane is given by

$$G_{yz} = \frac{\tau_{yz}}{\gamma_{yz}} \qquad (2.188)$$

In a similar way, by applying shear stress in the zx- and xy-planes, respectively, we can obtain the following:

$$G_{zx} = \frac{\tau_{zx}}{\gamma_{zx}} \qquad (2.189)$$

$$G_{xy} = \frac{\tau_{xy}}{\gamma_{xy}} \qquad (2.190)$$

Using Equation 2.186 in Equation 2.188 and extending the procedure to the other two planes, we finally get

$$G_{yz} = \frac{1}{S_{44}} \quad \text{and} \quad G_{zx} = \frac{1}{S_{55}} \quad \text{and} \quad G_{xy} = \frac{1}{S_{66}} \qquad (2.191)$$

Now, from Equations 2.181 through 2.183 and 2.191, we obtain the expressions for the compliance matrix components for an orthotropic material as follows:

$$S_{11} = \frac{1}{E_x} \tag{2.192}$$

$$S_{22} = \frac{1}{E_y} \tag{2.193}$$

$$S_{33} = \frac{1}{E_z} \tag{2.194}$$

$$S_{44} = \frac{1}{G_{yz}} \tag{2.195}$$

$$S_{55} = \frac{1}{G_{zx}} \tag{2.196}$$

$$S_{66} = \frac{1}{G_{xy}} \tag{2.197}$$

$$S_{12} = -\frac{\nu_{xy}}{E_x} = -\frac{\nu_{yx}}{E_y} \tag{2.198}$$

$$S_{13} = -\frac{\nu_{xz}}{E_x} = -\frac{\nu_{zx}}{E_z} \tag{2.199}$$

$$S_{23} = -\frac{\nu_{yz}}{E_y} = -\frac{\nu_{zy}}{E_z} \tag{2.200}$$

Thus, for an orthotropic material, the stress–strain relation can be written as

$$\begin{Bmatrix} \varepsilon_{xx} \\ \varepsilon_{yy} \\ \varepsilon_{zz} \\ \gamma_{yz} \\ \gamma_{zx} \\ \gamma_{xy} \end{Bmatrix} = \begin{bmatrix} \frac{1}{E_x} & -\frac{\nu_{xy}}{E_x} & -\frac{\nu_{xz}}{E_x} & 0 & 0 & 0 \\ -\frac{\nu_{xy}}{E_x} & \frac{1}{E_y} & -\frac{\nu_{yz}}{E_y} & 0 & 0 & 0 \\ -\frac{\nu_{xz}}{E_x} & -\frac{\nu_{yz}}{E_y} & \frac{1}{E_z} & 0 & 0 & 0 \\ 0 & 0 & 0 & \frac{1}{G_{yz}} & 0 & 0 \\ 0 & 0 & 0 & 0 & \frac{1}{G_{zx}} & 0 \\ 0 & 0 & 0 & 0 & 0 & \frac{1}{G_{xy}} \end{bmatrix} \begin{Bmatrix} \sigma_{xx} \\ \sigma_{yy} \\ \sigma_{zz} \\ \tau_{yz} \\ \tau_{zx} \\ \tau_{xy} \end{Bmatrix} \tag{2.201}$$

Basic Solid Mechanics

2.7.3.6 Transversely Isotropic Materials

A transversely isotropic material is an orthotropic material that exhibits isotropic behavior in one plane of symmetry. Taking the plane yz- as the material plane of symmetry possessing isotropic properties, it can be seen that

$$C_{22} = C_{33} \quad \text{and} \quad C_{12} = C_{13} \quad \text{and} \quad C_{55} = C_{66}$$

Further, it can be shown that

$$C_{44} = \frac{C_{22} - C_{23}}{2}$$

Thus, the number of independent elastic constants reduces to 5. The independent engineering constants for a transversely isotropic material with yz- as the plane of isotropy are

- E_x Young's modulus in the x-direction
- E_y Young's modulus in the y-direction
- ν_{xy} Major Poisson's ratio in the xy-plane
- G_{xy} In-plane shear modulus in the xy-plane
- ν_{yz} Major Poisson's ratio in the yz-plane

In-plane shear modulus, G_{yz} in the plane of isotropy is related to the major Poisson's ratio, ν_{yz} in the same plane as

$$G_{yz} = \frac{E_y}{2(1+\nu_{yz})} \tag{2.202}$$

Thus, G_{yz} can also be considered as the fifth independent engineering constant in place of ν_{yz}. Note that $E_y = E_z$, $\nu_{xz} = \nu_{xy}$, and $G_{zx} = G_{xy}$.

The stiffness matrix for this material is given by

$$[C] = \begin{bmatrix} C_{11} & C_{12} & C_{12} & 0 & 0 & 0 \\ C_{12} & C_{22} & C_{23} & 0 & 0 & 0 \\ C_{12} & C_{23} & C_{22} & 0 & 0 & 0 \\ 0 & 0 & 0 & \dfrac{C_{22}-C_{23}}{2} & 0 & 0 \\ 0 & 0 & 0 & 0 & C_{55} & 0 \\ 0 & 0 & 0 & 0 & 0 & C_{55} \end{bmatrix} \tag{2.203}$$

The compliance matrix for this material is given by

$$[S] = \begin{bmatrix} S_{11} & S_{12} & S_{12} & 0 & 0 & 0 \\ S_{12} & S_{22} & S_{23} & 0 & 0 & 0 \\ S_{12} & S_{23} & S_{22} & 0 & 0 & 0 \\ 0 & 0 & 0 & 2(S_{22}-S_{23}) & 0 & 0 \\ 0 & 0 & 0 & 0 & S_{55} & 0 \\ 0 & 0 & 0 & 0 & 0 & S_{55} \end{bmatrix} \tag{2.204}$$

Then, the stress–strain relation for a transversely isotropic material with yz as the plane of symmetry can be written as

$$\begin{Bmatrix} \varepsilon_{xx} \\ \varepsilon_{yy} \\ \varepsilon_{zz} \\ \gamma_{yz} \\ \gamma_{zx} \\ \gamma_{xy} \end{Bmatrix} = \begin{bmatrix} \dfrac{1}{E_x} & -\dfrac{\nu_{xy}}{E_x} & -\dfrac{\nu_{xy}}{E_x} & 0 & 0 & 0 \\ -\dfrac{\nu_{xy}}{E_x} & \dfrac{1}{E_y} & -\dfrac{\nu_{yz}}{E_y} & 0 & 0 & 0 \\ -\dfrac{\nu_{xy}}{E_x} & -\dfrac{\nu_{yz}}{E_y} & \dfrac{1}{E_y} & 0 & 0 & 0 \\ 0 & 0 & 0 & \dfrac{2(1+\nu_{yz})}{E_y} & 0 & 0 \\ 0 & 0 & 0 & 0 & \dfrac{1}{G_{xy}} & 0 \\ 0 & 0 & 0 & 0 & 0 & \dfrac{1}{G_{xy}} \end{bmatrix} \begin{Bmatrix} \sigma_{xx} \\ \sigma_{yy} \\ \sigma_{zz} \\ \tau_{yz} \\ \tau_{zx} \\ \tau_{xy} \end{Bmatrix} \quad (2.205)$$

2.7.3.7 Cubic Symmetry

An orthotropic material, in which all three material planes of symmetry are identical, is known as a material with cubic symmetry. The number of independent elastic constants reduces to three. Note that the three orthogonal planes are identical but they do not have isotropy. Equation 2.206 shows the elastic stiffness matrix.

$$[C] = \begin{bmatrix} C_{11} & C_{12} & C_{12} & 0 & 0 & 0 \\ C_{12} & C_{11} & C_{12} & 0 & 0 & 0 \\ C_{12} & C_{12} & C_{11} & 0 & 0 & 0 \\ 0 & 0 & 0 & C_{44} & 0 & 0 \\ 0 & 0 & 0 & 0 & C_{44} & 0 \\ 0 & 0 & 0 & 0 & 0 & C_{44} \end{bmatrix} \quad (2.206)$$

2.7.3.8 Isotropic Materials

Further reduction in the number of independent elastic constants from three to two is possible when one of the three orthogonal planes in a cubic symmetric material is isotropic. Such a material is called isotropic. It can be seen that in an isotropic material, there are infinite numbers of material planes of symmetry. Thus, material properties are not dependent on the directions and the elastic stiffness matrix takes the following form:

$$[C] = \begin{bmatrix} C_{11} & C_{12} & C_{12} & 0 & 0 & 0 \\ C_{12} & C_{11} & C_{12} & 0 & 0 & 0 \\ C_{12} & C_{12} & C_{11} & 0 & 0 & 0 \\ 0 & 0 & 0 & \dfrac{C_{11}-C_{12}}{2} & 0 & 0 \\ 0 & 0 & 0 & 0 & \dfrac{C_{11}-C_{12}}{2} & 0 \\ 0 & 0 & 0 & 0 & 0 & \dfrac{C_{11}-C_{12}}{2} \end{bmatrix} \quad (2.207)$$

Basic Solid Mechanics

The compliance matrix reduces to

$$[S] = \begin{bmatrix} S_{11} & S_{12} & S_{12} & 0 & 0 & 0 \\ S_{12} & S_{11} & S_{12} & 0 & 0 & 0 \\ S_{12} & S_{12} & S_{11} & 0 & 0 & 0 \\ 0 & 0 & 0 & 2(S_{11}-S_{12}) & 0 & 0 \\ 0 & 0 & 0 & 0 & 2(S_{11}-S_{12}) & 0 \\ 0 & 0 & 0 & 0 & 0 & 2(S_{11}-S_{12}) \end{bmatrix} \quad (2.208)$$

Then, the stress–strain relation for an isotropic material can be written as

$$\begin{Bmatrix} \varepsilon_{xx} \\ \varepsilon_{yy} \\ \varepsilon_{zz} \\ \gamma_{yz} \\ \gamma_{zx} \\ \gamma_{xy} \end{Bmatrix} = \begin{bmatrix} \frac{1}{E} & \frac{-\nu}{E} & \frac{-\nu}{E} & 0 & 0 & 0 \\ \frac{-\nu}{E} & \frac{1}{E} & \frac{-\nu}{E} & 0 & 0 & 0 \\ \frac{-\nu}{E} & \frac{-\nu}{E} & \frac{1}{E} & 0 & 0 & 0 \\ 0 & 0 & 0 & \frac{2(1+\nu)}{E} & 0 & 0 \\ 0 & 0 & 0 & 0 & \frac{2(1+\nu)}{E} & 0 \\ 0 & 0 & 0 & 0 & 0 & \frac{2(1+\nu)}{E} \end{bmatrix} \begin{Bmatrix} \sigma_{xx} \\ \sigma_{yy} \\ \sigma_{zz} \\ \tau_{yz} \\ \tau_{zx} \\ \tau_{xy} \end{Bmatrix} \quad (2.209)$$

It may be noted that we have three engineering elastic constants describing an isotropic material, namely Young's modulus E, shear modulus G, and Poisson's ratio ν. It may further be noted that out of these three constants only two are independent as the third one can be expressed in terms of the first two as follows:

$$G = \frac{E}{2(1+\nu)} \quad (2.210)$$

2.8 PLANE ELASTICITY PROBLEMS

In the preceding sections, we framed the governing equations for a 3D linear elasticity problem. The boundary conditions are provided in the form of forces, displacements, or both specified on the boundary. In certain boundary value problems, owing to their particular way of loading, geometry and boundary conditions, it is possible to ignore some of the stress or strain components. Considerable computational efficiency is achieved by idealizing these problems as 2D. Two such important idealized problems are plane stress problem and plane strain problem [2]. In this section, we shall briefly discuss these two plane problems.

2.8.1 Plane Stress

Let us consider a thin plate in the xy-plane (Figure 2.18). The thickness of the plate is small compared to the other two dimensions and the applied forces act only in the

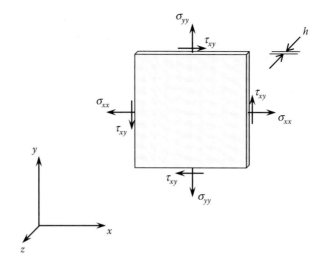

FIGURE 2.18 A thin plate—pictorial representation of plane stress.

plane of the plate, that is, in the *xy*-plane. A problem like this can be idealized as one in which the stresses in the thickness direction are zero. Also, the nonzero stresses are functions of *x* and *y* only. This is called a plane stress problem, which is mathematically described as

$$\sigma_{zz} = \tau_{zx} = \tau_{yz} = 0 \tag{2.211}$$

$$\sigma_{xx} = \sigma_{xx}(x,y) \quad \text{and} \quad \sigma_{yy} = \sigma_{yy}(x,y) \quad \text{and} \quad \tau_{xy} = \tau_{xy}(x,y) \tag{2.212}$$

2.8.1.1 Plane Stress Problem in Orthotropic Materials

Under plane stress condition, the strains in orthotropic materials can be obtained by using Equation 2.201 as

$$\varepsilon_{xx} = \frac{\sigma_{xx}}{E_x} - \frac{\nu_{xy}\sigma_{yy}}{E_x} \tag{2.213}$$

$$\varepsilon_{yy} = \frac{\sigma_{yy}}{E_y} - \frac{\nu_{xy}\sigma_{xx}}{E_x} \tag{2.214}$$

$$\varepsilon_{zz} = -\left(\frac{\nu_{xz}\sigma_{xx}}{E_x} + \frac{\nu_{yz}\sigma_{yy}}{E_y}\right) \tag{2.215}$$

$$\gamma_{yz} = 0 \tag{2.216}$$

$$\gamma_{zx} = 0 \tag{2.217}$$

$$\gamma_{xy} = \frac{\tau_{xy}}{G_{xy}} \tag{2.218}$$

Basic Solid Mechanics 71

By solving the above equations, the nonzero stresses can be obtained and the results can be written in the matrix form as

$$\begin{Bmatrix} \sigma_{xx} \\ \sigma_{yy} \\ \tau_{xy} \end{Bmatrix} = \begin{bmatrix} \dfrac{E_x}{1-\nu_{xy}\nu_{yx}} & \dfrac{\nu_{xy}E_y}{1-\nu_{xy}\nu_{yx}} & 0 \\ \dfrac{\nu_{xy}E_y}{1-\nu_{xy}\nu_{yx}} & \dfrac{E_y}{1-\nu_{xy}\nu_{yx}} & 0 \\ 0 & 0 & G_{xy} \end{bmatrix} \begin{Bmatrix} \varepsilon_{xx} \\ \varepsilon_{yy} \\ \gamma_{xy} \end{Bmatrix} \qquad (2.219)$$

Equation 2.219 gives the stress state in an orthotropic material under plane stress condition. Note that there are three nonzero stress components, but, there are four nonzero strain components.

2.8.1.2 Plane Stress Problem in Isotropic Materials

Under plane stress condition, the strains in isotropic materials can be obtained by using Equation 2.209 as

$$\varepsilon_{xx} = \frac{\sigma_{xx}}{E} - \frac{\nu \sigma_{yy}}{E} \qquad (2.220)$$

$$\varepsilon_{yy} = \frac{\sigma_{yy}}{E} - \frac{\nu \sigma_{xx}}{E} \qquad (2.221)$$

$$\varepsilon_{zz} = -\nu \left(\frac{\sigma_{xx}}{E} + \frac{\sigma_{yy}}{E} \right) \qquad (2.222)$$

$$\gamma_{yz} = 0 \qquad (2.223)$$

$$\gamma_{zx} = 0 \qquad (2.224)$$

$$\gamma_{xy} = \frac{2(1+\nu)}{E}\tau_{xy} \qquad (2.225)$$

By solving the above equations, the nonzero stresses can be obtained and the results can be written in the matrix form as

$$\begin{Bmatrix} \sigma_{xx} \\ \sigma_{yy} \\ \tau_{xy} \end{Bmatrix} = \begin{bmatrix} \dfrac{E}{1-\nu^2} & \dfrac{\nu E}{1-\nu^2} & 0 \\ \dfrac{\nu E}{1-\nu^2} & \dfrac{E}{1-\nu^2} & 0 \\ 0 & 0 & G \end{bmatrix} \begin{Bmatrix} \varepsilon_{xx} \\ \varepsilon_{yy} \\ \gamma_{xy} \end{Bmatrix} \qquad (2.226)$$

Equation 2.226 gives the stress state in an isotropic material under plane stress condition. Note that, as in the case of orthotropic materials, there are three nonzero stress components, but, there are four nonzero strain components.

2.8.2 Plane Strain

Let us consider a long thick cylinder under internal pressure (Figure 2.19). For the sake of simplicity, let us orient the axis of the cylinder along the z-axis such that the xy-plane is normal to the axis of the cylinder. Some characteristic features of this problem are

- Geometry—the cross-sectional dimensions are small compared to the longitudinal dimension.
- Loading—the applied forces are normal to the longitudinal axis. They are functions of x and y only, and independent of z.
- Boundary conditions—the ends are restrained such that displacement gradients are zero at the ends.

Let us consider a cross section far from the ends. The displacements along x- and y-directions are functions of x and y only. On the other hand, the displacement along the z-direction is zero. Mathematically,

$$u_x = u_x(x,y) \tag{2.227}$$

$$u_y = u_y(x,y) \tag{2.228}$$

$$u_z = 0 \tag{2.229}$$

Thus,

$$\varepsilon_{xx} \equiv \frac{\partial u_x}{\partial x} = \varepsilon_{xx}(x,y) \tag{2.230}$$

$$\varepsilon_{yy} \equiv \frac{\partial u_y}{\partial y} = \varepsilon_{yy}(x,y) \tag{2.231}$$

$$\varepsilon_{zz} \equiv \frac{\partial u_z}{\partial z} = 0 \tag{2.232}$$

$$\gamma_{yz} \equiv \frac{\partial u_z}{\partial y} + \frac{\partial u_y}{\partial z} = 0 \tag{2.233}$$

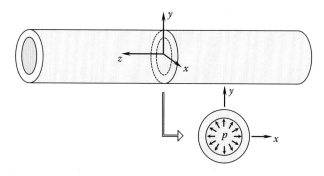

FIGURE 2.19 A long cylinder under internal pressure—pictorial representation of an example of plane strain.

Basic Solid Mechanics

$$\gamma_{zx} \equiv \frac{\partial u_x}{\partial z} + \frac{\partial u_z}{\partial x} = 0 \qquad (2.234)$$

$$\gamma_{xy} \equiv \frac{\partial u_y}{\partial x} + \frac{\partial u_x}{\partial y} = \gamma_{xy}(x, y) \qquad (2.235)$$

A problem with these characteristics is known as a plane strain problem. Some common examples of plane strain problems are thick pipe under internal or external pressure, dam, tunnel, etc.

2.8.2.1 Plane Strain Problem in Orthotropic Materials

By substituting the strains from Equations 2.230 through 2.235 in Equation 2.201, after some arithmetic manipulation, we get the following:

$$\varepsilon_{xx} = \left(\frac{1}{E_x} - \frac{\nu_{zx}^2}{E_z}\right)\sigma_{xx} - \left(\frac{\nu_{xy}}{E_x} + \frac{\nu_{yz}\nu_{zx}}{E_y}\right)\sigma_{yy} \qquad (2.236)$$

$$\varepsilon_{yy} = -\left(\frac{\nu_{xy}}{E_x} + \frac{\nu_{yz}\nu_{zx}}{E_y}\right)\sigma_{xx} + \left(\frac{1}{E_y} - \frac{\nu_{yz}^2 E_z}{E_y^2}\right)\sigma_{yy} \qquad (2.237)$$

$$\gamma_{xy} = \frac{\tau_{xy}}{G_{xy}} \qquad (2.238)$$

We can solve the above equations for σ_{xx}, σ_{yy}, and τ_{xy}, and express them in terms of the nonzero strain components. Further, σ_{zz} can be expressed in terms of σ_{xx} and σ_{yy} as

$$\sigma_{zz} = \nu_{zx}\sigma_{xx} + \frac{E_z \nu_{yz}}{E_y}\sigma_{yy} \qquad (2.239)$$

Thus, we see that there are four nonzero (three independent) stress components.

2.8.2.2 Plane Strain Problem in Isotropic Materials

For isotropic materials, we put $E_x = E_y = E_z = E$, $\nu_{xy} = \nu_{yz} = \nu_{zx} = \nu$, and $G_{xy} = G$ and obtain the following:

$$\varepsilon_{xx} = \left(\frac{1-\nu^2}{E}\right)\sigma_{xx} - \left(\frac{\nu + \nu^2}{E}\right)\sigma_{yy} \qquad (2.240)$$

$$\varepsilon_{yy} = \left(\frac{1-\nu^2}{E}\right)\sigma_{yy} - \left(\frac{\nu + \nu^2}{E}\right)\sigma_{xx} \qquad (2.241)$$

$$\gamma_{xy} = \frac{\tau_{xy}}{G} \qquad (2.242)$$

We can solve the above equations for σ_{xx}, σ_{yy}, and τ_{xy}, and express them in terms of the nonzero strain components in the matrix form as follows:

$$\begin{Bmatrix} \sigma_{xx} \\ \sigma_{yy} \\ \tau_{xy} \end{Bmatrix} = \begin{bmatrix} \dfrac{E(1-\nu)}{(1+\nu)(1-2\nu)} & \dfrac{\nu E}{(1+\nu)(1-2\nu)} & 0 \\ \dfrac{\nu E}{(1+\nu)(1-2\nu)} & \dfrac{E(1-\nu)}{(1+\nu)(1-2\nu)} & 0 \\ 0 & 0 & G \end{bmatrix} \begin{Bmatrix} \varepsilon_{xx} \\ \varepsilon_{yy} \\ \gamma_{xy} \end{Bmatrix} \qquad (2.243)$$

Further, σ_{zz} can be expressed in terms of σ_{xx} and σ_{yy} as

$$\sigma_{zz} = \nu(\sigma_{xx} + \sigma_{yy}) \qquad (2.244)$$

Thus, we see that there are four nonzero stress components. However, it may be noted that only three stress components are independent.

2.9 SUMMARY

The basic concepts of solid mechanics have been reviewed in this chapter. Solid mechanics is based on the fundamental concept of continuum, which is a state of a continuously distributed material without any crack or flaw. The governing equations belonging to four broad areas—kinematics, kinetics, constitutive relations, and thermodynamics—based on certain fundamental principles govern the behavior of a material; concepts in the first three areas are discussed in detail, whereas an introductory remark is made in respect of thermodynamic principles as applied to solid mechanics.

Kinematics is the subject of geometric changes in a body without any concern for the factors that cause such changes. The variables involved in kinematics are the displacements and strains and, the primary output of kinematics are the strain–displacement relations. Kinetics is the study of forces and moments acting on a body in static or dynamic equilibrium and it is based on conservation of momenta, which gives us the equilibrium equations or equations of motion. Experimental observations together with thermodynamic considerations provide the constitutive relations. The constitutive relations are developed for different idealized materials and they relate kinematics to kinetics. The strain–displacement relations, equations of motion, and constitutive models provide the necessary equations to determine the required parameters such as displacements, strains, and stresses.

Computational efforts can be greatly reduced by the plane stress and plane strain approximations. These approximations are valid in certain boundary conditions, loading, and geometry such that a 3D problem can be treated as 2D with reduced numbers of parameters.

EXERCISE PROBLEMS

2.1 If a bar of length 400 mm elongates under an axial tensile force by 0.4 mm, determine the following strains in the bar: (i) engineering strain, (ii) true strain, (iii) Green strain, and (iv) Almansi strain.

Basic Solid Mechanics

2.2 The displacement vector (coordinate axes O-XYZ and o-xyz are superimposed) at a point is given by

$$\boldsymbol{u} = (4x^2 + 3y + z + 2)\boldsymbol{e}_x + (x + 2y^2 + 2z + 5)\boldsymbol{e}_y + (2x + 3y + 5z^2 + 3)\boldsymbol{e}_z$$

If the initial coordinate of a point is $(1,0,-2)$, determine its coordinates in the deformed configuration.

2.3 The displacement vector (coordinate axes O-XYZ and o-xyz are superimposed) at a point is given by

$$\boldsymbol{u} = (4x^2 + 3y)\boldsymbol{e}_x + (2y^2 + 5)\boldsymbol{e}_y + (5z^2 + 3)\boldsymbol{e}_z$$

Determine the deformation gradient and displacement gradient. Verify that

$$[\boldsymbol{F}] = [\boldsymbol{I}] + [\boldsymbol{D}]$$

2.4 Consider the displacement field given in Exercise 2.3. Determine the infinitesimal strains at a point whose initial coordinates are $(1,-2,4)$.

2.5 The displacement vector (coordinate axes O-XYZ and o-xyz are superimposed) at a point is given by

$$\boldsymbol{u} = (4x + 3y)\boldsymbol{e}_x + (2y - 5)\boldsymbol{e}_y + (5z + 6)\boldsymbol{e}_z$$

If the coordinates of a point in the deformed configuration are $(1, 4, 0)$, determine the original coordinates and the infinitesimal strains at the point.

2.6 Consider the displacement field given in Exercise 2.5. Determine the change in length of the line segment joining two points whose original coordinates are $(4, 0, 2)$ and $(2, 1, 0)$.

2.7 The stress tensor at a point is given by

$$[\boldsymbol{\sigma}] = \begin{bmatrix} 3 & 1 & 0 \\ 1 & 3 & 0 \\ 0 & 0 & 1 \end{bmatrix} \text{MPa}$$

Determine the principal stresses and associated principal planes.

2.8 The stress tensor at a point is given by

$$[\boldsymbol{\sigma}] = \begin{bmatrix} 5 & 1 & 0 \\ 1 & 4 & 0 \\ 0 & 0 & 2 \end{bmatrix} \text{MPa}$$

Determine the stress vectors on the positive and negative xy-, yz-, and zx-planes.

2.9 The stress tensor at a point is given by

$$[\boldsymbol{\sigma}] = \begin{bmatrix} 6 & -2 & 1 \\ -2 & 4 & 0 \\ 1 & 0 & 2 \end{bmatrix} \text{MPa}$$

Determine the shear stress on the plane associated with the normal given by

$$n = \frac{1}{\sqrt{3}}(e_x + e_y + e_z)$$

2.10 The stress tensor at a point is given by

$$[\sigma] = \begin{bmatrix} 2x & y & -z \\ y & 4y & 0 \\ -z & 0 & 6z \end{bmatrix} \text{MPa}$$

Determine the body forces acting on the body if it is at rest.

2.11 The stress tensor at a point is given by

$$[\sigma] = \begin{bmatrix} 250 & 40 & 75 \\ 40 & 200 & 0 \\ 75 & 0 & 200 \end{bmatrix} \text{MPa}$$

Write the stresses as a (6 × 1) vector in the contracted notation.

2.12 The strain tensor at a point is given by

$$[\varepsilon] = \begin{bmatrix} 1.2 & 0.8 & 0.4 \\ 0.8 & 1 & 0.2 \\ 0.4 & 0.2 & 1.1 \end{bmatrix} \times 10^{-4}$$

Write the strains as a (6 × 1) vector in the contracted notation.

2.13 For an anisotropic material, the generalized Hooke's law states that there are 81 elastic constants. Work out systematically and show that under various criteria, the number of these constants can be reduced leading to five and two independent elastic constants, respectively, for transversely isotropic material and isotropic material.

2.14 Young's modulus and shear modulus of aluminum are given as 70 and 26 GPa, respectively. Determine its Poisson's ratio.

2.15 The stress tensor at a point is given as

$$[\sigma] = \begin{bmatrix} 150 & 25 & 0 \\ 25 & 140 & 0 \\ 0 & 0 & 160 \end{bmatrix} \text{MPa}$$

Determine the strain tensor if Young's modulus and shear modulus are 70 and 26 GPa, respectively.

2.16 Derive the elastic stiffness matrix for an orthotropic material in terms of its elastic constants.

Hint: Compliance matrix is given by Equation 2.201. $[C] = [S]^{-1}$.

2.17 Given the compliance matrix of an orthotropic material by Equation 2.201, derive the compliance matrix expression of an isotropic material in terms of its elastic constants.

Hint: For an isotropic material $E_x = E_y = E_z = E$, and so on.

Basic Solid Mechanics

2.18 The material properties of an orthotropic material are as follows:

$E_x = 150\,\text{GPa}, E_y = 12\,\text{GPa}, E_z = 12\,\text{GPa}, G_{yz} = 4.5\,\text{GPa}, G_{xy} = 8\,\text{GPa},$
$G_{zx} = 8\,\text{GPa}, \nu_{xy} = 0.2, \nu_{xz} = 0.2, \text{ and } \nu_{yz} = 0.3$

Determine the stiffness matrix and the compliance matrix.

2.19 For an orthotropic material, the elastic constants are given as

$E_x = 150\,\text{GPa}, E_y = 12\,\text{GPa}, E_z = 12\,\text{GPa}, G_{yz} = 4.5\,\text{GPa}, G_{xy} = 8\,\text{GPa},$
$G_{zx} = 8\,\text{GPa}, \nu_{xy} = 0.2, \nu_{xz} = 0.2, \text{ and } \nu_{yz} = 0.3$

The stress tensor at a point is given as

$$[\sigma] = \begin{bmatrix} 1200 & 30 & 40 \\ 30 & 25 & 0 \\ 40 & 0 & 20 \end{bmatrix} \text{MPa}$$

Determine the strain tensor.

2.20 In a thin plate (in the xy-plane) under in-plane loads, the strains are given as

$$\varepsilon_{xx} = \varepsilon_{yy} = 1.2 \times 10^{-4} \quad \text{and} \quad \gamma_{xy} = 0.$$

Determine the stresses if $E = 200$ GPa and $\nu = 0.3$.
What is the normal and shear strains in the z-direction and xz-/yz-planes?

Hint: Use plane stress idealization.

2.21 In a thin plate (in the xy-plane) under in-plane loads, the following stresses are applied:

$$\sigma_{xx} = 1400\,\text{MPa}, \sigma_{yy} = 250\,\text{MPa}, \text{ and } \tau_{xy} = 0$$

If the out-of-plane stresses are zero, determine the strains in the plate. The orthotropic material properties are as follows:

$E_x = 160\,\text{GPa}, E_y = 8\,\text{GPa}, E_z = 8\,\text{GPa}, G_{yz} = 3\,\text{GPa}, G_{xy} = 6\,\text{GPa},$
$G_{zx} = 6\,\text{GPa}, \nu_{xy} = 0.2, \nu_{xz} = 0.2, \text{ and } \nu_{yz} = 0.33$

Solve it first by using 3D orthotropic constitutive relation and then verify the results by using plane stress idealization.

2.22 A long tube of internal diameter 80 mm and thickness 8 mm is pressurized to 120 MPa. Determine the stresses and strains in the pipe.
Following data are given:

$$E = 200\,GPa, \nu = 0.3$$

Hint: Use force equilibrium to determine the membrane stresses. Use plane strain idealization for strains.

2.23 Solve the problem in Exercise 2.22 if the material is changed to orthotropic with the following data:

$$E_x = 60\,\text{GPa}, E_y = 60\,\text{GPa}, G_{xy} = 4\,\text{GPa}, \nu_{xy} = 0.2$$

2.24 Consider two Cartesian coordinate systems $O\text{-}xyz$ and $O\text{-}x'y'z'$, where the second system is obtained by rotating the first one about x-axis by 30°. The stress at a point in the first coordinate system by

$$[\sigma] = \begin{bmatrix} 1200 & 30 & 40 \\ 30 & 25 & 0 \\ 40 & 0 & 20 \end{bmatrix} \text{MPa}$$

What is the stress tensor in the second coordinate system?

2.25 Consider the problem given in Exercise 2.24. Determine the strains in the first coordinate system and then get the strains by transformation. Verify the results by first transformation of stress to the second coordinate system followed by determination of strains. Assume the following isotropic material data:

$$E = 200\,\text{GPa}, \nu = 0.3$$

REFERENCES AND SUGGESTED READING

1. J. N. Reddy, *An Introduction to Continuum Mechanics—With Applications*, Cambridge University Press, Cambridge, 2010.
2. S. P. Timoshenko and J. N. Goodier, *Theory of Elasticity*, third edition, McGraw-Hill, New York, 1970.
3. A. K. Singh, *Mechanics of Solids*, PHI Learning, New Delhi, 2011.
4. G. E. Mase, *Theory and Problems of Continuum Mechanics*, McGraw-Hill, New York, 1970.
5. E. H. Dill, *Continuum Mechanics: Elasticity, Plasticity and Viscoelasticity*, CRC Press, Boca Raton, FL, 2007.
6. A.-R. Ragab and S. E. Bayoumi, *Engineering Solid Mechanics—Fundamentals and Applications*, CRC Press, Boca Raton, FL, 1999.
7. Y. C. Fung and P. Tong, *Classical and Computational Solid Mechanics*, World Scientific Publishing, Singapore, 2001.
8. Y. C. Fung, *Foundations of Solid Mechanics*, Prentice Hall Inc., Englewood Cliffs, NJ, 1965.
9. I. H. Shames and F. A. Cozzereli, *Elastic and Inelastic Stress Analysis*, Taylor & Francis, London, 1997.

3

Micromechanics of a Lamina

3.1 CHAPTER ROAD MAP

A laminate is a laminated composite structural element, and laminate design is a crucial aspect in the overall design of a composite structure. As mentioned in Chapter 1, laminae are the building blocks in a composite structure; knowledge of lamina behavior is essential for the design of a composite structure and analysis of a lamina is the starting point. Figure 3.1 presents a schematic representation of the process of composite laminate analysis (and design) at different levels. A lamina is a multiphase element and its behavior can be studied at two levels—micro level and macro level. For micromechanical analysis of a lamina, the necessary input data are obtained from the experimental study of its constituents, viz. reinforcements and matrix, and lamina behavior is estimated as functions of the constituent properties. The lamina characteristics are then used in the analysis of the lamina at the macro level and subsequent laminate design and analysis. Alternatively, the input data for the macro-level analysis of a lamina and subsequent laminate design and analysis can be directly obtained from an experimental study of the lamina. Thus, in the context of product design, the micromechanics of a lamina can be considered as an alternative to the experimental study of the lamina.

In this chapter, we provide an introductory remark followed by a brief review of the basic micromechanics concepts. There are many micromechanics models in the literature. Our focus is not a review of these models; instead, we dwell on the formulations of some mechanics of materials-based models for the evaluation of lamina thermoelastic parameters and briefly touch upon the elasticity-based models and semiempirical models.

3.2 PRINCIPAL NOMENCLATURE

A	Area of cross section of a representative volume element
A_c, A_f, A_m	Areas of cross section of composite, fibers, and matrix, respectively, in a representative volume element
b_c, b_f, b_m	Widths of composite, fibers, and matrix, respectively, in a representative volume element
d	Fiber diameter
E_c	Young's modulus of isotropic composite
E_{1c}, E_{2c}	Young's moduli in the longitudinal and transverse directions, respectively, of transversely isotropic composite
E_f	Young's modulus of isotropic fibers
E_{1f}, E_{2f}	Young's moduli in the longitudinal and transverse directions, respectively, of transversely isotropic fibers
E_m	Young's modulus of matrix
F_c	Total force on composite (representative volume element)
F_f, F_m	Forces shared by the fibers and matrix, respectively
G_f	Shear modulus of isotropic fibers

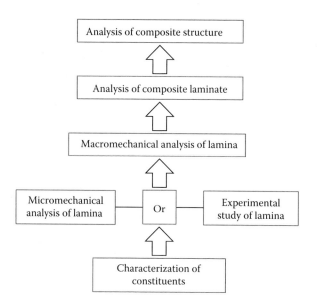

FIGURE 3.1 Schematic representation of composite laminate analysis process.

G_{12f}, G_{23f}	Shear moduli in the longitudinal and transverse planes, respectively, of transversely isotropic fibers
G_m	Shear modulus of matrix
l, b, t	Length, width, and thickness, respectively, of a representative volume element
l_c, l_f, l_m	Lengths of composite, fibers, and matrix, respectively, in a representative volume element
s	Fiber spacing
t_c, t_f, t_m	Thicknesses of composite, fibers, and matrix, respectively, in a representative volume element
V_f, V_m, V_v	Fiber volume fraction, matrix volume fraction, and voids volume fraction, respectively
$(V_f)_{cri}, (V_f)_{min}$	Critical fiber volume fraction and minimum fiber volume fraction, respectively
v_c	Total volume of composite
v_f, v_m, v_v	Volumes of fibers, matrix, and voids, respectively
W_f, W_m	Mass fraction of fibers and mass fraction of matrix, respectively
w_c	Total weight of composite
w_f, w_m	Mass of fibers and mass of matrix, respectively
α_c	Coefficient of thermal expansion of isotropic composite
α_{1c}, α_{2c}	Longitudinal and transverse coefficients of thermal expansion, respectively, of transversely isotropic composite
α_{1f}, α_{2f}	Longitudinal and transverse coefficients of thermal expansion, respectively, of transversely isotropic fibers
α_m	Coefficient of thermal expansion of matrix
β_c	Coefficient of moisture expansion of isotropic composite
β_{1c}, β_{2c}	Longitudinal and transverse coefficients of moisture expansion, respectively, of transversely isotropic composite
β_{1f}, β_{2f}	Longitudinal and transverse coefficients of moisture expansion, respectively, of transversely isotropic fibers
β_m	Coefficient of moisture expansion of matrix
$\gamma_{12c}, \gamma_{23c}$	Longitudinal (in a longitudinal plane) and transverse (in a transverse plane) shear strains, respectively, in composite

Micromechanics of a Lamina

$(\gamma_c)_{ult}$	Ultimate shear strain in isotropic composite
$(\gamma_{12c})_{ult}, (\gamma_{23c})_{ult}$	Ultimate longitudinal (in a longitudinal plane) and transverse (in a transverse plane) shear strains, respectively, in transversely isotropic composite
$\gamma_{12f}, \gamma_{23f}$	Longitudinal (in a longitudinal plane) and transverse (in a transverse plane) shear strains, respectively, in fibers
$(\gamma_f)_{ult}$	Ultimate shear strain in isotropic fibers
$(\gamma_{12f})_{ult}, (\gamma_{23f})_{ult}$	Ultimate longitudinal (in a longitudinal plane) and transverse (in a transverse plane) shear strains, respectively, in transversely isotropic fibers
$\gamma_{12m}, \gamma_{23m}$	Longitudinal (in a longitudinal plane) and transverse (in a transverse plane) shear strain, respectively, in matrix
$(\gamma_m)_{ult}$	Ultimate shear strain in matrix
$\Delta_c, \Delta_f, \Delta_m$	Deformations in composite, fibers, and matrix, respectively
$\Delta C_c, \Delta C_f, \Delta C_m$	Changes in moisture content in composite, fibers, and matrix, respectively
Δl	Change in length of a representative volume element
$\Delta l_c, \Delta l_f, \Delta l_m$	Changes in length of composite, fibers, and matrix, respectively, in a representative volume element
ΔT	Change in temperature
$\varepsilon_{1c}^T, \varepsilon_{2c}^T$	Longitudinal and transverse tensile strains, respectively, in composite
$\varepsilon_{1c}^C, \varepsilon_{2c}^C$	Longitudinal and transverse compressive strains, respectively, in composite
$(\varepsilon_c^T)_{ult}$	Ultimate tensile strain in isotropic composite
$(\varepsilon_{1c}^T)_{ult}, (\varepsilon_{2c}^T)_{ult}$	Ultimate longitudinal and transverse tensile strains, respectively, in transversely isotropic composite
$(\varepsilon_{1c}^C)_{ult}, (\varepsilon_{2c}^C)_{ult}$	Ultimate longitudinal and transverse compressive strains, respectively, in transversely isotropic composite
$\varepsilon_{1f}^T, \varepsilon_{2f}^T$	Longitudinal and transverse tensile strains, respectively, in fibers
$\varepsilon_{1f}^C, \varepsilon_{2f}^C$	Longitudinal and transverse compressive strains, respectively, in fibers
$(\varepsilon_f^T)_{ult}$	Ultimate tensile strain in isotropic fibers
$(\varepsilon_{1f}^T)_{ult}, (\varepsilon_{2f}^T)_{ult}$	Ultimate longitudinal and transverse tensile strains, respectively, in transversely isotropic fibers
$(\varepsilon_{1f}^C)_{ult}, (\varepsilon_{2f}^C)_{ult}$	Ultimate longitudinal and transverse compressive strains, respectively, in transversely isotropic fibers
$\varepsilon_{1m}^T, \varepsilon_{2m}^T$	Longitudinal and transverse tensile strains, respectively, in matrix
$\varepsilon_{1m}^C, \varepsilon_{2m}^C$	Longitudinal and transverse compressive strains, respectively, in matrix
$(\varepsilon_m^T)_{ult}$	Ultimate tensile strain in matrix
η	Fiber packing factor (in Halpin–Tsai equations)
ν_f	Poisson's ratio of isotropic fibers
ν_{12f}, ν_{23f}	Major Poisson's ratios (in the longitudinal plane and transverse plane, respectively) of transversely isotropic fibers
ν_m	Poisson's ratio of matrix
ξ	Reinforcing factor (in Halpin–Tsai equations)
ρ_c, ρ_f, ρ_m	Density of composite, fibers, and matrix, respectively
$\sigma_{1c}^T, \sigma_{2c}^T$	Longitudinal and transverse tensile stresses, respectively, in composite
$\sigma_{1c}^C, \sigma_{2c}^C$	Longitudinal and transverse compressive stresses, respectively, in composite
$(\sigma_c^T)_{ult}$	Ultimate tensile stress in isotropic composite (i.e., tensile strength of isotropic composite)

$(\sigma_{1c}^T)_{ult}, (\sigma_{2c}^T)_{ult}$	Ultimate longitudinal and transverse tensile stresses, respectively, in transversely isotropic composite (i.e., longitudinal and transverse tensile strengths of transversely isotropic composite)
$(\sigma_{1c}^C)_{ult}, (\sigma_{2c}^C)_{ult}$	Ultimate longitudinal and transverse compressive stresses, respectively, in transversely isotropic composite (i.e., longitudinal and transverse compressive strengths of transversely isotropic composite)
$\sigma_{1f}^T, \sigma_{2f}^T$	Longitudinal and transverse tensile stresses in fibers
$\sigma_{1f}^C, \sigma_{2f}^C$	Longitudinal and transverse compressive stresses in fibers
$(\sigma_f^T)_{ult}$	Ultimate tensile stress in isotropic fibers (i.e., tensile strength of isotropic fibers)
$(\sigma_{1f}^T)_{ult}, (\sigma_{2f}^T)_{ult}$	Ultimate longitudinal and transverse tensile stresses, respectively, in transversely isotropic fibers (i.e., longitudinal and transverse tensile strengths of transversely isotropic fibers)
$(\sigma_{1f}^C)_{ult}, (\sigma_{2f}^C)_{ult}$	Ultimate longitudinal and transverse compressive stresses, respectively, in transversely isotropic fibers (i.e., longitudinal and transverse compressive strengths of transversely isotropic fibers)
$\sigma_{1m}^T, \sigma_{2m}^T$	Longitudinal and transverse tensile stresses, respectively, in matrix
$\sigma_{1m}^C, \sigma_{2m}^C$	Longitudinal and transverse compressive stresses, respectively, in matrix
$(\sigma_m^T)_{ult}, (\sigma_m^C)_{ult}$	Ultimate tensile and compressive stresses, respectively, in matrix (i.e., tensile and compressive strengths of matrix)
τ_{12c}, τ_{23c}	Longitudinal (in a longitudinal plane) and transverse (in a transverse plane) shear stresses, respectively, in composite
$(\tau_c)_{ult}$	Ultimate shear stress (i.e., shear strength) of isotropic composite
$(\tau_{12c})_{ult}, (\tau_{23c})_{ult}$	Ultimate longitudinal and transverse shear stresses, respectively, in transversely isotropic composite (i.e., longitudinal and transverse shear strengths)
τ_{12f}, τ_{23f}	Longitudinal (in a longitudinal plane) and transverse (in a transverse plane) shear stresses, respectively, in fibers
$(\tau_f)_{ult}$	Ultimate shear stress (i.e., shear strength) of isotropic fibers
$(\tau_{12f})_{ult}, (\tau_{23f})_{ult}$	Ultimate longitudinal and transverse shear stresses, respectively, in transversely isotropic fibers (i.e., longitudinal and transverse shear strength)
τ_{12m}, τ_{23m}	Longitudinal (in a longitudinal plane) and transverse (in a transverse plane) shear stresses, respectively, in matrix
$(\tau_m)_{ult}$	Ultimate shear stress in matrix (i.e., shear strength of matrix)

3.3 INTRODUCTION

A composite lamina is made up of two constituents—reinforcements and matrix. As we know, these constituents combine together and act in unison as a single entity. Micromechanics is the study in which the interaction of the reinforcements and the matrix is considered and their effect on the gross behavior of the lamina is determined. Toward this, we need to determine several thermoelastic parameters of the lamina in terms of constituent properties. These parameters include

- Elastic moduli
- Strength parameters
- Coefficients of thermal expansion (CTEs)
- Coefficients of moisture expansion (CMEs)

Extensive work, as reflected by numerous research papers available in the literature, has been done in the field of micromechanics. The subject is also discussed at different levels of treatment in many texts on the mechanics of composites [1–5]. Micromechanics models have been of keen research interest and several approaches have been adopted to develop models for the prediction of various parameters, especially elastic moduli, of a unidirectional lamina. A detailed survey of various approaches is provided by Chamis and Sendeckyj [6]; these approaches are netting analysis, mechanics of materials, self-consistent models, bounding techniques based on variational principles, exact solutions, statistical methods, finite element methods, microstructure theories, and semiempirical models. The netting models and mechanics of materials-based models involve grossly simplifying assumptions. The rest of the approaches are based on the principles of elasticity and they, barring the semiempirical models, are typically associated with rigorous treatment and complex mathematical and graphical expressions. Thus, for the sake of convenience of discussion, the micromechanics models can be put into a simple classification as follows:

- Netting models
- Mechanics of materials-based models
- Elasticity-based models
- Semiempirical models

Netting models are highly simplified models in which the bond between the fibers and the matrix is ignored for estimating the longitudinal stiffness and strength of a unidirectional lamina; it is assumed that longitudinal stiffness and strength are provided completely by the fibers. On the other hand, transverse and shear stiffness and Poisson's effect are assumed to be provided by the matrix. These models typically underestimate the properties of a lamina but due to their simplicity they are still used in the preliminary ply design of pressure vessels [7].

The mechanics of materials-based models too involve grossly simplifying assumptions (see, for instance, References 8–10). Averaged stresses and strains are used in force and energy balance in a representative volume element (RVE) to derive the desired expressions for elastic parameters. Typically, the continuity of displacement across the interface between the constituents is maintained. Some of the common assumptions in micromechanics (see Section 3.4.1) are relaxed/modified suitably and a number of mechanics of materials-based models have been proposed in the past. Several of these models relate to different assumed geometrical array of fibers (square, rectangular, hexagonal, etc.), fiber alignment, inclusion of voids, etc.

Elasticity-based models involve more rigorous treatment of the lamina behavior (see, for instance, References 11–20). In an exact method, an elasticity problem within the general frame of assumptions (see Section 3.4.1) is formulated and solved by various techniques, including numerical methods such as the finite element method. A variation of the exact method is the self-consistent model. Variational principles are employed to obtain bounds on the elastic parameters. In the statistical methods, the restrictions of aligned fibers in regular array are relaxed and the elastic parameters are allowed to vary randomly with position. All these models, however, are somewhat complex and they have limited utility in the design of a product. Also, many variables that actually influence the lamina elastic behavior are ignored, leading to unreliable estimates. In semiempirical models, the mathematical complexity is reduced and the effects of process-related variables are taken into account by incorporating empirical factors [21].

An exhaustive discussion of the models available in the literature is beyond the scope of this book; for in-depth reviews, interested readers can refer to References 6, 22, and 23 and the bibliographies provided therein. In this chapter, we shall attempt to provide an overall idea required in a product design environment. With this in mind, we shall discuss the mechanics of materials models in detail for all the parameters listed above.

A brief discussion is also provided on the elasticity approach and the semiempirical approach for the elastic moduli.

3.4 BASIC MICROMECHANICS

3.4.1 Assumptions and Restrictions

Micromechanics models are based on a number of simplifying assumptions and restrictions in respect of lamina, its constituents, that is, fibers and matrix, and the interface. These assumptions and restrictions are as follows:

- The lamina is (i) macroscopically homogeneous, (ii) macroscopically orthotropic, (iii) linearly elastic, and (iv) initially stress-free.
- The fibers are (i) homogeneous, (ii) linearly elastic, (iii) isotropic, (iv) regularly spaced, (v) perfectly aligned, and (vi) void-free.
- The matrix is (i) homogeneous, (ii) isotropic, (iii) linearly elastic, and (iv) void-free.
- The interface between fibers and matrix has (i) perfect bond, (ii) no voids, and (iii) no interphase, that is, fiber–matrix interaction zone.

Some of the restrictions are not realistic and some of them are relaxed in the derivations of various models. For example, glass fibers are isotropic, but carbon and aramid fibers are highly anisotropic. They can be considered as transversely isotropic and their elastic moduli and strengths are direction-dependent. As we shall see in the next section, the mechanics of materials-based models discussed here can accommodate anisotropic (transversely isotropic) fibers. Fibers are generally randomly spaced and their alignment is not perfect. Similarly, the matrix can have voids and the lamina can have initial stresses. Also, an interphase is present at the interface between the fibers and the matrix.

3.4.2 Micromechanics Variables

The general procedure, irrespective of the micromechanics model used, is to express the desired parameter in terms of a number of basic micromechanics variables. These variables are as follows:

- Elastic moduli of fibers and matrix
- Strengths of fibers and matrix
- Densities of fibers and matrix
- Volume fractions of fibers, matrix, and voids
- Mass fractions of fibers and matrix

3.4.2.1 Elastic Moduli and Strengths of Fibers and Matrix

The elastic moduli and strengths of fibers and matrix are determined experimentally. The number of these parameters to be determined experimentally for use in micromechanics would depend on the restriction in respect of behaviors of fibers and matrix. Certain fibers such as carbon are highly anisotropic and they can be considered as transversely isotropic. For these fibers, we need five stiffness parameters: E_{1f}, E_{2f}, G_{12f}, ν_{12f}, and ν_{23f}. For isotropic fibers such as glass, the number of stiffness parameters reduces to three—E_f, G_f, and ν_f. On the other hand, all common matrix materials are isotropic for which we need the three stiffness parameters—E_m, G_m, and ν_m. Further, under the restriction of homogeneousness, all of these parameters are uniform across the fibers or matrix.

3.4.2.2 Volume Fractions

As we know, a composite material is made up of primarily two constituents—fibers and matrix. However, during the manufacture of a composite laminate, deviations do

Micromechanics of a Lamina

occur and voids are introduced. Thus, the total volume of a composite material consists of three parts—fibers, matrix, and voids. Fiber volume fraction is defined as the ratio of the volume of fibers in the composite material to the total volume of composite. Similarly, matrix volume fraction is defined as the ratio of the volume of matrix to the total volume of composite, and voids volume fraction is defined as the ratio of the volume of voids to the total volume of composite. Thus,

$$V_f = \frac{v_f}{v_c}, \quad V_m = \frac{v_m}{v_c}, \quad \text{and} \quad V_v = \frac{v_v}{v_c} \quad (3.1)$$

where
- V_f fiber volume fraction
- V_m matrix volume fraction
- V_v voids volume fraction
- v_f volume of fibers
- v_m volume of matrix
- v_v volume of voids
- v_c total volume of composite material

It is clear that

$$v_f + v_m + v_v = v_c \quad (3.2)$$

Dividing both the sides by v_c, we get

$$V_f + V_m + V_v = 1 \quad (3.3)$$

For an ideal composite material, $v_v = V_v = 0$ and we get

$$V_f + V_m = 1 \quad (3.4)$$

We shall see in the subsequent sections that fiber volume fraction is a key parameter that greatly influences lamina properties such as longitudinal modulus and major Poisson's ratio. It is useful to know the theoretical maximum fiber volume fraction of a lamina. In a composite material, fibers are packed in a random fashion. However, with a view to determining the maximum theoretical fiber volume fraction, as shown in Figure 3.2, let us consider two regular arrays of fibers—square array and triangular array. Fiber volume fractions can be expressed as

$$\text{For square array,} \quad V_f = \frac{\pi d^2}{4s^2} \quad (3.5)$$

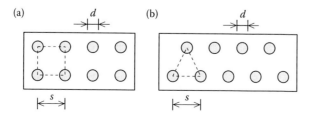

FIGURE 3.2 Schematic representation of fiber packing. (a) Square array. (b) Triangular array.

and

$$\text{For triangular array,} \quad V_f = \frac{\pi d^2}{2\sqrt{3}s^2} \tag{3.6}$$

where d and s are fiber diameter and fiber spacing, respectively.

For maximum fiber packing, $d = s$. Thus, theoretical maximum fiber volume fractions with fibers of circular cross section are

$$\text{For square array,} \quad (V_f)_{max} = \frac{\pi}{4} = 0.79 \tag{3.7}$$

$$\text{For triangular array,} \quad (V_f)_{max} = \frac{\pi}{2\sqrt{3}} = 0.91 \tag{3.8}$$

where $(V_f)_{max}$ is the theoretical maximum fiber volume fraction.

3.4.2.3 Mass Fractions

Fiber mass fraction is defined as the ratio of the mass of fibers to the total mass of composite material. Similarly, matrix mass fraction is defined as the ratio of the mass of matrix to the total mass of composite. Thus,

$$W_f = \frac{w_f}{w_c} \tag{3.9}$$

$$W_m = \frac{w_m}{w_c} \tag{3.10}$$

where
- W_f fiber mass fraction
- W_m matrix mass fraction
- w_f mass of fibers
- w_m mass of matrix
- w_c mass of the composite material

It is clear that

$$w_c = w_f + w_m \tag{3.11}$$

Now, we know that the product of density and volume is the mass contained in that volume. Then, for the composite, fibers, and matrix, we can write the following:

$$w_c = \rho_c v_c \tag{3.12}$$

$$w_f = \rho_f v_f \tag{3.13}$$

$$w_m = \rho_m v_m \tag{3.14}$$

where ρ_c, ρ_f, and ρ_m are densities of composite, fibers, and matrix, respectively.

Substituting Equations 3.12 through 3.14 in Equation 3.11, we get

$$\rho_c v_c = \rho_f v_f + \rho_m v_m \tag{3.15}$$

Dividing both the sides by v_c and using Equation 3.1, we get

$$\rho_c = \rho_f V_f + \rho_m V_m \tag{3.16}$$

Equation 3.16 is the rule of mixtures expression for density of composite.

Now, substituting Equations 3.12 through 3.14 in Equations 3.9 and 3.10, we get the following:

$$W_f = \frac{\rho_f}{\rho_c} V_f \tag{3.17}$$

$$W_m = \frac{\rho_m}{\rho_c} V_m \tag{3.18}$$

Then, substituting Equation 3.16 in Equations 3.17 and 3.18, with simple manipulation, we get the expressions for mass fractions for fibers and matrix as

$$W_f = \frac{(\rho_f/\rho_m)V_f}{(\rho_f/\rho_m)V_f + V_m} \tag{3.19}$$

Taking voids fraction as zero, $V_m = 1 - V_f$, and we get the following:

$$W_f = \frac{(\rho_f/\rho_m)V_f}{1 + (\rho_f/\rho_m - 1)V_f} \tag{3.20}$$

3.4.3 Representative Volume Element

An RVE is considered for obtaining expressions of the various elastic moduli and strengths. Figure 3.3a shows the schematic representation of a unidirectional lamina. The fibers are taken as straight and regularly aligned. Let the fiber spacings be b_c and t_c in the width and thickness directions, respectively. Then, we take an RVE of size $l_c \times b_c \times t_c$ as shown in Figure 3.3b such that by placing the RVEs repeatedly next to each other, we can obtain the complete lamina. Further, it is presumed that the responses of the RVEs to applied loads are identical and thus the analysis of an RVE

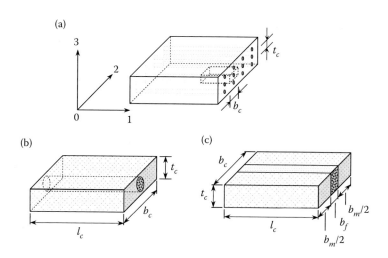

FIGURE 3.3 (a) Schematic representation of a unidirectional lamina. (b) Representative volume element. (c) Idealized volume element. (Adapted with permission from A. K. Kaw, *Mechanics of Composite Materials*, CRC Press, Boca Raton, FL, 2006.)

is sufficient for determining the characteristics of the complete lamina. The RVE is further simplified as shown in Figure 3.3c.

Now, the total cross-sectional area of composite in the RVE, A_c, the cross-sectional area of the fibers, A_f, and the cross-sectional area of the matrix, A_m, are, respectively, given by

$$A_c = b_c t_c \tag{3.21}$$

$$A_f = b_f t_c \tag{3.22}$$

$$A_m = b_m t_c \tag{3.23}$$

It is easy to see that for zero voids fraction,

$$A_c = A_f + A_m \tag{3.24}$$

3.5 MECHANICS OF MATERIALS-BASED MODELS

3.5.1 Evaluation of Elastic Moduli

A unidirectional lamina (Figure 3.3a) is an orthotropic body characterized by four elastic constants—longitudinal modulus (E_{1c}) along the fiber direction, transverse modulus (E_{2c}) normal to the fiber direction, shear modulus (G_{12c}) in the plane of the lamina, and major Poisson's ratio (ν_{12c}).

Notes:

- We have used a Cartesian coordinate system O-123 usually known as the material coordinate system. Here, 1-direction is the longitudinal direction, which is along the fibers, 2-direction is the transverse direction, which is normal to the fibers in the plane of the lamina, and 3-direction is normal to the plane of the lamina.
- In the general nomenclature, composite elastic moduli are represented by E_1, E_2, G_{12}, etc. However, in this chapter, we shall add an additional suffix "c" to stress on the fact that the parameter belongs to the composite. Similarly, suffixes "f" and "m" are used for fibers and matrix, respectively. Thus, E_{1c} is the longitudinal Young's modulus of composite, E_{2f} is the transverse Young's modulus of fibers, E_m is the Young's modulus of matrix, and so on.

3.5.1.1 Longitudinal Modulus (E_{1c})

Let us consider a unidirectional lamina under uniaxial load in the fiber direction. An RVE under this loading condition is shown in Figure 3.4a. The RVE can be compared with a system of springs with different stiffnesses in parallel. This springs-in-parallel analogy is shown in Figure 3.4b.

Now, the total force taken by the volume element is shared by the fibers and the matrix. Thus,

$$F_c = F_f + F_m \tag{3.25}$$

where

F_c total force on the representative volume element
F_f force shared by the fibers
F_m force shared by the matrix

Micromechanics of a Lamina

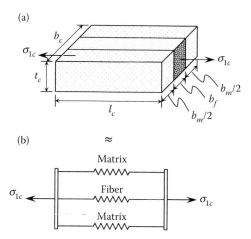

FIGURE 3.4 (a) Representative volume element under uniaxial stress in the fiber direction. (b) Springs-in-parallel analogy. (Adapted in parts with permission from R. M. Jones, *Mechanics of Composite Materials*, second edition, Taylor & Francis, New York, 1999; A. K. Kaw, *Mechanics of Composite Materials*, CRC Press, Boca Raton, FL, 2006.)

From Equation 3.25, we obtain

$$\sigma_{1c} A_c = \sigma_{1f} A_f + \sigma_{1m} A_m \tag{3.26}$$

where

- σ_{1c} longitudinal stress in the composite material
- σ_{1f} longitudinal stress in the fibers
- σ_{1m} longitudinal stress in the matrix

Now, under the restriction that the composite, fibers, and matrix are elastic, we bring in Hooke's law and write Equation 3.26:

$$E_{1c} \varepsilon_{1c} A_c = E_{1f} \varepsilon_{1f} A_f + E_m \varepsilon_{1m} A_m \tag{3.27}$$

The fibers and matrix are perfectly bonded, and thus, the longitudinal strains in the composite, fibers, and matrix are equal, that is, $\varepsilon_{1c} = \varepsilon_{1f} = \varepsilon_{1m}$. Then, from Equation 3.27, we obtain

$$E_{1c} = E_{1f} \frac{A_f}{A_c} + E_m \frac{A_m}{A_c} \tag{3.28}$$

In the above equation, we can multiply the numerator and the denominator in the area fractions by the length l_c of the RVE and see that the area fractions are equal to the corresponding volume fractions. Thus, we obtain the expression for the longitudinal modulus as follows:

$$E_{1c} = E_{1f} V_f + E_m V_m \tag{3.29}$$

Equation 3.29 is a very popular one and it is referred to as the "rule of mixtures" for the longitudinal modulus of a unidirectional composite. Under the restriction that there is no void in the composite, we can also write it as

$$E_{1c} = E_{1f} V_f + E_m (1 - V_f) \tag{3.30}$$

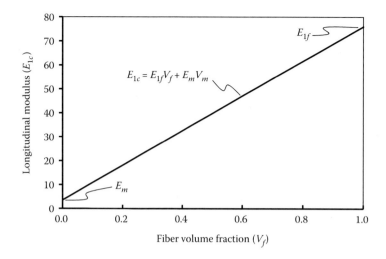

FIGURE 3.5 Longitudinal modulus by mechanics of materials approach (constituent material data from Example 3.1).

Figure 3.5 shows the variation of the longitudinal modulus w.r.t. the fiber volume fraction for the data given in Example 3.1. As seen from the figure, the rule of mixtures gives a simple linear relation in terms of the constituent moduli and volume fractions. It is widely used in design and analysis; it is not only simple but also reliable as predictions made for the longitudinal modulus by the rule of mixtures tally well with experimental results. For most advanced polymeric matrix composite materials, the fiber modulus is far higher than the matrix modulus. In these materials, changes in the matrix modulus do not have any appreciable impact on the composite modulus. Further, as we mentioned before, the RVE can be compared with a system of springs-in-parallel. From the springs-in-parallel analogy (Figure 3.4b) of the RVE, it can be seen that the resultant stiffness of the three springs is controlled by the stiffer spring, viz. the fibers. Thus, we may conclude that the longitudinal modulus of a unidirectional lamina is a fiber-dominated property.

3.5.1.2 Transverse Modulus (E_{2c})

An RVE stressed in the transverse direction as shown in Figure 3.6a is considered next. Under the load as shown in the figure, the RVE undergoes gross extension in the transverse direction. Owing to Poisson's effect, it undergoes contraction in the longitudinal

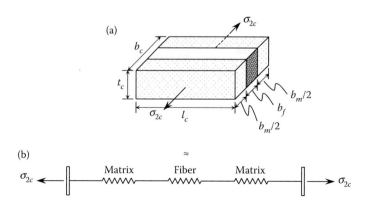

FIGURE 3.6 (a) Representative volume element under transverse stress. (b) Springs-in-series analogy. (Adapted in parts with permission from R. M. Jones, *Mechanics of Composite Materials*, second edition, Taylor & Francis, New York, 1999; A. K. Kaw, *Mechanics of Composite Materials*, CRC Press, Boca Raton, FL, 2006.)

Micromechanics of a Lamina

direction. The RVE can be compared with a system of springs with different stiffnesses in series. This springs-in-series analogy is shown in Figure 3.6b. The gross transverse extension in the transverse is the sum total of transverse extensions in the fibers and matrix. Thus,

$$\Delta_c = \Delta_f + \Delta_m \tag{3.31}$$

where
- Δ_c gross transverse extension in the composite
- Δ_f transverse extension in the fibers
- Δ_m transverse extension in the matrix

Bringing in the definition of normal strains, Equation 3.31 can be written as

$$\varepsilon_{2c} b_c = \varepsilon_{2f} b_f + \varepsilon_{2m} b_m \tag{3.32}$$

where
- ε_{2c} transverse strain in the composite
- ε_{2f} transverse strain in the fibers
- ε_{2m} transverse strain in the matrix

Dividing both the sides by b_c, Equation 3.32 can be written as

$$\varepsilon_{2c} = \varepsilon_{2f} \frac{b_f}{b_c} + \varepsilon_{2m} \frac{b_m}{b_c} \tag{3.33}$$

Now, multiplying the numerator and denominator, in the width fractions in the right-hand side of the above equation, by the product of length and thickness of the RVE, $l_c t_c$, we see that the width fractions are nothing but fiber volume fraction and matrix volume fraction, respectively. Thus,

$$\frac{b_f}{b_c} = V_f \tag{3.34}$$

$$\frac{b_m}{b_c} = V_m \tag{3.35}$$

Further, transverse strains in composite, fibers, and matrix are related to the respective moduli as

$$\varepsilon_{2c} = \frac{\sigma_{2c}}{E_{2c}} \tag{3.36}$$

$$\varepsilon_{2f} = \frac{\sigma_{2f}}{E_{2f}} \tag{3.37}$$

$$\varepsilon_{2m} = \frac{\sigma_{2m}}{E_m} \tag{3.38}$$

Then, substituting Equations 3.34 through 3.38 in Equation 3.33, we get

$$\frac{\sigma_{2c}}{E_{2c}} = \frac{\sigma_{2f}}{E_{2f}} V_f + \frac{\sigma_{2m}}{E_m} V_m \tag{3.39}$$

Now, we look at the RVE under transverse stress in Figure 3.6, and notice that the cross-sectional area normal to the transverse stress is the same for the composite as a whole as well as the fibers and matrix. Thus,

$$\sigma_{2c} = \sigma_{2f} = \sigma_{2m} \tag{3.40}$$

Using Equation 3.40 in Equation 3.39, we get

$$\frac{1}{E_{2c}} = \frac{V_f}{E_{2f}} + \frac{V_m}{E_m} \tag{3.41}$$

or

$$E_{2c} = \frac{E_{2f} E_m}{E_m V_f + E_{2f} V_m} \tag{3.42}$$

Taking void content as zero, Equation 3.42 can be written as

$$E_{2c} = \frac{E_{2f} E_m}{E_m V_f + E_{2f}(1 - V_f)} \tag{3.43}$$

The variation of E_{2c} with V_f for the data given in Example 3.1, based on Equation 3.43, is shown in Figure 3.7. The variation in the transverse modulus is rather sharp at high fiber volume fractions. Such high fiber volume fractions, however, are unrealistic. On the other hand, E_{2c} rises at a very low rate up to a fiber volume fraction of about 0.8 and it is very close to the matrix modulus. Further, as mentioned earlier, the representative volume under transverse stress can be with a system of springs-in-series. From the springs-in-series analogy (Figure 3.6b), we can see that the resultant stiffness of the springs is influenced heavily by the weak springs (matrix). In a unidirectional composite lamina under transverse stress, gross deformation of the lamina is primarily dependent on the matrix deformations. Thus, we may conclude that the transverse modulus of a unidirectional lamina is a matrix-dominated property.

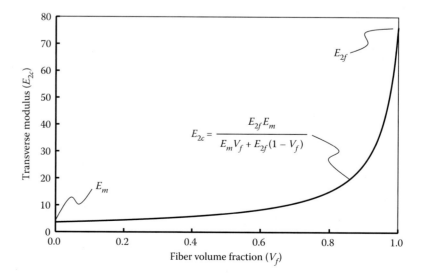

FIGURE 3.7 Transverse modulus by mechanics of materials approach (constituent material data from Example 3.1).

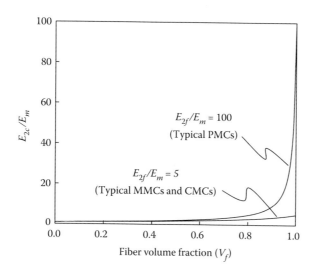

FIGURE 3.8 Variation of transverse modulus with different fiber-to-matrix modulus ratios.

Another way to express the composite transverse modulus is in the nondimensionalized form as follows:

$$\frac{E_{2c}}{E_m} = \frac{1}{1 + (E_m/E_{2f} - 1)V_f} \tag{3.44}$$

From the above equation, we see that, if $E_{2f}/E_m = 1$ or $E_{2f} = E_m$, irrespective of the fiber volume fraction, $E_{2c}/E_m = 1$ or $E_{2c} = E_{2f} = E_m$. In other words, in a unidirectional lamina, if the fiber and matrix moduli are equal, the transverse modulus of the composite is equal to the modulus of the fibers or matrix. For MMCs and CMCs, fiber and matrix moduli are of similar order, and E_{2f}/E_m values are typically small. On the other hand, fiber-to-matrix modulus ratios are very large in PMCs. Typical E_{2c}/E_m plots for these two cases are shown in Figure 3.8. The mechanics of materials-based model for E_{2c} is a simple one, but it does not compare well with experimental results. In general, this approach leads to underestimate of the transverse modulus.

3.5.1.3 Major Poisson's Ratio (ν_{12c})

The major Poisson's ratio is defined as the negative ratio of transverse normal strain to longitudinal normal strain under uniaxial loading in the fiber direction. Thus,

$$\nu_{12c} = -\frac{\varepsilon_{2c}}{\varepsilon_{1c}} \quad \text{with } \sigma_{1c} \neq 0 \quad \text{and all others zero} \tag{3.45}$$

The model for the major Poisson's ratio is similar to that for the longitudinal modulus and we consider an RVE under uniaxial force in the longitudinal direction as shown in Figure 3.9. The lamina deforms in the longitudinal direction due to direct stress and in the transverse direction due to Poisson's effect.

Now, the total transverse deformation is the sum of transverse deformations in the fibers and matrix. (Note that transverse deformations are negative.) Thus,

$$\Delta_c^T = \Delta_f^T + \Delta_m^T \tag{3.46}$$

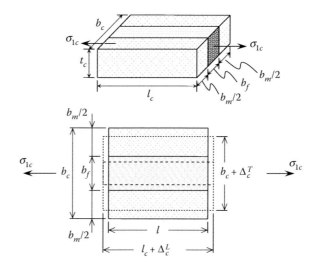

FIGURE 3.9 Representative volume element under uniaxial stress in the fiber direction for the determination of the major Poisson's ratio. (Adapted with permission from A. K. Kaw, *Mechanics of Composite Materials*, CRC Press, Boca Raton, FL, 2006.)

where

Δ_c^T total transverse deformation in composite
Δ_f^T transverse deformation in the fibers
Δ_m^T transverse deformation in the matrix

Deformations in the composite and the constituents can be related to the respective strains and we can write Equation 3.46 as

$$b_c \varepsilon_{2c} = b_f \varepsilon_{2f} + b_m \varepsilon_{2m} \tag{3.47}$$

where

ε_{2c} transverse strain in the composite
ε_{2f} transverse strain in the fibers
ε_{2m} transverse strain in the matrix

Now, under the restriction that the fibers and matrix are perfectly bonded, the longitudinal strains in the composite, fibers, and matrix are all equal, that is, $\varepsilon_{1c} = \varepsilon_{1f} = \varepsilon_{1m}$. Then, dividing both the sides of Equation 3.47 with the width of the RVE, b_c, and longitudinal strain, ε_{1c} (or ε_{1f} or ε_{1m}), we get the following:

$$\frac{\varepsilon_{2c}}{\varepsilon_{1c}} = \frac{b_f}{b_c} \frac{\varepsilon_{2f}}{\varepsilon_{1f}} + \frac{b_m}{b_c} \frac{\varepsilon_{2m}}{\varepsilon_{1m}} \tag{3.48}$$

Now, by definition

$$\nu_{12c} = -\frac{\varepsilon_{2c}}{\varepsilon_{1c}} \tag{3.49}$$

$$\nu_{12f} = -\frac{\varepsilon_{2f}}{\varepsilon_{1f}} \tag{3.50}$$

$$\nu_m = -\frac{\varepsilon_{2m}}{\varepsilon_{1m}} \tag{3.51}$$

Micromechanics of a Lamina

Substituting the above in Equation 3.48 and noting that the width fractions are equal to the corresponding volume fractions, we get

$$\nu_{12c} = \nu_{12f} V_f + \nu_m V_m \tag{3.52}$$

For zero void content,

$$\nu_{12c} = \nu_{12f} V_f + \nu_m (1 - V_f) \tag{3.53}$$

Equations 3.52 and 3.53 are the rule of mixtures expressions for the major Poisson's ratio. We had seen before that the longitudinal modulus is a fiber-dominated property whereas the transverse modulus is matrix-dominated. Fiber and matrix Poisson's ratios are not much different from each other and thus, composite Poisson's ratio is neither fiber-dominated nor matrix-dominated.

3.5.1.4 In-Plane Shear Modulus (G_{12c})

For developing a model for the in-plane shear modulus, an RVE is subjected to in-plane shear stress as shown in Figure 3.10. The total shear deformation in the volume element is the sum of shear deformations in the fibers and the matrix. Thus,

$$\Delta_c = \Delta_f + \Delta_m \tag{3.54}$$

where

- Δ_c shear deformation in the composite
- Δ_f shear deformation in the fibers
- Δ_m shear deformation in the matrix

Shear deformations are related to the shear strains and shear strains can in turn be related to the shear stresses. Thus, we can express the shear deformations as follows:

$$\Delta_c = \gamma_{12c} b_c = \frac{\tau_{12c}}{G_{12c}} b_c \tag{3.55}$$

$$\Delta_f = \gamma_{12f} b_f = \frac{\tau_{12f}}{G_{12f}} b_f \tag{3.56}$$

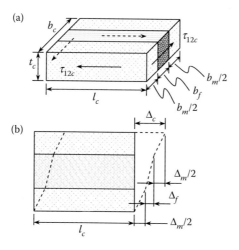

FIGURE 3.10 (a) Representative volume element under shear stress. (b) Shear deformation. (Adapted in parts with permission from R. M. Jones, *Mechanics of Composite Materials*, second edition, Taylor & Francis, New York, 1999; A. K. Kaw, *Mechanics of Composite Materials*, CRC Press, Boca Raton, FL, 2006.)

$$\Delta_m = \gamma_{12m} b_m = \frac{\tau_{12m}}{G_m} b_m \tag{3.57}$$

We may note here that the shear stresses in composite, fibers, and matrix are all equal, that is, $\tau_{12c} = \tau_{12f} = \tau_{12m}$. Then, substituting Equations 3.55 through 3.57 in Equation 3.54, we get

$$\frac{b_c}{G_{12c}} = \frac{b_f}{G_{12f}} + \frac{b_m}{G_m} \tag{3.58}$$

Dividing both the sides of the above equation with b_c, and noting that $b_f/b_c = V_f$ and $b_m/b_c = V_m$, we get the following relation for the in-plane shear modulus of a unidirectional composite:

$$\frac{1}{G_{12c}} = \frac{V_f}{G_{12f}} + \frac{V_m}{G_m} \tag{3.59}$$

or

$$G_{12c} = \frac{G_{12f} G_m}{G_m V_f + G_{12f} V_m} \tag{3.60}$$

Under the restriction that there is no void,

$$G_{12c} = \frac{G_{12f} G_m}{G_m V_f + G_{12f}(1 - V_f)} \tag{3.61}$$

Equations 3.60 and 3.61 are the models by the mechanics of materials-based approach for the in-plane shear modulus of a unidirectional lamina. These equations are very similar to those for the transverse modulus. As with E_{2c}, G_{12c} is also a matrix-dominated property. A typical variation of G_{12c} with V_f is shown in Figure 3.11.

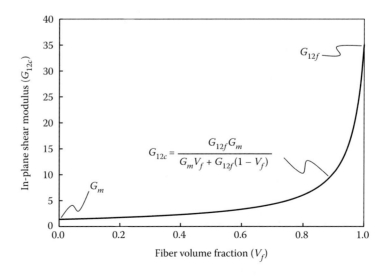

FIGURE 3.11 In-plane shear modulus by mechanics of materials approach (constituent material data from Example 3.1).

Micromechanics of a Lamina

EXAMPLE 3.1

For a unidirectional glass/epoxy lamina, the constituent material properties are as follows: $E_f = 76$ GPa, $\nu_f = 0.2$, $G_f = 35$ GPa, $E_m = 3.6$ GPa, $\nu_m = 0.3$, $G_m = 1.4$ GPa. Consider zero void content and a fiber volume fraction of 0.6.

(a) Determine the composite longitudinal modulus, transverse modulus, major Poisson's ratio, and in-plane shear modulus. (b) Apply a longitudinal force on the lamina and determine the ratio of axial forces shared by fibers and matrix. (c) Consider the cross section of fibers as circular and determine the maximum possible composite longitudinal modulus, transverse modulus, major Poisson's ratio, and in-plane shear modulus.

Solution

Glass fiber is isotropic and we can replace E_{1f} and E_{2f} with E_f, G_{12f} with G_f, and ν_{12f} with ν_f. Then, using Equations 3.30, 3.43, 3.53, and 3.6), respectively, the longitudinal modulus, transverse modulus, major Poisson's ratio, and in-plane shear modulus are obtained as

$$E_{1c} = 0.6 \times 76 + (1-0.6) \times 3.6 = 47.04 \text{ GPa}$$

$$E_{2c} = \frac{76 \times 3.6}{0.6 \times 3.6 + (1-0.6) \times 76} = 8.40 \text{ GPa}$$

$$\nu_{12c} = 0.6 \times 0.2 + (1-0.6) \times 0.3 = 0.24$$

$$G_{12c} = \frac{35 \times 1.4}{0.6 \times 1.4 + (1-0.6) \times 35} = 3.30 \text{ GPa}$$

Let us apply a longitudinal force F_{1c} on the composite. The ratio in which load sharing takes place is as follows:

$$\frac{F_{1f}}{F_{1m}} = \frac{E_f \varepsilon_{1f} A_f}{E_m \varepsilon_{1m} A_m}$$

We know under uniaxial longitudinal loads, longitudinal strains in fibers and matrix are equal to each other. Also, note that $A_f/A_c = V_f$ and $A_m/A_c = V_m$. Then, dividing the numerator and denominator by A_c, we obtain the desired load sharing ratio as

$$\frac{F_{1f}}{F_{1m}} = \frac{E_f V_f}{E_m V_m} = \frac{76 \times 0.6}{3.6 \times 0.4} = 31.67$$

We see that the fibers take 31.67 times the axial load taken by the matrix. In other words, the fibers take about 97% of the total axial load on the composite, whereas the matrix takes only about 3%.

For fibers of circular cross section, the maximum theoretical volume fraction of fibers is given by

$$(V_f)_{max} = \frac{\pi}{2\sqrt{3}} = 0.9069$$

The corresponding elastic moduli for this fiber volume fraction are given by

$$E_{1c} = 0.9069 \times 76 + (1 - 0.9069) \times 3.6 = 69.26 \text{ GPa}$$

$$E_{2c} = \frac{76 \times 3.6}{0.9069 \times 3.6 + (1 - 0.9069) \times 76} = 26.46 \text{ GPa}$$

$$\nu_{12c} = 0.9069 \times 0.2 + (1 - 0.9069) \times 0.3 = 0.209$$

$$G_{12c} = \frac{35 \times 1.4}{0.9069 \times 1.4 + (1 - 0.9069) \times 35} = 10.82 \text{ GPa}$$

EXAMPLE 3.2

For a unidirectional carbon/epoxy lamina, the constituent material properties are given as follows: $E_{1f} = 240$ GPa, $E_{2f} = 24$ GPa, $\nu_{12f} = 0.3$, $G_{12f} = 22$ GPa, $E_m = 3.6$ GPa, $\nu_m = 0.3$, $G_m = 1.4$ GPa.

a. Determine the composite longitudinal modulus, transverse modulus, major Poisson's ratio, and in-plane shear modulus.
b. Apply a longitudinal force on the composite and determine the ratio of axial forces shared by fibers and matrix.
c. Consider circular cross section of fibers and determine the maximum possible composite longitudinal modulus, transverse modulus, major Poisson's ratio, and in-plane shear modulus.
d. Compare the elastic moduli of the carbon/epoxy lamina with those of glass/epoxy lamina in Example 3.1.

Take a fiber volume fraction of 0.6 and zero void content.

Solution

Using Equations 3.30, 3.43, 3.53, and 3.61, respectively, the longitudinal modulus, transverse modulus, major Poisson's ratio, and in-plane shear modulus are obtained as

$$E_{1c} = 0.6 \times 240 + (1 - 0.6) \times 3.6 = 145.44 \text{ GPa}$$

$$E_{2c} = \frac{24 \times 3.6}{0.6 \times 3.6 + (1 - 0.6) \times 24} = 7.35 \text{ GPa}$$

$$\nu_{12c} = 0.6 \times 0.3 + (1 - 0.6) \times 0.3 = 0.3$$

$$G_{12c} = \frac{22 \times 1.4}{0.6 \times 1.4 + (1 - 0.6) \times 22} = 3.20 \text{ GPa}$$

Under a longitudinal force on the composite, the ratio in which load sharing takes place is calculated as follows:

$$\frac{F_{1f}}{F_{1m}} = \frac{E_{1f} V_f}{E_m V_m} = \frac{240 \times 0.6}{3.6 \times 0.4} = 100$$

We see that the fibers take 100 times the axial load taken by the matrix. In other words, the fibers take about 99% of the total axial load on the composite, whereas the matrix takes only about 1%.

Micromechanics of a Lamina

TABLE 3.1

Comparison of Elastic Moduli (Example 3.2)

Elastic Modulus	UD Glass/Epoxy Lamina		UD Carbon/Epoxy Lamina	
	Absolute Value	As a Ratio w.r.t. Matrix Property	Absolute Value	As a Ratio w.r.t. Matrix Property
E_{1c}	47.0	13.1	145.4	40.4
E_{2c}	8.4	2.3	7.4	2.1
G_{12c}	3.3	2.4	3.2	2.3
ν_{12c}	0.24	0.8	0.3	1.0

For fibers of circular cross section, the maximum theoretical volume fraction of fibers is given by

$$(V_f)_{max} = \frac{\pi}{2\sqrt{3}} = 0.9069$$

The corresponding elastic moduli for this fiber volume fraction are given by

$$E_{1c} = 0.9069 \times 240 + (1-0.9069) \times 3.6 = 217.99 \text{ GPa}$$
$$E_{2c} = \frac{24 \times 3.6}{0.9069 \times 3.6 + (1-0.9069) \times 24} = 15.71$$
$$\nu_{12c} = 0.9069 \times 0.3 + (1-0.9069) \times 0.3 = 0.3$$
$$G_{12c} = \frac{22 \times 1.4}{0.9069 \times 1.4 + (1-0.9069) \times 22} = 9.28 \text{ GPa}$$

A comparison of the elastic moduli of the carbon/epoxy lamina with those of the glass/epoxy lamina in the previous example is given in Table 3.1.

Note: From the comparison made above, we find that w.r.t. the matrix modulus, the longitudinal modulus of the lamina is greatly increased by the reinforcements. The increase is more prominent in the case of carbon/epoxy lamina as the longitudinal modulus of carbon fiber is higher than the glass fiber modulus. The transverse modulus and the in-plane shear modulus are increased by the reinforcements only marginally. On the other hand, the major Poisson's ratio remains largely uninfluenced. In other words, the longitudinal modulus of a unidirectional lamina is fiber-dominated, the transverse and in-plane shear moduli are matrix-dominated, and the major Poisson's ratio is neutral to fibers or matrix.

3.5.2 Evaluation of Strengths

The strength of a material is the maximum stress that it can be subjected to before failure. There are five strength parameters (Table 3.2) to be evaluated for complete characterization of strength of a unidirectional composite lamina. Each of these strength parameters corresponds to a specific combination of loading direction and nature of load.

The fibers and matrix have their own failure characteristics as individual entities and in the form of composite material as well. Consequently, in a unidirectional lamina

TABLE 3.2
Strength Parameters of a Unidirectional Lamina

Strength Parameter	Nature of Load Applied	Loading Direction
Longitudinal tensile strength	Tensile force	Along the fiber direction
Longitudinal compressive strength	Compressive force	Along the fiber direction
Transverse tensile strength	Tensile force	Normal to the fiber direction (in the plane of the lamina)
Transverse compressive strength	Compressive force	Normal to the fiber direction (in the plane of the lamina)
In-plane shear strength	Shear force	In the plane of the lamina

under different loading conditions, different failure modes can be found. The failure of a lamina is highly sensitive to local imperfections such as voids, fiber kink, etc. These imperfections, however, do not affect the stiffness characteristics to the same extent. Stiffness may be considered as a global parameter with a smoothening effect as far as local imperfections are concerned. As a result, the models for the evaluation of strengths of a lamina are more complex than those for moduli.

3.5.2.1 Longitudinal Tensile Strength $(\sigma_{1c}^T)_{ult}$

The failure characteristics of a composite lamina depend on the failure characteristics of its constituents—fibers and matrix and the interface between the two. The possible failure modes in a unidirectional lamina under longitudinal tensile load are fiber fracture, fiber fracture with fiber pullout, fiber debond, and matrix cracking. Fibers, matrix, and the interface have their own individual failure characteristics. As a result, the failure characteristics of a composite lamina can be quite involved. However, simplifying assumptions are made for the development of models for predicting the strength of a lamina. We made a number of simplifying assumptions and restrictions for the evaluation of elastic moduli. In addition to those assumptions and restrictions, we assume that individual fibers are of equal strengths. As per our restriction, the fiber is linearly elastic till failure. Thus, as we apply gradually increasing tensile stress in a fiber, its strain increases linearly till failure. The strain at which fiber fracture takes place is referred to as the maximum fiber strain or fiber failure strain and it would be denoted as $(\varepsilon_{1f}^T)_{ult}$. The corresponding stress in the fiber is the longitudinal tensile strength of fiber $(\sigma_{1f}^T)_{ult}$. Similarly, the matrix is linearly elastic till failure and under a gradually increasing tensile stress in the matrix, the strain increases linearly till failure. The strain at which matrix failure takes place is referred to as the maximum matrix strain or matrix failure strain and it would be denoted as $(\varepsilon_m^T)_{ult}$ and the corresponding stress in the matrix is the tensile strength of matrix, $(\sigma_m^T)_{ult}$. The strengths of the constituents are related to the limiting strains as follows:

$$\left(\sigma_{1f}^T\right)_{ult} = \left(\varepsilon_{1f}^T\right)_{ult} E_{1f} \tag{3.62}$$

and

$$\left(\sigma_m^T\right)_{ult} = \left(\varepsilon_m^T\right)_{ult} E_m \tag{3.63}$$

The mechanics of materials-based model for the longitudinal strength of a unidirectional lamina is governed by the failure strains (and strengths) of fibers and matrix together with the elastic moduli of fibers and matrix and fiber volume fraction.

Micromechanics of a Lamina

Now, there are two possible cases of maximum fiber strain relative to maximum matrix strain: (i) $(\varepsilon_{1f}^T)_{ult} < (\varepsilon_m^T)_{ult}$ and (ii) $(\varepsilon_{1f}^T)_{ult} > (\varepsilon_m^T)_{ult}$. Let us consider these cases separately.

Case 1: $(\varepsilon_{1f}^T)_{ult} < (\varepsilon_m^T)_{ult}$ (Figure 3.12)

Let us check the longitudinal tensile strength of a unidirectional lamina under different fiber volume fractions. When the fiber volume fraction is zero, the composite is nothing but pure matrix and the longitudinal tensile strength of the lamina is equal to the tensile strength of the matrix. At this point, the longitudinal tensile strength of the composite is given by

$$\left(\sigma_{1c}^T\right)_{ult} = \left(\sigma_m^T\right)_{ult} \tag{3.64}$$

As we gradually increase the fiber volume fraction, initially, the fibers hardly contribute to the strength of the lamina. At a very low fiber volume fraction, under small tensile load, the fiber tensile strain exceeds its failure strain and fiber fracture occurs. Fiber fracture implies a decrease in effective cross-sectional area of the lamina and an instantaneous increase in matrix strain. However, this increase in matrix strain at the same composite stress does not necessarily imply failure of the composite. The fractured fibers are like holes in the cross section of the composite lamina and the total load is taken by the matrix alone. The longitudinal strength of the composite lamina is then given by

$$\left(\sigma_{1c}^T\right)_{ult} = \left(\sigma_m^T\right)_{ult}(1 - V_f) \tag{3.65}$$

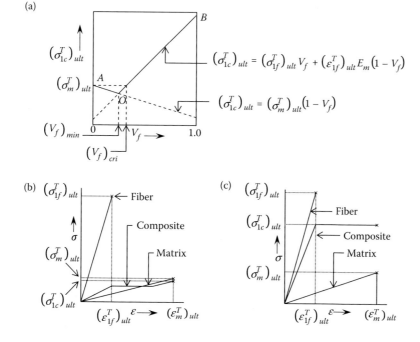

FIGURE 3.12 Mechanics of materials model for the longitudinal strength of a unidirectional lamina, $(\varepsilon_{1f}^T)_{ult} < (\varepsilon_m^T)_{ult}$. (a) Strength of a unidirectional lamina. (b) Stress–strain curves for a unidirectional lamina, $V_f < (V_f)_{min}$. (c) Stress–strain curves for a unidirectional lamina, $V_f > (V_f)_{min}$. (Adapted with permission in parts from R. M. Jones, *Mechanics of Composite Materials*, second edition, Taylor & Francis, New York, 1999; A. Kelly and G. J. Davies, *Metallurgical Reviews*, 10(37), 1965, 1–77.)

Equation 3.65 implies that the addition of fibers actually reduces lamina strength! Quite obviously, this contradicts the very principle of composites where reinforcements are provided for better properties. Thus, Equation 3.65 has to be valid for zero fiber volume fraction (i.e., pure matrix) and at low fiber volume fractions. At this point, let us introduce the concept of minimum fiber volume fraction, $(V_f)_{min}$, below which Equation 3.65 is valid. Note that in this region of fiber volume fractions, the longitudinal strength of the composite lamina is entirely contributed by the matrix alone. At fiber volume fractions above $(V_f)_{min}$ too, once the fiber tensile strain exceeds its failure strain, fiber fracture occurs and the matrix strain increases instantaneously. However, at high fiber volume fractions, this increase in matrix strain is beyond its failure strain and fiber fracture leads to complete failure of the composite lamina. Thus, the fiber failure strain can be considered as the failure strain of the composite lamina as well and its longitudinal tensile strength is then given by

$$\left(\sigma_{1c}^T\right)_{ult} = \left(\sigma_{1f}^T\right)_{ult} V_f + \left(\varepsilon_{1f}^T\right)_{ult} E_m (1 - V_f) \tag{3.66}$$

In Equation 3.66, the first term is the contribution from the fibers to the longitudinal tensile strength of the composite lamina, whereas the second term is the contribution from the matrix. Note that stress in the matrix at the point of maximum fiber strain is $(\varepsilon_{1f}^T)_{ult} E_m$.

Equations 3.65 and 3.66 represent the micromechanics-based model for the longitudinal strength of a unidirectional lamina. As indicated above, these equations are not valid for all fiber volume fractions; for volume fractions lower than $(V_f)_{min}$, Equation 3.65 is applicable, and for volume fractions above $(V_f)_{min}$, Equation 3.66 is applicable. Mathematically, $(V_f)_{min}$ is obtained by solving Equations 3.65 and 3.66 as

$$(V_f)_{min} = \frac{\left(\sigma_m^T\right)_{ult} - \left(\varepsilon_{1f}^T\right)_{ult} E_m}{\left(\sigma_m^T\right)_{ult} + \left(\sigma_{1f}^T\right)_{ult} - \left(\varepsilon_{1f}^T\right)_{ult} E_m} \tag{3.67}$$

Note that the lamina strength at $(V_f)_{min}$ is lower than the strength of the matrix. Thus, for the fibers to be effective in increasing the lamina strength above that of the matrix, we introduce another parameter referred to as the critical fiber volume fraction, $(V_f)_{cri}$. $(V_f)_{cri}$ is the fiber volume fraction above which the lamina strength is more than that of the matrix. Then, by replacing the lamina strength with matrix strength in Equation 3.66, one obtains the expression for critical fiber volume fraction as

$$(V_f)_{cri} = \frac{\left(\sigma_m^T\right)_{ult} - \left(\varepsilon_{1f}^T\right)_{ult} E_m}{\left(\sigma_{1f}^T\right)_{ult} - \left(\varepsilon_{1f}^T\right)_{ult} E_m} \tag{3.68}$$

The model for the longitudinal tensile strength of a unidirectional lamina, for the case $(\varepsilon_{1f}^T)_{ult} < (\varepsilon_m^T)_{ult}$, is pictorially explained in Figure 3.12. The line segments AO and OB represent the lamina strength at fiber volume fractions below and above $(V_f)_{min}$, respectively. Irrespective of the fiber volume fraction, the fiber fails first. At this point, there is readjustment in the load sharing and longitudinal strain increases at the same stress in the composite. Beyond this point, for $V_f < (V_f)_{min}$, the matrix continues to take load and the composite finally fails due to matrix failure. In this case, the strength of the composite and matrix strength are very close to each other. On the other hand, for $V_f > (V_f)_{min}$, the readjustment of load sharing immediately after fiber failure increases

Micromechanics of a Lamina

the strain sharply beyond the matrix failure strain and the composite fails at the same load. Also, in this case, the composite strength is far higher than that of the matrix.

Case 2: $(\varepsilon_{1f}^T)_{ult} > (\varepsilon_m^T)_{ult}$ (Figure 3.13)

The procedure for the development of the model in this case is similar to the first one. Thus, we check the longitudinal tensile strength of a unidirectional lamina under different fiber volume fractions. As in the first case, when fiber volume fraction is zero, the composite is nothing but pure matrix and the longitudinal tensile strength of the lamina is equal to the tensile strength of the matrix and the longitudinal tensile strength of the lamina is given by

$$\left(\sigma_{1c}^T\right)_{ult} = \left(\sigma_m^T\right)_{ult} \tag{3.69}$$

As we gradually increase the fiber volume fraction, at a low fiber volume fraction, under small tensile load, the matrix tensile strain exceeds its failure strain and matrix cracking occurs. Matrix cracking implies a decrease in the effective cross-sectional area of the lamina and an instantaneous increase in fiber strain. At a small fiber volume fraction, this increase in strain is very steep; the fiber strain exceeds its ultimate failure strain and the composite fails. Thus, the longitudinal strength of the composite lamina is then given by

$$\left(\sigma_{1c}^T\right)_{ult} = \left(\varepsilon_m^T\right)_{ult} E_{1f} V_f + \left(\sigma_m^T\right)_{ult} (1 - V_f) \tag{3.70}$$

Note the similarity of Equation 3.70 with Equation 3.66. At fiber volume fractions higher than a certain minimum value, $(V_f)_{min}$, after matrix cracking, the fibers continue to take loads till the strain reaches the fiber ultimate failure strain, that is, the

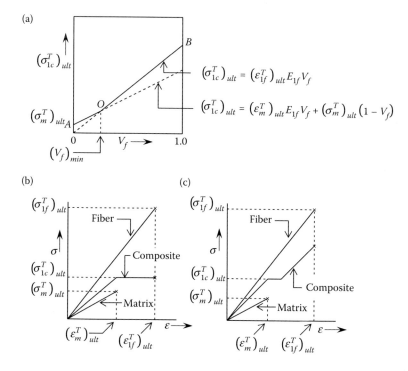

FIGURE 3.13 Mechanics of materials-based model for the longitudinal strength of a unidirectional lamina, $(\varepsilon_{1f}^T)_{ult} > (\varepsilon_m^T)_{ult}$. (a) Strength of the lamina. (b) Stress–strain curves for the lamina, $V_f < (V_f)_{min}$. (c) Stress–strain curves for the lamina, $V_f > (V_f)_{min}$.

fiber stress exceeds the ultimate fiber stress. The longitudinal strength of the composite lamina is then given by

$$\left(\sigma_{1c}^T\right)_{ult} = \left(\sigma_{1f}^T\right)_{ult} V_f \qquad (3.71)$$

Equation 3.70 is applicable at fiber volume fractions lower than $(V_f)_{min}$. Note at zero fiber volume fraction (i.e., pure matrix), it reduces to Equation 3.69. Now, $(V_f)_{min}$ is obtained by solving Equations 3.70 and 3.71 as

$$(V_f)_{min} = \frac{\left(\varepsilon_m^T\right)_{ult} E_m}{\left[\left(\varepsilon_{1f}^T\right)_{ult} - \left(\varepsilon_m^T\right)_{ult}\right] E_{1f} + \left(\varepsilon_m^T\right)_{ult} E_m} \qquad (3.72)$$

The model for the longitudinal tensile strength of a unidirectional lamina, for the case $(\varepsilon_{1f}^T)_{ult} > (\varepsilon_m^T)_{ult}$, is pictorially explained in Figure 3.13. The line segments AO and OB represent the lamina strength at fiber volume fractions below and above $(V_f)_{min}$, respectively. Irrespective of the fiber volume fraction, the matrix fails first. At this point, there is readjustment in the load sharing and longitudinal strain increases at the same stress in the composite. For $V_f < (V_f)_{min}$, the readjustment of load sharing immediately after matrix failure increases the strain sharply beyond the fiber failure strain and the composite fails at the same load. On the other hand, for $V_f > (V_f)_{min}$, the fiber continues to take load and the composite finally fails due to fiber failure.

EXAMPLE 3.3

For a unidirectional carbon/epoxy lamina, the constituent material properties are as follows: $E_{1f} = 375$ GPa, $(\sigma_{1f}^T)_{ult} = 3000$ MPa, $E_m = 3.6$ GPa, $(\sigma_m^T)_{ult} = 72$ MPa.

a. Determine the minimum fiber volume fraction and the critical fiber volume fraction.
b. Study the stress, strain, and load-sharing characteristics at a fiber volume fraction of 0.01.
c. Study the stress, strain, and load-sharing characteristics at a fiber volume fraction of 0.6.

Solution

First, we find the failure strains of fibers and matrix as follows:

$$\left(\varepsilon_{1f}^T\right)_{ult} = \frac{3000}{375,000} = 0.008$$

$$\left(\varepsilon_m^T\right)_{ult} = \frac{72}{3600} = 0.02$$

We see that the matrix failure strain is higher than fiber failure strain.

Using Equations 3.67 and 3.68, the minimum and critical fiber volume fractions are readily calculated as

$$(V_f)_{min} = \frac{72 - 0.008 \times 3600}{72 + 3000 - 0.008 \times 3600} = 0.0142$$

and

$$(V_f)_{cri} = \frac{72 - 0.008 \times 3600}{3000 - 0.008 \times 3600} = 0.0145$$

Thus, at a fiber volume fraction less than 1.42%, the longitudinal tensile strength of the composite is lower than the matrix tensile strength. Also, any additional increase in fiber volume fraction would actually reduce the composite tensile strength.

At a fiber volume fraction between 1.42% and 1.45%, the longitudinal tensile strength of the composite is lower than the matrix tensile strength. However, any additional increase in fiber volume fraction would increase the composite tensile strength.

At a fiber volume fraction higher than 1.45%, the longitudinal tensile strength of the composite is higher than the matrix tensile strength and any additional increase in fiber volume fraction would further increase the composite tensile strength.

In a unidirectional carbon/epoxy composite, the fiber volume fraction is generally around 50% to 60%. Composite strength is invariably much higher than that of the matrix and further increase in the fiber volume fraction would increase the composite strength. Very low fiber volume fraction such as 1% is impractical. However, for the sake of illustration, let us consider $V_f = 0.01$.

Let us first consider an RVE of unit cross-sectional area. Then, at fiber volume fraction, $V_f = 0.01$, the cross-sectional areas of composite, fibers, and matrix are

$$A_c = 1 \text{ mm}^2$$
$$A_f = 0.01 \text{ mm}^2$$
$$A_m = 0.99 \text{ mm}^2$$

Let us apply a tensile force on the RVE and gradually increase its magnitude. Fiber failure takes place when the longitudinal strain is 0.008.

Just before fiber failure, the longitudinal stresses in the fibers, matrix, and composite are calculated as follows:

$$\sigma_{1c}^T = 0.008 \times (0.01 \times 375{,}000 + 0.99 \times 3600) = 58.512 \text{ MPa}$$
$$\sigma_{1f}^T = 0.008 \times 375{,}000 = 3000 \text{ MPa}$$
$$\sigma_{1m}^T = 0.008 \times 3600 = 28.8 \text{ MPa}$$

Loads shared by the fibers, matrix, and composite are calculated by multiplying the stresses with the corresponding cross-sectional areas as follows:

$$F_{1c} = 58.512 \times 1.0 = 58.512 \text{ N} (=100\%)$$
$$F_{1f} = 3000 \times 0.01 = 30 \text{ N} (=51.27\%)$$
$$F_{1m} = 28.8 \times 0.99 = 28.512 \text{ N} (=48.73\%)$$

Immediately after fiber failure, load sharing goes through an instantaneous change and the total load is shared by the matrix alone, that is,

$$F_{1c} = 58.512 \text{ N} (=100\%)$$
$$F_{1f} = 0 (=0\%)$$
$$F_{1m} = 58.512 \text{ N} (=100\%)$$

and the corresponding stresses are

$$\sigma_{1c}^T = \frac{58.512}{1} = 58.512 \text{ MPa}$$
$$\sigma_{1f}^T = 0$$
$$\sigma_{1m}^T = \frac{58.512}{0.99} = 59.103 \text{ MPa}$$

The strains corresponding to these stresses are

$$\varepsilon_{1c}^T = \frac{58.512}{0.01 \times 0 + 0.99 \times 3600} = 0.0164$$
$$\varepsilon_{1f}^T = 0$$
$$\varepsilon_{1m}^T = \frac{59.103}{3600} = 0.0164$$

We find that when fiber failure takes place, the longitudinal strain increases at the same load. However, this increased strain is still lower than the failure strain of the matrix. Thus, the lamina has the capacity to take additional loads. On further loading, finally, the matrix fails when the strain reaches matrix failure strain. At this point, the stresses are as follows:

$$\sigma_{1c}^T = 0.02 \times (0.01 \times 0 + 0.99 \times 3600) = 71.28 \text{ MPa}$$
$$\sigma_{1f}^T = 0$$
$$\sigma_{1m}^T = 0.02 \times 3600 = 72 \text{ MPa}$$

No further loading is possible as the matrix fails at this load level. So, the stress in the composite is the strength of the composite, that is, the longitudinal strength of the composite at a fiber volume fraction of 1% is 71.28 MPa. We can also use Equation 3.65 and get the composite strength as

$$\left(\sigma_{1c}^T\right)_{ult} = 72 \times (1 - 0.01) = 71.28 \text{ MPa}$$

Let us now consider a fiber volume fraction of 0.6.

As in the previous case, let us consider an RVE of unit cross-sectional area. Then, at fiber volume fraction, $V_f = 0.6$, the cross-sectional areas of composite, fibers, and matrix are

$$A_c = 1 \text{ mm}^2$$
$$A_f = 0.6 \text{ mm}^2$$
$$A_m = 0.4 \text{ mm}^2$$

Let us apply a tensile force on the RVE and gradually increase its magnitude. Fiber failure takes place when the longitudinal strain is 0.008.

Just before fiber failure, the longitudinal stresses in the fibers, matrix, and composite are calculated as follows:

$$\sigma_{1c}^T = 0.008 \times (0.6 \times 375{,}000 + 0.4 \times 3600) = 1811.52 \text{ MPa}$$
$$\sigma_{1f}^T = 0.008 \times 375{,}000 = 3000 \text{ MPa}$$
$$\sigma_{1m}^T = 0.008 \times 3600 = 28.8 \text{ MPa}$$

Micromechanics of a Lamina

Loads shared by the fibers, matrix, and composite are calculated by multiplying the stresses with the corresponding cross-sectional areas as follows:

$$F_{1c} = 1811.52 \times 1.0 = 1811.52 \text{ N} (=100\%)$$
$$F_{1f} = 3000 \times 0.6 = 1800 \text{ N} (=99.36\%)$$
$$F_{1m} = 28.8 \times 0.4 = 11.52 \text{ N} (=0.64\%)$$

To check whether further loading is possible, let us consider an instant immediately after fiber failure. At this point, the total load is required to be shared by the matrix alone. That is,

$$F_{1c} = 1811.52 \text{ N} (=100\%)$$
$$F_{1f} = 0 (=0\%)$$
$$F_{1m} = 1811.52 \text{ N} (=100\%)$$

and the corresponding stresses are

$$\sigma_{1c}^T = \frac{1811.52}{1} = 1811.52 \text{ MPa}$$
$$\sigma_{1f}^T = 0$$
$$\sigma_{1m}^T = \frac{1811.52}{0.4} = 4528.8 \text{ MPa}$$

Note that the stress in the matrix is too high. The strains corresponding to these stresses are

$$\varepsilon_{1c}^T = \frac{1811.52}{0.6 \times 0 + 0.4 \times 3600} = 1.258$$
$$\varepsilon_{1f}^T = 0$$
$$\varepsilon_{1m}^T = \frac{4528.8}{3600} = 1.258$$

Note that the strain in the matrix is too high. We find that when fiber failure takes place, the matrix stress and strain increase instantaneously far beyond their limits. Thus, at this fiber volume fraction, the composite fails immediately after fiber failure and the corresponding composite stress is the composite strength, that is, the longitudinal tensile strength of the composite is 1811.52 MPa. We can also use Equation 3.66 and get the composite strength as

$$\left(\sigma_{1c}^T\right)_{ult} = 3000 \times 0.6 + 0.008 \times 3600 \times (1-0.6) = 1811.52 \text{ MPa}$$

3.5.2.2 Longitudinal Compressive Strength $(\sigma_{1c}^C)_{ult}$

The strength characteristics of a unidirectional lamina under longitudinal compression are different from and more complex than those under longitudinal tension. Typical failure modes associated with a unidirectional lamina under longitudinal compression are [2,25–27]

- Microbuckling of fibers in extension
- Microbuckling of fibers in shear
- Transverse tensile failure of matrix and/or interface
- Shear failure

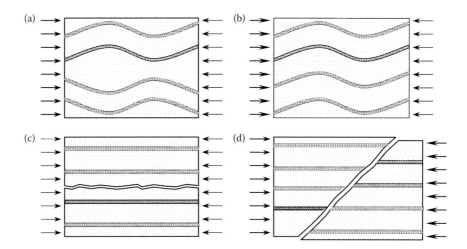

FIGURE 3.14 Typical failure modes in a unidirectional lamina under longitudinal compression. (a) Microbuckling of fibers in extension. (b) Microbuckling of fibers in shear. (c) Transverse tensile failure of matrix. (d) Shear failure. (Adapted with permission from A. K. Kaw, *Mechanics of Composite Materials*, CRC Press, Boca Raton, FL, 2006.)

These basic failure modes in a unidirectional lamina under longitudinal compression are schematically shown in Figure 3.14. In a unidirectional composite lamina under longitudinal compression, the fibers act like tiny columns and they tend to buckle. Local fiber buckling may take place either out-of-phase or in-phase. Out-of-phase fiber microbuckling (Figure 3.14a) is the extensional buckling mode in which the matrix undergoes extension and compression in the transverse direction. This type of failure is associated with low fiber volume fraction and the following approximate expression can be used for longitudinal compressive strength [1]:

$$\left(\sigma_{1c}^{C}\right)_{ult} \approx 2V_f \sqrt{\frac{V_f E_{1f} E_m}{3(1-V_f)}} \tag{3.73}$$

At high fiber volume fraction (typically, $V_f > 0.4$), fiber microbuckling occurs in-phase or the shear mode (Figure 3.14b) in which the matrix undergoes shear. This type of fiber microbuckling is more common and the longitudinal compressive strength of the composite in this mode can be approximated by [1]

$$\left(\sigma_{1c}^{C}\right)_{ult} \approx \frac{G_m}{1-V_f} \tag{3.74}$$

Transverse tensile failure of the matrix and/or interface (Figure 3.14c) takes place when the transverse tensile strain due to Poisson's effect under longitudinal compression exceeds the ultimate tensile strain of the matrix or the fiber–matrix interface. Now, transverse tensile strain due to Poisson's effect under a longitudinal compressive stress is given by

$$\varepsilon_{2c}^{T} = \nu_{12c}\varepsilon_{1c}^{C} = \frac{\nu_{12c}\sigma_{1c}^{C}}{E_{1c}} \tag{3.75}$$

Thus, in this mode of failure, the longitudinal compressive strength of a unidirectional lamina is given by

Micromechanics of a Lamina

$$\left(\sigma_{1c}^{C}\right)_{ult} = \frac{E_{1c}\left(\varepsilon_{2c}^{T}\right)_{ult}}{\nu_{12c}} \tag{3.76}$$

The composite elastic moduli are given by the mechanics of materials formulae given in Equations 3.30 and 3.53. The ultimate transverse tensile strain of the composite will be discussed in Section 3.5.2.3, and we can use the mechanics of materials relation from Equation 3.83. Thus, the longitudinal composite strength can be expressed as

$$\left(\sigma_{1c}^{C}\right)_{ult} = \left[\frac{E_{1f}V_f + E_m(1-V_f)}{\nu_{12f}V_f + \nu_m(1-V_f)}\right]\left[1 + \left(\frac{E_m}{E_{2f}} - 1\right)V_f\right]\left(\varepsilon_{m}^{T}\right)_{ult} \tag{3.77}$$

The fourth basic compression failure mode in a unidirectional lamina is the shear failure (Figure 3.14d) that is associated with high fiber volume fractions. In this case, the composite fails due to direct shear failure of the fibers and the composite strength is dictated by the shear strength of the fibers. By applying the rule of mixtures, longitudinal compressive stress in a unidirectional composite lamina can be shown as

$$\sigma_{1c}^{C} = \sigma_{1f}^{C}V_f + \sigma_{m}^{C}(1-V_f) \tag{3.78}$$

where
σ_{1c}^{C} longitudinal compressive stress in the composite
σ_{1f}^{C} longitudinal compressive stress in the fibers
σ_{1m}^{C} longitudinal compressive stress in the matrix

We know that the maximum shear stress under a longitudinal load is half the longitudinal stress and it occurs on a plane at 45° to the longitudinal axis. Thus, the longitudinal compressive stress in the fiber corresponding to fiber shear failure is $2(\tau_{12f})_{ult}$. Here, we put a restriction that the matrix failure strain is higher than that of the fiber. Then, at the point of fiber shear failure, the longitudinal compressive stress in the matrix is given by $\sigma_{1m}^{C} = 2(\tau_{12f})_{ult}E_m/E_{1f}$. Thus, we get the longitudinal compressive strength as

$$\left(\sigma_{1c}^{C}\right)_{ult} = 2(\tau_{12f})_{ult}\left[V_f + \frac{E_m}{E_{1f}}(1-V_f)\right] \tag{3.79}$$

Note: We presented in this section models for predicting the longitudinal compressive strength of a unidirectional lamina in different compression failure modes. The failure modes in compression are rather complex and theoretically predicted values do not have good match with experimental results. Thus, these relations are useful only in preliminary design calculations.

EXAMPLE 3.4

For a unidirectional carbon/epoxy lamina, the constituent material properties are given as follows: $E_{1f} = 240$ GPa, $E_{2f} = 22$ GPa, $G_{12f} = 22$ GPa, $\nu_{12f} = 0.3$, $E_m = 3.6$ GPa, $G_m = 1.4$ GPa, $\nu_m = 0.3$, $(\tau_{12f})_{ult} = 36$ MPa, and $(\sigma_{m}^{T})_{ult} = 72$ MPa. Determine the longitudinal compressive strength of the lamina. Take the fiber volume fraction as 0.6.

Solution

We shall first determine the longitudinal compressive strength as per different failure modes.

Fiber microbuckling in extensional mode: Using Equation 3.73, the longitudinal compressive strength of the composite is obtained as

$$\left(\sigma_{1c}^{C}\right)_{ult} = 2 \times 0.6 \times \sqrt{\frac{0.6 \times 240,000 \times 3600}{3 \times (1-0.6)}} = 24,942\,\text{MPa}$$

Fiber microbuckling in shear mode: Using Equation 3.74, the longitudinal compressive strength of the composite is obtained as

$$\left(\sigma_{1c}^{C}\right)_{ult} = \frac{1400}{1-0.6} = 3500\,\text{MPa}$$

Transverse tensile failure of matrix: The ultimate tensile strain of matrix is obtained as

$$\left(\varepsilon_{m}^{T}\right)_{ult} = \frac{72}{3600} = 0.02$$

Then, using Equation 3.77, longitudinal compressive strength of composite is obtained as

$$\left(\sigma_{1c}^{C}\right)_{ult} = \left[\frac{240,000 \times 0.6 + 3600 \times (1-0.6)}{0.3 \times 0.6 + 0.3 \times (1-0.6)}\right] \times \left[1 + \left(\frac{3600}{22,000} - 1\right) \times 0.6\right] \times 0.02$$
$$= 4830.4\,\text{MPa}$$

Shear failure mode: Using Equation 3.79, the longitudinal compressive strength of the composite is obtained as

$$\left(\sigma_{1c}^{C}\right)_{ult} = 2 \times 36 \times \left[0.6 + \frac{3600}{240,000}(1-0.6)\right] = 43.6\,\text{MPa}$$

Note: The variation of the longitudinal compressive strength of the unidirectional carbon/epoxy composite in Example 3.4 w.r.t. the fiber volume fraction is pictorially shown in Figure 3.15. As found in the above calculations as well as in the figure, there is hardly any comparison among the results as per different failure modes. As reported in the literature, experimental results also do not match well with theoretical predictions. Thus, in a practical design scenario, one would rather depend on experimental material data. However, in the absence of any experimental data, the designer may use the minimum of the above results. In this example, direct shear failure mode gives the minimum longitudinal compressive strength; as seen in the figure, the graph for this failure mode is almost coincident with the horizontal axis.

3.5.2.3 Transverse Tensile Strength $(\sigma_{2c}^{T})_{ult}$

The transverse tensile strength of a unidirectional composite lamina is a critical parameter. The "first ply failure" of a laminate is generally due to the transverse tensile failure of a lamina. As we know, suitable reinforcements greatly enhance the longitudinal

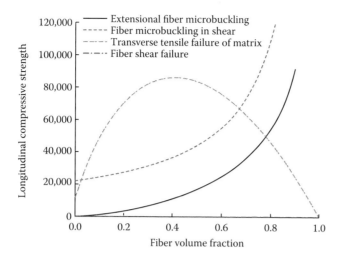

FIGURE 3.15 Variation of longitudinal compressive strength with fiber volume fraction as per various failure modes (Example 3.4).

properties of a unidirectional composite. However, it is not the case for the transverse properties. In fact, the transverse tensile strength of a unidirectional lamina can be lower than the tensile strength of the parent matrix! The fibers can be considered as discontinuities and stress/strain concentrations develop in the matrix around the fibers. It has been found from experimental observations that crack initiation usually takes place from the matrix with dense fiber packing indicating stress concentration in the matrix around the fibers. Theoretical analysis shows that this stress concentration factor is about two. However, stress or strain concentration is not the only factor that influences the transverse tensile strength. Among the several factors responsible for influencing the transverse tensile strength are strength of the matrix, strength of the fiber-to-matrix interface/interphase, fiber strength, fiber volume fraction, fiber packing and voids, etc. Theoretical models have been developed using these parameters. The effects of these factors on the transverse tensile strength are rather complex and the resulting models are also complex. Here, however, we would adopt a simplistic approach and discuss a mechanics of materials-based model in line with the model for the transverse modulus.

Let us consider the RVE used for deriving the expression for the transverse modulus (Figure 3.6). The gross transverse extension in the composite is the sum total of transverse extensions in the fiber and matrix, that is, $\Delta_c = \Delta_f + \Delta_m$. Expressing these transverse extensions in terms of transverse strains, we can show that

$$\varepsilon_{2c}^T = \varepsilon_{2f}^T V_f + \varepsilon_{2m}^T (1 - V_f) \tag{3.80}$$

Under transverse load, the transverse stresses in the fiber and matrix are equal, which gives us

$$\varepsilon_{2f}^T E_{2f} = \varepsilon_{2m}^T E_m \tag{3.81}$$

Using Equation 3.81 in Equation 3.80, we get

$$\varepsilon_{2c}^T = \left[1 + \left(\frac{E_m}{E_{2f}} - 1\right) V_f\right] \varepsilon_{2m}^T \tag{3.82}$$

In this mode of failure, failure takes place when the transverse tensile strain in the matrix becomes equal to the ultimate tensile strain of the matrix, that is, $\varepsilon_{2m}^T = (\varepsilon_m^T)_{ult}$. Then, we get the ultimate tensile strain of the unidirectional composite as

$$(\varepsilon_{2c}^T)_{ult} = \left[1 + \left(\frac{E_m}{E_{2f}} - 1\right) V_f\right] (\varepsilon_m^T)_{ult} \tag{3.83}$$

And the transverse tensile strength of the unidirectional composite is given by

$$(\sigma_{2c}^T)_{ult} = E_{2c} \left[1 + \left(\frac{E_m}{E_{2f}} - 1\right) V_f\right] (\varepsilon_m^T)_{ult} \tag{3.84}$$

Equation 3.84 represents a simple model for the transverse tensile strength of a unidirectional lamina using an RVE shown in Figure 3.6. By substituting the mechanics of materials expression for E_{2c} in Equation 3.84, it can be shown that the right-hand side of the equation is nothing but matrix tensile strength. In other words, in this model, the transverse tensile strength of the unidirectional composite is the same as the tensile strength of the matrix. Other models can also be developed making appropriate simplifying approximations. For example, let us consider a square array of fibers with circular cross section (Figure 3.16). Here, we make an approximation that the fiber-to-matrix interface is ineffective such that the total transverse stress is borne by the matrix alone. Thus,

$$\sigma_{2c}^T l_c t_c = \sigma_{2m}^T l_c (t_c - d) \tag{3.85}$$

where the terms l_c, t_c, and d are given in Figure 3.16. The composite fails when the transverse tensile stress in the matrix reaches the ultimate tensile stress of the matrix. Thus,

$$(\sigma_{2c}^T)_{ult} = (\sigma_m^T)_{ult} \left(\frac{t_c - d}{t_c}\right) \tag{3.86}$$

We can see that the fiber volume fraction for the RVE in Figure 3.16 (considering a square array of fibers, i.e., $b_c = t_c$) is given by

$$V_f = \frac{\pi}{4} \left(\frac{d}{t_c}\right)^2 \tag{3.87}$$

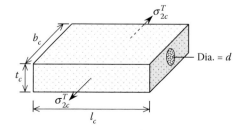

FIGURE 3.16 Representative volume element for transverse tensile strength with fibers of circular cross section in a square array.

Micromechanics of a Lamina

Substituting Equation 3.87 in Equation 3.86, we get

$$\left(\sigma_{2c}^T\right)_{ult} = \left(\sigma_m^T\right)_{ult}\left(1 - 2\sqrt{\frac{V_f}{\pi}}\right) \qquad (3.88)$$

EXAMPLE 3.5

For a unidirectional carbon/epoxy lamina, the constituent material properties are given as follows: $E_{1f} = 240$ GPa, $E_{2f} = 22$ GPa, $E_m = 3.6$ GPa, $(\sigma_{2f}^T)_{ult} = 36$ MPa, $(\sigma_m^T)_{ult} = 72$ MPa. Verify that the transverse tensile strength of the lamina is the same as the tensile strength of the matrix. Take the fiber volume fraction as 0.6.

Solution

We shall first determine the transverse modulus of the unidirectional composite and the ultimate tensile strain of the matrix.

Using Equation 3.43, the transverse modulus of the unidirectional composite is obtained as

$$E_{2c} = \frac{22 \times 3.6}{0.6 \times 3.6 + (1 - 0.6) \times 22} = 7.226 \text{ GPa}$$

The ultimate tensile strain of the matrix is obtained as

$$\left(\varepsilon_m^T\right)_{ult} = \frac{72}{3600} = 0.02$$

Then, using Equation 3.84, the transverse tensile strength of the unidirectional composite is obtained as

$$\left(\sigma_{2c}^T\right)_{ult} = 7226 \times \left[1 + \left(\frac{3.6}{22} - 1\right) \times 0.6\right] \times 0.02 = 72 \text{ MPa}$$

which is identically the same as the tensile strength of the matrix. (See note at the end of Section 3.5.2.5.)

We can also use Equation 3.88 and obtain the transverse tensile strength of the unidirectional composite as

$$\left(\sigma_{2c}^T\right)_{ult} = 72 \times \left(1 - 2 \times \sqrt{\frac{0.6}{\pi}}\right) = 9.1 \text{ MPa}$$

3.5.2.4 Transverse Compressive Strength $(\sigma_{2c}^C)_{ult}$

A number of failure modes are possible in a unidirectional lamina under transverse compression. They are compression failure of matrix, shear failure of matrix, fiber crushing, and fiber-to-matrix interface failure. These failure modes may occur independently or in combination with one another. Clearly, the final failure mechanism is complex. A simplistic approach in line with that for transverse tensile strength can be adopted and we can obtain a relation similar to Equation 3.84 as follows:

$$\left(\sigma_{2c}^C\right)_{ult} = E_{2c}\left[1 + \left(\frac{E_m}{E_{2f}} - 1\right)V_f\right]\left(\varepsilon_m^C\right)_{ult} \qquad (3.89)$$

where

$(\varepsilon_m^C)_{ult}$ ultimate compressive strain of the matrix

As in the case of the transverse tensile strength of the unidirectional composite, here too, the right-hand side of Equation 3.89 is actually the same as the matrix strength, which is compressive. Alternatively, considering fibers of circular cross section in a square array and assuming debond at the fiber-to-matrix interface, we can obtain a relation similar to Equation 3.88 as follows:

$$\left(\sigma_{2c}^C\right)_{ult} = \left(\sigma_m^C\right)_{ult}\left(1 - 2\sqrt{\frac{V_f}{\pi}}\right) \tag{3.90}$$

EXAMPLE 3.6

For a unidirectional carbon/epoxy lamina, the constituent material properties are given as follows: $E_{1f} = 240$ GPa, $E_{2f} = 22$ GPa, $E_m = 3.6$ GPa, $(\sigma_{2f}^T)_{ult} = 36$ MPa, $(\sigma_m^C)_{ult} = 108$ MPa. Verify that the transverse compressive strength of the lamina is the same as the compressive strength of the matrix. Take the fiber volume fraction as 0.6.

Solution

We shall first determine the transverse modulus of the unidirectional composite and the ultimate compressive strain of the matrix.

From Example 3.5, the transverse modulus of the unidirectional composite is

$$E_{2c} = 7.226 \text{ GPa}$$

The ultimate compressive strain of the matrix is obtained as

$$\left(\varepsilon_m^C\right)_{ult} = \frac{108}{3600} = 0.03$$

Then, using Equation 3.89, the transverse compressive strength of the unidirectional composite is obtained as

$$\left(\sigma_{2c}^C\right)_{ult} = 7226 \times \left[1 + \left(\frac{3.6}{22} - 1\right) \times 0.6\right] \times 0.03 = 108 \text{ MPa}$$

which is identically equal to the compressive strength of the matrix. (See note at the end of Section 3.5.2.5.)

Assuming fibers of circular cross section in a square array, we can also use Equation 3.90 and obtain the transverse compressive strength of the unidirectional composite as

$$\left(\sigma_{2c}^T\right)_{ult} = 108 \times \left(1 - 2 \times \sqrt{\frac{0.6}{\pi}}\right) = 13.6 \text{ MPa}$$

3.5.2.5 In-Plane Shear Strength $(\tau_{12c})_{ult}$

Under in-plane shear stress, the matrix shear strength and the fiber-to-matrix interface shear strength are critical. The fibers too come under shear stress, but fibers possess far higher shear strength and the possible failure modes are shear failure of matrix and the interface. Like in the case of transverse strengths, the failure mechanism in in-plane shear is also complex. Here, we adopt a simplistic approach in line with that for the

Micromechanics of a Lamina

in-plane shear modulus. Let us consider the RVE subjected to in-plane shear stress as shown in Figure 3.10. Total shear deformation in the volume element is the sum of shear deformations in the fiber and the matrix, that is, $\Delta_c = \Delta_f + \Delta_m$. Shear deformations are related to the shear strains and, with simple manipulation, we can express the shear strain in composite as follows:

$$\gamma_{12c} = \gamma_{12f}\frac{b_f}{b_c} + \gamma_{12m}\frac{b_m}{b_c} \tag{3.91}$$

or

$$\gamma_{12c} = \gamma_{12f}V_f + \gamma_{12m}(1-V_f) \tag{3.92}$$

The shear stresses in fiber and matrix are equal, that is, $\tau_{12f} = \tau_{12m}$, which gives us

$$\gamma_{12f}G_{12f} = \gamma_{12m}G_m \tag{3.93}$$

Using Equation 3.93 in Equation 3.92, after simple manipulations, we get

$$\gamma_{12c} = \left[1 + \left(\frac{G_m}{G_{12f}} - 1\right)V_f\right]\gamma_{12m} \tag{3.94}$$

Taking the mode of failure as the shear failure of the matrix, at failure, the shear strain in the matrix becomes equal to the ultimate shear strain of the matrix, that is, $\gamma_{12m} = (\gamma_m)_{ult}$. Thus,

$$(\gamma_{12c})_{ult} = \left[1 + \left(\frac{G_m}{G_{12f}} - 1\right)V_f\right](\gamma_m)_{ult} \tag{3.95}$$

Then, multiplying the above with the in-plane shear modulus, the in-plane shear strength of the composite is obtained as

$$(\tau_{12c})_{ult} = G_{12c}\left[1 + \left(\frac{G_m}{G_{12f}} - 1\right)V_f\right](\gamma_m)_{ult} \tag{3.96}$$

It can be shown that the right-hand side of Equation 3.96 is the same as the shear strength of the matrix.

EXAMPLE 3.7

For a unidirectional carbon/epoxy lamina, the constituent material properties are given as follows: $E_{1f} = 240$ GPa, $G_{12f} = 22$ GPa, $G_m = 1.4$ GPa, $(\sigma_{2f}^T)_{ult} = 36$ MPa, $(\tau_m)_{ult} = 35$ MPa. Verify that the in-plane shear strength of the lamina is the same as the shear strength of the matrix. Take the fiber volume fraction as 0.6.

Solution

We shall first determine the in-plane shear modulus of the unidirectional composite and the ultimate in-plane shear strain of the matrix.

Using Equation 3.61, the in-plane shear modulus of the unidirectional composite is obtained as

$$G_{12c} = \frac{22 \times 1.4}{0.6 \times 1.4 + (1 - 0.6) \times 22} = 3.195 \text{ GPa}$$

and the ultimate in-plane shear strain of the matrix is obtained as

$$(\gamma_m)_{ult} = \frac{35}{1400} = 0.025$$

Then, using Equation 3.96, the in-plane shear strength of the unidirectional composite is obtained as

$$(\tau_c)_{ult} = 3195 \times \left[1 + \left(\frac{1400}{22,000} - 1\right) \times 0.6\right] \times 0.025 = 35 \text{ MPa}$$

which is the same as the matrix shear strength. (See note below.)

Note on the mechanics of materials-based models for strengths: In this section, we discussed the mechanics of materials-based models for the strengths of a unidirectional lamina. We also worked out the strengths of a hypothetical unidirectional lamina using these models. Failure modes are complex and strength characteristics are far more complicated than stiffness. These models are based on highly simplified approximations, resulting in simple relations. As a result, these models are not reliable for use in any practical design and analysis exercise. It is always advisable that the designer utilizes experimentally determined strength data. The mechanics of materials-based models for strengths, at best, can be used in very preliminary design calculations. Further, the models for transverse tensile strength, transverse compressive strength, and in-plane shear strength are based on the inherent assumption that the composite failure is due to the matrix failure. The springs-in-series analogy is applicable in these cases and the failure of the composite is by the weakest link in the series, and as already found, the composite strengths are the same as the corresponding matrix strengths.

3.5.3 Evaluation of Thermal Coefficients

The CTE is a measure of relative change in dimensions w.r.t. change in temperature. For an isotropic material, it is defined as

$$\alpha = \frac{\Delta l}{l \Delta T} \qquad (3.97)$$

where

- α CTE of the material
- Δl change in length
- l original length
- ΔT change in temperature

For a unidirectional lamina, CTE is a direction-dependent parameter. Longitudinal CTE gives a relative change in dimension of the lamina in the fiber direction. Let us now derive an expression for the longitudinal thermal coefficient of expansion based

Micromechanics of a Lamina

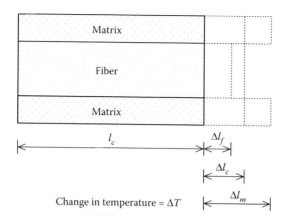

FIGURE 3.17 Representative volume element for longitudinal coefficient of thermal expansion—schematic representation of longitudinal deformations due to change in temperature.

on simple mechanics of materials approach. Let us consider an RVE and subject it to a temperature change of ΔT. If we consider the fibers and matrix to be free, that is, no bond at the interface, due to mismatch between the CTEs, they would undergo different thermal deformations, Δl_f and Δl_m, respectively (Figure 3.17). However, the bond between them restrains them from differential deformation and the net deformation is the same, that is, Δl_c, which is the thermal deformation of the composite. As a result, thermal stresses are generated in the fibers and the matrix although the net stress in the lamina is zero as there is no structural load. Thus, adding the longitudinal stresses in the fibers and matrix, we can write the following:

$$\sigma_{1f} A_f + \sigma_{1m} A_m = 0 \tag{3.98}$$

where
- A_f cross-sectional area of the fibers
- A_m cross-sectional area of the matrix

Dividing both the sides of the above equation with the cross-sectional area of the composite, A_c, we get

$$\sigma_{1f} V_f + \sigma_{1m}(1 - V_f) = 0 \tag{3.99}$$

We can bring in the thermal deformations and rewrite the equation above as follows:

$$E_{1f}\left(\frac{\Delta l_c - \Delta l_f}{l_c}\right) V_f + E_m\left(\frac{\Delta l_c - \Delta l_m}{l_c}\right)(1 - V_f) = 0 \tag{3.100}$$

or

$$E_{1f}(\Delta l_c - \Delta l_f) V_f + E_m(\Delta l_c - \Delta l_m)(1 - V_f) = 0 \tag{3.101}$$

Now, thermal deformations are related to the temperature change as follows:

$$\Delta l_c = \alpha_{1c} \Delta T l_c \tag{3.102}$$

$$\Delta l_f = \alpha_{1f} \Delta T l_c \tag{3.103}$$

$$\Delta l_m = \alpha_m \Delta T l_c \tag{3.104}$$

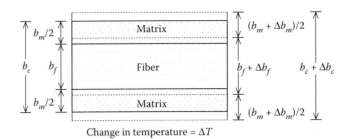

Change in temperature = ΔT

FIGURE 3.18 Representative volume element for transverse coefficient of thermal expansion—schematic representation of transverse deformations due to change in temperature.

Substituting the above in Equation 3.101 and dividing both the sides by $\Delta T l_c$, we get

$$E_{1f}(\alpha_{1c} - \alpha_{1f})V_f + E_m(\alpha_{1c} - \alpha_m)(1 - V_f) = 0 \qquad (3.105)$$

Rearranging the terms, we can show that the longitudinal CTE is given by

$$\alpha_{1c} = \frac{\alpha_{1f}E_{1f}V_f + \alpha_m E_m(1 - V_f)}{E_{1f}V_f + E_m(1 - V_f)} \qquad (3.106)$$

In the case of the transverse CTE, ignoring the Poisson's effect, a highly simplified mechanics of materials-based relation can be obtained by equating the total transverse thermal expansion to the sum of the thermal expansions of the fibers and matrix (Figure 3.18), that is, $\Delta b_c = \Delta b_f + \Delta b_m$
or

$$\alpha_{2c}\Delta T b_c = \alpha_{2f}\Delta T b_f + \alpha_m \Delta T b_m \qquad (3.107)$$

Dividing both the sides by $\Delta T b_c$, we get an expression for transverse thermal coefficient of expansion as

$$\alpha_{2c} = \alpha_{2f}V_f + \alpha_m(1 - V_f) \qquad (3.108)$$

Rigorous methods have been employed for both longitudinal CTE as well as transverse CTE. The expression for longitudinal CTE from rigorous analysis is the same as that from the mechanics of materials approach; for transverse CTE, it can be stated as given below [28]:

$$\alpha_{2c} = (1 + \nu_f)\alpha_{2f}V_f + (1 + \nu_m)\alpha_m V_m - \alpha_{1f}\nu_{12c} \qquad (3.109)$$

In the above relation, for ν_{12c}, the mechanics of materials expression can be used.

EXAMPLE 3.8

For a unidirectional carbon/epoxy lamina, the constituent material properties are given as follows: $E_{1f} = 240$ GPa, $E_m = 3.6$ GPa, $\alpha_{1f} = -0.5 \times 10^{-6}$ m/m/°C, $\alpha_{2f} = 6 \times 10^{-6}$ m/m/°C, $\alpha_m = 60 \times 10^{-6}$ m/m/°C, and $\nu_{12f} = 0.28$, $\nu_m = 0.3$. Determine the longitudinal and transverse CTEs. Take the fiber volume fraction as 0.6.

Solution

The longitudinal CTE is given by Equation 3.106 as follows:

$$\alpha_{1c} = \frac{-0.5 \times 240 \times 0.6 + 60 \times 3.6 \times (1 - 0.6)}{240 \times 0.6 + 3.6 \times (1 - 0.6)} \times 10^{-6} = 0.099 \times 10^{-6} \text{ m/m/°C}$$

Micromechanics of a Lamina

The transverse CTE is given by Equation 3.108 as follows:

$$\alpha_{2c} = [6 \times 0.6 + 60 \times (1 - 0.6)] \times 10^{-6} = 27.6 \times 10^{-6} \text{ m/m/°C}$$

The longitudinal CTE as per rigorous analysis is the same as per the mechanics of materials approach. The transverse CTE as per rigorous analysis is given by Equation 3.109. Toward this, we first determine the major Poisson's ratio as follows:

$$\nu_{12c} = 0.28 \times 0.6 + 0.3 \times (1 - 0.6) = 0.288$$

Transverse CTE is then obtained as follows:

$$\alpha_{2c} = [(1 + 0.28) \times 6 \times 0.6 + (1 + 0.3) \times 60 \times (1 - 0.6) - (-0.5) \times 0.288] \times 10^{-6}$$
$$= 35.952 \times 10^{-6} \text{ m/m/°C}$$

3.5.4 Evaluation of Moisture Coefficients

When a body absorbs moisture, it expands in size. The CME is a measure of relative change in dimension w.r.t. change in moisture content in the body. For an isotropic material, it is defined as

$$\beta = \frac{\Delta l}{l \Delta C} \tag{3.110}$$

where
- β CME of the material (m/m/kg/kg)
- Δl change in length (m)
- l original length (m)
- ΔC change in moisture content per unit mass of the body (kg/kg)

For a unidirectional lamina, the CME is a direction-dependent parameter. Thus, the longitudinal CME is defined as the change in linear dimension in the fiber direction per unit length per unit change in mass of moisture content per unit mass of the body. On the other hand, the transverse CME gives us a measure of relative change in dimension in the transverse direction. The unit of CME is m/m/kg/kg.

Let us now derive an expression for the longitudinal CME based on the simple mechanics of materials approach. Let us consider an RVE (Figure 3.19). Let the

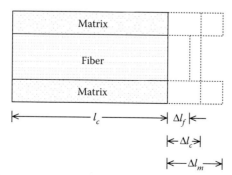

Change in moisture content per unit mass of composite = ΔC_c

FIGURE 3.19 Representative volume element for longitudinal coefficient of moisture expansion—schematic representation of longitudinal deformations due to change in moisture content.

volume element absorb a certain quantity of moisture. If we consider the fibers and matrix to be free, that is, no bond at the interface, due to mismatch between the CMEs, as shown in Figure 3.19, they would undergo different deformations, Δl_f and Δl_m, respectively. However, the bond between them restrains them from differential deformation and the net deformation is the same, that is, Δl_c, which is the net deformation of the composite due to moisture absorption. As a result, stresses are generated in the fibers and the matrix although net stress in the lamina is zero as there is no structural load. Thus, adding the longitudinal stresses in the fibers and matrix, we can write the following:

$$\sigma_{1f} A_f + \sigma_{1m} A_m = 0 \tag{3.111}$$

where
- A_f cross-sectional area of the fibers
- A_m cross-sectional area of the matrix

Dividing both the sides of the above equation with the cross-sectional area of the composite, A_f, we get

$$\sigma_{1f} V_f + \sigma_{1m}(1 - V_f) = 0 \tag{3.112}$$

We can bring in the deformations due to moisture absorption and rewrite the equation above as follows:

$$E_{1f}\left(\frac{\Delta l_c - \Delta l_f}{l_c}\right) V_f + E_m\left(\frac{\Delta l_c - \Delta l_m}{l_c}\right)(1 - V_f) = 0 \tag{3.113}$$

or

$$E_{1f}(\Delta l_c - \Delta l_f) V_f + E_m(\Delta l_c - \Delta l_m)(1 - V_f) = 0 \tag{3.114}$$

Note that till this point, the procedure for derivation is very similar to that for longitudinal CTE. However, the deformations are due to moisture expansion and they are related to the change in moisture content as follows:

$$\Delta l_c = \beta_{1c} \Delta C_c l_c \tag{3.115}$$

$$\Delta l_f = \beta_{1f} \Delta C_f l_c \tag{3.116}$$

$$\Delta l_m = \beta_m \Delta C_m l_c \tag{3.117}$$

Substituting of the above in Equation 3.114 and dividing both the sides by l_c, we get

$$E_{1f}(\beta_{1c} \Delta C_c - \beta_{1f} \Delta C_f) V_f + E_m(\beta_{1c} \Delta C_c - \beta_m \Delta C_m)(1 - V_f) = 0 \tag{3.118}$$

Rearranging the terms, we get

$$\beta_{1c} = \frac{\beta_{1f} E_{1f} \Delta C_f V_f + \beta_m E_m \Delta C_m (1 - V_f)}{E_{1c} \Delta C_c} \tag{3.119}$$

Micromechanics of a Lamina

Now, the total moisture content in the composite is equal to the sum of moisture contents in the fibers and matrix. Thus,

$$\Delta C_c \rho_c l_c b_c t_c = \Delta C_f \rho_f l_c b_f t_c + \Delta C_m \rho_m l_c b_m t_c \tag{3.120}$$

Dividing both the sides by $\rho_c l_c b_c t_c$, we get

$$\Delta C_c = \frac{1}{\rho_c}[\Delta C_f \rho_f V_f + \Delta C_m \rho_m (1-V_f)] \tag{3.121}$$

or

$$\Delta C_c = \frac{\Delta C_f \rho_f V_f + \Delta C_m \rho_m (1-V_f)}{\rho_f V_f + \rho_m (1-V_f)} \tag{3.122}$$

Substituting Equation 3.122 in Equation 3.119, we get the expression for the longitudinal CME as

$$\beta_{1c} = \left[\frac{\beta_{1f} E_{1f} \Delta C_f V_f + \beta_m E_m \Delta C_m (1-V_f)}{\Delta C_f \rho_f V_f + \Delta C_m \rho_m (1-V_f)}\right]\left[\frac{\rho_f V_f + \rho_m (1-V_f)}{E_{1f} V_f + E_m (1-V_f)}\right] \tag{3.123}$$

In the case of transverse CME, ignoring the Poisson's effect, a highly simplified mechanics of materials-based relation is derived here. The methodology is similar to that for the transverse CTE and we equate the total transverse expansion to the sum of the expansions of the fibers and matrix (Figure 3.20), that is, $\Delta b_c = \Delta b_f + \Delta b_m$. These deformations are related to the transverse CMEs and we get the following:

$$\beta_{2c} \Delta C_c b_c = \beta_{2f} \Delta C_f b_f + \beta_m \Delta C_m b_m \tag{3.124}$$

where
- β_{2c} transverse CME of the composite
- β_{2f} transverse CME of the fiber
- β_m CME of the matrix
- ΔC_c change in moisture content per unit mass of the composite
- ΔC_f change in moisture content per unit mass of the fibers
- ΔC_m change in moisture content per unit mass of the matrix

Dividing both the sides by $\Delta C_c b_c$, we get

$$\beta_{2c} = \frac{\beta_{2f} \Delta C_f V_f + \beta_m \Delta C_m (1-V_f)}{\Delta C_c} \tag{3.125}$$

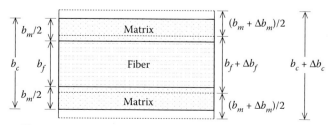

Change in moisture content per unit mass of composite = ΔC_c

FIGURE 3.20 Representative volume element for transverse coefficient of moisture expansion—schematic representation of transverse deformations due to change in moisture content.

Substituting ΔC_c from Equation 3.122, we can obtain an expression for transverse CME as

$$\beta_{2c} = \left[\frac{\beta_{2f}\Delta C_f V_f + \beta_m \Delta C_m(1-V_f)}{\Delta C_f \rho_f V_f + \Delta C_m \rho_m(1-V_f)}\right][\rho_f V_f + \rho_m(1-V_f)] \quad (3.126)$$

We have developed expressions for both longitudinal as well as transverse CMEs based on simplistic assumptions. In the case of the transverse coefficient, the Poisson's effect has been ignored. More rigorous methods have also been employed. The expression for the longitudinal CME from more rigorous analysis is the same as that from the mechanics of materials approach; for the transverse CME, it can be stated as follows [4]:

$$\beta_{2c} = \left[\frac{\beta_{2f}\Delta C_f(1+\nu_{12f})V_f + \beta_m \Delta C_m(1+\nu_m)(1-V_f)}{\Delta C_f \rho_f V_f + \Delta C_m \rho_m(1-V_f)}\right]\rho_c - \beta_{1c}\nu_{12c} \quad (3.127)$$

EXAMPLE 3.9

For a unidirectional carbon/epoxy lamina, the constituent material properties are given as follows: $E_{1f} = 240$ GPa, $E_m = 3.6$ GPa, $\rho_f = 1.8$, $\rho_m = 1.1$, $\beta_m = 0.35$ m/m/kg/kg, $\nu_{12f} = 0.28$, and $\nu_m = 0.3$. Determine the longitudinal and transverse CMEs. Take the fiber volume fraction as 0.6. Assume carbon fiber does not absorb moisture.

If a unidirectional laminate of size 400 mm × 300 mm × 8 mm absorbs 50 g moisture, determine the changed dimensions of the laminate.

Solution

Longitudinal CME is given by Equation 3.123. Now, under the given assumption that carbon fiber does not absorb moisture, $\Delta C_f = \beta_{1f} = \beta_{2f} = 0$. Then,

$$\beta_{1c} = \left[\frac{0 + 0.35 \times 3.6 \times \Delta C_m \times (1-0.6)}{0 + \Delta C_m \times 1.1 \times (1-0.6)}\right] \times \left[\frac{1.8 \times 0.6 + 1.1 \times (1-0.6)}{240 \times 0.6 + 3.6 \times (1-0.6)}\right]$$
$$= 0.012 \text{ m/m/kg/kg}$$

The transverse CME as per the mechanics of materials-based approach is given by Equation 3.126 as follows:

$$\beta_{2c} = \left[\frac{0 + 0.35 \times \Delta C_m \times (1-0.6)}{0 + \Delta C_m \times 1.1 \times (1-0.6)}\right] \times [1.8 \times 0.6 + 1.1 \times (1-0.6)]$$
$$= 0.484 \text{ m/m/kg/kg}$$

The longitudinal CME as per the rigorous analysis is the same as per the mechanics of materials approach. The transverse CTE as per the rigorous analysis is given by Equation 3.127. The density of the composite is given by

$$\rho_c = 1.8 \times 0.6 + 1.1 \times (1-0.6) = 1.52$$

And the longitudinal Poisson's ratio of the composite is given by

$$\nu_{12c} = 0.28 \times 0.6 + 0.3 \times (1-0.6) = 0.288$$

Then the transverse CME as per the rigorous approach is obtained as

$$\beta_{2c} = \left[\frac{0 + 0.35 \times \Delta C_m \times (1+0.3) \times (1-0.6)}{0 + \Delta C_m \times 1.1 \times (1-0.6)} \right] \times 1.52 - 0.012 \times 0.288$$
$$= 0.625 \text{ m/m/kg/kg}$$

The mass of the laminate is obtained as

$$w_c = \frac{40 \times 30 \times 0.8 \times 1.52}{1000} = 1.4592 \text{ kg}$$

The mass of the matrix is

$$w_m = \frac{(40 \times 30 \times 0.8) \times (1-0.6) \times 1.1}{1000} = 0.4224 \text{ kg}$$

The changes in the dimensions of the laminate (from the mechanics of materials approach) are obtained as follows:

$$\Delta l = 0.012 \times \frac{400}{1000} \times \frac{50}{1000} \times 1.4592 \times 1000 = 0.35 \text{ mm}$$
$$\Delta b = 0.484 \times \frac{300}{1000} \times \frac{50}{1000} \times 1.4592 \times 1000 = 10.59 \text{ mm}$$
$$\Delta h = 0.484 \times \frac{8}{1000} \times \frac{50}{1000} \times 1.4592 \times 1000 = 0.28 \text{ mm}$$

Thus, the changed dimensions of the unidirectional laminate are

$$400.35 \text{ mm} \times 310.59 \text{ mm} \times 8.28 \text{ mm}.$$

3.6 ELASTICITY-BASED MODELS

The mechanics of materials equations for stiffness, strength, and hygrothermal parameters presented in this chapter are derived by utilizing simple forms of equilibrium conditions. Simplifying assumptions such as uniform stress distribution are made. Compatibility and 3D stress–strain relations may not be satisfied at each point in the RVE. In the elasticity approach, too, the concept of RVE is adopted. However, this approach is based on more rigorous treatment, including stress–strain relations in three dimensions, equilibrium conditions, and compatibility conditions.

As we had mentioned earlier, there are several types of models that can be considered as elasticity models. Also, the classification of these models is a matter of convenience of discussion. One way to classify the elasticity-based models is to categorize them into three subcategories—models based on bounding techniques, models with exact solutions, and self-consistent models [2].

Bounding techniques are associated with finding the upper and lower bounds for the elastic moduli. Principles of minimum complementary energy and minimum potential energy have been used for lower bound and upper bound, respectively. The following bounds for the Young's modulus of an isotropic composite material can be derived [1,14]:

$$E_{\text{lower}} = \frac{E_m E_d}{V_d E_m + V_m E_d} \tag{3.128}$$

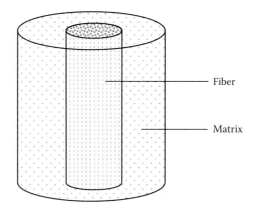

FIGURE 3.21 Representative volume element in the composite cylinder assemblage (CCA) approach. (Adapted with permission from A. K. Kaw, *Mechanics of Composite Materials*, CRC Press, Boca Raton, FL, 2006.)

and

$$E_{\text{upper}} = E_d V_d + E_m V_m \qquad (3.129)$$

where

E_{lower} lower bound for Young's modulus
E_{upper} upper bound for Young's modulus
E_m Young's modulus of the matrix
E_d Young's modulus of the dispersed material
V_m matrix volume fraction
V_d volume fraction of the dispersed matter

These bounds can also be interpreted as bounds for the transverse modulus of a unidirectional composite. However, the bounds are too far apart for most practical fiber volume fractions. One of the reasons that the bounds are far apart is that fiber packing geometry is not specified. In the exact solutions approach, a possible solution is assumed. The assumed solution involves stress, strain, or displacement components and it is verified whether the governing differential equations are satisfied by the assumed solutions. In the composite cylindrical assemblage (CCA) approach [15], the fibers are taken as circular in cross section and arranged either in a regular array or at random. The RVE is a combination of two cylinders—an inner cylinder that represents the fiber and an outer cylinder for the matrix (Figure 3.21). Appropriate boundary conditions are applied on the composite cylinder and the desired elastic modulus is obtained by analyzing the response of the composite cylinder to the applied boundary conditions.

3.7 SEMIEMPIRICAL MODELS

We had noted before that the mechanics of materials-based models for the transverse modulus and the in-plane shear modulus are not very reliable as they do not have good match with experimental results. Elasticity-based models are generally complicated, and in some cases, their applicability is also restricted to a rather narrow range of design variables. Empirical models, on the other hand, are simple and easy to use in a design environment. Halpin–Tsai equations [1,21] are the most commonly used empirical models and these models are briefly discussed in this section. These models were developed by curve fitting of experimental and elasticity-based model data. The parameters used in the curve fitting have physical significance and thus these models are called semiempirical.

3.7.1 General Form of Halpin–Tsai Equations

The general form of the Halpin–Tsai equations for the elastic moduli of a lamina can be expressed as follows:

$$M = \left(\frac{1 + \xi \eta V_f}{1 - \eta V_f}\right) M_m \qquad (3.130)$$

in which, the coefficient η is given by

$$\eta = \frac{(M_f/M_m) - 1}{(M_f/M_m) + \xi} \qquad (3.131)$$

and

- M desired composite modulus, that is, E_{1c}, E_{2c}, ν_{12c}, or G_{12c}
- M_f corresponding fiber modulus
- M_m corresponding matrix modulus
- V_f fiber volume fraction

The parameter ξ is called the reinforcing factor and it depends on fiber geometry, fiber packing geometry, and the loading condition. Halpin–Tsai equations are very simple to apply; however, their accuracy depends on the choice of the parameter ξ. It is determined by a procedure of curve fitting and comparing Equations 3.130 and 3.131 with elasticity solutions or reliable experimental data. The recommended values of ξ are given in Table 3.3.

Using the values for ξ given in Table 3.3, we can obtain the expressions for different elastic moduli.

3.7.2 Halpin–Tsai Equations for Elastic Moduli

3.7.2.1 Longitudinal Modulus

$$\xi = \infty \qquad (3.132)$$

Thus, from Equation 3.131

$$\eta = 0 \qquad (3.133)$$

TABLE 3.3
Recommended Reinforcing Factors

Desired Modulus	ξ	Remarks
Longitudinal modulus, E_1	∞	–
Transverse modulus, E_2	2	For fibers of circular cross section in a square array (Figure 3.22a)
	$\dfrac{2a}{b}$	For fibers of rectangular cross section in a triangular array (Figure 3.22b)
Major Poisson's ratio, ν_{12}	∞	–
In-plane shear modulus, G_{12}	1	For fibers of circular cross section in a square array (Figure 3.22a)
	$\sqrt{3}\ln\left(\dfrac{a}{b}\right)$	For fibers of rectangular cross section in a triangular array (Figure 3.22b)

Source: R. M. Jones, *Mechanics of Composite Materials*, second edition, Taylor & Francis, New York, 1999; S. W. Tsai and H. Thomas Hahn, *Introduction to Composite Materials*, Technomic Publishing, Lancaster, 1980; B. Paul, *Transactions of the Metallurgical Society of AIME*, 100, 1960, 36–41.

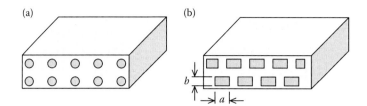

FIGURE 3.22 Fiber cross section and fiber packing in Halpin–Tsai equations. (a) Fibers of circular cross section in a square array. (b) Fibers of rectangular cross section in a triangular array. (Adapted with permission from A. K. Kaw, *Mechanics of Composite Materials*, CRC Press, Boca Raton, FL, 2006.)

and

$$\xi\eta = \frac{E_{1f}}{E_m} - 1 \tag{3.134}$$

On substitution in Equation 3.130, we get

$$E_{1c} = E_{1f}V_f + E_m(1-V_f) \tag{3.135}$$

Thus, we find that the Halpin–Tsai equation for the longitudinal modulus is the same as that by the mechanics of materials approach with zero void content.

3.7.2.2 Transverse Modulus

The Halphin–Tsai equation for the transverse modulus of a lamina is given by

$$E_{2c} = \left(\frac{1+\xi\eta V_f}{1-\eta V_f}\right) E_m \tag{3.136}$$

where

$$\eta = \frac{(E_{2f}/E_m)-1}{(E_{2f}/E_m)+\xi} \tag{3.137}$$

And $\xi = 2$ or $\xi = 2a/b$ as per Table 3.3.

3.7.2.3 Major Poisson's Ratio

As in the case of the longitudinal modulus, $\xi = 0$, and following a similar procedure, we can show that

$$\nu_{12c} = \nu_{12f}V_f + \nu_m(1-V_f) \tag{3.138}$$

Thus, we find that the Halpin–Tsai equation for the major Poisson's ratio is the same as that by the mechanics of materials approach with zero void content.

Micromechanics of a Lamina

3.7.2.4 In-Plane Shear Modulus

The Halphin–Tsai equation for the in-plane shear modulus of a unidirectional lamina is similar to that for the transverse modulus and it is given by

$$G_{12c} = \left(\frac{1+\xi\eta V_f}{1-\eta V_f}\right) G_m \qquad (3.139)$$

where

$$\eta = \frac{(G_{12f}/G_m) - 1}{(G_{12f}/G_m) + \xi} \qquad (3.140)$$

And $\xi = 1$ or $\xi = \sqrt{3}\ln(a/b)$ as per Table 3.3.

Halpin–Tsai equations for the elastic moduli of a lamina are simple to apply in a wide range of fiber volume fractions. For the longitudinal modulus and the major Poisson's ratio, Halpin–Tsai equations provide identical results as with the mechanics of materials approach and there is a good match with experimental data as well. The mechanics of materials approach gives underestimates of transverse and in-plane shear moduli, whereas Halpin–Tsai equations yield closer match.

EXAMPLE 3.10

For the unidirectional carbon/epoxy lamina in Example 3.2, determine the composite longitudinal modulus, transverse modulus, major Poisson's ratio, and in-plane shear modulus by using Halpin–Tsai equations. Take a fiber volume fraction of 0.6. Also, take the fiber cross section as circular in a square array. Compare the results with those obtained in Example 3.2.

Solution

The constituent material properties are as follows: $E_{1f} = 240$ GPa, $E_{2f} = 24$, $\nu_{12f} = 0.3$, $G_{12f} = 22$ GPa, $E_m = 3.6$ GPa, $\nu_m = 0.3$, and $G_m = 1.4$ GPa.

Using Equation 3.135, the longitudinal modulus of the lamina is obtained as

$$E_{1c} = 0.6 \times 240 + (1-0.6) \times 3.6 = 145.44 \text{ GPa}$$

For the transverse modulus, $\xi = 2$ for circular fibers in a square array. Then, using Equation 3.137, η is obtained as

$$\eta = \frac{(24/3.6) - 1}{(24/3.6) + 2} = 0.6538$$

Then, using Equation 3.136, we get the transverse modulus as

$$E_{2c} = \left(\frac{1 + 2 \times 0.6538 \times 0.6}{1 - 0.6538 \times 0.6}\right) \times 3.6 = 10.57 \text{ GPa}$$

Using Equation 3.138, the major Poisson's ratio is obtained as

$$\nu_{12c} = 0.6 \times 0.3 + (1-0.6) \times 0.3 = 0.3$$

For the in-plane shear modulus, $\xi = 1$ for circular fibers in a square array. Then, using Equation 3.140, η is obtained as

$$\eta = \frac{(22/1.4) - 1}{(22/1.4) + 1} = 0.8803$$

Then, using Equation 3.139, we get the in-plane shear modulus as

$$G_{12c} = \left(\frac{1 + 1 \times 0.8803 \times 0.6}{1 - 0.8803 \times 0.6}\right) \times 1.4 = 4.53 \text{ GPa}$$

The comparison of the results for the elastic moduli of the carbon/epoxy lamina by Halpin–Tsai equations with the mechanics of materials approach is given below:

Elastic Modulus	Mechanics of Materials Approach	Halpin–Tsai Equations
E_{1c}	145.4	145.4
E_{2c}	7.4	10.57
G_{12c}	3.2	4.53
ν_{12c}	0.3	0.3

Note: From the comparison made above, we find that the longitudinal modulus and the major Poisson's ratio are identical. This is a straightforward outcome as the expressions are identical in both the methods. On the other hand, the transverse modulus and the in-plane shear modulus as per Halpin–Tsai equations are larger than those by the mechanics of materials approach. We had mentioned before that the mechanics of materials approach, when compared with experimental data, underestimates the transverse and in-plane shear moduli. Thus, it is seen that Halpin–Tsai equations are better suited than the mechanics of materials approach for the evaluation of transverse and in-plane shear moduli.

3.8 SUMMARY

In this chapter, we reviewed the basic concepts and tools available for the analysis of a lamina at the micro level. At the micro level, the study of a composite material revolves around the determination of the composite lamina parameters, viz. elastic moduli, strengths and coefficients of thermal and moisture expansion from the knowledge of certain basic variables, viz. elastic moduli, strengths, densities, and volume fractions and mass fractions of the fibers and matrix. A number of models are available for the determination of the lamina parameters; they are of different types, including netting models, mechanics of materials-based models, elasticity-based models, and semiempirical models.

Typically, an RVE is considered in the micromechanics models. Several basic assumptions are also made, of which some are relaxed in some models. The mechanics of materials-based models are simple tools that are based on simple equilibrium considerations involving averaged stresses and strains. Elasticity-based models are based on more rigorous treatment of lamina behavior. They are generally complicated and they have limited applicability. Semiempirical models, on the other hand, are simple and easy to use in a design environment.

Micromechanics of a Lamina

Many of the assumptions made in micromechanics are unrealistic. As a result, there is generally a gap between micromechanics predictions and experimental results. Thus, while the study of micromechanics gives a good insight into how composites work, the models for the prediction of elastic moduli and strengths can at best be considered as an alternative to experimental work only in the preliminary design of a product. In this respect, the mechanics of materials-based models as well as Halpin–Tsai models are the most commonly used models.

EXERCISE PROBLEMS

3.1 Consider the data given below:

	Diameter (μm)	Density (g/cm^3)
Carbon fiber	7	1.80
Glass fiber	16	2.58
Kevlar fiber	12	1.45
Epoxy matrix	–	1.1

Consider the fiber cross section as circular. (a) Determine the fiber volume fraction in (i) square array of fiber packing and (ii) triangular array of fiber packing for fiber spacing-to-fiber diameter ratios (s/d) of 1.0, 1.25, 1.5, 1.75, and 2.0. Plot the v_f versus s/d curves. Are the fiber volume fractions dependent on fiber type/diameter? (b) What is the theoretical maximum fiber volume fraction for each of the three composites? Are they dependent on the fiber type/diameter?

3.2 Consider the data in Exercise 3.1. (a) Determine the fiber mass fraction in (i) square array of fiber packing and (ii) triangular array of fiber packing for fiber spacing-to-fiber diameter ratios (s/d) of 1.0, 1.25, 1.5, 1.75, and 2.0. Plot the W_f versus s/d curves. Are the fiber mass fractions dependent on fiber type? (b) What is the theoretical maximum fiber mass fraction for each of the three composites? Are they dependent on the fiber type?

3.3 In a matrix digestion test (see Chapter 11 for details) of carbon/epoxy sample, the following were recorded:

Mass of empty crucible = 30.1525 g
Mass of crucible with sample before matrix removal = 30.5903 g
Mass of crucible with sample after matrix removal = 30.4590 g
Density of fiber = 1.80 g/cm^3
Density of matrix 1.1 g/cm^3
Determine the (a) fiber mass fraction and (b) fiber volume fraction. Assume zero void content.

Hint: Use the rule of mixtures for composite density.

3.4 Consider the data given in Exercise 3.3. If the density of the sample is experimentally found as 1.48 g/cm^3, determine the (a) fiber mass fraction, (b) fiber volume fraction, (c) matrix volume fraction, and (d) voids volume fraction.

3.5 Consider a unidirectional carbon/epoxy and a unidirectional glass/epoxy lamina, each subjected to a uniaxial tension. E_{1f} = 240 GPa (carbon fiber), E_f = 76 GPa (glass fiber), and E_m = 3.5 GPa (epoxy matrix). Starting with small fiber volume fraction of 0.1, increase it gradually to 0.9 in steps of 0.1. Tabulate and plot the percentage share of load by the fibers in each lamina w.r.t. fiber volume fractions. Comment on the trend in load sharing in the two materials.

3.6 If the mass fraction of fiber in a unidirectional glass/epoxy lamina is 0.67, determine the void content. Assume the following data: $\rho_f = 2.54$ g/cm³, $\rho_m = 1.1$ g/cm³, and $\rho_c = 1.75$ g/cm³.

3.7 Consider a unidirectional glass/epoxy lamina with the following constituent material properties: $E_f = 76$ GPa, $\nu_f = 0.2$, $G_f = 35$ GPa, $E_m = 3.6$ GPa, $\nu_m = 0.3$, and $G_m = 1.4$ GPa. If the voids volume fraction is 2% and the fiber volume fraction is 60%. (a) Determine the composite longitudinal modulus, transverse modulus, major Poisson's ratio, and in-plane shear modulus. (b) Apply a longitudinal force on the lamina and determine the ratio of axial forces shared by fibers and matrix. (c) Consider the circular cross section of fibers and determine the maximum possible composite longitudinal modulus, transverse modulus, major Poisson's ratio, and in-plane shear modulus.

Compare the results with those in Example 3.1. Discuss the effect of voids on the composite mechanical properties.

3.8 Consider the following data for a unidirectional carbon/epoxy lamina:

$$E_{1f} = 240 \text{ GPa}, E_m = 3.5 \text{ GPa}, V_f = 0.55, \left(\sigma_{1f}^T\right)_{ult}$$
$$= 4500 \text{ MPa, and } \left(\sigma_m^T\right)_{ult} = 72 \text{ MPa}$$

(a) Determine the longitudinal tensile strength of the lamina. (b) If at an elevated temperature, the fiber and matrix moduli reduce by 10% and 20%, respectively, determine the change in the longitudinal strength of the lamina.

3.9 Consider a hybrid unidirectional carbon–glass/epoxy lamina with the following constituent characteristics:

Carbon fiber: $E_{1f} = 240$ GPa, $E_{2f} = 24$ GPa, $G_{12f} = 20$ GPa, $\nu_{12f} = 0.25$, and $V_f = 0.5$

Glass fiber: $E_f = 76$ GPa, $G_f = 36$ GPa, $\nu_f = 0.2$, and $V_f = 0.1$

Epoxy matrix: $E_m = 3.5$ GPa, $G_m = 1.4$ GPa, $\nu_m = 0.3$, and $V_f = 0.4$

Determine the (a) longitudinal modulus, (b) transverse modulus, (c) in-plane shear modulus, and (d) major Poisson's ratio of the lamina.

Hint: Replace the two reinforcement phases with an equivalent one. For example,

$$(E_{1f})_{\text{equivalent}} = (E_{1f})_{\text{carbon}} \times \frac{(V_f)_{\text{carbon}}}{(V_f)_{\text{carbon}} + (V_f)_{\text{glass}}} + (E_f)_{\text{glass}}$$
$$\times \frac{(V_f)_{\text{glass}}}{(V_f)_{\text{carbon}} + (V_f)_{\text{glass}}}$$

3.10 Consider a unidirectional glass/epoxy lamina with the following constituent material properties:

$$E_f = 76 \text{ GPa}, \nu_f = 0.2, G_f = 35 \text{ GPa}, E_m = 3.6 \text{ GPa}, \nu_m = 0.3, G_m = 1.4 \text{ GPa}$$

Assume zero void content and a fiber volume fraction of 0.6. Determine the composite longitudinal modulus, transverse modulus, major Poisson's ratio, and in-plane shear modulus using the Halpin–Tsai formulations. Compare the results with those in Example 3.1.

3.11 Determine the (a) longitudinal tensile strength, (b) longitudinal compressive strength, (c) transverse tensile strength, (d) transverse compressive strength, and (e) in-plane shear strength of the lamina in Exercise 3.7. Use the following additional data:

$$\left(\sigma_{1f}^T\right)_{ult}\Big|_{carbon} = 4500\,\text{MPa}, \left(\sigma_f^T\right)_{ult}\Big|_{glass} = 3500\,\text{MPa},$$

$$\left(\sigma_m^T\right)_{ult} = 72\,\text{MPa} \text{ and } (\tau_{12f})_{ult}\Big|_{carbon} = 36\,\text{MPa}$$

3.12 For a unidirectional carbon/epoxy lamina, the constituent material properties are given as follows: $E_{1f} = 240$ GPa, $E_m = 3.6$ GPa, $\alpha_{1f} = -0.5 \times 10^{-6}$ m/m/°C, $\alpha_{2f} = 6 \times 10^{-6}$ m/m/°C, $\alpha_m = 60 \times 10^{-6}$ m/m/°C, and $\nu_{12f} = 0.28$. Determine the change in dimensions w.r.t. room temperature dimensions for $V_f = 0.5$ and $V_f = 0.6$ if the lamina is subjected to an elevated temperature of 25°C above room temperature. The original dimensions are 300 mm × 300 mm. Comment on the effect of V_f.

3.13 Is it possible to design a unidirectional lamina such that its longitudinal dimension is temperature invariant? What should be the fiber volume fraction of the carbon/epoxy lamina in Exercise 3.12 if its length (dimension along the fiber direction) should not change when subjected to temperature change?

3.14 For a unidirectional carbon/epoxy lamina, the constituent material properties are given as follows: $E_{1f} = 240$ GPa, $E_m = 3.6$ GPa, $\rho_f = 1.8$, $\rho_m = 1.1$, $\beta_m = 0.35$ m/m/kg/kg, $\nu_{12f} = 0.28$, $\nu_m = 0.3$. If $V_f = 0.6$ and $V_v = 0.02$, determine the longitudinal and transverse CMEs. Assume carbon fiber does not absorb moisture. Compare the results with those in Example 3.9.

3.15 Given the material data in Exercise 3.14, plot the percentage change in volume of a unidirectional laminate of size 400 mm × 300 mm × 8 mm w.r.t. V_f if it absorbs 50 g moisture. Take void content as zero.

REFERENCES AND SUGGESTED READING

1. R. M. Jones, *Mechanics of Composite Materials*, second edition, Taylor & Francis, New York, 1999.
2. B. D. Agarwal, L. J. Broutman, and K. Chandrashekhara, *Analysis and Performance of Fiber Composites*, third edition, John Wiley & Sons, New York, 2006.
3. A. K. Kaw, *Mechanics of Composite Materials*, CRC Press, Boca Raton, FL, 2006.
4. S. W. Tsai and H. Thomas Hahn, *Introduction to Composite Materials*, Technomic Publishing, Lancaster, 1980.
5. E. J. Barbero, *Introduction to Composite Materials Design*, second edition, CRC Press, Boca Raton, FL, 2011.
6. C. C. Chamis and G. P. Sendeckyj, Critique on theories predicting thermoelastic properties of fibrous composites, *Journal of Composite Materials*, 2(3), 1968, 332–358.
7. B. W. Tew, Preliminary design of tubular composite structures using netting theory and composite degradation factors, *Journal of Pressure Vessel Technology*, 117(4), 1995, 390–394.
8. J. C. Ekvall, Elastic properties of orthotropic monofilament laminates, ASME Paper 61-AV-56, *Aviation Conference*, Los Angeles, California, March 12–16, 1961.
9. B. W. Shaffer, Stress–strain relations of reinforced plastics parallel and normal to the internal filaments, *AIAA Journal*, 2, 1964, 348.
10. Z. Hashin, *Theory of Fiber Reinforced Materials*, NASA-CR-1974, National Aeronautics and Space Administration, 1972.
11. R. Hill, Theory of mechanical properties of fiber-strengthened materials—III. Self-consistent model, *Journal of Mechanics and Physics of Solids*, 13, 1965, 189–198.

12. J. M. Whitney and M. B. Riley, Elastic properties of fiber reinforced composite materials, *AIAA Journal*, 4(9), 1966, 1537–1542.
13. J. M. Whitney, Elastic moduli of unidirectional composites with anisotropic filaments, *Journal of Composite Materials*, 1, 1967, 188–193.
14. B. Paul, Prediction of elastic constants of multiphase materials, *Transactions of the Metallurgical Society of AIME*, 100, 1960, 36–41.
15. Z. Hashin and S. Shtrikman, A variational approach to the theory of the elastic behavior of multiphase materials, *Journal of the Mechanics and Physics of Solids*, 100, 1963, 127–140.
16. Z. Hashin and B. W. Rosen, The elastic moduli of fibre reinforced materials, *ASME Journal of Applied Mechanics*, 100, 1964, 223–232.
17. D. F. Adams and S. W. Tsai, The influence of random filament packing on the transverse stiffness of unidirectional composites, *Journal of Composite Materials*, 3, 1969, 368–381.
18. R. Luciano and E. J. Barbero, Formulas for the stiffness of composites with periodic microstructure, *International Journal of Solids and Structures*, 31(21), 1994, 2933–2944.
19. E. J. Barbero, T. M. Damiani, and J. Trovillion, Micromechanics of fabric reinforced composites with periodic microstructure, *International Journal of Solids and Structures*, 42(9–10), 2005, 2489–2504.
20. S. G. Mogilevskaya, V. I. Kushch, and D. Nikolskiy, Evaluation of some approximate estimates for the effective tetragonal elastic moduli of two-phase fiber-reinforced composites, *Journal of Composite Materials*, 48(19), 2014, 2349–2362.
21. J. C. Halpin, *Effects of Environmental Factors on Composite Materials*, AFML-TR-67-423, June 1969.
22. Z. Hashin, Analysis of composite materials—A survey, *Journal of Applied Mechanics*, 50(3), 1983, 481–505.
23. N. Charalambakis, Homogenization techniques and micromechanics. A survey and perspectives, *Applied Mechanics Reviews*, 63, 2010.
24. A. Kelly and G. J. Davies, The principles of the fiber reinforcements of metals, *Metallurgical Reviews*, 10(37), 1965, 1–77.
25. L. B. Greszczuk, Microbuckling failure of circular fiber reinforced composites, *AIAA Journal*, 100, 1975, 1311–1318.
26. N. F. Dow and B. W. Rosen, *Evaluations of Filament Reinforced Composites for Aerospace Structural Applications*, NASA CR-207, April 1965.
27. H. Schuerch, Prediction of compressive strength in unidirectional boron fiber metal matrix composites, *AIAA Journal*, 4, 102, 1966.
28. R. A. Schapery, Thermal expansion co-efficients of composite materials based on energy principles, *Journal of Composite Materials*, 2(3), 1968, 380–404.

4

Macromechanics of a Lamina

4.1 CHAPTER ROAD MAP

In the chapter road map for Chapter 3, we stated that the analysis of a composite lamina is done at two levels—micro level and macro level. We also noted that, in the context of product design, the study of a lamina at the micro level is an alternative to the experimental study of the lamina. Effective hygro-thermo-mechanical parameters of a lamina are the typical output of micro-level analysis or experimental study of the lamina. These hygro-thermo-mechanical parameters are the input to a macro-level analysis of a lamina and this is an essential step in the analysis of a laminate leading to product design and development.

The objective of this chapter is to acquaint the reader with the behavior of a lamina at a macro level. We shall begin our study of macromechanics with an introductory discussion on lamina and the parameters that are required to describe a lamina. The mechanical behavior of a lamina is discussed in terms of the constitutive relations, engineering constants, and strength. Fiber orientation in a lamina is an important aspect and it is addressed in the discussion on topics in generally orthotropic lamina. Then, strength parameters of an orthotropic lamina are introduced and failure criteria are presented. Finally, the hygrothermal behavior of a lamina is discussed and the effects of temperature and moisture on specially orthotropic as well as generally orthotropic laminae are addressed.

Exposure to the introductory concepts of composites (Chapter 1) and basic solid mechanics (Chapter 2) is a prerequisite for effective assimilation of the concepts presented in this chapter; familiarity with micromechanics of lamina (Chapter 3) is desirable but not essential.

4.2 PRINCIPAL NOMENCLATURE

E_1, E_2	Young's moduli of a lamina in the material coordinates
E_x, E_y	Young's moduli of a lamina in the global coordinates
G_{12}	In-plane shear modulus of a lamina in the material coordinates
G_{xy}	In-plane shear modulus of a lamina in the global coordinates
$[Q]$	Reduced stiffness matrix of a lamina
$Q_{11}, Q_{12}, ..., Q_{66}$	Elements of the reduced stiffness matrix
$[\bar{Q}]$	Transformed reduced stiffness matrix of a lamina
$\bar{Q}_{11}, \bar{Q}_{12}, ..., \bar{Q}_{66}$	Elements of the transformed reduced stiffness matrix
$[R]$	Reuter matrix
$(R)_{11}^T, (R)_{22}^T$	Strength ratio/ultimate strain ratio for normal tensile stress in the directions -1 and -2, respectively
$(R)_{11}^C, (R)_{22}^C$	Strength ratio/ultimate strain ratio for normal compressive stress in the directions -1 and -2, respectively
$(R)_{12}^S$	Strength ratio for in-plane shear stress

$[S]$	Reduced compliance matrix of a lamina
$[\bar{S}]$	Transformed reduced compliance matrix of a lamina
$S_{11}, S_{12}, S_{22}, S_{66}$	Elements of the reduced compliance matrix
$\bar{S}_{11}, \bar{S}_{12}, ..., \bar{S}_{66}$	Elements of the transformed reduced compliance matrix
$[T]$	Transformation matrix
α_1, α_2	CTEs of a lamina in the material coordinates
$\alpha_x, \alpha_y, \alpha_{xy}$	CTEs of a lamina in the global coordinates
β_1, β_2	Coefficients of moisture expansion of a lamina in the material coordinates
$\beta_x, \beta_y, \beta_{xy}$	Coefficients of moisture expansion of a lamina in the global coordinates
γ_{12}	In-plane shear strains in a lamina in the material coordinates
γ_{xy}	In-plane shear strains in a lamina in the global coordinates
ΔT	Change in temperature
ΔC	Change in moisture content per unit mass of the lamina (kg/kg)
$\varepsilon_{11}, \varepsilon_{22}$	Normal strains in a lamina in the material coordinates
$\varepsilon_{xx}, \varepsilon_{yy}$	Normal strains in a lamina in the global coordinates
$\eta_{x,xy}, \eta_{y,xy}, \eta_{xy,x}, \eta_{xy,y}$	Shear coupling ratios
ν_{12}	Major Poisson's ratio of a lamina in the material coordinates
ν_{xy}	Major Poisson's ratio of a lamina in the global coordinates
σ_{11}, σ_{22}	Normal stresses in a lamina in the material coordinates
σ_{xx}, σ_{yy}	Normal stresses in a lamina in the global coordinates
$(\sigma_{11}^T)_{ult}, (\sigma_{22}^T)_{ult}$	Longitudinal and transverse tensile strengths, respectively, of a lamina
$(\sigma_{11}^C)_{ult}, (\sigma_{22}^C)_{ult}$	Longitudinal and transverse compressive strengths, respectively, of a lamina
τ_{12}	In-plane shear stress in a lamina in the material coordinates
τ_{xy}	In-plane shear stress in a lamina in the global coordinates
$(\tau_{12})_{ult}$	In-plane shear strength
θ	Orientation of a lamina w.r.t. x-axis

4.3 INTRODUCTION TO LAMINA

A lamina is a single layer or ply in a laminated composite material (Figure 4.1). It can be either flat or curved and it is made up of unidirectional, bidirectional, multidirectional, or randomly oriented fibers in the matrix material. At the micro level, as

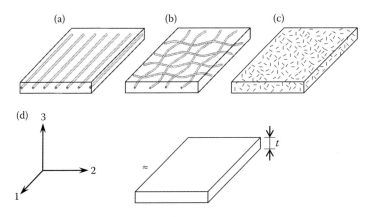

FIGURE 4.1 (a) Unidirectional composite ply. (b) Bidirectional composite ply. (c) Random fiber composite ply. (d) Composite lamina.

discussed in Chapter 3, the composite properties are determined from the interaction between the constituents. As noted in the chapter road map, macro-level analysis of a lamina is an essential step in the overall product design and analysis and it is presented in different styles and varying depth in many texts on mechanics of composites [1–6]. In macromechanics, a lamina is represented by its gross hygro-thermo-mechanical properties. A lamina is basically 2D and to describe it we need to know the ply thickness and the gross properties in the length and breadth directions. Taking the case of a unidirectional lamina, the direction along the fibers is the longitudinal direction or direction −1. However, the direction across the fibers in the plane of the lamina is the transverse direction or direction −2. Clearly, direction −3 is along the thickness of the lamina. The coordinate system O-123 is called the material coordinate system (or the local coordinate system) and it should be clearly differentiated from the laminate coordinate system O-xyz (or the global coordinate system).

The hygro-thermo-mechanical properties required in the macro-level analysis of a lamina are primarily as follows:

- Stiffness properties
  - Longitudinal modulus, E_1
  - Transverse modulus, E_2
  - In-plane shear modulus, G_{12}
  - Poisson's ratios, ν_{12} and ν_{23} (ν_{23} is needed if analysis involves ε_{33})

Note: In a transversely isotropic material, there are five independent elastic constants. Under the plane stress idealization, the Poisson's ratio ν_{23} is associated only with the out-of-plane normal strain ε_{33}. Thus, for lamina analysis involving in-plane stresses and strains, the remaining four stiffness parameters are sufficient.

- Strength parameters
  - Longitudinal tensile strength, $(\sigma_{11}^T)_{ult}$
  - Longitudinal compressive strength, $(\sigma_{11}^C)_{ult}$
  - Transverse tensile strength, $(\sigma_{22}^T)_{ult}$
  - Transverse compressive strength, $(\sigma_{22}^C)_{ult}$
  - In-plane shear strength, $(\tau_{12})_{ult}$
- Hygrothermal properties
  - Longitudinal CTE, α_1
  - Transverse CTE, α_2
  - Longitudinal CME, β_1
  - Transverse CME, β_2

As mentioned earlier, a lamina may be unidirectional, bidirectional, multidirectional, or randomly oriented. However, thankfully, at the macro level, the general methodology of analysis remains the same for all these different types of laminae. We only need to take care in using suitable values for the various stiffness and other parameters. For example, for a unidirectional lamina, $E_1 \neq E_2$, whereas, for a bidirectional fiber lamina, $E_1 \approx E_2$, and so on.

4.4 CONSTITUTIVE EQUATIONS OF A LAMINA

The structural analysis of a lamina involves the determination of the response (primarily strains and deformations) of the lamina to applied loads. Given the loads and the geometry of the lamina, the stresses are readily determined. Then, using the

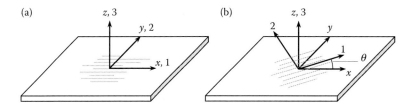

FIGURE 4.2 (a) Specially orthotropic lamina. (b) Generally orthotropic lamina.

stress–strain relations, the strains and deformations are determined. The stress–strain relations for different types of materials were discussed in Chapter 2. These relations contain a number of elastic constants and the number of independent elastic constants depends on the type of the material. A lamina, whether unidirectional or bidirectional, is orthotropic in nature. It may be recalled that there are nine independent elastic constants in an orthotropic material.

There is another aspect that needs attention—the orientation of the lamina. An orthotropic lamina whose material coordinates are aligned with the laminate coordinates is referred to as a specially orthotropic lamina (Figure 4.2a). On the other hand, an orthotropic lamina whose in-plane material coordinate axes are at some nonzero angle to the in-plane laminate coordinate axes is referred to as a generally orthotropic lamina (Figure 4.2b).

4.4.1 Specially Orthotropic Lamina

4.4.1.1 Constitutive Relation

Let us recall Equation 2.201 and rewrite the stress–strain relation for an orthotropic material in the material coordinate system as follows:

$$\begin{Bmatrix} \varepsilon_{11} \\ \varepsilon_{22} \\ \varepsilon_{33} \\ \gamma_{23} \\ \gamma_{31} \\ \gamma_{12} \end{Bmatrix} = \begin{bmatrix} \dfrac{1}{E_1} & -\dfrac{\nu_{12}}{E_1} & -\dfrac{\nu_{13}}{E_1} & 0 & 0 & 0 \\ -\dfrac{\nu_{12}}{E_1} & \dfrac{1}{E_2} & -\dfrac{\nu_{23}}{E_2} & 0 & 0 & 0 \\ -\dfrac{\nu_{13}}{E_1} & -\dfrac{\nu_{23}}{E_2} & \dfrac{1}{E_3} & 0 & 0 & 0 \\ 0 & 0 & 0 & \dfrac{1}{G_{23}} & 0 & 0 \\ 0 & 0 & 0 & 0 & \dfrac{1}{G_{31}} & 0 \\ 0 & 0 & 0 & 0 & 0 & \dfrac{1}{G_{12}} \end{bmatrix} \begin{Bmatrix} \sigma_{11} \\ \sigma_{22} \\ \sigma_{33} \\ \tau_{23} \\ \tau_{31} \\ \tau_{12} \end{Bmatrix} \quad (4.1)$$

where

$\varepsilon_{11}, \varepsilon_{22}, \varepsilon_{33}$ Normal strains
$\gamma_{23}, \gamma_{31}, \gamma_{12}$ Shear strains
$\sigma_{11}, \sigma_{22}, \sigma_{33}$ Normal stresses
$\tau_{23}, \tau_{31}, \tau_{12}$ Shear stresses
E_1, E_2, E_3 Young's moduli

Macromechanics of a Lamina

G_{23}, G_{31}, G_{12} Shear moduli
$\nu_{12}, \nu_{23}, \nu_{13}$ Poisson's ratios

Equation 4.1 provides the 3D stress–strain relations for an orthotropic composite material in the material coordinate system. Under the applied loads, the gross or apparent stress components are determined from the geometry and loading information. Note that the six strain components are then determined from the six stress components using the nine engineering constants. It is illustrated with the help of an example as given below.

EXAMPLE 4.1

Consider a unidirectional lamina made as shown in Figure 4.3a. The engineering constants for the material are as follows: $E_1 = 40$ GPa, $E_2 = 8$ GPa, $\nu_{12} = 0.25$, $\nu_{23} = 0.3$, and $G_{12} = 4$ GPa.

1. Apply a tensile force in the fiber direction (Figure 4.3b) and determine the strains and deformations.
2. Apply a tensile force in the transverse direction (Figure 4.3c) and determine the strains and deformations.
3. Apply a shear force in the plane of the lamina (Figure 4.3d) and determine the strain components.
4. Sketch the deformed lamina in each of the above cases.

Solution

Unidirectional lamina is transversely isotropic and thus

$E_3 = E_2 = 8$ GPa
$\nu_{13} = \nu_{12} = 0.25$
$G_{31} = G_{12} = 4$ GPa

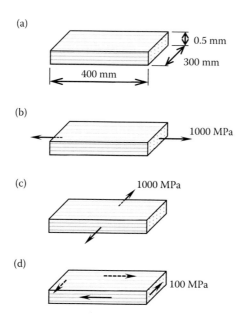

FIGURE 4.3 Lamina analysis in Example 4.1. (a) Unidirectional lamina. (b) UD lamina under longitudinal tensile stress. (c) UD lamina under transverse tensile stress. (d) UD lamina under in-plane shear stress.

Also,

$$G_{23} = \frac{E_2}{2(1+\nu_{23})} = \frac{8}{2\times(1+0.25)} = 3.2\,\text{GPa}$$

Then, using Equation 4.1, strain components under different loading conditions are obtained as follows:

1. Analysis of lamina under longitudinal tensile stress

$$\begin{Bmatrix} \varepsilon_{11} \\ \varepsilon_{22} \\ \varepsilon_{33} \\ \gamma_{23} \\ \gamma_{31} \\ \gamma_{12} \end{Bmatrix} = \begin{bmatrix} \dfrac{1}{40} & -\dfrac{0.25}{40} & -\dfrac{0.25}{40} & 0 & 0 & 0 \\ -\dfrac{0.25}{40} & \dfrac{1}{8} & -\dfrac{0.3}{8} & 0 & 0 & 0 \\ -\dfrac{0.25}{40} & -\dfrac{0.3}{8} & \dfrac{1}{8} & 0 & 0 & 0 \\ 0 & 0 & 0 & \dfrac{1}{3.2} & 0 & 0 \\ 0 & 0 & 0 & 0 & \dfrac{1}{4} & 0 \\ 0 & 0 & 0 & 0 & 0 & \dfrac{1}{4} \end{bmatrix} \begin{Bmatrix} 1000 \\ 0 \\ 0 \\ 0 \\ 0 \\ 0 \end{Bmatrix} \times 10^{-3} = \begin{Bmatrix} 0.025 \\ -0.00625 \\ -0.00625 \\ 0 \\ 0 \\ 0 \end{Bmatrix}$$

The corresponding deformations are

$$\begin{Bmatrix} u_1 \\ u_2 \\ u_3 \\ \phi_{23} \\ \phi_{31} \\ \phi_{12} \end{Bmatrix} = \begin{Bmatrix} 0.025 \times 400 \\ -0.00625 \times 300 \\ -0.00625 \times 0.5 \\ 0 \\ 0 \\ 0 \end{Bmatrix} = \begin{Bmatrix} 10 \\ -1.875 \\ -0.003 \\ 0 \\ 0 \\ 0 \end{Bmatrix}$$

2. Analysis of lamina under transverse tensile stress

$$\begin{Bmatrix} \varepsilon_{11} \\ \varepsilon_{22} \\ \varepsilon_{33} \\ \gamma_{23} \\ \gamma_{31} \\ \gamma_{12} \end{Bmatrix} = \begin{bmatrix} \dfrac{1}{40} & -\dfrac{0.25}{40} & -\dfrac{0.25}{40} & 0 & 0 & 0 \\ -\dfrac{0.25}{40} & \dfrac{1}{8} & -\dfrac{0.3}{8} & 0 & 0 & 0 \\ -\dfrac{0.25}{40} & -\dfrac{0.3}{8} & \dfrac{1}{8} & 0 & 0 & 0 \\ 0 & 0 & 0 & \dfrac{1}{3.2} & 0 & 0 \\ 0 & 0 & 0 & 0 & \dfrac{1}{4} & 0 \\ 0 & 0 & 0 & 0 & 0 & \dfrac{1}{4} \end{bmatrix} \begin{Bmatrix} 0 \\ 1000 \\ 0 \\ 0 \\ 0 \\ 0 \end{Bmatrix} \times 10^{-3} = \begin{Bmatrix} -0.00625 \\ 0.125 \\ -0.0375 \\ 0 \\ 0 \\ 0 \end{Bmatrix}$$

Macromechanics of a Lamina

The corresponding deformations are

$$\begin{Bmatrix} u_1 \\ u_2 \\ u_3 \\ \phi_{23} \\ \phi_{31} \\ \phi_{12} \end{Bmatrix} = \begin{Bmatrix} -0.00625 \times 400 \\ 0.125 \times 300 \\ -0.0375 \times 0.5 \\ 0 \\ 0 \\ 0 \end{Bmatrix} = \begin{Bmatrix} -2.5 \\ 37.5 \\ -0.019 \\ 0 \\ 0 \\ 0 \end{Bmatrix}$$

3. Analysis of lamina under in-plane shear stress

$$\begin{Bmatrix} \varepsilon_{11} \\ \varepsilon_{22} \\ \varepsilon_{33} \\ \gamma_{23} \\ \gamma_{31} \\ \gamma_{12} \end{Bmatrix} = \begin{bmatrix} \dfrac{1}{40} & -\dfrac{0.25}{40} & -\dfrac{0.25}{40} & 0 & 0 & 0 \\ -\dfrac{0.25}{40} & \dfrac{1}{8} & -\dfrac{0.3}{8} & 0 & 0 & 0 \\ -\dfrac{0.25}{40} & -\dfrac{0.3}{8} & \dfrac{1}{8} & 0 & 0 & 0 \\ 0 & 0 & 0 & \dfrac{1}{3.2} & 0 & 0 \\ 0 & 0 & 0 & 0 & \dfrac{1}{4} & 0 \\ 0 & 0 & 0 & 0 & 0 & \dfrac{1}{4} \end{bmatrix} \begin{Bmatrix} 0 \\ 0 \\ 0 \\ 0 \\ 0 \\ 100 \end{Bmatrix} \times 10^{-3} = \begin{Bmatrix} 0 \\ 0 \\ 0 \\ 0 \\ 0 \\ 0.025 \end{Bmatrix}$$

The corresponding deformations are

$$\begin{Bmatrix} u_1 \\ u_2 \\ u_3 \\ \phi_{23} \\ \phi_{31} \\ \phi_{12} \end{Bmatrix} = \begin{Bmatrix} 0 \\ 0 \\ 0 \\ 0 \\ 0 \\ 0.025 \times 180/\pi \end{Bmatrix} = \begin{Bmatrix} 0 \\ 0 \\ 0 \\ 0 \\ 0 \\ 1.43 \end{Bmatrix}$$

4. Deformed shapes of the lamina in each of the above cases are given in Figure 4.4.

The illustration above shows how an orthotropic lamina under in-plane loads can be analyzed. We have found that the assumption of transverse isotropy reduces the number of independent engineering constants. However, this simplification is basically on the material properties only and the number of stress–strain equations and their general structure remains the same. We have also found that under the in-plane loads for a thin lamina, there is negligible out-of-plane deformation.

On the other hand, gross reduction in the computational effort can be achieved by utilizing plane stress idealization. Let us then apply the concept of plane stress idealization (refer Chapter 2) in the analysis of a lamina. The thickness of a lamina is small compared to the other two dimensions. Also, only in-plane forces are applied on the lamina.

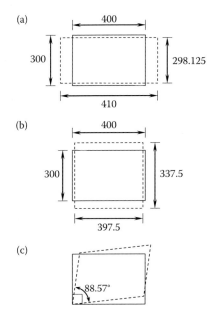

FIGURE 4.4 Deformation of unidirectional lamina (Example 4.1). (a) Deformation under longitudinal tensile stress. (b) Deformation under transverse tensile stress. (c) Deformation under in-plane shear stress.

Given this, a lamina can be analyzed as a plane stress problem. In the material coordinate system, from Chapter 2, we can rewrite Equations 2.213 through 2.218 as follows:

$$\begin{Bmatrix} \varepsilon_{11} \\ \varepsilon_{22} \\ \gamma_{12} \end{Bmatrix} = \begin{bmatrix} \dfrac{1}{E_1} & -\dfrac{\nu_{12}}{E_1} & 0 \\ -\dfrac{\nu_{12}}{E_1} & \dfrac{1}{E_2} & 0 \\ 0 & 0 & \dfrac{1}{G_{12}} \end{bmatrix} \begin{Bmatrix} \sigma_{11} \\ \sigma_{22} \\ \tau_{12} \end{Bmatrix} \quad (4.2)$$

and

$$\varepsilon_{33} = -\left(\dfrac{\nu_{13}\sigma_{11}}{E_1} + \dfrac{\nu_{23}\sigma_{22}}{E_2} \right) \quad (4.3)$$

$$\gamma_{23} = \gamma_{31} = 0 \quad (4.4)$$

The square matrix on the right-hand side of Equation 4.2, denoted by $[S]$, is called the reduced compliance matrix. Thus,

$$\begin{Bmatrix} \varepsilon_{11} \\ \varepsilon_{22} \\ \gamma_{12} \end{Bmatrix} = [S] \begin{Bmatrix} \sigma_{11} \\ \sigma_{22} \\ \tau_{12} \end{Bmatrix} \quad (4.5)$$

in which,

$$[S] = \begin{bmatrix} S_{11} & S_{12} & 0 \\ S_{12} & S_{22} & 0 \\ 0 & 0 & S_{66} \end{bmatrix} \quad (4.6)$$

and the elements of [S] are given by

$$S_{11} = \frac{1}{E_1} \tag{4.7}$$

$$S_{12} = -\frac{\nu_{12}}{E_1} \tag{4.8}$$

$$S_{22} = \frac{1}{E_2} \tag{4.9}$$

$$S_{66} = \frac{1}{G_{12}} \tag{4.10}$$

$$S_{16} = S_{26} = 0 \tag{4.11}$$

Notes:

1. In the compliance matrix, there are no terms with suffix containing digits 3, 4, or 5. For example, we do not have a term S_{13} and so on. In our alternate or contracted notation, σ_{11}, σ_{22}, and τ_{12} are represented, respectively, by σ_1, σ_2, and σ_6. Similarly, ε_{11}, ε_{22}, and γ_{12} are represented, respectively, by ε_1, ε_2, and ε_6. Thus, in the case of a lamina, the third, fourth, and fifth rows and columns of the full six by six compliance matrix are omitted.
2. For a specially orthotropic lamina, as we have seen, $S_{16} = S_{26} = 0$ and there is no coupling between normal stresses and shear strain. Similarly, there is no coupling between shear stress and normal strains.

The inverse form of Equation 4.2 is as follows:

$$\begin{Bmatrix} \sigma_{11} \\ \sigma_{22} \\ \tau_{12} \end{Bmatrix} = \begin{bmatrix} \dfrac{E_1}{1-\nu_{12}\nu_{21}} & \dfrac{\nu_{12}E_2}{1-\nu_{12}\nu_{21}} & 0 \\ \dfrac{\nu_{12}E_2}{1-\nu_{12}\nu_{21}} & \dfrac{E_2}{1-\nu_{12}\nu_{21}} & 0 \\ 0 & 0 & G_{12} \end{bmatrix} \begin{Bmatrix} \varepsilon_{11} \\ \varepsilon_{22} \\ \gamma_{12} \end{Bmatrix} \tag{4.12}$$

and

$$\sigma_{33} = \tau_{31} = \tau_{23} = 0 \tag{4.13}$$

The square matrix on the right-hand side of Equation 4.12, denoted by [**Q**], is called the reduced stiffness matrix. Thus,

$$\begin{Bmatrix} \sigma_{11} \\ \sigma_{22} \\ \tau_{12} \end{Bmatrix} = [\boldsymbol{Q}] \begin{Bmatrix} \varepsilon_{11} \\ \varepsilon_{22} \\ \gamma_{12} \end{Bmatrix} \tag{4.14}$$

in which

$$[Q] = \begin{bmatrix} Q_{11} & Q_{12} & 0 \\ Q_{12} & Q_{22} & 0 \\ 0 & 0 & Q_{66} \end{bmatrix} \quad (4.15)$$

and the elements of $[Q]$ are given by

$$Q_{11} = \frac{E_1}{1 - \nu_{12}\nu_{21}} \quad (4.16)$$

$$Q_{12} = \frac{\nu_{12} E_2}{1 - \nu_{12}\nu_{21}} \quad (4.17)$$

$$Q_{22} = \frac{E_2}{1 - \nu_{12}\nu_{21}} \quad (4.18)$$

$$Q_{66} = G_{12} \quad (4.19)$$

$$Q_{16} = Q_{26} = 0 \quad (4.20)$$

EXAMPLE 4.2

Consider the problem in Example 4.1 and solve it adopting plane stress idealization.

Solution

Using Equation 4.2, strain components under different loading conditions are obtained as follows:

1. Analysis of lamina under longitudinal tensile stress

$$\begin{Bmatrix} \varepsilon_{11} \\ \varepsilon_{22} \\ \gamma_{12} \end{Bmatrix} = \begin{bmatrix} \dfrac{1}{40} & -\dfrac{0.25}{40} & 0 \\ -\dfrac{0.25}{40} & \dfrac{1}{8} & 0 \\ 0 & 0 & \dfrac{1}{4} \end{bmatrix} \begin{Bmatrix} 1000 \\ 0 \\ 0 \end{Bmatrix} \times 10^{-3} = \begin{Bmatrix} 0.025 \\ -0.00625 \\ 0 \end{Bmatrix}$$

and

$$\varepsilon_{33} = -\left(\frac{0.25 \times 1000}{40} + \frac{0.3 \times 0}{8} \right) \times 10^{-3} = -0.00625$$

2. Analysis of lamina under transverse tensile stress

$$\begin{Bmatrix} \varepsilon_{11} \\ \varepsilon_{22} \\ \gamma_{12} \end{Bmatrix} = \begin{bmatrix} \dfrac{1}{40} & -\dfrac{0.25}{40} & 0 \\ -\dfrac{0.25}{40} & \dfrac{1}{8} & 0 \\ 0 & 0 & \dfrac{1}{4} \end{bmatrix} \begin{Bmatrix} 0 \\ 1000 \\ 0 \end{Bmatrix} \times 10^{-3} = \begin{Bmatrix} -0.00625 \\ 0.125 \\ 0 \end{Bmatrix}$$

Macromechanics of a Lamina

and

$$\varepsilon_{33} = -\left(\frac{0.25 \times 0}{40} + \frac{0.3 \times 1000}{8}\right) \times 10^{-3} = -0.0375$$

3. Analysis of lamina under in-plane shear stress

$$\begin{Bmatrix} \varepsilon_{11} \\ \varepsilon_{22} \\ \gamma_{12} \end{Bmatrix} = \begin{bmatrix} \frac{1}{40} & -\frac{0.25}{40} & 0 \\ -\frac{0.25}{40} & \frac{1}{8} & 0 \\ 0 & 0 & \frac{1}{4} \end{bmatrix} \begin{Bmatrix} 0 \\ 0 \\ 100 \end{Bmatrix} \times 10^{-3} = \begin{Bmatrix} 0 \\ 0 \\ 0.025 \end{Bmatrix}$$

and

$$\varepsilon_{33} = -\left(\frac{0.25 \times 0}{40} + \frac{0.3 \times 0}{8}\right) \times 10^{-3} = 0$$

4.4.1.2 Restrictions on Elastic Constants

The constitutive relation in the material coordinates (Equation 4.1) in three dimensions for an orthotropic material involves nine independent elastic constants. These elastic constants are bound by some physical principles, which lead to certain mathematical restrictions. In an isotropic material, such restrictions are simple and they can be stated as

$$E > 0 \tag{4.21}$$

$$G > 0 \tag{4.22}$$

$$-1 \leq \nu \leq 0.5 \tag{4.23}$$

in which the restriction on Poisson's ratio can be obtained from the restrictions that Young's modulus and bulk modulus are both positive.

In an orthotropic material, these restrictions are relatively complex; they can be derived from natural physical reasoning [1,2,7]. Let us consider a load case of a non-zero positive normal stress ($\sigma_{11} \neq 0$, $\sigma_{22} = \sigma_{33} = \tau_{23} = \tau_{31} = \tau_{12} = 0$); it should result in positive strain ε_{11}, which shows that S_{11} is positive. Extending the logic to the other possible load cases, we can see that the diagonal elements in the compliance matrix and thus normal and shear moduli are all positive.

$$S_{11} > 0 \quad \text{or} \quad E_1 > 0 \tag{4.24}$$

$$S_{22} > 0 \quad \text{or} \quad E_2 > 0 \tag{4.25}$$

$$S_{33} > 0 \quad \text{or} \quad E_3 > 0 \tag{4.26}$$

$$S_{44} > 0 \quad \text{or} \quad G_{23} > 0 \tag{4.27}$$

$$S_{55} > 0 \quad \text{or} \quad G_{31} > 0 \tag{4.28}$$

$$S_{66} > 0 \quad \text{or} \quad G_{12} > 0 \tag{4.29}$$

Next, it is possible under suitable conditions to simulate a case of deformation wherein $\varepsilon_{11} \neq 0$, $\varepsilon_{22} = \varepsilon_{33} = \gamma_{23} = \gamma_{31} = \gamma_{12} = 0$. The corresponding work done is positive, which shows C_{11} is positive. Extending the logic to such other possible cases of deformation, it can be seen that the diagonal elements in the 3D stiffness matrix are all positive. Then, inverting [S] from the first three diagonal elements, it can be shown that

$$C_{11} > 0 \quad \text{or} \quad 1 - \nu_{23}\nu_{32} > 0 \tag{4.30}$$

$$C_{22} > 0 \quad \text{or} \quad 1 - \nu_{13}\nu_{31} > 0 \tag{4.31}$$

$$C_{33} > 0 \quad \text{or} \quad 1 - \nu_{12}\nu_{21} > 0 \tag{4.32}$$

Noting that $\nu_{21} = \nu_{12}(E_2/E_1)$, $\nu_{32} = \nu_{23}(E_3/E_2)$, and $\nu_{31} = \nu_{13}(E_3/E_1)$, we can readily show that

$$|\nu_{23}| < \sqrt{\frac{E_2}{E_3}} \tag{4.33}$$

$$|\nu_{32}| < \sqrt{\frac{E_3}{E_2}} \tag{4.34}$$

$$|\nu_{13}| < \sqrt{\frac{E_1}{E_3}} \tag{4.35}$$

$$|\nu_{31}| < \sqrt{\frac{E_3}{E_1}} \tag{4.36}$$

$$|\nu_{12}| < \sqrt{\frac{E_1}{E_2}} \tag{4.37}$$

$$|\nu_{21}| < \sqrt{\frac{E_2}{E_1}} \tag{4.38}$$

Further, the stiffness and compliance matrices are positive definite [7]; the determinants are positive and it can be shown that

$$1 - \nu_{12}\nu_{21} - \nu_{23}\nu_{32} - \nu_{31}\nu_{13} - 2\nu_{21}\nu_{32}\nu_{13} > 0 \tag{4.39}$$

4.4.2 Generally Orthotropic Lamina

Unidirectional laminae are poor in transverse mechanical properties; to overcome this, usually they are placed in different orientations in a laminate. Bidirectional laminae

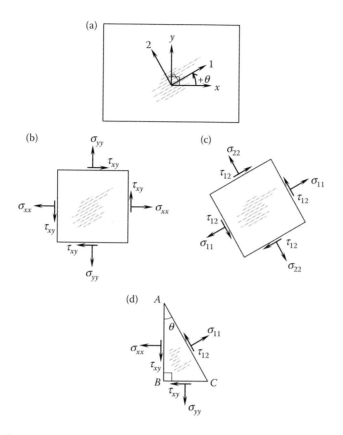

FIGURE 4.5 Stress transformation. (a) Generally orthotropic lamina. (b) Stress element in the global coordinates. (c) Stress element in the material coordinates. (d) Free body diagram of a triangular element.

are also placed in different orientations so as to achieve the desired laminate properties. Thus, it is necessary to know the stress–strain relations for a generally orthotropic lamina.

Let us consider a lamina as shown in Figure 4.5a. It is generally orthotropic as the material coordinates and the global coordinates of the lamina are not aligned. Let 1-direction be at an angle θ w.r.t. the x-direction. The sign convention adopted is clockwise positive as per the right-hand rule. Stress components in the global coordinate system and the material coordinate system are shown in Figure 4.5b and c, respectively. We shall first arrive at a transformation of stresses in the global coordinates to those in the material coordinates. Let us then consider a triangular element as shown in Figure 4.5d such that sides AB and BC are normal to the x- and y-directions, respectively, and side AC is normal to the 1-direction. Also, angle $\angle ABC$ is a right angle. Now, let the length of side AC be l such that the lengths of sides AB and BC are $l \cos \theta$ and $l \sin \theta$, respectively.

Now, by considering the equilibrium of forces in the 1-direction and after simple manipulation, we obtain

$$\sigma_{11} = \sigma_{xx} \cos^2 \theta + \sigma_{yy} \sin^2 \theta + 2\tau_{xy} \sin \theta \cos \theta \qquad (4.40)$$

Similarly, by considering the equilibrium of forces in the 2-direction and after simple manipulation, we obtain

$$\tau_{12} = -\sigma_{xx} \sin \theta \cos \theta + \sigma_{yy} \sin \theta \cos \theta + \tau_{xy}(\cos^2 \theta - \sin^2 \theta) \qquad (4.41)$$

We can consider another right-angled triangular element with its hypotenuse normal to the 2-direction and the other two sides normal to the x- and y-directions. Considering force equilibrium in the 2-direction in this element, we can obtain

$$\sigma_{22} = \sigma_{xx} \sin^2 \theta + \sigma_{yy} \cos^2 \theta - 2\tau_{xy} \sin \theta \cos \theta \tag{4.42}$$

Equations 4.40 through 4.42 can be written in the matrix form as

$$\begin{Bmatrix} \sigma_{11} \\ \sigma_{22} \\ \tau_{12} \end{Bmatrix} = \begin{bmatrix} \cos^2 \theta & \sin^2 \theta & 2\sin\theta\cos\theta \\ \sin^2 \theta & \cos^2 \theta & -2\sin\theta\cos\theta \\ -\sin\theta\cos\theta & \sin\theta\cos\theta & \cos^2\theta - \sin^2\theta \end{bmatrix} \begin{Bmatrix} \sigma_{xx} \\ \sigma_{yy} \\ \tau_{xy} \end{Bmatrix} \tag{4.43}$$

or

$$\begin{Bmatrix} \sigma_{11} \\ \sigma_{22} \\ \tau_{12} \end{Bmatrix} = [T] \begin{Bmatrix} \sigma_{xx} \\ \sigma_{yy} \\ \tau_{xy} \end{Bmatrix} \tag{4.44}$$

where the transformation matrix $[T]$ in Equation 4.44 is given by

$$[T] = \begin{bmatrix} \cos^2 \theta & \sin^2 \theta & 2\sin\theta\cos\theta \\ \sin^2 \theta & \cos^2 \theta & -2\sin\theta\cos\theta \\ -\sin\theta\cos\theta & \sin\theta\cos\theta & \cos^2\theta - \sin^2\theta \end{bmatrix} \tag{4.45}$$

The transformation matrix is usually written as

$$[T] = \begin{bmatrix} c^2 & s^2 & 2sc \\ s^2 & c^2 & -2sc \\ -sc & sc & c^2 - s^2 \end{bmatrix} \tag{4.46}$$

where $c = \cos \theta$ and $s = \sin \theta$.

Note: The stress transformation given above can also be directly obtained from Equation 2.122 discussed in Chapter 2.

Strain transformation equation is similar to that for stress transformation. Care, however, must be taken to use tensorial shear strain in the transformation equation. Thus, for the generally orthotropic lamina shown in Figure 4.5a, strain transformation is given by

$$\begin{Bmatrix} \varepsilon_{11} \\ \varepsilon_{22} \\ \gamma_{12}/2 \end{Bmatrix} = [T] \begin{Bmatrix} \varepsilon_{xx} \\ \varepsilon_{yy} \\ \gamma_{xy}/2 \end{Bmatrix} \tag{4.47}$$

Here, we introduce the Reuter matrix $[R]$ and rewrite Equation 4.47 as [8]

Macromechanics of a Lamina

$$\begin{Bmatrix} \varepsilon_{11} \\ \varepsilon_{22} \\ \gamma_{12} \end{Bmatrix} = [R][T][R]^{-1} \begin{Bmatrix} \varepsilon_{xx} \\ \varepsilon_{yy} \\ \gamma_{xy} \end{Bmatrix} \quad (4.48)$$

where

$$[R] = \begin{bmatrix} 1 & 0 & 0 \\ 0 & 1 & 0 \\ 0 & 0 & 2 \end{bmatrix} \quad (4.49)$$

The global stresses are obtained from the local stresses by inverting Equation 4.44 as

$$\begin{Bmatrix} \sigma_{xx} \\ \sigma_{yy} \\ \tau_{xy} \end{Bmatrix} = [T]^{-1} \begin{Bmatrix} \sigma_{11} \\ \sigma_{22} \\ \tau_{12} \end{Bmatrix} \quad (4.50)$$

Using Equation 4.14, then, we can write

$$\begin{Bmatrix} \sigma_{xx} \\ \sigma_{yy} \\ \tau_{xy} \end{Bmatrix} = [T]^{-1}[Q] \begin{Bmatrix} \varepsilon_{11} \\ \varepsilon_{22} \\ \gamma_{12} \end{Bmatrix} \quad (4.51)$$

Then, using Equation 4.48, finally, we can relate global stresses to global strains as

$$\begin{Bmatrix} \sigma_{xx} \\ \sigma_{yy} \\ \tau_{xy} \end{Bmatrix} = [T]^{-1}[Q][R][T][R]^{-1} \begin{Bmatrix} \varepsilon_{xx} \\ \varepsilon_{yy} \\ \gamma_{xy} \end{Bmatrix} \quad (4.52)$$

Equation 4.52 can be written in a simplified way as

$$\begin{Bmatrix} \sigma_{xx} \\ \sigma_{yy} \\ \tau_{xy} \end{Bmatrix} = [\bar{Q}] \begin{Bmatrix} \varepsilon_{xx} \\ \varepsilon_{yy} \\ \gamma_{xy} \end{Bmatrix} \quad (4.53)$$

The matrix $[\bar{Q}]$ is called the transformed reduced stiffness matrix and it is obtained by multiplying the five 3×3 matrices on the right-hand side of Equation 4.52, that is,

$$[\bar{Q}] = [T]^{-1}[Q][R][T][R]^{-1} \quad (4.54)$$

By carrying out the necessary multiplication and simplification, it can be shown that the elements of $[\bar{Q}]$ are given by

$$\bar{Q}_{11} = Q_{11} \cos^4 \theta + Q_{22} \sin^4 \theta + 2(Q_{12} + 2Q_{66}) \sin^2 \theta \cos^2 \theta \quad (4.55)$$

$$\bar{Q}_{12} = (Q_{11} + Q_{22} - 4Q_{66}) \sin^2 \theta \cos^2 \theta + Q_{12}(\sin^4 \theta + \cos^4 \theta) \quad (4.56)$$

$$\bar{Q}_{16} = (Q_{11} - Q_{12} - 2Q_{66})\sin\theta\cos^3\theta - (Q_{22} - Q_{12} - 2Q_{66})\sin^3\theta\cos\theta \quad (4.57)$$

$$\bar{Q}_{22} = Q_{11}\sin^4\theta + Q_{22}\cos^4\theta + 2(Q_{12} + 2Q_{66})\sin^2\theta\cos^2\theta \quad (4.58)$$

$$\bar{Q}_{26} = (Q_{11} - Q_{12} - 2Q_{66})\sin^3\theta\cos\theta - (Q_{22} - Q_{12} - 2Q_{66})\sin\theta\cos^3\theta \quad (4.59)$$

$$\bar{Q}_{66} = (Q_{11} - 2Q_{12} + Q_{22} - 2Q_{66})\sin^2\theta\cos^2\theta + Q_{66}(\sin^4\theta + \cos^4\theta) \quad (4.60)$$

The transformed reduced stiffness matrix $[\bar{Q}]$ has six generally nonzero elements as compared to the reduced stiffness matrix $[Q]$, which has four. The elements of $[\bar{Q}]$ are functions of stiffness elements Q_{11}, Q_{12}, Q_{22}, and Q_{66} and the lamina angle θ. Since the elements of $[Q]$ are functions of four engineering constants, we can say that the elements of $[\bar{Q}]$ are functions of four engineering constants and the lamina angle θ. Note that when the global coordinates and local coordinates are aligned, $\theta = 0$ and $[\bar{Q}]$ reduces to $[Q]$.

Note further that $Q_{16} = Q_{26} = 0$, but $\bar{Q}_{16} \neq 0$ and $\bar{Q}_{26} \neq 0$. As a result, as mentioned earlier, there is no shear-extension coupling in a specially orthotropic lamina, but in a generally orthotropic lamina, shear-extension coupling does exist. Thus, in a generally orthotropic lamina, normal stresses result in shear strains and shear stresses result in normal strain. Such a lamina, though orthotropic from material characteristics, looks like anisotropic and hence the name "generally orthotropic."

Given the global strains in a generally orthotropic lamina, we can obtain the global stresses from Equation 4.53 by inverting it as follows:

$$\begin{Bmatrix} \varepsilon_{xx} \\ \varepsilon_{yy} \\ \gamma_{xy} \end{Bmatrix} = [\bar{S}] \begin{Bmatrix} \sigma_{xx} \\ \sigma_{yy} \\ \tau_{xy} \end{Bmatrix} \quad (4.61)$$

in which $[\bar{S}]$ is the transformed reduced compliance matrix given by the inverse of the transformed reduced stiffness matrix, that is, $[\bar{S}] = [\bar{Q}]^{-1}$. $[\bar{S}]$ can also be obtained from the transformation of stress–strain relations in the local coordinates. Let us reproduce Equation 4.5 as

$$\begin{Bmatrix} \varepsilon_{11} \\ \varepsilon_{22} \\ \gamma_{12} \end{Bmatrix} = [S] \begin{Bmatrix} \sigma_{11} \\ \sigma_{22} \\ \tau_{12} \end{Bmatrix} \quad (4.62)$$

The inverse form of Equation 4.47 gives us

$$\begin{Bmatrix} \varepsilon_{xx} \\ \varepsilon_{yy} \\ \gamma_{xy}/2 \end{Bmatrix} = [T]^{-1} \begin{Bmatrix} \varepsilon_{11} \\ \varepsilon_{22} \\ \gamma_{12}/2 \end{Bmatrix} \quad (4.63)$$

Incorporating the Reuter matrix, we get

$$\begin{Bmatrix} \varepsilon_{xx} \\ \varepsilon_{yy} \\ \gamma_{xy} \end{Bmatrix} = [R][T]^{-1}[R]^{-1} \begin{Bmatrix} \varepsilon_{11} \\ \varepsilon_{22} \\ \gamma_{12} \end{Bmatrix} \quad (4.64)$$

Then, using Equation 4.62 in Equation 4.64, we obtain

$$\begin{Bmatrix} \varepsilon_{xx} \\ \varepsilon_{yy} \\ \gamma_{xy} \end{Bmatrix} = [R][T]^{-1}[R]^{-1}[S] \begin{Bmatrix} \sigma_{11} \\ \sigma_{22} \\ \tau_{12} \end{Bmatrix} \quad (4.65)$$

Finally, using Equation 4.44 in Equation 4.65, we obtain

$$\begin{Bmatrix} \varepsilon_{xx} \\ \varepsilon_{yy} \\ \gamma_{xy} \end{Bmatrix} = [R][T]^{-1}[R]^{-1}[S][T] \begin{Bmatrix} \sigma_{xx} \\ \sigma_{yy} \\ \tau_{xy} \end{Bmatrix} \quad (4.66)$$

Comparing Equation 4.61 with Equation 4.66, we get

$$[\bar{S}] = [R][T]^{-1}[R]^{-1}[S][T] \quad (4.67)$$

Matrix multiplication as given in the above equation can be carried out and it can be shown that the elements of $[\bar{S}]$ are given by

$$\bar{S}_{11} = S_{11} \cos^4 \theta + S_{22} \sin^4 \theta + (2S_{12} + S_{66}) \sin^2 \theta \cos^2 \theta \quad (4.68)$$

$$\bar{S}_{12} = (S_{11} + S_{22} - S_{66}) \sin^2 \theta \cos^2 \theta + S_{12}(\sin^4 \theta + \cos^4 \theta) \quad (4.69)$$

$$\bar{S}_{16} = (2S_{11} - 2S_{12} - S_{66}) \sin \theta \cos^3 \theta - (2S_{22} - 2S_{12} - S_{66}) \sin^3 \theta \cos \theta \quad (4.70)$$

$$\bar{S}_{22} = S_{11} \sin^4 \theta + S_{22} \cos^4 \theta + (2S_{12} + S_{66}) \sin^2 \theta \cos^2 \theta \quad (4.71)$$

$$\bar{S}_{26} = (2S_{11} - 2S_{12} - S_{66}) \sin^3 \theta \cos \theta - (2S_{22} - 2S_{12} - S_{66}) \sin \theta \cos^3 \theta \quad (4.72)$$

$$\bar{S}_{66} = 2(2S_{11} - 4S_{12} + 2S_{22} - S_{66}) \sin^2 \theta \cos^2 \theta + S_{66}(\sin^4 \theta + \cos^4 \theta) \quad (4.73)$$

EXAMPLE 4.3

Consider a unidirectional lamina as shown in Figure 4.6. The engineering constants for the material are as follows: $E_1 = 40$ GPa, $E_2 = 8$ GPa, $\nu_{12} = 0.25$, and $G_{12} = 4$ GPa. Determine the local stresses in the lamina.

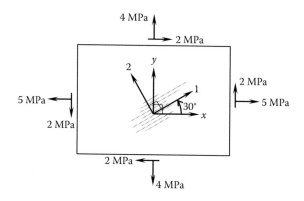

FIGURE 4.6 Generally orthotropic lamina under in-plane loading (Example 4.3).

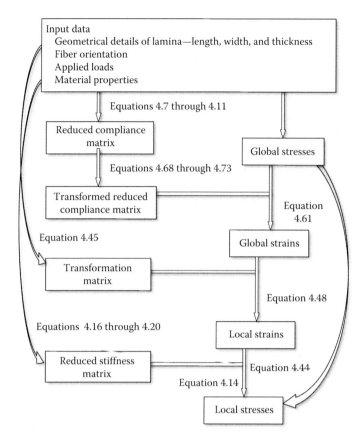

FIGURE 4.7 Analysis of a generally orthotropic lamina under in-plane loading.

Solution

The steps involved in the determination of local stresses in a generally orthotropic lamina are pictorially depicted in Figure 4.7.

Let us first use Equations 4.7 through 4.11 and determine the reduced compliance matrix as follows:

$$S_{11} = \frac{1}{E_1} = \frac{1}{40,000} = 0.000025 \, \text{MPa}^{-1}$$

$$S_{12} = -\frac{\nu_{12}}{E_1} = -\frac{0.25}{40,000} = -0.00000625 \, \text{MPa}^{-1}$$

$$S_{22} = \frac{1}{E_2} = \frac{1}{8000} = 0.000125 \, \text{MPa}^{-1}$$

$$S_{66} = \frac{1}{G_{12}} = \frac{1}{4000} = 0.00025 \, \text{MPa}^{-1}$$

Thus, the reduced compliance matrix is

$$[S] = \begin{bmatrix} 25 & -6.25 & 0 \\ -6.25 & 125 & 0 \\ 0 & 0 & 250 \end{bmatrix} \times 10^{-6} \, \text{MPa}^{-1}$$

Macromechanics of a Lamina

Next, the transformed reduced compliance matrix is determined by substituting the reduced compliance matrix elements from above in Equations 4.68 through 4.73. Thus,

$$\bar{S}_{11} = 66.4 \times 10^{-6} \text{ MPa}^{-1}$$

$$\bar{S}_{12} = -22.7 \times 10^{-6} \text{ MPa}^{-1}$$

$$\bar{S}_{16} = -62.2 \times 10^{-6} \text{ MPa}^{-1}$$

$$\bar{S}_{22} = 116.4 \times 10^{-6} \text{ MPa}^{-1}$$

$$\bar{S}_{26} = -24.4 \times 10^{-6} \text{ MPa}^{-1}$$

$$\bar{S}_{66} = 184.4 \times 10^{-6} \text{ MPa}^{-1}$$

Then,

$$[\bar{S}] = \begin{bmatrix} 66.4 & -22.7 & -62.2 \\ -22.7 & 116.4 & -24.4 \\ -62.2 & -24.4 & 184.4 \end{bmatrix} \times 10^{-6} \text{ MPa}^{-1}$$

Next, global strains are determined by using Equation 4.61 as follows:

$$\begin{Bmatrix} \varepsilon_{xx} \\ \varepsilon_{yy} \\ \gamma_{xy} \end{Bmatrix} = \begin{bmatrix} 66.4 & -22.7 & -62.2 \\ -22.7 & 116.4 & -24.4 \\ -62.2 & -24.4 & 184.4 \end{bmatrix} \times \begin{Bmatrix} 5 \\ 4 \\ 2 \end{Bmatrix} \times 10^{-6} = \begin{Bmatrix} 116.9 \\ 303.6 \\ -39.9 \end{Bmatrix} \times 10^{-6}$$

The transformation matrix is obtained from Equation 4.45. Thus,

$$[T] = \begin{bmatrix} \dfrac{3}{4} & \dfrac{1}{4} & \dfrac{\sqrt{3}}{2} \\ \dfrac{1}{4} & \dfrac{3}{4} & -\dfrac{\sqrt{3}}{2} \\ -\dfrac{\sqrt{3}}{4} & \dfrac{\sqrt{3}}{4} & \dfrac{1}{2} \end{bmatrix}$$

and the local strains are determined using Equation 4.48 as follows:

$$\begin{Bmatrix} \varepsilon_{11} \\ \varepsilon_{22} \\ \gamma_{12} \end{Bmatrix} = \begin{bmatrix} 1 & 0 & 0 \\ 0 & 1 & 0 \\ 0 & 0 & 2 \end{bmatrix} \times \begin{bmatrix} \dfrac{3}{4} & \dfrac{1}{4} & \dfrac{\sqrt{3}}{2} \\ \dfrac{1}{4} & \dfrac{3}{4} & -\dfrac{\sqrt{3}}{2} \\ -\dfrac{\sqrt{3}}{4} & \dfrac{\sqrt{3}}{4} & \dfrac{1}{2} \end{bmatrix} \times \begin{bmatrix} 1 & 0 & 0 \\ 0 & 1 & 0 \\ 0 & 0 & 2 \end{bmatrix}^{-1} \times \begin{Bmatrix} 116.9 \\ 303.6 \\ -39.9 \end{Bmatrix} \times 10^{-6} = \begin{Bmatrix} 146.3 \\ 274.2 \\ 141.7 \end{Bmatrix} \times 10^{-6}$$

Finally, for the determination of the local stresses, the reduced stiffness matrix elements are obtained from Equations 4.16 through 4.19 as follows:

$$Q_{11} = \frac{E_1}{1 - \nu_{12}\nu_{21}} = \frac{40,000}{1 - 0.25 \times 0.05} = 40,506 \, \text{MPa}$$

$$Q_{12} = \frac{\nu_{12} E_2}{1 - \nu_{12}\nu_{21}} = \frac{0.25 \times 8000}{1 - 0.25 \times 0.05} = 2025 \, \text{MPa}$$

$$Q_{22} = \frac{8000}{1 - 0.25 \times 0.05} = 8101 \, \text{MPa}$$

$$Q_{66} = G_{12} = 4000 \, \text{MPa}$$

Thus,

$$[Q] = \begin{bmatrix} 40,506 & 2025 & 0 \\ 2025 & 8101 & 0 \\ 0 & 0 & 4000 \end{bmatrix} \text{MPa}$$

Using Equation 4.14, local stresses are then determined as follows:

$$\begin{Bmatrix} \sigma_{11} \\ \sigma_{22} \\ \tau_{12} \end{Bmatrix} = \begin{bmatrix} 40,506 & 2025 & 0 \\ 2025 & 8101 & 0 \\ 0 & 0 & 4000 \end{bmatrix} \times \begin{Bmatrix} 146.3 \\ 274.2 \\ 141.7 \end{Bmatrix} \times 10^{-6} = \begin{Bmatrix} 6.48 \\ 2.52 \\ 0.57 \end{Bmatrix} \text{MPa}$$

The local stresses can also be determined directly using Equation 4.44 as follows:

$$\begin{Bmatrix} \sigma_{11} \\ \sigma_{22} \\ \tau_{12} \end{Bmatrix} = \begin{bmatrix} \frac{3}{4} & \frac{1}{4} & \frac{\sqrt{3}}{2} \\ \frac{1}{4} & \frac{3}{4} & -\frac{\sqrt{3}}{2} \\ -\frac{\sqrt{3}}{4} & \frac{\sqrt{3}}{4} & \frac{1}{2} \end{bmatrix} \times \begin{Bmatrix} 5 \\ 4 \\ 2 \end{Bmatrix} = \begin{Bmatrix} 6.48 \\ 2.52 \\ 0.57 \end{Bmatrix} \text{MPa}$$

4.5 ENGINEERING CONSTANTS OF A GENERALLY ORTHOTROPIC LAMINA

Laminae are stacked in different orientations in a laminate. The designer decides on the angle of orientation based primarily on structural considerations. The structural response of a lamina depends on the engineering constants. The engineering constants of a lamina in the material coordinates remain constants irrespective of the ply angle. However, in the global coordinates, it can be expected that the engineering constants would depend on the ply angle. Transformation relations for the engineering constants are presented in this section.

The engineering constants of an orthotropic material were introduced in Chapter 2. Recall that these constants were related to the compliance matrix by a process of applying specific nonzero stresses on a cubic stress element. Under plane stress idealization, the 3D problem has been simplified to a 2D one. Now, in the case of a generally orthotropic lamina, we adopt a similar process to determine the engineering constants in the global coordinates.

A point worth mentioning here is that in the case of a generally orthotropic lamina, owing to the presence of shear coupling, we have four additional parameters—the shear coupling ratios. As we know, shear coupling is the effect of either normal stress on shear strain or shear stress on normal strain. The shear coupling ratios are dimensionless parameters. We need to consider three different load cases for the determination of the engineering constants of a generally orthotropic lamina. Table 4.1 presents these load cases and the corresponding engineering constants. The load cases are pictorially shown in Figure 4.8.

Let us consider the first load case and apply a nonzero normal stress in the x-direction, that is,

$$\sigma_{xx} \neq 0, \quad \sigma_{yy} = \tau_{xy} = 0 \tag{4.74}$$

Then, from Equation 4.61, we get

$$\varepsilon_{xx} = \bar{S}_{11} \sigma_{xx} \tag{4.75}$$

TABLE 4.1
Determination of Engineering Constants of a Generally Orthotropic Lamina

Load Case	Description	Engineering Constants
1	Nonzero normal stress in the x-direction, that is, $\sigma_{xx} \neq 0, \sigma_{yy} = \tau_{xy} = 0$	Modulus of elasticity in the x-direction (E_x) Major Poisson's ratio in the xy-plane (ν_{xy}) Shear coupling ratio associated with normal stress in the x-direction and in-plane shear strain ($\eta_{xy,x}$)
2	Nonzero normal stress in the y-direction, that is, $\sigma_{yy} \neq 0, \sigma_{xx} = \tau_{xy} = 0$	Modulus of elasticity in the y-direction (E_y) Minor Poisson's ratio in the xy-plane (ν_{yx}) Shear coupling ratio associated with normal stress in the y-direction and in-plane shear strain ($\eta_{xy,y}$)
3	Nonzero in-plane shear stress in the xy-plane, that is, $\tau_{xy} \neq 0, \sigma_{xx} = \sigma_{yy} = 0$	Shear coupling ratios associated with shear stress in the xy-plane and normal strains in the x- and y-directions ($\eta_{x,xy}$ and $\eta_{y,xy}$) In-plane shear modulus (G_{xy})

Notes:
1. Table 4.1 indicates the presence of a total of nine engineering constants in a generally orthotropic lamina. There are two extensional moduli of elasticity (E_x and E_y), two Poisson's ratios (ν_{xy} and ν_{yx}), one in-plane shear modulus (G_{xy}), and four shear coupling ratios ($\eta_{xy,x}$, $\eta_{xy,y}$, $\eta_{x,xy}$, and $\eta_{y,xy}$). It will be seen that minor Poisson's ratio can be related to the major Poisson's ratio. Similarly, of the four shear coupling ratios, $\eta_{x,xy}$ is related to $\eta_{xy,x}$ and $\eta_{y,xy}$ is related to $\eta_{xy,y}$. Thus, there are six independent off-axis engineering constants in a generally orthotropic lamina. Of course, these six off-axis engineering constants are not truly independent as they, in turn, are related to the four material axis engineering constants.
2. The shear coupling ratios ($\eta_{x,xy}$ and $\eta_{y,xy}$) are associated with normal strains caused by in-plane shear stresses and they are called as the coefficients of mutual influence of the first kind. On the other hand, the shear coupling ratios ($\eta_{xy,x}$ and $\eta_{xy,y}$) are associated with in-plane shear strains caused by normal stresses and they are called the coefficients of mutual influence of the second kind. Also, note the difference in convention. In the Poisson's ratios, the two characters in the subscript correspond to "cause" and "effect," respectively. For example, in the case of ν_{xy}, normal stress in the x-direction is the cause and normal strain in the y-direction is the effect. In the shear coupling ratios, the characters in the subscript correspond to "effect" and "cause," respectively. Thus, in $\eta_{x,xy}$, in-plane shear strain in the xy-plane is the cause and normal stress in the x-direction is the effect, and so on.

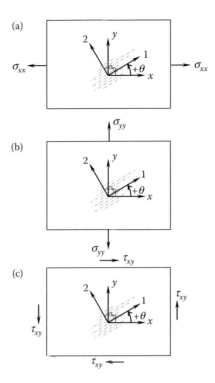

FIGURE 4.8 Determination of engineering constants of a generally orthotropic lamina—three load cases. (a) Lamina under uniaxial stress in the *x*-direction. (b) Lamina under uniaxial stress in the *y*-direction. (c) Lamina under in-plane shear. (Adapted with permission from A. K. Kaw, *Mechanics of Composite Materials*, CRC Press, Boca Raton, FL, 2006.)

$$\varepsilon_{yy} = \overline{S}_{12} \sigma_{xx} \tag{4.76}$$

$$\gamma_{xy} = \overline{S}_{16} \sigma_{xx} \tag{4.77}$$

The modulus of elasticity in the *x*-direction is defined as

$$E_x = \frac{\sigma_{xx}}{\varepsilon_{xx}} \tag{4.78}$$

Using Equation 4.75, we get

$$E_x = \frac{1}{\overline{S}_{11}} \tag{4.79}$$

Substituting Equation 4.68 together with Equations 4.7 through 4.10 in the equation above, we get the expression for the off-axis elastic modulus in the *x*-direction as follows:

$$E_x = \left[\frac{\cos^4 \theta}{E_1} + \frac{\sin^4 \theta}{E_2} + \left(\frac{1}{G_{12}} - \frac{2\nu_{12}}{E_1} \right) \sin^2 \theta \cos^2 \theta \right]^{-1} \tag{4.80}$$

The major Poisson's ratio in the *xy*-plane is defined as

$$\nu_{xy} = -\frac{\varepsilon_{yy}}{\varepsilon_{xx}} \tag{4.81}$$

Using Equations 4.75 and 4.76, we get

$$\nu_{xy} = -\frac{\bar{S}_{12}}{\bar{S}_{11}} \qquad (4.82)$$

Substituting Equation 4.69 together with Equations 4.7 through 4.10 and 4.79 in the equation above, we get the expression for off-axis major Poisson's ratio in the xy-plane as follows:

$$\nu_{xy} = E_x \left[\frac{\nu_{12}}{E_1}(\sin^4\theta + \cos^4\theta) - \left(\frac{1}{E_1} + \frac{1}{E_2} - \frac{1}{G_{12}}\right)\sin^2\theta\cos^2\theta \right] \qquad (4.83)$$

Shear coupling ratio associated with the normal stress in the x-direction is defined as

$$\eta_{xy,x} = \frac{\gamma_{xy}}{\varepsilon_{xx}} \qquad (4.84)$$

Using Equations 4.75 and 4.77, we get

$$\eta_{xy,x} = \frac{\bar{S}_{16}}{\bar{S}_{11}} \qquad (4.85)$$

Substituting Equation 4.70 together with Equations 4.7 through 4.10 and 4.79 in the equation above, we get the expression for shear coupling ratio associated with normal stress in the x-direction.

$$\eta_{xy,x} = E_x \left[\left(\frac{2}{E_1} + \frac{2\nu_{12}}{E_1} - \frac{1}{G_{12}}\right)\sin\theta\cos^3\theta - \left(\frac{2}{E_2} + \frac{2\nu_{12}}{E_1} - \frac{1}{G_{12}}\right)\sin^3\theta\cos\theta \right] \qquad (4.86)$$

Next, let us consider the second load case and apply a nonzero normal stress in the y-direction, that is,

$$\sigma_{yy} \neq 0, \quad \sigma_{xx} = \tau_{xy} = 0 \qquad (4.87)$$

Then, from Equation 4.61, we get

$$\varepsilon_{xx} = \bar{S}_{12}\sigma_{yy} \qquad (4.88)$$

$$\varepsilon_{yy} = \bar{S}_{22}\sigma_{yy} \qquad (4.89)$$

$$\gamma_{xy} = \bar{S}_{26}\sigma_{yy} \qquad (4.90)$$

The modulus of elasticity in the y-direction is defined as

$$E_y = \frac{\sigma_{yy}}{\varepsilon_{yy}} \qquad (4.91)$$

Using Equation 4.89, we get

$$E_y = \frac{1}{\overline{S}_{22}} \quad (4.92)$$

Substituting Equation 4.71 together with Equations 4.7 through 4.10 in the equation above, we get the expression for the off-axis elastic modulus in the y-direction as follows:

$$E_y = \left[\frac{\sin^4 \theta}{E_1} + \frac{\cos^4 \theta}{E_2} + \left(\frac{1}{G_{12}} - \frac{2\nu_{12}}{E_1} \right) \sin^2 \theta \cos^2 \theta \right]^{-1} \quad (4.93)$$

Next, the minor Poisson's ratio in the xy-plane is defined as

$$\nu_{yx} = -\frac{\varepsilon_{xx}}{\varepsilon_{yy}} \quad (4.94)$$

Using Equations 4.88 and 4.89, we get

$$\nu_{yx} = -\frac{\overline{S}_{12}}{\overline{S}_{22}} \quad (4.95)$$

Substituting Equation 4.69 together with Equations 4.7 through 4.10 and 4.92 in the equation above, we get the expression for off-axis minor Poisson's ratio in the xy-plane as follows:

$$\nu_{yx} = E_y \left[\frac{\nu_{12}}{E_1} (\sin^4 \theta + \cos^4 \theta) - \left(\frac{1}{E_1} + \frac{1}{E_2} - \frac{1}{G_{12}} \right) \sin^2 \theta \cos^2 \theta \right] \quad (4.96)$$

It can be seen from Equations 4.83 and 4.96 that the major and minor Poisson's ratios are related as

$$\frac{\nu_{xy}}{E_x} = \frac{\nu_{yx}}{E_y} \quad (4.97)$$

Shear coupling ratio associated with the normal stress in the y-direction is defined as

$$\eta_{xy,y} = \frac{\gamma_{xy}}{\varepsilon_{yy}} \quad (4.98)$$

Using Equations 4.89 and 4.90, we get

$$\eta_{xy,y} = \frac{\overline{S}_{26}}{\overline{S}_{22}} \quad (4.99)$$

Macromechanics of a Lamina

Substituting Equation 4.72 together with Equations 4.7 through 4.10 and 4.92 in the equation above, we get the expression for shear coupling ratio associated with normal stress in the y-direction.

$$\eta_{xy,y} = E_y \left[\left(\frac{2}{E_1} + \frac{2\nu_{12}}{E_1} - \frac{1}{G_{12}} \right) \sin^3\theta \cos\theta - \left(\frac{2}{E_2} + \frac{2\nu_{12}}{E_1} - \frac{1}{G_{12}} \right) \sin\theta \cos^3\theta \right] \quad (4.100)$$

Finally, let us consider the third load case and apply a nonzero in-plane shear stress in the xy-plane, that is,

$$\tau_{xy} \neq 0, \quad \sigma_{xx} = \sigma_{yy} = 0 \quad (4.101)$$

Then, from Equation 4.61, we get

$$\varepsilon_{xx} = \bar{S}_{16} \tau_{xy} \quad (4.102)$$

$$\varepsilon_{yy} = \bar{S}_{26} \tau_{xy} \quad (4.103)$$

$$\gamma_{xy} = \bar{S}_{66} \tau_{xy} \quad (4.104)$$

In this load case, first let us consider the in-plane shear modulus. It is defined as

$$G_{xy} = \frac{\tau_{xy}}{\gamma_{xy}} \quad (4.105)$$

Using Equation 4.104, we get

$$G_{xy} = \frac{1}{\bar{S}_{66}} \quad (4.106)$$

Substituting Equation 4.73 together with Equations 4.7 through 4.10 in the equation above, we get the expression for the in-plane shear modulus as follows:

$$G_{xy} = \left[2 \left(\frac{2}{E_1} + \frac{4\nu_{12}}{E_1} + \frac{2}{E_2} - \frac{1}{G_{12}} \right) \sin^2\theta \cos^2\theta + \frac{1}{G_{12}} (\sin^4\theta + \cos^4\theta) \right]^{-1} \quad (4.107)$$

In-plane shear stress results in normal strains in two directions—x- and y-directions. Thus, there are two shear coupling ratios associated with the in-plane shear stress. They are defined as

$$\eta_{x,xy} = \frac{\varepsilon_{xx}}{\gamma_{xy}} \quad (4.108)$$

and

$$\eta_{y,yx} = \frac{\varepsilon_{yy}}{\gamma_{xy}} \quad (4.109)$$

Using Equations 4.102 and 4.104, we get

$$\eta_{x,xy} = \frac{\overline{S}_{16}}{\overline{S}_{66}} \qquad (4.110)$$

and using Equations 4.103 and 4.104, we get

$$\eta_{y,xy} = \frac{\overline{S}_{26}}{\overline{S}_{66}} \qquad (4.111)$$

Substituting Equations 4.70 and 4.106 together with Equations 4.7 through 4.10 in Equation 4.110 above, we get

$$\eta_{x,xy} = G_{xy}\left[\left(\frac{2}{E_1} + \frac{2\nu_{12}}{E_1} - \frac{1}{G_{12}}\right)\sin\theta\cos^3\theta - \left(\frac{2}{E_2} + \frac{2\nu_{12}}{E_1} - \frac{1}{G_{12}}\right)\sin^3\theta\cos\theta\right] \qquad (4.112)$$

Similarly, substituting Equations 4.72 and 4.106 together with Equations 4.7 through 4.10 in Equation 4.111 above, we get

$$\eta_{y,xy} = G_{xy}\left[\left(\frac{2}{E_1} + \frac{2\nu_{12}}{E_1} - \frac{1}{G_{12}}\right)\sin^3\theta\cos\theta - \left(\frac{2}{E_2} + \frac{2\nu_{12}}{E_1} - \frac{1}{G_{12}}\right)\sin\theta\cos^3\theta\right] \qquad (4.113)$$

We mentioned earlier that $\eta_{xy,x}$ is related to $\eta_{x,xy}$. Similarly, $\eta_{xy,y}$ is related to $\eta_{y,xy}$. To verify this, let us consider Equation 4.110 and write the following:

$$\eta_{x,xy} = \frac{\overline{S}_{16}}{\overline{S}_{66}} = \frac{\overline{S}_{11}}{\overline{S}_{66}}\frac{\overline{S}_{16}}{\overline{S}_{11}} \qquad (4.114)$$

Now, from Equations 4.79, 4.85, and 4.106, respectively, we have $E_x = 1/\overline{S}_{11}$, $\eta_{xy,x} = \overline{S}_{16}/\overline{S}_{11}$, and $G_{xy} = 1/\overline{S}_{66}$. Thus, Equation 4.114 results in

$$\eta_{x,xy} = \frac{G_{xy}}{E_x}\eta_{xy,x} \qquad (4.115)$$

In a similar way, it can be shown that

$$\eta_{y,xy} = \frac{G_{xy}}{E_y}\eta_{xy,y} \qquad (4.116)$$

Pictorial representation of variation of the engineering constants with ply angle is helpful in the study of the effect of ply angle on the off-axis engineering constants. The example below is provided to present plots of variation of engineering constants with ply angle for carbon/epoxy and glass/epoxy laminae.

Macromechanics of a Lamina

EXAMPLE 4.4

The engineering constants in the material coordinates for carbon/epoxy and glass/epoxy laminae are given below:

Carbon/epoxy:

$$E_1 = 140 \text{ GPa}, E_2 = 10 \text{ GPa}, \nu_{12} = 0.28, G_{12} = 6 \text{ GPa}$$

Glass/epoxy:

$$E_1 = 40 \text{ GPa}, E_2 = 8 \text{ GPa}, \nu_{12} = 0.25, G_{12} = 4 \text{ GPa}$$

Determine the engineering constants at a ply angle of 30°. Plot the variation of the engineering constants with ply angle.

Solution

$\sin 30° = 0.5$ and $\cos 30° = 0.866$.

Using Equation 4.80, the elastic modulus in the x-direction is determined as follows:

Carbon/epoxy:

$$E_x = \left[\frac{0.866^4}{140} + \frac{0.5^4}{10} + \left(\frac{1}{6} - \frac{2 \times 0.28}{140}\right) \times 0.5^2 \times 0.866^2\right]^{-1} = 24.53 \text{ GPa}$$

Glass/epoxy:

$$E_x = \left[\frac{0.866^4}{40} + \frac{0.5^4}{8} + \left(\frac{1}{4} - \frac{2 \times 0.25}{40}\right) \times 0.5^2 \times 0.866^2\right]^{-1} = 15.06 \text{ GPa}$$

Using Equation 4.93, the elastic modulus in the y-direction is determined as follows:

Carbon/epoxy:

$$E_y = \left[\frac{0.5^4}{140} + \frac{0.866^4}{10} + \left(\frac{1}{6} - \frac{2 \times 0.28}{140}\right) \times 0.5^2 \times 0.866^2\right]^{-1} = 11.47 \text{ GPa}$$

Glass/epoxy:

$$E_y = \left[\frac{0.5^4}{40} + \frac{0.866^4}{8} + \left(\frac{1}{4} - \frac{2 \times 0.25}{40}\right) \times 0.5^2 \times 0.866^2\right]^{-1} = 8.59 \text{ GPa}$$

Using Equation 4.83, the major Poisson's ratio is determined as follows:

Carbon/epoxy:

$$\nu_{xy} = 24.53 \times \left[\frac{0.28}{140}(0.5^4 + 0.866^4) - \left(\frac{1}{140} + \frac{1}{10} - \frac{1}{6}\right) \times 0.5^2 \times 0.866^2\right] = 0.3$$

Glass/epoxy:

$$V_{xy} = 15.06 \times \left[\frac{0.25}{40}(0.5^4 + 0.866^4) - \left(\frac{1}{40} + \frac{1}{8} - \frac{1}{4} \right) \times 0.5^2 \times 0.866^2 \right] = 0.34$$

Using Equation 4.107, the in-plane shear modulus is determined as follows:

Carbon/epoxy:

$$G_{xy} = \left[2 \times \left(\frac{2}{140} + \frac{4 \times 0.28}{140} + \frac{2}{10} - \frac{1}{6} \right) \times 0.5^2 \times 0.866^2 + \frac{1}{6} \times (0.5^4 + 0.866^4) \right]^{-1}$$
$$= 8.0 \text{ GPa}$$

Glass/epoxy:

$$G_{xy} = \left[2 \times \left(\frac{2}{40} + \frac{4 \times 0.25}{40} + \frac{2}{8} - \frac{1}{4} \right) \times 0.5^2 \times 0.866^2 + \frac{1}{4} \times (0.5^4 + 0.866^4) \right]^{-1}$$
$$= 5.42 \text{ GPa}$$

Using Equation 4.86, the shear coupling ratio ($\eta_{x,xy}$) is determined as follows:

Carbon/epoxy:

$$\eta_{xy,x} = 24.53 \times \left[\left(\frac{2}{140} + \frac{2 \times 0.28}{140} - \frac{1}{6} \right) \times 0.5 \times 0.866^3 \right.$$
$$\left. - \left(\frac{2}{10} + \frac{2 \times 0.28}{140} - \frac{1}{6} \right) \times 0.5^3 \times 0.866 \right] = -1.28$$

Glass/epoxy:

$$\eta_{xy,x} = 15.06 \times \left[\left(\frac{2}{40} + \frac{2 \times 0.25}{40} - \frac{1}{4} \right) \times 0.5 \times 0.866^3 \right.$$
$$\left. - \left(\frac{2}{8} + \frac{2 \times 0.25}{40} - \frac{1}{4} \right) \times 0.5^3 \times 0.866 \right] = -0.94$$

Using Equation 4.100, the shear coupling ratio ($\eta_{xy,y}$) is determined as follows:

Carbon/epoxy:

$$\eta_{xy,y} = 11.47 \times \left[\left(\frac{2}{140} + \frac{2 \times 0.28}{140} - \frac{1}{6} \right) \times 0.5^3 \times 0.866 \right.$$
$$\left. - \left(\frac{2}{10} + \frac{2 \times 0.28}{140} - \frac{1}{6} \right) \times 0.5 \times 0.866^3 \right] = -0.32$$

Glass/epoxy:

$$\eta_{xy,y} = 8.59 \times \left[\left(\frac{2}{40} + \frac{2 \times 0.25}{40} - \frac{1}{4}\right) \times 0.5^3 \times 0.866 \right.$$
$$\left. - \left(\frac{2}{8} + \frac{2 \times 0.25}{40} - \frac{1}{4}\right) \times 0.5 \times 0.866^3\right] = -0.21$$

Variations of the off-axis engineering constants with lamina angle are shown in Figures 4.9 through 4.14. These plots given in Figures 4.9 through 4.14 are for the specific carbon/epoxy and glass/epoxy composites in Example 4.4. These are not generalized plots. However, certain observations, helpful in product design, can be made from these plots.

First, the extensional moduli in the x- and y-directions move in opposite directions as the lamina angle increases; the variations of E_x and E_y are symmetric w.r.t. $\theta = 45°$. E_x is the maximum when $\theta = 0$. As θ increases gradually, initially it sharply decreases; at higher values of θ, it decreases at a low rate and eventually reaches its minimum when $\theta = 90°$. On the other hand, E_y is the minimum when $\theta = 0$. With gradually increasing θ, it increases at a low rate initially; at high values of θ, it increases sharply and finally it reaches its maximum at $\theta = 90°$.

Second, the in-plane shear modulus is the maximum at a lamina angle of 45°. It shows that a 45° lamina offers the maximum resistance to shear deformation.

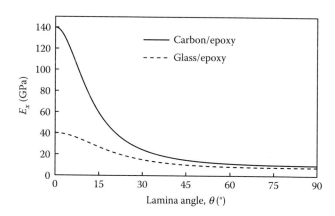

FIGURE 4.9 Variation of elastic modulus in the x-direction with lamina angle (Example 4.4).

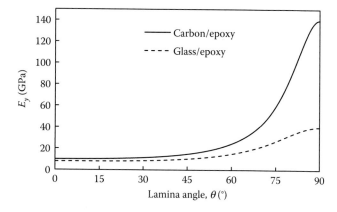

FIGURE 4.10 Variation of elastic modulus in the y-direction with lamina angle (Example 4.4).

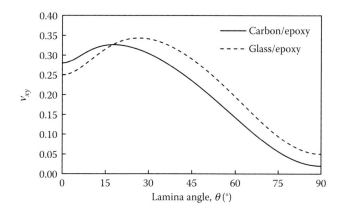

FIGURE 4.11 Variation of major Poisson's ratio with lamina angle (Example 4.4).

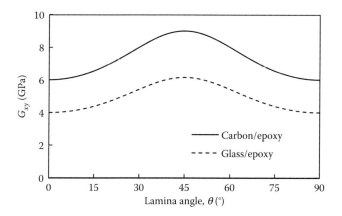

FIGURE 4.12 Variation of in-plane shear modulus with lamina angle (Example 4.4).

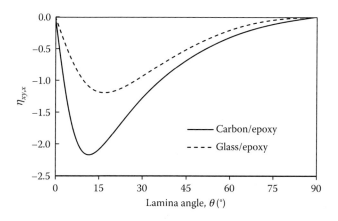

FIGURE 4.13 Variation of shear coupling ratio ($\eta_{xy,x}$) with lamina angle (Example 4.4).

Third, shear coupling is zero at either $\theta = 0°$ or $\theta = 90°$. The variations of the shear coupling ratios are too symmetric w.r.t. $\theta = 45°$.

The observations made above in respect of the off-axis elastic constants are quite generic in nature. Also, the equations for reduced transformed stiffness matrix or reduced transformed compliance matrix are rather complicated for any direct use in design calculations. Invariant forms of stiffness and compliance matrices were

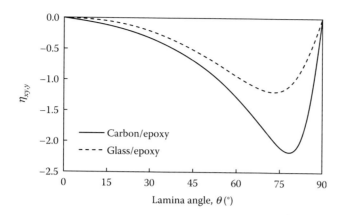

FIGURE 4.14 Variation of shear coupling ratio ($\eta_{xy,y}$) with lamina angle (Example 4.4).

introduced by Tsai and Pagano [9]. Using the invariants, the reduced transformed stiffness matrix elements can be written as

$$\bar{Q}_{11} = U_1 + U_2 \cos 2\theta + U_3 \cos 4\theta \tag{4.117}$$

$$\bar{Q}_{12} = U_4 - U_3 \cos 4\theta \tag{4.118}$$

$$\bar{Q}_{16} = \frac{U_2 \sin 2\theta}{2} + U_3 \sin 4\theta \tag{4.119}$$

$$\bar{Q}_{22} = U_1 - U_2 \cos 2\theta + U_3 \cos 4\theta \tag{4.120}$$

$$\bar{Q}_{26} = \frac{U_2 \sin 2\theta}{2} - U_3 \sin 4\theta \tag{4.121}$$

$$\bar{Q}_{66} = \frac{U_1 - U_4}{2} - U_3 \cos 4\theta \tag{4.122}$$

In the above equations, the four invariants are given by

$$U_1 = \frac{1}{8}(3Q_{11} + 2Q_{12} + 3Q_{22} + 4Q_{66}) \tag{4.123}$$

$$U_2 = \frac{1}{2}(Q_{11} - Q_{22}) \tag{4.124}$$

$$U_3 = \frac{1}{8}(Q_{11} - 2Q_{12} + Q_{22} - 4Q_{66}) \tag{4.125}$$

$$U_4 = \frac{1}{8}(Q_{11} + 6Q_{12} + Q_{22} - 4Q_{66}) \tag{4.126}$$

Note that the invariants are functions only of Q_{11}, Q_{12}, Q_{22}, and Q_{66}; they are not functions of θ.

Similar to the reduced transformed stiffness matrix, the reduced transformed compliance matrix elements can also be expressed in terms of invariants as follows:

$$\overline{S}_{11} = V_1 + V_2 \cos 2\theta + V_3 \cos 4\theta \tag{4.127}$$

$$\overline{S}_{12} = V_4 - V_3 \cos 4\theta \tag{4.128}$$

$$\overline{S}_{16} = V_2 \sin 2\theta + 2V_3 \sin 4\theta \tag{4.129}$$

$$\overline{S}_{22} = V_1 - V_2 \cos 2\theta + V_3 \cos 4\theta \tag{4.130}$$

$$\overline{S}_{26} = V_2 \sin 2\theta - 2V_3 \sin 4\theta \tag{4.131}$$

$$\overline{S}_{66} = 2(V_1 - V_4) - 4V_3 \cos 4\theta \tag{4.132}$$

The invariants in the above equations are given by

$$V_1 = \frac{1}{8}(3S_{11} + 2S_{12} + 3S_{22} + S_{66}) \tag{4.133}$$

$$V_2 = \frac{1}{2}(S_{11} - S_{22}) \tag{4.134}$$

$$V_3 = \frac{1}{8}(S_{11} - 2S_{12} + S_{22} - S_{66}) \tag{4.135}$$

$$V_4 = \frac{1}{8}(S_{11} + 6S_{12} + S_{22} - S_{66}) \tag{4.136}$$

As we know, a laminate contains a number of laminae arranged in different orientations. Working out the details of laminate ply orientations, so as to achieve the desired end characteristics, is a major part of composite part design. Invariant forms of stiffness and compliance matrices are helpful in the design of a laminate as these invariants help the designer to directly check the effects of change in lamina angle on the reduced transformed stiffness and compliance matrices.

4.6 STRENGTH

One of the essential aspects of design and analysis of a structural part is to ensure that stresses in the part are less than the corresponding strengths of the material. Strength is the maximum stress that a material can take before failure. It is an experimentally determined parameter for which standard test specimens have been devised. During the loading of a test specimen, simple stress field is generated in the test specimen and based on the information of failure load and specimen geometry, stress at failure, that is, strength is determined. Unlike the standard test specimen, stress field in a practical structural part, however, is not a simple one and stress data cannot be directly compared with the strength data. As a consequence, a number of failure criteria have been developed for isotropic as well as orthotropic materials [10].

Macromechanics of a Lamina

In an isotropic material, there are three strength parameters:

- Tensile strength
- Compressive strength
- Shear strength

Tensile strength is the maximum normal stress in tension that a material can be subjected to before failure. Similarly, compressive strength and shear strength are the maximum normal stress in compression and maximum shear stress, respectively. Principal stresses are the maximum stresses and, in an isotropic material, strengths are not direction-dependent. Thus, the failure criteria for isotropic materials are based on principal stresses.

4.6.1 Strength of an Orthotropic Lamina

The concept of strengths in orthotropic materials is a little complex as the strengths, like stiffness, are direction-dependent. In principle, we can have infinite numbers of strengths corresponding to each possible fiber orientation. In a practical design problem, however, it is not a feasible proposition. Instead, in an orthotropic lamina, the following five basic strength parameters are used:

- Longitudinal tensile strength, $(\sigma_{11}^T)_{ult}$
- Longitudinal compressive strength, $(\sigma_{11}^C)_{ult}$
- Transverse tensile strength, $(\sigma_{22}^T)_{ult}$
- Transverse compressive strength, $(\sigma_{22}^C)_{ult}$
- In-plane shear strength, $(\tau_{12})_{ult}$

The ultimate longitudinal tensile strength of an orthotropic lamina is the maximum normal stress in tension that the lamina can be subjected to in the 1-direction before failure. Similarly, the ultimate longitudinal compressive strength is the maximum normal stress in compression that the lamina can be subjected to in the 1-direction before failure. Transverse strengths, on the other hand, are the corresponding maximum stresses in the 2-direction. Finally, the ultimate shear strength is the maximum in-plane shear stress in the 1-2 plane. (Note that for a unidirectional lamina, 1- and 2-directions are along the fibers and across the fibers, respectively, whereas for a bidirectional lamina, they are along the weft and warp directions, respectively.)

A number of failure criteria are available for orthotropic materials, of which four popular failure criteria are presented here; they are (i) maximum stress failure criterion, (ii) maximum strain failure criterion, (iii) Tsai–Hill failure criterion, and (iv) Tsai–Wu failure criterion. Before discussing these failure criteria, we shall dwell upon a few useful points.

Sign convention for stresses and strengths (Figure 4.15): The standard sign convention for stresses and strengths is that all normal stresses are positive in tension and negative in compression. Similarly, normal strengths are positive in tension and negative in compression. Positive and negative shear stresses can have different effects on a lamina. For example, a 45° lamina under positive and negative shear stresses is shown in Figure 4.16. However, maximum shear stress at failure, irrespective of whether it is positive or negative, remains the same. Thus, shear strength is considered as positive and, as we shall see shortly, in the failure criteria, the absolute value of the shear stress is compared with the shear strength.

Stress transformation: We had seen before that stresses in one coordinate system can be easily transformed to another by utilizing suitable transformation matrix.

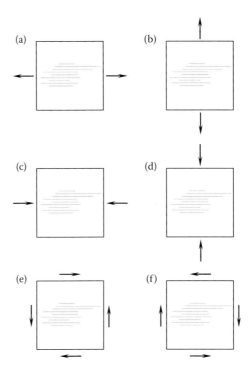

FIGURE 4.15 Sign convention for stresses. (a) Positive normal stress in the longitudinal direction (longitudinal tensile stress). (b) Positive normal stress in the transverse direction (transverse tensile stress). (c) Negative normal stress in the longitudinal direction (longitudinal compressive stress). (d) Negative normal stress in the transverse direction (transverse compressive stress). (e) Positive shear stress. (f) Negative shear stress.

The transformation matrix can be constructed with the angles of orientation of one coordinate system w.r.t. the other. On the other hand, strengths are influenced by many local factors and simple transformations similar to stresses are not possible. Thus, comparison of stresses with strengths is generally made in the material axes.

Principal stresses versus material axis strengths: Principal stresses are the maximum stresses at a point in a body whether it is made up of isotropic or orthotropic material. However, in an orthotropic material, principal stresses are not of primary use as the strength data are available in the material axes and in the failure criteria, the stresses in the material axes are compared with the strengths that are in the material axes.

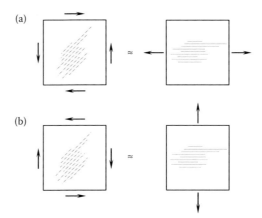

FIGURE 4.16 Shear stress on a 45° lamina. (a) Positive shear stress. (b) Negative shear stress. (Adapted with permission from R. M. Jones, *Mechanics of Composite Materials*, Taylor & Francis, New York, 1999.)

Macromechanics of a Lamina

4.6.2 Failure Criteria

4.6.2.1 Maximum Stress Failure Criterion

In this failure criterion, five different modes of failure, each related to excessive stress, are identified. They are

- Failure of the lamina due to excessive tensile stress in the 1-direction
- Failure of the lamina due to excessive compressive stress in the 1-direction
- Failure of the lamina due to excessive tensile stress in the 2-direction
- Failure of the lamina due to excessive compressive stress in the 2-direction
- Failure of the lamina due to excessive in-plane shear stress in the 1-2 plane

As per the maximum stress failure criterion, a lamina is considered to have failed if any one of the stresses in the material axes exceeds the corresponding strength. Thus, for the lamina to be safe, the following need to be satisfied:

$$\left(\sigma_{11}^C\right)_{ult} < \sigma_{11} < \left(\sigma_{11}^T\right)_{ult} \tag{4.137}$$

$$\left(\sigma_{22}^C\right)_{ult} < \sigma_{22} < \left(\sigma_{22}^T\right)_{ult} \tag{4.138}$$

$$|\tau_{12}| < (\tau_{12})_{ult} \tag{4.139}$$

In the above inequalities, the sign convention is as stated in the previous paragraphs. It can be seen that in this criterion, comparison of stresses with strengths is made in the respective direction only and there is no interaction between different modes of failure. Further, the inequalities only indicate whether the lamina is safe or not; they do not provide any information on the available factor of safety. In this connection, the concept of strength ratio is useful. Strength ratios corresponding to the five strength parameters are defined as

$$(R)_{11}^T = \frac{\left(\sigma_{11}^T\right)_{ult}}{\sigma_{11}} \tag{4.140}$$

$$(R)_{11}^C = \frac{\left(\sigma_{11}^C\right)_{ult}}{\sigma_{11}} \tag{4.141}$$

$$(R)_{22}^T = \frac{\left(\sigma_{22}^T\right)_{ult}}{\sigma_{22}} \tag{4.142}$$

$$(R)_{22}^C = \frac{\left(\sigma_{22}^C\right)_{ult}}{\sigma_{22}} \tag{4.143}$$

$$(R)_{12}^S = \frac{(\tau_{12})_{ult}}{|\tau_{12}|} \tag{4.144}$$

where $(R)_{11}^T$ is the strength ratio for normal tensile stress in the 1-direction and so on. Note that strength ratios are always positive. $(R)_{11}^T < 1$ indicates that the lamina has failed due to excessive tensile load in 1-direction and the load needs to be reduced by a factor of $(R)_{11}^T$. $(R)_{11}^T > 1$ indicates that the lamina is safe and the applied tensile load in 1-direction can be increased by a factor of $(R)_{11}^T$. Finally, $(R)_{11}^T = 1$ indicates that the lamina has just failed. Similar conclusions can be drawn in respect of the remaining strength ratios as well.

EXAMPLE 4.5

Consider a carbon/epoxy lamina with its fiber orientation at 30° to the x-direction. It is subjected to the following stresses: $\sigma_{xx} = 1200$ MPa, $\sigma_{yy} = 350$ MPa, and $\tau_{xy} = 800$ MPa. Check whether the lamina is safe under these stresses. The following strength data are given: $(\sigma_{11}^T)_{ult} = 2000$ MPa, $(\sigma_{11}^C)_{ult} = 800$ MPa, $(\sigma_{22}^T)_{ult} = 40$ MPa, $(\sigma_{22}^C)_{ult} = 150$ MPa, and $(\tau_{12})_{ult} = 70$ MPa.

Solution

First, we note that $\sin 30° = 0.5$ and $\cos 30° = 0.866$. Then, using Equation 4.43, the stresses are readily transformed from the global coordinates to local coordinates as follows:

$$\begin{Bmatrix} \sigma_{11} \\ \sigma_{22} \\ \tau_{12} \end{Bmatrix} = \begin{bmatrix} 0.75 & 0.25 & 0.866 \\ 0.25 & 0.75 & -0.866 \\ -0.433 & 0.433 & 0.5 \end{bmatrix} \begin{Bmatrix} 1200 \\ 350 \\ 800 \end{Bmatrix} = \begin{Bmatrix} 1680.3 \\ -130.3 \\ -32.0 \end{Bmatrix} \text{MPa}$$

Corresponding strength ratios are calculated as follows:

$$(R)_{11}^T = \frac{2000}{1680.3} = 1.19$$

$$(R)_{22}^C = \frac{-150}{-130.3} = 1.15$$

$$(R)_{12}^S = \frac{70}{32.0} = 2.19$$

From the above, it is clear that the lamina is safe under the given loads. While the normal stresses along the fibers and across can be increased by factors of 1.19 and 1.15, respectively, the shear stress can be increased by a factor of 2.19 before the lamina fails. Note, however, that the stresses in the local coordinates cannot be directly increased or decreased. Given the lamina geometry, it is the global stresses that can be varied which, in turn, cause the stresses in the local coordinates to change. For an efficient design these strength ratios would be equal to one. For the given strength data and lamina angle, these global stresses can be obtained by solving the following eight sets of simultaneous equations:

$$1. \begin{bmatrix} 0.75 & 0.25 & 0.866 \\ 0.25 & 0.75 & -0.866 \\ -0.433 & 0.433 & 0.5 \end{bmatrix} \begin{Bmatrix} \sigma_{xx} \\ \sigma_{yy} \\ \tau_{xy} \end{Bmatrix} = \begin{Bmatrix} 2000 \\ 40 \\ 70 \end{Bmatrix}$$

$$2. \begin{bmatrix} 0.75 & 0.25 & 0.866 \\ 0.25 & 0.75 & -0.866 \\ -0.433 & 0.433 & 0.5 \end{bmatrix} \begin{Bmatrix} \sigma_{xx} \\ \sigma_{yy} \\ \tau_{xy} \end{Bmatrix} = \begin{Bmatrix} 2000 \\ 40 \\ -70 \end{Bmatrix}$$

Macromechanics of a Lamina

3. $\begin{bmatrix} 0.75 & 0.25 & 0.866 \\ 0.25 & 0.75 & -0.866 \\ -0.433 & 0.433 & 0.5 \end{bmatrix} \begin{Bmatrix} \sigma_{xx} \\ \sigma_{yy} \\ \tau_{xy} \end{Bmatrix} = \begin{Bmatrix} 2000 \\ -150 \\ 70 \end{Bmatrix}$

4. $\begin{bmatrix} 0.75 & 0.25 & 0.866 \\ 0.25 & 0.75 & -0.866 \\ -0.433 & 0.433 & 0.5 \end{bmatrix} \begin{Bmatrix} \sigma_{xx} \\ \sigma_{yy} \\ \tau_{xy} \end{Bmatrix} = \begin{Bmatrix} 2000 \\ -150 \\ -70 \end{Bmatrix}$

5. $\begin{bmatrix} 0.75 & 0.25 & 0.866 \\ 0.25 & 0.75 & -0.866 \\ -0.433 & 0.433 & 0.5 \end{bmatrix} \begin{Bmatrix} \sigma_{xx} \\ \sigma_{yy} \\ \tau_{xy} \end{Bmatrix} = \begin{Bmatrix} -800 \\ 40 \\ 70 \end{Bmatrix}$

6. $\begin{bmatrix} 0.75 & 0.25 & 0.866 \\ 0.25 & 0.75 & -0.866 \\ -0.433 & 0.433 & 0.5 \end{bmatrix} \begin{Bmatrix} \sigma_{xx} \\ \sigma_{yy} \\ \tau_{xy} \end{Bmatrix} = \begin{Bmatrix} -800 \\ 40 \\ -70 \end{Bmatrix}$

7. $\begin{bmatrix} 0.75 & 0.25 & 0.866 \\ 0.25 & 0.75 & -0.866 \\ -0.433 & 0.433 & 0.5 \end{bmatrix} \begin{Bmatrix} \sigma_{xx} \\ \sigma_{yy} \\ \tau_{xy} \end{Bmatrix} = \begin{Bmatrix} -800 \\ -150 \\ 70 \end{Bmatrix}$

8. $\begin{bmatrix} 0.75 & 0.25 & 0.866 \\ 0.25 & 0.75 & -0.866 \\ -0.433 & 0.433 & 0.5 \end{bmatrix} \begin{Bmatrix} \sigma_{xx} \\ \sigma_{yy} \\ \tau_{xy} \end{Bmatrix} = \begin{Bmatrix} -800 \\ -150 \\ -70 \end{Bmatrix}$

On solving the above equations, we get the following, respectively:

1. $\begin{Bmatrix} \sigma_{xx} \\ \sigma_{yy} \\ \tau_{xy} \end{Bmatrix} = \begin{Bmatrix} 1449.4 \\ 590.6 \\ 883.7 \end{Bmatrix}$ MPa

2. $\begin{Bmatrix} \sigma_{xx} \\ \sigma_{yy} \\ \tau_{xy} \end{Bmatrix} = \begin{Bmatrix} 1570.6 \\ 469.4 \\ 813.7 \end{Bmatrix}$ MPa

3. $\begin{Bmatrix} \sigma_{xx} \\ \sigma_{yy} \\ \tau_{xy} \end{Bmatrix} = \begin{Bmatrix} 1401.9 \\ 448.1 \\ 966.0 \end{Bmatrix}$ MPa

4. $\begin{Bmatrix} \sigma_{xx} \\ \sigma_{yy} \\ \tau_{xy} \end{Bmatrix} = \begin{Bmatrix} 1523.1 \\ 326.9 \\ 896.0 \end{Bmatrix}$ MPa

5. $\begin{Bmatrix} \sigma_{xx} \\ \sigma_{yy} \\ \tau_{xy} \end{Bmatrix} = \begin{Bmatrix} -650.6 \\ -109.4 \\ -328.7 \end{Bmatrix}$ MPa

6. $\begin{Bmatrix} \sigma_{xx} \\ \sigma_{yy} \\ \tau_{xy} \end{Bmatrix} = \begin{Bmatrix} -529.4 \\ -230.6 \\ -398.7 \end{Bmatrix}$ MPa

7. $\begin{Bmatrix} \sigma_{xx} \\ \sigma_{yy} \\ \tau_{xy} \end{Bmatrix} = \begin{Bmatrix} -698.1 \\ -251.9 \\ -246.5 \end{Bmatrix}$ MPa

8. $\begin{Bmatrix} \sigma_{xx} \\ \sigma_{yy} \\ \tau_{xy} \end{Bmatrix} = \begin{Bmatrix} -576.9 \\ -373.1 \\ -316.5 \end{Bmatrix}$ MPa

Note that strength ratios corresponding to all the eight solutions are identically one, which is desirable from an efficient design point of view. (We are not discussing concepts like factor of safety, margin of safety, etc. yet!) Note, further, that global loads are generally a design input and in a real design scenario, given the design loads, we would rather stack a number of laminae in a laminate and vary lamina angles and thicknesses. We shall discuss the details of these issues in subsequent chapters.

4.6.2.2 Maximum Strain Failure Criterion

The maximum strain failure criterion is similar to the maximum stress failure criterion. Here, too, five different modes of failure are identified. However, unlike the previous failure criterion, here, the modes of failure are related to the strains and not stresses. They are

- Failure of the lamina due to excessive tensile strain in the 1-direction
- Failure of the lamina due to excessive compressive strain in the 1-direction
- Failure of the lamina due to excessive tensile strain in the 2-direction
- Failure of the lamina due to excessive compressive strain in the 2-direction
- Failure of the lamina due to excessive in-plane shear strain in the 1-2 plane

As per this criterion, a lamina is considered to have failed if any one of the strains in the material axes exceeds the corresponding ultimate strain. Thus, for the lamina to be safe, the following need to be satisfied:

$$\left(\varepsilon_{11}^C\right)_{ult} < \varepsilon_{11} < \left(\varepsilon_{11}^T\right)_{ult} \tag{4.145}$$

$$\left(\varepsilon_{22}^C\right)_{ult} < \varepsilon_{22} < \left(\varepsilon_{22}^T\right)_{ult} \tag{4.146}$$

$$|\gamma_{12}| < (\gamma_{12})_{ult} \tag{4.147}$$

in which

$(\varepsilon_{11}^T)_{ult}$ ultimate longitudinal tensile strain
$(\varepsilon_{11}^C)_{ult}$ ultimate longitudinal compressive strain
$(\varepsilon_{22}^T)_{ult}$ ultimate transverse tensile strain
$(\varepsilon_{22}^C)_{ult}$ ultimate transverse compressive strain
$(\gamma_{12})_{ult}$ ultimate in-plane shear strain

In the above inequalities, the sign convention is similar to that in stresses and strengths. Thus, normal strains are positive in tension and negative in compression. Similarly, normal ultimate strains are positive in tension and negative in compression.

Macromechanics of a Lamina

Similar to the in-plane shear stresses, in-plane shear strains can be either positive or negative, whereas the ultimate in-plane shear strain is positive. As in the case of the maximum stress criterion, in this failure criterion too, there is no interaction between different failure modes. Comparison of strains with ultimate strains is made in the respective direction only. Further, the inequalities only indicate whether the lamina is safe or not; they do not provide any information on the available factor of safety. In this connection, the following ultimate strain ratios can be defined:

$$(R)_{11}^T = \frac{\left(\varepsilon_{11}^T\right)_{ult}}{\varepsilon_{11}} \qquad (4.148)$$

$$(R)_{11}^C = \frac{\left(\varepsilon_{11}^C\right)_{ult}}{\varepsilon_{11}} \qquad (4.149)$$

$$(R)_{22}^T = \frac{\left(\varepsilon_{22}^T\right)_{ult}}{\varepsilon_{22}} \qquad (4.150)$$

$$(R)_{22}^C = \frac{\left(\varepsilon_{22}^C\right)_{ult}}{\varepsilon_{22}} \qquad (4.151)$$

$$(R)_{12}^S = \frac{(\gamma_{12})_{ult}}{|\gamma_{12}|} \qquad (4.152)$$

where $(R)_{11}^T$ is the ultimate strain ratio for normal tensile strain in the 1-direction and so on. Similar to the strength ratios, the ultimate strain ratios are always positive. $(R)_{11}^T < 1$ indicates that the lamina has failed due to excessive tensile load in the 1-direction and the load needs to be reduced by a factor of $(R)_{11}^T$. $(R)_{11}^T > 1$ indicates that the lamina is safe and the applied tensile load in the 1-direction can be increased by a factor of $(R)_{11}^T$. Finally, $(R)_{11}^T = 1$ indicates that the lamina has just failed. Similar conclusions can be drawn in respect of the remaining ultimate strain ratios as well.

EXAMPLE 4.6

Consider the carbon/epoxy lamina in Example 4.5. Analyze whether the lamina is safe under these stresses. Employ the maximum strain failure criterion. Other data remain the same as in that example. The elastic constants are $E_1 = 140$ GPa, $E_2 = 10$ GPa, $G_{12} = 6$ GPa, and $\nu_{12} = 0.28$.

Solution

Given $\sigma_{xx} = 1200$ MPa, $\sigma_{yy} = 350$ MPa, and $\tau_{xy} = 800$ MPa. Then, the stresses in the local coordinates are (refer Example 4.5)

$$\begin{Bmatrix} \sigma_{11} \\ \sigma_{22} \\ \tau_{12} \end{Bmatrix} = \begin{Bmatrix} 1680.3 \\ -130.3 \\ -32.0 \end{Bmatrix} \text{MPa}$$

Using Equation 4.2, the strains in the local coordinates are obtained as follows:

$$\begin{Bmatrix} \varepsilon_{11} \\ \varepsilon_{22} \\ \gamma_{12} \end{Bmatrix} = \begin{bmatrix} \dfrac{1}{140} & -\dfrac{0.28}{140} & 0 \\ -\dfrac{0.28}{140} & \dfrac{1}{10} & 0 \\ 0 & 0 & \dfrac{1}{6} \end{bmatrix} \begin{Bmatrix} 1680.3 \\ -130.3 \\ -32.0 \end{Bmatrix} \times 10^{-3} = \begin{Bmatrix} 12262 \\ -16391 \\ -5333 \end{Bmatrix} \times 10^{-6}$$

Next, given the strength data and elastic moduli, we find the ultimate strains as follows:

$$\left(\varepsilon_{11}^T\right)_{ult} = \dfrac{2000}{140} \times 10^{-3} = 14{,}286 \times 10^{-6}$$

$$\left(\varepsilon_{11}^C\right)_{ult} = \dfrac{-800}{140} \times 10^{-3} = -5714 \times 10^{-6}$$

$$\left(\varepsilon_{22}^T\right)_{ult} = \dfrac{40}{10} \times 10^{-3} = 4000 \times 10^{-6}$$

$$\left(\varepsilon_{22}^C\right)_{ult} = \dfrac{-150}{10} \times 10^{-3} = -15{,}000 \times 10^{-6}$$

$$(\gamma_{12})_{ult} = \dfrac{70}{6} \times 10^{-3} = 11{,}667 \times 10^{-6}$$

Then, the ultimate strain ratios are determined as follows:

$$(R)_{11}^T = \dfrac{14{,}286 \times 10^{-6}}{12{,}262 \times 10^{-6}} = 1.17$$

$$(R)_{22}^T = \dfrac{-15{,}000 \times 10^{-6}}{-16{,}391 \times 10^{-6}} = 0.92$$

$$(R)_{12}^S = \dfrac{11{,}667 \times 10^{-6}}{5333 \times 10^{-6}} = 2.19$$

Two points may be noted. First, the lamina is marginally unsafe in the transverse direction. Second, the ultimate strain ratios are pretty similar to the strength ratios obtained in Example 4.5 for the same lamina under the same loads. In fact, if we put $\nu_{12} = 0$, the ultimate strain ratios in the maximum strain criterion can be found to be identically the same as the strength ratios in the maximum stress criterion. In other words, but for the Poisson's effect, maximum strain criterion, and maximum stress criterion give the same results.

4.6.2.3 Tsai–Hill Failure Criterion

Unlike the maximum stress and maximum strain criteria, wherein there is no interaction between different failure modes, the Tsai–Hill failure criterion [11,12] considers

Macromechanics of a Lamina

the interaction between different strength parameters. This criterion is based on the von Mises distortional energy criterion for isotropic materials. As per the von Mises criterion, a body fails when the energy of distortion becomes more than a certain critical level or the distortion energy of failure. Hill extended the von Mises yield criterion to orthotropic materials and then Tsai adapted it to predict the failure of a unidirectional lamina.

The Hill's criterion for anisotropic material in three dimension can be stated as

$$A\sigma_{11}^2 + B\sigma_{22}^2 + C\sigma_{33}^2 + D\sigma_{11}\sigma_{22} + E\sigma_{22}\sigma_{33} + F\sigma_{33}\sigma_{11} + G\tau_{23}^2 + H\tau_{31}^2 + I\tau_{12}^2 = 1$$

(4.153)

In a plane stress problem, $\sigma_{33} = \tau_{31} = \tau_{23} = 0$ and thus, the Hill's criterion for a lamina reduces to

$$A\sigma_{11}^2 + B\sigma_{22}^2 + D\sigma_{11}\sigma_{22} + I\tau_{12}^2 = 1$$

(4.154)

When a body undergoes deformation, strain energy is stored in the body. Strain energy has two components—energy of dilation and energy of distortion. The first component is associated with change in size and the second with change in shape. The von Mises yield criterion for isotropic materials is based on energy of distortion, that is, the energy associated with changing the shape but not the volume of the body. On the other hand, in an orthotropic material, distortion and dilation cannot be separated and the Hill's yield criterion is not based on distortion energy; rather the coefficients in Equations 4.153 and 4.154 are based on orthotropic failure strengths. Restricting our focus on the plane stress case in Equation 4.154, the coefficients are determined by applying the following load cases.

First, $\sigma_{11} = (\sigma_{11}^T)_{ult}$ and $\sigma_{22} = \tau_{12} = 0$. At this point, the lamina fails due to excessive stress in the longitudinal direction and from Equation 4.154, we get

$$A = \frac{1}{\left(\sigma_{11}^T\right)_{ult}^2}$$

(4.155)

Second, $\sigma_{22} = (\sigma_{22}^T)_{ult}$ and $\sigma_{11} = \tau_{12} = 0$. Then, from Equation 4.154, we get

$$B = \frac{1}{\left(\sigma_{22}^T\right)_{ult}^2}$$

(4.156)

Third, $\tau_{12} = (\tau_{12})_{ult}$ and $\sigma_{11} = \sigma_{22} = 0$, which, along with Equation 4.154, gives us

$$I = \frac{1}{(\tau_{12})_{ult}^2}$$

(4.157)

Coefficient D is associated with two stress components and for its determination we have to have a biaxial stress field such that the following holds good at lamina failure:

$$\sigma_{11} \neq 0, \sigma_{22} \neq 0, \quad \text{and} \quad \tau_{12} = 0$$

A convenient way to achieve this is to consider lamina under a stress field given by

$$\sigma_{xx} = \sigma_{yy} = \sigma \neq 0 \quad \text{and} \quad \tau_{xy} = 0$$

From this, it can be readily found that

$$\sigma_{11} = \sigma_{22} = \sigma \quad \text{and} \quad \tau_{12} = 0$$

Transverse strength is normally far lower than the longitudinal strength and we can presume that failure takes place at $\sigma = (\sigma_{22}^T)_{ult}$. Substituting this in Equation 4.154 together with Equations 4.155 and 4.156, we get

$$D = -\frac{1}{\left(\sigma_{11}^T\right)_{ult}^2} \tag{4.158}$$

Now, by substituting the expressions of the coefficients from Equations 4.155 through 4.158 in Equation 4.154, and noting that the left-hand side of Equation 4.154 has to be less than 1, we can obtain an expression for the Tsai–Hill failure criterion as

$$\left(\frac{\sigma_{11}}{\left(\sigma_{11}^T\right)_{ult}}\right)^2 + \left(\frac{\sigma_{22}}{\left(\sigma_{22}^T\right)_{ult}}\right)^2 + \left(\frac{\tau_{12}}{(\tau_{12})_{ult}}\right)^2 - \left(\frac{\sigma_{11}}{\left(\sigma_{11}^T\right)_{ult}}\right)\left(\frac{\sigma_{22}}{\left(\sigma_{11}^T\right)_{ult}}\right) < 1 \tag{4.159}$$

Note that we considered the normal stresses as tensile only. When compressive stresses are also considered, the Tsai–Hill failure criterion can be suitably modified as follows:

$$\left(\frac{\sigma_{11}}{F_1}\right)^2 + \left(\frac{\sigma_{22}}{F_2}\right)^2 + \left(\frac{\tau_{12}}{F_3}\right)^2 - \left(\frac{\sigma_{11}}{F_1}\right)\left(\frac{\sigma_{22}}{F_1}\right) < 1 \tag{4.160}$$

where
$F_1 = (\sigma_{11}^T)_{ult}$ if $\sigma_{11} > 0$, that is, σ_{11} is tensile
$ = (\sigma_{11}^C)_{ult}$ if $\sigma_{11} < 0$, that is, σ_{11} is compressive
$F_2 = (\sigma_{22}^T)_{ult}$ if $\sigma_{22} > 0$, that is, σ_{22} is tensile
$ = (\sigma_{22}^C)_{ult}$ if $\sigma_{22} < 0$, that is, σ_{22} is compressive
$F_3 = (\tau_{12})_{ult}$

The Tsai–Hill failure criterion considers the interaction between different strength parameters and it indicates the failure of a lamina in a combined way based on all the applied stresses and strength parameters. A gross strength ratio can be calculated by taking the square root of the left-hand side of Equation 4.160. However, this failure criterion does not indicate the mode of failure of a lamina.

EXAMPLE 4.7
Consider the carbon/epoxy lamina in Example 4.5. Check whether the lamina is safe under these stresses. Employ the Tsai–Hill failure criterion. Other data remain the same as in that example.

Solution
Stresses applied on the lamina are $\sigma_{xx} = 1200$ MPa, $\sigma_{yy} = 350$ MPa, and $\tau_{xy} = 800$ MPa, that is, the stresses in the global coordinates are

$$\begin{Bmatrix} \sigma_{xx} \\ \sigma_{yy} \\ \tau_{xy} \end{Bmatrix} = \begin{Bmatrix} 1200 \\ 350 \\ 800 \end{Bmatrix} \text{MPa}$$

Macromechanics of a Lamina

And the stresses in the local coordinates are

$$\begin{Bmatrix} \sigma_{11} \\ \sigma_{22} \\ \tau_{12} \end{Bmatrix} = \begin{Bmatrix} 1680.3 \\ -130.3 \\ -32.0 \end{Bmatrix} \text{MPa}$$

σ_{11} is tensile, whereas σ_{22} is compressive. Thus, the constants F_1, F_2, and F_3 in Equation 4.160 are the tensile strength, compressive strength, and in-plane shear strength, respectively. That is,

$F_1 = 2000$ MPa
$F_2 = -150$ MPa
$F_3 = 70$ MPa

Then,

$$\left(\frac{\sigma_{11}}{F_1}\right)^2 + \left(\frac{\sigma_{22}}{F_2}\right)^2 + \left(\frac{\tau_{12}}{F_3}\right)^2 - \left(\frac{\sigma_{11}}{F_1}\right)\left(\frac{\sigma_{22}}{F_1}\right) = \left(\frac{1680.3}{2000}\right)^2 + \left(\frac{-130.3}{-150}\right)^2$$
$$+ \left(\frac{-32}{70}\right)^2 - \left(\frac{1680.3}{2000}\right)\left(\frac{-130.3}{-800}\right) = 1.53$$

which is greater than one and thus, the lamina is not safe.

4.6.2.4 Tsai–Wu Failure Criterion

In its most general form, using indicial notation, the Tsai–Wu criterion [13] can be expressed as

$$F_i \sigma_i + F_{ij} \sigma_i \sigma_j = 1 \tag{4.161}$$

In the above equation, the contracted notation is used for stress components and $i, j = 1, 2, \ldots, 6$. For an orthotropic lamina in plane stress condition, the Tsai–Wu criterion is of the following form:

$$F_1 \sigma_1 + F_2 \sigma_2 + F_6 \sigma_6 + F_{11} \sigma_1^2 + F_{22} \sigma_2^2 + F_{66} \sigma_6^2 + (F_{12} + F_{21}) \sigma_1 \sigma_2$$
$$+ (F_{16} + F_{61}) \sigma_1 \sigma_6 + (F_{26} + F_{62}) \sigma_2 \sigma_6 = 1 \tag{4.162}$$

Let us substitute $H_{12} = F_{12} + F_{21}$, $H_{16} = F_{16} + F_{61}$, and $H_{26} = F_{26} + F_{62}$ and rewrite Equation 4.162, using tensor notation for stress components, as

$$F_1 \sigma_{11} + F_2 \sigma_{22} + F_6 \tau_{12} + F_{11} \sigma_{11}^2 + F_{22} \sigma_{22}^2 + F_{66} \tau_{12}^2$$
$$+ H_{12} \sigma_{11} \sigma_{22} + H_{16} \sigma_{11} \tau_{12} + H_{26} \sigma_{22} \tau_{12} = 1 \tag{4.163}$$

To determine the coefficients, we adopt a similar procedure as in the case of the Tsai–Hill criterion and apply some load cases as follows:

The first and second load cases applied are a uniaxial longitudinal stress equal to the ultimate longitudinal tensile stress and a uniaxial longitudinal stress equal to the ultimate longitudinal compressive stress, respectively, that is,

$$\sigma_{11} = (\sigma_{11}^T)_{ult} \quad \text{and} \quad \sigma_{22} = \tau_{12} = 0$$

and

$$\sigma_{11} = (\sigma_{11}^C)_{ult} \text{ and } \sigma_{22} = \tau_{12} = 0$$

On substitution of the first two load cases in Equation 4.163, we obtain, respectively

$$F_1\left(\sigma_{11}^T\right)_{ult} + F_{11}\left(\sigma_{11}^T\right)_{ult}^2 = 1 \qquad (4.164)$$

$$F_1\left(\sigma_{11}^C\right)_{ult} + F_{11}\left(\sigma_{11}^C\right)_{ult}^2 = 1 \qquad (4.165)$$

Solving Equations 4.164 and 4.165, the coefficients F_1 and F_{11} are obtained as

$$F_1 = \frac{1}{\left(\sigma_{11}^T\right)_{ult}} + \frac{1}{\left(\sigma_{11}^C\right)_{ult}} \qquad (4.166)$$

$$F_{11} = -\frac{1}{\left(\sigma_{11}^T\right)_{ult}\left(\sigma_{11}^C\right)_{ult}} \qquad (4.167)$$

The third and fourth load cases applied are a uniaxial transverse stress equal to the ultimate transverse tensile stress and a uniaxial transverse stress equal to the ultimate transverse compressive stress, respectively, that is,

$$\sigma_{22} = (\sigma_{22}^T)_{ult} \quad \text{and} \quad \sigma_{11} = \tau_{12} = 0$$

and

$$\sigma_{22} = (\sigma_{22}^C)_{ult} \quad \text{and} \quad \sigma_{11} = \tau_{12} = 0$$

On substitution of the third and fourth two load cases in Equation 4.163, we obtain, respectively

$$F_2\left(\sigma_{22}^T\right)_{ult} + F_{22}\left(\sigma_{22}^T\right)_{ult}^2 = 1 \qquad (4.168)$$

$$F_2\left(\sigma_{22}^C\right)_{ult} + F_{22}\left(\sigma_{22}^C\right)_{ult}^2 = 1 \qquad (4.169)$$

Solving Equations 4.168 and 4.169, the coefficients F_2 and F_{22} are obtained as

$$F_2 = \frac{1}{\left(\sigma_{22}^T\right)_{ult}} + \frac{1}{\left(\sigma_{22}^C\right)_{ult}} \qquad (4.170)$$

$$F_{22} = -\frac{1}{\left(\sigma_{22}^T\right)_{ult}\left(\sigma_{22}^C\right)_{ult}} \qquad (4.171)$$

The fifth load case applied is an in-plane shear stress equal to the ultimate in-plane shear stress together with nonzero normal stresses, that is,

$$\tau_{12} = (\tau_{12})_{ult}, \sigma_{11} = \sigma_1, \sigma_{22} = \sigma_2$$

On substitution of the fifth load case in Equation 4.163, we obtain,

$$F_1\sigma_1 + F_2\sigma_2 + F_6(\tau_{12})_{ult} + F_{11}\sigma_1^2 + F_{22}\sigma_2^2 + F_{66}(\tau_{12})_{ult}^2$$
$$+ H_{12}\sigma_1\sigma_2 + H_{16}\sigma_1(\tau_{12})_{ult} + H_{26}\sigma_2(\tau_{12})_{ult} = 1 \qquad (4.172)$$

Macromechanics of a Lamina

Note here that in-plane shear stress has the same ultimate strength whether it is positive or negative. Thus, Equation 4.172 can also be written as

$$F_1\sigma_1 + F_2\sigma_2 - F_6(\tau_{12})_{ult} + F_{11}\sigma_1^2 + F_{22}\sigma_2^2 + F_{66}(\tau_{12})_{ult}^2 \\ + H_{12}\sigma_1\sigma_2 - H_{16}\sigma_1(\tau_{12})_{ult} - H_{26}\sigma_2(\tau_{12})_{ult} = 1 \quad (4.173)$$

The existence of Equations 4.172 and 4.173 is possible only when the coefficients in the linear terms in $(\tau_{12})_{ult}$ vanish, that is,

$$F_6 = H_{16} = H_{26} = 0 \quad (4.174)$$

Substituting the above in Equation 4.172, and taking $\sigma_1 = \sigma_2 = 0$, we get

$$F_{66} = \frac{1}{(\tau_{12})_{ult}^2} \quad (4.175)$$

There are nine coefficients in Equation 4.163, of which we have obtained the expressions for eight coefficients by the procedure discussed above. The coefficient H_{12} is left but it cannot be obtained directly from the strength parameters. One of the experimental methods proposed for the determination of this coefficient involves the application of a tensile stress in the x-direction on a 45° lamina. It can be readily seen that

$$\sigma_{11} = \sigma_{22} = -\tau_{12} = \frac{(\sigma_{xx})_{ult}}{2} \quad (4.176)$$

where $(\sigma_{xx})_{ult}$ is the applied uniaxial tension at failure of the lamina. From Equation 4.163, it can be shown that

$$H_{12} = \frac{4}{(\sigma_{xx})_{ult}^2} - \frac{2}{(\sigma_{xx})_{ult}} \left[\frac{1}{(\sigma_{11}^T)_{ult}} + \frac{1}{(\sigma_{11}^C)_{ult}} + \frac{1}{(\sigma_{22}^T)_{ult}} + \frac{1}{(\sigma_{22}^C)_{ult}} \right] \\ + \left[\frac{1}{(\sigma_{11}^T)_{ult}(\sigma_{11}^C)_{ult}} + \frac{1}{(\sigma_{22}^T)_{ult}(\sigma_{22}^C)_{ult}} \right] \quad (4.177)$$

The effect of H_{12} on the final strength assessment is not significant. Further, the expression given by Equation 4.177 is not very elegant. Some empirical expressions have been suggested. One such empirical expression is as follows:

$$H_{12} = -\frac{1}{2\sqrt{(\sigma_{11}^T)_{ult}(\sigma_{11}^C)_{ult}(\sigma_{22}^T)_{ult}(\sigma_{22}^C)_{ult}}} \quad (4.178)$$

For the lamina to be safe, the left-hand side of Equation 4.163 has to be less than one and the Tsai–Wu failure criterion for an orthotropic lamina in plane stress condition can be expressed as

$$F_1\sigma_{11} + F_2\sigma_{22} + F_{11}\sigma_{11}^2 + F_{22}\sigma_{22}^2 + F_{66}\tau_{12}^2 + H_{12}\sigma_{11}\sigma_{22} < 1 \quad (4.179)$$

in which, the coefficients are given by Equations 4.166, 4.167, 4.170, 4.171, 4.175 and 4.177 or 4.178.

EXAMPLE 4.8

Consider the carbon/epoxy lamina in Example 4.5. Analyze whether the lamina is safe under these stresses. Employ the Tsai–Hill failure criterion. Other data remain the same as in that example. Elastic constants are $E_1 = 140$ GPa, $E_2 = 10$ GPa, $\nu_{12} = 0.28$, and $G_{12} = 6$ GPa.

Solution

The stresses applied on the lamina are $\sigma_{xx} = 1200$ MPa, $\sigma_{yy} = 350$ MPa, and $\tau_{xy} = 800$ MPa, that is, the stresses in the global coordinates are

$$\begin{Bmatrix} \sigma_{xx} \\ \sigma_{yy} \\ \tau_{xy} \end{Bmatrix} = \begin{Bmatrix} 1200 \\ 350 \\ 800 \end{Bmatrix} \text{MPa}$$

and the stresses in the local coordinates are

$$\begin{Bmatrix} \sigma_{11} \\ \sigma_{22} \\ \tau_{12} \end{Bmatrix} = \begin{Bmatrix} 1680.3 \\ -130.3 \\ -32.0 \end{Bmatrix} \text{MPa}$$

The coefficients in Equation 4.179 are obtained as follows:

$$F_1 = \frac{1}{2000} + \frac{1}{-800} = -7.5 \times 10^{-4}$$

$$F_{11} = \frac{-1}{2000 \times (-800)} = 6.25 \times 10^{-7}$$

$$F_2 = \frac{1}{40} + \frac{1}{-150} = 1.833 \times 10^{-2}$$

$$F_{22} = \frac{-1}{40 \times (-150)} = 1.667 \times 10^{-4}$$

$$F_{66} = \frac{1}{70^2} = 2.041 \times 10^{-4}$$

$$H_{12} = -\frac{1}{2\sqrt{2000 \times (-800) \times 40 \times (-150)}} = -5.103 \times 10^{-6}$$

Macromechanics of a Lamina

Then, the left-hand side of Equation 4.179 can be calculated as

$$F_1\sigma_{11} + F_2\sigma_{22} + F_{11}\sigma_{11}^2 + F_{22}\sigma_{22}^2 + F_{66}\tau_{12}^2 + H_{12}\sigma_{11}\sigma_{22}$$
$$= -7.5 \times 10^{-4} \times 1680.3 + 1.833 \times 10^{-2} \times (-130.3) + 6.25 \times 10^{-7} \times 1680.3^2$$
$$+ 1.667 \times 10^{-4} \times (-130.3)^2 + 2.041 \times 10^{-4} \times 32.0^2 - 5.103 \times 10^{-6}$$
$$\times 1680.3 \times (-130.3) = 2.273$$

which is greater than one and thus the lamina is not safe.

4.6.2.5 Discussion on Failure Criteria

In the preceding sections, we presented four failure criteria commonly used in the design and analysis of composite structures. A natural question arises as to which criterion should be used in a specific design case. With a view to getting an answer to this, in this section, we shall have a brief comparative study of these failure criteria. Let us consider the following example.

EXAMPLE 4.9

Consider a carbon/epoxy lamina with the following strength parameters and elastic constants:

$(\sigma_{11}^T)_{ult} = 2000\,\text{MPa}$, $(\sigma_{11}^C)_{ult} = -800\,\text{MPa}$, $(\sigma_{22}^T)_{ult} = 40\,\text{MPa}$,
$(\sigma_{22}^C)_{ult} = -150\,\text{MPa}$, and $(\tau_{12})_{ult} = 70\,\text{MPa}$

$E_1 = 140\,\text{GPa}$, $E_2 = 10\,\text{GPa}$, $\nu_{12} = 0.28$, and $G_{12} = 6\,\text{GPa}$

Consider the following load case and determine the failure loads as per different failure criteria for varying lamina angles:

$\sigma_{xx} \neq 0$ and $\sigma_{yy} = \tau_{xy} = 0$

Solution

Using Equation 4.43, the stresses in the local coordinate axes can be written as

$$\begin{Bmatrix} \sigma_{11} \\ \sigma_{22} \\ \tau_{12} \end{Bmatrix} = \begin{Bmatrix} \cos^2\theta \\ \sin^2\theta \\ -\cos\theta\sin\theta \end{Bmatrix} \sigma_{xx}$$

Then, using Equation 4.2, the strains in the local coordinates can be written as

$$\begin{Bmatrix} \varepsilon_{11} \\ \varepsilon_{22} \\ \gamma_{12} \end{Bmatrix} = \begin{Bmatrix} \dfrac{\cos^2\theta - \nu_{12}\sin^2\theta}{E_1} \\ \dfrac{\sin^2\theta - (E_2/E_1)\nu_{12}\cos^2\theta}{E_2} \\ -\dfrac{\cos\theta\sin\theta}{G_{12}} \end{Bmatrix} \sigma_{xx}$$

Under the given uniaxial load case, two subcases are possible: (i) tensile stress along the x-direction and (ii) compressive stress along the x-direction. Let us first consider tensile stress, that is, $\sigma_{xx} > 0$.

Maximum Stress Failure Criterion

For $\sigma_{xx} > 0$ and $0 \leq \theta \leq 90°$, $0 \leq \sigma_{xx} \cos^2 \theta$. Thus, σ_{11} is positive, that is, tensile. In a similar way, σ_{22} is also tensile. Hence, as per the maximum stress criterion, we compare these local stresses with the corresponding ultimate tensile stresses. On the other hand, τ_{12} is negative. However, as discussed before, in-plane shear strength is the same irrespective of its sign and we need to compare the absolute value of the shear stress with the corresponding strength. Thus, the maximum values of σ_{xx} are obtained as follows:

Failure due to excessive longitudinal tensile stress

$$(\sigma_{xx})_{max} = \frac{2000}{\cos^2 \theta}$$

Failure due to excessive transverse tensile stress

$$(\sigma_{xx})_{max} = \frac{40}{\sin^2 \theta}$$

Failure due to excessive in-plane shear stress

$$(\sigma_{xx})_{max} = \frac{70}{\cos \theta \sin \theta}$$

Maximum Strain Criterion

We note that the ultimate strains are given by

$$\left(\varepsilon_{11}^T\right)_{ult} = \frac{\left(\sigma_{11}^T\right)_{ult}}{E_1}$$

$$\left(\varepsilon_{11}^C\right)_{ult} = \frac{\left(\sigma_{11}^C\right)_{ult}}{E_1}$$

$$\left(\varepsilon_{22}^T\right)_{ult} = \frac{\left(\sigma_{22}^T\right)_{ult}}{E_2}$$

$$\left(\varepsilon_{22}^C\right)_{ult} = \frac{\left(\sigma_{22}^C\right)_{ult}}{E_2}$$

$$(\gamma_{12})_{ult} = \frac{(\tau_{12})_{ult}}{G_{12}}$$

Then, comparing the local strains with the ultimate strains, we obtain the maximum values of σ_{xx} as follows:

Failure due to excessive longitudinal tensile strain (for $\cos^2 \theta > 0.28 \sin^2 \theta$)

$$(\sigma_{xx})_{max} = \frac{2000}{\cos^2 \theta - 0.28 \sin^2 \theta}$$

Macromechanics of a Lamina

Failure due to excessive longitudinal compressive strain (for $\cos^2 \theta < 0.28 \sin^2 \theta$)

$$(\sigma_{xx})_{max} = \frac{-800}{\cos^2 \theta - 0.28 \sin^2 \theta}$$

Failure due to excessive transverse tensile strain (for $\sin^2 \theta > 0.02 \cos^2 \theta$)

$$(\sigma_{xx})_{max} = \frac{40}{\sin^2 \theta - 0.02 \cos^2 \theta}$$

Failure due to excessive transverse compressive strain (for $\sin^2 \theta < 0.02 \cos^2 \theta$)

$$(\sigma_{xx})_{max} = \frac{-150}{\sin^2 \theta - 0.02 \cos^2 \theta}$$

Failure due to excessive in-plane shear strain

$$(\sigma_{xx})_{max} = \frac{70}{\cos \theta \sin \theta}$$

Tsai–Hill Criterion

The coefficients in Equation 4.160 are given by

$$F_1 = 2000 \, \text{MPa}$$

$$F_2 = 40 \, \text{MPa}$$

$$F_3 = 70 \, \text{MPa}$$

Note that the tensile strengths that are used for F_1 and F_2 as σ_{11} and σ_{22} are positive quantities. Then, using Equation 4.160, we get

$$\left[\left(\frac{\cos^2 \theta}{2000} \right)^2 + \left(\frac{\sin^2 \theta}{40} \right)^2 + \left(\frac{\cos \theta \sin \theta}{70} \right)^2 - \left(\frac{\cos \theta}{2000} \right) \left(\frac{\sin \theta}{2000} \right) \right] (\sigma_{xx})^2_{max} = 1$$

or

$$(\sigma_{xx})_{max} = \frac{10}{\sqrt{\left[\frac{\cos^4 \theta}{4 \times 10^4} + \frac{\sin^4 \theta}{16} + \frac{\cos^2 \theta \sin^2 \theta}{49} - \frac{\cos \theta \sin \theta}{4 \times 10^4} \right]}}$$

Tsai–Wu Criterion

The coefficients in Equation 4.179 are given by

$$F_1 = \frac{1}{2000} + \frac{1}{-800} = -7.5 \times 10^{-4}$$

$$F_{11} = \frac{-1}{2000 \times (-800)} = 6.25 \times 10^{-7}$$

$$F_2 = \frac{1}{40} + \frac{1}{-150} = 1.833 \times 10^{-2}$$

$$F_{22} = \frac{-1}{40 \times (-150)} = 1.667 \times 10^{-4}$$

$$F_{66} = \frac{1}{70^2} = 2.041 \times 10^{-4}$$

$$H_{12} = -\frac{1}{2\sqrt{2000 \times (-800) \times 40 \times (-150)}} = -5.103 \times 10^{-6}$$

Then, the Tsai Wu criterion in Equation 4.179 can be written as

$$[F_{11} \cos^4 \theta + F_{22} \sin^4 \theta + (F_{66} + H_{12}) \cos^2 \theta \sin^2 \theta] \sigma_{xx}^2 + [F_1 \cos^2 \theta + F_2 \sin^2 \theta] \sigma_{xx} = 1$$

Substituting the values of the coefficients from above, after simple manipulation, we get

$$[6.25 \cos^4 \theta + 1667 \sin^4 \theta + 1990 \cos^2 \theta \sin^2 \theta] \sigma_{xx}^2$$
$$+ [(1.833 \times 10^5) \sin^2 \theta - 7500 \cos^2 \theta] \sigma_{xx} - 10^7 = 0$$

By solving the quadratic equation, we get two roots. The positive root is the maximum tensile $(\sigma_{xx})_{max}$ as per the Tsai–Wu criterion.

We have obtained expressions for $(\sigma_{xx})_{max}$ in terms of θ as per all the four failure criteria. θ is varied from 0° to 90° and the values of maximum uniaxial stress $(\sigma_{xx})_{max}$ are tabulated in Table 4.2. The results are pictorially presented in Figures 4.17 through 4.19.

Next, let us consider the uniaxial compressive load, that is, $\sigma_{xx} < 0$ and determine the variation of $(\sigma_{xx})_{min}$.

Maximum Stress Criterion

Following a similar procedure as in the case of uniaxial tensile loading, we get the maximum uniaxial compressive stress as follows:

Failure due to excessive longitudinal compressive stress

$$(\sigma_{xx})_{min} = \frac{-800}{\cos^2 \theta}$$

Failure due to excessive transverse compressive stress

$$(\sigma_{xx})_{min} = \frac{-150}{\sin^2 \theta}$$

TABLE 4.2

Maximum Uniaxial Tensile Stress, $(\sigma_{xx})_{max}$, on a Carbon/Epoxy Lamina (Example 4.9) as Per Different Failure Criteria

Lamina Angle (°)	Maximum Stress Criterion	Maximum Strain Criterion	Tsai–Hill Criterion	Tsai–Wu Criterion
0	2000.0	2000.0	2000.0	2000.0
5	806.2	806.2	745.6	842.9
10	409.3	409.3	385.5	404.0
15	280.0	280.0	252.3	251.3
20	217.8	217.8	183.3	177.7
25	182.8	182.8	141.5	135.4
30	160.0	161.7	113.7	108.3
35	121.6	126.8	94.2	89.9
40	96.8	99.6	80.0	76.7
45	80.0	81.6	69.5	67.0
50	68.2	69.1	61.5	59.7
55	59.6	60.2	55.4	54.1
60	53.3	53.7	50.7	49.8
65	48.7	48.9	47.1	46.5
70	45.3	45.4	44.4	44.0
75	42.9	42.9	42.4	42.2
80	41.2	41.3	41.0	41.0
85	40.3	40.3	40.3	40.2
90	40.0	40.0	40.0	40.0

Failure due to excessive in-plane shear stress

$$(\sigma_{xx})_{min} = \frac{-70}{\cos\theta \sin\theta}$$

Maximum Strain Criterion

We obtain the maximum values of σ_{xx} as follows:

Failure due to excessive longitudinal compressive strain (for $\cos^2\theta > 0.28\sin^2\theta$)

$$(\sigma_{xx})_{min} = \frac{-800}{\cos^2\theta - 0.28\sin^2\theta}$$

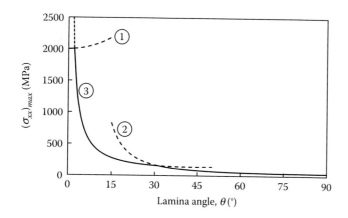

FIGURE 4.17 Failure load as per the maximum stress failure criterion of a lamina (Example 4.9) under uniaxial tensile stress σ_{xx}. (1) Failure due to excessive longitudinal tensile stress. (2) Failure due to excessive transverse tensile stress. (3) Failure due to excessive in-plane shear stress.

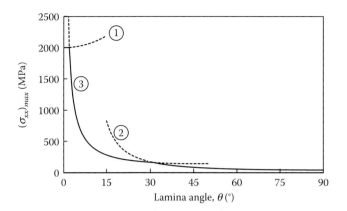

FIGURE 4.18 Failure load as per the maximum strain failure criterion of a lamina (Example 4.9) under uniaxial tensile stress σ_{xx}. (1) Failure due to excessive longitudinal tensile strain. (2) Failure due to excessive transverse tensile strain. (3) Failure due to excessive in-plane shear strain.

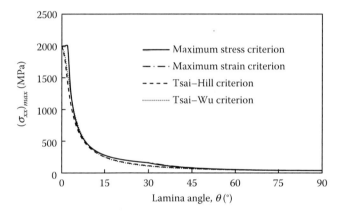

FIGURE 4.19 Comparison of failure load by different failure criteria of a lamina (Example 4.9) under uniaxial tensile stress σ_{xx}.

Failure due to excessive longitudinal tensile strain (for $\cos^2\theta < 0.28\sin^2\theta$)

$$(\sigma_{xx})_{min} = \frac{2000}{\cos^2\theta - 0.28\sin^2\theta}$$

Failure due to excessive transverse compressive stress (for $\sin^2\theta > 0.02\cos^2\theta$)

$$(\sigma_{xx})_{min} = \frac{-150}{\sin^2\theta - 0.02\cos^2\theta}$$

Failure due to excessive transverse tensile strain (for $\sin^2\theta < 0.02\cos^2\theta$)

$$(\sigma_{xx})_{min} = \frac{40}{\sin^2\theta - 0.02\cos^2\theta}$$

Failure due to excessive in-plane shear strain

$$(\sigma_{xx})_{min} = -\frac{70}{\cos\theta\sin\theta}$$

Tsai–Hill Criterion

The coefficients in Equation 4.160 are given by

$$F_1 = -800 \, \text{MPa}$$

$$F_2 = -150 \, \text{MPa}$$

$$F_3 = 70 \, \text{MPa}$$

Note that compressive strengths are used for F_1 and F_2 as σ_{11} and σ_{22} are negative quantities. Then, using Equation 4.160, we get

$$\left[\left(\frac{\cos^2\theta}{-800}\right)^2 + \left(\frac{\sin^2\theta}{-150}\right)^2 + \left(\frac{\cos\theta\sin\theta}{70}\right)^2 - \left(\frac{\cos\theta}{-800}\right)\left(\frac{\sin\theta}{-800}\right)\right](\sigma_{xx})^2_{min} = 1$$

or

$$(\sigma_{xx})_{min} = -\frac{10}{\sqrt{\left[\dfrac{\cos^4\theta}{6400} + \dfrac{\sin^4\theta}{225} + \dfrac{\cos^2\theta\sin^2\theta}{49} - \dfrac{\cos\theta\sin\theta}{6400}\right]}}$$

Tsai–Wu Criterion

We discussed the procedure for the determination of $(\sigma_{xx})_{max}$ as per the Tsai–Wu criterion in the subcase $\sigma_{xx} > 0$. We found that the solution gives us two roots. The negative root is the maximum compressive stress (algebraically minimum) as per the Tsai–Wu criterion.

The results in the case compressive σ_{xx} are tabulated in Table 4.3 and plotted in Figures 4.20 through 4.22.

Based on the results in Example 4.9 and the discussions on the failure criteria, the following points may be noted:

The maximum stress failure criterion indicates that the mode of failure and its pictorial representation are composed of three curves. In Figures 4.17, the first curve corresponds to the failure of the lamina due to excessive tensile stress in the longitudinal direction. Owing to normally very high longitudinal modulus as compared to transverse and in-plane shear moduli, this mode of failure is prevalent only at very low lamina angles. The failure of the lamina at high lamina angles is due to excessive tensile stress in the transverse direction. At low to medium lamina angles, the mode of failure is in-plane shear failure.

The maximum strain failure criterion is similar to the maximum stress criterion and it also indicates the mode of failure. On the other hand, Tsai–Hill and Tsai–Wu criteria are unified criteria and thus in these failure criteria, mode of failure cannot be found directly.

Maximum stress and maximum strain criteria are associated with cusps formed at the junctions between curves representing different failure modes. These cusps are not found in the Tsai–Hill and Tsai–Wu criteria, wherein variation of failure stress w.r.t. lamina angle is rather smooth.

The results of experimental works [14,15] indicate good comparison with Tsai–Hill and Tsai–Wu criteria. On the other hand, the maximum stress and maximum strain

TABLE 4.3

Maximum Uniaxial Compressive Stress, $(\sigma_{xx})_{max}$, on a Carbon/Epoxy Lamina (Example 4.9) as Per Different Failure Criteria

Lamina Angle (°)	Maximum Stress Criterion	Maximum Strain Criterion	Tsai–Hill Criterion	Tsai–Wu Criterion
0	−800.0	−800.0	−800.0	−800.0
5	−806.1	−806.2	−582.8	−558.2
10	−409.3	−409.3	−372.4	−377.4
15	−280.0	−280.0	−268.0	−289.8
20	−217.8	−217.8	−211.3	−241.3
25	−182.8	−182.8	−177.3	−211.5
30	−161.7	−161.7	−155.7	−192.0
35	−149.0	−149.0	−141.7	−178.7
40	−142.2	−142.2	−132.6	−169.5
45	−140.0	−140.0	−127.3	−163.0
50	−142.2	−142.2	−124.7	−158.5
55	−149.0	−149.0	−124.5	−155.3
60	−161.7	−161.7	−126.3	−153.2
65	−182.6	−182.8	−129.8	−151.8
70	−169.9	−170.3	−134.5	−151.0
75	−160.8	−161.0	−139.9	−150.5
80	−154.7	−154.8	−145.1	−150.2
85	−151.1	−151.2	−148.8	−150.0
90	−150.0	−150.0	−150.0	−150.0

failure criteria cannot estimate failure stress so well especially near the cusps, where the experimental values are typically lower than the predicted ones.

Given the local stress/strain fields, the maximum stress and maximum strain criteria involve hardly any calculation other than simple comparison with the corresponding ultimate strength or strain parameters. Thus, these criteria are extremely simple to use in a design case.

Tsai–Hill and Tsai–Wu failure criteria are similar in respect of their approach and performance. The Tsai–Wu criterion has a broader scope and it differentiates compressive stress from tensile stress better than the Tsai–Hill criterion. However, the Tsai–Wu criterion is more complicated to use.

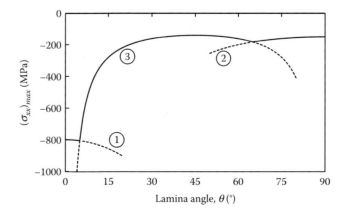

FIGURE 4.20 Failure load as per the maximum stress failure criterion of a lamina (Example 4.9) under uniaxial compressive stress σ_{xx}. (1) Failure due to excessive longitudinal compressive stress. (2) Failure due to excessive transverse compressive stress. (3) Failure due to excessive in-plane shear stress.

Macromechanics of a Lamina

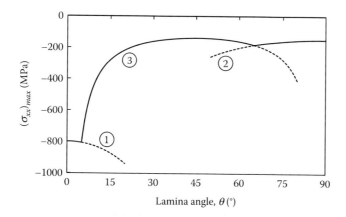

FIGURE 4.21 Failure load as per the maximum strain failure criterion of a lamina (Example 4.9) under uniaxial compressive stress σ_{xx}. (1) Failure due to excessive longitudinal compressive strain. (2) Failure due to excessive transverse compressive strain. (3) Failure due to excessive in-plane shear strain.

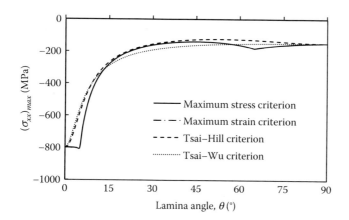

FIGURE 4.22 Comparison of failure load by different failure criteria of a lamina (Example 4.9) under uniaxial compressive stress σ_{xx}.

The four failure criteria have their own individual characteristics. From the point of view of simplicity, the maximum stress or maximum strain criterion can be used especially in the development phase of a design exercise. These criteria help to identify the mode of failure and thereby the designer can make appropriate modifications to the ply sequence. On the other hand, although the Tsai–Hill and Tsai–Wu criteria are more complicated, they are found to have better match with experimental data and can be utilized in the final analysis. In other words, the failure criteria may be taken as complement to each other.

4.7 HYGROTHERMAL EFFECTS

The constitutive relations presented in Section 4.4 do not consider any aspect of the environment that the composite lamina is in. Among the many environmental factors, the effects of change in temperature and change in moisture content on PMCs are the most significant.

Thermal stresses can occur in two ways. First, residual thermal stresses can develop due to temperature change during processing. Most PMCs are cured at high temperatures. Owing to mismatch in their CTEs, fibers and matrix undergo differential expansion during the temperature rise phase. At an elevated temperature, the matrix cross-links and fibers and matrix unite. During the cooling phase, the fibers and matrix tend to shrink differentially but the bond at the interface between the two restrains them. As a result, on cooling, the composite develops residual stresses.

Second, thermal stresses can develop due to temperature change during product service life. This is due to mismatch between CTEs in the longitudinal and transverse directions of a unidirectional lamina. When a single lamina is subjected to a change in temperature, it undergoes differential deformation in the two directions and clearly the thermal strains are different in the two directions. However, no thermal stresses develop unless the lamina is restrained from deformation. In most practical composite structures, a number of laminae are stacked in different directions to make a laminate. Owing to the presence of bond between the laminae, the individual laminae in a laminate, when subjected to a temperature change, restrain each other from free thermal deformations and thermal stresses develop.

The next major environmental factor to be considered in the constitutive modeling is moisture absorption. Most PMCs can absorb or deabsorb moisture. Owing to moisture absorption, the composite swells. Similar to the thermal strains, owing to mismatch in the CMEs of a unidirectional lamina, the swelling strains are different in the longitudinal and transverse directions and stresses can develop in a laminate with laminae at different angles.

Hygrothermal effects on a laminate shall be discussed in Chapter 5. Here, we discuss the effects on laminae—a specially orthotropic lamina and a generally orthotropic lamina.

4.7.1 Hygrothermal Effects in Specially Orthotropic Lamina

As we just mentioned above, when subjected to a change in temperature or moisture content, a unidirectional lamina undergoes different deformations in the longitudinal and transverse directions. However, the lamina does not undergo any shear deformation in the local material coordinate system. Thus, the total strains in a unidirectional lamina, subjected to in-plane loads and change in temperature as well as moisture content, are given by

$$\begin{Bmatrix} \varepsilon_{11} \\ \varepsilon_{22} \\ \gamma_{12} \end{Bmatrix} = [S] \begin{Bmatrix} \sigma_{11} \\ \sigma_{22} \\ \tau_{12} \end{Bmatrix} + \Delta T \begin{Bmatrix} \alpha_1 \\ \alpha_2 \\ 0 \end{Bmatrix} + \Delta C \begin{Bmatrix} \beta_1 \\ \beta_2 \\ 0 \end{Bmatrix} \quad (4.180)$$

where
- $[S]$ Compliance matrix given by Equations 4.6 through 4.11
- ΔT Change in temperature
- ΔC Change in moisture content per unit mass of the lamina (kg/kg)
- $\sigma_{11}, \sigma_{22}, \tau_{12}$ In-plane hygrothermal stresses in the lamina
- α_1, α_2 CTEs in the longitudinal and transverse directions, respectively
- β_1, β_2 CMEs in the longitudinal and transverse directions, respectively

Then, the stresses are obtained by inverting Equation 4.180 as

$$\begin{Bmatrix} \sigma_{11} \\ \sigma_{22} \\ \tau_{12} \end{Bmatrix} = [Q] \left(\begin{Bmatrix} \varepsilon_{11} \\ \varepsilon_{22} \\ \gamma_{12} \end{Bmatrix} - \Delta T \begin{Bmatrix} \alpha_1 \\ \alpha_2 \\ 0 \end{Bmatrix} - \Delta C \begin{Bmatrix} \beta_1 \\ \beta_2 \\ 0 \end{Bmatrix} \right) \quad (4.181)$$

where $[Q]\ (=[S]^{-1})$ is the reduced stiffness matrix given by Equations 4.15 through 4.20. When the lamina is not constrained from any deformation during hygrothermal expansion, no hygrothermal stresses develop. Thus, for an unconstrained lamina,

$$\begin{Bmatrix} \varepsilon_{11} \\ \varepsilon_{22} \\ \gamma_{12} \end{Bmatrix} = \Delta T \begin{Bmatrix} \alpha_1 \\ \alpha_2 \\ 0 \end{Bmatrix} + \Delta C \begin{Bmatrix} \beta_1 \\ \beta_2 \\ 0 \end{Bmatrix} \quad (4.182)$$

On the other hand, the hygrothermal stresses are the maximum when the lamina is fully constrained. In such a case,

$$\begin{Bmatrix} \varepsilon_{11} \\ \varepsilon_{22} \\ \gamma_{12} \end{Bmatrix} = 0 \quad (4.183)$$

and the hygrothermal stresses are

$$\begin{Bmatrix} \sigma_{11} \\ \sigma_{22} \\ \tau_{12} \end{Bmatrix} = -[Q] \left(\Delta T \begin{Bmatrix} \alpha_1 \\ \alpha_2 \\ 0 \end{Bmatrix} + \Delta C \begin{Bmatrix} \beta_1 \\ \beta_2 \\ 0 \end{Bmatrix} \right) \quad (4.184)$$

4.7.2 Hygrothermal Effects in Generally Orthotropic Lamina

A specially orthotropic lamina, when subjected to hygrothermal loading, undergoes normal deformations along the local material coordinates but not in-plane shear deformation. However, an off-axis lamina or a generally orthotropic lamina undergoes in-plane shear deformation in addition to normal deformations (Figure 4.23). Thus, the total strains in a generally orthotropic unidirectional lamina, subjected to in-plane loads and change in temperature as well as moisture content, are given by

$$\begin{Bmatrix} \varepsilon_{xx} \\ \varepsilon_{yy} \\ \gamma_{xy} \end{Bmatrix} = [\bar{S}] \begin{Bmatrix} \sigma_{xx} \\ \sigma_{yy} \\ \tau_{xy} \end{Bmatrix} + \Delta T \begin{Bmatrix} \alpha_x \\ \alpha_y \\ \alpha_{xy} \end{Bmatrix} + \Delta C \begin{Bmatrix} \beta_x \\ \beta_y \\ \beta_{xy} \end{Bmatrix} \quad (4.185)$$

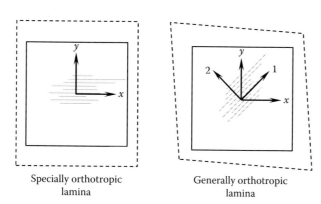

Specially orthotropic lamina   Generally orthotropic lamina

FIGURE 4.23 Hygrothermal deformations in a specially orthotropic lamina and a generally orthotropic lamina (dotted lines for deformed shapes).

where

- $[\bar{S}]$ Transformed reduced compliance matrix given by Equations 4.68 through 4.73
- ΔT Change in temperature
- ΔC Change in moisture content per unit mass of the lamina (kg/kg)
- $\sigma_{xx}, \sigma_{yy}, \tau_{xy}$ In-plane hygrothermal stresses in the global coordinate axes
- $\alpha_x, \alpha_y, \alpha_{xy}$ CTEs in the global coordinate axes
- $\beta_x, \beta_y, \beta_{xy}$ CMEs in the global coordinate axes

By inverting Equation 4.185, the global stresses are obtained as

$$\begin{Bmatrix} \sigma_{xx} \\ \sigma_{yy} \\ \tau_{xy} \end{Bmatrix} = [\bar{Q}] \left(\begin{Bmatrix} \varepsilon_{xx} \\ \varepsilon_{yy} \\ \gamma_{xy} \end{Bmatrix} - \Delta T \begin{Bmatrix} \alpha_x \\ \alpha_y \\ \alpha_{xy} \end{Bmatrix} - \Delta C \begin{Bmatrix} \beta_x \\ \beta_y \\ \beta_{xy} \end{Bmatrix} \right) \quad (4.186)$$

where $[\bar{Q}]$ is the transformed reduced stiffness matrix. The CTEs and CMEs in the global coordinates are related to those in the local coordinates in the same way as the global strains are related to the local strains. Then, these coefficients in the global coordinates are obtained by transformation from the local material coordinates as follows:

$$\begin{Bmatrix} \alpha_x \\ \alpha_y \\ \alpha_{xy}/2 \end{Bmatrix} = [T]^{-1} \begin{Bmatrix} \alpha_1 \\ \alpha_2 \\ 0 \end{Bmatrix} \quad (4.187)$$

$$\begin{Bmatrix} \beta_x \\ \beta_y \\ \beta_{xy}/2 \end{Bmatrix} = [T]^{-1} \begin{Bmatrix} \beta_1 \\ \beta_2 \\ 0 \end{Bmatrix} \quad (4.188)$$

in which the transformation matrix $[T]$ is given by Equation 4.45.

Using the Reuter's matrix, Equations 4.187 and 4.188 can be written as

$$\begin{Bmatrix} \alpha_x \\ \alpha_y \\ \alpha_{xy} \end{Bmatrix} = [R][T]^{-1}[R]^{-1} \begin{Bmatrix} \alpha_1 \\ \alpha_2 \\ 0 \end{Bmatrix} \quad (4.189)$$

$$\begin{Bmatrix} \beta_x \\ \beta_y \\ \beta_{xy} \end{Bmatrix} = [R][T]^{-1}[R]^{-1} \begin{Bmatrix} \beta_1 \\ \beta_2 \\ 0 \end{Bmatrix} \quad (4.190)$$

Similar to the case of a lamina in the material coordinates, when a specially orthotropic lamina is not constrained from any deformation during hygrothermal expansion, no hygrothermal stresses develop. Thus, for an unconstrained generally orthotropic lamina,

$$\begin{Bmatrix} \varepsilon_{xx} \\ \varepsilon_{yy} \\ \gamma_{xy} \end{Bmatrix} = \Delta T \begin{Bmatrix} \alpha_x \\ \alpha_y \\ \alpha_{xy} \end{Bmatrix} + \Delta C \begin{Bmatrix} \beta_x \\ \beta_y \\ \beta_{xy} \end{Bmatrix} \quad (4.191)$$

Macromechanics of a Lamina

and for fully constrained generally orthotropic lamina,

$$\begin{Bmatrix} \varepsilon_{xx} \\ \varepsilon_{yy} \\ \gamma_{xy} \end{Bmatrix} = 0 \qquad (4.192)$$

and

$$\begin{Bmatrix} \sigma_{xx} \\ \sigma_{yy} \\ \tau_{xy} \end{Bmatrix} = -[\bar{Q}]\left(\Delta T \begin{Bmatrix} \alpha_x \\ \alpha_y \\ \alpha_{xy} \end{Bmatrix} + \Delta C \begin{Bmatrix} \beta_x \\ \beta_y \\ \beta_{xy} \end{Bmatrix}\right) \qquad (4.193)$$

EXAMPLE 4.10

Consider a unidirectional glass/epoxy lamina with the following properties:

$\alpha_1 = 8.0 \times 10^{-6}$ m/m/°C, $\alpha_2 = 20.0 \times 10^{-6}$ m/m/°C
$\beta_1 = 0.01$ m/m/kg/kg, $\beta_2 = 0.06$ m/m/kg/kg
$E_1 = 40$ GPa, $E_2 = 6$ GPa, $\nu_{12} = 0.25$, and $G_{12} = 4$ GPa

If the lamina is subjected to a temperature rise of 50°C and it absorbs moisture @ 0.01 kg/kg, determine the hygrothermal stresses and strains. Consider the following two cases:

1. Specially orthotropic lamina fully constrained in the longitudinal direction
2. Generally orthotropic lamina at a lamina angle of 30°, fully constrained in the x-direction.

Solution

1. The reduced stiffness matrix is given by

$$[Q] = \begin{bmatrix} 40.379 & 1.514 & 0 \\ 1.514 & 6.057 & 0 \\ 0 & 0 & 4 \end{bmatrix} \text{GPa}$$

For unconstrained lamina, the hygrothermal strains are given by Equation 4.182 as follows:

$$\begin{Bmatrix} \varepsilon_{11} \\ \varepsilon_{22} \\ \gamma_{12} \end{Bmatrix} = 50 \times \begin{Bmatrix} 8.0 \\ 20.0 \\ 0 \end{Bmatrix} \times 10^{-6} + 0.01 \times \begin{Bmatrix} 0.01 \\ 0.6 \\ 0 \end{Bmatrix} = \begin{Bmatrix} 5.0 \\ 70.0 \\ 0 \end{Bmatrix} \times 10^{-4}$$

However, the lamina is constrained in the longitudinal direction. Thus, $\varepsilon_{11} = 0$ and for the given end conditions, the hygrothermal stresses are obtained as follows:

$$\begin{Bmatrix} \sigma_{11} \\ \sigma_{22} \\ \tau_{12} \end{Bmatrix} = \begin{bmatrix} 40,379 & 1514 & 0 \\ 1514 & 6057 & 0 \\ 0 & 0 & 4000 \end{bmatrix} \times \left(\begin{Bmatrix} 0 \\ 70.0 \\ 0 \end{Bmatrix} \times 10^{-4} - 50 \times \begin{Bmatrix} 8.0 \\ 20.0 \\ 0 \end{Bmatrix} \times 10^{-6} - 0.01 \times \begin{Bmatrix} 0.01 \\ 0.6 \\ 0 \end{Bmatrix} \right)$$

$$= \begin{bmatrix} 40,379 & 1514 & 0 \\ 1514 & 6057 & 0 \\ 0 & 0 & 4000 \end{bmatrix} \left(\begin{Bmatrix} -5.0 \\ 0 \\ 0 \end{Bmatrix} \times 10^{-4} \right) = - \begin{Bmatrix} 20.2 \\ 0.8 \\ 0 \end{Bmatrix} \text{MPa}$$

In the case of the generally orthotropic lamina, first let us find the reduced transformed stiffness matrix elements. Using Equation 4.55 through 4.60, we get

$$\bar{Q}_{11} = 40,379 \times \cos^4 30° + 6057 \times \sin^4 30° + 2 \times (1514 + 2 \times 4000) \\ \times \sin^2 30° \cos^2 30° = 26660 \text{ MPa}$$

$$\bar{Q}_{12} = (40,379 + 6057 - 4 \times 4000) \times \sin^2 30° \times \cos^2 30° \\ + 1514 \times (\sin^4 30° + \cos^4 30°) = 6653 \text{ MPa}$$

$$\bar{Q}_{16} = (40,379 - 1514 - 2 \times 4000) \times \sin 30° \times \cos^3 30° - (6057 - 1514 - 2 \times 4000) \\ \times \sin^3 30° \times \cos 30° = 10398 \text{ MPa}$$

$$\bar{Q}_{22} = 40,379 \times \sin^4 30° + 6057 \times \cos^4 30° + 2 \times (1514 + 2 \times 4000) \\ \times \sin^2 30° \cos^2 30° = 9499 \text{ MPa}$$

$$\bar{Q}_{26} = (40,379 - 1514 - 2 \times 4000) \times \sin^3 30° \cos 30° - (6057 - 1514 - 2 \times 4000) \\ \times \sin 30° \cos^3 30° = 4464 \text{ MPa}$$

$$\bar{Q}_{66} = (40,379 - 2 \times 1514 + 6057 - 2 \times 4000) \times \sin^2 30° \cos^2 30° + 4000 \\ \times (\sin^4 30° + \cos^4 30°) = 9139 \text{ MPa}$$

Thus, the reduced transformed stiffness matrix is

$$[\bar{Q}] = \begin{bmatrix} 26,660 & 6653 & 10,398 \\ 6653 & 9499 & 4464 \\ 10,398 & 4464 & 9139 \end{bmatrix}$$

Using Equation 4.45, for a lamina angle of 30°, the transformation matrix is obtained as

$$[T] = \begin{bmatrix} 0.75 & 0.25 & 0.866 \\ 0.25 & 0.75 & -0.866 \\ -0.433 & 0.433 & 0.5 \end{bmatrix}$$

Using Equations 4.189 and 4.190, the off-axis CTEs and CMEs are determined as follows. (Note that $[R][T]^{-1}[R]^{-1} = [T]^T$.)

$$\begin{Bmatrix} \alpha_x \\ \alpha_y \\ \alpha_{xy} \end{Bmatrix} = \begin{bmatrix} 0.75 & 0.25 & -0.433 \\ 0.25 & 0.75 & 0.433 \\ 0.866 & -0.866 & 0.5 \end{bmatrix} \begin{Bmatrix} 8.0 \\ 20.0 \\ 0 \end{Bmatrix} \times 10^{-6} = \begin{Bmatrix} 11.0 \\ 17.0 \\ -10.39 \end{Bmatrix} \times 10^{-6}$$

$$\begin{Bmatrix} \beta_x \\ \beta_y \\ \beta_{xy} \end{Bmatrix} = \begin{bmatrix} 0.75 & 0.25 & -0.433 \\ 0.25 & 0.75 & 0.433 \\ 0.866 & -0.866 & 0.5 \end{bmatrix} \begin{Bmatrix} 0.01 \\ 0.6 \\ 0 \end{Bmatrix} = \begin{Bmatrix} 0.1575 \\ 0.4525 \\ -0.5110 \end{Bmatrix}$$

Then, for an unconstrained lamina, the global hygrothermal strains would be given by

$$\begin{Bmatrix} \varepsilon_{xx} \\ \varepsilon_{yy} \\ \gamma_{xy} \end{Bmatrix} = 50 \times \begin{Bmatrix} 11.0 \\ 17.0 \\ -10.39 \end{Bmatrix} \times 10^{-6} + 0.01 \times \begin{Bmatrix} 0.1575 \\ 0.4525 \\ -0.5110 \end{Bmatrix} = \begin{Bmatrix} 2.125 \\ 5.375 \\ -5.6295 \end{Bmatrix} \times 10^{-3}$$

However, as given in the problem, the lamina is constrained in the x-direction. Thus, $\varepsilon_{xx} = 0$ and we can obtain the hygrothermal stresses as follows:

$$\begin{Bmatrix} \sigma_{xx} \\ \sigma_{yy} \\ \tau_{xy} \end{Bmatrix} = \begin{bmatrix} 26{,}660 & 6653 & 10{,}398 \\ 6653 & 9499 & 4464 \\ 10398 & 4464 & 9139 \end{bmatrix} \times \left(\begin{Bmatrix} 0.0 \\ 5.375 \\ -5.6295 \end{Bmatrix} \times 10^{-3} - 50 \times \begin{Bmatrix} 11.0 \\ 17.0 \\ -10.39 \end{Bmatrix} \times 10^{-6} - 0.01 \right.$$

$$\left. \times \begin{Bmatrix} 0.1575 \\ 0.4525 \\ -0.5110 \end{Bmatrix} \right) = \begin{bmatrix} 26{,}660 & 6653 & 10398 \\ 6653 & 9499 & 4464 \\ 10398 & 4464 & 9139 \end{bmatrix} \times \begin{Bmatrix} -2.125 \\ 0 \\ 0 \end{Bmatrix} \times 10^{-3} = \begin{Bmatrix} -56.65 \\ -14.13 \\ -22.10 \end{Bmatrix} \text{MPa}$$

Note that the generally orthotropic lamina, owing to its constraint in one direction, develops compressive hygrothermal normal stresses in both the directions and a negative in-plane shear stress. Note, further, that the specially orthotropic lamina develops only hygrothermal normal stresses but no in-plane shear stress.

4.8 SUMMARY

The behavior of a lamina is studied at a macro level in this chapter. Toward this, the lamina is introduced as a 2D ply described in terms of various gross or apparent parameters related to the geometry, stiffness, strength, and hygrothermal state of the lamina. The concepts of material coordinates and global coordinates are presented.

Constitutive relations have been developed based on plane stress idealization for specially orthotropic lamina as well as generally orthotropic lamina. For a specially orthotropic lamina, these relations involve the material elastic constants, whereas for a generally orthotropic lamina, the orientation of the lamina comes into consideration as an additional parameter.

Unlike in isotropic materials, the strengths of an orthotropic lamina are direction-dependent properties and five strength parameters can be identified. Several failure criteria involving strengths or failure strains are presented.

Among the many environmental factors, the effects of change in temperature and change in moisture content on polymeric matrix composites are the most significant. The constitutive relations are modified to take into account these environmental changes by introducing certain coefficients, viz. CTEs and CMEs.

EXERCISE PROBLEMS

4.1 Consider a specially orthotropic lamina of size 400 mm × 300 mm with the following engineering constants: $E_1 = 140$ GPa, $E_2 = 6$ GPa, $\nu_{12} = 0.25$, $\nu_{23} = 0.3$, $G_{12} = 4$ GPa, and $G_{23} = 4$ GPa. Determine the strains and deformations in the following load cases:
   a. Tensile stress of 1600 MPa in the longitudinal direction
   b. Tensile stress of 30 MPa in the transverse direction
   c. Positive shear stress of 60 MPa in the plane of the lamina
   Sketch the deformed configuration of the lamina in each of the above cases. Use 3D orthotropic constitutive relations given by Equation 4.1.

4.2 Consider the lamina in Exercise 4.1 and apply all the three load cases simultaneously. Determine the strains and deformations and compare the results with those in Exercise 4.1. What conclusion can we draw in respect of the combined strains and deformations vis-à-vis the strains and deformations when the three stresses are applied separately? Use 3D orthotropic constitutive relations given by Equation 4.1.

4.3 Determine the reduced stiffness matrix $[Q]$ of a unidirectional glass/epoxy lamina with the following material data: $E_1 = 40$ GPa, $E_2 = 8$ GPa, $\nu_{12} = 0.25$, and $G_{12} = 4$ GPa.

4.4 Solve the problem in Exercise 4.1 using plane stress idealization.

4.5 Solve the problem in Exercise 4.2 using plane stress idealization.

4.6 Consider the lamina in Exercise 4.3. If the lamina is subjected to only a uniaxial tension, determine its maximum value (maximum tensile force) so as to limit the longitudinal deformation within 8 mm.

4.7 Consider a unidirectional carbon/epoxy laminate of size 500 mm × 500 mm × 5 mm subjected to in-plane normal stresses. Determine the ratio of applied longitudinal stress to transverse stress so as to achieve equal deformation in both the directions. Assume the following data: $E_1 = 150$ GPa, $E_2 = 8$ GPa, $\nu_{12} = 0.25$, $\nu_{23} = 0.3$, and $G_{12} = 4$ GPa.

4.8 Derive the restrictions on Poisson's ratio in terms of elastic moduli, that is, show that

$$|\nu_{12}| < \sqrt{\frac{E_1}{E_2}}, \quad |\nu_{21}| < \sqrt{\frac{E_2}{E_1}}, \quad |\nu_{23}| < \sqrt{\frac{E_2}{E_3}}, \quad |\nu_{32}| < \sqrt{\frac{E_3}{E_2}},$$

$$|\nu_{31}| < \sqrt{\frac{E_3}{E_1}}, \quad \text{and} \quad |\nu_{13}| < \sqrt{\frac{E_1}{E_3}}$$

Hint: Diagonal elements of the stiffness matrix are all positive.

4.9 Determine the transformed reduced stiffness matrix $[\bar{Q}]$ of the unidirectional glass/epoxy lamina in Exercise 4.3 for each of the following ply orientations: (i) 0°, (ii) 45°, and (iii) 90°. Do you see any specialty in $[\bar{Q}]_{0°}$ and $[\bar{Q}]_{90°}$?

4.10 Write a code in MATLAB®/C/C++ for the determination of (i) reduced compliance matrix $[S]$, (ii) transformed reduced compliance matrix $[\bar{S}]$, (iii) transformation matrix $[T]$, (iv) reduced stiffness matrix $[Q]$, (v) transformed reduced stiffness matrix $[\bar{Q}]$, (vi) global stresses, (vii) global strains, (viii) local strains, and (v) local stresses. Use geometrical details of lamina, fiber orientation, and applied loads and material properties as input variables.

Hint: Refer Figure 4.7.

4.11 A generally orthotropic square lamina with fibers oriented at 30° to the global x-axis is subjected to a pure shear stress of 150 MPa. Determine the change in shape of the lamina. What happens if the sign of the applied shear stress is reversed? Draw neat sketches to indicate the changes. Assume the following material data: $E_1 = 160$ GPa, $E_2 = 10$ GPa, $\nu_{12} = 0.15$, and $G_{12} = 8$ GPa.

4.12 Consider the lamina in Exercise 4.11 and subject it to the following in-plane stresses: $\sigma_{xx} = 1000$ MPa, $\sigma_{yy} = 200$ MPa, and $\tau_{xy} = 150$ MPa. Determine the local stresses in the lamina.

4.13 Consider a unidirectional carbon/epoxy laminate of size 400 mm × 300 mm × 8 mm. The laminate is subjected to loads as shown in Figure 4.24. Determine the global strains, local strains, and local stresses. Assume the following material data: $E_1 = 150$ GPa, $E_2 = 8$ GPa, $\nu_{12} = 0.25$, $\nu_{23} = 0.3$, and $G_{12} = 4$ GPa.

4.14 Consider a globally orthotropic lamina. If the global strains are equal ($\varepsilon_{xx} = \varepsilon_{yy}$), determine the fiber orientation angle (θ) when the lamina is subjected to equal global normal stresses ($\sigma_{xx} = \sigma_{yy}$) and zero in-plane shear stress ($\tau_{xy} = 0$).

4.15 Plot the variations of elastic moduli E_x, E_y, ν_{xy}, and G_{xy} w.r.t. ply angle. Assume the following material data: $E_1 = 180$ GPa, $E_2 = 12$ GPa, $\nu_{12} = 0.2$, and $G_{12} = 8$ GPa.

4.16 Consider the unidirectional laminate in Exercise 4.13. The following strength data are given: $(\sigma_{11}^T)_{ult} = 2400$ MPa, $(\sigma_{11}^C)_{ult} = 800$ MPa, $(\sigma_{22}^T)_{ult} = 40$ MPa, $(\sigma_{22}^C)_{ult} = 200$ MPa, and $(\tau_{12})_{ult} = 75$ MPa. Check if the lamina is safe under the applied loads using the maximum stress failure criterion. Comment on the mode of failure.

4.17 Solve the problem in Exercise 4.16 using the maximum strain failure criterion. Comment on the mode of failure.

4.18 Solve the problem in Exercise 4.16 using the Tsai–Hill failure criterion. Comment on the mode of failure.

4.19 Solve the problem in Exercise 4.16 using the Tsai–Wu failure criterion. Comment on the mode of failure.

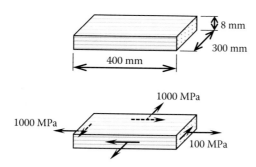

FIGURE 4.24 Unidirectional laminate (Exercise 4.13).

4.20 Consider the unidirectional laminate in Exercise 4.13 together with the strength data given in Exercise 4.16. Vary the ply orientation angle from 0° to 90° and generate the failure load versus angle plots as per (a) the maximum stress failure criterion, (b) the maximum strain failure criterion, (c) the Tsai–Hill failure criterion, and (d) the Tsai–Wu failure criterion. Mark in each of the four cases the ranges of ply angles clearly indicating the mode(s) of failure.

4.21 For a unidirectional glass/epoxy lamina with 30° ply orientation, determine the off-axis CTEs and CMEs. Assume the following data: $\alpha_1 = 8 \times 10^{-6}$ m/m/°C, $\alpha_2 = 20 \times 10^{-6}$ m/m/°C, $\beta_1 = 0.01$ m/m/kg/kg, $\beta_2 = 0.6$ m/m/kg/kg.

4.22 Consider a unidirectional glass/epoxy laminate of size 500 mm × 500 mm × 5 mm. The fibers are oriented at an angle of 30° to the x-direction. If the temperature is raised by 25°C, determine the changes in dimensions. Assume the following data: $\alpha_1 = 8 \times 10^{-6}$ m/m/°C and $\alpha_2 = 20 \times 10^{-6}$ m/m/°C

4.23 Consider the laminate in Exercise 4.22. If the laminate is subjected simultaneously to a humid environment and elevated temperature such that it absorbs 20 g of moisture and the temperature increases by 25°C, determine the changes in dimensions. Assume the following additional data: $\beta_1 = 0.01$ m/m/kg/kg and $\beta_2 = 0.6$ m/m/kg/kg

4.24 Consider the laminate in Exercise 4.22 together with the data on CMEs given in Exercise 4.23. The laminate is fully constrained in both x- and y-directions. If it is subjected to an increase in temperature of 25°C and it absorbs 20 g moisture, determine the hygrothermal stresses and strains in the laminate. Assume the following data: $\alpha_1 = 8 \times 10^{-6}$ m/m/°C, $\alpha_2 = 20 \times 10^{-6}$ m/m/°C, $\beta_1 = 0.01$ m/m/kg/kg, $\beta_2 = 0.6$ m/m/kg/kg, $E_1 = 40$ GPa, $E_2 = 6$ GPa, $\nu_{12} = 0.25$, $G_{12} = 4$ GPa, and $\rho_c = 1.45$ g/cm³.

REFERENCES AND SUGGESTED READING

1. R. M. Jones, *Mechanics of Composite Materials*, Taylor & Francis, New York, 1999.
2. B. D. Agarwal, L. J. Broutman, and K. Chandrashekhara, *Analysis and Performance of Fiber Composites*, John Wiley & Sons, New York, 2006.
3. A. K. Kaw, *Mechanics of Composite Materials*, CRC Press, Boca Raton, FL, 2006.
4. J. N. Reddy, *Mechanics of Laminated Composite Plates and Shells: Theory and Analysis*, CRC Press, Boca Raton, FL, 2004.
5. M. Mukhopadhyay, *Mechanics of Composite Materials and Structures*, Universities Press, Hyderabad, India, 2009.
6. V. V. Vasiliev and E. V. Morozov, *Mechanics and Analysis of Composite Materials*, Elsevier Science, Amsterdam, 2001.
7. B. M. Lempriere, Poisson's ratio in orthotropic materials, *AIAA Journal*, 6(11), 1968, 2226–2227.
8. R. C. Reuter, Jr., Concise property transformation relations for an anisotropic lamina, *Journal of Composite Materials*, 5, 1971, 270–272.
9. S. W. Tsai and N. J. Pagano, Invariant properties of composite materials, *Composite Materials Workshop* (S. W. Tsai, J. C. Halpin, and N. J. Pagano, eds.), St. Louis, Missouri, July 13–21, 1967, Technomic, Westport, Connecticut, 1968, pp. 233–253.
10. M. N. Nahas, Survey of failure and post-failure theories of laminated fiber reinforced composites, *Journal of Composites, Technology and Research*, 8(4), 1986, 138–153.
11. S. W. Tsai, Strength theories of filamentary structures, *Fundamental Aspects of Fiber Reinforced Plastic Composites* (R. T. Schwartz and H. S. Schwartz, eds.), Interscience, New York, 1968, Chapter 1, pp. 3–11.
12. R. Hill, *Mathematical Theory of Plasticity*, Oxford University Press, Oxford, UK, 1950.
13. S. W. Tsai and E. M. Wu, A general theory of strength for anisotropic materials, *Journal of Composite Materials*, 5(1), 1971, 58–80.
14. S. W. Tsai and H. T. Hahn, *Introduction to Composite Materials*, Technomic Publishing Company, Lancester, PA, 1980.
15. R. Byron Pipes and B. W. Cole, On the off-axis strength test for anisotropic materials, *Journal of Composite Materials*, 7, 1973, 246–256.

5

Macromechanics of a Laminate

5.1 CHAPTER ROAD MAP

We addressed a single layer or a lamina in Chapters 3 and 4. However, individual laminae are not of any direct practical use. Typically, a single lamina is very thin, for example, approximately 0.5 mm, and it is not usually capable of providing appropriate strength and stiffness required of any real-life structure. Further, a unidirectional lamina is exceptionally strong and stiff in the longitudinal direction but very poor in the transverse direction. Thus, a single lamina is not suitable in a bidirectional loading environment. As a result, it is essential, in almost all practical cases, to stack the laminae and bond them in the form of a laminate.

Analysis of a laminate—an essential part in the overall design of a product—is done at a macro level based on the gross hygro-thermo-mechanical characteristics of the laminae comprising the laminate. The objective of this chapter is to address the various aspects of laminate analysis at a macro level. Laminae can be stacked in different ways and several special laminate configurations can be identified. We shall see in this chapter that the general response of a laminate depends greatly on the type of stacking sequence. Several theories have been proposed by researchers for analysis of a laminate. A review of these theories is not intended here. Classical laminate theory (CLT) is the most popular theory in the analysis of a composite laminate and topics related to basic assumptions and restrictions, kinematics, kinetics, and constitutive relations are discussed. Brief introductory remarks are made in respect of shear deformation and layerwise theories. Hygrothermal effects on laminate behavior are discussed next; constitutive relations are suitably modified; and CTEs and CMEs are introduced. Special cases of laminate stacking sequences have great significance in laminate behavior and several such cases are addressed. Finally, the general failure behavior of a laminate is presented.

A thorough understanding of the macro-level analysis of a lamina (Chapter 4) is a prerequisite for effective assimilation of the concepts of laminate analysis discussed in this chapter. As mentioned in the road map of Chapter 4, exposure to the introductory concepts of composites (Chapter 1) and basic solid mechanics (Chapter 2), in turn, is essential whereas, familiarity with micromechanics of lamina (Chapter 3) is desirable but not essential.

5.2 PRINCIPAL NOMENCLATURE

A, B	Lame' parameters
$[A], [B], [D]$	Laminate stiffness matrices—extensional stiffness matrix, extension–bending coupling stiffness matrix, and bending stiffness matrix, respectively

$[A^*]$, $[B^*]$, $[C^*]$, $[D^*]$	Laminate compliance matrices—extensional compliance matrix, extension–bending coupling compliance matrices, and bending compliance matrix, respectively
E_1, E_2, G_{12}, ν_{12}	Material elastic constants
$\{M\}$	Vector of moment resultants
$\{M^{HT}\}$	Vector of fictitious hygrothermal moment resultants
M_{xx}, M_{yy}, M_{xy}	Bending and twisting moment per unit length, that is, moment resultants
$\{N\}$	Vector of force resultants
$\{N^{HT}\}$	Vector of fictitious hygrothermal force resultants
N_{xx}, N_{yy}, N_{xy}	Normal and shear force per unit length, that is, force resultants
$[Q]$	Reduced stiffness matrix for a lamina
$[\bar{Q}]$	Transformed reduced stiffness matrix for a lamina
R_1, R_2	Principal radii of curvature
$[R]$	Reuter matrix
$[T]$	Transformation matrix
t	Laminate thickness
u, v, w	Displacements in the x-, y-, and z-directions, respectively
u_0, v_0, w_0	Middle surface displacements
x, y, z	Cartesian coordinates
α, β, ξ	Curvilinear orthogonal coordinates
ε_{11}, ε_{22}, γ_{12}	Local strains
ε_{xx}, ε_{yy}, γ_{xy}	Global strains
$\{\varepsilon^0\}$	Vector of middle surface strains
ε_{xx}^0, ε_{yy}^0, γ_{xy}^0	Middle surface strains
$\{\kappa\}$	Vector of middle surface curvatures
κ_{xx}, κ_{yy}, κ_{xy}	Middle surface curvatures
σ_{11}, σ_{22}, τ_{12}	Local stresses
σ_{xx}, σ_{yy}, τ_{xy}	Global stresses

5.3 LAMINATE CODES

A laminate is an integral composite structural element that is made by bonding together a number of laminae. (Lamina, plural laminae, is a term more common in the context of mechanics of laminated composites. The term ply is more common in the context of processing of composites. We shall use both these terms as synonymous to each other.) Figure 5.1 shows a schematic representation of a laminate made up of six unidirectional laminae. Note that details of the laminae, other than the angle of orientation, have not been provided. Obviously, more details, such as ply thickness, ply material (carbon/epoxy, glass/epoxy, etc.), ply type (unidirectional, bidirectional, etc.), ply sequence, etc. are needed for complete description of a laminate. A number of ply combinations are possible that gives us different types of laminate configurations. Laminate description is generally given using codes. The usual practice for writing a laminate code is to write the angles separated by slashes inside a pair of square brackets. For example, let us consider the laminate shown below:

0°
90°
0°
90°

Macromechanics of a Laminate

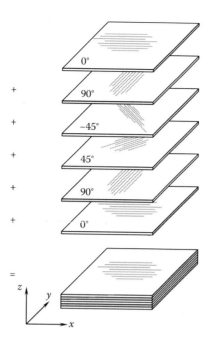

FIGURE 5.1 Schematic representation of a laminate with exploded views of the laminae.

The topmost ply is at 0° w.r.t. the global *x*-direction and the subsequent plies are as shown. The code for this laminate would be

$$[0°/90°/0°/90°]$$

The code above indicates only the lamina angles and nothing else. In such a case, material system and lamina thickness and lamina type are generally the same for all the laminae. Also, these lamina details are required to be mentioned separately. In the case of hybrid laminates, it is convenient to provide details in the code itself. For example, in the above laminate, if the material systems for the 0° and 90° plies are carbon/epoxy and glass/epoxy, respectively, we can write the laminate code as follows:

$$[0°_{CE}/90°_{GE}/0°_{CE}/90°_{GE}]$$

We can introduce more generalizations. For example, if the thicknesses of carbon/epoxy and glass/epoxy plies are 0.5 and 0.6 mm, respectively, the code can be written as follows:

$$[0°_{(CE,0.5)}/90°_{(GE,0.6)}/0°_{(CE,0.5)}/90°_{(GE,0.6)}]$$

Now, let us consider the following laminate:

45°
45°
−45°
−45°
45°
45°

The code for this laminate can be

$$[45°/45°/-45°/-45°/45°/45°]$$

Ply sequence as above containing adjacent laminae with the same angle can be described in simplified code as follows:

$$[45°_2/-45°_2/45°_2]$$

In the code above, the suffix 2 indicates that the corresponding lamina repeats twice. We also notice that the laminate is symmetric w.r.t. the midplane. Such laminate symmetry can be utilized and the code can be further simplified as follows:

$$[45°_2/-45°]_s$$

The suffix s indicates that the laminate is symmetric w.r.t. the midplane. In the above example of a symmetric laminate the number of plies is six, which is even and in such a case the midplane is between two plies. If the number of plies in a symmetric laminate is odd, the midplane passes through the mid-ply and a bar is put on the mid-ply. For example, consider the following laminate:

90°
45°
−45°
60°
−45°
45°
90°

This laminate has an odd number of plies; it is symmetric and the midplane passes through the 60° ply. The code for this laminate is

$$[90°/45°/-45°/\overline{60°}]_s$$

5.4 CLASSIFICATION OF LAMINATE ANALYSIS THEORIES

A number of theories have been proposed for analysis of laminated composite plates and shells (see, for instance, References 1 and 2 for a review of these theories). Broadly, these theories can be classified into two classes—2D theories and 3D theories. This classification can be stated as follows:

1. 2D equivalent single-layer theories
    a. Classical laminate theory
        i. Classical laminated plate theory (CLPT)
        ii. Classical laminated shell theory (CLST)
    b. Shear deformation theory
        i. First-order shear deformation theory (FSDT)
        ii. Third-order shear deformation theory (TSDT)

2. 3D elasticity theories
   a. Classical 3D elasticity formulations
   b. Layerwise theories

A large number of laminated composite structures are thin wherein the length and breadth dimensions are far larger than the thickness dimension. These structures can be either plates or shells. Analysis of such a thin structure is grossly simplified by reducing a 3D problem into a 2D one. In the first category of laminate analysis theories, a 2D approach is adopted. Invariably suitable assumptions are made in respect of the kinematic deformation or state of stress through the thickness of the laminate. CLT, perhaps the most popular composite laminate analysis theory, belongs to the 2D approach. CLT for laminated composite plates is called the CLPT. In the case of shells, it takes marginally modified shape and is called the CLST. The other major class of laminate analysis theories in the 2D approach is the shear deformation theories.

The second approach is the 3D approach, in which there are two major classes. The first one is the classical 3D elasticity theories, in which 3D kinematic deformations and equilibrium equations are considered. Similarly, constitutive relations are also in three dimensions. In layerwise theories, the laminate is described in terms of a number of mathematical layers with discrete layerwise displacement fields.

5.5 CLASSICAL LAMINATED PLATE THEORY

5.5.1 Basic Assumptions

CLPT is an extension of the classical plate theory of isotropic materials to laminated composite materials. It is assumed that *Kirchhoff hypothesis* for plates is applicable. In addition to Kirchhoff hypothesis, several other assumptions are made and similar restrictions are placed in the formulation of CLPT. They are tabulated in Table 5.1.

TABLE 5.1
Assumptions and Restrictions in CLPT

Description	Remarks/Implications
1. The plies are perfectly bonded together with infinitely thin bond	Assumption The plies do not slip over each other and displacements and strains are continuous across the interfaces between plies
2. The laminate is thin, that is, the laminate thickness is much smaller than the other two dimensions. Also, the laminate is loaded only in its plane	Restriction Plane stress condition holds
3. Straight lines perpendicular to the middle surface of the laminate before deformation (i.e., transverse normals) remain straight and perpendicular to the middle surface	Kirchhoff hypothesis In-plane displacements u and v are linear functions of z coordinate Transverse shear strains are negligible, that is, $\gamma_{xz} = \gamma_{yz} = 0$
4. The transverse normals do not undergo any change in lengths	Kirchhoff hypothesis Normal strain in the thickness direction is zero ($\varepsilon_{zz} = 0$)
5. The strains and displacements are small	Restriction $\varepsilon_{xx}, \varepsilon_{yy}, \gamma_{xy} \ll 1$
6. Each ply is of uniform thickness	Restriction
7. The material of each ply is homogeneous, orthotropic, and linearly elastic	Restriction

Note: The terms hypothesis, assumption, and restriction have similar meanings and often they are used loosely as synonymous to each other. This may not matter as, for all practical purposes, the final mathematical formulations remain unaffected. However, to have a proper understanding of the subject, it is important to make distinctions between these terms. Hypothesis is a proposition. It is a statement or a collection of statements that is put forward to explain a phenomenon. A hypothesis becomes a theory when it is proved by some means. An assumption is something that is believed to be true. It is related to something unknown and it is required for the development of a mathematical model. An assumption in engineering cannot be arbitrary and there should be reasons for assuming something. On the other hand, a restriction is a bounding condition within which a theory is valid; it is not essential for the development of the theory.

5.5.2 Kinematics of CLPT: Strain–Displacement Relations

Assumption 1 in Table 5.1 forms a base for analysis of a laminate. It is further assumed that the bond is not shear deformable. Thus, the plies do not slip over one another and the strains and displacements are continuous across the lamina interfaces. As a consequence, the laminae act as a laminate.

Let us now consider a laminate in the Cartesian coordinate system (Figure 5.2). The deformed and undeformed configurations of a transverse normal in the xz-plane are shown. The origin of the coordinate system is on the middle surface of the laminate such that $z = 0$ at any point on the middle surface. The displacements along x-, y-, and z-directions of any point in the laminate are u, v, and w, respectively. (These displacements are denoted by u_x, u_y, and u_z, respectively, in Chapter 2.) We use a subscript nought to indicate middle surface displacements. Thus, u_0, v_0, and w_0 are the middle surface displacements. Now, under the restriction that the laminate is thin, we assume that Kirchhoff hypothesis holds. Let us consider a transverse normal, that is, a straight line perpendicular to the middle surface of the laminate in the undeformed configuration (line segment AB). As per our assumption 3, the transverse normal remains straight and perpendicular to the middle surface in the deformed configuration. Thus, we can conclude that the in-plane displacements u and v are linear functions of z. Then, as shown in the figure, in the xz-plane, the displacement along x-direction of a point P on the transverse normal is given by

$$u_P = u_0 - \alpha z_P \tag{5.1}$$

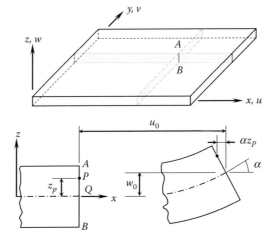

FIGURE 5.2 Geometry of deformation of a transverse normal in the xz-plane.

Macromechanics of a Laminate

where α is the slope of the laminate middle surface in the x-direction and it is given by

$$\alpha = \frac{\partial w_0}{\partial x} \tag{5.2}$$

Thus, the displacement u at any point at a distance z from the middle surface in the laminate is given by

$$u = u_0 - z\frac{\partial w_0}{\partial x} \tag{5.3}$$

Following similar considerations in the yz-plane, we can relate the displacement v at any point to that of the middle surface as function of z. On the other hand, the displacement w is independent of z. Thus, the total displacement field at any point (x, y, z) is expressed in terms of the middle surface displacements as follows:

$$\begin{Bmatrix} u(x,y,z) \\ v(x,y,z) \\ w(x,y,z) \end{Bmatrix} = \begin{Bmatrix} u_0(x,y) - z\dfrac{\partial w_0(x,y)}{\partial x} \\ v_0(x,y) - z\dfrac{\partial w_0(x,y)}{\partial y} \\ w_0(x,y) \end{Bmatrix} \tag{5.4}$$

or, simply,

$$\begin{Bmatrix} u \\ v \\ w \end{Bmatrix} = \begin{Bmatrix} u_0 - z\dfrac{\partial w_0}{\partial x} \\ v_0 - z\dfrac{\partial w_0}{\partial y} \\ w_0 \end{Bmatrix} \tag{5.5}$$

Now, let us go back to assumption 3. Since the transverse normal remains straight and perpendicular to the middle surface in the deformed configuration, the shear strains in the xz- and yz-planes (i.e., the planes perpendicular to the middle surface of the laminate) are zero. Also, the transverse normal remains unchanged in length. Accordingly, normal strain in the z-direction vanishes. Thus,

$$\gamma_{xz} = \gamma_{yz} = \varepsilon_{zz} = 0 \tag{5.6}$$

Thus, the nonzero laminate strains are reduced to ε_{xx}, ε_{yy}, and γ_{xy}. Under the restriction 5, these strains are small and we can use the small strain expressions, presented in Chapter 2, as follows:

$$\varepsilon_{xx} = \frac{\partial u}{\partial x} \tag{5.7}$$

$$\varepsilon_{yy} = \frac{\partial v}{\partial y} \tag{5.8}$$

$$\gamma_{xy} = \frac{\partial u}{\partial y} + \frac{\partial v}{\partial x} \tag{5.9}$$

Then, using Equations 5.3 and 5.5, we get the following:

$$\varepsilon_{xx} = \frac{\partial u_0}{\partial x} - z\frac{\partial^2 w_0}{\partial x^2} \tag{5.10}$$

$$\varepsilon_{yy} = \frac{\partial v_0}{\partial y} - z\frac{\partial^2 w_0}{\partial y^2} \tag{5.11}$$

$$\gamma_{xy} = \frac{\partial u_0}{\partial y} + \frac{\partial v_0}{\partial x} - 2z\frac{\partial^2 w_0}{\partial x \partial y} \tag{5.12}$$

The strain–displacement relations given above can be expressed in the matrix form as follows:

$$\begin{Bmatrix}\varepsilon_{xx}\\ \varepsilon_{yy}\\ \gamma_{xy}\end{Bmatrix} = \begin{Bmatrix}\dfrac{\partial u_0}{\partial x}\\ \dfrac{\partial v_0}{\partial y}\\ \dfrac{\partial u_0}{\partial y}+\dfrac{\partial v_0}{\partial x}\end{Bmatrix} + z\begin{Bmatrix}-\dfrac{\partial^2 w_0}{\partial x^2}\\ -\dfrac{\partial^2 w_0}{\partial y^2}\\ -2\dfrac{\partial^2 w_0}{\partial x \partial y}\end{Bmatrix} \tag{5.13}$$

which is further simplified as follows:

$$\begin{Bmatrix}\varepsilon_{xx}\\ \varepsilon_{yy}\\ \gamma_{xy}\end{Bmatrix} = \begin{Bmatrix}\varepsilon_{xx}^0\\ \varepsilon_{yy}^0\\ \gamma_{xy}^0\end{Bmatrix} + z\begin{Bmatrix}\kappa_{xx}\\ \kappa_{yy}\\ \kappa_{xy}\end{Bmatrix} \tag{5.14}$$

where

$$\begin{Bmatrix}\varepsilon_{xx}^0\\ \varepsilon_{yy}^0\\ \gamma_{xy}^0\end{Bmatrix} = \begin{Bmatrix}\dfrac{\partial u_0}{\partial x}\\ \dfrac{\partial v_0}{\partial y}\\ \dfrac{\partial u_0}{\partial y}+\dfrac{\partial v_0}{\partial x}\end{Bmatrix} \tag{5.15}$$

are the middle surface strains and

$$\begin{Bmatrix}\kappa_{xx}\\ \kappa_{yy}\\ \kappa_{xy}\end{Bmatrix} = \begin{Bmatrix}-\dfrac{\partial^2 w_0}{\partial x^2}\\ -\dfrac{\partial^2 w_0}{\partial y^2}\\ -2\dfrac{\partial^2 w_0}{\partial x \partial y}\end{Bmatrix} \tag{5.16}$$

Macromechanics of a Laminate

are the middle surface curvatures. Equation 5.14 is the strain–displacement relation for a laminated plate. It relates the strains at any point in the laminate to the middle surface strains and curvatures. The middle surface strains and curvatures, in turn, are related to the middle surface displacements by Equations 5.15 and 5.16.

5.5.3 Kinetics of CLPT: Force and Moment Resultants

Let us go back to the assumptions and restrictions in Table 5.1. As mentioned therein, the laminate is loaded only in its plane. The resultant force on any cross section can be obtained by integrating the stresses over the surface. (Similar is the case with the moments.) However, instead of the resultant force (and moment) over the complete surface, it is more convenient to consider the resultant force (and moment) per unit length of the cross section. These are referred to as the force and moment resultants (Figure 5.3) and the nomenclature followed is

N_{xx}, N_{yy}: normal forces per unit length
N_{xy}: in-plane shear force per unit length
M_{xx}, M_{yy}: bending moments per unit length
M_{xy}: twisting moment per unit length

Force resultants are obtained by integrating the stresses, whereas the moment resultants are obtained by integrating the products of stresses and corresponding z-coordinates. In both the cases, integration is done across the thickness. Thus,

$$N_{xx} = \int_{-h/2}^{h/2} \sigma_{xx} dz \tag{5.17}$$

$$N_{yy} = \int_{-h/2}^{h/2} \sigma_{yy} dz \tag{5.18}$$

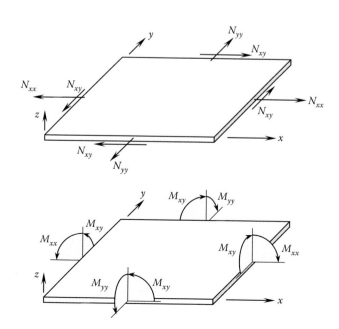

FIGURE 5.3 Force and moment resultants on a plate. (Adapted with permission from R. M. Jones, *Mechanics of Composite Materials*, second edition, Taylor & Francis, New York, 1999; A. K. Kaw, *Mechanics of Composite Materials*, CRC Press, Boca Raton, FL, 2006.)

$$N_{xy} = \int_{-h/2}^{h/2} \tau_{xy} dz \qquad (5.19)$$

$$M_{xx} = \int_{-h/2}^{h/2} z\sigma_{xx} dz \qquad (5.20)$$

$$M_{yy} = \int_{-h/2}^{h/2} z\sigma_{yy} dz \qquad (5.21)$$

$$M_{xy} = \int_{-h/2}^{h/2} z\tau_{xy} dz \qquad (5.22)$$

In the above expressions for force and moment resultants, h is the thickness of the laminate. Note that the integration is done across the total thickness of the laminate. However, we shall see below that stresses are stepwise continuous functions of z and the integrations indicated above are actually done for each lamina and summed up across all the laminae in a laminate. To see this, let us go back to the constitutive relations for a lamina.

The global stresses in each lamina are given by Equation 4.53 as follows:

$$\begin{Bmatrix} \sigma_{xx} \\ \sigma_{yy} \\ \tau_{xy} \end{Bmatrix} = [\bar{Q}] \begin{Bmatrix} \varepsilon_{xx} \\ \varepsilon_{yy} \\ \gamma_{xy} \end{Bmatrix} \qquad (5.23)$$

Substituting Equation 5.14, we get

$$\begin{Bmatrix} \sigma_{xx} \\ \sigma_{yy} \\ \tau_{xy} \end{Bmatrix} = [\bar{Q}] \begin{Bmatrix} \varepsilon_{xx}^0 \\ \varepsilon_{yy}^0 \\ \gamma_{xy}^0 \end{Bmatrix} + z[\bar{Q}] \begin{Bmatrix} \kappa_{xx} \\ \kappa_{yy} \\ \kappa_{xy} \end{Bmatrix} \qquad (5.24)$$

Equation 5.24 gives the global stresses at any point in the laminate. $[\bar{Q}]$ is the reduced transformed stiffness matrix of the lamina at the point, where the stresses are being determined, is located. For a given loading on a laminate, following points may be noted:

- Middle surface strains and curvatures are functions of x and y (Equations 5.15 and 5.16).
- For a given set of values of x and y, strains are linear functions of z alone (Equation 5.14). Thus, the strains vary linearly through the thickness of the whole laminate.
- For a given set of values of x and y, stresses are linear functions of z and they depend on $[\bar{Q}]$ of the corresponding lamina (Equation 5.23). Thus, the stresses vary linearly through the thickness of a lamina. However, the variations of stresses in different laminae within the same laminate are likely to be different. In other words, the stresses are step-linear functions of z.

At this point, let us consider the following example:

EXAMPLE 5.1

Consider the glass/epoxy laminate [90°/0°/90°] shown in Figure 5.4. Each ply is of equal thickness and the following material properties are given

$$E_1 = 40 \text{ GPa}, E_2 = 6 \text{ GPa}, \nu_{12} = 0.25, \text{ and } G_{12} = 4 \text{ GPa}$$

The variation of normal strain in the x-direction is shown in the same figure. If all other strains are zero, determine the normal stresses in the x-direction.

Solution

The reduced stiffness matrix (see Chapter 4) is given by

$$[Q] = \begin{bmatrix} \dfrac{40}{1-0.25\times 0.0375} & \dfrac{0.25\times 6}{1-0.25\times 0.0375} & 0 \\ \dfrac{0.25\times 6}{1-0.25\times 0.0375} & \dfrac{6}{1-0.25\times 0.0375} & 0 \\ 0 & 0 & 4 \end{bmatrix} = \begin{bmatrix} 40.379 & 1.514 & 0 \\ 1.514 & 6.057 & 0 \\ 0 & 0 & 4 \end{bmatrix} \text{GPa}$$

Noting that $\sin 0° = \cos 90° = 0$ and, $\sin 90° = \cos 0° = 1$ the transformed reduced stiffness matrices (see Chapter 4) can be readily obtained as

$$[\bar{Q}]_{0°} = \begin{bmatrix} 40.379 & 1.514 & 0 \\ 1.514 & 6.057 & 0 \\ 0 & 0 & 4 \end{bmatrix} \text{GPa}$$

and

$$[\bar{Q}]_{90°} = \begin{bmatrix} 6.057 & 1.514 & 0 \\ 1.514 & 40.379 & 0 \\ 0 & 0 & 4 \end{bmatrix} \text{GPa}$$

The global stresses in a ply are given by

$$\begin{Bmatrix} \sigma_{xx} \\ \sigma_{yy} \\ \tau_{xy} \end{Bmatrix} = [\bar{Q}] \begin{Bmatrix} \varepsilon_{xx} \\ \varepsilon_{yy} \\ \gamma_{xy} \end{Bmatrix}$$

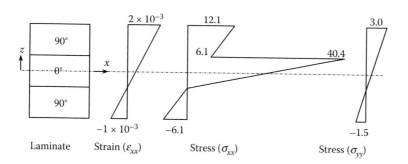

FIGURE 5.4 Typical stress variations in a laminate (Example 5.1).

Given that $\varepsilon_{yy} = \gamma_{xy} = 0$ and the variation of ε_{xx} as shown in the figure, the stresses can be readily determined.

Top 90° ply (top point):

$$\begin{Bmatrix} \sigma_{xx} \\ \sigma_{yy} \\ \tau_{xy} \end{Bmatrix} = \begin{bmatrix} 6057 & 1514 & 0 \\ 1514 & 40{,}379 & 0 \\ 0 & 0 & 4 \end{bmatrix} \begin{Bmatrix} 2 \times 10^{-3} \\ 0 \\ 0 \end{Bmatrix} = \begin{Bmatrix} 12.1 \\ 3.0 \\ 0 \end{Bmatrix} \text{MPa}$$

Top 90° ply (bottom point):

$$\begin{Bmatrix} \sigma_{xx} \\ \sigma_{yy} \\ \tau_{xy} \end{Bmatrix} = \begin{bmatrix} 6057 & 1514 & 0 \\ 1514 & 40{,}379 & 0 \\ 0 & 0 & 4 \end{bmatrix} \begin{Bmatrix} 1 \times 10^{-3} \\ 0 \\ 0 \end{Bmatrix} = \begin{Bmatrix} 6.1 \\ 1.5 \\ 0 \end{Bmatrix} \text{MPa}$$

0° ply (top point):

$$\begin{Bmatrix} \sigma_{xx} \\ \sigma_{yy} \\ \tau_{xy} \end{Bmatrix} = \begin{bmatrix} 40{,}379 & 1514 & 0 \\ 1514 & 6057 & 0 \\ 0 & 0 & 4 \end{bmatrix} \begin{Bmatrix} 1 \times 10^{-3} \\ 0 \\ 0 \end{Bmatrix} = \begin{Bmatrix} 40.4 \\ 1.5 \\ 0 \end{Bmatrix} \text{MPa}$$

0° ply (bottom point):

$$\begin{Bmatrix} \sigma_{xx} \\ \sigma_{yy} \\ \tau_{xy} \end{Bmatrix} = \begin{bmatrix} 40{,}379 & 1514 & 0 \\ 1514 & 6057 & 0 \\ 0 & 0 & 4 \end{bmatrix} \begin{Bmatrix} 0 \\ 0 \\ 0 \end{Bmatrix} = \begin{Bmatrix} 0 \\ 0 \\ 0 \end{Bmatrix}$$

Bottom 90° ply (top point):

$$\begin{Bmatrix} \sigma_{xx} \\ \sigma_{yy} \\ \tau_{xy} \end{Bmatrix} = \begin{bmatrix} 6057 & 1514 & 0 \\ 1514 & 40{,}379 & 0 \\ 0 & 0 & 4 \end{bmatrix} \begin{Bmatrix} 0 \\ 0 \\ 0 \end{Bmatrix} = \begin{Bmatrix} 0 \\ 0 \\ 0 \end{Bmatrix}$$

Bottom 90° ply (bottom point):

$$\begin{Bmatrix} \sigma_{xx} \\ \sigma_{yy} \\ \tau_{xy} \end{Bmatrix} = \begin{bmatrix} 6057 & 1514 & 0 \\ 1514 & 40{,}379 & 0 \\ 0 & 0 & 4 \end{bmatrix} \begin{Bmatrix} -1 \times 10^{-3} \\ 0 \\ 0 \end{Bmatrix} = \begin{Bmatrix} -6.1 \\ -1.5 \\ 0 \end{Bmatrix} \text{MPa}$$

Variations of σ_{xx} and σ_{yy} are shown in Figure 5.4. (Note that τ_{xy} is zero everywhere.)

From the above example and the preceding discussions, it is clear that the integrations involved in Equations 5.17 through 5.22 have to be stepwise. Thus, stresses are integrated for each lamina and summed up for all the laminae in a laminate to get the force and moment resultants. Toward this, a coordinate system as shown in Figure 5.5 is adopted and the expressions for force and moment resultants are obtained as follows:

Macromechanics of a Laminate

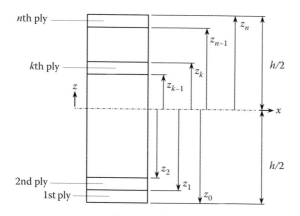

FIGURE 5.5 Coordinates of plies in a laminate.

$$N_{xx} = \sum_{k=1}^{n} \int_{z_{k-1}}^{z_k} \sigma_{xx} dz \qquad (5.25)$$

$$N_{yy} = \sum_{k=1}^{n} \int_{z_{k-1}}^{z_k} \sigma_{yy} dz \qquad (5.26)$$

$$N_{xy} = \sum_{k=1}^{n} \int_{z_{k-1}}^{z_k} \tau_{xy} dz \qquad (5.27)$$

$$M_{xx} = \sum_{k=1}^{n} \int_{z_{k-1}}^{z_k} z\sigma_{xx} dz \qquad (5.28)$$

$$M_{yy} = \sum_{k=1}^{n} \int_{z_{k-1}}^{z_k} z\sigma_{yy} dz \qquad (5.29)$$

$$M_{xy} = \sum_{k=1}^{n} \int_{z_{k-1}}^{z_k} z\tau_{xy} dz \qquad (5.30)$$

Equations 5.25 through 5.30 can be expressed in the matrix form as follows:

$$\begin{Bmatrix} N_{xx} \\ N_{yy} \\ N_{xy} \end{Bmatrix} = \sum_{k=1}^{n} \int_{z_{k-1}}^{z_k} \begin{Bmatrix} \sigma_{xx} \\ \sigma_{yy} \\ \tau_{xy} \end{Bmatrix} dz \qquad (5.31)$$

$$\begin{Bmatrix} M_{xx} \\ M_{yy} \\ M_{xy} \end{Bmatrix} = \sum_{k=1}^{n} \int_{z_{k-1}}^{z_k} z \begin{Bmatrix} \sigma_{xx} \\ \sigma_{yy} \\ \tau_{xy} \end{Bmatrix} dz \qquad (5.32)$$

5.5.4 Constitutive Relations in CLPT

In this section, we relate the force and moment resultants to the strains. Let us substitute Equation 5.24 in Equations 5.31 and 5.32. Then,

$$\begin{Bmatrix} N_{xx} \\ N_{yy} \\ N_{xy} \end{Bmatrix} = \sum_{k=1}^{n} \int_{z_{k-1}}^{z_k} [\bar{Q}]_k \begin{Bmatrix} \varepsilon_{xx}^0 \\ \varepsilon_{yy}^0 \\ \gamma_{xy}^0 \end{Bmatrix} dz + \sum_{k=1}^{n} \int_{z_{k-1}}^{z_k} z [\bar{Q}]_k \begin{Bmatrix} \kappa_{xx} \\ \kappa_{yy} \\ \kappa_{xy} \end{Bmatrix} dz \qquad (5.33)$$

$$\begin{Bmatrix} M_{xx} \\ M_{yy} \\ M_{xy} \end{Bmatrix} = \sum_{k=1}^{n} \int_{z_{k-1}}^{z_k} z [\bar{Q}]_k \begin{Bmatrix} \varepsilon_{xx}^0 \\ \varepsilon_{yy}^0 \\ \gamma_{xy}^0 \end{Bmatrix} dz + \sum_{k=1}^{n} \int_{z_{k-1}}^{z_k} z^2 [\bar{Q}]_k \begin{Bmatrix} \kappa_{xx} \\ \kappa_{yy} \\ \kappa_{xy} \end{Bmatrix} dz \qquad (5.34)$$

In the equations above, middle surface strains and curvatures are independent of z and $[\bar{Q}]_k$ is constant for each lamina. (The suffix k is put to indicate that $[\bar{Q}]_k$ is associated with the kth ply.) Thus, the middle surface strains and curvatures and the reduced transformed stiffness matrix can be brought outside the integration and Equations 5.33 and 5.34 can be written as

$$\begin{Bmatrix} N_{xx} \\ N_{yy} \\ N_{xy} \end{Bmatrix} = \left(\sum_{k=1}^{n} [\bar{Q}]_k \int_{z_{k-1}}^{z_k} dz \right) \begin{Bmatrix} \varepsilon_{xx}^0 \\ \varepsilon_{yy}^0 \\ \gamma_{xy}^0 \end{Bmatrix} + \left(\sum_{k=1}^{n} [\bar{Q}]_k \int_{z_{k-1}}^{z_k} z\,dz \right) \begin{Bmatrix} \kappa_{xx} \\ \kappa_{yy} \\ \kappa_{xy} \end{Bmatrix} \qquad (5.35)$$

$$\begin{Bmatrix} M_{xx} \\ M_{yy} \\ M_{xy} \end{Bmatrix} = \left(\sum_{k=1}^{n} [\bar{Q}]_k \int_{z_{k-1}}^{z_k} z\,dz \right) \begin{Bmatrix} \varepsilon_{xx}^0 \\ \varepsilon_{yy}^0 \\ \gamma_{xy}^0 \end{Bmatrix} + \left(\sum_{k=1}^{n} [\bar{Q}]_k \int_{z_{k-1}}^{z_k} z^2\,dz \right) \begin{Bmatrix} \kappa_{xx} \\ \kappa_{yy} \\ \kappa_{xy} \end{Bmatrix} \qquad (5.36)$$

The terms inside the small brackets in the above equations are the laminate stiffness matrices and we can write

$$\begin{Bmatrix} N_{xx} \\ N_{yy} \\ N_{xy} \end{Bmatrix} = [A] \begin{Bmatrix} \varepsilon_{xx}^0 \\ \varepsilon_{yy}^0 \\ \gamma_{xy}^0 \end{Bmatrix} + [B] \begin{Bmatrix} \kappa_{xx} \\ \kappa_{yy} \\ \kappa_{xy} \end{Bmatrix} \qquad (5.37)$$

$$\begin{Bmatrix} M_{xx} \\ M_{yy} \\ M_{xy} \end{Bmatrix} = [B] \begin{Bmatrix} \varepsilon_{xx}^0 \\ \varepsilon_{yy}^0 \\ \gamma_{xy}^0 \end{Bmatrix} + [D] \begin{Bmatrix} \kappa_{xx} \\ \kappa_{yy} \\ \kappa_{xy} \end{Bmatrix} \qquad (5.38)$$

where

$$[A] = \sum_{k=1}^{n} [\bar{Q}]_k \int_{z_{k-1}}^{z_k} dz = \sum_{k=1}^{n} [\bar{Q}]_k (z_k - z_{k-1}) \qquad (5.39)$$

Macromechanics of a Laminate

$$[B] = \sum_{k=1}^{n} [\bar{Q}]_k \int_{z_{k-1}}^{z_k} z\, dz = \frac{1}{2} \sum_{k=1}^{n} [\bar{Q}]_k \left(z_k^2 - z_{k-1}^2\right) \qquad (5.40)$$

$$[D] = \sum_{k=1}^{n} [\bar{Q}]_k \int_{z_{k-1}}^{z_k} z^2\, dz = \frac{1}{3} \sum_{k=1}^{n} [\bar{Q}]_k \left(z_k^3 - z_{k-1}^3\right) \qquad (5.41)$$

Equations 5.37 and 5.38 can be combined and we can write in the explicit matrix notation as follows:

$$\begin{Bmatrix} N_{xx} \\ N_{yy} \\ N_{xy} \\ M_{xx} \\ M_{yy} \\ M_{xy} \end{Bmatrix} = \begin{bmatrix} A_{11} & A_{12} & A_{16} & B_{11} & B_{12} & B_{16} \\ A_{12} & A_{22} & A_{26} & B_{12} & B_{22} & B_{26} \\ A_{16} & A_{26} & A_{66} & B_{16} & B_{26} & B_{66} \\ B_{11} & B_{12} & B_{16} & D_{11} & D_{12} & D_{16} \\ B_{12} & B_{22} & B_{26} & D_{12} & D_{22} & D_{26} \\ B_{16} & B_{26} & B_{66} & D_{16} & D_{26} & D_{66} \end{bmatrix} \begin{Bmatrix} \varepsilon_{xx}^0 \\ \varepsilon_{yy}^0 \\ \gamma_{xy}^0 \\ \kappa_{xx} \\ \kappa_{yy} \\ \kappa_{xy} \end{Bmatrix} \qquad (5.42)$$

In a more contracted form, Equation 5.42 can be written as

$$\begin{Bmatrix} N \\ M \end{Bmatrix} = \begin{bmatrix} A & B \\ B & D \end{bmatrix} \begin{Bmatrix} \varepsilon^0 \\ \kappa \end{Bmatrix} \qquad (5.43)$$

where

$$\{N\} = \begin{Bmatrix} N_{xx} \\ N_{yy} \\ N_{xy} \end{Bmatrix} \qquad (5.44)$$

$$\{M\} = \begin{Bmatrix} M_{xx} \\ M_{yy} \\ M_{xy} \end{Bmatrix} \qquad (5.45)$$

$$[A] = \begin{bmatrix} A_{11} & A_{12} & A_{16} \\ A_{12} & A_{22} & A_{26} \\ A_{16} & A_{26} & A_{66} \end{bmatrix} \qquad (5.46)$$

$$[B] = \begin{bmatrix} B_{11} & B_{12} & B_{16} \\ B_{12} & B_{22} & B_{26} \\ B_{16} & B_{26} & B_{66} \end{bmatrix} \qquad (5.47)$$

$$[D] = \begin{bmatrix} D_{11} & D_{12} & D_{16} \\ D_{12} & D_{22} & D_{26} \\ D_{16} & D_{26} & D_{66} \end{bmatrix} \qquad (5.48)$$

$$\{\varepsilon^0\} = \begin{Bmatrix} \varepsilon_{xx}^0 \\ \varepsilon_{yy}^0 \\ \gamma_{xy}^0 \end{Bmatrix} \tag{5.49}$$

$$\{\kappa\} = \begin{Bmatrix} \kappa_{xx} \\ \kappa_{yy} \\ \kappa_{xy} \end{Bmatrix} \tag{5.50}$$

[*A*], [*B*], and [*D*] are the laminate stiffness matrices. Note that [*A*] associates the force resultants with the middle surface strains and [*D*] associates the moment resultants with the middle surface curvatures. On the other hand, [*B*] associates force resultants with middle surface curvatures and moment resultants with middle surface strains. Thus, [*A*], [*B*], and [*D*], respectively, are called the extensional, coupling, and bending stiffness matrices of a laminate.

In mechanics of composites, we often come across problems wherein, given the force and moment resultants on a laminate, we are required to find the strains in the laminate. In such a case, the inverse of Equation 5.43 is useful. Then, we can express the middle surface strains and curvatures as follows:

$$\begin{Bmatrix} \varepsilon^0 \\ \kappa \end{Bmatrix} = \begin{bmatrix} A & B \\ B & D \end{bmatrix}^{-1} \begin{Bmatrix} N \\ M \end{Bmatrix} \tag{5.51}$$

The equation above involves inversion of a 6 × 6 matrix, which is obviously not a simple task without the use of a computer. To reduce the computational effort, the following process can be adopted.

Equation 5.43 can be split into two equations as follows:

$$\{N\} = [A]\{\varepsilon^0\} + [B]\{\kappa\} \tag{5.52}$$

$$\{M\} = [B]\{\varepsilon^0\} + [D]\{\kappa\} \tag{5.53}$$

After simple manipulation, Equation 5.52 gives us

$$\{\varepsilon^0\} = [A]^{-1}\{N\} - [A]^{-1}[B]\{\kappa\} \tag{5.54}$$

Substituting Equation 5.54 in Equation 5.53, we get

$$\{M\} = [B][A]^{-1}\{N\} - [B][A]^{-1}[B]\{\kappa\} + [D]\{\kappa\} \tag{5.55}$$

Equations 5.54 and 5.55 can be combined as follows:

$$\begin{Bmatrix} \varepsilon^0 \\ M \end{Bmatrix} = \begin{bmatrix} A' & B' \\ C' & D' \end{bmatrix} \begin{Bmatrix} N \\ \kappa \end{Bmatrix} \tag{5.56}$$

where

$$[A'] = [A]^{-1} \tag{5.57}$$

Macromechanics of a Laminate

$$[B'] = -[A]^{-1}[B] \tag{5.58}$$

$$[C'] = [B][A]^{-1} = -[B']^T \tag{5.59}$$

$$[D'] = [D] - [B][A]^{-1}[B] \tag{5.60}$$

Note that $[A']$ and $[D']$ are symmetric, whereas $[B']$ and $[C']$ are not necessarily symmetric.

Equation 5.56 actually contains six equations, of which the first three equations correspond to $\{\varepsilon^0\}$ and the remaining to $\{\kappa\}$. From the second part, we can express $\{\kappa\}$ as

$$\{\kappa\} = [D']^{-1}\{M\} - [D']^{-1}[C']\{N\} \tag{5.61}$$

Substituting Equation 5.61 in the first part of Equation 5.56, we get

$$\{\varepsilon^0\} = ([A'] - [B'][D']^{-1}[C'])\{N\} + [B'][D']^{-1}\{M\} \tag{5.62}$$

Combining Equations 5.61 and 5.62, we finally get

$$\begin{Bmatrix} \varepsilon^0 \\ \kappa \end{Bmatrix} = \begin{bmatrix} A^* & B^* \\ C^* & D^* \end{bmatrix} \begin{Bmatrix} N \\ M \end{Bmatrix} \tag{5.63}$$

where $[A^*]$, $[B^*]$, and $[D^*]$ are called the extensional, coupling, and bending compliance matrices, respectively, and is given by

$$[A^*] = [A'] - [B'][D']^{-1}[C'] \tag{5.64}$$

$$[B^*] = [B'][D']^{-1} \tag{5.65}$$

$$[C^*] = -[D']^{-1}[C'] \tag{5.66}$$

$$[D^*] = [D']^{-1} \tag{5.67}$$

Note that $\begin{bmatrix} A^* & B^* \\ C^* & D^* \end{bmatrix}$ is symmetric and thus, $[C^*] = [B^*]^T$. Note further that $[A^*]$ and $[D^*]$ are symmetric whereas $[B^*]$ and $[C^*]$ are not necessarily symmetric.

The procedure of laminate analysis might apparently look complex and tedious. However, a systematic approach (Table 5.2) can be adopted. The example below illustrates the process. (A good computing facility for matrix calculations is necessary.)

EXAMPLE 5.2

Consider the carbon/epoxy laminate [90°/60°/30°/0°] shown in Figure 5.6. Ply thicknesses and gross applied loads are given in the figure. Material properties are as follows:

$$E_1 = 125\,\text{GPa},\ E_2 = 10\,\text{GPa},\ \nu_{12} = 0.25,\ \text{and}\ G_{12} = 8\,\text{GPa}$$

Determine the local stresses and strains in each ply.

TABLE 5.2
Analysis of a Laminate

Step	Input Data
1. Determine the force and moment resultants {N} and {M}	Applied loads Laminate dimensions Equations 5.17 through 5.22
2. Determine the reduced stiffness matrix [Q] for the material(s)	Material properties—E_1, E_2, G_{12}, and ν_{12} Equations 4.16 through 4.20
3. Determine the transformed reduced stiffness matrix $[\overline{Q}]$ for the each ply	[Q] from step 2 above Ply angle Equations 4.45, 4.49, and 4.54 or Equations 4.55 through 4.60
4. Determine the z-coordinates of the plies	Details of laminate ply thicknesses Figure 5.5
5. Determine the laminate stiffness matrices—[A], [B], and [D]	$[\overline{Q}]$ from step 3 above z-coordinate details of each ply Equations 5.39 through 5.41
6. Determine the laminate compliance matrices—[A*], [B*], [C*], and [D*]	[A], [B], and [D] from step 5 above Equations 5.64 through 5.67
7. Determine the laminate middle surface strains and curvatures—{ε^0} and {κ}	[A*], [B*], [C*], and [D*] from step 6 above Equation 5.63
8. Determine the global strains	{ε^0} and {κ} from step 7 above z-coordinates from step 4 Equation 5.14
9. Determine the global stresses	ε_{xx}, ε_{yy} and γ_{xy} from step 8 above $[\overline{Q}]$ from step 3 above Equation 4.53
10. Determine the local strains in each ply	ε_{xx}, ε_{yy}, and γ_{xy} from step 8 above [T] from ply angle Equations 4.45 and 4.48
11. Determine the local stresses in each ply	{σ_{xx}}, {σ_{yy}}, and {τ_{xy}} from step 9 above [T] from ply angle Equations 4.44 and 4.45
12. Apply lamina failure criteria to each ply	ε_{11}, ε_{22}, and γ_{12} from step 10 above σ_{11}, σ_{22}, and τ_{12} from step 11 above Lamina failure criteria (Section 4.6.2)

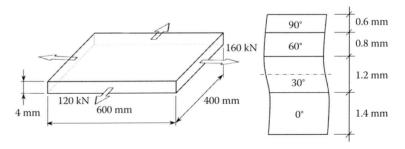

FIGURE 5.6 Analysis of a laminate (Example 5.2).

Solution

Step 1.

The force and moment resultants are given by

$$N_{xx} = \int_{-2}^{2} \left(\frac{160 \times 10^3}{400 \times 4}\right) dz = 400 \, \text{N/mm}$$

Macromechanics of a Laminate

$$N_{yy} = \int_{-2}^{2} \left(\frac{120 \times 10^3}{600 \times 4} \right) dz = 200 \, \text{N/mm}$$

$$N_{xy} = M_{xx} = M_{yy} = M_{xy} = 0$$

Step 2.

The reduced stiffness matrix for the material is given by (Equations 4.16 through 4.20)

$$[Q] = \begin{bmatrix} 125.628 & 2.513 & 0 \\ 2.513 & 10.050 & 0 \\ 0 & 0 & 8 \end{bmatrix} \times 10^3 \, \text{MPa}$$

Step 3.

The transformed reduced stiffness matrices can be readily obtained as (Equations 4.55 through 4.60)

$$[\bar{Q}]_{0°} = \begin{bmatrix} 125.628 & 2.513 & 0 \\ 2.513 & 10.050 & 0 \\ 0 & 0 & 8 \end{bmatrix} \times 10^3 \, \text{MPa}$$

$$[\bar{Q}]_{30°} = \begin{bmatrix} 78.236 & 21.010 & 35.703 \\ 21.010 & 20.447 & 14.344 \\ 35.703 & 14.344 & 26.498 \end{bmatrix} \times 10^3 \, \text{MPa}$$

$$[\bar{Q}]_{60°} = \begin{bmatrix} 20.447 & 21.010 & 14.344 \\ 21.010 & 78.236 & 35.703 \\ 14.344 & 35.703 & 26.498 \end{bmatrix} \times 10^3 \, \text{MPa}$$

$$[\bar{Q}]_{90°} = \begin{bmatrix} 10.050 & 2.513 & 0 \\ 2.513 & 125.628 & 0 \\ 0 & 0 & 8 \end{bmatrix} \times 10^3 \, \text{MPa}$$

Step 4.

z-coordinates of different plies are determined as follows (Figures 5.5 and 5.6):

$$z_0 = -2 \, \text{mm}$$

$$z_1 = -2 + 1.4 = -0.6 \, \text{mm}$$

$$z_2 = -0.6 + 1.2 = 0.6 \, \text{mm}$$

$$z_3 = 0.6 + 0.8 = 1.4 \, \text{mm}$$

$$z_4 = 1.4 + 0.6 = 2.0 \, \text{mm}$$

Step 5.

Laminate stiffness matrices are then determined as follows (Equations 5.39 through 5.41):

$$[A] = \begin{bmatrix} 125.628 & 2.513 & 0 \\ 2.513 & 10.050 & 0 \\ 0 & 0 & 8 \end{bmatrix} (2.0 - 0.6) \times 10^3$$

$$+ \begin{bmatrix} 78.236 & 21.010 & 35.703 \\ 21.010 & 20.447 & 14.344 \\ 35.703 & 14.344 & 26.498 \end{bmatrix} (0.6 + 0.6) \times 10^3$$

$$+ \begin{bmatrix} 20.447 & 21.010 & 14.344 \\ 21.010 & 78.236 & 35.703 \\ 14.344 & 35.703 & 26.498 \end{bmatrix} (1.4 - 0.6) \times 10^3$$

$$+ \begin{bmatrix} 10.050 & 2.513 & 0 \\ 2.513 & 125.628 & 0 \\ 0 & 0 & 8 \end{bmatrix} (2.0 - 1.4) \times 10^3$$

$$= \begin{bmatrix} 292.151 & 47.045 & 54.319 \\ 47.045 & 176.573 & 45.775 \\ 54.319 & 45.775 & 68.995 \end{bmatrix} \times 10^3 \, \text{MPa.mm}$$

$$[B] = \begin{bmatrix} 125.628 & 2.513 & 0 \\ 2.513 & 10.050 & 0 \\ 0 & 0 & 8 \end{bmatrix} \left(\frac{0.36 - 4.0}{2} \right) \times 10^3$$

$$+ \begin{bmatrix} 78.236 & 21.010 & 35.703 \\ 21.010 & 20.447 & 14.344 \\ 35.703 & 14.344 & 26.498 \end{bmatrix} \left(\frac{0.36 - 0.36}{2} \right) \times 10^3$$

$$+ \begin{bmatrix} 20.447 & 21.010 & 14.344 \\ 21.010 & 78.236 & 35.703 \\ 14.344 & 35.703 & 26.498 \end{bmatrix} \left(\frac{1.96 - 0.36}{2} \right) \times 10^3$$

$$+ \begin{bmatrix} 10.050 & 2.513 & 0 \\ 2.513 & 125.628 & 0 \\ 0 & 0 & 8 \end{bmatrix} \left(\frac{4.0 - 1.96}{2} \right) \times 10^3$$

$$= \begin{bmatrix} -202.034 & 14.798 & 11.475 \\ 14.798 & 172.438 & 28.562 \\ 11.475 & 28.562 & 14.798 \end{bmatrix} \times 10^3 \, \text{MPa.mm}^2$$

$$[D] = \begin{bmatrix} 125.628 & 2.513 & 0 \\ 2.513 & 10.050 & 0 \\ 0 & 0 & 8 \end{bmatrix} \left(\frac{8-0.216}{3}\right) \times 10^3$$

$$+ \begin{bmatrix} 78.236 & 21.010 & 35.703 \\ 21.010 & 20.447 & 14.344 \\ 35.703 & 14.344 & 26.498 \end{bmatrix} \left(\frac{0.216+0.216}{3}\right) \times 10^3$$

$$+ \begin{bmatrix} 20.447 & 21.010 & 14.344 \\ 21.010 & 78.236 & 35.703 \\ 14.344 & 35.703 & 26.498 \end{bmatrix} \left(\frac{2.744-0.216}{3}\right) \times 10^3$$

$$+ \begin{bmatrix} 10.050 & 2.513 & 0 \\ 2.513 & 125.628 & 0 \\ 0 & 0 & 8 \end{bmatrix} \left(\frac{8.0-2.744}{3}\right) \times 10^3$$

$$= \begin{bmatrix} 372.067 & 31.651 & 17.228 \\ 31.651 & 315.049 & 32.151 \\ 17.228 & 32.151 & 60.918 \end{bmatrix} \times 10^3 \; \text{MPa.mm}^3$$

Step 6.

Having determined the [A], [B], and [D] matrices, the [A*], [B*], [C*], and [D*] matrices are determined using Equations 5.57 through 5.60 and Equations 5.64 through 5.67. The final values of these matrices are

$$[A^*] = \begin{bmatrix} 7.925 & -1.218 & -6.185 \\ -1.218 & 14.939 & -5.508 \\ -6.185 & -5.508 & 23.313 \end{bmatrix} \times 10^{-6} \, (\text{MPa.mm})^{-1}$$

$$[B^*] = \begin{bmatrix} 4.545 & 0.497 & -0.967 \\ -0.388 & -7.438 & -1.402 \\ -3.928 & 1.763 & -1.735 \end{bmatrix} \times 10^{-6} \, (\text{MPa.mm}^2)^{-1}$$

$$[C^*] = \begin{bmatrix} 4.545 & -0.388 & -3.928 \\ 0.497 & -7.438 & 1.763 \\ -0.967 & -1.402 & -1.735 \end{bmatrix} \times 10^{-6} \, (\text{MPa.mm}^2)^{-1}$$

$$[D^*] = \begin{bmatrix} 5.353 & -0.060 & -1.202 \\ -0.060 & 7.149 & -0.791 \\ -1.202 & -0.791 & 18.434 \end{bmatrix} \times 10^{-6} \, (\text{MPa.mm}^3)^{-1}$$

Note: Instead of determining the [A*], [B*], [C*], and [D*] matrices, one can directly invert the $\begin{bmatrix} A & B \\ B & D \end{bmatrix}$ matrix, whose size is 6 × 6, in which case, of course, a computer would be usually essential.

Step 7.

Now, the laminate middle surface strains and curvatures are obtained using Equation 5.63

$$\begin{Bmatrix} \varepsilon_{xx}^0 \\ \varepsilon_{yy}^0 \\ \gamma_{xy}^0 \\ \kappa_{xx} \\ \kappa_{yy} \\ \kappa_{xy} \end{Bmatrix} = \begin{bmatrix} 7.925 & -1.218 & -6.185 & 4.545 & 0.497 & -0.967 \\ -1.218 & 14.939 & -5.508 & -0.388 & -7.438 & -1.402 \\ -6.185 & -5.508 & 23.313 & -3.928 & 1.763 & -1.735 \\ 4.545 & -0.388 & -3.928 & 5.353 & -0.060 & -1.202 \\ 0.497 & -7.438 & 1.763 & -0.060 & 7.149 & -0.791 \\ -0.967 & -1.402 & -1.735 & -1.202 & -0.791 & 18.434 \end{bmatrix} \times \begin{Bmatrix} 400 \\ 200 \\ 0 \\ 0 \\ 0 \\ 0 \end{Bmatrix} \times 10^{-6}$$

$$= \begin{Bmatrix} 2926.3 \\ 2500.5 \\ -3575.7 \\ 1740.3 \\ -1288.7 \\ -667.1 \end{Bmatrix} \times 10^{-6}$$

or the middle surface strains and curvatures are

$$\begin{Bmatrix} \varepsilon_{xx}^0 \\ \varepsilon_{yy}^0 \\ \gamma_{xy}^0 \end{Bmatrix} = \begin{Bmatrix} 2926.3 \\ 2500.5 \\ -3575.7 \end{Bmatrix} \times 10^{-6}$$

and

$$\begin{Bmatrix} \kappa_{xx} \\ \kappa_{yy} \\ \kappa_{xy} \end{Bmatrix} = \begin{Bmatrix} 1740.3 \\ -1288.7 \\ -667.1 \end{Bmatrix} \times 10^{-6}$$

Step 8.

The global strains at different locations are given by Equation 5.14 as follows:

Bottom of laminate (or, bottom of 0° ply):

$$\begin{Bmatrix} \varepsilon_{xx} \\ \varepsilon_{yy} \\ \gamma_{xy} \end{Bmatrix} = \left(\begin{Bmatrix} 2926.3 \\ 2500.5 \\ -3575.7 \end{Bmatrix} - 2.0 \times \begin{Bmatrix} 1740.3 \\ -1288.7 \\ -667.1 \end{Bmatrix} \right) \times 10^{-6} = \begin{Bmatrix} -554.3 \\ 5077.9 \\ -2241.5 \end{Bmatrix} \times 10^{-6}$$

Top of 0° ply (or, bottom of 30° ply):

$$\begin{Bmatrix} \varepsilon_{xx} \\ \varepsilon_{yy} \\ \gamma_{xy} \end{Bmatrix} = \left(\begin{Bmatrix} 2926.3 \\ 2500.5 \\ -3575.7 \end{Bmatrix} - 0.6 \times \begin{Bmatrix} 1740.3 \\ -1288.7 \\ -667.1 \end{Bmatrix} \right) \times 10^{-6} = \begin{Bmatrix} 1882.1 \\ 3273.7 \\ -3175.4 \end{Bmatrix} \times 10^{-6}$$

Top of 30° ply (or, bottom of 60° ply):

$$\begin{Bmatrix} \varepsilon_{xx} \\ \varepsilon_{yy} \\ \gamma_{xy} \end{Bmatrix} = \left(\begin{bmatrix} 2926.3 \\ 2500.5 \\ -3575.7 \end{bmatrix} + 0.6 \times \begin{bmatrix} 1740.3 \\ -1288.7 \\ -667.1 \end{bmatrix} \right) \times 10^{-6} = \begin{Bmatrix} 3970.5 \\ 1727.3 \\ -3976.0 \end{Bmatrix} \times 10^{-6}$$

Top of 60° ply (or, bottom of 90° ply):

$$\begin{Bmatrix} \varepsilon_{xx} \\ \varepsilon_{yy} \\ \gamma_{xy} \end{Bmatrix} = \left(\begin{bmatrix} 2926.3 \\ 2500.5 \\ -3575.7 \end{bmatrix} + 1.4 \times \begin{bmatrix} 1740.3 \\ -1288.7 \\ -667.1 \end{bmatrix} \right) \times 10^{-6} = \begin{Bmatrix} 5362.7 \\ 696.3 \\ -4509.6 \end{Bmatrix} \times 10^{-6}$$

Top of laminate (or, top of 90° ply):

$$\begin{Bmatrix} \varepsilon_{xx} \\ \varepsilon_{yy} \\ \gamma_{xy} \end{Bmatrix} = \left(\begin{bmatrix} 2926.3 \\ 2500.5 \\ -3575.7 \end{bmatrix} + 2.0 \times \begin{bmatrix} 1740.3 \\ -1288.7 \\ -667.1 \end{bmatrix} \right) \times 10^{-6} = \begin{Bmatrix} 6406.9 \\ -76.9 \\ -4909.9 \end{Bmatrix} \times 10^{-6}$$

Step 9.

Having determined the global strains, now we determine the global stresses, using Equation 4.53, as follows:

Bottom of laminate (or bottom of 0° ply):

$$\begin{Bmatrix} \sigma_{xx} \\ \sigma_{yy} \\ \tau_{xy} \end{Bmatrix} = \begin{bmatrix} 125.628 & 2.513 & 0 \\ 2.513 & 10.050 & 0 \\ 0 & 0 & 8 \end{bmatrix} \begin{Bmatrix} -554.3 \\ 5077.9 \\ -2241.5 \end{Bmatrix} \times 10^{-3} = \begin{Bmatrix} -56.9 \\ 49.6 \\ -17.9 \end{Bmatrix} \text{MPa}$$

Top of 0° ply:

$$\begin{Bmatrix} \sigma_{xx} \\ \sigma_{yy} \\ \tau_{xy} \end{Bmatrix} = \begin{bmatrix} 125.628 & 2.513 & 0 \\ 2.513 & 10.050 & 0 \\ 0 & 0 & 8 \end{bmatrix} \begin{Bmatrix} 1882.1 \\ 3273.7 \\ -3175.4 \end{Bmatrix} \times 10^{-3} = \begin{Bmatrix} 244.7 \\ 37.6 \\ -25.4 \end{Bmatrix} \text{MPa}$$

Bottom of 30° ply:

$$\begin{Bmatrix} \sigma_{xx} \\ \sigma_{yy} \\ \tau_{xy} \end{Bmatrix} = \begin{bmatrix} 78.236 & 21.010 & 35.703 \\ 21.010 & 20.447 & 14.344 \\ 35.703 & 14.344 & 26.498 \end{bmatrix} \begin{Bmatrix} 1882.1 \\ 3273.7 \\ -3175.4 \end{Bmatrix} \times 10^{-3} = \begin{Bmatrix} 102.7 \\ 60.9 \\ 30.0 \end{Bmatrix} \text{MPa}$$

Top of 30° ply:

$$\begin{Bmatrix} \sigma_{xx} \\ \sigma_{yy} \\ \tau_{xy} \end{Bmatrix} = \begin{bmatrix} 78.236 & 21.010 & 35.703 \\ 21.010 & 20.447 & 14.344 \\ 35.703 & 14.344 & 26.498 \end{bmatrix} \begin{Bmatrix} 3970.5 \\ 1727.3 \\ -3976.0 \end{Bmatrix} \times 10^{-3} = \begin{Bmatrix} 205.0 \\ 61.7 \\ 61.2 \end{Bmatrix} \text{MPa}$$

Bottom of 60° ply:

$$\begin{Bmatrix} \sigma_{xx} \\ \sigma_{yy} \\ \tau_{xy} \end{Bmatrix} = \begin{bmatrix} 20.447 & 21.010 & 14.344 \\ 21.010 & 78.236 & 35.703 \\ 14.344 & 35.703 & 26.498 \end{bmatrix} \begin{Bmatrix} 3970.5 \\ 1727.3 \\ -3976.0 \end{Bmatrix} \times 10^{-3} = \begin{Bmatrix} 60.4 \\ 76.6 \\ 13.3 \end{Bmatrix} \text{MPa}$$

Top of 60° ply:

$$\begin{Bmatrix} \sigma_{xx} \\ \sigma_{yy} \\ \tau_{xy} \end{Bmatrix} = \begin{bmatrix} 20.447 & 21.010 & 14.344 \\ 21.010 & 78.236 & 35.703 \\ 14.344 & 35.703 & 26.498 \end{bmatrix} \begin{Bmatrix} 5362.7 \\ 696.3 \\ -4509.6 \end{Bmatrix} \times 10^{-3} = \begin{Bmatrix} 59.6 \\ 6.1 \\ -17.7 \end{Bmatrix} \text{MPa}$$

Bottom of 90° ply:

$$\begin{Bmatrix} \sigma_{xx} \\ \sigma_{yy} \\ \tau_{xy} \end{Bmatrix} = \begin{bmatrix} 10.050 & 2.513 & 0 \\ 2.513 & 125.628 & 0 \\ 0 & 0 & 8 \end{bmatrix} \begin{Bmatrix} 5362.7 \\ 696.3 \\ -4509.6 \end{Bmatrix} \times 10^{-3} = \begin{Bmatrix} 55.6 \\ 101.0 \\ -36.1 \end{Bmatrix} \text{MPa}$$

Top of laminate (or top of 90° ply):

$$\begin{Bmatrix} \sigma_{xx} \\ \sigma_{yy} \\ \tau_{xy} \end{Bmatrix} = \begin{bmatrix} 10.050 & 2.513 & 0 \\ 2.513 & 125.628 & 0 \\ 0 & 0 & 8 \end{bmatrix} \begin{Bmatrix} 6406.9 \\ -76.9 \\ -4909.9 \end{Bmatrix} \times 10^{-3} = \begin{Bmatrix} 64.2 \\ 6.4 \\ -39.3 \end{Bmatrix} \text{MPa}$$

Step 10.

Now, we apply transformations of strains. The transformation is given by Equation 4.48 as

$$\begin{Bmatrix} \varepsilon_{11} \\ \varepsilon_{22} \\ \gamma_{12} \end{Bmatrix} = [R][T][R]^{-1} \begin{Bmatrix} \varepsilon_{xx} \\ \varepsilon_{yy} \\ \gamma_{xy} \end{Bmatrix}$$

Note that

$$[R] = \begin{bmatrix} 1 & 0 & 0 \\ 0 & 1 & 0 \\ 0 & 0 & 2 \end{bmatrix}$$

and

$$[T] = \begin{bmatrix} \cos^2\theta & \sin^2\theta & 2\sin\theta\cos\theta \\ \sin^2\theta & \cos^2\theta & -2\sin\theta\cos\theta \\ -\sin\theta\cos\theta & \sin\theta\cos\theta & \cos^2\theta - \sin^2\theta \end{bmatrix}$$

Macromechanics of a Laminate

First, the transformation matrices for different θ are determined as follows:

$$([R][T][R]^{-1})_{0°} = \begin{bmatrix} 1 & 0 & 0 \\ 0 & 1 & 0 \\ 0 & 0 & 2 \end{bmatrix} \begin{bmatrix} 1 & 0 & 0 \\ 0 & 1 & 0 \\ 0 & 0 & 1 \end{bmatrix} \begin{bmatrix} 1 & 0 & 0 \\ 0 & 1 & 0 \\ 0 & 0 & 2 \end{bmatrix}^{-1} = \begin{bmatrix} 1 & 0 & 0 \\ 0 & 1 & 0 \\ 0 & 0 & 1 \end{bmatrix}$$

$$([R][T][R]^{-1})_{30°} = \begin{bmatrix} 1 & 0 & 0 \\ 0 & 1 & 0 \\ 0 & 0 & 2 \end{bmatrix} \begin{bmatrix} 0.75 & 0.25 & 0.866 \\ 0.25 & 0.75 & -0.866 \\ -0.433 & 0.433 & 0.5 \end{bmatrix} \begin{bmatrix} 1 & 0 & 0 \\ 0 & 1 & 0 \\ 0 & 0 & 2 \end{bmatrix}^{-1}$$

$$= \begin{bmatrix} 0.75 & 0.25 & 0.433 \\ 0.25 & 0.75 & -0.433 \\ -0.866 & 0.866 & 0.5 \end{bmatrix}$$

$$([R][T][R]^{-1})_{60°} = \begin{bmatrix} 1 & 0 & 0 \\ 0 & 1 & 0 \\ 0 & 0 & 2 \end{bmatrix} \begin{bmatrix} 0.25 & 0.75 & 0.866 \\ 0.75 & 0.25 & -0.866 \\ -0.433 & 0.433 & -0.5 \end{bmatrix} \begin{bmatrix} 1 & 0 & 0 \\ 0 & 1 & 0 \\ 0 & 0 & 2 \end{bmatrix}^{-1}$$

$$= \begin{bmatrix} 0.25 & 0.75 & 0.433 \\ 0.75 & 0.25 & -0.433 \\ -0.866 & 0.866 & -0.5 \end{bmatrix}$$

$$([R][T][R]^{-1})_{90°} = \begin{bmatrix} 1 & 0 & 0 \\ 0 & 1 & 0 \\ 0 & 0 & 2 \end{bmatrix} \begin{bmatrix} 0 & 1 & 0 \\ 1 & 0 & 0 \\ 0 & 0 & -1 \end{bmatrix} \begin{bmatrix} 1 & 0 & 0 \\ 0 & 1 & 0 \\ 0 & 0 & 2 \end{bmatrix}^{-1} = \begin{bmatrix} 0 & 1 & 0 \\ 1 & 0 & 0 \\ 0 & 0 & -1 \end{bmatrix}$$

Using the above, the local strains are obtained as follows:
Bottom of 0° ply:

$$\begin{Bmatrix} \varepsilon_{11} \\ \varepsilon_{22} \\ \gamma_{12} \end{Bmatrix} = \begin{bmatrix} 1 & 0 & 0 \\ 0 & 1 & 0 \\ 0 & 0 & 1 \end{bmatrix} \begin{Bmatrix} -554.3 \\ 5077.9 \\ -2241.5 \end{Bmatrix} \times 10^{-6} = \begin{Bmatrix} -554.3 \\ 5077.9 \\ -2241.5 \end{Bmatrix} \times 10^{-6}$$

Top of 0° ply:

$$\begin{Bmatrix} \varepsilon_{11} \\ \varepsilon_{22} \\ \gamma_{12} \end{Bmatrix} = \begin{bmatrix} 1 & 0 & 0 \\ 0 & 1 & 0 \\ 0 & 0 & 1 \end{bmatrix} \begin{Bmatrix} 1882.1 \\ 3273.7 \\ -3175.4 \end{Bmatrix} \times 10^{-6} = \begin{Bmatrix} 1882.1 \\ 3273.7 \\ -3175.4 \end{Bmatrix} \times 10^{-6}$$

Bottom of 30° ply:

$$\begin{Bmatrix} \varepsilon_{11} \\ \varepsilon_{22} \\ \gamma_{12} \end{Bmatrix} = \begin{bmatrix} 0.75 & 0.25 & 0.433 \\ 0.25 & 0.75 & -0.433 \\ -0.866 & 0.866 & 0.5 \end{bmatrix} \begin{Bmatrix} 1882.1 \\ 3273.7 \\ -3175.4 \end{Bmatrix} \times 10^{-6} = \begin{Bmatrix} 855.1 \\ 4300.7 \\ -382.6 \end{Bmatrix} \times 10^{-6}$$

Top of 30° ply:

$$\begin{Bmatrix} \varepsilon_{11} \\ \varepsilon_{22} \\ \gamma_{12} \end{Bmatrix} = \begin{bmatrix} 0.75 & 0.25 & 0.433 \\ 0.25 & 0.75 & -0.433 \\ -0.866 & 0.866 & 0.5 \end{bmatrix} \begin{Bmatrix} 3970.5 \\ 1727.3 \\ -3976.0 \end{Bmatrix} \times 10^{-6} = \begin{Bmatrix} 1688.1 \\ 4009.7 \\ -3930.6 \end{Bmatrix} \times 10^{-6}$$

Bottom of 60° ply:

$$\begin{Bmatrix} \varepsilon_{11} \\ \varepsilon_{22} \\ \gamma_{12} \end{Bmatrix} = \begin{bmatrix} 0.25 & 0.75 & 0.433 \\ 0.75 & 0.25 & -0.433 \\ -0.866 & 0.866 & -0.5 \end{bmatrix} \begin{Bmatrix} 3970.5 \\ 1727.3 \\ -3976.0 \end{Bmatrix} \times 10^{-6} = \begin{Bmatrix} 566.5 \\ 5131.3 \\ 45.4 \end{Bmatrix} \times 10^{-6}$$

Top of 60° ply:

$$\begin{Bmatrix} \varepsilon_{11} \\ \varepsilon_{22} \\ \gamma_{12} \end{Bmatrix} = \begin{bmatrix} 0.25 & 0.75 & 0.433 \\ 0.75 & 0.25 & -0.433 \\ -0.866 & 0.866 & -0.5 \end{bmatrix} \begin{Bmatrix} 5362.7 \\ 696.3 \\ -4509.6 \end{Bmatrix} \times 10^{-6} = \begin{Bmatrix} -89.8 \\ 6148.8 \\ -1786.3 \end{Bmatrix} \times 10^{-6}$$

Bottom of 90° ply:

$$\begin{Bmatrix} \varepsilon_{11} \\ \varepsilon_{22} \\ \gamma_{12} \end{Bmatrix} = \begin{bmatrix} 0 & 1 & 0 \\ 1 & 0 & 0 \\ 0 & 0 & -1 \end{bmatrix} \begin{Bmatrix} 5362.7 \\ 696.3 \\ -4509.6 \end{Bmatrix} \times 10^{-6} = \begin{Bmatrix} 696.3 \\ 5362.7 \\ 4509.6 \end{Bmatrix} \times 10^{-6}$$

Top of 90° ply:

$$\begin{Bmatrix} \varepsilon_{11} \\ \varepsilon_{22} \\ \gamma_{12} \end{Bmatrix} = \begin{bmatrix} 0 & 1 & 0 \\ 1 & 0 & 0 \\ 0 & 0 & -1 \end{bmatrix} \begin{Bmatrix} 6406.9 \\ -76.9 \\ -4909.9 \end{Bmatrix} \times 10^{-6} = \begin{Bmatrix} -76.9 \\ 6406.9 \\ 4909.9 \end{Bmatrix} \times 10^{-6}$$

Step 11.

Finally, we carry out stress transformation using Equation 4.44. Then, the local stresses are as follows:

Bottom of laminate (or bottom of 0° ply):

$$\begin{Bmatrix} \sigma_{11} \\ \sigma_{22} \\ \tau_{12} \end{Bmatrix} = \begin{bmatrix} 1 & 0 & 0 \\ 0 & 1 & 0 \\ 0 & 0 & 1 \end{bmatrix} \begin{Bmatrix} -56.9 \\ 49.6 \\ -17.9 \end{Bmatrix} = \begin{Bmatrix} -56.9 \\ 49.6 \\ -17.9 \end{Bmatrix} \text{MPa}$$

Top of 0° ply:

$$\begin{Bmatrix} \sigma_{11} \\ \sigma_{22} \\ \tau_{12} \end{Bmatrix} = \begin{bmatrix} 1 & 0 & 0 \\ 0 & 1 & 0 \\ 0 & 0 & 1 \end{bmatrix} \begin{Bmatrix} 244.7 \\ 37.6 \\ -25.4 \end{Bmatrix} = \begin{Bmatrix} 244.7 \\ 37.6 \\ -25.4 \end{Bmatrix} \text{MPa}$$

Bottom of 30° ply:

$$\begin{Bmatrix}\sigma_{11}\\\sigma_{22}\\\tau_{12}\end{Bmatrix}=\begin{bmatrix}0.75 & 0.25 & 0.866\\0.25 & 0.75 & -0.866\\-0.433 & 0.433 & 0.5\end{bmatrix}\begin{Bmatrix}102.7\\60.9\\30.0\end{Bmatrix}=\begin{Bmatrix}118.2\\45.4\\-3.1\end{Bmatrix}\text{MPa}$$

Top of 30° ply:

$$\begin{Bmatrix}\sigma_{11}\\\sigma_{22}\\\tau_{12}\end{Bmatrix}=\begin{bmatrix}0.75 & 0.25 & 0.866\\0.25 & 0.75 & -0.866\\-0.433 & 0.433 & 0.5\end{bmatrix}\begin{Bmatrix}205.0\\61.7\\61.2\end{Bmatrix}=\begin{Bmatrix}222.2\\44.5\\-31.4\end{Bmatrix}\text{MPa}$$

Bottom of 60° ply:

$$\begin{Bmatrix}\sigma_{11}\\\sigma_{22}\\\tau_{12}\end{Bmatrix}=\begin{bmatrix}0.25 & 0.75 & 0.866\\0.75 & 0.25 & -0.866\\-0.433 & 0.433 & -0.5\end{bmatrix}\begin{Bmatrix}60.4\\76.6\\13.3\end{Bmatrix}=\begin{Bmatrix}84.1\\52.9\\0.4\end{Bmatrix}\text{MPa}$$

Top of 60° ply:

$$\begin{Bmatrix}\sigma_{11}\\\sigma_{22}\\\tau_{12}\end{Bmatrix}=\begin{bmatrix}0.25 & 0.75 & 0.866\\0.75 & 0.25 & -0.866\\-0.433 & 0.433 & -0.5\end{bmatrix}\begin{Bmatrix}59.6\\6.1\\-17.7\end{Bmatrix}=\begin{Bmatrix}4.1\\61.6\\-14.3\end{Bmatrix}\text{MPa}$$

Bottom of 90° ply:

$$\begin{Bmatrix}\sigma_{11}\\\sigma_{22}\\\tau_{12}\end{Bmatrix}=\begin{bmatrix}0 & 1 & 0\\1 & 0 & 0\\0 & 0 & -1\end{bmatrix}\begin{Bmatrix}55.6\\101.0\\-36.1\end{Bmatrix}=\begin{Bmatrix}101.0\\55.6\\36.1\end{Bmatrix}\text{MPa}$$

Top of laminate (or top of 90° ply):

$$\begin{Bmatrix}\sigma_{11}\\\sigma_{22}\\\tau_{12}\end{Bmatrix}=\begin{bmatrix}0 & 1 & 0\\1 & 0 & 0\\0 & 0 & -1\end{bmatrix}\begin{Bmatrix}64.2\\6.4\\-39.3\end{Bmatrix}=\begin{Bmatrix}6.4\\64.2\\39.3\end{Bmatrix}\text{MPa}$$

Step 12.

The local strains and stresses determined here can now be used in a suitable lamina failure criterion to estimate failure of the laminae. We shall discuss laminate failure in a subsequent section.

5.6 CLASSICAL LAMINATED SHELL THEORY

Shells are structural elements bounded by two curved surfaces. There are two classes of shells—thin shells and thick shells. When the largest ratio of shell thickness to radius of curvature is far small compared to unity, the shell is called a thin shell. In general, a shell is considered as thin, when

$$\left(\frac{h}{R}\right)_{max} \leq \frac{1}{20} \tag{5.68}$$

CLT, as applied to shells, is referred to as the CLST. A shell is similar to a plate except for one characteristic—presence of curvature. In fact, a plate can be considered as a special case of a shell whose principal radii of curvature are infinite. The assumptions, kinematics, kinetics, and constitutive relations in CLST are similar to those in CLPT with marginal differences.

5.6.1 Geometry of the Middle Surface

We will see in the subsequent sections that the governing equations for a shell are formulated in terms of Lame' parameters and principal radii of curvature. In this section, these terms in respect of the middle surface are introduced. Our discussion on geometric characteristics of the middle surface shall be limited primarily to these terms.

Figure 5.7a shows the middle surface of a shell in the orthogonal curvilinear coordinates. We can assign continually varying values to these coordinates such as $\alpha = \alpha_1$, $\alpha_2, \alpha_3, \ldots$ and $= \beta_1, \beta_2, \beta_3, \ldots$. The α- and β-coordinate lines are mutually perpendicular to each other at all intersection points. On the other hand, the ξ-coordinate is the linear axis such that, if we consider a pair of α- and β-coordinate lines, then the ξ-axis passing through their intersection point is normal to the curvilinear coordinate lines.

At any point on the curved surface, we can consider a section along some plane containing the normal at that point. Such a section is called a normal section. The plane cuts the curved surface along a plane curve. Let us consider a normal section through an arbitrary point P (Figure 5.7b). Let P' be a neighboring point on the normal section. The lines OP and OP' are the normals at these two points such that they

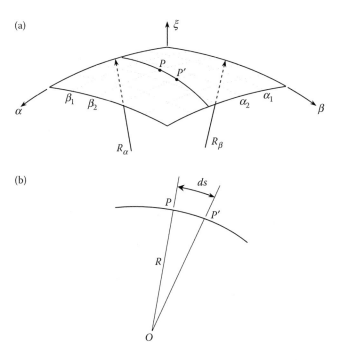

FIGURE 5.7 Geometry of the middle surface. (a) Orthogonal curvilinear coordinates. (b) Radius of curvature in an arbitrary normal section.

Macromechanics of a Laminate

intersect at point O. Then, as $ds \to 0$, the point O is defined as the center of curvature and the distance R between the points O and P is the radius of curvature at the point P. The curvature at any point is defined as the reciprocal of the radius of curvature, that is, $\kappa = 1/R$. Now, an infinite numbers of normal sections can be formed at the point P and corresponding radii of curvature and centers of curvature. However, there are two orthogonal normal sections that are of special interest such that one radius of curvature is the maximum of all possible radii of curvature and the other is the minimum. These are called the principal normal sections or principal directions and principal radii of curvature. The plane curves along which the principal normal sections cut the surface are called the principal lines of curvature or simply lines of curvature. We shall denote the principal radii by $R_1\ (= 1/\kappa_1)$ and $R_2\ (= 1/\kappa_2)$. Further, we shall align the orthogonal coordinates α and β along the principal lines of curvature. Thus, $R_1 = R_\alpha$ and $R_2 = R_\beta$.

We have defined principal radii of curvature. We need to know two more parameters before proceeding to the kinematics and kinetics of a shell. These are the Lame' parameters—A and B. Let us consider two arbitrary neighboring points P and P' on the curved surface (Figure 5.8). Denoting the position vector of the point P by $\mathbf{r}$, the square of the differential arc length can be shown as

$$(ds)^2 = |d\mathbf{r}|^2 = d\mathbf{r} \cdot d\mathbf{r} = \left(\frac{\partial \mathbf{r}}{\partial \alpha} d\alpha + \frac{\partial \mathbf{r}}{\partial \beta} d\beta\right) \cdot \left(\frac{\partial \mathbf{r}}{\partial \alpha} d\alpha + \frac{\partial \mathbf{r}}{\partial \beta} d\beta\right) \tag{5.69}$$

Now, let us introduce

$$A = \left(\frac{\partial \mathbf{r}}{\partial \alpha} \cdot \frac{\partial \mathbf{r}}{\partial \alpha}\right)^{-1/2} \quad \text{and} \quad B = \left(\frac{\partial \mathbf{r}}{\partial \beta} \cdot \frac{\partial \mathbf{r}}{\partial \beta}\right)^{-1/2} \tag{5.70}$$

Further, note that for orthogonal coordinates,

$$\frac{\partial \mathbf{r}}{\partial \alpha} \cdot \frac{\partial \mathbf{r}}{\partial \beta} = 0 \tag{5.71}$$

Then, Equation 5.69 can be written as

$$(ds)^2 = A^2 (d\alpha)^2 + B^2 (d\beta)^2 \tag{5.72}$$

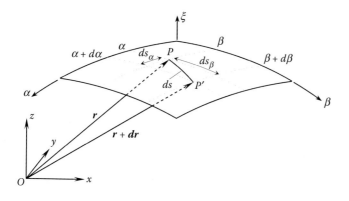

FIGURE 5.8 Differential arc.

The parameters A and B are called the Lamé parameters. The differential arc length in an arbitrary direction can be expressed in terms of the corresponding differential arc lengths in the orthogonal curvilinear coordinate directions as follows:

$$(ds)^2 = (ds_\alpha)^2 + (ds_\beta)^2 \tag{5.73}$$

where ds_α and ds_β are the differential arc lengths in the α- and β-directions, respectively.

Then, comparing Equations 5.72 and 5.73,

$$ds_\alpha = Ad\alpha \quad \text{and} \quad ds_\beta = Bd\beta \tag{5.74}$$

or

$$A = \frac{ds_\alpha}{d\alpha} \quad \text{and} \quad B = \frac{ds_\beta}{d\beta} \tag{5.75}$$

The Lamé parameters, principal radii of curvature, and the orthogonal curvilinear coordinates are interrelated as follows [5]:

$$\frac{\partial}{\partial \beta}\left(\frac{A}{R_1}\right) = \frac{1}{R_2}\frac{\partial A}{\partial \beta} \tag{5.76}$$

$$\frac{\partial}{\partial \alpha}\left(\frac{B}{R_2}\right) = \frac{1}{R_1}\frac{\partial B}{\partial \alpha} \tag{5.77}$$

$$\frac{\partial}{\partial \alpha}\left(\frac{1}{A}\frac{\partial B}{\partial \alpha}\right) + \frac{\partial}{\partial \beta}\left(\frac{1}{B}\frac{\partial A}{\partial \beta}\right) = -\frac{AB}{R_1 R_2} \tag{5.78}$$

Equations 5.76 and 5.77 are referred to as the *Codazzi conditions* and Equation 5.78 as the *Gauss condition*.

We will see in the next sections that kinematic, kinetic, and constitutive relations for a shell are formulated in terms of Lamé parameters and the principal radii of curvatures. So, with this brief introduction to the geometric characteristics of a general shell, we shall now proceed to the governing equations for a shell.

5.6.2 Kinematics of CLST: Strain–Displacement Relations

Let us consider a general shell of uniform thickness as shown in Figure 5.9. The strain–displacement relations in the orthogonal curvilinear coordinates for the shell in the middle surface are given by

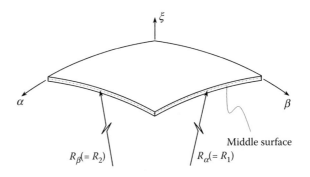

FIGURE 5.9 General shell in curvilinear coordinates.

$$\varepsilon_{\alpha\alpha} = \frac{1}{A}\frac{\partial u_0}{\partial \alpha} + \frac{1}{AB}\frac{\partial A}{\partial \beta}v_0 + \frac{w_0}{R_\alpha} \qquad (5.79)$$

$$\varepsilon_{\beta\beta} = \frac{1}{B}\frac{\partial v_0}{\partial \beta} + \frac{1}{AB}\frac{\partial B}{\partial \alpha}u_0 + \frac{w_0}{R_\beta} \qquad (5.80)$$

$$\gamma_{\alpha\beta} = \frac{1}{A}\frac{\partial v_0}{\partial \alpha} + \frac{1}{B}\frac{\partial u_0}{\partial \beta} - \frac{1}{AB}\left(\frac{\partial A}{\partial \beta}u_0 + \frac{\partial B}{\partial \alpha}v_0\right) \qquad (5.81)$$

where

$\varepsilon_{\alpha\alpha}, \varepsilon_{\beta\beta}$ Normal strains in the α- and β-directions, respectively
$\gamma_{\alpha\beta}$ In-plane shear strain in the α-β plane
A, B Lame' parameters
R_α, R_β Radii of curvature that are also the principal radii of curvature as the orthogonal curvilinear coordinates are aligned with the principal directions
u_0, v_0, w_0 Middle surface displacements

The detailed derivation of the above strain–displacement relation for a general shell is beyond the scope of this book. Interested reader may refer to Reference 5 among others. Here, we shall use these relations in the context of a circular cylindrical shell (Figure 5.10). Let us choose the coordinate system in such a way that

$$\alpha = x \quad \text{and} \quad \beta = y \qquad (5.82)$$

Then, square of an arbitrary differential arc length is given by

$$(ds)^2 = (dx)^2 + (dy)^2 \qquad (5.83)$$

Comparing Equation 5.83 with Equation 5.73 and taking Equations 5.75 and 5.82 into account, we find that

$$A = 1 \quad \text{and} \quad B = 1 \qquad (5.84)$$

Also, the principal radii of curvature are as follows:

$$R_1 = \infty \quad \text{and} \quad R_2 = R \qquad (5.85)$$

Substituting the above in Equations 5.79 through 5.81, the strain–displacement relations for a circular cylindrical shell are obtained as follows:

$$\varepsilon_{xx} = \frac{\partial u_0}{\partial x} \qquad (5.86)$$

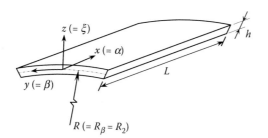

FIGURE 5.10 Circular cylindrical shell.

$$\varepsilon_{yy} = \frac{\partial v_0}{\partial y} + \frac{w_0}{R} \quad (5.87)$$

$$\gamma_{xy} = \frac{\partial v_0}{\partial x} + \frac{\partial u_0}{\partial y} \quad (5.88)$$

Note that the strain–displacement relations for a circular cylindrical shell become identically the same as those for a plate if we substitute $R \to \infty$.

5.6.3 Kinetics of CLST: Force and Moment Resultants

Figure 5.11 shows the force and moment resultants in a shell in the orthogonal curvilinear coordinates. Force and moment resultants in a shell are defined, respectively, as the forces and moments per unit length of the cross section of the shell. Note that the length is measured along the middle surface of the shell. Note further that the curvilinear lengths at different ξ from the middle surface are different and, as a result, the expressions for force and moment resultants are different in a shell from those in a plate.

Let us consider a shell element as shown in Figure 5.12. The curvilinear lengths along the middle surface are (refer Equation 5.74)

$$ds_\alpha = A d\alpha \quad \text{and} \quad ds_\beta = B d\beta \quad (5.89)$$

whereas the curvilinear lengths at distance ξ from the middle surface are

$$ds'_\alpha = \left(1 + \frac{\xi}{R_1}\right) A d\alpha \quad \text{and} \quad ds'_\beta = \left(1 + \frac{\xi}{R_2}\right) B d\alpha \quad (5.90)$$

Then, the force resultant on the differential shell element in the α-direction is given by

$$N_{\alpha\alpha} = \left(\int_{-h/2}^{h/2} \sigma_{\alpha\alpha} ds'_\beta d\xi\right) \Big/ ds_\beta \quad (5.91)$$

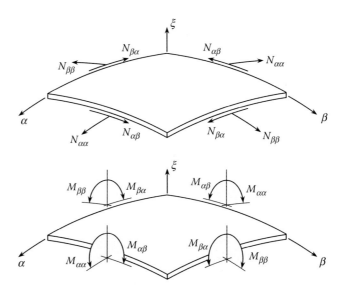

FIGURE 5.11 Force and moment resultants in a shell.

Macromechanics of a Laminate

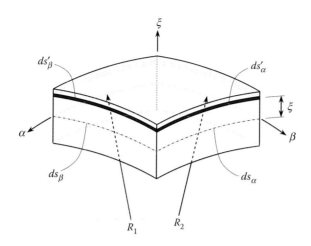

FIGURE 5.12 A shell element. (Adapted with permission from J. N. Reddy, *Mechanics of Laminated Composite Plates and Shells*, CRC Press, Boca Raton, FL, 2004.)

On simplification of the above using Equations 5.89 and 5.90 and extending the procedure to other cases, we get the expressions for the force and moment resultants as follows:

$$N_{\alpha\alpha} = \int_{-h/2}^{h/2} \sigma_{\alpha\alpha}\left(1 + \frac{\xi}{R_2}\right) d\xi \tag{5.92}$$

$$N_{\beta\beta} = \int_{-h/2}^{h/2} \sigma_{\beta\beta}\left(1 + \frac{\xi}{R_1}\right) d\xi \tag{5.93}$$

$$N_{\alpha\beta} = \int_{-h/2}^{h/2} \tau_{\alpha\beta}\left(1 + \frac{\xi}{R_2}\right) d\xi \tag{5.94}$$

$$N_{\beta\alpha} = \int_{-h/2}^{h/2} \tau_{\alpha\beta}\left(1 + \frac{\xi}{R_1}\right) d\xi \tag{5.95}$$

$$M_{\alpha\alpha} = \int_{-h/2}^{h/2} \xi\sigma_{\alpha\alpha}\left(1 + \frac{\xi}{R_2}\right) d\xi \tag{5.96}$$

$$M_{\beta\beta} = \int_{-h/2}^{h/2} \xi\sigma_{\beta\beta}\left(1 + \frac{\xi}{R_1}\right) d\xi \tag{5.97}$$

$$M_{\alpha\beta} = \int_{-h/2}^{h/2} \xi\tau_{\alpha\beta}\left(1 + \frac{\xi}{R_2}\right) d\xi \tag{5.98}$$

$$M_{\beta\alpha} = \int_{-h/2}^{h/2} \xi\tau_{\alpha\beta}\left(1 + \frac{\xi}{R_1}\right) d\xi \tag{5.99}$$

Note that, in the case of shells, $N_{\alpha\beta} \neq N_{\beta\alpha}$ and $M_{\alpha\beta} \neq M_{\beta\alpha}$. For thin shells, however, the terms ξ/R_1 and ξ/R_2 can be ignored as compared to unity and the expressions for force and moment resultants take identical forms as in plates. Also, for thin shells, $N_{\alpha\beta} = N_{\beta\alpha}$ and $M_{\alpha\beta} = M_{\beta\alpha}$.

For a circular cylindrical shell (Figure 5.10), we can substitute Equations 5.82 and 5.85 in Equations 5.92 through 5.99. Thus, for circular cylindrical shells,

$$N_{xx} = \int_{-h/2}^{h/2} \sigma_{xx}\left(1 + \frac{z}{R}\right)dz \tag{5.100}$$

$$N_{yy} = \int_{-h/2}^{h/2} \sigma_{yy}\,dz \tag{5.101}$$

$$N_{xy} = \int_{-h/2}^{h/2} \tau_{xy}\left(1 + \frac{z}{R}\right)dz \tag{5.102}$$

$$N_{yx} = \int_{-h/2}^{h/2} \tau_{xy}\,dz \tag{5.103}$$

$$M_{xx} = \int_{-h/2}^{h/2} z\sigma_{xx}\left(1 + \frac{z}{R}\right)dz \tag{5.104}$$

$$M_{yy} = \int_{-h/2}^{h/2} z\sigma_{yy}\,dz \tag{5.105}$$

$$M_{xy} = \int_{-h/2}^{h/2} z\tau_{xy}\left(1 + \frac{z}{R}\right)dz \tag{5.106}$$

$$M_{yx} = \int_{-h/2}^{h/2} z\tau_{xy}\,dz \tag{5.107}$$

Note that the expressions for a circular cylindrical shell become identically the same as those for a plate if we substitute $R \to \infty$.

5.6.4 Constitutive Relations in CLST

The final constitutive relations in CLST are the same as in CLPT. In terms of laminate stiffnesses, these relations are

$$\begin{Bmatrix} N \\ M \end{Bmatrix} = \begin{bmatrix} A & B \\ B & D \end{bmatrix} \begin{Bmatrix} \varepsilon^0 \\ \kappa \end{Bmatrix} \tag{5.108}$$

and, in terms of compliances,

$$\begin{Bmatrix} \varepsilon^0 \\ \kappa \end{Bmatrix} = \begin{bmatrix} A^* & B^* \\ C^* & D^* \end{bmatrix} \begin{Bmatrix} N \\ M \end{Bmatrix} \qquad (5.109)$$

5.7 HYGROTHERMAL EFFECTS IN A LAMINATE

Thermal stresses can occur in a laminate in two ways [6]. First, residual thermal stresses can develop due to temperature change during processing. The CTE of a lamina depends on the material as well as angle of orientation of the lamina. Owing to mismatch in their CTEs, the laminae in a laminate undergo differential expansion during temperature rise phase. At an elevated temperature, the laminae get bonded. During the cooling phase, the laminae at different orientations (and possibly of different materials) tend to shrink differentially but the bond at the interface between the laminae restrains them. As a result, on cooling the composite develops residual stresses.

Second, thermal stresses can develop due to temperature change during product service life. If individual laminae are subjected to a change in temperature, they undergo differential deformations and clearly the thermal strains are different in the different laminae even in the same direction. However, no thermal stresses develop unless the laminae are restrained from deformation. In a laminate, when subjected to a temperature change, the individual laminae restrain each other from free thermal deformations and, as a result, thermal stresses develop.

Similar to the thermal strains, owing to mismatch in the CMEs, the swelling strains are different in different laminae in a laminate and stresses can develop in a laminate with laminae at different angles. Response of a laminate to hygrothermal stresses [7] is important in the design of a product; hygrothermal constitutive relations and the effective coefficients are presented in the following sections.

5.7.1 Hygrothermal Constitutive Relations

Let us consider a laminate subjected to hygrothermal loads. Let us rewrite Equation 4.186 for the total hygrothermal stresses at a point in a generally orthotropic lamina in the laminate, as follows:

$$\begin{Bmatrix} \sigma_{xx} \\ \sigma_{yy} \\ \tau_{xy} \end{Bmatrix}_{HT} = [\bar{Q}] \left(\begin{Bmatrix} \varepsilon_{xx} \\ \varepsilon_{yy} \\ \gamma_{xy} \end{Bmatrix} - \Delta T \begin{Bmatrix} \alpha_x \\ \alpha_y \\ \alpha_{xy} \end{Bmatrix} - \Delta C \begin{Bmatrix} \beta_x \\ \beta_y \\ \beta_{xy} \end{Bmatrix} \right) \qquad (5.110)$$

Let us have a closer look at the terms inside the brackets on the right-hand side of Equation 5.110 (Figure 5.13). The second and third terms are the hygrothermal strains in the lamina, if it were unrestrained. Note that the lamina is a restrained one as it is part of a laminate. Thus, the total strains in the lamina are likely to be different from the hygrothermal strains. The first term inside the brackets is the vector of total strains in the lamina. The difference between the total strains and the hygrothermal strains is the vector of strains called the mechanical strains, that is,

$$\begin{Bmatrix} \varepsilon_{xx} \\ \varepsilon_{yy} \\ \gamma_{xy} \end{Bmatrix}_m = \begin{Bmatrix} \varepsilon_{xx} \\ \varepsilon_{yy} \\ \gamma_{xy} \end{Bmatrix} - \begin{Bmatrix} \varepsilon_{xx} \\ \varepsilon_{yy} \\ \gamma_{xy} \end{Bmatrix}_{HT} \qquad (5.111)$$

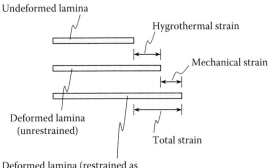

FIGURE 5.13 Schematic representation of total strain, hygrothermal strain, and mechanical strain in a lamina.

and

$$\left\{\begin{array}{c}\sigma_{xx}\\ \sigma_{yy}\\ \tau_{xy}\end{array}\right\}_{HT} = [\bar{Q}]\left\{\begin{array}{c}\varepsilon_{xx}\\ \varepsilon_{yy}\\ \gamma_{xy}\end{array}\right\}_{m} \tag{5.112}$$

where

$$\left\{\begin{array}{c}\varepsilon_{xx}\\ \varepsilon_{yy}\\ \gamma_{xy}\end{array}\right\}_{m}$$

are the mechanical strains,

$$\left\{\begin{array}{c}\varepsilon_{xx}\\ \varepsilon_{yy}\\ \gamma_{xy}\end{array}\right\}$$

are the total strains,

$$\left\{\begin{array}{c}\varepsilon_{xx}\\ \varepsilon_{yy}\\ \gamma_{xy}\end{array}\right\}_{HT} = \Delta T \left\{\begin{array}{c}\alpha_{x}\\ \alpha_{y}\\ \alpha_{xy}\end{array}\right\} + \Delta C \left\{\begin{array}{c}\beta_{x}\\ \beta_{y}\\ \beta_{xy}\end{array}\right\}$$

are the hygrothermal strains, and

$$\left\{\begin{array}{c}\sigma_{xx}\\ \sigma_{yy}\\ \tau_{xy}\end{array}\right\}_{HT}$$

are the hygrothermal stresses.

Macromechanics of a Laminate

The resultant hygrothermal force or moment on the laminate is zero. So, integration of the hygrothermal stresses across the laminate thickness is zero. Thus,

$$\int_{-h/2}^{h/2} \begin{Bmatrix} \sigma_{xx} \\ \sigma_{yy} \\ \tau_{xy} \end{Bmatrix}_{HT} dz = 0 \tag{5.113}$$

$$\int_{-h/2}^{h/2} \begin{Bmatrix} \sigma_{xx} \\ \sigma_{yy} \\ \tau_{xy} \end{Bmatrix}_{HT} z\, dz = 0 \tag{5.114}$$

The hygrothermal stress variation across the laminate thickness is discontinuous and a stepwise integration is carried out, as follows:

$$\sum_{k=1}^{n} \int_{z_{k-1}}^{z_k} \begin{Bmatrix} \sigma_{xx} \\ \sigma_{yy} \\ \tau_{xy} \end{Bmatrix}_{HT} dz = 0 \tag{5.115}$$

$$\sum_{k=1}^{n} \int_{z_{k-1}}^{z_k} \begin{Bmatrix} \sigma_{xx} \\ \sigma_{yy} \\ \tau_{xy} \end{Bmatrix}_{HT} z\, dz = 0 \tag{5.116}$$

Substituting Equation 5.112 in Equations 5.115 and 5.116, we get

$$\sum_{k=1}^{n} \int_{z_{k-1}}^{z_k} [\bar{Q}]_k \begin{Bmatrix} \varepsilon_{xx} \\ \varepsilon_{yy} \\ \gamma_{xy} \end{Bmatrix}_m dz = 0 \tag{5.117}$$

$$\sum_{k=1}^{n} \int_{z_{k-1}}^{z_k} [\bar{Q}]_k \begin{Bmatrix} \varepsilon_{xx} \\ \varepsilon_{yy} \\ \gamma_{xy} \end{Bmatrix}_m z\, dz = 0 \tag{5.118}$$

The mechanical strains are substituted with the total strains and the hygrothermal strains. Total strains are then further substituted with the middle surface strains and curvatures and we get the following:

$$\sum_{k=1}^{n} \int_{z_{k-1}}^{z_k} [\bar{Q}]_k \left(\begin{Bmatrix} \varepsilon_{xx}^0 \\ \varepsilon_{yy}^0 \\ \gamma_{xy}^0 \end{Bmatrix} + z \begin{Bmatrix} \kappa_{xx} \\ \kappa_{yy} \\ \kappa_{xy} \end{Bmatrix} - \begin{Bmatrix} \varepsilon_{xx} \\ \varepsilon_{yy} \\ \gamma_{xy} \end{Bmatrix}_{HT} \right) dz = 0 \tag{5.119}$$

$$\sum_{k=1}^{n} \int_{z_{k-1}}^{z_k} [\bar{Q}]_k \left(\begin{Bmatrix} \varepsilon_{xx}^0 \\ \varepsilon_{yy}^0 \\ \gamma_{xy}^0 \end{Bmatrix} + z \begin{Bmatrix} \kappa_{xx} \\ \kappa_{yy} \\ \kappa_{xy} \end{Bmatrix} - \begin{Bmatrix} \varepsilon_{xx} \\ \varepsilon_{yy} \\ \gamma_{xy} \end{Bmatrix}_{HT} \right) z\, dz = 0 \tag{5.120}$$

We can write the above two equations in the following form:

$$\left(\sum_{k=1}^{n}\int_{z_{k-1}}^{z_k}[\bar{Q}]_k\,dz\right)\begin{Bmatrix}\varepsilon_{xx}^0\\\varepsilon_{yy}^0\\\gamma_{xy}^0\end{Bmatrix} + \left(\sum_{k=1}^{n}\int_{z_{k-1}}^{z_k}[\bar{Q}]_k\,zdz\right)\begin{Bmatrix}\kappa_{xx}\\\kappa_{yy}\\\kappa_{xy}\end{Bmatrix} - \left(\sum_{k=1}^{n}\int_{z_{k-1}}^{z_k}[\bar{Q}]_k\begin{Bmatrix}\varepsilon_{xx}\\\varepsilon_{yy}\\\gamma_{xy}\end{Bmatrix}_{HT}dz\right) = 0 \quad (5.121)$$

$$\left(\sum_{k=1}^{n}\int_{z_{k-1}}^{z_k}[\bar{Q}]_k\,zdz\right)\begin{Bmatrix}\varepsilon_{xx}^0\\\varepsilon_{yy}^0\\\gamma_{xy}^0\end{Bmatrix} + \left(\sum_{k=1}^{n}\int_{z_{k-1}}^{z_k}[\bar{Q}]_k\,z^2dz\right)\begin{Bmatrix}\kappa_{xx}\\\kappa_{yy}\\\kappa_{xy}\end{Bmatrix} - \left(\sum_{k=1}^{n}\int_{z_{k-1}}^{z_k}[\bar{Q}]_k\begin{Bmatrix}\varepsilon_{xx}\\\varepsilon_{yy}\\\gamma_{xy}\end{Bmatrix}_{HT}zdz\right) = 0 \quad (5.122)$$

Note that the hygrothermal strains, unlike the middle surface strains and curvatures, cannot be taken outside the integrations. Using the definitions of $[A]$, $[B]$, and $[D]$ matrices from Equations 5.39 through 5.41, and taking the terms associated with total strains to the right-hand side, we can write Equations 5.121 and 5.122 as follows:

$$\sum_{k=1}^{n}\int_{z_{k-1}}^{z_k}[\bar{Q}]_k\begin{Bmatrix}\varepsilon_{xx}\\\varepsilon_{yy}\\\gamma_{xy}\end{Bmatrix}_{HT}dz = [A]\begin{Bmatrix}\varepsilon_{xx}^0\\\varepsilon_{yy}^0\\\gamma_{xy}^0\end{Bmatrix} + [B]\begin{Bmatrix}\kappa_{xx}\\\kappa_{yy}\\\kappa_{xy}\end{Bmatrix} \quad (5.123)$$

$$\sum_{k=1}^{n}\int_{z_{k-1}}^{z_k}[\bar{Q}]_k\begin{Bmatrix}\varepsilon_{xx}\\\varepsilon_{yy}\\\gamma_{xy}\end{Bmatrix}_{HT}zdz = [B]\begin{Bmatrix}\varepsilon_{xx}^0\\\varepsilon_{yy}^0\\\gamma_{xy}^0\end{Bmatrix} + [D]\begin{Bmatrix}\kappa_{xx}\\\kappa_{yy}\\\kappa_{xy}\end{Bmatrix} \quad (5.124)$$

The terms on the left-hand side of the above equations are the hygrothermal force and moment resultants. These force/moment resultants, in fact, are fictitious force/moment resultants. The fictitious hygrothermal force/moment resultants have two components—thermal and hygroscopic such that

$$\{N^{HT}\} = \{N^T\} + \{N^H\} \quad (5.125)$$

$$\{M^{HT}\} = \{M^T\} + \{M^H\} \quad (5.126)$$

where

$$\{N^{HT}\} = \sum_{k=1}^{n}[\bar{Q}]_k\begin{Bmatrix}\varepsilon_{xx}\\\varepsilon_{yy}\\\gamma_{xy}\end{Bmatrix}_{HT}(z_k - z_{k-1})$$

are the fictitious hygrothermal force resultants (from the left-hand side of Equation 5.123)

$$\{N^T\} = \Delta T\sum_{k=1}^{n}[\bar{Q}]_k\begin{Bmatrix}\alpha_x\\\alpha_y\\\alpha_{xy}\end{Bmatrix}(z_k - z_{k-1})$$

are the fictitious thermal force resultants (writing the thermal strains in terms of CTE)

$$\{N^H\} = \Delta C \sum_{k=1}^{n} [\bar{Q}]_k \begin{Bmatrix} \beta_x \\ \beta_y \\ \beta_{xy} \end{Bmatrix} (z_k - z_{k-1})$$

are the fictitious hygroscopic force resultants (writing the hygroscopic strains in terms of CME)

$$\{M^{HT}\} = \frac{1}{2} \sum_{k=1}^{n} [\bar{Q}]_k \begin{Bmatrix} \varepsilon_{xx} \\ \varepsilon_{yy} \\ \gamma_{xy} \end{Bmatrix}_{HT} (z_k^2 - z_{k-1}^2)$$

are the fictitious hygrothermal force resultants (from the left-hand side of Equation 5.124)

$$\{M^T\} = \frac{\Delta T}{2} \sum_{k=1}^{n} [\bar{Q}]_k \begin{Bmatrix} \alpha_x \\ \alpha_y \\ \alpha_{xy} \end{Bmatrix} (z_k^2 - z_{k-1}^2)$$

are the fictitious thermal moment resultants (writing the thermal strains in terms of CTE), and

$$\{M^H\} = \frac{\Delta C}{2} \sum_{k=1}^{n} [\bar{Q}]_k \begin{Bmatrix} \beta_x \\ \beta_y \\ \beta_{xy} \end{Bmatrix} (z_k^2 - z_{k-1}^2)$$

are the fictitious hygroscopic moment resultants (writing the hygroscopic strains in terms of CME)

Then, Equations 5.123 and 5.124 can be written as follows:

$$\{N^{HT}\} = \{N^T\} + \{N^H\} = [A] \begin{Bmatrix} \varepsilon_{xx}^0 \\ \varepsilon_{yy}^0 \\ \gamma_{xy}^0 \end{Bmatrix} + [B] \begin{Bmatrix} \kappa_{xx} \\ \kappa_{yy} \\ \kappa_{xy} \end{Bmatrix} \quad (5.127)$$

$$\{M^{HT}\} = \{M^T\} + \{M^H\} = [B] \begin{Bmatrix} \varepsilon_{xx}^0 \\ \varepsilon_{yy}^0 \\ \gamma_{xy}^0 \end{Bmatrix} + [D] \begin{Bmatrix} \kappa_{xx} \\ \kappa_{yy} \\ \kappa_{xy} \end{Bmatrix} \quad (5.128)$$

Combining the above, in a compact form, we can write

$$\begin{Bmatrix} N^{HT} \\ M^{HT} \end{Bmatrix} = \begin{Bmatrix} N^T \\ M^T \end{Bmatrix} + \begin{Bmatrix} N^H \\ M^H \end{Bmatrix} = \begin{bmatrix} A & B \\ B & D \end{bmatrix} \begin{Bmatrix} \varepsilon^0 \\ \kappa \end{Bmatrix} \quad (5.129)$$

or, in the explicit form,

$$\begin{Bmatrix} N_{xx} \\ N_{yy} \\ N_{xy} \\ M_{xx} \\ M_{yy} \\ M_{xy} \end{Bmatrix}^{HT} = \begin{Bmatrix} N_{xx} \\ N_{yy} \\ N_{xy} \\ M_{xx} \\ M_{yy} \\ M_{xy} \end{Bmatrix}^{T} + \begin{Bmatrix} N_{xx} \\ N_{yy} \\ N_{xy} \\ M_{xx} \\ M_{yy} \\ M_{xy} \end{Bmatrix}^{H} = \begin{bmatrix} A_{11} & A_{12} & A_{16} & B_{11} & B_{12} & B_{16} \\ A_{12} & A_{22} & A_{26} & B_{12} & B_{22} & B_{26} \\ A_{16} & A_{26} & A_{66} & B_{16} & B_{26} & B_{66} \\ B_{11} & B_{12} & B_{16} & D_{11} & D_{12} & D_{16} \\ B_{12} & B_{22} & B_{26} & D_{12} & D_{22} & D_{26} \\ B_{16} & B_{26} & B_{66} & D_{16} & D_{26} & D_{66} \end{bmatrix} \begin{Bmatrix} \varepsilon^{0}_{xx} \\ \varepsilon^{0}_{yy} \\ \gamma^{0}_{xy} \\ \kappa_{xx} \\ \kappa_{yy} \\ \kappa_{xy} \end{Bmatrix}$$

(5.130)

Equations 5.129 and 5.130 are the constitutive relations for a laminate under hygrothermal loads. In the compliance form, the middle surface strains and curvatures in a laminate under hygrothermal loads can be obtained as follows:

$$\begin{Bmatrix} \varepsilon^0 \\ \kappa \end{Bmatrix} = \begin{bmatrix} A^* & B^* \\ C^* & D^* \end{bmatrix} \begin{Bmatrix} N^{HT} \\ M^{HT} \end{Bmatrix} = \begin{bmatrix} A^* & B^* \\ C^* & D^* \end{bmatrix} \left(\begin{Bmatrix} N^T \\ M^T \end{Bmatrix} + \begin{Bmatrix} N^H \\ M^H \end{Bmatrix} \right) \quad (5.131)$$

5.7.2 Coefficients of Thermal Expansion and Coefficients of Moisture Expansion of a Laminate

The middle surface strains, in terms of CTE and CME, in a laminate under the action of pure hygrothermal loads are given by

$$\begin{Bmatrix} \varepsilon^0_{xx} \\ \varepsilon^0_{yy} \\ \gamma^0_{xy} \end{Bmatrix} = \Delta T \begin{Bmatrix} \alpha_x \\ \alpha_y \\ \alpha_{xy} \end{Bmatrix} + \Delta C \begin{Bmatrix} \beta_x \\ \beta_y \\ \beta_{xy} \end{Bmatrix} \quad (5.132)$$

The CTEs are defined as changes in length per unit length per unit change in temperature. Then, the CTEs are obtained by substituting $\Delta T = 1$ and $\Delta C = 0$ in Equation 5.132 as

$$\begin{Bmatrix} \alpha_x \\ \alpha_y \\ \alpha_{xy} \end{Bmatrix} = \begin{Bmatrix} \varepsilon^0_{xx} \\ \varepsilon^0_{yy} \\ \gamma^0_{xy} \end{Bmatrix}_{\substack{\Delta T=1 \\ \Delta C=0}} \quad (5.133)$$

Then, using the upper half of Equation 5.131, we obtain the CTEs as follows:

$$\begin{Bmatrix} \alpha_x \\ \alpha_y \\ \alpha_{xy} \end{Bmatrix} = \begin{bmatrix} A^*_{11} & A^*_{12} & A^*_{16} \\ A^*_{12} & A^*_{22} & A^*_{26} \\ A^*_{16} & A^*_{26} & A^*_{66} \end{bmatrix} \begin{Bmatrix} N^T_{xx} \\ N^T_{yy} \\ N^T_{xy} \end{Bmatrix}_{\substack{\Delta T=1 \\ \Delta C=0}} + \begin{bmatrix} B^*_{11} & B^*_{12} & B^*_{16} \\ B^*_{21} & B^*_{22} & B^*_{26} \\ B^*_{61} & B^*_{62} & B^*_{66} \end{bmatrix} \begin{Bmatrix} M^T_{xx} \\ M^T_{yy} \\ M^T_{xy} \end{Bmatrix}_{\substack{\Delta T=1 \\ \Delta C=0}} \quad (5.134)$$

In a similar way, the CMEs are obtained by substituting $\Delta T = 0$ and $\Delta C = 1$, as follows:

$$\begin{Bmatrix} \beta_x \\ \beta_y \\ \beta_{xy} \end{Bmatrix} = \begin{bmatrix} A_{11}^* & A_{12}^* & A_{16}^* \\ A_{12}^* & A_{22}^* & A_{26}^* \\ A_{16}^* & A_{26}^* & A_{66}^* \end{bmatrix} \begin{Bmatrix} N_{xx}^H \\ N_{yy}^H \\ N_{xy}^H \end{Bmatrix}_{\substack{\Delta T=0 \\ \Delta C=1}} + \begin{bmatrix} B_{11}^* & B_{12}^* & B_{16}^* \\ B_{21}^* & B_{22}^* & B_{26}^* \\ B_{61}^* & B_{62}^* & B_{66}^* \end{bmatrix} \begin{Bmatrix} M_{xx}^T \\ M_{yy}^T \\ M_{xy}^T \end{Bmatrix}_{\substack{\Delta T=0 \\ \Delta c=1}} \quad (5.135)$$

EXAMPLE 5.3

Consider a glass/epoxy laminate [45°/−45°/45°/−45°]. Each ply 1.25 mm in thickness and the following material properties are given.

$E_1 = 40$ GPa, $E_2 = 6$ GPa, $\nu_{12} = 0.25$, $G_{12} = 4$ GPa, $\alpha_1 = 8 \times 10^{-6}$ m/m/°C, $\alpha_2 = 20 \times 10^{-6}$ m/m/°C, and $\alpha_{12} = 0$.

The laminate is cured at a temperature of 125°C and then brought down to ambient temperature of 25°C. Estimate residual strains in the laminate.

Solution

For the given material properties, the reduced stiffness matrix is obtained as

$$[Q] = \begin{bmatrix} 40.379 & 1.514 & 0 \\ 1.514 & 6.057 & 0 \\ 0 & 0 & 4 \end{bmatrix} \times 10^3 \text{ MPa}$$

Noting that $\sin 45° = \cos 45° = \cos(-45°) = 0.7071$ and $\sin(-45°) = -0.7071$, the transformed reduced stiffness matrices can be readily obtained as

$$[\bar{Q}]_{45°} = \begin{bmatrix} 16.366 & 8.366 & 8.580 \\ 8.366 & 16.366 & 8.580 \\ 8.580 & 8.580 & 10.852 \end{bmatrix} \times 10^3 \text{ MPa}$$

and

$$[\bar{Q}]_{-45°} = \begin{bmatrix} 16.366 & 8.366 & -8.580 \\ 8.366 & 16.366 & -8.580 \\ -8.580 & -8.580 & 10.852 \end{bmatrix} \times 10^3 \text{ MPa}$$

The z coordinates for the four ply laminate are as follows:

$z_0 = -2.5$ mm, $z_1 = -1.25$ mm, $z_2 = 0$ mm, $z_3 = 1.25$ mm, and $z_4 = 2.5$ mm

Then, the laminate stiffness matrices are determined (Equations 5.39 through 5.41), as follows (details of calculations are not shown):

$$[A] = \begin{bmatrix} 81.830 & 41.830 & 0 \\ 41.830 & 81.830 & 0 \\ 0 & 0 & 54.259 \end{bmatrix} \times 10^3 \text{ MPa} \cdot \text{mm}$$

$$[B] = \begin{bmatrix} 0 & 0 & -26.814 \\ 0 & 0 & -26.814 \\ -26.814 & -26.814 & 0 \end{bmatrix} \times 10^3 \, \text{MPa} \cdot \text{mm}^2$$

$$[D] = \begin{bmatrix} 170.478 & 87.145 & 0 \\ 87.145 & 170.478 & 0 \\ 0 & 0 & 113.039 \end{bmatrix} \times 10^3 \, \text{MPa} \cdot \text{mm}^3$$

Next, the laminate compliance matrices are determined (details of calculations are not shown)

$$[A^*] = \begin{bmatrix} 17.007 & -7.993 & 0 \\ -7.993 & 17.007 & 0 \\ 0 & 0 & 20.544 \end{bmatrix} \times 10^{-6} (\text{MPa} \cdot \text{mm})^{-1}$$

$$[B^*] = \begin{bmatrix} 0 & 0 & 2.138 \\ 0 & 0 & 2.138 \\ 2.138 & 2.138 & 0 \end{bmatrix} \times 10^{-6} (\text{MPa} \cdot \text{mm}^2)^{-1}$$

$$[C^*] = \begin{bmatrix} 0 & 0 & 2.138 \\ 0 & 0 & 2.138 \\ 2.138 & 2.138 & 0 \end{bmatrix} \times 10^{-6} (\text{MPa} \cdot \text{mm}^2)^{-1}$$

$$[D^*] = \begin{bmatrix} 8.163 & -3.837 & 0 \\ -3.837 & 8.163 & 0 \\ 0 & 0 & 9.861 \end{bmatrix} \times 10^{-6} (\text{MPa} \cdot \text{mm}^3)^{-1}$$

The transformation matrices are (refer Equation 4.45)

$$[T]_{45°} = \begin{bmatrix} 0.5 & 0.5 & 1 \\ 0.5 & 0.5 & -1 \\ -0.5 & 0.5 & 0 \end{bmatrix}$$

$$[T]_{-45°} = \begin{bmatrix} 0.5 & 0.5 & -1 \\ 0.5 & 0.5 & 1 \\ 0.5 & -0.5 & 0 \end{bmatrix}$$

Then, the off-axis CTEs are given by (refer Equation 4.189)

$$\begin{Bmatrix} \alpha_x \\ \alpha_y \\ \alpha_{xy} \end{Bmatrix}_{45°} = \begin{bmatrix} 0.5 & 0.5 & -0.5 \\ 0.5 & 0.5 & 0.5 \\ 1 & -1 & 0 \end{bmatrix} \begin{Bmatrix} 8.0 \\ 20.0 \\ 0 \end{Bmatrix} \times 10^{-6} = \begin{Bmatrix} 14.0 \\ 14.0 \\ -12.0 \end{Bmatrix} \times 10^{-6}$$

Macromechanics of a Laminate

and

$$\begin{Bmatrix} \alpha_x \\ \alpha_y \\ \alpha_{xy} \end{Bmatrix}_{-45°} = \begin{bmatrix} 0.5 & 0.5 & 0.5 \\ 0.5 & 0.5 & -0.5 \\ -1 & 1 & 0 \end{bmatrix} \begin{Bmatrix} 8.0 \\ 20.0 \\ 0 \end{Bmatrix} \times 10^{-6} = \begin{Bmatrix} 14.0 \\ 14.0 \\ 12.0 \end{Bmatrix} \times 10^{-6}$$

Now, the fictitious thermal force resultants are given by (refer Equations 5.125 and 5.126)

$$\{N^T\} = -100 \times \left(\begin{bmatrix} 16.366 & 8.366 & 8.580 \\ 8.366 & 16.366 & 8.580 \\ 8.580 & 8.580 & 10.852 \end{bmatrix} \begin{Bmatrix} 14 \\ 14 \\ -12 \end{Bmatrix} \times 1.25 \right.$$

$$+ \begin{bmatrix} 16.366 & 8.366 & -8.580 \\ 8.366 & 16.366 & -8.580 \\ -8.580 & -8.580 & 10.852 \end{bmatrix} \begin{Bmatrix} 14 \\ 14 \\ 12 \end{Bmatrix} \times 1.25$$

$$+ \begin{bmatrix} 16.366 & 8.366 & 8.580 \\ 8.366 & 16.366 & 8.580 \\ 8.580 & 8.580 & 10.852 \end{bmatrix} \begin{Bmatrix} 14 \\ 14 \\ -12 \end{Bmatrix} \times 1.25$$

$$\left. + \begin{bmatrix} 16.366 & 8.366 & -8.5804 \\ 8.366 & 16.366 & -8.5804 \\ -8.5804 & -8.5804 & 10.852 \end{bmatrix} \begin{Bmatrix} 14 \\ 14 \\ 12 \end{Bmatrix} \times 1.25 \right) \times 10^{-3} = \begin{Bmatrix} -121.640 \\ -121.640 \\ 0 \end{Bmatrix} \text{N/mm}$$

and

$$\{M^T\} = -50 \times \left(\begin{bmatrix} 16.366 & 8.366 & 8.580 \\ 8.366 & 16.366 & 8.580 \\ 8.580 & 8.580 & 10.852 \end{bmatrix} \begin{Bmatrix} 14 \\ 14 \\ -12 \end{Bmatrix} \times (4.688) \right.$$

$$+ \begin{bmatrix} 16.366 & 8.366 & -8.580 \\ 8.366 & 16.366 & -8.580 \\ -8.580 & -8.580 & 10.852 \end{bmatrix} \begin{Bmatrix} 14 \\ 14 \\ 12 \end{Bmatrix} \times (1.563)$$

$$+ \begin{bmatrix} 16.366 & 8.366 & 8.580 \\ 8.366 & 16.366 & 8.580 \\ 8.580 & 8.580 & 10.852 \end{bmatrix} \begin{Bmatrix} 14 \\ 14 \\ -12 \end{Bmatrix} \times (-1.563)$$

$$\left. + \begin{bmatrix} 16.366 & 8.366 & -8.580 \\ 8.366 & 16.366 & -8.580 \\ -8.580 & -8.580 & 10.852 \end{bmatrix} \begin{Bmatrix} 14 \\ 14 \\ 12 \end{Bmatrix} \times (-4.688) \right) \times 10^{-3} = \begin{Bmatrix} 0 \\ 0 \\ -34.385 \end{Bmatrix} \text{N} \cdot \text{mm/mm}$$

Then, the middle surface strains and curvatures are determined as

$$\begin{Bmatrix} \varepsilon_{xx}^0 \\ \varepsilon_{yy}^0 \\ \gamma_{xy}^0 \end{Bmatrix} = \left(\begin{bmatrix} 17.007 & -7.993 & 0 \\ -7.993 & 17.007 & 0 \\ 0 & 0 & 20.544 \end{bmatrix} \times \begin{Bmatrix} -121.640 \\ -121.640 \\ 0 \end{Bmatrix} + \begin{bmatrix} 0 & 0 & 2.138 \\ 0 & 0 & 2.138 \\ 2.138 & 2.138 & 0 \end{bmatrix} \right.$$

$$\left. \times \begin{Bmatrix} 0 \\ 0 \\ 34.385 \end{Bmatrix} \right) \times 10^{-6} = \begin{Bmatrix} -1.17 \\ -1.17 \\ 0 \end{Bmatrix} \times 10^{-3}$$

and

$$\begin{Bmatrix} \kappa_{xx} \\ \kappa_{yy} \\ \kappa_{xy} \end{Bmatrix} = \left(\begin{bmatrix} 0 & 0 & 2.138 \\ 0 & 0 & 2.138 \\ 2.138 & 2.138 & 0 \end{bmatrix} \times \begin{Bmatrix} -121.640 \\ -121.640 \\ 0 \end{Bmatrix} + \begin{bmatrix} 8.163 & -3.837 & 0 \\ -3.837 & 8.163 & 0 \\ 0 & 0 & 9.861 \end{bmatrix} \right.$$

$$\left. \times \begin{Bmatrix} 0 \\ 0 \\ -34.385 \end{Bmatrix} \right) \times 10^{-6} = \begin{Bmatrix} 0 \\ 0 \\ -0.859 \end{Bmatrix} \times 10^{-3}$$

5.8 SPECIAL CASES OF LAMINATES

Stacking sequence of a laminate is a critical design parameter and it has been of great interest among researches as reflected by numerous papers related to topics such as effect of stacking sequence on general laminate performance, effect of stacking sequence on laminates with specific features such as holes, ply thickness, and clustering, optimization of stacking sequence, etc. (see References 8–12 among many others). Depending on the lamina angles, thicknesses, materials, and types of reinforcements, different types of laminate stacking sequences are possible. There are certain types of stacking sequences that have special significance in the design and analysis of laminated composite structures (see Reference 3 for a comprehensive discussion). The stacking sequence of a laminate has direct influence on the laminate stiffness matrices. In the subsequent sections, we shall see that, in some special cases of laminate stacking sequences, some of the terms in the laminate stiffness matrices vanish. As a result, in these laminate stacking sequences, computational effort is greatly reduced.

Further, each term in the laminate stiffness matrices has specific effect on the final performance of the laminate under different loading conditions. Thus, a good understanding of the terms in the [A], [B], and [D] matrices is essential in the design and analysis of a laminated composite structure.

5.8.1 Significance of Stiffness Matrix Terms

Let us rewrite the laminate constitutive relation in Equation 5.42 in a split form as follows:

$$\begin{Bmatrix} N_{xx} \\ N_{yy} \\ N_{xy} \end{Bmatrix} = \begin{bmatrix} A_{11} & A_{12} & A_{16} \\ A_{12} & A_{22} & A_{26} \\ A_{16} & A_{26} & A_{66} \end{bmatrix} \begin{Bmatrix} \varepsilon_{xx}^0 \\ \varepsilon_{yy}^0 \\ \gamma_{xy}^0 \end{Bmatrix} + \begin{bmatrix} B_{11} & B_{12} & B_{16} \\ B_{12} & B_{22} & B_{26} \\ B_{16} & B_{26} & B_{66} \end{bmatrix} \begin{Bmatrix} \kappa_{xx} \\ \kappa_{yy} \\ \kappa_{xy} \end{Bmatrix} \quad (5.136)$$

$$\begin{Bmatrix} M_{xx} \\ M_{yy} \\ M_{xy} \end{Bmatrix} = \begin{bmatrix} B_{11} & B_{12} & B_{16} \\ B_{12} & B_{22} & B_{26} \\ B_{16} & B_{26} & B_{66} \end{bmatrix} \begin{Bmatrix} \varepsilon_{xx}^0 \\ \varepsilon_{yy}^0 \\ \gamma_{xy}^0 \end{Bmatrix} + \begin{bmatrix} D_{11} & D_{12} & D_{16} \\ D_{12} & D_{22} & D_{26} \\ D_{16} & D_{26} & D_{66} \end{bmatrix} \begin{Bmatrix} \kappa_{xx} \\ \kappa_{yy} \\ \kappa_{xy} \end{Bmatrix} \quad (5.137)$$

The [A] matrix associates the force resultants with the middle surface strains and [D] matrix associates the moment resultants with the middle surface curvatures. On the other hand, [B] matrix associates force resultants with middle surface curvatures and moment resultants with middle surface strains. Let us have a more detailed look at Equations 5.136 and 5.137 and note that each term in the stiffness matrices has unique contribution in associating the force/moment resultants with the strains/curvatures. For example, the term A_{11} associates the force resultant N_{xx} with the middle surface strain ε_{xx}^0. There is no coupling involved in this case. All the terms are checked and the details of such associations are tabulated in Table 5.3.

Before we proceed further, let us understand the terms association and coupling used in the table. The diagonal terms in the [A] and [D] matrices are nonzero, that is, $A_{11} \neq 0, A_{22} \neq 0, A_{66} \neq 0, D_{11} \neq 0, D_{22} \neq 0$, and $D_{66} \neq 0$. Similarly, the Poisson's terms

TABLE 5.3
Laminate Stiffness Matrix Terms and Corresponding Association between Stress Resultants and Strains/Curvatures

Stiffness Matrix Term	Association between		Associated Coupling
	Force/Moment Resultant	Strain/Curvature	
A_{11}	N_{xx}	ε_{xx}^0	No coupling
A_{22}	N_{yy}	ε_{yy}^0	No coupling
A_{66}	N_{xy}	γ_{xy}^0	No coupling
A_{12}	N_{xx}	ε_{yy}^0	Extension Poisson coupling
	N_{yy}	ε_{xx}^0	Extension Poisson coupling
A_{16}	N_{xx}	γ_{xy}^0	Extension–shear coupling
	N_{xy}	ε_{xx}^0	Extension–shear coupling
A_{26}	N_{yy}	γ_{xy}^0	Extension–shear coupling
	N_{xy}	ε_{yy}^0	Extension–shear coupling
B_{11}	N_{xx}	κ_{xx}	Extension–bending coupling
	M_{xx}	ε_{xx}^0	Extension–bending coupling
B_{22}	N_{yy}	κ_{yy}	Extension–bending coupling
	M_{yy}	ε_{yy}^0	Extension–bending coupling
B_{66}	N_{xy}	κ_{xy}	Shear–twisting coupling
	M_{xy}	γ_{xy}^0	Shear–twisting coupling
B_{12}	N_{xx}	κ_{yy}	Extension–bending Poisson coupling
	N_{yy}	κ_{xx}	Extension–bending Poisson coupling
	M_{xx}	ε_{yy}^0	Extension–bending Poisson coupling
	M_{yy}	ε_{xx}^0	Extension–bending Poisson coupling
B_{16}	N_{xx}	κ_{xy}	Extension–twisting coupling
	N_{xy}	κ_{xx}	Bending–shear coupling
	M_{xx}	γ_{xy}^0	Bending–shear coupling
	M_{xy}	ε_{xx}^0	Extension–twisting coupling
B_{26}	N_{yy}	κ_{xy}	Extension–twisting coupling
	N_{xy}	κ_{yy}	Bending–shear coupling
	M_{yy}	γ_{xy}^0	Bending–shear coupling
	M_{xy}	ε_{yy}^0	Extension–twisting coupling
D_{11}	M_{xx}	κ_{xx}	No coupling
D_{22}	M_{yy}	κ_{yy}	No coupling
D_{66}	M_{xy}	κ_{xy}	No coupling
D_{12}	M_{xx}	κ_{yy}	Bending Poisson coupling
	M_{yy}	κ_{xx}	Bending Poisson coupling
D_{16}	M_{xx}	κ_{xy}	Bending–twisting coupling
	M_{xy}	κ_{xx}	Bending–twisting coupling
D_{26}	M_{yy}	κ_{xy}	Bending–twisting coupling
	M_{xy}	κ_{yy}	Bending–twisting coupling

in these two matrices are also nonzero for nonzero Poisson's ratio, that is, $A_{12} \neq 0$ and $D_{12} \neq 0$. On the other hand, all other terms in the laminate stiffness matrices can be zero for certain stacking sequences.

Consider the term A_{11}. Consider a loading state such that $N_{xx} \neq 0$ and all other force and moment resultants are zero. As we see, A_{11} associates N_{xx} with ε_{xx}^0. It merely signifies that, A_{11} being a nonzero quantity, the laminate would experience normal strain ε_{xx}^0 under the application of N_{xx}. To obtain the value of the strain, let us write Equation 5.63 in the explicit split form as follows:

$$\begin{Bmatrix} \varepsilon_{xx}^0 \\ \varepsilon_{yy}^0 \\ \gamma_{xy}^0 \end{Bmatrix} = \begin{bmatrix} A_{11}^* & A_{12}^* & A_{16}^* \\ A_{12}^* & A_{22}^* & A_{26}^* \\ A_{16}^* & A_{26}^* & A_{66}^* \end{bmatrix} \begin{Bmatrix} N_{xx} \\ N_{yy} \\ N_{xy} \end{Bmatrix} + \begin{bmatrix} B_{11}^* & B_{12}^* & B_{16}^* \\ B_{21}^* & B_{22}^* & B_{26}^* \\ B_{61}^* & B_{62}^* & B_{66}^* \end{bmatrix} \begin{Bmatrix} M_{xx} \\ M_{yy} \\ M_{xy} \end{Bmatrix} \quad (5.138)$$

$$\begin{Bmatrix} \kappa_{xx} \\ \kappa_{yy} \\ \kappa_{xy} \end{Bmatrix} = \begin{bmatrix} B_{11}^* & B_{12}^* & B_{16}^* \\ B_{21}^* & B_{22}^* & B_{26}^* \\ B_{61}^* & B_{62}^* & B_{66}^* \end{bmatrix} \begin{Bmatrix} N_{xx} \\ N_{yy} \\ N_{xy} \end{Bmatrix} + \begin{bmatrix} D_{11}^* & D_{12}^* & D_{16}^* \\ D_{12}^* & D_{22}^* & D_{26}^* \\ D_{16}^* & D_{26}^* & D_{66}^* \end{bmatrix} \begin{Bmatrix} M_{xx} \\ M_{yy} \\ M_{xy} \end{Bmatrix} \quad (5.139)$$

For the given loads, the normal strain in the x-direction is given by: $\varepsilon_{xx}^0 = A_{11}^* N_{xx}$. Note that A_{11}^* contains a number of terms of $[A]$, $[B]$, and $[D]$ matrices. Further, existence of A_{11} does not indicate any deformation in other directions. In other words, there is no coupling.

Let us now consider a stiffness term with coupling, say, A_{16}. Considering the same loading condition, that is, other force and moment resultants being zero and a nonzero N_{xx}, for a nonzero A_{16}, the laminate would undergo in-plane shear deformation. Note that this association is between force resultant and strains in two different directions and it is a case of extension–shear coupling. Now, consider a load case with $N_{xy} \neq 0$ and all others zero. A nonzero A_{16} would imply a nonzero extension ε_{xx}^0, which is also a case of extension-shear coupling.

As mentioned earlier, some of the laminate stiffness matrix terms can be made to vanish under certain laminate stacking sequences. These special stacking sequences have significance in laminate design and analysis from two angles. First, with the reduction of nonzero terms, computational effort can be grossly reduced. Second, with a number of coupling terms vanishing, undesired coupling effects can be avoided. In the following sections, we shall discuss some of these special stacking sequences.

5.8.2 Single-Ply Laminate

Three types of single-ply laminates are possible—single isotropic ply, single specially orthotropic ply, and single generally orthotropic ply. Irrespective of the type, for a single-ply laminate (laminate thickness $h =$ lamina thickness t),

$$[B] = [\bar{Q}]\left[\left(\frac{t}{2}\right)^2 - \left(-\frac{t}{2}\right)^2\right] = 0 \quad (5.140)$$

5.8.2.1 Single Isotropic Ply

Let us consider an isotropic material with elastic constants E and v. In this case, $E_1 = E_2 = E$, $v_{12} = v_{21} = v$, and $G_{12} = G = E/2(1 + v)$. Then, for a single isotropic ply, the $[\bar{Q}]$ ($= [Q]$) matrix is readily obtained as follows (Equations 4.16 through 4.20):

Macromechanics of a Laminate

$$[\bar{Q}] = [Q] = \begin{bmatrix} \dfrac{E}{1-\nu^2} & \dfrac{\nu E}{1-\nu^2} & 0 \\ \dfrac{\nu E}{1-\nu^2} & \dfrac{E}{1-\nu^2} & 0 \\ 0 & 0 & \dfrac{E}{2(1+\nu)} \end{bmatrix} \quad (5.141)$$

Then, using Equations 5.39 through 5.41, the extensional, coupling, and bending stiffness matrices can be readily obtained and it is found that $[A]$ and $[D]$ are partially populated and $[B] = 0$. Laminate thickness $h =$ lamina thickness t. Then, the constitutive relations can be written as

$$\begin{Bmatrix} N_{xx} \\ N_{yy} \\ N_{xy} \end{Bmatrix} = \begin{bmatrix} \dfrac{Eh}{1-\nu^2} & \dfrac{\nu Eh}{1-\nu^2} & 0 \\ \dfrac{\nu Eh}{1-\nu^2} & \dfrac{Eh}{1-\nu^2} & 0 \\ 0 & 0 & \dfrac{Eh}{2(1+\nu)} \end{bmatrix} \begin{Bmatrix} \varepsilon_{xx}^0 \\ \varepsilon_{yy}^0 \\ \gamma_{xy}^0 \end{Bmatrix} \quad (5.142)$$

$$\begin{Bmatrix} M_{xx} \\ M_{yy} \\ M_{xy} \end{Bmatrix} = \begin{bmatrix} \dfrac{Eh^3}{12(1-\nu^2)} & \dfrac{\nu Eh^3}{12(1-\nu^2)} & 0 \\ \dfrac{\nu Eh^3}{12(1-\nu^2)} & \dfrac{Eh^3}{12(1-\nu^2)} & 0 \\ 0 & 0 & \dfrac{Eh^3}{24(1+\nu)} \end{bmatrix} \begin{Bmatrix} \kappa_{xx} \\ \kappa_{yy} \\ \kappa_{xy} \end{Bmatrix} \quad (5.143)$$

5.8.2.2 Single Specially Orthotropic Ply

For a single specially orthotropic ply, the stiffness matrices can be expressed in terms of the reduced stiffness matrices and the laminate thickness h. $[A]$ and $[D]$ are partially populated and $[B] = 0$. Then, the constitutive relations can be written as

$$\begin{Bmatrix} N_{xx} \\ N_{yy} \\ N_{xy} \end{Bmatrix} = h \begin{bmatrix} Q_{11} & Q_{12} & 0 \\ Q_{12} & Q_{22} & 0 \\ 0 & 0 & Q_{66} \end{bmatrix} \begin{Bmatrix} \varepsilon_{xx}^0 \\ \varepsilon_{yy}^0 \\ \gamma_{xy}^0 \end{Bmatrix} \quad (5.144)$$

$$\begin{Bmatrix} M_{xx} \\ M_{yy} \\ M_{xy} \end{Bmatrix} = \dfrac{h^3}{12} \begin{bmatrix} Q_{11} & Q_{12} & 0 \\ Q_{12} & Q_{22} & 0 \\ 0 & 0 & Q_{66} \end{bmatrix} \begin{Bmatrix} \kappa_{xx} \\ \kappa_{yy} \\ \kappa_{xy} \end{Bmatrix} \quad (5.145)$$

The reduced stiffness matrix can be expressed in terms of the orthotropic material properties and Equations 5.144 and 5.145 can be written as

$$\begin{Bmatrix} N_{xx} \\ N_{yy} \\ N_{xy} \end{Bmatrix} = h \begin{bmatrix} \dfrac{E_1}{1-\nu_{12}\nu_{21}} & \dfrac{\nu_{12}E_2}{1-\nu_{12}\nu_{21}} & 0 \\ \dfrac{\nu_{12}E_2}{1-\nu_{12}\nu_{21}} & \dfrac{E_2}{1-\nu_{12}\nu_{21}} & 0 \\ 0 & 0 & G_{12} \end{bmatrix} \begin{Bmatrix} \varepsilon_{xx}^0 \\ \varepsilon_{yy}^0 \\ \gamma_{xy}^0 \end{Bmatrix} \quad (5.146)$$

$$\begin{Bmatrix} M_{xx} \\ M_{yy} \\ M_{xy} \end{Bmatrix} = \frac{h^3}{12} \begin{bmatrix} \dfrac{E_1}{1-\nu_{12}\nu_{21}} & \dfrac{\nu_{12}E_2}{1-\nu_{12}\nu_{21}} & 0 \\ \dfrac{\nu_{12}E_2}{1-\nu_{12}\nu_{21}} & \dfrac{E_2}{1-\nu_{12}\nu_{21}} & 0 \\ 0 & 0 & G_{12} \end{bmatrix} \begin{Bmatrix} \kappa_{xx} \\ \kappa_{yy} \\ \kappa_{xy} \end{Bmatrix} \quad (5.147)$$

5.8.2.3 Single Generally Orthotropic Ply

In the case of a generally orthotropic single ply, [A] and [D] are fully populated and [B] = 0. Then, the constitutive relations can be written in terms of the transformed reduced stiffness matrices and laminate thickness h, as follows:

$$\begin{Bmatrix} N_{xx} \\ N_{yy} \\ N_{xy} \end{Bmatrix} = h \begin{bmatrix} \bar{Q}_{11} & \bar{Q}_{12} & \bar{Q}_{16} \\ \bar{Q}_{12} & \bar{Q}_{22} & \bar{Q}_{26} \\ \bar{Q}_{16} & \bar{Q}_{26} & \bar{Q}_{66} \end{bmatrix} \begin{Bmatrix} \varepsilon^0_{xx} \\ \varepsilon^0_{yy} \\ \gamma^0_{xy} \end{Bmatrix} \quad (5.148)$$

$$\begin{Bmatrix} M_{xx} \\ M_{yy} \\ M_{xy} \end{Bmatrix} = \frac{h^3}{12} \begin{bmatrix} \bar{Q}_{11} & \bar{Q}_{12} & \bar{Q}_{16} \\ \bar{Q}_{12} & \bar{Q}_{22} & \bar{Q}_{26} \\ \bar{Q}_{16} & \bar{Q}_{26} & \bar{Q}_{66} \end{bmatrix} \begin{Bmatrix} \kappa_{xx} \\ \kappa_{yy} \\ \kappa_{xy} \end{Bmatrix} \quad (5.149)$$

5.8.3 Symmetric Laminate

If in a laminate, the laminae angles, thicknesses, and materials are symmetric w.r.t. the middle surface, then the laminate is called a symmetric laminate. An example of a symmetric laminate is

$$\left[0°_{(CE,0.8)} / 90°_{(GE,0.4)} \right]_s$$

or

0° (carbon/epoxy, 0.8 mm thick)
90° (glass/epoxy, 0.4 mm thick)
90° (glass/epoxy, 0.4 mm thick)
0° (carbon/epoxy, 0.8 mm thick)

A symmetric laminate can have either even or odd number of plies. In a symmetric laminate with even number of plies, for each lamina below the middle surface, there will be an identical lamina above. Thus, for a laminate with n laminae, there will be $n/2$ identical laminae. Let us consider the kth lamina below the middle surface (Figure 5.14a). Let us denote the corresponding lamina above the middle surface by k'. Then,

$$\left[\bar{Q} \right]_{k'} = \left[\bar{Q} \right]_k \quad \text{and} \quad z_{k'-1} = -z_k \quad \text{and} \quad z_{k'} = -z_{k-1} \quad (5.150)$$

Now, the extension–bending coupling matrix can be obtained as

$$[B] = \frac{1}{2} \sum_{k=1}^{n/2} \left(\left[\bar{Q} \right]_k \left(z_k^2 - z_{k-1}^2 \right) + \left[\bar{Q} \right]_{k'} \left(z_{k'}^2 - z_{k'-1}^2 \right) \right) = 0 \quad (5.151)$$

Macromechanics of a Laminate

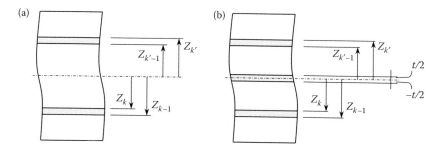

FIGURE 5.14 (a) Symmetric laminate with even number of plies. (b) Symmetric laminate with odd number of plies.

For a symmetric laminate with odd number of plies, there will be an extra ply in the middle such that the middle surface passes through this ply and the coupling matrix is given by

$$[B] = \frac{1}{2}\sum_{k=1}^{n-1/2}\left([\bar{Q}]_k\left(z_k^2 - z_{k-1}^2\right) + [\bar{Q}]_{k'}\left(z_{k'}^2 - z_{k'-1}^2\right)\right) + \frac{1}{2}[\bar{Q}]_{\text{midply}}\left((t/2)^2 - (-t/2)^2\right) = 0$$

(5.152)

Thus, for a symmetric laminate, irrespective of even or odd number of plies, the extension–bending coupling stiffness matrix $[B] = 0$. As a result, there is no coupling between extension and bending in a symmetric laminate. In other words, if a laminate is subjected to only in-plane forces, it will not have any middle surface curvature. Similarly, if it is subjected to only moments, it will have zero extensions in the middle surface. The constitutive relations for a general symmetric laminate take the following form:

$$\begin{Bmatrix} N_{xx} \\ N_{yy} \\ N_{xy} \end{Bmatrix} = \begin{bmatrix} A_{11} & A_{12} & A_{16} \\ A_{12} & A_{22} & A_{26} \\ A_{16} & A_{26} & A_{66} \end{bmatrix} \begin{Bmatrix} \varepsilon_{xx}^0 \\ \varepsilon_{yy}^0 \\ \gamma_{xy}^0 \end{Bmatrix}$$

(5.153)

$$\begin{Bmatrix} M_{xx} \\ M_{yy} \\ M_{xy} \end{Bmatrix} = \begin{bmatrix} D_{11} & D_{12} & D_{16} \\ D_{12} & D_{22} & D_{26} \\ D_{16} & D_{26} & D_{66} \end{bmatrix} \begin{Bmatrix} \kappa_{xx} \\ \kappa_{yy} \\ \kappa_{xy} \end{Bmatrix}$$

(5.154)

Absence of extension–bending coupling in a symmetric laminate is advantageous in more than one way. First, computations during design and analysis are grossly simplified. Since $[B] = 0$, we can see that $[B^*] = [C^*] = 0$ and the inverse forms of Equations 5.153 and 5.154 are as follows:

$$\begin{Bmatrix} \varepsilon_{xx}^0 \\ \varepsilon_{yy}^0 \\ \gamma_{xy}^0 \end{Bmatrix} = \begin{bmatrix} A_{11}^* & A_{12}^* & A_{16}^* \\ A_{12}^* & A_{22}^* & A_{26}^* \\ A_{16}^* & A_{26}^* & A_{66}^* \end{bmatrix} \begin{Bmatrix} N_{xx} \\ N_{yy} \\ N_{xy} \end{Bmatrix}$$

(5.155)

$$\begin{Bmatrix} \kappa_{xx} \\ \kappa_{yy} \\ \kappa_{xy} \end{Bmatrix} = \begin{bmatrix} D_{11}^* & D_{12}^* & D_{16}^* \\ D_{12}^* & D_{22}^* & D_{26}^* \\ D_{16}^* & D_{26}^* & D_{66}^* \end{bmatrix} \begin{Bmatrix} M_{xx} \\ M_{yy} \\ M_{xy} \end{Bmatrix}$$

(5.156)

in which, the elements of the compliance matrices are given directly in terms of the stiffness matrix elements, as follows:

$$A_{11}^* = \frac{A_{22}A_{66} - A_{26}^2}{\text{Det}[A]} \qquad (5.157)$$

$$A_{12}^* = \frac{A_{26}A_{16} - A_{12}A_{66}}{\text{Det}[A]} \qquad (5.158)$$

$$A_{16}^* = \frac{A_{12}A_{26} - A_{22}A_{16}}{\text{Det}[A]} \qquad (5.159)$$

$$A_{22}^* = \frac{A_{11}A_{66} - A_{16}^2}{\text{Det}[A]} \qquad (5.160)$$

$$A_{26}^* = \frac{A_{12}A_{16} - A_{11}A_{26}}{\text{Det}[A]} \qquad (5.161)$$

$$A_{66}^* = \frac{A_{11}A_{22} - A_{12}^2}{\text{Det}[A]} \qquad (5.162)$$

$$D_{11}^* = \frac{D_{22}D_{66} - D_{26}^2}{\text{Det}[D]} \qquad (5.163)$$

$$D_{12}^* = \frac{D_{26}D_{16} - D_{12}D_{66}}{\text{Det}[D]} \qquad (5.164)$$

$$D_{16}^* = \frac{D_{12}D_{26} - D_{22}D_{16}}{\text{Det}[D]} \qquad (5.165)$$

$$D_{22}^* = \frac{D_{11}D_{66} - D_{16}^2}{\text{Det}[D]} \qquad (5.166)$$

$$D_{26}^* = \frac{D_{12}D_{16} - D_{11}D_{26}}{\text{Det}[D]} \qquad (5.167)$$

$$D_{66}^* = \frac{D_{11}D_{22} - D_{12}^2}{\text{Det}[D]} \qquad (5.168)$$

Second, processing of symmetric laminate is simpler as undesirable warpage due to thermal loads during curing can be avoided. Third, during functional usage, a symmetric laminate does not exhibit undesirable bending or twisting deformations.

5.8.4 Antisymmetric Laminate

If in a laminate the plies are symmetric in respect of material and ply thickness but the ply angles at the same distance above and below the middle surface are negative of each other, then the laminate is called antisymmetric. The total number of plies in an antisymmetric laminate is always even. An example of an antisymmetric laminate is

$$\left[60°_{(CE,0.8)}/45°_{(GE,0.4)}/-45°_{(GE,0.4)}/-60°_{(CE,0.8)}\right]$$

or

60° (carbon/epoxy, 0.8 mm thick)
45° (glass/epoxy, 0.4 mm thick)
−45° (glass/epoxy, 0.4 mm thick)
−60° (carbon/epoxy, 0.8 mm thick)

In an antisymmetric laminate, the extension–shear coupling and bending–twisting coupling terms in the [A] and [D] matrices are zero, that is, $A_{16} = A_{26} = D_{16} = D_{26} = 0$ and the constitutive relations for an antisymmetric laminate can be written as follows:

$$\begin{Bmatrix} N_{xx} \\ N_{yy} \\ N_{xy} \end{Bmatrix} = \begin{bmatrix} A_{11} & A_{12} & 0 \\ A_{12} & A_{22} & 0 \\ 0 & 0 & A_{66} \end{bmatrix} \begin{Bmatrix} \varepsilon_{xx}^0 \\ \varepsilon_{yy}^0 \\ \gamma_{xy}^0 \end{Bmatrix} + \begin{bmatrix} B_{11} & B_{12} & B_{16} \\ B_{12} & B_{22} & B_{26} \\ B_{16} & B_{26} & B_{66} \end{bmatrix} \begin{Bmatrix} \kappa_{xx} \\ \kappa_{yy} \\ \kappa_{xy} \end{Bmatrix} \quad (5.169)$$

$$\begin{Bmatrix} M_{xx} \\ M_{yy} \\ M_{xy} \end{Bmatrix} = \begin{bmatrix} B_{11} & B_{12} & B_{16} \\ B_{12} & B_{22} & B_{26} \\ B_{16} & B_{26} & B_{66} \end{bmatrix} \begin{Bmatrix} \varepsilon_{xx}^0 \\ \varepsilon_{yy}^0 \\ \gamma_{xy}^0 \end{Bmatrix} + \begin{bmatrix} D_{11} & D_{12} & 0 \\ D_{12} & D_{22} & 0 \\ 0 & 0 & D_{66} \end{bmatrix} \begin{Bmatrix} \kappa_{xx} \\ \kappa_{yy} \\ \kappa_{xy} \end{Bmatrix} \quad (5.170)$$

5.8.5 Balanced Laminate

A balanced laminate is one in which for each $+\theta$ ply there is one $-\theta$ ply in the laminate. The plies in a balanced laminate are in pairs and the total number of plies is even. The location of the plies in a pair can be anywhere in the laminate but the materials and thicknesses of the plies in a pair are the same. An example of a balanced laminate is

$$\left[-45°_{(GE,0.4)} / 60°_{(CE,0.8)} / -60°_{(CE,0.8)} / 45°_{(GE,0.4)} \right]$$

or

−45° (glass/epoxy, 0.4 mm thick)
60° (carbon/epoxy, 0.8 mm thick)
−60° (carbon/epoxy, 0.8 mm thick)
45° (glass/epoxy, 0.4 mm thick)

In a balanced laminate, the extension–shear coupling terms in the [A] matrix are zero, that is, $A_{16} = A_{26} = 0$. Thus, the constitutive relations for an antisymmetric laminate can be written as follows:

$$\begin{Bmatrix} N_{xx} \\ N_{yy} \\ N_{xy} \end{Bmatrix} = \begin{bmatrix} A_{11} & A_{12} & 0 \\ A_{12} & A_{22} & 0 \\ 0 & 0 & A_{66} \end{bmatrix} \begin{Bmatrix} \varepsilon_{xx}^0 \\ \varepsilon_{yy}^0 \\ \gamma_{xy}^0 \end{Bmatrix} + \begin{bmatrix} B_{11} & B_{12} & B_{16} \\ B_{12} & B_{22} & B_{26} \\ B_{16} & B_{26} & B_{66} \end{bmatrix} \begin{Bmatrix} \kappa_{xx} \\ \kappa_{yy} \\ \kappa_{xy} \end{Bmatrix} \quad (5.171)$$

$$\begin{Bmatrix} M_{xx} \\ M_{yy} \\ M_{xy} \end{Bmatrix} = \begin{bmatrix} B_{11} & B_{12} & B_{16} \\ B_{12} & B_{22} & B_{26} \\ B_{16} & B_{26} & B_{66} \end{bmatrix} \begin{Bmatrix} \varepsilon_{xx}^0 \\ \varepsilon_{yy}^0 \\ \gamma_{xy}^0 \end{Bmatrix} + \begin{bmatrix} D_{11} & D_{12} & D_{16} \\ D_{12} & D_{22} & D_{26} \\ D_{16} & D_{26} & D_{66} \end{bmatrix} \begin{Bmatrix} \kappa_{xx} \\ \kappa_{yy} \\ \kappa_{xy} \end{Bmatrix} \quad (5.172)$$

5.8.6 Cross-Ply Laminate

A laminate is called a cross-ply laminate if it contains only 0° and 90° plies, that is, only specially orthotropic plies. Note that a cross-ply laminate is not necessarily symmetric and it has no restriction on the materials and ply thicknesses. An example of a cross-ply laminate is

$$\left[90°_{(GE,0.4)}/0°_{(CE,0.8)}/90_{(CE,0.8)}/90°_{(GE,0.4)}\right]$$

or

90° (glass/epoxy, 0.4 mm thick)
0° (carbon/epoxy, 0.8 mm thick)
90° (carbon/epoxy, 0.8 mm thick)
90° (glass/epoxy, 0.4 mm thick)

For 0° and 90° plies, the terms $\bar{Q}_{16}$ and $\bar{Q}_{26}$ in the transformed reduced stiffness matrix are zero. Thus, $A_{16} = A_{26} = B_{16} = B_{26} = D_{16} = D_{26} = 0$ and the constitutive relations can be written as follows:

$$\begin{Bmatrix} N_{xx} \\ N_{yy} \\ N_{xy} \end{Bmatrix} = \begin{bmatrix} A_{11} & A_{12} & 0 \\ A_{12} & A_{22} & 0 \\ 0 & 0 & A_{66} \end{bmatrix} \begin{Bmatrix} \varepsilon^0_{xx} \\ \varepsilon^0_{yy} \\ \gamma^0_{xy} \end{Bmatrix} + \begin{bmatrix} B_{11} & B_{12} & 0 \\ B_{12} & B_{22} & 0 \\ 0 & 0 & B_{66} \end{bmatrix} \begin{Bmatrix} \kappa_{xx} \\ \kappa_{yy} \\ \kappa_{xy} \end{Bmatrix} \quad (5.173)$$

$$\begin{Bmatrix} M_{xx} \\ M_{yy} \\ M_{xy} \end{Bmatrix} = \begin{bmatrix} B_{11} & B_{12} & 0 \\ B_{12} & B_{22} & 0 \\ 0 & 0 & B_{66} \end{bmatrix} \begin{Bmatrix} \varepsilon^0_{xx} \\ \varepsilon^0_{yy} \\ \gamma^0_{xy} \end{Bmatrix} + \begin{bmatrix} D_{11} & D_{12} & 0 \\ D_{12} & D_{22} & 0 \\ 0 & 0 & D_{66} \end{bmatrix} \begin{Bmatrix} \kappa_{xx} \\ \kappa_{yy} \\ \kappa_{xy} \end{Bmatrix} \quad (5.174)$$

A cross-ply laminate can be symmetric and in a symmetric cross-ply laminate, the constitutive relations become

$$\begin{Bmatrix} N_{xx} \\ N_{yy} \\ N_{xy} \end{Bmatrix} = \begin{bmatrix} A_{11} & A_{12} & 0 \\ A_{12} & A_{22} & 0 \\ 0 & 0 & A_{66} \end{bmatrix} \begin{Bmatrix} \varepsilon^0_{xx} \\ \varepsilon^0_{yy} \\ \gamma^0_{xy} \end{Bmatrix} \quad (5.175)$$

$$\begin{Bmatrix} M_{xx} \\ M_{yy} \\ M_{xy} \end{Bmatrix} = \begin{bmatrix} D_{11} & D_{12} & 0 \\ D_{12} & D_{22} & 0 \\ 0 & 0 & D_{66} \end{bmatrix} \begin{Bmatrix} \kappa_{xx} \\ \kappa_{yy} \\ \kappa_{xy} \end{Bmatrix} \quad (5.176)$$

5.8.7 Angle-Ply Laminate

An angle-ply laminate is one in which at least one ply angle is other than 0° or 90°. In a general case, in an angle-ply laminate, all the stiffness matrices are fully populated. Specific cases such as symmetric angle-ply, antisymmetric angle-ply, etc. can be made, in which cases some of the terms in the stiffness matrices vanish.

5.8.8 Quasi-Isotropic Laminate

A quasi-isotropic laminate is isotropic in the xy-plane. In such a laminate, the extensional stiffness matrix behaves like that of an isotropic material, which implies

Macromechanics of a Laminate

$$A_{11} = A_{22} \tag{5.177}$$

$$A_{16} = A_{26} = 0 \tag{5.178}$$

$$A_{66} = \frac{A_{11} - A_{12}}{2} \tag{5.179}$$

Note that the other two stiffness matrices [B] and [D] may not behave like isotropic materials. Some examples of quasi-isotropic laminates are

$$[0°/30°/60°/90°], [0°/60°/-60°], [0°/45°/90°/-45°], \text{etc.}$$

EXAMPLE 5.4

Consider the problem in Example 5.3. If the ply sequence is altered to: [45°/−45°/−45°/45°], estimate residual strains in the laminate. Other details remain the same.

Solution

Since the material properties are the same as in Example 5.3, the reduced stiffness matrix remains unchanged as follows:

$$[Q] = \begin{bmatrix} 40.379 & 1.514 & 0 \\ 1.514 & 6.057 & 0 \\ 0 & 0 & 4 \end{bmatrix} \times 10^3 \text{ MPa}$$

The transformed reduced stiffness matrices are

$$[\overline{Q}]_{45°} = \begin{bmatrix} 16.366 & 8.366 & 8.580 \\ 8.366 & 16.366 & 8.580 \\ 8.580 & 8.580 & 10.852 \end{bmatrix} \times 10^3 \text{ MPa}$$

and

$$[\overline{Q}]_{-45°} = \begin{bmatrix} 16.366 & 8.366 & -8.580 \\ 8.366 & 16.366 & -8.580 \\ -8.580 & -8.580 & 10.852 \end{bmatrix} \times 10^3 \text{ MPa}$$

The z coordinates for the four ply laminate are as follows:

$$z_0 = -2.5 \text{ mm}, z_1 = -1.25 \text{ mm}, z_2 = 0 \text{ mm}, z_3 = 1.25 \text{ mm, and } z_4 = 2.5 \text{ mm}$$

Since the ply sequence is symmetric, there is no extension–bending coupling, that is, [B] = 0. The laminate stiffness matrices are determined (Equations 5.39 through 5.41), as follows (details of calculations are not shown):

$$[A] = \begin{bmatrix} 81.830 & 41.830 & 0 \\ 41.830 & 81.830 & 0 \\ 0 & 0 & 54.259 \end{bmatrix} \times 10^3 \text{ MPa} \cdot \text{mm}$$

$$[\boldsymbol{B}] = \begin{bmatrix} 0 & 0 & 0 \\ 0 & 0 & 0 \\ 0 & 0 & 0 \end{bmatrix}$$

$$[\boldsymbol{D}] = \begin{bmatrix} 170.478 & 87.145 & 67.035 \\ 87.145 & 170.478 & 67.035 \\ 67.035 & 67.035 & 113.039 \end{bmatrix} \times 10^3 \, \text{MPa} \cdot \text{mm}^3$$

Next, the laminate compliance matrices are determined (details of calculations are not shown)

$$[\boldsymbol{A}^*] = \begin{bmatrix} 16.543 & -8.457 & 0 \\ -8.457 & 16.543 & 0 \\ 0 & 0 & 18.430 \end{bmatrix} \times 10^{-6} \, (\text{MPa} \cdot \text{mm})^{-1}$$

$$[\boldsymbol{B}^*] = \begin{bmatrix} 0 & 0 & 0 \\ 0 & 0 & 0 \\ 0 & 0 & 0 \end{bmatrix}$$

$$[\boldsymbol{C}^*] = \begin{bmatrix} 0 & 0 & 0 \\ 0 & 0 & 0 \\ 0 & 0 & 0 \end{bmatrix}$$

$$[\boldsymbol{D}^*] = \begin{bmatrix} 8.807 & -3.193 & -3.329 \\ -3.193 & 8.807 & -3.329 \\ -3.329 & -3.329 & 12.795 \end{bmatrix} \times 10^{-6} \, (\text{MPa} \cdot \text{mm}^3)^{-1}$$

The transformation matrices are

$$[\boldsymbol{T}]_{45°} = \begin{bmatrix} 0.5 & 0.5 & 1 \\ 0.5 & 0.5 & -1 \\ -0.5 & 0.5 & 0 \end{bmatrix}$$

$$[\boldsymbol{T}]_{-45°} = \begin{bmatrix} 0.5 & 0.5 & -1 \\ 0.5 & 0.5 & 1 \\ 0.5 & -0.5 & 0 \end{bmatrix}$$

Then, the off-axis CTEs are obtained from Example 5.3 as

$$\begin{Bmatrix} \alpha_x \\ \alpha_y \\ \alpha_{xy} \end{Bmatrix}_{45°} = \begin{Bmatrix} 14.0 \\ 14.0 \\ -12.0 \end{Bmatrix} \times 10^{-6}$$

Macromechanics of a Laminate

and

$$\begin{Bmatrix} \alpha_x \\ \alpha_y \\ \alpha_{xy} \end{Bmatrix}_{-45°} = \begin{Bmatrix} 14.0 \\ 14.0 \\ 12.0 \end{Bmatrix} \times 10^{-6}$$

Now, the fictitious thermal force resultants are given by

$$\{N^T\} = -100 \times \left(\begin{bmatrix} 16.366 & 8.366 & 8.580 \\ 8.366 & 16.366 & 8.580 \\ 8.580 & 8.580 & 10.852 \end{bmatrix} \begin{Bmatrix} 14 \\ 14 \\ -12 \end{Bmatrix} \times 1.25 \right.$$

$$+ \begin{bmatrix} 16.366 & 8.366 & -8.580 \\ 8.366 & 16.366 & -8.580 \\ -8.580 & -8.580 & 10.852 \end{bmatrix} \begin{Bmatrix} 14 \\ 14 \\ 12 \end{Bmatrix} \times 1.25$$

$$+ \begin{bmatrix} 16.366 & 8.366 & -8.580 \\ 8.366 & 16.366 & -8.580 \\ -8.580 & -8.580 & 10.852 \end{bmatrix} \begin{Bmatrix} 14 \\ 14 \\ 12 \end{Bmatrix} \times 1.25$$

$$\left. + \begin{bmatrix} 16.366 & 8.366 & 8.580 \\ 8.366 & 16.366 & 8.580 \\ 8.580 & 8.580 & 10.852 \end{bmatrix} \begin{Bmatrix} 14 \\ 14 \\ -12 \end{Bmatrix} \times 1.25 \right) \times 10^{-3} = \begin{Bmatrix} -121.640 \\ -121.640 \\ 0 \end{Bmatrix} \text{N/mm}$$

and

$$\{M^T\} = -50 \times \left(\begin{bmatrix} 16.366 & 8.366 & 8.580 \\ 8.366 & 16.366 & 8.580 \\ 8.580 & 8.580 & 10.852 \end{bmatrix} \begin{Bmatrix} 14 \\ 14 \\ -12 \end{Bmatrix} \times (-4.688) \right.$$

$$+ \begin{bmatrix} 16.366 & 8.366 & -8.580 \\ 8.366 & 16.366 & -8.580 \\ -8.580 & -8.580 & 10.852 \end{bmatrix} \begin{Bmatrix} 14 \\ 14 \\ 12 \end{Bmatrix} \times (-1.563)$$

$$+ \begin{bmatrix} 16.366 & 8.366 & -8.580 \\ 8.366 & 16.366 & -8.580 \\ -8.580 & -8.580 & 10.852 \end{bmatrix} \begin{Bmatrix} 14 \\ 14 \\ 12 \end{Bmatrix} \times (1.563)$$

$$\left. + \begin{bmatrix} 16.366 & 8.366 & 8.580 \\ 8.366 & 16.366 & 8.580 \\ 8.580 & 8.580 & 10.852 \end{bmatrix} \begin{Bmatrix} 14 \\ 14 \\ -12 \end{Bmatrix} \times (4.688) \right) \times 10^{-3} = \begin{Bmatrix} 0 \\ 0 \\ 0 \end{Bmatrix} \text{N} \cdot \text{mm/mm}$$

Then, the middle surface strains and curvatures are determined as

$$\begin{Bmatrix} \varepsilon_{xx}^0 \\ \varepsilon_{yy}^0 \\ \gamma_{xy}^0 \end{Bmatrix} = \begin{bmatrix} 16.543 & -8.457 & 0 \\ -8.457 & 16.543 & 0 \\ 0 & 0 & 18.430 \end{bmatrix} \times \begin{Bmatrix} -121.640 \\ -121.640 \\ 0 \end{Bmatrix} \times 10^{-6} = \begin{Bmatrix} -0.984 \\ -0.984 \\ 0 \end{Bmatrix} \times 10^{-3}$$

and

$$\begin{Bmatrix} \kappa_{xx} \\ \kappa_{yy} \\ \kappa_{xy} \end{Bmatrix} = \begin{bmatrix} 8.807 & -3.193 & -3.329 \\ -3.193 & 8.8071 & -3.329 \\ -3.329 & -3.329 & 12.795 \end{bmatrix} \times \begin{Bmatrix} 0 \\ 0 \\ 0 \end{Bmatrix} \times 10^{-6} = \begin{Bmatrix} 0 \\ 0 \\ 0 \end{Bmatrix}$$

5.9 FAILURE ANALYSIS OF A LAMINATE

5.9.1 First Ply Failure and Last Ply Failure

Failure analysis of a laminate is the study, in which we study the stress and strain behavior of the laminate under increasing load until failure occurs. The goal of failure analysis is to determine the strength of the laminate, which is the maximum load that it can take before failure. Before we proceed further, let us note the following points:

- Failure of a laminate depends on failure of the individual laminae
- In general, failures of all the laminae do not occur simultaneously
- Failure of a single ply need not necessarily trigger failures of all the plies

As we know, laminae are the building blocks in a laminate, and it is natural that failure of a laminate depends on failure of the individual laminae. However, failure of the laminate can also occur without the failure of the individual laminae. Sometimes, the laminae get separated from each other due to interlaminar shear failure even though the individual laminae may still be intact. The laminae are stacked in different orientations. Also, the materials of the laminae may be different. As a result, strength and stiffness characteristics of different laminae are different. Each lamina responds to the applied loads as per its own strength/stiffness characteristics. Quite naturally, it can be expected that each lamina will have its own failure loads and failure of all the laminae in the laminate will not occur simultaneously.

When a lamina fails, the applied loads will have to be shared by the remaining laminae and the stresses and strains increase. If the revised stresses and strains in the remaining laminae are still within safe limits, the laminate, as a whole, continues to take higher loads until ultimate failure of the laminate due to failure of all the laminae.

From the above discussions, it is clear that the failure process in a laminate is not catastrophic; rather it is a gradual process and failure may be concluded to have occurred at the point of initiation of failure, at the point of completion, or in between. Thus, failure load of a laminate depends on the adopted philosophy on laminate failure. Broadly, there are two ways to define laminate failure—first ply failure (FPF) and last ply failure (LPF). FPF is defined as the failure of a laminate when ply failure process starts with the failure of the first ply. The FPF is a rather conservative approach to designing a composite product and it is used generally in primary structures. LPF, also known as ultimate laminate failure (ULF), is defined as the failure of a laminate when all the plies fail. LPF signifies the ultimate load-carrying capacity of a laminate.

5.9.2 Progressive Failure Analysis

The failure phenomenon of an individual lamina was discussed in Chapter 4. As we saw there, a lamina has basically five direction-dependent failure modes—longitudinal tensile failure, longitudinal compressive failure, transverse tensile failure, transverse compressive failure, and in-plane shear failure. While longitudinal failures are controlled by the reinforcement, transverse and in-plane shear failures are controlled by the matrix.

A step-by-step procedure for laminate analysis is given in Table 5.2. The main output of laminate analysis are as follows: local stresses (σ_{11}, σ_{22}, and τ_{12}) and local strains (ε_{11}, ε_{22}, and γ_{12}) in each ply. (Local stresses and strains are in the principal material directions.) Having obtained the local stresses and strains, a suitable failure criterion is applied to each lamina. Generally, in a failure analysis process, a small load is applied on the laminate such that all the plies are safe. Each ply is checked for available factor of safety or margin of safety and the ply with the minimum available factor is identified. Then, the applied load is extrapolated to determine the load at which the first ply fails.

When a ply fails, a redistribution of load sharing takes place. This redistribution depends on the mode of failure of the ply. Thus, it is necessary to identify the mode of failure of each lamina. For example, a ply may fail by cracks parallel to the fibers due to high transverse tensile stress. The ply is incapable of providing transverse/in-plane shear stiffness/strength while the longitudinal stiffness/strength remains unaffected. In the progressive failure analysis process, such a ply is degraded. There are two methods for ply degradation—*total ply degradation* and *partial ply degradation*.

In the total ply degradation method, when a ply fails, irrespective of the mode of failure, all the stiffnesses are made zero. (Note that in actual practice, a ply is degraded by making the stiffnesses near zero but not zero as otherwise mathematical singularity problems arise.) Keeping the degraded plies geometrically unchanged, laminate stiffness matrices are determined again. The redistribution of stresses lead to higher stresses in the remaining plies and applied loads are increased further, if possible, till next failure takes place. The process is then continued till ultimate failure occurs.

In the partial ply degradation method, when a ply fails, the stiffnesses corresponding to the mode of failure are made zero. Thus, if a ply fails by cracks parallel to the fibers due to high transverse tensile stress, the stiffnesses E_2 and G_{12} are made zero, keeping E_1 unchanged. Similarly, if a ply fails due to high longitudinal stress, E_1 is degraded.

Note: Often, failure of a ply due to high transverse normal stress/strain or in-plane shear stress/strain is not considered as ply failure. For example, in a composite pressure vessel, where the stress state is predominantly biaxial tensile, a ply is considered to have failed only when it fails due to tensile stress in the fiber direction. Also, accuracy of failure theories in predicting laminate failure under all loading and boundary conditions is still a subject of debate (see, for instance, Reference 13 for a review of failure theories). Given above is a procedure that is found convenient and acceptable in a wide range of design cases (also see References 4, 14, and 15).

EXAMPLE 5.5

Consider a glass/epoxy cylindrical composite pressure vessel of diameter 400 mm. If the ply sequence is [90°/60°/30°]$_s$, determine the maximum internal pressure that the pressure vessel can contain. Each ply is 1.25 mm in thickness and the material properties are as follows:

$$E_1 = 40 \text{ GPa}, E_2 = 8 \text{ GPa}, \nu_{12} = 0.25, \text{ and } G_{12} = 4 \text{ GPa}$$

The strength data are as follows:

$(\sigma_{11}^T)_{ult} = 1000\,\text{MPa}, (\sigma_{11}^C)_{ult} = 600\,\text{MPa}, (\sigma_{22}^T)_{ult} = 30\,\text{MPa}, (\sigma_{22}^C)_{ult} = 120\,\text{MPa},$
$(\tau_{12})_{ult} = 70\,\text{MPa}$

Solution

In a progressive failure analysis, as we know, the applied loads are gradually increased and, at different load levels, specific plies are either partially or totally degraded. While the geometrical details remain unchanged, the characteristics related to material and loads change at different load levels.

The transformation matrices for the plies are

$$[T]_{90°} = \begin{bmatrix} 0 & 1 & 0 \\ 1 & 0 & 0 \\ 0 & 0 & -1 \end{bmatrix}$$

$$[T]_{60°} = \begin{bmatrix} 0.25 & 0.75 & 0.866 \\ 0.75 & 0.25 & -0.866 \\ -0.433 & 0.433 & -0.5 \end{bmatrix}$$

$$[T]_{30°} = \begin{bmatrix} 0.75 & 0.25 & 0.866 \\ 0.25 & 0.75 & -0.866 \\ -0.433 & 0.433 & 0.5 \end{bmatrix}$$

and the z coordinates for the six ply laminate are as follows:

$z_0 = -3.75\,\text{mm}, z_1 = -2.50\,\text{mm}, z_2 = -1.25\,\text{mm}, z_3 = 0.0\,\text{mm},$
$z_4 = 1.25\,\text{mm}, z_5 = 2.50\,\text{mm}, \text{and } z_6 = 3.75\,\text{mm}.$

Let us denote internal pressure by p. The internal pressure would result in a state of membrane stress in the pressure vessel shell.

Let x-axis be along the axis of the pressure vessel and let y-axis along the circumference. Also, let z-axis be outward normal axis. Then, from basic strength of materials approach, if R is the radius of the cylinder, the force resultants can be determined as follows:

$$\begin{Bmatrix} N_{xx} \\ N_{yy} \\ N_{xy} \end{Bmatrix} = \begin{Bmatrix} \dfrac{pR}{2} \\ pR \\ 0 \end{Bmatrix}$$

and

$$\begin{Bmatrix} M_{xx} \\ M_{yy} \\ M_{xy} \end{Bmatrix} = \begin{Bmatrix} 0 \\ 0 \\ 0 \end{Bmatrix}$$

Macromechanics of a Laminate

Load step 1: $p = 1$ MPa

Force and moment resultants are

$$\begin{Bmatrix} N_{xx} \\ N_{yy} \\ N_{xy} \end{Bmatrix} = \begin{Bmatrix} 100 \\ 200 \\ 0 \end{Bmatrix}$$

and

$$\begin{Bmatrix} M_{xx} \\ M_{yy} \\ M_{xy} \end{Bmatrix} = \begin{Bmatrix} 0 \\ 0 \\ 0 \end{Bmatrix}$$

The reduced stiffness matrix is the same for all the plies and it is given by

$$[\boldsymbol{Q}]_{90°} = [\boldsymbol{Q}]_{60°} = [\boldsymbol{Q}]_{30°} = \begin{bmatrix} 40.506 & 2.025 & 0 \\ 2.025 & 8.101 & 0 \\ 0 & 0 & 4 \end{bmatrix} \times 10^3 \text{ MPa}$$

The transformed reduced stiffness matrices are

$$[\bar{\boldsymbol{Q}}]_{90°} = \begin{bmatrix} 8.101 & 2.025 & 0 \\ 2.025 & 40.506 & 0 \\ 0 & 0 & 4 \end{bmatrix} \times 10^3 \text{ MPa}$$

$$[\bar{\boldsymbol{Q}}]_{60°} = \begin{bmatrix} 10.848 & 7.380 & 3.925 \\ 7.380 & 27.051 & 10.107 \\ 3.925 & 10.107 & 9.354 \end{bmatrix} \times 10^3 \text{ MPa}$$

$$[\bar{\boldsymbol{Q}}]_{30°} = \begin{bmatrix} 27.051 & 7.380 & 10.107 \\ 7.380 & 10.848 & 3.925 \\ 10.107 & 3.925 & 9.354 \end{bmatrix} \times 10^3 \text{ MPa}$$

The laminate stiffness matrices are

$$[\boldsymbol{A}] = \begin{bmatrix} 115.000 & 41.962 & 35.080 \\ 41.962 & 196.013 & 35.080 \\ 35.080 & 35.080 & 56.772 \end{bmatrix} \times 10^3 \text{ MPa} \cdot \text{mm}$$

$$[\boldsymbol{B}] = \begin{bmatrix} 0 & 0 & 0 \\ 0 & 0 & 0 \\ 0 & 0 & 0 \end{bmatrix}$$

$$[\boldsymbol{D}] = \begin{bmatrix} 334.520 & 126.978 & 48.931 \\ 126.978 & 1262.790 & 97.234 \\ 48.931 & 97.234 & 196.400 \end{bmatrix} \times 10^3 \text{ MPa} \cdot \text{mm}^3$$

Next, the laminate compliance matrices are determined (details of calculations are not shown)

$$[A^*] = \begin{bmatrix} 10.993 & -1.279 & -6.002 \\ -1.279 & 5.885 & -2.846 \\ -6.002 & -2.846 & 23.082 \end{bmatrix} \times 10^{-6} \ (\text{MPa} \cdot \text{mm})^{-1}$$

$$[B^*] = \begin{bmatrix} 0 & 0 & 0 \\ 0 & 0 & 0 \\ 0 & 0 & 0 \end{bmatrix}$$

$$[C^*] = \begin{bmatrix} 0 & 0 & 0 \\ 0 & 0 & 0 \\ 0 & 0 & 0 \end{bmatrix}$$

$$[D^*] = \begin{bmatrix} 3.188 & -0.270 & -0.661 \\ -0.270 & 0.846 & -0.352 \\ -0.661 & -0.352 & 5.430 \end{bmatrix} \times 10^{-6} \ (\text{MPa} \cdot \text{mm}^3)^{-1}$$

Then, using Equation 5.63, the middle surface strains are obtained as follows:

$$\begin{Bmatrix} \varepsilon_{xx}^0 \\ \varepsilon_{yy}^0 \\ \gamma_{xy}^0 \end{Bmatrix} = \begin{bmatrix} 10.993 & -1.279 & -6.002 \\ -1.279 & 5.885 & -2.846 \\ -6.002 & -2.846 & 23.082 \end{bmatrix} \times \begin{Bmatrix} 100 \\ 200 \\ 0 \end{Bmatrix} \times 10^{-6} = \begin{Bmatrix} 843.5 \\ 1049.1 \\ -1169.4 \end{Bmatrix} \times 10^{-6}$$

and

$$\begin{Bmatrix} \kappa_{xx} \\ \kappa_{yy} \\ \kappa_{xy} \end{Bmatrix} = \begin{Bmatrix} 0 \\ 0 \\ 0 \end{Bmatrix}$$

Now, using Equation 5.14, the global strains in each ply can be determined. In general, the ply strains would depend on their z-locations w.r.t. the middle surface. However, in this case, the middle surface curvatures are zero. Thus, the global strains in the plies are all equal and they are equal to those of the middle surface, that is,

$$\begin{Bmatrix} \varepsilon_{xx} \\ \varepsilon_{yy} \\ \gamma_{xy} \end{Bmatrix}_{90°} = \begin{Bmatrix} \varepsilon_{xx} \\ \varepsilon_{yy} \\ \gamma_{xy} \end{Bmatrix}_{60°} = \begin{Bmatrix} \varepsilon_{xx} \\ \varepsilon_{yy} \\ \gamma_{xy} \end{Bmatrix}_{30°} = \begin{Bmatrix} 843.5 \\ 1049.1 \\ -1169.4 \end{Bmatrix} \times 10^{-6}$$

Then, using Equation 4.53, the global stresses in each ply are determined as follows:

$$\begin{Bmatrix} \sigma_{xx} \\ \sigma_{yy} \\ \tau_{xy} \end{Bmatrix}_{90°} = \begin{bmatrix} 8.101 & 2.025 & 0 \\ 2.025 & 40.506 & 0 \\ 0 & 0 & 4.000 \end{bmatrix} \times \begin{Bmatrix} 843.5 \\ 1049.1 \\ -1169.4 \end{Bmatrix} \times 10^{-3} = \begin{Bmatrix} 8.96 \\ 44.20 \\ -4.68 \end{Bmatrix} \text{MPa}$$

Macromechanics of a Laminate

$$\begin{Bmatrix} \sigma_{xx} \\ \sigma_{yy} \\ \tau_{xy} \end{Bmatrix}_{60°} = \begin{bmatrix} 10.848 & 7.380 & 3.925 \\ 7.380 & 27.051 & 10.107 \\ 3.925 & 10.107 & 9.354 \end{bmatrix} \times \begin{Bmatrix} 843.5 \\ 1049.1 \\ -1169.4 \end{Bmatrix} \times 10^{-3} = \begin{Bmatrix} 12.30 \\ 22.78 \\ 2.97 \end{Bmatrix} \text{MPa}$$

$$\begin{Bmatrix} \sigma_{xx} \\ \sigma_{yy} \\ \tau_{xy} \end{Bmatrix}_{30°} = \begin{bmatrix} 27.051 & 7.380 & 10.107 \\ 7.380 & 10.848 & 3.925 \\ 10.107 & 3.925 & 9.354 \end{bmatrix} \times \begin{Bmatrix} 843.5 \\ 1049.1 \\ -1169.4 \end{Bmatrix} \times 10^{-3} = \begin{Bmatrix} 18.74 \\ 13.02 \\ 1.70 \end{Bmatrix} \text{MPa}$$

Having determined the global stresses in each ply, the local stresses in the plies are obtained by transformation; these are

$$\begin{Bmatrix} \sigma_{11} \\ \sigma_{22} \\ \tau_{12} \end{Bmatrix}_{90°} = \begin{Bmatrix} 44.20 \\ 8.96 \\ 4.68 \end{Bmatrix} \text{MPa}$$

$$\begin{Bmatrix} \sigma_{11} \\ \sigma_{22} \\ \tau_{12} \end{Bmatrix}_{60°} = \begin{Bmatrix} 22.74 \\ 12.35 \\ 3.05 \end{Bmatrix} \text{MPa}$$

$$\begin{Bmatrix} \sigma_{11} \\ \sigma_{22} \\ \tau_{12} \end{Bmatrix}_{30°} = \begin{Bmatrix} 18.78 \\ 12.97 \\ -1.63 \end{Bmatrix} \text{MPa}$$

We can now apply a suitable failure criterion. Let us choose maximum stress failure criterion for our analysis. (Alternatively, we can also determine the local strains and apply maximum strain failure criterion.) The strength ratios as per maximum stress failure criterion are calculated, as follows:

90° ply:

$$(R)_{11} = \frac{1000}{44.20} = 22.62$$

$$(R)_{22} = \frac{30}{8.96} = 3.35$$

$$(R)_{12} = \frac{70}{4.68} = 14.96$$

60° ply:

$$(R)_{11} = \frac{1000}{22.74} = 43.98$$

$$(R)_{22} = \frac{30}{12.35} = 2.43$$

TABLE 5.4

Local Stresses at Different Internal Pressures (Example 5.5)

Load Step	Pressure (MPa)	Description	90° Ply Stresses (MPa)			60° Ply Stresses (MPa)			30° Ply Stresses (MPa)		
			σ_{11}	σ_{22}	τ_{12}	σ_{11}	σ_{22}	τ_{12}	σ_{11}	σ_{22}	τ_{12}
1	1.00	Start of loading	44.2	9.0	4.7	22.7	12.4	3.1	18.8	13.0	−1.6
2	2.31	Before FPF	102.1	20.8	10.9	52.4	28.6	7.2	43.4	30.0	−3.7
3	2.31	30° plies partially degraded	120.5	28.4	16.9	49.6	39.6	8.5	39.0	–	–
4	2.31	30° and 60° plies partially degraded	155.7	48.4	31.6	21.6	–	–	51.4	–	–
5	2.31	30°, 60°, and 90° plies partially degraded	277.2	–	–	−184.8	–	–	184.8	–	–
6	7.51	30°, 60°, and 90° plies partially degraded	900.9	–	–	−600.6	–	–	600.6	–	–

$$(R)_{12} = \frac{70}{3.05} = 22.95$$

30° ply:

$$(R)_{11} = \frac{1000}{18.78} = 53.25$$

$$(R)_{22} = \frac{30}{12.97} = 2.31$$

$$(R)_{12} = \frac{70}{1.63} = 42.94$$

The local stresses in the plies and the strength ratios at 1 MPa pressure are tabulated in Tables 5.4 and 5.5, respectively.

The minimum strength ratio from above is found as 2.31. We know that a ply fails when the strength ratio is 1. Thus, the pressure can be increased by 2.31 times the currently applied pressure before FPF takes place. Since the currently applied pressure is 1 MPa, the internal pressure can be increased to 2.31 MPa.

Load step 2: $p = 2.31$ MPa.

Local stresses in the plies are obtained by linear extrapolation and tabulated in Table 5.4. The strength ratios are also calculated in a similar way and tabulated in Table 5.5. These stresses and strength ratios are just before FPF. We can see that

TABLE 5.5

Strength Ratios at Different Internal Pressures (Example 5.5)

Load Step	Pressure (MPa)	Description	90° Ply Strength Ratios			60° Ply Strength Ratios			30° Ply Strength Ratios		
			R_{11}	R_{22}	R_{12}	R_{11}	R_{22}	R_{12}	R_{11}	R_{22}	R_{12}
1	1	Start of loading	22.62	3.35	14.96	43.98	2.43	22.95	53.25	2.31	42.94
2	2.31	Before FPF	9.79	1.45	6.48	19.04	1.05	9.94	23.05	**1.00**	18.45
3	2.31	30° plies partially degraded	8.30	1.06	4.14	20.16	**0.76**	8.24	25.64	–	–
4	2.31	30° and 60° plies partially degraded	6.42	**0.62**	2.22	46.30	–	–	19.46	–	–
5	2.31	30°, 60°, and 90° plies partially degraded	3.61	–	–	3.25	–	–	5.41	–	–
6	7.51	30°, 60°, and 90° plies partially degraded	1.11	–	–	**1.00**	–	–	1.67	–	–

FPF takes place in the form of failure of the 30° plies due to excessive transverse tensile stress. Now, we can degrade the 30° plies. Let us adopt the partial ply degradation method.

Load step 3: $p = 2.31$ MPa.

Put $E_2 = G_{12} \approx 0$ in the 30° plies, that is, these plies are partially degraded keeping the geometry intact.

The revised reduced stiffness matrices are

$$[Q]_{90°} = [Q]_{60°} = \begin{bmatrix} 40.506 & 2.025 & 0 \\ 2.025 & 8.101 & 0 \\ 0 & 0 & 4 \end{bmatrix} \times 10^3 \text{ MPa}$$

$$[Q]_{30°} = \begin{bmatrix} 40 & 0 & 0 \\ 0 & 0 & 0 \\ 0 & 0 & 0 \end{bmatrix} \times 10^3 \text{ MPa}$$

Note that the reduced stiffness matrix for the 30° plies is different from the others.

The revised transformed reduced stiffness matrices are

$$[\bar{Q}]_{90°} = \begin{bmatrix} 8.101 & 2.025 & 0 \\ 2.025 & 40.506 & 0 \\ 0 & 0 & 4 \end{bmatrix} \times 10^3 \text{ MPa}$$

$$[\bar{Q}]_{60°} = \begin{bmatrix} 10.848 & 7.380 & 3.925 \\ 7.380 & 27.051 & 10.107 \\ 3.925 & 10.107 & 9.354 \end{bmatrix} \times 10^3 \text{ MPa}$$

$$[\bar{Q}]_{30°} = \begin{bmatrix} 22.500 & 7.500 & 12.990 \\ 7.500 & 2.500 & 4.330 \\ 12.990 & 4.330 & 7.500 \end{bmatrix} \times 10^3 \text{ MPa}$$

Note that the transformed reduced stiffness matrix for the 30° plies is different from what it was before the ply degradation.

The laminate stiffness matrices are then determined followed by determination of middle surface strains and finally the local ply stresses and strength ratios. For the sake of brevity, we would not show all these steps. The final local ply stresses and the strength ratios are tabulated in Tables 5.4 and 5.5. On partial degradation of the 30° plies, as we can see, redistribution of loads takes place; the transverse tensile stress in the 60° plies rise beyond its strength and the corresponding strength ratio becomes smaller than unity. This redistribution takes place instantaneously, that is, without any increase in the pressure. Thus, we degrade the 60° plies partially in respect of the transverse and shear stiffnesses.

Load step 4: $p = 2.31$ MPa.

Put $E_2 = G_{12} \approx 0$ in the 60° plies in addition to the 30° plies. The procedure followed is similar to what we had done in the previous load step and the local stresses and the strength ratios are tabulated.

Now, we see that the 90° plies, as indicated by the lower than unity strength ratio, too have failed in respect of the transverse and shear stiffnesses.

Load step 5: $p = 2.31$ MPa

Put $E_2 = G_{12} \approx 0$ in the 90° plies in addition to the 60° and 30° plies. In other words, all the plies are degraded now. The strength ratios are all greater than unity, which indicates that the pressure can be increased before next ply failure. The minimum strength ratio at this stage is 3.25, which means that the pressure can be increased by 3.25 times, that is, to 7.51 MPa.

Load step 6: $p = 7.51$ MPa

Put $E_2 = G_{12} \approx 0$ in the 90°, 60°, and 30° plies, that is, all the plies. The strength ratio corresponding to the longitudinal stress in the 90° plies is unity. Thus, at this pressure, the 60° plies have failed fully.

We can conclude that final failure takes place at $p = 7.51$ MPa by fiber fracture in the 60° plies.

5.10 OTHER TOPICS IN A LAMINATE ANALYSIS

5.10.1 Interlaminar Stress

CLT assumes a plane stress state in the laminate. In other words, the out-of-plane stresses are considered as zero, that is,

$$\sigma_{zz} = \tau_{xz} = \tau_{yz} = 0 \tag{5.180}$$

The first stress in the above equation is the interlaminar normal stress, whereas the other two are the interlaminar shear stresses. These stresses are zero in the regions of laminate far away from the free edges. Near the free edges of a laminate (e.g., the sides of a laminate or holes and cutouts), the interlaminar stresses are rather large and in many cases, they are the causes for free edge delamination. Behavior of interlaminar stresses under different conditions including loads, laminate configurations, and stacking sequence is crucial in the design of composite products and different types of solutions have been proposed (see, for instance, References 16–18).

To see how interlaminar stresses are generated, let us consider a cross-ply laminate [90°/0°/90°] under transverse tensile force as shown in Figure 5.15a (also see Reference 19). Let us first have a physical explanation. Under the tensile force in the y-direction, the laminate undergoes contraction in the x-direction. As we know, for unidirectional composites, v_{12} is far greater than v_{21}. As a result, if the plies were unbonded, that is, free to deform without any constraint from adjacent plies, the 90° plies would undergo more contraction than the 0° ply. However, the bond between the plies does not allow such differential deformation. The 90° plies expand relative to free deformed configuration whereas the 0° ply contracts (Figure 5.15b). This means that the 90° plies and the 0° will be in tension and compression, respectively, in the x-direction (Figure 5.15c). As we can see in Figure 5.15c, if we consider one-half of the laminate, it will be in equilibrium. Now, let us look at the free body diagrams of the individual plies (Figure 5.15d). Considering the free body diagram of the top ply, we see that there is no external stress applied on the ply in the x-direction. For equilibrium to be achieved, the only possible way is to have an interlaminar shear stress τ_{zx} as shown. The in-plane normal stress σ_{xx} and the interlaminar shear stress τ_{zx}, however, generate a moment. Now, for moment equilibrium, the countering moment is provided by interlaminar normal stresses (Figure 5.15d). Note that the interlaminar normal stresses near the free edge are compressive in nature. The nature of these stresses, whether tensile or compressive,

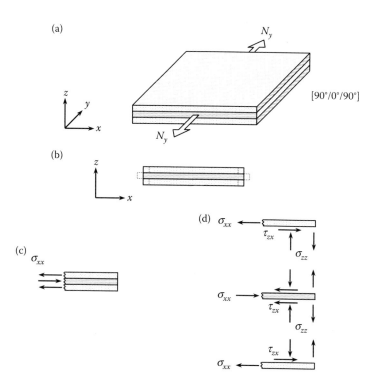

FIGURE 5.15 (a) A cross-ply laminate under transverse tensile load. (b) Cross section of the deformed laminate (dotted lines show the laminae, if they were unbounded). (c) Free body diagram of one-half of the laminate. (d) Free body diagrams of the individual laminae.

depends on the ply sequence. For example, if we change the ply sequence in our illustration from [90°/0°/90°] to [0°/90°/0°], with a little insight, we can readily see that the interlaminar normal stresses near the free edge will be tensile. Often, edge delamination is caused by interlaminar tensile normal stresses.

5.10.2 Shear Deformation Theories

The kinematic assumptions are marginally relaxed in the shear deformation theories [1,20–22]. As per Kirchhoff hypothesis, which is assumed to be valid in CLT, a transverse normal before deformation remains straight and perpendicular to the middle surface after deformation.

In the FSDT, as shown in Figure 5.16, the assumption of perpendicularity of a transverse normal is relaxed (also see Reference 1). Then, the displacement field of the FSDT is of the following form:

$$\begin{Bmatrix} u(x,y,z) \\ v(x,y,z) \\ w(x,y,z) \end{Bmatrix} = \begin{Bmatrix} u_0(x,y) + z\phi_x(x,y) \\ v_0(x,y) + z\phi_y(x,y) \\ w_0(x,y) \end{Bmatrix} \quad (5.181)$$

where (u, v, w) are the displacements at a distance of z from the middle surface and $(u_0, v_0, w_0, \phi_x, \phi_y)$ are the generalized displacements of the middle surface. Note that (ϕ_x, ϕ_y) are the rotations of the transverse normal about the y- and x-axes, respectively, and they are given by

$$\phi_x = \frac{\partial u}{\partial z} \quad \text{and} \quad \phi_y = \frac{\partial v}{\partial z} \quad (5.182)$$

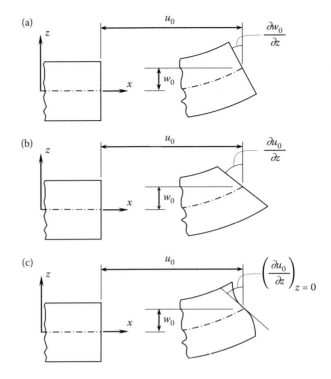

FIGURE 5.16 Kinematics of different laminate analysis theories: (a) CLT, (b) FSDT, and (c) TSDT.

In addition to the in-plane force/moment resultants, the transverse force resultants exist in FSDT and they are given by

$$Q_x = \int_{-h/2}^{h/2} k\tau_{zx} dz \qquad (5.183)$$

$$Q_y = \int_{-h/2}^{h/2} k\tau_{yz} dz \qquad (5.184)$$

where
- Q_x Transverse shear force resultant on the face normal to x-axis
- Q_y Transverse shear force resultant on the face normal to y-axis
- k Shear correction factor

The final constitutive relation for a plate as per FSDT is given by

$$\begin{Bmatrix} N \\ M \\ Q \end{Bmatrix} = \begin{bmatrix} A & B & 0 \\ B & D & 0 \\ 0 & 0 & S \end{bmatrix} \begin{Bmatrix} \varepsilon^0 \\ \kappa^0 \\ \gamma^0 \end{Bmatrix} \qquad (5.185)$$

where the new terms (other than the terms associated with CLT) are

$$\{Q\} = \begin{Bmatrix} Q_x \\ Q_y \end{Bmatrix}, \quad \text{transverse shear force resultants} \qquad (5.186)$$

$$\{\gamma^0\} = \begin{Bmatrix} \gamma_{yz}^0 \\ \gamma_{xz}^0 \end{Bmatrix}, \quad \text{transverse shear strains} \qquad (5.187)$$

Higher order shear deformations theories, which represent kinematics and interlaminar shear characteristics of a plate in a better way, have also been developed. Shear correction factor required in the FSDT [23] is not required in the higher order theories. However, these theories are computationally more involved.

In the TSDT, the assumption of regarding a transverse normal is further relaxed. Thus, the transverse normal is free to rotate and it need not remain straight (Figure 5.16). Then, the displacement field of the TSDT is of the following form:

$$\begin{Bmatrix} u(x,y,z) \\ v(x,y,z) \\ w(x,y,z) \end{Bmatrix} = \begin{Bmatrix} u_0(x,y) + z\phi_x(x,y) + z^2\theta_x(x,y) + z^3\psi_x(x,y) \\ v_0(x,y) + z\phi_y(x,y) + z^2\theta_y(x,y) + z^3\psi_y(x,y) \\ w_0(x,y) \end{Bmatrix} \quad (5.188)$$

where as in FSDT, (u, v, w) are the displacements at a distance of z from the middle surface and $(u_0, v_0, w_0, \phi_x, \phi_y)$ are the generalized displacements of the middle surface. Similarly, (ϕ_x, ϕ_y) are the rotations of the tangent to the transverse normal about the y- and x-axes, respectively, at $z = 0$. In addition, we have functions (θ_x, θ_y) and (ψ_x, ψ_y) to be determined. They are given by

$$\phi_x = \left(\frac{\partial u}{\partial z}\right)_{z=0} \quad \text{and} \quad \phi_y = \left(\frac{\partial v}{\partial z}\right)_{z=0} \quad (5.189)$$

$$\theta_x = \frac{1}{2}\left(\frac{\partial^2 u}{\partial z^2}\right)_{z=0} \quad \text{and} \quad \theta_y = \frac{1}{2}\left(\frac{\partial^2 v}{\partial z^2}\right)_{z=0} \quad (5.190)$$

$$\psi_x = \frac{1}{6}\left(\frac{\partial^3 u}{\partial z^3}\right)_{z=0} \quad \text{and} \quad \theta_y = \frac{1}{6}\left(\frac{\partial^3 v}{\partial z^3}\right)_{z=0} \quad (5.191)$$

5.10.3 Layerwise Theories

The 2D equivalent single-layer theories viz. CLT and shear deformation theories are adequate in the analysis of thin laminated composite plates and shells. However, for thick laminates and special structural elements such as sandwich and grid stiffened shell, etc., the equivalent single-layer theories provide erroneous solutions. Also, owing to the presence of interlaminar shear and normal stresses near the free edges of a laminate, CLT is not valid in these regions. In such cases, a fully 3D elasticity solution or a layerwise theory is useful.

Layerwise theories provide more realistic representation of the kinematic behavior of a moderately thick to thick composite laminate [1,2,24]. The laminate is considered to be made up of a number of mathematical layers (the number of mathematical layers need not be equal to the number of physical plies) and discrete layerwise displacement field is assumed. The layerwise displacements are such that the displacement components are continuous but the derivatives of the displacements w.r.t. the thickness coordinate may be discontinuous.

Layerwise theories can be classified into two classes—the partial layerwise theories and the full layerwise theories. The displacement field in a partial layerwise theory provides for layerwise in-plane deformations only. On the other hand, the displacement in a full layerwise theory provides for layerwise deformations in all the three dimensions.

Detailed discussion on the layerwise theories is beyond the scope of this book; interested reader may consult Reference 1 among others.

5.11 SUMMARY

Analysis of a laminate is presented in this chapter. A laminate is made by stacking a number of laminae and its analysis is carried out at a macro level based on the macro-level parameters of the laminae. Several theories have been proposed by researchers. CLT, as applied to plates and shells, is possibly the most popular theory in the analysis of a composite laminate. It is an equivalent single-layer theory; the kinematics, kinetics, and constitutive relations are based on certain assumptions and restrictions, notable among them is the Kirchhoff hypothesis.

Hygrothermal stresses develop in a laminate during processing as well as service life. Constitutive relations are suitably modified by incorporating certain fictitious hygrothermal force/moment resultants.

The laminate constitutive relations are written in terms of laminate stiffness matrices and each term in these matrices has certain significance in the final laminate behavior. By considering certain specific stacking sequences, some of these terms can be made to vanish and special cases of laminates are obtained. These special laminates are of great utility in design.

Composite failure behavior is different from metallic failure behavior. Laminate failure depends on individual laminae failures and in general, it is progressive in nature and not catastrophic. In the context of product design, different philosophies of failure loads exist; two broad ways of defining failure are FPF and LPF.

There are certain issues that cannot be addressed within the framework of CLT. For example, interlaminar stresses, which are zero as per CLT, do exist in reality especially near free edges and the same can be studied by adopting a 3D approach.

EXERCISE PROBLEMS

5.1 Write the following laminate stacking sequences in expanded and contracted codes.

(a)	(b)	(c)	(d)
0°	0°	0°	0° (CE)
45°	−45°	30°	30° (CE)
−45°	45°	30°	30° (GE)
90°	90°	60°	60° (CE)
90°	0°	60°	60° (GE)
−45°	90°	90°	90° (CE)
45°	45°	60°	60° (GE)
0°	−45°	60°	60° (CE)
	0°	30°	30° (GE)
		30°	30° (CE)
		0°	0° (CE)

Note: CE = carbon/epoxy, GE = glass/epoxy.

5.2 Consider the carbon/epoxy laminate with stacking sequence: $[0°/45°/-45°/90°]_s$ and material data: $E_1 = 160$ GPa, $E_2 = 8$ GPa, $\nu_{12} = 0.25$, and $G_{12} = 4$ GPa. Thickness of each ply is 0.5 mm. The laminate is loaded in such a way that the strains are as follows: $\varepsilon_{xx} = 4 \times 10^{-6}$ (top), $\varepsilon_{xx} = 0$ (bottom), and $\varepsilon_{yy} = \gamma_{xy} = 0$ (everywhere). Determine the stresses in the laminate and draw the stress distribution across the laminate thickness. Assume linear strain variation across the laminate thickness.

Hint: Determine the transformed reduced stiffness matrices for each ply. Use Equation 5.23.

5.3 Consider a carbon/epoxy laminate with stacking sequence: [90°/0°/90°/0°] and the following material data: $E_1 = 160$ GPa, $E_2 = 8$ GPa, $\nu_{12} = 0.25$, and $G_{12} = 4$ GPa. Thickness of each play is 2.0 mm. Determine the laminate stiffness matrices [*A*], [*B*], and [*D*].

5.4 Consider the laminate in Exercise 5.3. Stacking sequence is changed to [90°/0°/0°/90°]. Determine the laminate stiffness matrices [*A*], [*B*], and [*D*]. What striking change can you notice?

5.5 Determine the laminate stiffness matrices [*A*], [*B*], and [*D*] of a laminate laid-up with unidirectional carbon/epoxy laminae as per the following stacking sequence: [0°/45°/0°]. Assume the following material data: $E_1 = 140$ GPa, $E_2 = 6$ GPa, $G_{12} = 4$ GPa, and $\nu_{12} = 0.2$. Thickness of each ply = 1 mm.

5.6 Consider the laminate in Exercise 5.5. If the unidirectional carbon/epoxy lamina is replaced with a bidirectional carbon/epoxy lamina, determine the laminate stiffness matrices [*A*], [*B*], and [*D*]. Assume the following additional material data for the bidirectional lamina: $E_1 = 40$ GPa, $E_2 = 40$ GPa, $G_{12} = 10$ GPa, and $\nu_{12} = 0.25$. Compare the results with those of Exercise 5.5.

5.7 Write a code in MATLAB/C/C++ for the determination of
  1. Force and moment resultants {*N*} and {*M*}
  2. Reduced stiffness matrix of the material [*Q*]
  3. Transformed reduced stiffness matrix [$\bar{Q}$] of each ply
  4. Laminate stiffness matrices [*A*], [*B*], and [*D*]
  5. Laminate compliance matrices [*A**], [*B**], [*C**], and [*D**]
  6. Laminate middle surface strains and curvatures {ε^0} and {κ}
  7. Global strains ε_{xx}, ε_{yy}, and γ_{xy} at any specified point in the laminate thickness
  8. Global stresses σ_{xx}, σ_{yy}, and τ_{xy} at any specified point in the laminate thickness
  9. Local strains ε_{11}, ε_{22}, and γ_{12} in each ply
  10. Local stresses σ_{11}, σ_{22}, and τ_{12} in each ply
  11. Consider the following as input data:
     a. Laminate dimensions
     b. Total number of plies and thickness and orientation of each ply
     c. Applied loads
     d. Material properties E_1, E_2, ν_{12}, and G_{12}

5.8 Consider an antisymmetric laminate [90°/60°/30°/−30°/−60°/90°] with material data $E_1 = 150$ GPa, $E_2 = 6$ GPa, $G_{12} = 4$ GPa, and $\nu_{12} = 0.2$. Determine the laminate stiffness matrices [*A*], [*B*], and [*D*]. Verify that $A_{16} = A_{26} = D_{16} = D_{26} = 0$.

5.9 Consider a balanced laminate [90°/60°/−60°/−30°/30°/90°] with material data $E_1 = 150$ GPa, $E_2 = 6$ GPa, $G_{12} = 4$ GPa, and $\nu_{12} = 0.2$. Determine the laminate stiffness matrices [*A*], [*B*], and [*D*]. Verify that $A_{16} = A_{26} = 0$.

5.10 Consider a cross-ply laminate [90°/0°/90°/90°/90°/0°] with material data $E_1 = 40$ GPa, $E_2 = 8$ GPa, $G_{12} = 4$ GPa, and $\nu_{12} = 0.2$. Determine the laminate stiffness matrices [*A*], [*B*], and [*D*]. Verify that $A_{16} = A_{26} = B_{16} = B_{26} = D_{16} = D_{26} = 0$.

5.11 Consider a quasi-isotropic laminate [0°/30°/60°/90°] with material data $E_1 = 150$ GPa, $E_2 = 6$ GPa, $G_{12} = 4$ GPa, and $\nu_{12} = 0.2$. Determine the laminate stiffness matrices [*A*], [*B*], and [*D*]. Verify that $A_{11} = A_{22}$, $A_{16} = A_{26} = 0$, and $A_{66} = (A_{11} - A_{12})/2$.

5.12 Consider a carbon/epoxy laminate with ply sequence: $[90°/45°/-45°/0°]_s$ each ply being 0.5 mm thick and material data: $E_1 = 160$ GPa, $E_2 = 10$ GPa, $\nu_{12} = 0.2$, and $G_{12} = 4$ GPa. The middle surface strains and curvatures are given as

$$\{\varepsilon^0\} = \begin{Bmatrix} 3.0 \\ 2.0 \\ -4.0 \end{Bmatrix} \times 10^{-3} \text{ mm/mm} \quad \text{and} \quad \{\kappa\} = \begin{Bmatrix} 1.8 \\ -1.2 \\ -0.6 \end{Bmatrix} \times 10^{-3} \text{ mm}^{-1}$$

Determine the global strains, local strains, and local stresses in the outermost ply.

5.13 Consider a carbon/epoxy laminate of size 400 mm × 400 mm. Following data are given:

Ply sequence: $[0°/30°/-30°/60°/-60°/90°]$
Material data: $E_1 = 160$ GPa, $E_2 = 10$ GPa, $\nu_{12} = 0.2$, and $G_{12} = 4$ GPa

Apply a tensile force of 1200 kN in the longitudinal direction (in the direction of the 0° ply) and determine the middle surface strains and curvatures.

5.14 Consider the laminate in Exercise 5.13 and apply the tensile force now in the transverse direction (in direction of the 90° ply). Determine the middle surface strains and curvatures. Do you see any similarity/difference in the laminate behavior in this exercise vis-à-vis the laminate behavior in the previous exercise?

5.15 Consider a cylindrical pressure vessel of 300 mm inner diameter under an internal pressure of 2 MPa. Following data are given:

Ply sequence: $[90°/30°/-30°/90°]$
Material data: $E_1 = 140$ GPa, $E_2 = 8$ GPa, $\nu_{12} = 0.2$, and $G_{12} = 4$ GPa
Determine the local stresses in each ply.

Hint: Consider force equilibrium to find N_{xx} and N_{yy}. $N_{xy} = M_{xx} = M_{yy} = M_{xy} = 0$. (Take x-direction along the axis of the pressure vessel and y-direction along the circumference.)

5.16 Consider the pressure vessel in Exercise 5.15. Apply maximum stress criterion and check whether the pressure vessel is safe. Increase the pressure from zero till LPF and study the failure behavior. Draw plots to show variations of local stresses in each ply w.r.t. pressure. Indicate clearly events such as matrix cracking, FPF, and LPF. Comment on load sharing by the plies. Following strength data are given:

$$\left(\sigma_{11}^T\right)_{ult} = 1600 \text{ MPa}, \left(\sigma_{11}^C\right)_{ult} = 1000 \text{ MPa}, \left(\sigma_{22}^T\right)_{ult} = 50 \text{ MPa},$$
$$\left(\sigma_{22}^C\right)_{ult} = 200 \text{ MPa}, (\tau_{12})_{ult} = 40 \text{ MPa}$$

5.17 Solve Exercise 5.16 by using maximum strain criterion.

5.18 Consider a carbon/epoxy laminate with the following data:

Ply sequence: $[0°/90°/0°/90°]$ each ply being 0.5 mm thick
Material data: $E_1 = 150$ GPa, $E_2 = 8$ GPa, $\nu_{12} = 0.25$, $G_{12} = 6$ GPa, $\alpha_1 = 0.02 \times 10^{-6}$ m/m/°C, and $\alpha_2 = 22.5 \times 10^{-6}$ m/m/°C

The laminate was cured at 160°C and cooled down to ambient temperature of 25°C. Determine the residual strains in the laminate.

5.19 If the stacking of the laminate in Exercise 5.18 is changed to [0°/90°/90°/0°], other data remaining unchanged, determine the residual strains. Is there any change in the results vis-à-vis the results in Exercise 5.18?

5.20 Consider a glass/epoxy laminate of size 500 mm × 500 mm with the following data:

Ply sequence: [0°/45°/−45°/90°] each ply being 1.0 mm thick
Material data: $E_1 = 40$ GPa, $E_2 = 8$ GPa, $\nu_{12} = 0.25$, $G_{12} = 4$ GPa, $\alpha_1 = 8.0 \times 10^{-6}$ m/m/°C, and $\alpha_2 = 20.0 \times 10^{-6}$ m/m/°C

If the temperature is raised by 25°C, determine the changed dimensions of the laminate. Is there any warpage? Assume the laminate to be initially stress free.

5.21 Solve the problem in Exercise 5.19, if the ply sequence is changed to [0°/45°/45°/0°]. Is there any change in the laminate behavior?

REFERENCES AND SUGGESTED READING

1. J. N. Reddy, *Mechanics of Laminated Composite Plates and Shells*, CRC Press, Boca Raton, FL, 2004.
2. J. N. Reddy, An evaluation of equivalent single-layer and layerwise theories of composite laminates, *Composite Structures*, 25(1–4), 1993, 21–35.
3. R. M. Jones, *Mechanics of Composite Materials*, second edition, Taylor & Francis, New York, 1999.
4. A. K. Kaw, *Mechanics of Composite Materials*, CRC Press, Boca Raton, FL, 2006.
5. E. Ventsel and T. Krauthammer, *Thin Plates and Shells—Theory, Analysis and Applications*, Marcel Dekker, New York, 2001.
6. H. Thomas Hahn, Residual stresses in polymer matrix composite laminates, *Journal of Composite Materials*, 10(4), 1976, 266–278.
7. R. Byron Pipes, J. R. Vinson, and T.-W. Chou, On the hygrothermal response of laminated composite systems, *Journal of Composite Materials*, 10(2), 1076, 129–148.
8. N. J. Pagano and R. Byron Pipes, The influence of stacking sequence on laminate strength, *Journal of Composite Materials*, 1971, 5, 50–57.
9. H. Ghiasi, D. Pasini, and L. Lessard, Optimal stacking sequence design of composite materials, Part-I: Constant stiffness design, *Composite Structures*, 90(1), 2009, 1–11.
10. H. Ghiasi, K. Fayazbakhsh, D. Pasini, and L. Lessard, Optimal stacking sequence design of composite materials, Part-II: Variable stiffness design, *Composite Structures*, 93(1), 2010, 1–13.
11. C. T. Herakovich, Influence of layer thickness of angle-ply laminates, *Journal of Composite Materials*, 16(3), 1982, 216–227.
12. S. R. Hallett and M. R. Wisnom, Effect of stacking sequence on open-hole tensile strength of composite laminates, *AIAA Journal*, 47(7), 2009, 1692–1699.
13. U. Icardi, S. Locatto, and A. Longo, Assessment of recent theories for predicting failure of composite laminates, *Applied Mechanics Reviews*, 60(2), 2007, 76–86.
14. B. D. Agarwal, L. J. Broutman, and K. Chandrashekhara, *Analysis and Performance of Fiber Composites*, John Wiley & Sons, New York, 2006.
15. V. V. Vasiliev and E. V. Morozov, *Mechanics and Analysis of Composite Materials*, Elsevier Science, Amsterdam, 2001.
16. N. J. Salamon, An assessment of the interlaminar stress problem in laminated composites, *Journal of Composite Materials*, 14(1), 1980, 177–194.
17. R. Byron and N. J. Pagano, Interlaminar stresses in composite laminates under uniform axial extension, *Journal of Composite Materials*, 4, 1970, 538–548.
18. C. Kassapoglou and P. A. Lagace, Closed form solutions for the interlaminar stress field in angle-ply and cross-ply laminates, *Journal of Composite Materials*, 21(4), 1987, 292–308.
19. M. Mukhopadhyay, *Mechanics of Composite Materials and Structures*, Universities Press, Hyderabad, India, 2009.
20. A. K. Noor and W. Scott Burton, Assessment of shear deformation theories for multilayered composite plates, *Applied Mechanics Reviews*, 42(1), 1989, 1–13.
21. J. M. Whitney and N. J. Pagano, Shear deformation in heterogeneous anisotropic plates, *Journal of Applied Mechanics*, 37(4), 1970, 1031–1036.
22. J. N. Reddy, A simple higher order theory for laminated composite plates, *Journal of Applied Mechanics*, 51(4), 1984, 745–752.
23. J. M. Whitney, Shear correction factors for orthotropic laminates under static load, *Journal of Applied Mechanics*, 40(1), 1973, 302–304.
24. J. N. Reddy, A generalization of two-dimensional theories of laminated composite plates, *Communications in Applied Numerical Methods*, 3(3), 1987, 173–180.

6

Analysis of Laminated Beams, Columns, and Rods

6.1 CHAPTER ROAD MAP

A logical end to the study of the analytical theories of laminated composites is a discussion on methods for solution to problems in laminated composite structures. A number of solution methods have been developed and these methods can be broadly classified into two classes—(i) analytical and semianalytical methods and (ii) numerical methods. Analytical methods result in either closed-form solutions or infinite series solutions and they may provide exact solutions to the governing equations. (An exact solution to a problem is one that satisfies the governing equations at all the points in the domain. Further, it satisfies the initial and boundary conditions.) In this chapter, we extend the basic principles of solid mechanics and macromechanical theories in laminated composites, discussed in Chapters 2, 4, and 5, to obtain analytical solutions of 1D problems in bending, vibration, and buckling.

In general structural engineering parlance, beam, column, and rod are 1D structural elements. A beam is a 1D structural element under bending due to transverse loading. Analytical modeling of laminated beams with different cross sections is presented and solutions are obtained for in-plane and interlaminar stresses as well as displacements under bending loads. When a 1D structural element is subjected to axial compression, it is referred to as a column. Buckling is an important phenomenon in a column and governing equations and solutions for buckling load are presented. For problems in beam vibration, solutions are obtained for natural frequencies. Specific cases, described by end support conditions and applied loads, are considered in each of these three 1D problems.

The solution methods presented in this chapter are a direct extension of the topics on analysis of lamina and laminate. Thus, a thorough understanding of the concepts presented in Chapters 4 and 5 together with basic solid mechanics in Chapter 2 is a prerequisite for effective assimilation of the concepts discussed in this chapter.

6.2 PRINCIPAL NOMENCLATURE

$[A], [B], [D]$	Laminate stiffness matrices—extensional stiffness matrix, extension–bending coupling stiffness matrix, and bending stiffness matrix, respectively
$A_{11}, A_{12}, ..., A_{66}$	Elements of the extensional stiffness matrix
$B_{11}, B_{12}, ..., B_{66}$	Elements of the extension–bending coupling stiffness matrix
$D_{11}, D_{12}, ..., D_{66}$	Elements of the bending stiffness matrix
$[A^*], [B^*], [C^*], [D^*]$	Laminate compliance matrices—extensional compliance matrix, extension–bending coupling compliance matrices, and bending compliance matrix, respectively

$D_{11}^*, D_{12}^*, \ldots, D_{66}^*$	Elements of the bending compliance matrix
E_{xx}^b, E_{xx}^{ex}	Effective bending modulus and extensional modulus, respectively, of a beam
$(F)^{(1)}, (F)^{(2)}, \ldots$	Total force on individual sectional elements
I_{yy}	Area moment of inertia
l, b, h	Length, breadth, and height of a beam
M	Total bending moment
M_{xx}, M_{yy}, M_{xy}	Bending and twisting moment per unit length, that is, moment resultants
N_{xx}, N_{yy}, N_{xy}	Normal and shear force per unit length, that is, force resultants
P	Applied concentrated force
P_{cr}	Critical buckling load
p	Applied pressure
$[Q]$	Reduced stiffness matrix of a lamina
$[\bar{Q}]$	Transformed reduced stiffness matrix of a lamina
$\bar{Q}_{11}, \bar{Q}_{12}, \ldots, \bar{Q}_{66}$	Elements of the transformed reduced stiffness matrix
$q(x)$	Transverse loads
R_A, R_B	Support reactions
u, v, w	Displacements in the x-, y-, and z-directions, respectively
u_0, v_0, w_0	Middle surface displacements
W	Buckling displacement
w_0^e	Prebuckling equilibrium displacement
V, V_{xx}	Total shear force and shear force per unit width of a beam, respectively
x, y, z	Cartesian coordinates
z_{c1}, z_{c2}	Distances of centroids of sectional elements from neutral axis
$\varepsilon_{xx}^0, \varepsilon_{yy}^0, \gamma_{xy}^0$	Middle surface strains
$\kappa_{xx}, \kappa_{yy}, \kappa_{xy}$	Middle surface curvatures
$\sigma_{xx}, \sigma_{yy}, \tau_{xy}$	In-plane stresses
$\sigma_{zz}, \tau_{zx}, \tau_{yz}$	Interlaminar stresses

6.3 INTRODUCTION

Figure 6.1 shows the configuration of a 1D laminated structural element. For a structural element to be considered as 1D, the following conditions have to be satisfied:

- The width and thickness are small compared to the length, that is, $b/l \ll 1$ and $h/l \ll 1$.
- The loads and displacements are functions of x only, that is, they are independent of y.

Note that all the three dimensions of a beam are finite.

The other case of 1D laminated plate problem is cylindrical bending of a plate strip (Figure 6.2). In this case, the y-dimension b' of the laminated plate is large and the applied transverse loads $q(x)$, displacements $u_0(x)$, $v_0(x)$, and $w_0(x)$ are functions of x only. Such a problem can be solved by considering a strip of width b such that $b/b' \ll 1$.

The cross-sectional shape and ply details greatly influence the overall performance of a 1D structural element. Cross-sectional configurations in common use are as follows:

- Solid cross sections
  - Rectangular
  - Circular

Analysis of Laminated Beams, Columns, and Rods

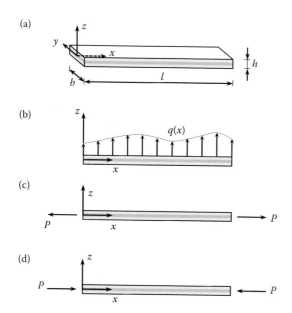

FIGURE 6.1 One-dimensional laminated structural elements. (a) Geometry. (b) Beam. (c) Rod. (d) Column.

- Thin-walled cross sections
  - Open ended, for example, T-section, I-section
  - Closed ended, for example, box-section

Solid rectangular cross sections are usually used in beams as well as columns. In the case of beams, ply orientation w.r.t. the loading direction is critical and we have two distinct subtypes of rectangular cross sections as follows:

- Rectangular (plies normal to the loading direction)
- Rectangular (plies parallel to the loading direction)

Solid circular cross sections are usually used in applications with axial compression or tension. Thin-walled cross sections are most commonly used as beams and sometimes as compression members as well.

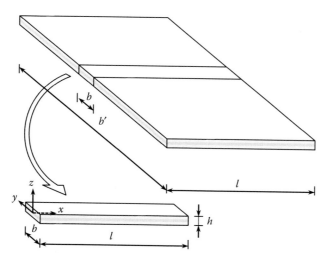

FIGURE 6.2 Cylindrical bending of laminated plate. (Adapted with permission from J. N. Reddy, *Mechanics of Laminated Composite Plates and Shells*, CRC Press, 2004.)

In the following sections, we consider the first type of laminated 1D problems (Figure 6.1). For bending problems, we shall consider (i) solid rectangular cross section (plies normal to the loading direction), (ii) solid rectangular cross section (plies parallel to the loading direction), (iii) T-section, (iv) I-section, and (v) box-section. For axial compression and vibration problems, we shall consider only solid rectangular cross section. Further, we shall limit our discussion to symmetric ply laminates only.

6.4 BENDING OF A LAMINATED BEAM (SOLID RECTANGULAR CROSS SECTION: PLIES NORMAL TO LOADING DIRECTION)

A laminated composite beam of solid rectangular cross section is shown in Figure 6.1a and b. In the case of plies normal to the loading direction, the plies are laid-up in the xy-plane. For clarity, a zoomed view is given in Figure 6.3.

6.4.1 Basic Assumptions and Restrictions

The analytical procedure for beam bending is an extension of CLPT [1–3] and the assumptions of CLPT hold good. In addition, certain restrictions are also placed. These assumptions and restrictions are enumerated below:

1. The plies are perfectly bonded with infinitely thin bond. (Assumption.)
2. Straight lines perpendicular to the middle surface of the laminate before deformation (i.e., transverse normals) remain straight and perpendicular to the middle surface. (Kirchhoff hypothesis. $\gamma_{zx} = \gamma_{yz} = 0$.)
3. The transverse normals do not undergo any change in lengths. (Kirchhoff hypothesis. $\varepsilon_{zz} = 0$.)
4. The strains and displacements are small. (Restriction.)
5. Each ply is of uniform thickness. (Restriction.)
6. The material of each ply is homogeneous, orthotropic, and linearly elastic. (Restriction.)
7. The thickness h and width b of the beam are small compared to the length l. (Restriction. Further, Poisson's effect and shear coupling are negligible.)
8. The laminate is symmetric. (Restriction.)
9. Only transverse loads $q(x)$ act on the beam. (Restriction.)

The assumptions and restrictions from serial numbers 1 to 6 are the same as in CLPT. These assumptions and restrictions form the basis for the development of the CLPT, and their implications were discussed before (Section 5.5, Chapter 5). The restriction in

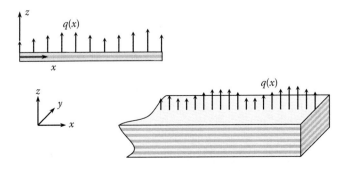

FIGURE 6.3 Laminated composite beam of solid rectangular cross section with plies normal to loading direction.

Analysis of Laminated Beams, Columns, and Rods

respect of thickness of the beam (at serial no. 7 above) is also as per CLPT. However, in addition to the thickness, width of the beam is also small compared to the length. In such a case, Poisson's effect and shear coupling are negligible.

As per the restriction at serial no. 8, there is no extension–bending coupling, which implies that

$$[B] = 0 \quad (6.1)$$

Finally, we need to look at the restriction on applied loads. Only transverse loads $q(x)$ act on the beam and as a result all in-plane normal and shear force resultants are zero. The transverse loads are functions of only x, that is, $q(x)$ is uniform along y. Further, no torque is applied. Thus, only nonzero stress resultant is M_{xx}. In other words,

$$M_{xx} \neq 0 \quad (6.2)$$

and

$$N_{xx} = N_{yy} = N_{xy} = M_{yy} = M_{xy} = 0 \quad (6.3)$$

6.4.2 Governing Equations

Let us go back to the laminate constitutive equations as per CLPT (Equations 5.42 in Chapter 5). Let us write these equations in the explicit form as follows:

$$\begin{Bmatrix} N_{xx} \\ N_{yy} \\ N_{xy} \\ M_{xx} \\ M_{yy} \\ M_{xy} \end{Bmatrix} = \begin{bmatrix} A_{11} & A_{12} & A_{16} & B_{11} & B_{12} & B_{16} \\ A_{12} & A_{22} & A_{26} & B_{12} & B_{22} & B_{26} \\ A_{16} & A_{26} & A_{66} & B_{16} & B_{26} & B_{66} \\ B_{11} & B_{12} & B_{16} & D_{11} & D_{12} & D_{16} \\ B_{12} & B_{22} & B_{26} & D_{12} & D_{22} & D_{26} \\ B_{16} & B_{26} & B_{66} & D_{16} & D_{26} & D_{66} \end{bmatrix} \begin{Bmatrix} \varepsilon_{xx}^0 \\ \varepsilon_{yy}^0 \\ \gamma_{xy}^0 \\ \kappa_{xx} \\ \kappa_{yy} \\ \kappa_{xy} \end{Bmatrix} \quad (6.4)$$

Substituting Equations 6.1 through 6.3 in Equation 6.4, we get

$$\begin{Bmatrix} M_{xx} \\ 0 \\ 0 \end{Bmatrix} = \begin{bmatrix} D_{11} & D_{12} & D_{16} \\ D_{12} & D_{22} & D_{26} \\ D_{16} & D_{26} & D_{66} \end{bmatrix} \begin{Bmatrix} \kappa_{xx} \\ \kappa_{yy} \\ \kappa_{xy} \end{Bmatrix} \quad (6.5)$$

The middle surface curvatures are given by Equation 5.16 and we rewrite it as follows:

$$\begin{Bmatrix} \kappa_{xx} \\ \kappa_{yy} \\ \kappa_{xy} \end{Bmatrix} = \begin{Bmatrix} -\dfrac{\partial^2 w_0}{\partial x^2} \\ -\dfrac{\partial^2 w_0}{\partial y^2} \\ -2\dfrac{\partial^2 w_0}{\partial x \partial y} \end{Bmatrix} \quad (6.6)$$

Substituting Equation 6.6 in Equation 6.5 and then inverting it we get

$$\begin{Bmatrix} \dfrac{\partial^2 w_0}{\partial x^2} \\ \dfrac{\partial^2 w_0}{\partial y^2} \\ 2\dfrac{\partial^2 w_0}{\partial x \partial y} \end{Bmatrix} = -\begin{bmatrix} D_{11}^* & D_{12}^* & D_{16}^* \\ D_{12}^* & D_{22}^* & D_{26}^* \\ D_{16}^* & D_{26}^* & D_{66}^* \end{bmatrix} \begin{Bmatrix} M_{xx} \\ 0 \\ 0 \end{Bmatrix} \quad (6.7)$$

Equation 6.7 can be written in an explicit form as follows:

$$\dfrac{\partial^2 w_0}{\partial x^2} = -D_{11}^* M_{xx} \quad (6.8)$$

$$\dfrac{\partial^2 w_0}{\partial y^2} = -D_{12}^* M_{xx} \quad (6.9)$$

$$2\dfrac{\partial^2 w_0}{\partial x \partial y} = -D_{16}^* M_{xx} \quad (6.10)$$

The above three partial differential equations are satisfied by the following general solution:

$$w_0 = Ax^2 + By^2 + Cxy + Dx + Ey + F \quad (6.11)$$

where A, B, C, D, E, and F are constants. Apparently, the transverse displacement of the middle surface w_0 is not independent of y. However, as noted in the previous section, for beams, the Poisson's effect (D_{12}^*) and shear coupling effect (D_{16}^*) can be neglected. In such a case, the constants associated with y^2 and xy must vanish, that is, $B = C = 0$. Then, by applying boundary conditions, it can be shown that $E = 0$. In other words, the transverse displacement of the middle surface w_0 is a function of x alone, that is,

$$w_0 = w_0(x) \quad (6.12)$$

The moment resultant M_{xx} is the bending moment on the beam per unit width. Denoting the total bending moment by M, we see that

$$M = bM_{xx} \quad (6.13)$$

For a beam of rectangular cross section (Figure 6.1), the area moment of inertia is given by

$$I_{yy} = \dfrac{bh^3}{12} \quad (6.14)$$

Now, we introduce a term called effective bending modulus of the beam E_{xx}^b, given by

$$E_{xx}^b = \dfrac{12}{h^3 D_{11}^*} \quad (6.15)$$

Analysis of Laminated Beams, Columns, and Rods

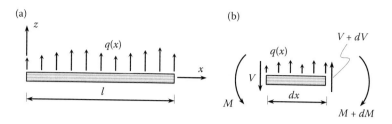

FIGURE 6.4 (a) Transverse loads on a beam. (b) Positive bending moment and shear force on a beam.

From Equations 6.14 and 6.15, the effective bending stiffness is given by

$$E^b_{xx} I_{yy} = \frac{b}{D^*_{11}} \qquad (6.16)$$

Substituting Equation 6.16 in Equation 6.8, with the help of Equation 6.13, we get

$$\frac{\partial^2 w_0}{\partial x^2} = -\frac{M}{E^b_{xx} I_{yy}} \qquad (6.17)$$

The transverse loads q result not only in bending moment but also shear force on the beam (Figure 6.4). Let us denote shear force per unit width of the beam by V_{xx} and total shear force on a cross section of the beam by V. Similarly, the applied load q is the load per unit length of the beam; it can be related to an applied pressure on the beam. Let us denote the applied pressure by p. Note that for a beam, both p and q are independent of y. Clearly,

$$V = bV_{xx} \qquad (6.18)$$

and

$$q = bp \qquad (6.19)$$

Now, by considering static moment equilibrium (Figure 6.4b), it can be shown that bending moment and shear force are related as

$$V_{xx} = \frac{\partial M_{xx}}{\partial x} \quad \text{or} \quad V = \frac{\partial M}{\partial x} \qquad (6.20)$$

On the other hand, shear force and the applied loads are related as

$$p = -\frac{\partial V_{xx}}{\partial x} \quad \text{or} \quad q = -\frac{\partial V}{\partial x} \qquad (6.21)$$

Then, differentiating Equation 6.17 twice and using Equations 6.20 and 6.21, we get

$$\frac{\partial^4 w_0}{\partial x^4} = \frac{q}{E^b_{xx} I_{yy}} \qquad (6.22)$$

For a beam, the applied loads q, bending moment M, and middle surface displacement w_0 are all functions of x alone and they are independent of y. Thus, Equations 6.17 and 6.22 can, respectively, be written as

$$\frac{d^2 w_0}{dx^2} = -\frac{M}{E^b_{xx} I_{yy}} \quad \text{or} \quad \frac{d^2 w_0}{dx^2} = -\frac{D^*_{11} M}{b} \tag{6.23}$$

and

$$\frac{d^4 w_0}{dx^4} = \frac{q}{E^b_{xx} I_{yy}} \quad \text{or} \quad \frac{d^4 w_0}{dx^4} = -\frac{D^*_{11} q}{b} \tag{6.24}$$

Equations 6.23 and 6.24 are the governing equations for beam bending under lateral loads. Solutions to these equations, in terms of values of $w_0(x)$, $q(x)$, and $M(x)$, can be obtained by direct integration and application of appropriate boundary conditions. The general solutions are

$$w_0(x) = -\frac{D^*_{11}}{b} \iint M(x)\, dx\, dx + C_1 x + C_2 \tag{6.25}$$

and

$$w_0(x) = -\frac{D^*_{11}}{b} \iiiint q(x)\, dx\, dx\, dx\, dx + C_3 x^3 + C_4 x^2 + C_5 x + C_6 \tag{6.26}$$

We shall consider some specific cases of beam. However, before that, let us dwell upon determination of stresses in a beam.

6.4.3 In-Plane Stresses

From the solution of the governing beam equations, $w_0(x)$ is obtained as a function of x and curvatures are obtained by differentiation of $w_0(x)$. Then, from Equation 5.24 in Chapter 5 and, noting that the in-plane strains are zero, the in-plane stresses at any point are obtained as follows:

$$\begin{Bmatrix} \sigma_{xx} \\ \sigma_{yy} \\ \tau_{xy} \end{Bmatrix}^{(k)} = z \begin{bmatrix} \bar{Q}_{11} & \bar{Q}_{12} & \bar{Q}_{16} \\ \bar{Q}_{12} & \bar{Q}_{22} & \bar{Q}_{26} \\ \bar{Q}_{16} & \bar{Q}_{26} & \bar{Q}_{66} \end{bmatrix}^{(k)} \begin{Bmatrix} \kappa_{xx} \\ \kappa_{yy} \\ \kappa_{xy} \end{Bmatrix} \tag{6.27}$$

that is,

$$\begin{Bmatrix} \sigma_{xx} \\ \sigma_{yy} \\ \tau_{xy} \end{Bmatrix}^{(k)} = -z \begin{bmatrix} \bar{Q}_{11} & \bar{Q}_{12} & \bar{Q}_{16} \\ \bar{Q}_{12} & \bar{Q}_{22} & \bar{Q}_{26} \\ \bar{Q}_{16} & \bar{Q}_{26} & \bar{Q}_{66} \end{bmatrix}^{(k)} \begin{Bmatrix} \dfrac{\partial^2 w_0}{\partial x^2} \\ \dfrac{\partial^2 w_0}{\partial y^2} \\ 2\dfrac{\partial^2 w_0}{\partial x \partial y} \end{Bmatrix} \tag{6.28}$$

where the superscript k refers to the kth ply corresponding to the point at which the stresses are being determined. Then, using Equations 6.7 and 6.13,

$$\begin{Bmatrix} \sigma_{xx} \\ \sigma_{yy} \\ \tau_{xy} \end{Bmatrix}^{(k)} = \frac{z}{b} \begin{bmatrix} \bar{Q}_{11} & \bar{Q}_{12} & \bar{Q}_{16} \\ \bar{Q}_{12} & \bar{Q}_{22} & \bar{Q}_{26} \\ \bar{Q}_{16} & \bar{Q}_{26} & \bar{Q}_{66} \end{bmatrix}^{(k)} \begin{bmatrix} D_{11}^* & D_{12}^* & D_{16}^* \\ D_{12}^* & D_{22}^* & D_{26}^* \\ D_{16}^* & D_{26}^* & D_{66}^* \end{bmatrix} \begin{Bmatrix} M \\ 0 \\ 0 \end{Bmatrix} \quad (6.29)$$

In the explicit form, the in-plane stresses are

$$\sigma_{xx}^{(k)}(x,z) = \frac{z}{b} \left(\bar{Q}_{11}^{(k)} D_{11}^* + \bar{Q}_{12}^{(k)} D_{12}^* + \bar{Q}_{16}^{(k)} D_{16}^* \right) M(x) \quad (6.30)$$

$$\sigma_{yy}^{(k)}(x,z) = \frac{z}{b} \left(\bar{Q}_{12}^{(k)} D_{11}^* + \bar{Q}_{22}^{(k)} D_{12}^* + \bar{Q}_{26}^{(k)} D_{16}^* \right) M(x) \quad (6.31)$$

$$\tau_{xy}^{(k)}(x,z) = \frac{z}{b} \left(\bar{Q}_{16}^{(k)} D_{11}^* + \bar{Q}_{26}^{(k)} D_{12}^* + \bar{Q}_{66}^{(k)} D_{16}^* \right) M(x) \quad (6.32)$$

Equations 6.29 or 6.30 through 6.32 give the in-plane stresses at any point in the beam under bending.

For a beam of homogeneous isotropic material, the elements of $[\bar{Q}]$ and $[D]$ matrices can be expressed in terms of isotropic material constants and thickness of the beam (refer Section 5.8.2.1, Chapter 5). Then, the expressions for elements of $[D^*]$ matrix can be obtained. Finally, it can be shown that Equations 6.30 through 6.32 can be reduced to

$$\sigma_{xx}^{(k)}(x,z) = \frac{zM(x)}{I_{yy}} \quad (6.33)$$

$$\sigma_{yy}^{(k)}(x,z) = 0 \quad (6.34)$$

$$\tau_{xy}^{(k)}(x,z) = 0 \quad (6.35)$$

Note that Equations 6.33 through 6.35 are the beam bending equations as per classical beam theory for homogeneous isotropic material.

6.4.4 Interlaminar Stresses

Interlaminar stresses are of great significance near free edges of a laminate [4,5]. The extent of the region where interlaminar stresses are large is of the order of the laminate thickness. In a beam, the width being small, interlaminar stresses are rather large [3]. 3D elasticity-based approach is required for the determination of these stresses. Let us recollect the equilibrium equations given by Equations 2.135, Chapter 2 and rewrite for a static case with zero body forces as follows:

$$\frac{\partial \sigma_{xx}}{\partial x} + \frac{\partial \tau_{xy}}{\partial y} + \frac{\partial \tau_{zx}}{\partial z} = 0 \quad (6.36)$$

$$\frac{\partial \tau_{xy}}{\partial x}+\frac{\partial \sigma_{yy}}{\partial y}+\frac{\partial \tau_{yz}}{\partial z}=0 \tag{6.37}$$

$$\frac{\partial \tau_{zx}}{\partial x}+\frac{\partial \tau_{yz}}{\partial y}+\frac{\partial \sigma_{zz}}{\partial z}=0 \tag{6.38}$$

The interlaminar normal stress σ_{zz} and interlaminar shear stresses τ_{zx} and τ_{yz} in any ply are determined by integration of the above three partial differential equations in that ply. Thus, the interlaminar stresses in the kth ply are obtained as

$$\tau_{zx}^{(k)} = -\int_{z_{k-1}}^{z}\left(\frac{\partial \sigma_{xx}}{\partial x}+\frac{\partial \tau_{xy}}{\partial y}\right)dz + C_1^{(k)} \tag{6.39}$$

$$\tau_{yz}^{(k)} = -\int_{z_{k-1}}^{z}\left(\frac{\partial \tau_{xy}}{\partial x}+\frac{\partial \sigma_{yy}}{\partial y}\right)dz + C_2^{(k)} \tag{6.40}$$

$$\sigma_{zz}^{(k)} = -\int_{k-1}^{z}\left(\frac{\partial \tau_{zx}}{\partial x}+\frac{\partial \tau_{yz}}{\partial y}\right)dz + C_3^{(k)} \tag{6.41}$$

Note that for the kth ply, $z_{k-1} \leq z \leq z_k$. Note further that in a beam, the variables being independent of y, their derivatives w.r.t. y are zero. Then, using Equations 6.30 and 6.32 in Equations 6.39 and 6.40, respectively, we get $\tau_{zx}^{(k)}$ and $\tau_{yz}^{(k)}$. $\tau_{zx}^{(k)}$ and $\tau_{yz}^{(k)}$, in turn, are substituted in Equation 6.41. Thus, we get

$$\tau_{zx}^{(k)} = -\frac{1}{b}\frac{\partial M}{\partial x}\left(\bar{Q}_{11}^{(k)}D_{11}^* + \bar{Q}_{12}^{(k)}D_{12}^* + \bar{Q}_{16}^{(k)}D_{16}^*\right)\int_{z_{k-1}}^{z} z\,dz + C_1^{(k)} \tag{6.42}$$

$$\tau_{yz}^{(k)} = -\frac{1}{b}\frac{\partial M}{\partial x}\left(\bar{Q}_{16}^{(k)}D_{11}^* + \bar{Q}_{26}^{(k)}D_{12}^* + \bar{Q}_{66}^{(k)}D_{16}^*\right)\int_{z_{k-1}}^{z} z\,dz + C_2^{(k)} \tag{6.43}$$

$$\sigma_{zz}^{(k)} = \frac{1}{b}\frac{\partial}{\partial x}\left(\frac{\partial M}{\partial x}\right)\left(\bar{Q}_{11}^{(k)}D_{11}^* + \bar{Q}_{12}^{(k)}D_{12}^* + \bar{Q}_{16}^{(k)}D_{16}^*\right)\int_{z_{k-1}}^{z}\int_{z_{k-1}}^{z} z\,dz\,dz + C_3^{(k)} \tag{6.44}$$

Now, $\partial M/\partial x = V$, $V/b = V_x$, and $\partial V/\partial x = -q$. Then,

$$\tau_{zx}^{(k)} = -\frac{V}{b}\left(\bar{Q}_{11}^{(k)}D_{11}^* + \bar{Q}_{12}^{(k)}D_{12}^* + \bar{Q}_{16}^{(k)}D_{16}^*\right)\left(\frac{z^2 - z_{k-1}^2}{2}\right) + C_1^{(k)} \tag{6.45}$$

$$\tau_{yz}^{(k)} = -\frac{V}{b}\left(\bar{Q}_{16}^{(k)}D_{11}^* + \bar{Q}_{26}^{(k)}D_{12}^* + \bar{Q}_{66}^{(k)}D_{16}^*\right)\left(\frac{z^2 - z_{k-1}^2}{2}\right) + C_2^{(k)} \tag{6.46}$$

Analysis of Laminated Beams, Columns, and Rods

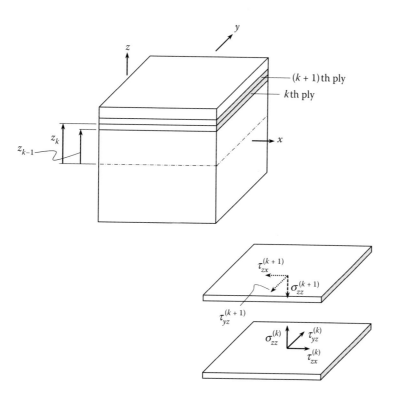

FIGURE 6.5 Interlaminar stresses at the ply interfaces.

$$\sigma_{zz}^{(k)} = -\frac{q}{b}\left(\bar{Q}_{11}^{(k)}D_{11}^* + \bar{Q}_{12}^{(k)}D_{12}^* + \bar{Q}_{16}^{(k)}D_{16}^*\right)\left(\frac{z^3 - z_{k-1}^3}{6}\right) + C_3^{(k)} \qquad (6.47)$$

Equations 6.45 through 6.47 give us the interlaminar stresses at a point in the kth ply. The constants of integration for the bottom most ply are determined by equating the interlaminar stresses at $z = 0$ to the applied shear and normal stresses on the bottom face of the beam. For the remaining plies from second ply upward, the following interface continuity conditions are utilized (Figure 6.5):

$$\tau_{zx}^{(k)}(x, z_k) = \tau_{zx}^{(k+1)}(x, z_k) \qquad (6.48)$$

$$\tau_{yz}^{(k)}(x, z_k) = \tau_{yz}^{(k+1)}(x, z_k) \qquad (6.49)$$

$$\sigma_{zz}^{(k)}(x, z_k) = \sigma_{zz}^{(k+1)}(x, z_k) \qquad (6.50)$$

6.4.5 Specific Cases of Beam Bending

6.4.5.1 Simply Supported Beam under Point Load

A simply supported beam AB under a single point load is shown in Figure 6.6. The moment boundary conditions, $M(x) = 0$ at $x = 0$ and $x = l$, are utilized to determine the support reactions, as follows:

$$R_A = \frac{P(l-a)}{l} \qquad (6.51)$$

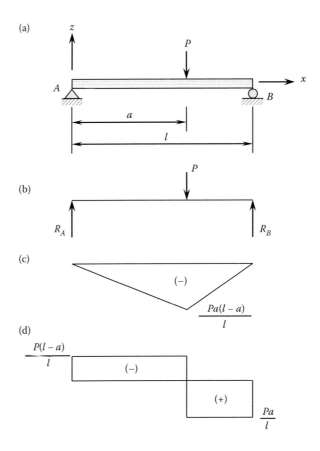

FIGURE 6.6 (a) Simply supported beam under a single point load. (b) Support reactions. (c) Bending moment distribution. (d) Shear force distribution.

and

$$R_B = \frac{Pa}{l} \tag{6.52}$$

The bending moment at a distance x is given by

$$M(x) = -\frac{P(l-a)x}{l} \quad \text{for } 0 \leq x \leq a \tag{6.53}$$

$$M(x) = -\frac{P(l-x)a}{l} \quad \text{for } a \leq x \leq l \tag{6.54}$$

Shear force, $V = dM/dx$, is given by

$$V(x) = -\frac{P(l-a)}{l} \quad \text{for } 0 \leq x \leq a \tag{6.55}$$

$$V(x) = \frac{Pa}{l} \quad \text{for } a \leq x \leq l \tag{6.56}$$

Note that the bending moment is negative everywhere, which implies that $\sigma_{xx}^{(k)}(x,z)$ is tensile at the bottom half of the beam and compressive at the top half. By substituting

Analysis of Laminated Beams, Columns, and Rods

the expressions of bending moments in Equations 6.30 through 6.32, the in-plane stresses can be readily obtained. Of the three in-plane stresses, longitudinal bending stress $\sigma_{xx}^{(k)}(x,z)$ is of special interest in analysis of beam and it is given by

$$\sigma_{xx}^{(k)}(x,z) = -\frac{P(l-a)xz}{bl}\left(\bar{Q}_{11}^{(k)}D_{11}^* + \bar{Q}_{12}^{(k)}D_{12}^* + \bar{Q}_{16}^{(k)}D_{16}^*\right) \quad \text{for } 0 \leq x \leq a \quad (6.57)$$

$$\sigma_{xx}^{(k)}(x,z) = -\frac{Pa(l-x)z}{bl}\left(\bar{Q}_{11}^{(k)}D_{11}^* + \bar{Q}_{12}^{(k)}D_{12}^* + \bar{Q}_{16}^{(k)}D_{16}^*\right) \quad \text{for } a \leq x \leq l \quad (6.58)$$

The transverse displacements are given by Equation 6.25, in which the constants of integration are determined by utilizing the displacement boundary conditions and continuity conditions, as follows:

$$w_0(x) = \frac{P(l-a)x^3}{6E_{xx}^b I_{yy} l} + C_1 x + C_2 \quad \text{for } 0 \leq x \leq a \quad (6.59)$$

$$w_0(x) = \frac{Pa(3lx^2 - x^3)}{6E_{xx}^b I_{yy} l} + C_3 x + C_4 \quad \text{for } a \leq x \leq l \quad (6.60)$$

The boundary conditions related to displacement are

$$(w_0)_{x=0} = 0 \quad (6.61)$$

and

$$(w_0)_{x=l} = 0 \quad (6.62)$$

Using Equation 6.61 in Equation 6.59 and Equation 6.62 in Equation 6.60, we get, respectively,

$$C_2 = 0 \quad (6.63)$$

and

$$C_4 = -\frac{Pal^2}{3E_{xx}^b I_{yy}} - C_3 l \quad (6.64)$$

Differentiating Equations 6.59 and 6.60, we get the slopes in the two regions of x, as follows:

$$\frac{dw_0}{dx} = \frac{P(l-a)x^2}{2E_{xx}^b I_{yy} l} + C_1 \quad \text{for } 0 \leq x \leq a \quad (6.65)$$

$$\frac{dw_0}{dx} = \frac{Pa(2lx - x^2)}{2E_{xx}^b I_{yy} l} + C_3 \quad \text{for } a \leq x \leq l \quad (6.66)$$

As we can see there are two expressions for slope in the two regions for x. The slope, however, is continuous across the point ($x = a$). Then, substituting $x = a$ in Equations 6.65 and 6.66, and equating them, we get

$$C_3 = -\frac{Pa^2}{2E_{xx}^b I_{yy}} + C_1 \qquad (6.67)$$

Similar to slope, displacement is also continuous across the point ($x = a$). Then, substituting $x = a$ in Equations 6.59 and 6.60, and equating them, we get

$$C_1 = -\frac{Pa(2l-a)(l-a)}{6E_{xx}^b I_{yy} l} \qquad (6.68)$$

Then, utilizing Equations 6.63, 6.64, 6.67, and 6.68, in Equations 6.59 and 6.60, we get

$$w_0(x) = \frac{P(l-a)[x^3 - a(2l-a)x]}{6E_{xx}^b I_{yy} l} \quad \text{for } 0 \leq x \leq a \qquad (6.69)$$

$$w_0(x) = \frac{Pa(l-x)[(l-x)^2 - (l^2 - a^2)]}{6E_{xx}^b I_{yy} l} \quad \text{for } a \leq x \leq l \qquad (6.70)$$

Note that the displacements are negative, that is, opposite to positive z-direction or downward. Maximum bending moment, bending stress, and displacement occur when the applied load is centrally located and the corresponding values are obtained by substituting $x = a = l/2$, as follows:

$$(M)_{max} = -\frac{Pl}{4} \qquad (6.71)$$

$$\left(\sigma_{xx}^{(k)}\right)_{max} = \pm \frac{Plh}{8b}\left(\bar{Q}_{11}^{(k)} D_{11}^* + \bar{Q}_{12}^{(k)} D_{12}^* + \bar{Q}_{16}^{(k)} D_{16}^*\right) \qquad (6.72)$$

$$(w_0)_{max} = -\frac{Pl^3}{48 E_{xx}^b I_{yy}} \qquad (6.73)$$

Note: Bending stress is the maximum in the top and bottom plies.

EXAMPLE 6.1

Consider a carbon/epoxy simply supported beam with the following dimensions: $l = 500$ mm, $b = 20$ mm, and $h = 6$ mm. The ply sequence of the beam laminate is [0°/90°/0°], each ply being 2 mm in thickness. Determine the maximum displacement, longitudinal in-plane stress σ_{xx} and interlaminar normal stress σ_{zz} at the center of the beam. The beam is under a central point load of 100 N. The point load is applied over an area of 20 mm × 20 mm. Material properties are as follows:

$$E_1 = 125 \text{ GPa}, E_2 = 10 \text{ GPa}, \nu_{12} = 0.25, \text{ and } G_{12} = 8 \text{ GPa}$$

Analysis of Laminated Beams, Columns, and Rods

Solution

For the given material properties and ply sequence, the transformed reduced stiffness matrix and the laminate bending compliance matrix are obtained as follows (detailed calculations are not shown):

$$[\bar{Q}]^{(1)} = [\bar{Q}]^{(3)} = \begin{bmatrix} 125.6281 & 2.5126 & 0 \\ 2.5126 & 10.0503 & 0 \\ 0 & 0 & 8 \end{bmatrix} \times 10^3 \text{ MPa}$$

$$[\bar{Q}]^{(2)} = \begin{bmatrix} 10.0503 & 2.5126 & 0 \\ 2.5126 & 125.6281 & 0 \\ 0 & 0 & 8 \end{bmatrix} \times 10^3 \text{ MPa}$$

$$[D^*] = \begin{bmatrix} 0.4595 & -0.0806 & 0 \\ -0.0806 & 3.8907 & 0 \\ 0 & 0 & 6.9444 \end{bmatrix} \times 10^{-6} \text{ (MPa} \cdot \text{mm}^3)^{-1}$$

Effective bending stiffness of the beam is

$$E_{xx}^b I_{yy} = \frac{b}{D_{11}^*} = \frac{20}{0.4595 \times 10^{-6}} = 43.5256 \times 10^6 \text{ N} \cdot \text{mm}^2$$

z-coordinates of different plies are as follows:

$$z_0 = -3 \text{ mm}, z_1 = -1 \text{ mm}, z_2 = 1 \text{ mm, and } z_3 = 3 \text{ mm}$$

Displacement under the point load is the maximum displacement and it is given by (Equation 6.73)

$$(w_0)_{max} = -\frac{100 \times 500^3}{48 \times 43.5256 \times 10^6} = -5.98 \text{ mm}$$

Maximum bending stresses occur at the bottom and top of the beam at the center. They are given by (Equation 6.72):

At the bottom of the beam under the central point load,

$$\left(\sigma_{xx}^{(1)}\right)_{max} = \frac{100 \times 500 \times 6}{8 \times 20 \times 10^3} \times (125.6281 \times 0.4595 - 2.5126 \times 0.0806)$$
$$= 107.86 \text{ MPa (tensile)}$$

At the top of the beam under the central point load,

$$\left(\sigma_{xx}^{(3)}\right)_{max} = -\frac{100 \times 500 \times 6}{8 \times 20 \times 10^3} \times (125.6281 \times 0.4595 - 2.5126 \times 0.0806)$$
$$= -107.86 \text{ MPa (compressive)}$$

For determining the interlaminar normal stresses in the beam under the point load, we need to consider the local applied load distribution. (Note that under a strictly pointed force, the local normal stress would be infinite!) Then, we see that for the given "pointed load" of 100 N over a local beam length 20 mm, the "uniformly distributed load" is $q = 5$ N/mm.

Now, we utilize Equation 6.47 and determine the interlaminar normal stresses as follows:

At the bottom face of the beam, there is no applied load and as a result, the interlaminar normal stress at the bottom face in the first ply is zero, that is,

$$\sigma_{zz}^{(1)}(250,-3) = 0$$

Comparing this with Equation 6.47 and substituting $z = -3 = z_0$ for the first ply, we readily find

$$C_3^{(1)} = 0$$

Then, at the top of the first ply,

$$\sigma_{zz}^{(1)}(250,-1) = -\frac{5}{20} \times (125.6281 \times 0.4595 - 2.5126 \times 0.0806) \times \left(\frac{-1+27}{6}\right) \times 10^{-3}$$
$$= -0.0623 \, \text{MPa (compressive)}$$

Equating stresses at the interface between first and second plies, the interlaminar normal stress at the bottom of the second ply is

$$\sigma_{zz}^{(2)}(250,-1) = -0.0623 \, \text{MPa (compressive)}$$

Comparing this with Equation 6.47 and substituting $z = -1 = z_1$ for the second ply, we readily find

$$C_3^{(2)} = -0.0623 \, \text{MPa}$$

Then, at the top of the second ply,

$$\sigma_{zz}^{(2)}(250,1) = -\frac{5}{20} \times (10.0503 \times 0.4595 - 2.5126 \times 0.0806) \times \left(\frac{1+1}{6}\right)$$
$$\times 10^{-3} - 0.0623 = -0.0628 \, \text{MPa (compressive)}$$

Comparing this with Equation 6.47 and substituting $z = 1 = z_1$ for the third ply, we readily find

$$C_3^{(3)} = -0.0628 \, \text{MPa}$$

Then, at the top of the third ply,

$$\sigma_{zz}^{(3)}(250,3) = -\frac{5}{20} \times (125.6281 \times 0.4595 - 2.5126 \times 0.0806) \times \left(\frac{27-1}{6}\right)$$
$$\times 10^{-3} - 0.0628 = -0.125 \, \text{MPa (compressive)}$$

Note that the interlaminar normal stress at the top of the third ply under the pointed load, as expected, is equal in magnitude to the local applied stress.

6.4.5.2 Simply Supported Beam under Uniformly Distributed Load

A simply supported beam *AB* under a uniformly distributed load is shown in Figure 6.7. The moment boundary conditions, $M(x) = 0$ at $x = 0$ and $x = l$, are utilized to determine the support reactions, as follows:

$$R_A = R_B = \frac{ql}{2} \tag{6.74}$$

The bending moment at a distance x is given by

$$M(x) = -\frac{qx(l-x)}{2} \tag{6.75}$$

Shear force, $V = dM/dx$, is given by

$$V(x) = -\frac{q(l-2x)}{2} \tag{6.76}$$

Similar to the beam under a pointed load, in this case too, the bending moment is negative everywhere, which implies that $\sigma_{xx}^{(k)}(x,z)$ is tensile at the bottom half of the beam and compressive at the top half. By substituting the expression of bending moment in Equation 6.30, the longitudinal bending stress $\sigma_{xx}^{(k)}(x,z)$ is obtained as

$$\sigma_{xx}^{(k)}(x,z) = -\frac{qx(l-x)z}{2b}\left(\bar{Q}_{11}^{(k)}D_{11}^* + \bar{Q}_{12}^{(k)}D_{12}^* + \bar{Q}_{16}^{(k)}D_{16}^*\right) \tag{6.77}$$

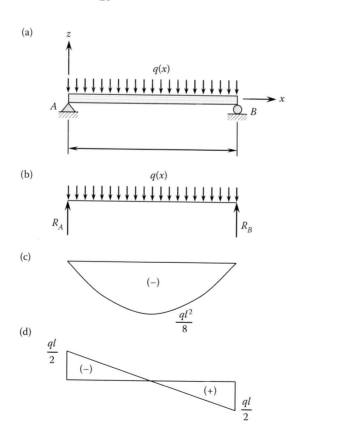

FIGURE 6.7 (a) Simply supported beam under uniformly distributed load. (b) Support reactions. (c) Bending moment distribution. (d) Shear force distribution.

The transverse displacements are given by Equation 6.25, in which the constants of integration are determined by utilizing the displacement boundary conditions, as follows:

$$w_0(x) = \frac{q(2lx^3 - x^4)}{24E_{xx}^b I_{yy}} + C_1 x + C_2 \tag{6.78}$$

Utilizing the displacement boundary conditions, $w_0(x) = 0$ at $x = 0$ and $x = l$, in Equation 6.78, we get

$$C_2 = 0 \tag{6.79}$$

and

$$C_1 = -\frac{ql^3}{24E_{xx}^b I_{yy}} \tag{6.80}$$

Then, substituting Equations 6.79 and 6.80 in Equation 6.78, we get

$$w_0(x) = -\frac{qx(x^3 - 2lx^2 + l^3)}{24E_{xx}^b I_{yy}} \tag{6.81}$$

Note that the displacements are negative, that is, opposite to positive z-direction or downward. Maximum bending moment, bending stress, and displacement occur at the midpoint of the beam and the corresponding values are obtained by substituting $x = l/2$ as follows:

$$(M)_{max} = -\frac{ql^2}{8} \tag{6.82}$$

$$\left(\sigma_{xx}^{(k)}\right)_{max} = \pm \frac{ql^2 h}{16b}\left(\bar{Q}_{11}^{(k)} D_{11}^* + \bar{Q}_{12}^{(k)} D_{12}^* + \bar{Q}_{16}^{(k)} D_{16}^*\right) \tag{6.83}$$

$$(w_0)_{max} = -\frac{5ql^4}{384 E_{xx}^b I_{yy}} \tag{6.84}$$

Note: Bending stress is the maximum in the top and bottom plies.

6.4.5.3 Fixed Beam under Point Load

A fixed beam AB under a point load is shown in Figure 6.8. Owing to the presence of end support moments, the support reactions cannot be directly found by employing boundary conditions. Instead, the following procedure is adopted.

Let us denote the end support moments and reactions as shown in the figure. Then, the bending moment at a distance x is given by

$$M(x) = M_A - R_A x \quad \text{for } 0 \leq x \leq a \tag{6.85}$$

$$M(x) = M_A - (R_A - P)x - Pa \quad \text{for } a \leq x \leq l \tag{6.86}$$

Analysis of Laminated Beams, Columns, and Rods

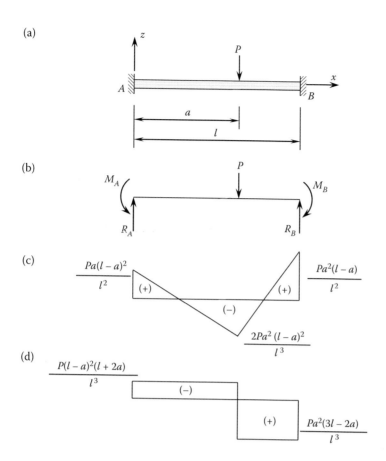

FIGURE 6.8 (a) Fixed beam under a point load. (b) End support moments and reactions. (c) Bending moment distribution. (d) Shear force distribution.

Shear force, $V = dM/dx$, is given by

$$V(x) = -R_A \quad \text{for } 0 \leq x \leq a \quad (6.87)$$

$$V(x) = P - R_A \quad \text{for } a \leq x \leq l \quad (6.88)$$

Note that the bending moment is positive near the supports and negative under the load, which implies that $\sigma_{xx}^{(k)}(x,z)$ is tensile at the top half of the beam near the supports and compressive at the bottom half. On the other hand, it is tensile at the bottom half and compressive at the top under the load. The longitudinal bending stress $\sigma_{xx}^{(k)}(x,z)$ is given by

$$\sigma_{xx}^{(k)}(x,z) = \frac{(M_A - R_A x)z}{b}\left(\bar{Q}_{11}^{(k)} D_{11}^* + \bar{Q}_{12}^{(k)} D_{12}^* + \bar{Q}_{16}^{(k)} D_{16}^*\right) \quad \text{for } 0 \leq x \leq a \quad (6.89)$$

$$\sigma_{xx}^{(k)}(x,z) = \frac{[M_A - (R_A - P)x - Pa]z}{b}\left(\bar{Q}_{11}^{(k)} D_{11}^* + \bar{Q}_{12}^{(k)} D_{12}^* + \bar{Q}_{16}^{(k)} D_{16}^*\right) \quad \text{for } a \leq x \leq l$$

$$(6.90)$$

The transverse displacements are given by Equation 6.25, in which the constants of integration are determined by utilizing the boundary conditions and continuity conditions, as follows:

$$w_0(x) = \frac{R_A x^3 - 3M_A x^2}{6E^b_{xx} I_{yy}} + C_1 x + C_2 \quad \text{for } 0 \leq x \leq a \tag{6.91}$$

$$w_0(x) = \frac{(R_A - P)x^3 - 3(M_A - Pa)x^2}{6E^b_{xx} I_{yy}} + C_3 x + C_4 \quad \text{for } a \leq x \leq l \tag{6.92}$$

The slope is given by

$$\frac{dw_0}{dx} = \frac{R_A x^2 - 2M_A x}{2E^b_{xx} I_{yy}} + C_1 \quad \text{for } 0 \leq x \leq a \tag{6.93}$$

$$\frac{dw_0}{dx} = \frac{(R_A - P)x^2 - 2(M_A - Pa)x}{2E^b_{xx} I_{yy}} + C_3 \quad \text{for } a \leq x \leq l \tag{6.94}$$

Let us apply boundary conditions, $w_0 = dw_0/dx = 0$ at $x = 0$. Then, from Equations 6.91 and 6.93, we get

$$C_1 = C_2 = 0 \tag{6.95}$$

Equations 6.93 and 6.94 give two expressions for slope for the two regions of x. The slope, however, is continuous across the point ($x = a$). Then, substituting $x = a$ in Equations 6.93 and 6.94, and equating them, we get after some arithmetic manipulations

$$C_3 = -\frac{Pa^2}{2E^b_{xx} I_{yy}} \tag{6.96}$$

Similar to slope, displacement is also continuous across the point ($x = a$). Then, substituting $x = a$ in Equations 6.91 and 6.92, and equating them, together with Equations 6.93 and 6.94, we get

$$C_4 = \frac{Pa^3}{6E^b_{xx} I_{yy}} \tag{6.97}$$

We have got all the four constants of integration; but the end support moments and reactions are not known yet.

Let us now apply the boundary condition $dw_0/dx = 0$ at $x = l$. Then, from Equation 6.94, together with Equation 6.96, we get

$$P(l-a)^2 - R_A l^2 + 2M_A l = 0 \tag{6.98}$$

Then, let us now apply the boundary condition $w_0(x) = 0$ at $x = l$. Then, from Equation 6.92, together with Equations 6.96 and 6.97, we get

$$P(l-a)^3 - R_A l^3 + 3M_A l^2 = 0 \tag{6.99}$$

Solving Equations 6.98 and 6.99, we get

$$M_A = \frac{Pa(l-a)^2}{l^2} \tag{6.100}$$

$$R_A = \frac{P(l-a)^2(l+2a)}{l^3} \tag{6.101}$$

Considering static equilibrium, it can be readily seen that

$$M_B = \frac{Pa^2(l-a)}{l^2} \tag{6.102}$$

$$R_B = \frac{Pa^2(3l-2a)}{l^3} \tag{6.103}$$

We can substitute the values/expressions of C_1, C_2, C_3, C_4, M_A, and R_A in corresponding equations and obtain the expressions for bending moment, longitudinal bending stress, and displacement. Resulting expressions, however, are involved and not convenient. We would rather be more interested in the maximum values of these parameters. Maximum bending moment, bending stress occur at the end supports as well as the midpoint when the applied load is centrally located. On the other hand, maximum displacement occurs under the load. These values are

$$(M)_{max} = \pm \frac{Pl}{8} \tag{6.104}$$

$$\left(\sigma_{xx}^{(k)}\right)_{max} = \pm \frac{Plh}{16b}\left(\bar{Q}_{11}^{(k)}D_{11}^* + \bar{Q}_{12}^{(k)}D_{12}^* + \bar{Q}_{16}^{(k)}D_{16}^*\right) \tag{6.105}$$

$$(w_0)_{max} = -\frac{Pl^3}{192 E_{xx}^b I_{yy}} \tag{6.106}$$

Note: Bending stress is the maximum on the top and bottom plies.

6.4.5.4 Fixed Beam under Uniformly Distributed Load

A fixed beam AB under uniformly distributed load is shown in Figure 6.9. Owing to the symmetry of the beam, the reactions are readily obtained as

$$R_A = R_B = \frac{ql}{2} \tag{6.107}$$

Bending moment at a distance x is given by

$$M(x) = M_A - \frac{qlx}{2} + \frac{qx^2}{2} \tag{6.108}$$

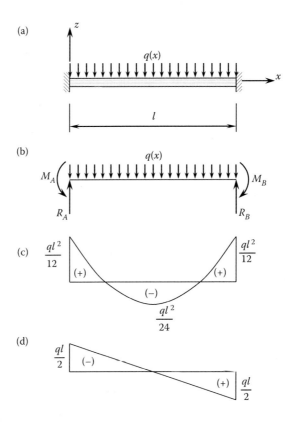

FIGURE 6.9 (a) Fixed beam under uniformly distributed load. (b) End support moments and reactions. (c) Bending moment distribution. (d) Shear force distribution.

Shear force, $V = dM/dx$, is given by

$$V(x) = -\frac{q(l-2x)}{2} \tag{6.109}$$

The transverse displacements are given by Equation 6.25

$$w_0(x) = \frac{-qx^4 + 2qlx^3 - 12M_A x^2}{24 E_{xx}^b I_{yy}} + C_1 x + C_2 \tag{6.110}$$

Differentiating Equation 6.110, we get the slope as

$$\frac{dw_0}{dx} = \frac{-2qx^3 + 3qlx^2 - 12M_A x}{12 E_{xx}^b I_{yy}} + C_1 \tag{6.111}$$

Utilizing the boundary conditions, $w_0 = dw_0/dx = 0$ at $x = 0$, from Equations 6.110 and 6.111, we get

$$C_1 = C_2 = 0 \tag{6.112}$$

Next, owing to symmetry, we see that the slope is zero at the midpoint of the beam. Thus,

$$\frac{dw_0}{dx}\left(x = \frac{l}{2}\right) = 0 \tag{6.113}$$

Analysis of Laminated Beams, Columns, and Rods

Then, from Equations 6.111 and 6.112, we get

$$M_A = \frac{ql^2}{12} \tag{6.114}$$

From Equation 6.108, bending moment is obtained as

$$M(x) = \frac{q(l^2 + 6x^2 - 6lx)}{12} \tag{6.115}$$

Substituting Equation 6.115 in Equation 6.30, the longitudinal bending stress is given by

$$\sigma_{xx}^{(k)}(x,z) = \frac{q(l^2 + 6x^2 - 6lx)z}{12b}\left(\bar{Q}_{11}^{(k)}D_{11}^* + \bar{Q}_{12}^{(k)}D_{12}^* + \bar{Q}_{16}^{(k)}D_{16}^*\right) \tag{6.116}$$

Utilizing Equations 6.112 and 6.114 in Equation 6.110, we get

$$w_0(x) = \frac{-qx^2(l-x)^2}{24E_{xx}^b I_{yy}} \tag{6.117}$$

Bending moment is the maximum at the end supports. Thus, longitudinal bending stress is also the maximum at the end supports. However, maximum displacement occurs at the midpoint of the beam. They are

$$(M)_{max} = \frac{ql^2}{12} \tag{6.118}$$

$$\left(\sigma_{xx}^{(k)}\right)_{max} = \pm\frac{ql^2 h}{24b}\left(\bar{Q}_{11}^{(k)}D_{11}^* + \bar{Q}_{12}^{(k)}D_{12}^* + \bar{Q}_{16}^{(k)}D_{16}^*\right) \tag{6.119}$$

$$(w_0)_{max} = -\frac{ql^4}{384 E_{xx}^b I_{yy}} \tag{6.120}$$

Note: Bending stress is the maximum on the top and bottom faces near the ends.

6.4.5.5 Cantilever Beam under Point Load

A cantilever beam AB under a pointed tip load is shown in Figure 6.10. Under static equilibrium conditions, the end reaction and support moment are readily obtained as

$$R_A = P \tag{6.121}$$

and

$$M_A = Pl \tag{6.122}$$

Bending moment at a distance x is given by

$$M(x) = M_A - R_A x = P(l - x) \tag{6.123}$$

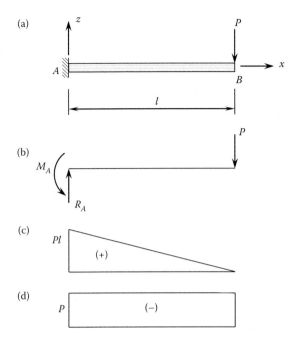

FIGURE 6.10 (a) Cantilever beam under tip load. (b) End support moment and reaction. (c) Bending moment distribution. (d) Shear force distribution.

Shear force, $V = dM/dx$, is given by

$$V(x) = -P \tag{6.124}$$

The transverse displacements are given by Equation 6.25, in which the constants of integration are determined by utilizing the boundary conditions, $w_0 = dw_0/dx = 0$ at $x = 0$, and it can be shown that

$$w_0(x) = -\frac{Px^2(3l-x)}{6E_{xx}^b I_{yy}} \tag{6.125}$$

Substituting Equation 6.123 in Equation 6.30, the longitudinal bending stress is obtained as

$$\sigma_{xx}^{(k)}(x,z) = \frac{P(l-x)z}{b}\left(\bar{Q}_{11}^{(k)}D_{11}^* + \bar{Q}_{12}^{(k)}D_{12}^* + \bar{Q}_{16}^{(k)}D_{16}^*\right) \tag{6.126}$$

Bending moment and longitudinal bending stress are the maximum at the end support. However, maximum displacement occurs at the free end of the beam. Maximum bending moment, bending stress, and displacements are

$$(M)_{max} = Pl \tag{6.127}$$

$$\left(\sigma_{xx}^{(k)}\right)_{max} = \pm\frac{Plh}{2b}\left(\bar{Q}_{11}^{(k)}D_{11}^* + \bar{Q}_{12}^{(k)}D_{12}^* + \bar{Q}_{16}^{(k)}D_{16}^*\right) \tag{6.128}$$

$$(w_0)_{max} = -\frac{Pl^3}{3E_{xx}^b I_{yy}} \tag{6.129}$$

6.4.5.6 Cantilever Beam under Uniformly Distributed Load

A cantilever beam *AB* under a uniformly distributed load is shown in Figure 6.11. Under static equilibrium conditions, the end reaction and support moment are readily obtained as

$$R_A = ql \tag{6.130}$$

and

$$M_A = \frac{ql^2}{2} \tag{6.131}$$

Bending moment at a distance x is given by

$$M(x) = M_A - R_A x + \frac{qx^2}{2} = \frac{q(l-x)^2}{2} \tag{6.132}$$

Shear force, $V = dM/dx$, is given by

$$V(x) = -q(l-x) \tag{6.133}$$

The transverse displacements are given by Equation 6.25, in which the constants of integration are determined by utilizing the boundary conditions, $w_0 = dw_0/dx = 0$ at $x = 0$, and it can be shown that

$$w_0(x) = -\frac{qx^2(x^2 - 4lx + 6l^2)}{24 E^b_{xx} I_{yy}} \tag{6.134}$$

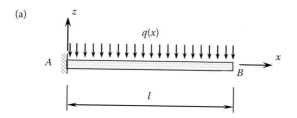

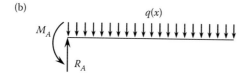

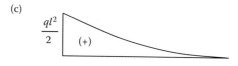

FIGURE 6.11 (a) Cantilever beam under uniformly distributed load. (b) End support moment and reaction. (c) Bending moment distribution. (d) Shear force distribution.

Substituting Equation 6.132 in Equation 6.30, the longitudinal bending stress is given by

$$\sigma_{xx}^{(k)}(x,z) = \frac{q(l-x)^2 z}{2b} \left(\bar{Q}_{11}^{(k)} D_{11}^* + \bar{Q}_{12}^{(k)} D_{12}^* + \bar{Q}_{16}^{(k)} D_{16}^* \right) \qquad (6.135)$$

Bending moment and longitudinal bending stress are the maximum at the end support. However, maximum displacement occurs at the free end of the beam. Maximum bending moment, bending stress, and displacements are

$$(M)_{max} = \frac{ql^2}{2} \qquad (6.136)$$

$$\left(\sigma_{xx}^{(k)} \right)_{max} = \pm \frac{ql^2 h}{4b} \left(\bar{Q}_{11}^{(k)} D_{11}^* + \bar{Q}_{12}^{(k)} D_{12}^* + \bar{Q}_{16}^{(k)} D_{16}^* \right) \qquad (6.137)$$

$$(w_0)_{max} = -\frac{ql^4}{8 E_{xx}^b I_{yy}} \qquad (6.138)$$

6.5 BENDING OF A LAMINATED BEAM (SOLID RECTANGULAR CROSS SECTION: PLIES PARALLEL TO LOADING DIRECTION)

Figure 6.12 shows a laminated composite beam of rectangular cross section with its plies parallel to the loading plane, which in this case is the xy-plane. Note carefully the subtle change in the nomenclature of beam cross-sectional dimensions. Now, the beam width and height are denoted by h and b, respectively. Note further that h also denotes the overall laminate thickness.

As we are dealing with only symmetric ply sequence, for the ply orientation parallel to the loading plane, the beam-bending behavior is similar to that of an isotropic beam. Under the action of the transverse loads $q = q(x)$, bending moment $M = M(x)$ is generated. (M may also include applied pure bending moment.) The beam bends about the z-axis in such a way that the xz-plane, which divides the beam cross section at the midheight, is the neutral plane. The longitudinal strain in the midplane at a

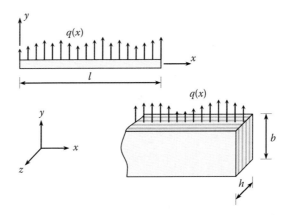

FIGURE 6.12 Laminated composite beam of solid rectangular cross section with plies parallel to loading plane.

Analysis of Laminated Beams, Columns, and Rods

distance y from the neutral plane is the same as that in any ply at the same distance and is given by

$$\varepsilon_{xx}^0 = \varepsilon_{xx} = \frac{y}{R} \tag{6.139}$$

where R is the radius of curvature of the longitudinal axis in the xy-plane.

At this point, before proceeding any further, let us bring in the concept of effective extensional modulus [3]. For a symmetric laminate, it can be shown that it is given by

$$E_{xx}^{ex} = \frac{1}{hA_{11}^*} \tag{6.140}$$

Also, the area moment of inertia of the beam cross section about the bending axis is given by

$$I_{zz} = \frac{hb^3}{12} \tag{6.141}$$

Thus, the effective extensional stiffness of the beam is

$$E_{xx}^{ex} I_{zz} = \frac{b^3}{12 A_{11}^*} \tag{6.142}$$

Now, for a symmetric laminate, the constitutive relation in terms of the compliance matrix is (refer Equation 5.155, Chapter 5)

$$\begin{Bmatrix} \varepsilon_{xx}^0 \\ \varepsilon_{yy}^0 \\ \gamma_{xy}^0 \end{Bmatrix} = \begin{bmatrix} A_{11}^* & A_{12}^* & A_{16}^* \\ A_{12}^* & A_{22}^* & A_{26}^* \\ A_{16}^* & A_{26}^* & A_{66}^* \end{bmatrix} \begin{Bmatrix} N_{xx} \\ N_{yy} \\ N_{xy} \end{Bmatrix} \tag{6.143}$$

N_{xx} being the only nonzero stress resultant, we get

$$\varepsilon_{xx}^0 = A_{11}^* N_{xx} \tag{6.144}$$

Using Equation 6.139 in Equation 6.144, we get

$$N_{xx} = \frac{y}{R A_{11}^*} \tag{6.145}$$

The externally applied moment M must be balanced by the moment due to internally generated stress resultants. Then,

$$M = \int_{-b/2}^{b/2} N_{xx} y \, dy \tag{6.146}$$

Substituting N_{xx} from Equation 6.145 in Equation 6.146, we get

$$M = \int_{-b/2}^{b/2} \frac{y^2}{RA_{11}^*} dy = \frac{b^3}{12RA_{11}^*} \quad (6.147)$$

Using Equations 6.142 and 6.147, we find a relation between the radius of curvature and the applied moment as follows:

$$MR = E_{xx}^{ex} I_{zz} \quad (6.148)$$

Next, we need to determine the stresses. It can be seen that for a given cross section, that is, given x, curvature is a constant and the longitudinal strain depends only on y. In other words, all the points in a given cross section at a given distance from the neutral plane are strained equally. However, the stresses depend on one more parameter, viz. the individual ply moduli and follow a stepwise variation across the laminate thickness, that is, beam width. Thus, we introduce a term effective longitudinal stress σ_{xx}^{eff} such that the axial stress resultant $N_{xx} = h\sigma_{xx}^{eff}$. Then, using Equation 6.145, we get an expression for effective longitudinal stress as follows:

$$\sigma_{xx}^{eff} = \frac{N_{xx}}{h} = \frac{y}{hRA_{11}^*} \quad (6.149)$$

Using Equations 6.141 and 6.147 in Equation 6.149, we find the following:

$$\sigma_{xx}^{eff} = \frac{yM}{I_{zz}} \quad (6.150)$$

Equations 6.148 and 6.150 can be combined and expressed as follows:

$$\frac{M}{I_{zz}} = \frac{\sigma_{xx}^{eff}}{y} = \frac{E_{xx}^{ex}}{R} \quad (6.151)$$

Note that Equation 6.151 is very similar to the classical isotropic beam-bending equation [6].

Note further that σ_{xx}^{eff} is not the actual stress. To determine the actual ply stresses, we need to find N_{xx} and then apply classical laminate analysis steps as described earlier (refer Table 5.2, Chapter 5).

Beam displacements under different cases can be determined by various expressions described in Section 6.4.5. However, care must be taken to replace $E_{xx}^b I_{yy}$ with $E_{xx}^{ex} I_{zz}$.

EXAMPLE 6.2

Consider the carbon/epoxy simply supported beam discussed in Example 6.1. For ready reference, we repeat the details here.

Dimensions: $l = 500$ mm, $b = 20$ mm, and $h = 6$ mm
Ply sequence: [0°/90°/0°], each ply being 2 mm in thickness

Analysis of Laminated Beams, Columns, and Rods

Determine the maximum displacement and longitudinal in-plane stress σ_{xx} at the center of the beam. The beam is under a central point load of 100 N applied parallel to the plies. The point load is applied over an area of 6 mm × 20 mm. Material properties are as follows:

$$E_1 = 125 \text{ GPa}, E_2 = 10 \text{ GPa}, \nu_{12} = 0.25, \text{ and } G_{12} = 8 \text{ GPa}$$

Solution

For the given material properties and ply sequence, the transformed reduced stiffness matrix and the laminate compliance matrices are as follows:

$$[\bar{Q}]^{(1)} = [\bar{Q}]^{(3)} = \begin{bmatrix} 125.6281 & 2.5126 & 0 \\ 2.5126 & 10.0503 & 0 \\ 0 & 0 & 8 \end{bmatrix} \times 10^3 \text{ MPa}$$

$$[\bar{Q}]^{(2)} = \begin{bmatrix} 10.0503 & 2.5126 & 0 \\ 2.5126 & 125.6281 & 0 \\ 0 & 0 & 8 \end{bmatrix} \times 10^3 \text{ MPa}$$

$$[A^*] = \begin{bmatrix} 1.9163 & -0.0991 & 0 \\ -0.0991 & 3.4362 & 0 \\ 0 & 0 & 20.8333 \end{bmatrix} \times 10^{-6} \text{ (MPa} \cdot \text{mm)}^{-1}$$

$$[D^*] = \begin{bmatrix} 0.4595 & -0.0806 & 0 \\ -0.0806 & 3.8907 & 0 \\ 0 & 0 & 6.9444 \end{bmatrix} \times 10^{-6} \text{ (MPa} \cdot \text{mm}^3)^{-1}$$

$$[B^*] = [C^*] = 0$$

Effective extensional stiffness of the beam is given by Equation 6.142 as follows:

$$E_{xx}^{ex} I_{zz} = \frac{b^3}{12 A_{11}^*} = \frac{20^3}{12 \times 1.9163 \times 10^{-6}} = 347.8926 \times 10^6 \text{ N} \cdot \text{mm}^2$$

z-coordinates of different plies are as follows:

$$z_0 = -3 \text{ mm}, z_1 = -1 \text{ mm}, z_2 = 1 \text{ mm}, \text{ and } z_3 = 3 \text{ mm}$$

Displacement under the point load is the maximum displacement, and it is obtained by using effective extensional stiffness in Equation 6.73 as

$$(w_0)_{max} = -\frac{100 \times 500^3}{48 \times 347.8926 \times 10^6} = -0.75 \text{ mm}$$

Maximum bending moment is $M = 12{,}500 \text{ N} \cdot \text{mm}$

Maximum effective bending stress occurs at the bottom and top of the beam at the center. They are given by Equation 6.150:

At the bottom of the beam under the central point load,

$$\left(\sigma_{xx}^{eff}\right)_{max} = \frac{10 \times 12500}{6 \times 20^3/12} = 31.25\,\text{MPa}$$

which implies $N_{xx} = 31.25 \times 6 = 187.5$ N/mm. N_{xx} is positive (tensile) at the bottom of the beam and negative (compressive) at the top.

To determine the in-plane stresses in the plies at the outermost face of the beam, we proceed as follows. The midplane strains and curvatures are given by

$$\begin{Bmatrix} \varepsilon_{xx}^0 \\ \varepsilon_{yy}^0 \\ \gamma_{xy}^0 \end{Bmatrix} = \begin{bmatrix} 1.9163 & -0.0991 & 0 \\ -0.0991 & 3.4362 & 0 \\ 0 & 0 & 20.8333 \end{bmatrix} \times \begin{Bmatrix} 187.5 \\ 0 \\ 0 \end{Bmatrix} \times 10^{-6} = \begin{Bmatrix} 3.593 \\ -0.186 \\ 0 \end{Bmatrix} \times 10^{-4}$$

$$\begin{Bmatrix} \kappa_{xx} \\ \kappa_{yy} \\ \kappa_{xy} \end{Bmatrix} = \begin{Bmatrix} 0 \\ 0 \\ 0 \end{Bmatrix}$$

Then, the global strains

$$\begin{Bmatrix} \varepsilon_{xx} \\ \varepsilon_{yy} \\ \varepsilon_{xy} \end{Bmatrix} = \begin{Bmatrix} 3.593 \\ -0.186 \\ 0 \end{Bmatrix} \times 10^{-4}$$

Note that the strains are the same in all the plies. Global stresses in the outermost plies (0°) are obtained as

$$\begin{Bmatrix} \sigma_{xx} \\ \sigma_{yy} \\ \tau_{xy} \end{Bmatrix}^{(1)} = \begin{bmatrix} 125.6281 & 2.5126 & 0 \\ 2.5126 & 10.0503 & 0 \\ 0 & 0 & 8 \end{bmatrix} \times \begin{Bmatrix} 3.593 \\ -0.186 \\ 0 \end{Bmatrix} \times 10^{-1} = \begin{Bmatrix} 45.1 \\ 0.7 \\ 0 \end{Bmatrix} \text{MPa}$$

The local stresses are obtained by transformation as follows:

$$\begin{Bmatrix} \sigma_{11} \\ \sigma_{22} \\ \tau_{12} \end{Bmatrix}^{(1)} = \begin{bmatrix} 1 & 0 & 0 \\ 0 & 1 & 0 \\ 0 & 0 & 1 \end{bmatrix} \begin{Bmatrix} 45.1 \\ 0.7 \\ 0 \end{Bmatrix} = \begin{Bmatrix} 45.1 \\ 0.7 \\ 0 \end{Bmatrix} \text{MPa}$$

Similarly, the global and local stresses in the middle ply (90°) are obtained as

$$\begin{Bmatrix} \sigma_{xx} \\ \sigma_{yy} \\ \tau_{xy} \end{Bmatrix}^{(2)} = \begin{bmatrix} 10.0503 & 2.5126 & 0 \\ 2.5126 & 125.6281 & 0 \\ 0 & 0 & 8 \end{bmatrix} \times \begin{Bmatrix} 3.593 \\ -0.186 \\ 0 \end{Bmatrix} \times 10^{-1} = \begin{Bmatrix} 3.6 \\ -1.4 \\ 0 \end{Bmatrix} \text{MPa}$$

$$\begin{Bmatrix} \sigma_{11} \\ \sigma_{22} \\ \tau_{12} \end{Bmatrix}^{(2)} = \begin{bmatrix} 0 & 1 & 0 \\ 1 & 0 & 0 \\ 0 & 0 & -1 \end{bmatrix} \begin{Bmatrix} 3.6 \\ -1.4 \\ 0 \end{Bmatrix} = \begin{Bmatrix} -1.4 \\ 3.6 \\ 0 \end{Bmatrix} \text{MPav}$$

6.6 BENDING OF A LAMINATED COMPOSITE BEAM (THIN-WALLED CROSS SECTION)

In an isotropic beam, the expressions for displacement involve the term EI, which is the flexural rigidity of the beam. In the analysis of laminated beam of rectangular cross-section, as discussed in Section 6.5, we need to determine the effective or equivalent flexural rigidity of the laminated beam. In the analysis of a laminated composite beam of thin-walled cross section too, we need to determine its equivalent flexural rigidity [3]. However, in this case, owing to the presence of more than one element in the cross-section, the analysis procedure is a little complex. The level of complexity increases particularly with nonsymmetric cross sections such as T-section and L-section. Note that unequal thickness in the top and bottom flanges (or left and right webs) can also make an otherwise symmetric cross section nonsymmetric.

The first step in the analysis process is to determine the locations of the centroids of the flange(s) and web(s). Next, the total moment supported by the beam is expressed in terms of the beam cross-sectional details and compliance matrix elements. Then, by comparing the total moment expression with the classical beam equation, we obtain the expression for the equivalent flexural rigidity.

Note: Thin-walled sections are composed of at least two elements—flanges(s) and web(s). During analysis, we need to consider an axis system and a set of stiffness/compliance properties for each of these elements as well as the beam as a whole. Before we move further, it is important to note carefully the convention for the axes and other parameters.

The cross-sectional elements are assigned a number as indicated in the respective figures. This number is put within small bracket as superscript to indicate to which element the stiffness/compliance parameter belongs. For example, in the case of an I-section (Figure 6.15 in Section 6.6.2), $(A_{11}^*)^{(3)}$ is the extensional compliance matrix element of the bottom flange, whereas, in the case of a box-section (Figure 6.16 in Section 6.6.3), it refers to the same compliance matrix element of the right web.

Any parameter without any superscript within small bracket belongs to the beam as a whole.

For the beam, x-axis is in the axial direction, y-axis is in the horizontal plane, and z-axis is in the vertical plane. The x-, y-, and z-axes are orthogonal to each other and they all pass through the centroid of the beam cross section. For the flange(s) and web(s), the axes pass through the respective centroid. The local x-axis (i.e., that of a flange or web x-axis) is parallel to the x-axis of the beam. However, the local y-axis is always parallel to the local plies and the local z-axis is normal to it. Thus, for a flange, y-axis is horizontal and z-axis is vertical, but for a web, y- and z-axes are in the vertical and horizontal planes, respectively.

6.6.1 T-Section

T-section consists of a flange and a web. As shown in Figure 6.13a–c, laminated beam construction can be achieved either by integral lay-up of the flange and web or by adhesive bonding or a combination of both. The exact ply orientation depends on the manufacturing scheme; however, for simplicity of analysis, as indicated in Figure 6.13d, the

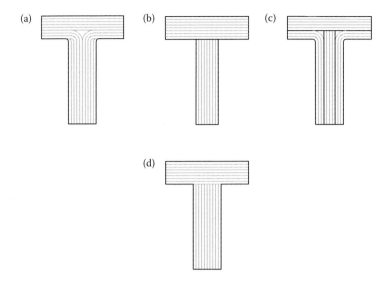

FIGURE 6.13 T-section laminated composite beam. (a) Integral lay-up of flange and web. (b) Adhesive bonding of flange and web. (c) Combination of adhesive bonding and integral lay-up. (d) Ply orientation for analysis.

flange and web are assumed to consist of all continuous horizontal and vertical plies with perfect bond between them.

Figure 6.14 shows details of a T-section laminated composite beam. Under the action of a pure bending moment, the beam bends about the y-axis. The axial strain distribution w.r.t. z is linear. We take the axis system in such a way as to align the xy-plane with the neutral plane.

Let us now find the locations of the centroids, toward which we find the axial force component supported by each element. First, we consider the flange. The axial strain at any point in the beam is given by

$$\varepsilon_{xx} = \frac{z}{R} \tag{6.152}$$

where R is the radius of curvature of the neutral axis in the xz-plane.

Then, the flange midplane axial strain is given by

$$\left(\varepsilon_{xx}^0\right)^{(1)} = \frac{z_{c1}}{R} \tag{6.153}$$

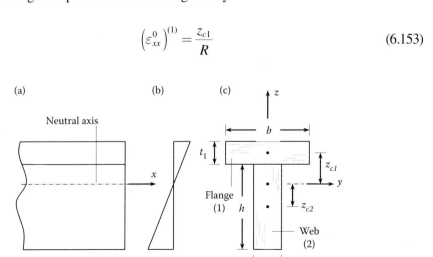

FIGURE 6.14 Laminated composite beam of T-cross section. (a) Beam in front view. (b) Typical axial strain distribution. (c) Cross-sectional details.

Since, we have restricted our analysis to symmetric ply sequence, the midplane strain is also given by

$$\left(\varepsilon_{xx}^{0}\right)^{(1)} = \left(A_{11}^{*}\right)^{(1)} \left(N_{xx}\right)^{(1)} \quad (6.154)$$

where $(A_{11}^{*})^{(1)}$ is the extensional compliance matrix element for the flange ply sequence and $(N_{xx})^{(1)}$ is the axial force resultant in the flange.

Then, comparing Equations 6.153 and 6.154, we get

$$(N_{xx})^{(1)} = \frac{z_{c1}}{R\left(A_{11}^{*}\right)^{(1)}} \quad (6.155)$$

Note that $(N_{xx})^{(1)}$ is a fixed quantity for a given beam configuration and ply sequence. Then, the total axial force in the flange can be obtained as

$$(F)^{(1)} = \frac{bz_{c1}}{R\left(A_{11}^{*}\right)^{(1)}} \quad (6.156)$$

Now, we turn our attention to the web. The plies are parallel to the loading plane and, as we had seen before, the web acts more like an isotropic beam. The midplane axial strain varies with z and is given by

$$\left(\varepsilon_{xx}^{0}\right)^{(2)} = \frac{z}{R} \quad (6.157)$$

The web ply sequence is also symmetric and accordingly the midplane strains are given by

$$\left(\varepsilon_{xx}^{0}\right)^{(2)} = \left(A_{11}^{*}\right)^{(2)} (N_{xx})^{(2)} \quad (6.158)$$

Comparing Equations 6.157 and 6.158, we get

$$(N_{xx})^{(2)} = \frac{z}{R\left(A_{11}^{*}\right)^{(2)}} \quad (6.159)$$

Clearly, $(N_{xx})^{(2)}$ depends linearly on z and the total axial force in the web is obtained by integration as follows:

$$(F)^{(2)} = \int_{-(h/2+z_{c2})}^{h/2-z_{c2}} \frac{z}{R\left(A_{11}^{*}\right)^{(2)}} dz \quad (6.160)$$

After few simple steps of manipulation, we obtain the following:

$$(F)^{(2)} = -\frac{hz_{c2}}{R\left(A_{11}^{*}\right)^{(2)}} \quad (6.161)$$

Now, under pure bending net axial force in the beam is zero, that is,

$$(F)^{(1)} + (F)^{(2)} = 0 \tag{6.162}$$

Also, from the cross-sectional details, we get

$$z_{c1} + z_{c2} = \frac{h}{2} + \frac{t_1}{2} \tag{6.163}$$

Using Equations 6.156, 6.161, and 6.163 in Equation 6.162, it can be shown that

$$z_{c1} = \frac{h+t_1}{2\left[1+(b/h)\left((A^*_{11})^{(2)}/(A^*_{11})^{(1)}\right)\right]} \tag{6.164}$$

and

$$z_{c2} = \frac{h+t_1}{2\left[1+(h/b)\left((A^*_{11})^{(1)}/(A^*_{11})^{(2)}\right)\right]} \tag{6.165}$$

Equations 6.164 and 6.165 give the locations of the centroids of the flange and web, respectively. Note that we have expressed the distances to the centroids from the neutral axis in terms of cross-sectional details and compliance matrix elements, which are all known.

Let us now find the moment components supported by the flange and web. There are two parts in the moment component supported by the flange—one due to the force resultant and the other due to the moment resultant. The moment resultant in the flange is given by

$$(M_{xx})^{(1)} = \frac{(\kappa_{xx})^{(1)}}{(D^*_{11})^{(1)}} \tag{6.166}$$

By definition, curvature is given by

$$(\kappa_{xx})^{(1)} = \frac{1}{R} \tag{6.167}$$

Thus,

$$(M_{xx})^{(1)} = \frac{1}{R(D^*_{11})^{(1)}} \tag{6.168}$$

Then, using Equations 6.156 and 6.168, the total moment component supported by the flange is obtained as

$$(M)^{(1)} = \frac{1}{R}\left[\frac{bz_{c1}^2}{(A^*_{11})^{(1)}} + \frac{b}{(D^*_{11})^{(1)}}\right] \tag{6.169}$$

Analysis of Laminated Beams, Columns, and Rods

The web does not undergo any curvature in the plane normal to its plies and its moment resultant is zero. Thus, the moment component supported by the web is only due to its force resultant and it is obtained by integration as follows:

$$(M)^{(2)} = \int_{-(h/2+z_{c2})}^{h/2-z_{c2}} \frac{z^2}{R\left(A_{11}^*\right)^{(2)}} dz \qquad (6.170)$$

which after simple manipulation leads to

$$(M)^{(2)} = \frac{h}{12R\left(A_{11}^*\right)^{(2)}} \left(h^2 + 12z_{c2}^2\right) \qquad (6.171)$$

Then, using Equations 6.169 and 6.171, the total moment supported by the beam is obtained as follows:

$$M = \frac{1}{R}\left[\frac{bz_{c1}^2}{\left(A_{11}^*\right)^{(1)}} + \frac{b}{\left(D_{11}^*\right)^{(1)}} + \frac{h}{12\left(A_{11}^*\right)^{(2)}}\left(h^2 + 12z_{c2}^2\right)\right] \qquad (6.172)$$

Now, from the classical beam equation, the effective bending rigidity of the beam is

$$E_{xx}^b I_{yy} = MR \qquad (6.173)$$

in which, E_{xx}^b and I_{yy} are the effective bending modulus and moment of inertia, respectively. Substituting Equation 6.172 in Equation 6.173, we get

$$E_{xx}^b I_{yy} = \frac{bz_{c1}^2}{\left(A_{11}^*\right)^{(1)}} + \frac{b}{\left(D_{11}^*\right)^{(1)}} + \frac{h}{12\left(A_{11}^*\right)^{(2)}}\left(h^2 + 12z_{c2}^2\right) \qquad (6.174)$$

Equation 6.174 gives us an expression for the effective or equivalent bending rigidity of the T-section laminated composite beam. Note that the expression is in terms of cross-sectional details and compliance matrix elements, which are all known. Note further that having determined the effective bending rigidity of the beam, the maximum displacement of the beam can be found for different loading and end conditions.

Let us now consider the stresses and strains in the beam.

Using Equation 6.173 in Equations 6.155 and 6.168, the axial force resultant and moment resultant in the flange are obtained as follows:

$$(N_{xx})^{(1)} = \frac{z_{c1}M}{\left(A_{11}^*\right)^{(1)} E_{xx}^b I_{yy}} \qquad (6.175)$$

$$(M_{xx})^{(1)} = \frac{M}{\left(D_{11}^*\right)^{(1)} E_{xx}^b I_{yy}} \qquad (6.176)$$

Note that other stress resultants are all zero, that is, $(N_{yy})^{(1)} = (N_{xy})^{(1)} = (M_{yy})^{(1)} = (M_{xy})^{(1)} = 0$, which implies

$$\{N\}^{(1)} = \left\{ \begin{array}{c} \dfrac{z_{c1}M}{\left(A_{11}^*\right)^{(1)} E_{xx}^b I_{yy}} \\ 0 \\ 0 \end{array} \right\} \quad \text{and} \quad \{M\}^{(1)} = \left\{ \begin{array}{c} \dfrac{M}{\left(D_{11}^*\right)^{(1)} E_{xx}^b I_{yy}} \\ 0 \\ 0 \end{array} \right\} \quad (6.177)$$

Then, the strains and stresses are determined by following standard procedure as explained in Table 5.2, Chapter 5.

In the case of the web, the only nonzero stress resultant is $(N_{xx})^{(2)}$ and it is obtained by using Equations 6.159 and 6.173 as

$$(N_{xx})^{(2)} = \dfrac{zM}{\left(A_{11}^*\right)^{(2)} E_{xx}^b I_{yy}} \quad (6.178)$$

Thus, the stress resultant vectors for the web are

$$\{N\}^{(1)} = \left\{ \begin{array}{c} \dfrac{zM}{\left(A_{11}^*\right)^{(2)} E_{xx}^b I_{yy}} \\ 0 \\ 0 \end{array} \right\} \quad \text{and} \quad \{M\}^{(1)} = \left\{ \begin{array}{c} 0 \\ 0 \\ 0 \end{array} \right\} \quad (6.179)$$

Then, the strains and stresses in the web are determined by following classical laminate analysis procedure.

EXAMPLE 6.3

Consider a carbon/epoxy cantilevered T-section beam with the following dimensions: length $l = 500$ mm, flange width $b = 20$ mm, web height $h = 30$ mm, and flange thickness $t = 4$ mm. Thickness of the web is the same as that of the flange. The flange and the web are composed of $0°$ plies, each ply being 0.5 mm in thickness. Determine the maximum displacement if the beam is under a tip point load of 100 N. Material properties are as follows:

$$E_1 = 125 \text{ GPa}, \; E_2 = 10 \text{ GPa}, \; \nu_{12} = 0.25, \; \text{and} \; G_{12} = 8 \text{ GPa}$$

Solution

For the given material properties and ply sequence, the transformed reduced stiffness matrix and the laminate compliance matrix are obtained as follows (detailed calculations are not shown):

$$[\bar{Q}] = \begin{bmatrix} 125.6281 & 2.5126 & 0 \\ 2.5126 & 10.0503 & 0 \\ 0 & 0 & 8 \end{bmatrix} \times 10^3 \text{ MPa (same for all the plies)}$$

Analysis of Laminated Beams, Columns, and Rods

$$[A^*] = \begin{bmatrix} 2 & -0.5 & 0 \\ -0.5 & 25 & 0 \\ 0 & 0 & 31.25 \end{bmatrix} \times 10^{-6} (\text{MPa} \cdot \text{mm})^{-1}$$

$$[D^*] = \begin{bmatrix} 1.5000 & -0.3750 & 0 \\ -0.3750 & 18.7500 & 0 \\ 0 & 0 & 23.4375 \end{bmatrix} \times 10^{-6} (\text{MPa} \cdot \text{mm}^3)^{-1}$$

Distances to the flange and web centroids from the neutral axis are obtained as

$$z_{c1} = \frac{h+t_1}{2\left[1+(b/h)\left((A_{11}^*)^{(2)}/(A_{11}^*)^{(1)}\right)\right]} = \frac{30+4}{2[1+(20/30)]} = 10.2 \text{ mm}$$

$$z_{c2} = \frac{h+t_1}{2\left[1+(h/b)\left((A_{11}^*)^{(1)}/(A_{11}^*)^{(2)}\right)\right]} = \frac{30+4}{2[1+(30/20)]} = 6.8 \text{ mm}$$

Effective bending stiffness of the beam is

$$E_{xx}^b I_{yy} = \frac{bz_{c1}^2}{(A_{11}^*)^{(1)}} + \frac{b}{(D_{11}^*)^{(1)}} + \frac{h}{12(A_{11}^*)^{(2)}}\left(h^2 + 12z_{c2}^2\right)$$

$$= \frac{20 \times 10.2^2}{2.0 \times 10^{-6}} + \frac{20}{1.5 \times 10^{-6}} + \frac{30}{12 \times 2.0 \times 10^{-6}}[30^2 + 12 \times 6.8^2]$$

$$= 2.872 \times 10^9 \text{ MPa.mm}^4$$

Displacement under the point load is the maximum displacement and it is given by (Equation 6.129)

$$(w_0)_{max} = -\frac{100 \times 500^3}{3 \times 2.872 \times 10^9} = -1.45 \text{ mm}$$

The stresses and strains vary along the length. Bending moment is maximum at the fixed end of the cantilevered beam and we shall find the stresses and strains at the top and bottom faces of the beam at the fixed end.

In order to find the bending stresses in the flange, we determine the nonzero stress resultants as follows (refer Equations 6.175 and 6.176):

$$(N_{xx})^{(1)} = \frac{10.2 \times (100 \times 500)}{2 \times 10^{-6} \times 2.872 \times 10^9} = 88.788 \text{ N/mm}$$

$$(M_{xx})^{(1)} = \frac{100 \times 500}{1.5 \times 10^{-6} \times 2.872 \times 10^9} = 11.606 \text{ N} \cdot \text{mm/mm}$$

Midplane strains and curvatures in the flange are

$$\begin{Bmatrix} \varepsilon_{xx}^0 \\ \varepsilon_{yy}^0 \\ \gamma_{xy}^0 \end{Bmatrix}^{(1)} = \begin{bmatrix} 2 & -0.5 & 0 \\ -0.5 & 25 & 0 \\ 0 & 0 & 31.25 \end{bmatrix} \times \begin{Bmatrix} 88.788 \\ 0 \\ 0 \end{Bmatrix} \times 10^{-6} = \begin{Bmatrix} 177.576 \\ -44.394 \\ 0 \end{Bmatrix} \times 10^{-6}$$

and

$$\begin{Bmatrix} \kappa_{xx}^0 \\ \kappa_{yy}^0 \\ \kappa_{xy}^0 \end{Bmatrix}^{(1)} = \begin{bmatrix} 1.5000 & -0.3750 & 0 \\ -0.3750 & 18.7500 & 0 \\ 0 & 0 & 23.4375 \end{bmatrix} \times \begin{Bmatrix} 11.606 \\ 0 \\ 0 \end{Bmatrix} \times 10^{-6} = \begin{Bmatrix} 17.409 \\ -4.352 \\ 0 \end{Bmatrix} \times 10^{-6}$$

Global strains at the top face of the flange are

$$\begin{Bmatrix} \varepsilon_{xx} \\ \varepsilon_{yy} \\ \gamma_{xy} \end{Bmatrix}^{(1)} = \begin{Bmatrix} 177.576 \\ -44.394 \\ 0 \end{Bmatrix} \times 10^{-6} + 2 \times \begin{Bmatrix} 17.409 \\ -4.352 \\ 0 \end{Bmatrix} \times 10^{-6} = \begin{Bmatrix} 212.394 \\ -53.098 \\ 0 \end{Bmatrix} \times 10^{-6}$$

Global stresses at the top face of the flange are

$$\begin{Bmatrix} \sigma_{xx} \\ \sigma_{yy} \\ \tau_{xy} \end{Bmatrix}^{(1)} = \begin{bmatrix} 125.6281 & 2.5126 & 0 \\ 2.5126 & 10.0503 & 0 \\ 0 & 0 & 8 \end{bmatrix} \times \begin{Bmatrix} 212.394 \\ -53.098 \\ 0 \end{Bmatrix} \times 10^{-3} = \begin{Bmatrix} 26.5 \\ 0 \\ 0 \end{Bmatrix} \text{MPa}$$

It can be seen that the local stresses are the same as the global stresses.

Then, we turn our attention to the web stresses, for which we find the only nonzero stress resultant in the web as follows (refer Equation 6.178):

$$(N_{xx})^{(2)} = \frac{(10.2 - 4/2) \times (100 \times 500)}{2 \times 10^{-6} \times 2.872 \times 10^9} = 71.379 \text{ N/mm}$$

(at the web-to-flange interface)

$$(N_{xx})^{(2)} = \frac{-(30/2 + 6.8) \times (100 \times 500)}{2 \times 10^{-6} \times 2.872 \times 10^9} = -189.763 \text{ N/mm}$$

(at the bottom of the beam)

Midplane strains and curvatures in the web (bottom face of the beam) are

$$\begin{Bmatrix} \varepsilon_{xx}^0 \\ \varepsilon_{yy}^0 \\ \gamma_{xy}^0 \end{Bmatrix}^{(2)} = \begin{bmatrix} 2 & -0.5 & 0 \\ -0.5 & 25 & 0 \\ 0 & 0 & 31.25 \end{bmatrix} \times \begin{Bmatrix} -189.763 \\ 0 \\ 0 \end{Bmatrix} \times 10^{-6} = \begin{Bmatrix} -379.526 \\ 94.882 \\ 0 \end{Bmatrix} \times 10^{-6}$$

and

$$\left\{\begin{matrix}\kappa_{xx}^0\\ \kappa_{yy}^0\\ \kappa_{xy}^0\end{matrix}\right\}^{(2)} = \left\{\begin{matrix}0\\ 0\\ 0\end{matrix}\right\}$$

Global strains in the web (bottom face of the beam) are

$$\left\{\begin{matrix}\varepsilon_{xx}\\ \varepsilon_{yy}\\ \gamma_{xy}\end{matrix}\right\}^{(2)} = \left\{\begin{matrix}-379.526\\ 94.882\\ 0\end{matrix}\right\} \times 10^{-6}$$

Global stresses in the web (bottom face of the beam) are

$$\left\{\begin{matrix}\sigma_{xx}\\ \sigma_{yy}\\ \tau_{xy}\end{matrix}\right\}^{(2)} = \begin{bmatrix}125.6281 & 2.5126 & 0\\ 2.5126 & 10.0503 & 0\\ 0 & 0 & 8\end{bmatrix} \times \left\{\begin{matrix}-379.526\\ 94.882\\ 0\end{matrix}\right\} \times 10^{-3} = \left\{\begin{matrix}-47.44\\ 0\\ 0\end{matrix}\right\} \text{MPa}$$

Note that the stresses in the beam are predominantly uniaxial—tensile at the top face of the beam and compressive at the bottom.

6.6.2 I-Section

An I-section consists of a top flange, a web, and a bottom flange. As in the case of T-section, for simplicity of analysis, the flanges are assumed to consist of all continuous horizontal plies and the web vertical plies. Further, the flanges and the web are assumed to have perfect bond between them.

The principle and procedure involved in the analysis of an I-section are similar to those for a T-section. Accordingly, we shall avoid details of analysis steps and rather concentrate on the procedure. Figure 6.15 shows details of an I-section laminated composite beam. We need to first find the centroids of the three elements—top flange, web,

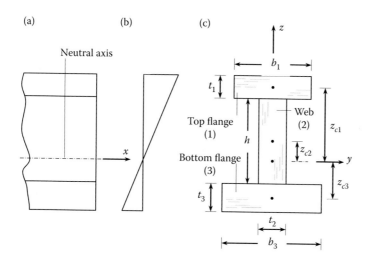

FIGURE 6.15 Laminated composite beam of I-cross section. (a) Beam in front view. (b) Typical axial strain distribution. (c) Cross-sectional details.

and bottom flange. Toward this, we need the axial forces acting in the three elements. By following a procedure similar to the one discussed in the previous section, we can show that the total forces in the top flange, web, and bottom flange, respectively, are given by

$$(F)^{(1)} = \frac{b_1 z_{c1}}{R\left(A_{11}^*\right)^{(1)}} \quad (6.180)$$

$$(F)^{(2)} = \frac{h z_{c2}}{R\left(A_{11}^*\right)^{(2)}} \quad (6.181)$$

$$(F)^{(3)} = -\frac{b_3 z_{c3}}{R\left(A_{11}^*\right)^{(3)}} \quad (6.182)$$

Note that we have considered the beam to be in tension above the neutral plane and in compression below it. The opposite can also be considered; final results will be the same. Under the action of a pure bending moment, the net axial force is zero, that is, $(F)^{(1)} + (F)^{(2)} + (F)^{(3)} = 0$, which results in

$$\frac{b_1 z_{c1}}{\left(A_{11}^*\right)^{(1)}} + \frac{h z_{c2}}{\left(A_{11}^*\right)^{(2)}} - \frac{b_3 z_{c3}}{\left(A_{11}^*\right)^{(3)}} = 0 \quad (6.183)$$

Further, from Figure 6.15, we get the following two relations:

$$z_{c1} + z_{c3} = h + \frac{t_1 + t_3}{2} \quad (6.184)$$

$$z_{c1} - z_{c2} = \frac{h + t_1}{2} \quad (6.185)$$

From Equations 6.183, 6.184, and 6.185, we obtain the following expressions for the distances to the centroids from the neutral axis:

$$z_{c1} = \frac{\left(A_{11}^*\right)^{(1)}\left[h(h+t_1)\left(A_{11}^*\right)^{(3)} + b_3(2h + t_1 + t_3)\left(A_{11}^*\right)^{(2)}\right]}{2\left[b_1\left(A_{11}^*\right)^{(2)}\left(A_{11}^*\right)^{(3)} + h\left(A_{11}^*\right)^{(1)}\left(A_{11}^*\right)^{(3)} + b_3\left(A_{11}^*\right)^{(1)}\left(A_{11}^*\right)^{(2)}\right]} \quad (6.186)$$

$$z_{c2} = z_{c1} - \frac{h + t_1}{2} \quad (6.187)$$

$$z_{c3} = h + \frac{t_1 + t_3}{2} - z_{c1} \quad (6.188)$$

Note that, for I-section with identical top and bottom flanges, the cross section is symmetric about the y-axis and the centroid of the beam coincides with that of the web. In such a case, $z_{c1} = (h + t)/2$, t being flange thickness and $z_{c2} = 0$.

Analysis of Laminated Beams, Columns, and Rods

Next, we need to find the moment components supported by the flanges and web. There are two parts in the moment component supported by each flange—one due to the force resultant and the other due to the moment resultant. Then, by following a procedure similar to that used in Section 6.6.1, we can show that the total moment components supported by the top and bottom flanges are as follows:

$$(M)^{(1)} = \frac{1}{R}\left[\frac{b_1 z_{c1}^2}{\left(A_{11}^*\right)^{(1)}} + \frac{b_1}{\left(D_{11}^*\right)^{(1)}}\right] \tag{6.189}$$

$$(M)^{(3)} = \frac{1}{R}\left[-\frac{b_3 z_{c3}^2}{\left(A_{11}^*\right)^{(3)}} + \frac{b_3}{\left(D_{11}^*\right)^{(3)}}\right] \tag{6.190}$$

On the other hand, the moment component supported by the web is obtained by integration as follows:

$$(M)^{(2)} = \frac{h}{12R\left(A_{11}^*\right)^{(2)}}\left(h^2 + 12 z_{c2}^2\right) \tag{6.191}$$

Then, the total moment supported by the beam is

$$M = \frac{1}{R}\left[\frac{b_1 z_{c1}^2}{\left(A_{11}^*\right)^{(1)}} + \frac{b_1}{\left(D_{11}^*\right)^{(1)}} + \frac{h}{12\left(A_{11}^*\right)^{(2)}}\left(h^2 + 12 z_{c2}^2\right) - \frac{b_3 z_{c3}^2}{\left(A_{11}^*\right)^{(3)}} + \frac{b_3}{\left(D_{11}^*\right)^{(3)}}\right] \tag{6.192}$$

Now, comparing the above expression with classical beam-bending equation, we can readily obtain the expression for effective bending rigidity as follows:

$$E_{xx}^b I_{yy} = \frac{b_1 z_{c1}^2}{\left(A_{11}^*\right)^{(1)}} + \frac{b_1}{\left(D_{11}^*\right)^{(1)}} + \frac{h}{12\left(A_{11}^*\right)^{(2)}}\left(h^2 + 12 z_{c2}^2\right) - \frac{b_3 z_{c3}^2}{\left(A_{11}^*\right)^{(3)}} + \frac{b_3}{\left(D_{11}^*\right)^{(3)}} \tag{6.193}$$

6.6.3 Box-Section

A box-section consists of a top flange, a left web, a right web, and a bottom flange. As in the previous cases, for simplicity of analysis, the flanges are assumed to consist of all continuous horizontal plies and the webs vertical plies. Further, the flanges and the webs are assumed to have perfect bond between them. Also, we shall restrict our study to identical webs, that is, the cross section is symmetric about z-axis. Figure 6.16 shows details of a box-section laminated composite beam. Note that we have considered the centroid of the beam below the centroids of the web. However, the final results remain unchanged provided we take care for the signs of the axial forces. Like in the previous cases, we need to first find the centroids of the cross-sectional elements, which are four in number in this case. Toward this, we need the axial forces acting in the top flange, left web, right web, and the bottom flange, which are, respectively, given by

$$(F)^{(1)} = \frac{b z_{c1}}{R\left(A_{11}^*\right)^{(1)}} \tag{6.194}$$

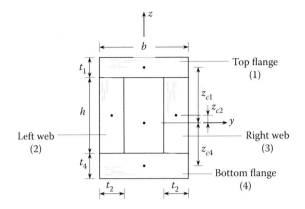

FIGURE 6.16 Cross-sectional details of box-section.

$$(F)^{(2)} = \frac{hz_{c2}}{R(A_{11}^*)^{(2)}} \quad (6.195)$$

$$(F)^{(3)} = \frac{hz_{c2}}{R(A_{11}^*)^{(2)}} \quad (6.196)$$

$$(F)^{(4)} = -\frac{bz_{c4}}{R(A_{11}^*)^{(4)}} \quad (6.197)$$

Like in the other two cases, here too we have considered the beam to be in tension above the neutral plane and in compression below it. For pure bending, the net axial force is zero, that is, $(F)^{(1)} + (F)^{(2)} + (F)^{(3)} + (F)^{(4)} = 0$, which results in

$$\frac{bz_{c1}}{(A_{11}^*)^{(1)}} + \frac{2hz_{c2}}{(A_{11}^*)^{(2)}} - \frac{bz_{c4}}{(A_{11}^*)^{(4)}} = 0 \quad (6.198)$$

Further, from Figure 6.16, we get the following two relations:

$$z_{c1} + z_{c4} = h + \frac{t_1 + t_4}{2} \quad (6.199)$$

$$z_{c1} - z_{c2} = \frac{h + t_1}{2} \quad (6.200)$$

From Equations 6.198 through 6.200, we obtain the following expressions for the distances to the centroids from the neutral axis:

$$z_{c1} = \frac{(A_{11}^*)^{(1)}\left[2h(h+t_1)(A_{11}^*)^{(4)} + b(2h+t_1+t_4)(A_{11}^*)^{(2)}\right]}{2\left[b(A_{11}^*)^{(2)}(A_{11}^*)^{(4)} + 2h(A_{11}^*)^{(1)}(A_{11}^*)^{(4)} + b(A_{11}^*)^{(1)}(A_{11}^*)^{(2)}\right]} \quad (6.201)$$

Analysis of Laminated Beams, Columns, and Rods

$$z_{c2} = z_{c1} - \frac{h+t_1}{2} \tag{6.202}$$

$$z_{c4} = h + \frac{t_1+t_2}{2} - z_{c1} \tag{6.203}$$

Note that, for box-section with identical top and bottom flanges, the cross section is symmetric about the y-axis and the centroid of the beam coincides with that of the web. In such a case, $z_{c1} = (h+t_1)/2$, $z_{c2} = 0$, and $z_{c4} = (h+t_4)/2$.

Next, we need to find the moment components supported by the flanges and webs. There are two parts in the moment component supported by each flange—one due to the force resultant and the other due to the moment resultant. Then, by following a procedure similar to that used in Section 6.6.1, we can show that the total moment components supported by the top and bottom flanges are as follows:

$$(M)^{(1)} = \frac{1}{R}\left[\frac{bz_{c1}^2}{(A_{11}^*)^{(1)}} + \frac{b}{(D_{11}^*)^{(1)}}\right] \tag{6.204}$$

$$(M)^{(4)} = \frac{1}{R}\left[-\frac{bz_{c3}^2}{(A_{11}^*)^{(4)}} + \frac{b}{(D_{11}^*)^{(4)}}\right] \tag{6.205}$$

On the other hand, the moment component supported by each web is obtained by integration as follows:

$$(M)^{(2)} = \frac{h}{12R(A_{11}^*)^{(2)}}\left(h^2 + 12z_{c2}^2\right) \tag{6.206}$$

Then, the total moment supported by the beam is

$$M = \frac{1}{R}\left[\frac{bz_{c1}^2}{(A_{11}^*)^{(1)}} + \frac{b}{(D_{11}^*)^{(1)}} + \frac{h}{6(A_{11}^*)^{(2)}}\left(h^2 + 12z_{c2}^2\right) - \frac{bz_{c3}^2}{(A_{11}^*)^{(4)}} + \frac{b}{(D_{11}^*)^{(4)}}\right] \tag{6.207}$$

Now, comparing the above expression with classical beam-bending equation, we can readily obtain the expression for effective bending rigidity as follows:

$$E_{xx}^b I_{yy} = \frac{bz_{c1}^2}{(A_{11}^*)^{(1)}} + \frac{b}{(D_{11}^*)^{(1)}} + \frac{h}{6(A_{11}^*)^{(2)}}\left(h^2 + 12z_{c2}^2\right) - \frac{bz_{c3}^2}{(A_{11}^*)^{(4)}} + \frac{b}{(D_{11}^*)^{(4)}} \tag{6.208}$$

6.7 BUCKLING OF A COLUMN

6.7.1 Concept of Buckling

When a structure is subjected to compressive loading, depending on the load level, it responds in two ways—(i) stable elastic deformation and (ii) buckling. Buckling is

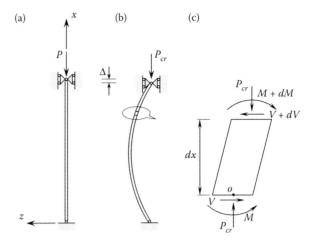

FIGURE 6.17 (a) Elastic shortening of a column under axial compressive force. (b) Buckled column. (c) Free body diagram of a differential element.

essentially an instability-related phenomenon that occurs at loads above certain critical level, when the structure becomes unstable. A stable structural configuration is an equilibrium configuration, in which small disturbances result in small response such that on removal of the disturbance, the structure regains its original equilibrium configuration. On the other hand, an unstable configuration is an equilibrium configuration, in which a small disturbance can result in excessive deformation such that, even after removal of the disturbance, the structure does not come back to its original configuration. At low levels of compressive loading, the structure deforms elastically and remains stable. As the loads are increased gradually, at the critical level of loads referred above, it becomes unstable and with small disturbance excessive deformations result. This sudden excessive deformation is called buckling and the load at which buckling occurs is the buckling load. We can define buckling load as the load at which the stable equilibrium configuration of the structure suddenly becomes unstable and the structure takes another stable configuration usually accompanied by large deformation or displacement.

Let us consider a column under an axial compressive force P (Figure 6.17) and gradually increase the load from zero. At low values of P, the column undergoes elastic axial shortening. The applied force being axial, it does not create any moment on the column. Let us now apply a small lateral force causing a small lateral displacement. (The word *small* implies that the displacement is in the immediate vicinity of the equilibrium straight configuration.) Now, owing to the lateral displacement, the applied axial compressive force creates a disturbing moment, which is resisted by a restoring moment caused by the bending stiffness of the column. At small loads, the restoring moment is higher than the disturbing moment and, after removal of the small lateral displacement, the column regains its original straight equilibrium configuration. However, at large enough load, the disturbing moment exceeds the restoring moment and the column becomes unstable. The load, corresponding to which the disturbing moment is just becomes equal to the restoring moment, is the buckling load.

6.7.2 Governing Equations

Before we proceed to the derivation of the governing equations, a discussion on buckling displacement is in order [1,7,8]. During the prebuckling phase, an axially loaded column undergoes marginal transverse displacement w_0^e due to imperfections. Note that for a straight column under axial load without imperfections, the prebuckling equilibrium displacement w_0^e is zero. The buckling displacement, denoted by W, is measured from onset of buckling. Thus, the total transverse displacement w_0 includes both w_0^e and W, that is,

$$w_0 = w_0^e + W \tag{6.209}$$

In other words, in general, total transverse displacement of the buckled column is not identical with the buckling displacement.

Let us consider a free body diagram of a differential element of the column (Figure 6.17c). Considering equilibrium of moments about point O, it can be shown that

$$V = \frac{dM}{dx} - P\frac{dw_0}{dx} \tag{6.210}$$

Utilizing Equation 6.23 in the above, we get

$$V = -\left(E_{xx}^b I_{yy}\frac{d^3 w_0}{dx^3} + P\frac{dw_0}{dx}\right) \tag{6.211}$$

Differentiating both sides w.r.t. x, and noting that $dV/dx = -q$, we get

$$E_{xx}^b I_{yy}\frac{d^4 w_0}{dx^4} + P\frac{d^2 w_0}{dx^2} - q = 0 \tag{6.212}$$

For a column, in the absence of transverse load, $q = 0$. Thus,

$$E_{xx}^b I_{yy}\frac{d^4 w_0}{dx^4} + P\frac{d^2 w_0}{dx^2} = 0 \tag{6.213}$$

It can be seen that in the prebuckling equilibrium configuration,

$$E_{xx}^b I_{yy}\frac{d^4 w_0^e}{dx^4} + P\frac{d^2 w_0^e}{dx^2} = 0 \tag{6.214}$$

Then, from Equations 6.209, 6.213, and 6.214, denoting the axial load at the onset of buckling by P_{cr}, the buckling equilibrium equation can be written as

$$E_{xx}^b I_{yy}\frac{d^4 W}{dx^4} + P_{cr}\frac{d^2 W}{dx^2} = 0 \tag{6.215}$$

Equation 6.215 is the governing equation for column buckling. It should be noted that this is an eigenvalue problem. (An eigenvalue problem is one that is of the mathematical form $A(u) + \lambda B(u) = 0$, in which A and B are differential operators, u is the eigenvector, and λ is the eigenvalue.) The eigenvalues and eigenvectors in a column buckling problem are, respectively, the buckling loads and buckling mode shapes. The objective is to determine the buckling loads and the corresponding mode shapes. The minimum buckling load is of great importance from design point of view and is called the critical buckling load.

The general solution of Equation 6.215 is

$$W(x) = A\sin\lambda x + B\cos\lambda x + Cx + D \tag{6.216}$$

where

$$\lambda^2 = \frac{P_{cr}}{E_{xx}^b I_{yy}} \tag{6.217}$$

The four constants A, B, C, and D are determined by using the boundary conditions. The governing differential equation is a fourth-order differential equation and the boundary conditions can involve derivatives up to the third-order. The terms W and dW/dx are the displacement and slope, and the corresponding boundary conditions are called *geometric boundary conditions*. On the other hand, the boundary conditions involving d^2W/dx^2 and d^3W/dx^3 (along with dW/dx) are related, respectively, to moment and shear and called *natural boundary conditions*.

Note: Buckling displacement is actually a variation from the prebuckled equilibrium configuration. More on this shall be addressed in the next chapter.

6.7.3 Specific Cases of Column Buckling

6.7.3.1 Simply Supported Column

For a simply supported column (Figure 6.18), displacements and moments at both the ends are zero. Thus,

$$(w_0)_{x=0} = (w_0)_{x=l} = 0 \tag{6.218}$$

and

$$\left(\frac{d^2w_0}{dx^2}\right)_{x=0} = \left(\frac{d^2w_0}{dx^2}\right)_{x=l} = 0 \tag{6.219}$$

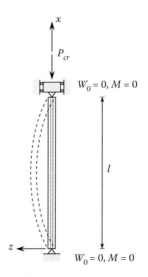

FIGURE 6.18 Simply supported column.

Analysis of Laminated Beams, Columns, and Rods

These conditions (Equations 6.218 and 6.219) would hold good for buckling displacements as well. Thus,

$$(W)_{x=0} = (W)_{x=l} = 0 \qquad (6.220)$$

and

$$\left(\frac{d^2W}{dx^2}\right)_{x=0} = \left(\frac{d^2W}{dx^2}\right)_{x=l} = 0 \qquad (6.221)$$

Using the boundary conditions from Equations 6.220 and 6.221 in Equation 6.216, we get the following:

$$B + D = 0 \qquad (6.222)$$

$$A \sin \lambda l + B \cos \lambda l + Cl + D = 0 \qquad (6.223)$$

$$B = 0 \qquad (6.224)$$

$$A \sin \lambda l + B \cos \lambda l = 0 \qquad (6.225)$$

From the above four equations, we find that

$$B = C = D = 0 \qquad (6.226)$$

and

$$A \sin \lambda l = 0 \qquad (6.227)$$

which implies

$$\lambda = \frac{n\pi}{l}, \quad n = 1, 2, \ldots \qquad (6.228)$$

Then, the buckling modes are

$$W(x) = \Delta \sin \frac{n\pi x}{l} \qquad (6.229)$$

where $\Delta = A \neq 0$ is the buckle amplitude, which is indeterminate.

The critical buckling load is ($n = 1$)

$$P_{cr} = \pi^2 \left(\frac{E_{xx}^b I_{yy}}{l^2}\right) \qquad (6.230)$$

and the corresponding buckling mode is

$$W(x) = \Delta \sin \frac{\pi x}{l} \qquad (6.231)$$

6.7.3.2 Fixed-Free Column

For a fixed-free column (Figure 6.19), displacement and slope are zero at the fixed end ($x = 0$) and moment and shear force are zero at the free end ($x = l$). Thus,

$$(w_0)_{x=0} = \left(\frac{dw_0}{dx}\right)_{x=0} = 0 \tag{6.232}$$

and

$$\left(\frac{d^2 w_0}{dx^2}\right)_{x=l} = E_{xx}^b I_{yy} \left(\frac{d^3 w_0}{dx^3}\right)_{x=l} + P\left(\frac{dw_0}{dx}\right)_{x=l} = 0 \tag{6.233}$$

Equations 6.232 and 6.233, in turn, imply that

$$(W)_{x=0} = \left(\frac{dW}{dx}\right)_{x=0} = 0 \tag{6.234}$$

and

$$\left(\frac{d^2 W}{dx^2}\right)_{x=l} = E_{xx}^b I_{yy} \left(\frac{d^3 W}{dx^3}\right)_{x=l} + P\left(\frac{dW}{dx}\right)_{x=l} = 0 \tag{6.235}$$

Using the boundary conditions from Equations 6.234 and 6.235 in Equation 6.216, we get the following:

$$B + D = 0 \tag{6.236}$$

$$A\lambda + C = 0 \tag{6.237}$$

$$A \sin \lambda l + B \cos \lambda l = 0 \tag{6.238}$$

$$PC = 0 \tag{6.239}$$

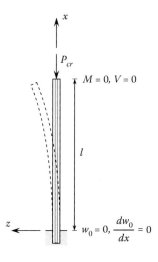

FIGURE 6.19 Fixed-free column.

Analysis of Laminated Beams, Columns, and Rods

From the above four equations, we find that

$$A = C = 0 \tag{6.240}$$

and

$$B \cos \lambda l = 0 \tag{6.241}$$

which implies

$$\lambda = \frac{\pi(2n-1)}{2l}, \quad n = 1, 2, \ldots \tag{6.242}$$

Then, the buckling modes are

$$W(x) = \Delta \cos \frac{\pi(2n-1)x}{2l} \tag{6.243}$$

where $\Delta = B \neq 0$ is the buckle amplitude, which is indeterminate.

The critical buckling load is ($n = 1$)

$$P_{cr} = \frac{\pi^2}{4} \left(\frac{E_{xx}^b I_{yy}}{l^2} \right) \tag{6.244}$$

and the corresponding buckling mode is

$$W(x) = \Delta \cos \frac{\pi x}{2l} \tag{6.245}$$

6.7.3.3 Fixed-Fixed Column

For a fixed-fixed column (Figure 6.20), displacement and slope are zero at both the ends $x = 0$ and $x = l$. Thus,

$$(w_0)_{x=0} = \left(\frac{dw_0}{dx} \right)_{x=0} = 0 \tag{6.246}$$

and

$$(w_0)_{x=l} = \left(\frac{dw_0}{dx} \right)_{x=l} = 0 \tag{6.247}$$

Equations 6.246 and 6.247, in turn, imply that

$$(W)_{x=0} = \left(\frac{dW}{dx} \right)_{x=0} = 0 \tag{6.248}$$

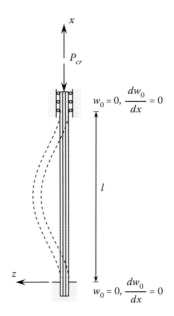

FIGURE 6.20 Fixed-fixed column.

and

$$(W)_{x=l} = \left(\frac{dW}{dx}\right)_{x=l} = 0 \qquad (6.249)$$

Using the boundary conditions from Equations 6.248 and 6.249 in Equation 6.216, we get the following:

$$B + D = 0 \qquad (6.250)$$

$$A\lambda + C = 0 \qquad (6.251)$$

$$A\sin\lambda l + B\cos\lambda l + Cl + D = 0 \qquad (6.252)$$

$$A\lambda\cos\lambda l - B\lambda\sin\lambda l + C = 0 \qquad (6.253)$$

From the above four equations, after some manipulation, it can be shown that

$$A\sin\frac{\lambda l}{2}\left(2\sin\frac{\lambda l}{2} - \lambda l\cos\frac{\lambda l}{2}\right) = 0 \qquad (6.254)$$

which is the buckling criterion. Equation 6.254 has two parts; the first part gives us

$$\lambda = \frac{2n\pi}{l}, \quad n = 1, 2, \ldots \qquad (6.255)$$

Then, the buckling modes from this part are

$$W(x) = \Delta\sin\frac{2n\pi x}{l} \qquad (6.256)$$

where $\Delta = A \neq 0$ is the buckle amplitude, which is indeterminate.

The critical buckling load is ($n = 1$)

$$P_{cr} = 4\pi^2 \left(\frac{E_{xx}^b I_{yy}}{l^2} \right) \quad (6.257)$$

and the corresponding buckling mode is

$$W(x) = \Delta \sin \frac{2\pi x}{l} \quad (6.258)$$

The second part of Equation 6.254 can also be used to obtain P_{cr}, which, it can be shown, is higher than what we get from Equation 6.257. For a complete solution, Equations 2.50 through 2.53 can be expressed in terms of the constants A and B, as follows:

$$\begin{bmatrix} \sin \lambda l - \lambda l & \cos \lambda l - 1 \\ \cos \lambda l - 1 & -\sin \lambda l \end{bmatrix} \begin{Bmatrix} A \\ B \end{Bmatrix} = 0 \quad (6.259)$$

For a nontrivial solution, the determinant of the coefficient matrix has to be zero that gives us the following characteristic equation:

$$\lambda l \sin \lambda l + 2 \cos \lambda l - 2 = 0 \quad (6.260)$$

The solution of this characteristic equation gives the buckling loads and buckling modes; the minimum is the critical buckling load, which can be shown to be the same as in Equation 6.257.

6.8 VIBRATION OF A BEAM

6.8.1 Concept of Vibration

The subject of vibration has been dealt in depth by many, for instance, References 9–11. Any oscillatory motion of a body is called vibration. An oscillatory motion that repeats itself at equal intervals of time is called periodic motion. The simplest form of periodic motion is a harmonic motion (Figure 6.21) and it can be represented as a sine function as follows:

$$x(t) = A \sin \omega t \quad (6.261)$$

where x, A, ω, and t are, respectively, the oscillatory motion, amplitude of motion, frequency of motion, and time.

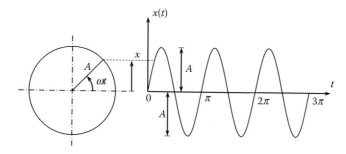

FIGURE 6.21 Simple harmonic motion.

There are two broad types of vibrations—free vibration and forced vibration. When a body vibrates due to an initial excitation at time $t = 0$ and it continues to vibrate under the action of forces inherent to itself and no external forces after time zero, it is called free vibration. The frequencies at which free vibration occurs are the natural frequencies of the body. On the other hand, if the body vibrates under the action of an external force, it is called forced vibration. If the frequency of the external excitation coincides with any of the natural frequencies of a body, resonance takes place accompanied by potentially dangerous large oscillations. In this section, our objective is to find the natural frequencies and corresponding mode shapes of laminated beams.

6.8.2 Governing Equations

Let us consider a laminated beam under transverse load as shown in Figure 6.22. Note that the transverse load $q(x, t)$, bending moment $M(x, t)$, shear force $V(x, t)$, and transverse displacement $w_0(x, t)$ are functions of not only x but also t. Also, note that $q = 0$ for free vibration. Considering dynamic equilibrium of forces in the z-direction, we get

$$q\,dx + dV = \rho A\,dx \frac{\partial^2 w_0}{\partial t^2} \tag{6.262}$$

where ρ and A are the density and area of cross section of the beam, respectively.

Noting, $dV = (\partial V/\partial x)dx$, from Equation 6.262, we get

$$\frac{\partial V}{\partial x} + q = \rho A \frac{\partial^2 w_0}{\partial t^2} \tag{6.263}$$

Next, considering moment equilibrium about y-axis through point O and neglecting higher-order terms of dx, it can be shown that

$$V = \frac{\partial M}{\partial x} \tag{6.264}$$

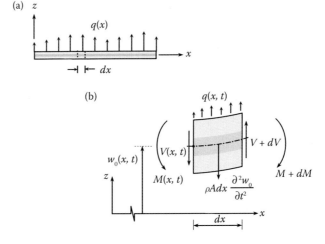

FIGURE 6.22 Transverse vibration of a beam. (a) Beam under transverse load. (b) Differential element at time t.

Analysis of Laminated Beams, Columns, and Rods

Then, Equation 6.263 becomes

$$\frac{\partial^2 M}{\partial x^2} + q = \rho A \frac{\partial^2 w_0}{\partial t^2} \qquad (6.265)$$

Then, utilizing Equation 6.23 and noting that for free vibration, $q = 0$, we get

$$E_{xx}^b I_{yy} \frac{\partial^4 w_0}{\partial x^4} + \rho A \frac{\partial^2 w_0}{\partial t^2} = 0 \qquad (6.266)$$

Equation 6.266 is the equation of motion for a laminated beam and it is an eigenvalue problem.

Note: During bending of a beam, in the CLT, a normal plane section remains plane and normal to the middle surface. Thus, a point undergoes rotary motion w.r.t. the y-axis and the equation of motion involves rotary inertia as well. In the derivation of the above equation of motion, we have considered the inertia force in the z-direction (the term associated with $\partial^2 w_0/\partial t^2$). However, rotary inertia has been neglected.

In Equation 6.266, E_{xx}^b, I_{yy}, ρ, and A are functions of x alone, whereas w_0 is a function of x and t. The solution can be found by adopting the method of separation, where we assume

$$w_0(x,t) = W(x)T(t) \qquad (6.267)$$

such that $W(x)$ and $T(t)$ are functions of x and t, respectively. $W(x)$ and $T(t)$ are, respectively, found by boundary conditions and initial conditions.

Utilizing Equation 6.267 in Equation 6.266, we get

$$\frac{1}{W}\left(\frac{E_{xx}^b I_{yy}}{\rho A}\right)\frac{d^4 W}{dx^4} + \frac{1}{T}\frac{d^2 T}{dt^2} = 0 \qquad (6.268)$$

Equation 6.268 can be expressed as

$$\frac{1}{W}\left(\frac{E_{xx}^b I_{yy}}{\rho A}\right)\frac{d^4 W}{dx^4} = -\frac{1}{T}\frac{d^2 T}{dt^2} = \omega^2 \qquad (6.269)$$

Equation 6.269 can be expressed as two equations as follows:

$$\frac{d^4 W(x)}{dx^4} - \left(\frac{\rho A \omega^2}{E_{xx}^b I_{yy}}\right) W(x) = 0 \qquad (6.270)$$

$$\frac{d^2 T(t)}{dt^2} + \omega^2 T(t) = 0 \qquad (6.271)$$

The general solution of Equation 6.271 is

$$T(t) = A \cos \omega t + B \sin \omega t \qquad (6.272)$$

where the constants A and B are obtained from initial conditions.

On the other hand, the general solution of Equation 6.270 is

$$W(x) = C\cos\beta x + D\sin\beta x + E\cosh\beta x + F\sinh\beta x \quad (6.273)$$

where

$$\beta^4 = \frac{\rho A \omega^2}{E_{xx}^b I_{yy}} \quad (6.274)$$

and the constants C, D, E, and F and the value of β are obtained by using the boundary conditions.

Now, by rearranging the terms in Equation 6.274, the natural frequency is obtained as

$$\omega = \beta^2 \sqrt{\frac{E_{xx}^b I_{yy}}{\rho A}} \quad (6.275)$$

There are, in fact, infinite numbers of normal modes given by $W(x)$, each associated with a natural frequency. Then, the total response of the beam is obtained by superposition as follows:

$$w_0(x,t) = \sum_{n=1}^{\infty} W_n(x)(A_n \cos\omega_n t + B_n \sin\omega_n t) \quad (6.276)$$

Note: In the above equation, $w_0(x, t)$ is the total response of the middle surface of the laminated beam, the subscript 0 being used for middle surface. On the right-hand side, the subscript n, $n = 1, 2, ...$, is used for the nth normal mode.

6.8.3 Specific Cases

6.8.3.1 Simply Supported Beam

For a simply supported beam of length l, displacements and moments at both the ends are zero. Thus,

$$(w_0)_{x=0} = (w_0)_{x=l} = 0 \quad (6.277)$$

and

$$\left(\frac{d^2 w_0}{dx^2}\right)_{x=0} = \left(\frac{d^2 w_0}{dx^2}\right)_{x=l} = 0 \quad (6.278)$$

Then, from Equation 6.267,

$$(W)_{x=0} = (W)_{x=l} = 0 \quad (6.279)$$

and

$$\left(\frac{d^2 W}{dx^2}\right)_{x=0} = \left(\frac{d^2 W}{dx^2}\right)_{x=l} = 0 \quad (6.280)$$

Analysis of Laminated Beams, Columns, and Rods

Using the boundary conditions from Equations 6.279 and 6.280 at $x = 0$, in Equation 6.273, it can be readily shown that

$$C = E = 0 \tag{6.281}$$

Then, application of the boundary conditions at $x = l$ leads to

$$\begin{bmatrix} \sin \beta l & \sinh \beta l \\ -\sin \beta l & \sinh \beta l \end{bmatrix} \begin{Bmatrix} D \\ F \end{Bmatrix} = 0 \tag{6.282}$$

For a nontrivial solution, the determinant of the square matrix has to be zero, that is,

$$\sin \beta l \sinh \beta l = 0 \tag{6.283}$$

$\sinh \beta l = 0$ would mean $\beta = 0$ or $\omega = 0$, that is, no vibration. Thus, $\sinh \beta l \neq 0$ and $\sin \beta l = 0$. This implies

$$D \neq 0 \quad \text{and} \quad F = 0 \tag{6.284}$$

Next, $\sin \beta l = 0$ gives the roots as

$$\beta_n l = n\pi, \quad n = 1, 2, \ldots \tag{6.285}$$

or the natural frequencies are

$$\omega_n = \left(\frac{n\pi}{l}\right)^2 \sqrt{\frac{E_{xx}^b I_{yy}}{\rho A}}, \quad n = 1, 2, \ldots \tag{6.286}$$

and the normal modes are

$$W_n(x) = D_n \sin \frac{n\pi x}{l}, \quad n = 1, 2, \ldots \tag{6.287}$$

6.8.3.2 Fixed-Free Beam

For a beam of length l fixed at $x = 0$ and free at $x = l$, displacement and slope are zero at $x = 0$ and bending moment and shear force are zero at $x = l$. Thus,

$$(w_0)_{x=0} = \left(\frac{dw_0}{dx}\right)_{x=0} = 0 \tag{6.288}$$

and

$$\left(\frac{d^2 w_0}{dx^2}\right)_{x=l} = \left(\frac{d^3 w_0}{dx^3}\right)_{x=l} = 0 \tag{6.289}$$

Then, from Equation 6.267,

$$(W)_{x=0} = \left(\frac{dW}{dx}\right)_{x=0} = 0 \tag{6.290}$$

and

$$\left(\frac{d^2W}{dx^2}\right)_{x=l} = \left(\frac{d^3W}{dx^3}\right)_{x=l} = 0 \qquad (6.291)$$

Using the boundary conditions from Equations 6.290 at $x = 0$, in Equation 6.273, it can be readily shown that

$$C + E = 0 \qquad (6.292)$$

and

$$D + F = 0 \qquad (6.293)$$

Then, application of the boundary conditions at $x = l$, together with Equations 6.292 and 6.293, leads to

$$\begin{bmatrix} \cos\beta l + \cosh\beta l & \sin\beta l + \sinh\beta l \\ -\sin\beta l + \sinh\beta l & \cos\beta l + \cosh\beta l \end{bmatrix} \begin{Bmatrix} E \\ F \end{Bmatrix} = 0 \qquad (6.294)$$

For a nontrivial solution, the determinant of the square matrix has to be zero, that is,

$$(\cos\beta l + \cosh\beta l)^2 - (\sin\beta l + \sinh\beta l)(-\sin\beta l + \sinh\beta l) = 0 \qquad (6.295)$$

By simplifying Equation 6.295, it can be shown that

$$\cos\beta l \cosh\beta l + 1 = 0 \qquad (6.296)$$

Equation 6.296 is the frequency equation, from which the frequencies of a fixed-free beam can be determined.

6.8.3.3 Fixed-Fixed Beam

For a beam of length l fixed at $x = 0$ as well at $x = l$, displacement and slope are zero at both the ends. Thus,

$$(w_0)_{x=0} = \left(\frac{dw_0}{dx}\right)_{x=0} = 0 \qquad (6.297)$$

and

$$(w_0)_{x=l} = \left(\frac{dw_0}{dx}\right)_{x=l} = 0 \qquad (6.298)$$

Then, from Equation 6.267,

$$(W)_{x=0} = \left(\frac{dW}{dx}\right)_{x=0} = 0 \qquad (6.299)$$

and

$$(W)_{x=l} = \left(\frac{dW}{dx}\right)_{x=l} = 0 \qquad (6.300)$$

Using the boundary conditions from Equation 6.299 at $x = 0$, in Equation 6.273, it can be readily shown that

$$C + E = 0 \qquad (6.301)$$

and

$$D + F = 0 \qquad (6.302)$$

Then, application of the boundary conditions at $x = l$, together with Equations 6.301 and 6.302, leads to

$$\begin{bmatrix} -\cos\beta l + \cosh\beta l & -\sin\beta l + \sinh\beta l \\ \sin\beta l + \sinh\beta l & -\cos\beta l + \cosh\beta l \end{bmatrix} \begin{Bmatrix} E \\ F \end{Bmatrix} = 0 \qquad (6.303)$$

For a nontrivial solution, the determinant of the square matrix has to be zero, that is,

$$(-\cos\beta l + \cosh\beta l)^2 - (\sin\beta l + \sinh\beta l)(-\sin\beta l + \sinh\beta l) = 0 \qquad (6.304)$$

By simplifying Equation 6.304, it can be shown that

$$\cos\beta l \cosh\beta l - 1 = 0 \qquad (6.305)$$

Equation 6.305 is the frequency equation, from which the frequencies of a fixed-fixed beam can be determined.

6.9 SUMMARY

In this chapter, we studied the analytical issues in respect of 1D laminated composite structural elements. Basic principles of solid mechanics and macromechanics of laminated composites are suitably modified to account for 1D nature of beams and columns. Governing equations are developed for bending of a laminated beam, buckling of a column, and vibration of a beam and analytical solutions are obtained for (i) in-plane and interlaminar stresses and displacements in beam bending, (ii) critical buckling load in column buckling, and (iii) natural frequency in beam vibration. Some of the key points to be noted are

- Analytical treatment of the 1D elements is based on the basic assumptions and restrictions of CLPT with some added restrictions.
- A beam cross section can be either solid or thin-walled.
- For a beam of solid cross section, ply orientation w.r.t. loading direction is important. Depending on the ply orientation, either effective bending stiffness or effective extensional stiffness determines the bending behavior of a laminated beam.

- In the case of a beam with thin-walled cross section, the relative sizes and locations of the web(s) and flange(s) are of key importance.
- The governing equation for buckling of laminated composite column is expressed in terms of buckling displacement; it is an eigenvalue equation that gives us the critical buckling load and it is influenced by the effective bending stiffness of the column.
- The governing equation for vibration of laminated composite beam is expressed in terms of transverse displacement; it is an eigenvalue equation that gives us natural frequency. It is influenced by effective bending stiffness, density, and area of cross section of the beam.

EXERCISE PROBLEMS

6.1 Consider a carbon/epoxy simply supported beam of solid rectangular cross-section under a central point load of 200 N. The dimensions of the beam are $l = 600$ mm, $b = 18$ mm, and $h = 10$ mm and the ply sequence is $[0_2^\circ/90^\circ/0_2^\circ]_s$. The beam is laid-up with plies of equal thickness and the plies are stacked normal to the loading direction. Determine the maximum displacement and in-plane normal stress. Assume the following material data:

$$E_1 = 125 \text{ GPa}, E_2 = 8 \text{ GPa}, G_{12} = 6 \text{ GPa, and } \nu_{12} = 0.25.$$

6.2 Consider the beam in Exercise 6.1. If the ply sequence is changed to $[0_2^\circ/45^\circ/0_2^\circ]$ and the 45° ply is laid-up with bidirectional fabric with epoxy resin, determine the maximum displacement and in-plane axial normal stress. Assume the following material data for bidirectional lamina:

$$E_1 = 40 \text{ GPa}, E_2 = 40 \text{ GPa}, G_{12} = 10 \text{ GPa, and } \nu_{12} = 0.25.$$

6.3 Consider the beam in Exercise 6.1. If the point load is moved to a position 100 mm from one end, determine the maximum displacement and in-plane axial normal stress.

6.4 Consider a carbon/epoxy simply supported beam of solid rectangular cross section under a uniformly distributed load of 0.5 N/mm. The dimensions of the beam are $l = 500$ mm, $b = 10$ mm, and $h = 5$ mm and the ply sequence is $[0_2^\circ/45^\circ/0_2^\circ]$. The beam is laid-up with plies of equal thickness and the plies are stacked normal to the loading direction. Determine (i) deflections, (ii) axial stresses at the bottom face, and (iii) axial stresses at the top face at every 50 mm of the beam length and draw the plots showing variations of the three parameters along the length. What is the location of maximum displacement? Assume the following material data:

$$E_1 = 140 \text{ GPa}, E_2 = 8 \text{ GPa}, G_{12} = 6 \text{ GPa, and } \nu_{12} = 0.25.$$

6.5 Solve the problem in Exercise 6.4 if the beam support conditions are changed to fixed. Keep other data unchanged.

6.6 Solve the problem in Exercise 6.4 if the beam support conditions are changed to a cantilever. Keep other data unchanged.

6.7 Solve the problem in Exercise 6.4 if the beam support conditions are changed to a cantilever with a tip load of 80 N. Keep other data unchanged.

6.8 Consider a carbon/epoxy simply supported beam of solid rectangular cross section under a uniformly distributed load of 0.5 N/mm. The dimensions of the beam are $l = 500$ mm, $b = 10$ mm, and $h = 5$ mm and the ply sequence is $[0_2°/45°/0_2°]$. The beam is laid-up with unidirectional and bidirectional plies of equal thickness and the plies are stacked normal to the loading direction. Determine the interlaminar normal stress and shear stresses at the interfaces of the plies in the middle cross section of the beam. Plot the variations of stresses across the thickness. What is the maximum deflection? Assume the following material data:

For the 0° plies: $E_1 = 140$ GPa, $E_2 = 8$ GPa, $G_{12} = 6$ GPa, and $\nu_{12} = 0.25$.
For the 45° plies: $E_1 = 40$ GPa, $E_2 = 40$ GPa, $G_{12} = 10$ GPa, and $\nu_{12} = 0.20$.

6.9 Solve the problem in Exercise 6.8 if the ply sequence is changed to (i) [0°/45°/0°/45°/0°] and (ii) [45°/0°/45°/0°/45°]. Keep other data unchanged.

6.10 Determine the maximum uniformly distributed load that can be allowed on a cantilever beam of dimensions 400 mm × 12 mm × 4 mm, if the maximum deflection is to be limited to 2 mm. Cross section of the beam is solid rectangular and the plies are laid-up normal to the loading direction as per the following sequence: [0°/45°/0°/45°/0°]. 0° plies are unidirectional and 1 mm in thickness, whereas 45° plies are bidirectional and 0.5 mm in thickness. Assume the following material data:

For the 0° plies: $E_1 = 160$ GPa, $E_2 = 10$ GPa, $G_{12} = 6$ GPa, and $\nu_{12} = 0.25$.
For the 45° plies: $E_1 = 40$ GPa, $E_2 = 40$ GPa, $G_{12} = 10$ GPa, and $\nu_{12} = 0.20$.

6.11 Solve the problem in Exercise 6.10, if the following strength data are given:

For the 0° plies: $(\sigma_{11}^T)_{ult} = 2400$ MPa, $(\sigma_{11}^C)_{ult} = 800$ MPa, $(\sigma_{22}^T)_{ult} = 40$ MPa, $(\sigma_{22}^C)_{ult} = 200$ MPa, and $(\tau_{12})_{ult} = 75$ MPa
For the 45° plies: $(\sigma_{11}^T)_{ult} = 1000$ MPa, $(\sigma_{11}^C)_{ult} = 600$ MPa, $(\sigma_{22}^T)_{ult} = 1000$ MPa, $(\sigma_{22}^C)_{ult} = 600$ MPa, and $(\tau_{12})_{ult} = 50$ MPa

6.12 Determine the effective extensional stiffness of a beam with the data given below. Ply sequence: [0°/45°/0°/45°/0°]; 0° plies are unidirectional and 1.0 mm in thickness; 45° plies are bidirectional and 0.5 mm in thickness. Plies are stacked in the loading direction.

Beam cross section: solid rectangular, $b = 12$ mm and $h = 5$ mm

Material properties:

Unidirectional plies: $E_1 = 160$ GPa, $E_2 = 10$ GPa, $G_{12} = 6$ GPa, and $\nu_{12} = 0.25$
Bidirectional plies: $E_1 = 40$ GPa, $E_2 = 40$ GPa, $G_{12} = 10$ GPa, and $\nu_{12} = 0.20$

6.13 Determine the maximum deflection and in-plane stresses in the beam in Exercise 6.12. The length of the beam is 500 mm and consider a uniformly distributed load of 0.25 N/mm.

6.14 Consider a carbon/epoxy simply supported T-section beam with the following dimensions: length $l = 500$ mm, flange width $b = 20$ mm, web height $h = 30$ mm, and flange thickness $t = 4$ mm. Thickness of the web is the same as that of the flange. The flange and the web are composed of 0° plies, each ply being 0.5 mm in thickness. Apply a uniformly distributed load of 0.2 N/mm and determine the deflections of the beam and draw the variation in deflection along its length. Determine the in-plane stresses in the web and flange. Material properties are as follows:

$$E_1 = 150 \text{ GPa}, E_2 = 10 \text{ GPa}, \nu_{12} = 0.25, \text{ and } G_{12} = 8 \text{ GPa}$$

6.15 Consider a carbon/epoxy fixed beam of I-section with the following dimensions: length $l = 500$ mm, top flange width $b_1 = 20$ mm, bottom flange width $b_2 = 20$ mm, web height $h = 30$ mm, and flange thickness $t = 4$ mm. Thickness of the web is the same as that of the flange. The flange and the web are composed of 0° plies, each ply being 0.5 mm in thickness. Apply a uniformly distributed load of 0.2 N/mm and determine (i) effective flexural rigidity of the beam, (ii) maximum deflection of the beam, and (iii) material properties are as follows:

$$E_1 = 150 \text{ GPa}, E_2 = 10 \text{ GPa}, \nu_{12} = 0.25, \text{ and } G_{12} = 8 \text{ GPa}$$

6.16 Consider a carbon/epoxy fixed beam of box-section with the following dimensions: length $l = 800$ mm, flange width $b = 30$ mm, web height $h = 40$ mm, and flange thickness $t = 4$ mm. Thickness of the web is the same as that of the flange. The flange and the web are composed of 0° plies, each ply being 0.5 mm in thickness. Apply a uniformly distributed load of 0.5 N/mm and determine (i) effective flexural rigidity of the beam, (ii) maximum deflection of the beam, and (iii) material properties are as follows:

$$E_1 = 160 \text{ GPa}, E_2 = 10 \text{ GPa}, \nu_{12} = 0.25, \text{ and } G_{12} = 8 \text{ GPa}$$

6.17 Extend the procedure for a T-section beam to an I-beam and derive expressions for (i) force and moment resultants in the flanges, (ii) force and moment resultant in the web, and (iii) effective bending rigidity of the flanges, web, and the beam as a whole.

6.18 Extend the procedure for a T-section beam to a box-beam and derive expressions for (i) force and moment resultants in the flanges, (ii) force and moment resultant in the webs, and (iii) effective bending rigidity of the flanges, webs, and the beam as a whole.

6.19 Determine the critical buckling load of a column of length $l = 400$ mm fixed at one end. Following data are given:

Cross section: solid rectangular, $b = 12$ mm, $h = 5$ mm
Ply sequence: $[0_2°/45_2°/0_2°]$; 0° plies are unidirectional and 1.0 mm in thickness; 45° plies are bidirectional and 0.5 mm in thickness

Material data:

Unidirectional plies: $E_1 = 140$ GPa, $E_2 = 10$ GPa, $G_{12} = 6$ GPa, and $\nu_{12} = 0.25$
Bidirectional plies: $E_1 = 40$ GPa, $E_2 = 40$ GPa, $G_{12} = 10$ GPa, and $\nu_{12} = 0.20$

6.20 Determine the critical buckling load of a column of length $l = 500$ mm fixed at both ends. Following data are given:

Cross section: T-section, $b = 20$ mm, $h = 30$ mm, $t_1 = t_2 = 4$ mm
Ply sequence: unidirectional 0° plies of 0.5 mm thickness everywhere
Material data: $E_1 = 140$ GPa, $E_2 = 10$ GPa, $G_{12} = 6$ GPa, and $\nu_{12} = 0.25$

6.21 Determine the critical buckling load of a column of length $l = 800$ mm fixed at both ends. Following data are given:

Cross section: box-section, $b = 40$ mm, $h = 40$ mm, $t_1 = t_2 = t_3 = t_4 = 4$ mm
Ply sequence: unidirectional 0° plies of 0.5 mm thickness everywhere
Material data: $E_1 = 140$ GPa, $E_2 = 10$ GPa, $G_{12} = 6$ GPa, and $\nu_{12} = 0.25$

6.22 Determine the natural frequencies of a simply supported beam of solid rectangular cross section with the following data:

Dimensions: $l = 500$ mm, $b = 20$ mm, $h = 5$ mm

Ply sequence: $[0_2°/45_2°/0_2°]$; 0° plies are unidirectional and 1.0 mm in thickness; 45° plies are bidirectional and 0.5 mm in thickness;

Material data:

Unidirectional plies: $E_1 = 140$ GPa, $E_2 = 10$ GPa, $G_{12} = 6$ GPa, and $\nu_{12} = 0.25$
Bidirectional plies: $E_1 = 40$ GPa, $E_2 = 40$ GPa, $G_{12} = 10$ GPa, and $\nu_{12} = 0.20$
Density of composite: 1.5 g/cm³

6.23 Solve the problem in Exercise 6.22, if the support conditions of the beam are changed to fixed at one end and free at the other.

Hint: Solve Equation 6.296 for β and use in Equation 6.275 for ω.

REFERENCES AND SUGGESTED READING

1. J. N. Reddy, *Mechanics of Laminated Composite Plates and Shells*, CRC Press, Boca Raton, FL, 2004.
2. L. P. Kollar and G. S. Springer, *Mechanics of Composite Structures*, Cambridge University Press, Cambridge, 2003.
3. M. E. Tuttle, *Structural Analysis of Polymeric Composite Materials*, Marcel Dekker, New York, 2004.
4. R. Byron Pipes and N. J. Pagano, Interlaminar stresses in composite laminates under uniform axial extension, *Journal of Composite Materials*, 4(4), 1970, 538–546.
5. N. J. Pagano and R. Byron Pipes, The influence of stacking sequence on laminate strength, *Journal of Composite Materials*, 5, 1971, 50–57.
6. S. P. Timoshenko, *Strength of Materials: Part-I Elementary Theory and Problems*, third edition, Wadsworth Publishing Company, New Delhi, India, 1986.
7. R. M. Jones, *Buckling of Bars, Plates, and Shells*, Bull Ridge Publishing, Blacksburg, Virginia, 2006.
8. S. P. Timoshenko and J. M. Gere, *Theory of Elastic Stability*, second edition, McGraw-Hill Book Company, Singapore, 1963.
9. S. S. Rao, *Vibration of Continuous Systems*, John Wiley & Sons, Hoboken, NJ, 2007.
10. A. A. Khedir and J. N. Reddy, Free vibration of cross-ply laminated beams with arbitrary boundary condition, *International Journal of Engineering Science*, 32(12), 1994, 1971–1980.
11. K. Chandrashekhara and K. M. Bangera, Vibration of symmetrically laminated clamped-free beam with a mass at the free end, *Journal of Sound and Vibration*, 160(1), 1993, 93–101.

7 Analytical Solutions for Laminated Plates

7.1 CHAPTER ROAD MAP

In chapter 6, we discussed analytical solutions of 1D problems in laminated composites. We extend the discussion to two dimensions, that is, plate problems, and bending, buckling, and vibration of rectangular plates. Governing equations, viz. equilibrium equations, buckling equations, and vibration equations and associated boundary conditions are discussed.

Both exact and approximate analytical methods are available for solution of plate problems. Navier, Levy, and Ritz methods are presented and solutions are obtained for in-plane and interlaminar stresses as well as deflection in a plate under bending loads. Specific cases of boundary conditions and ply sequence are considered. For plate buckling and vibration problems, solutions for critical buckling load and natural frequency are presented.

A thorough understanding of the concepts presented in Chapters 4 and 5 preceded by a review of basic solid mechanics in Chapter 2 is a prerequisite for effective assimilation of the concepts discussed in this chapter. Familiarity with the solutions for beam problems presented in Chapter 6 is desirable as it gives a logical continuity.

7.2 PRINCIPAL NOMENCLATURE

$[A]$	Extensional stiffness matrix
$A_{11}, A_{12}, ..., A_{66}$	Elements of the laminate extensional stiffness matrix
a, b, h	Length, width, and thickness of a rectangular plate
$[B]$	Extension–bending coupling stiffness matrix
$B_{11}, B_{12}, ..., B_{66}$	Elements of the laminate extension–bending coupling stiffness matrix
C_1, C_2, C_3, C_4	Clamped boundary conditions
$[D]$	Bending stiffness matrix
$D_{11}, D_{12}, ..., D_{66}$	Elements of the laminate bending stiffness matrix
$\bar{K}_{yz}$	Equivalent Kirchhoff force
M_{xx}, M_{yy}, M_{xy}	In-plane moment resultants
N_{xx}, N_{yy}, N_{xy}	In-plane force resultants
$\bar{N}_{xx}, \bar{N}_{yy}, \bar{N}_{xy}$	Applied compressive in-plane normal and shear force resultants on the plate edges
Q_{mn}	Coefficients in the double Fourier series expansion of applied transverse load
Q_m	Coefficients in the single Fourier series expansion of applied transverse load
U	Strain energy

U_{mn}, V_{mn}, W_{mn}	Coefficients in the double Fourier series expansion of middle surface displacements
U_m, V_m, W_m	Coefficients in the single Fourier series expansion of middle surface displacements
u_0, v_0, w_0	Plate middle surface displacements in the x-, y-, and z-directions, respectively
$q(x, y)$	Lateral (out-of-plane) loads
S_1, S_2, S_3, S_4	Simply supported boundary conditions
V	Negative work done by applied loads
V_{xz}, V_{yz}	Out-of-plane force resultants
t	Time
x, y, z	Cartesian coordinates
α, β	Angles of slope of the middle surface at a point in the x–z and y–z planes, respectively
$\delta M_{xx}, \delta M_{yy}, \delta M_{xy}$	Variations of the induced in-plane moment resultants from their prebuckled equilibrium state
$\delta N_{xx}, \delta N_{yy}, \delta N_{xy}$	Variations of the induced in-plane force resultants from their prebuckled equilibrium state
$\delta u_0, \delta v_0, \delta w_0$	Variation of the displacements in the middle surface of the plate
$\varepsilon_{xx}^0, \varepsilon_{yy}^0, \gamma_{xy}^0$	In-plane strains at the middle surface
$\kappa_{xx}, \kappa_{yy}, \kappa_{xy}$	Laminate middle surface curvatures
Π	Total potential energy
ρ	Area density of the laminated plate
$\sigma_{xx}, \sigma_{yy}, \tau_{xy}$	In-plane normal and shear stresses
$\sigma_{zz}, \tau_{yz}, \tau_{zx}$	Out-of-plane normal and shear stresses

7.3 INTRODUCTION

Laminated composite plates are extensively used in diverse applications. In many structural applications, plates are used as primary load-bearing structural elements and efficient design and analysis of these structural elements is critical for overall acceptable performance of the structure. Design and analysis of a plate require determination of the responses of the plate to the applied loads. When a structural element is loaded, depending on its configuration, material stiffness characteristics, and nature and magnitude of the applied loads, it responds to the applied loads by undergoing deformations, which involves in-plane displacements, out-of-plane displacements, or both. Out-of-plane deflections of plate bending are mostly associated with lateral loads. In-plane displacements take place when the plate is subjected to in-plane tensile or compressive forces. In-plane compressive force can also result in out-of-plane deflections that are associated with buckling.

Analysis of laminated composite plates under in-plane loads is given in Chapter 5. In this chapter, bending and free vibration of laminate plates under transverse loads and buckling under in-plane compression are addressed. The analytical procedures involved in the laminated plates are an extension of the well-established isotropic plate theories, see for instance References 1 and 2. However, owing to the anisotropic nature of composites, plate theories in laminated plates are more complex [3–9]. Analysis of laminated composite plates is an extensively studied subject (see References 10–18 among others). An in-depth review of the solutions available is not intended here; for comprehensive review and specific information on analytical results, interested reader can consult References 3–6 and the bibliography provided therein. On the other hand, governing equations, boundary conditions, and solution methods are introduced in this section. Solutions for bending, buckling, and vibration in some specific cases are addressed for demonstration, subsequently in Sections 7.4 through 7.6.

7.3.1 Rectangular Laminated Plate under General Loading

The geometry and coordinates of a rectangular plate are shown in Figure 7.1a. All the three physical dimensions of a plate are finite; however, the thickness is small compared to the other physical dimensions, that is, $h/a \ll 1$ and $h/b \ll 1$. Also, the loads and displacements are functions of x as well as y, and the analytical procedures, which are essentially derived from 3D elasticity formulations, are 2D in nature.

Figure 7.1b and c shows the rectangular plate under force and moment resultants. The force and moment resultants were first introduced (Figure 5.3) in Chapter 5. Note that there is a subtle difference—here, we have introduced the transverse loads $q(x, y)$ and the transverse force resultants V_{xz} and V_{yz}. Apparently, this is a violation of the restriction that the plate is loaded only in its plane and resulting plane stress conditions. Obviously, the transverse loads $q(x, y)$ result in nonzero transverse stresses σ_{zz}, τ_{zx}, and τ_{yz}. For a thin plate, however, the magnitudes of the out-of-plane stresses are far lower than the in-plane stresses. Thus, plane stress conditions exist in an approximate way.

Now, as we can see that five force/moment resultants act on each edge—three force resultants and two moment resultants. For example, on the edges normal to the x-axis, the force and moment resultants are N_{xx}, N_{xy}, V_{xz}, M_{xx}, and M_{xy}. Similarly, on the edges normal to the y-axis, the force and moment resultants are N_{yy}, N_{yx}, V_{yz}, M_{yy}, and M_{yx}. We have got two new force resultants, defined as follows:

$$V_{xz} = \int_{-h/2}^{h/2} \tau_{xz} \, dz \qquad (7.1)$$

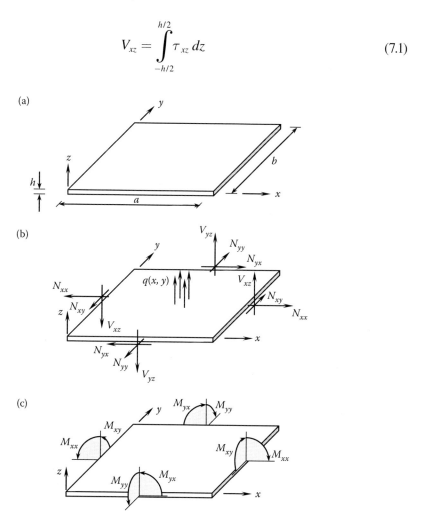

FIGURE 7.1 Rectangular plate. (a) Geometry and coordinates. (b) Force resultants. (c) Moment resultants. (Adapted with permission from R. M. Jones, *Mechanics of Composite Materials*, Taylor & Francis, New York, 1998.)

$$V_{yz} = \int_{-h/2}^{h/2} \tau_{yz}\, dz \tag{7.2}$$

It is important to distinguish between the applied stress resultants on the edges and the induced stress resultants at the interior points. The five applied stress resultants on an edge of a plate are all functions of the coordinate direction tangent to the edge. For example, the normal force resultant applied on the edge normal to the positive x-axis is a function of y, that is, $N_{xx} = N_{xx}(y)$. (Of course, it can be constant as a special case.) At interior points, the induced stress resultants are functions of both x and y, for example, $N_{xx} = N_{xx}(x, y)$. It follows that the magnitudes of stress resultants on opposite edges are not necessarily equal. For example, N_{xx} on the edge normal to the positive x-axis and that on the edge normal to the negative x-axis may be different in magnitude. Similarly, the applied stress resultants on the edges are likely to be different from those induced stress resultants at the interior points. Note, further, that N_{xy} and N_{yx} are not, in general, equal; however, at the corner points they are equal.

7.3.2 Governing Equations for Bending, Buckling, and Vibration of Laminated Plates

7.3.2.1 Equilibrium Equations for Laminated Plate Bending

The basic assumptions and restrictions, which form the basis for the development of the CLPT, and their implications are discussed in Section 5.5.1, Chapter 5. The same assumptions and restrictions hold good, based on which, in this section, we derive the governing equations for laminated plate bending as per CLT.

Let us consider an arbitrary point $P(x, y)$ in the middle surface of a rectangular plate as shown in Figure 7.2. Let us, then, consider a differential plate element around the point such that all faces in the undeformed state are equidistant from the point. Figures 7.3 and 7.4, respectively, show the force resultants and moment resultants on the differential element. Transverse loads, force and moment resultants all act simultaneously. However, for the sake of clarity, the transverse loads are not shown in these two figures. Also, again for the sake of clarity, the force and moment resultants are shown separately. Note that the transverse loads $q(x, y)$, which are functions of x and y are applied loads, whereas the force and moment resultants N_{xx}, N_{yy}, N_{xy}, V_{xz}, V_{yz}, M_{xx}, M_{yy}, and M_{xy} are induced at the point $P(x, y)$. Clearly, the force and moment resultants are all functions of (x, y). Then, the normal force resultant on the faces normal to the negative

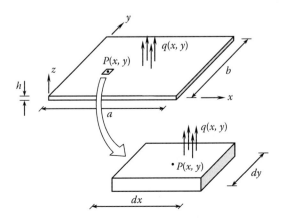

FIGURE 7.2 Differential plate element under transverse loading.

Analytical Solutions for Laminated Plates

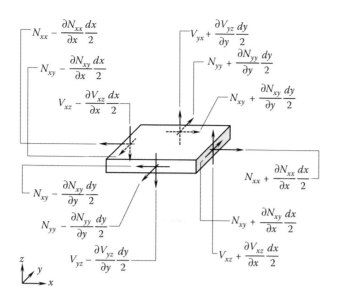

FIGURE 7.3 Force resultants on an undeformed differential plate element. Note: Transverse loads and moment resultants are not shown for clarity.

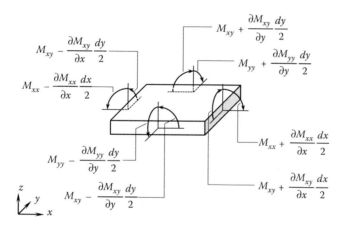

FIGURE 7.4 Moment resultants on an undeformed differential plate element. Note: Transverse loads and force resultants are not shown for clarity.

and positive x-directions, respectively, are $N_{xx} - (\partial N_{xx}/\partial x)(dx/2)$ and $N_{xx} + (\partial N_{xx}/\partial x)(dx/2)$. All other force and moment resultants can be obtained in a similar way.

The plate-bending equilibrium equations can be derived by considering force and moment equilibrium of the differential element in the deformed configuration. Figure 7.5 shows the deformed configurations of the differential plate element in the xz- and yz-planes. Let α and β denote the angles of slope of the middle surface at the point P in the xz- and yz-planes, respectively. Clearly, α and β are functions of x and y and their values on the faces of the plate element are as shown in the figure.

Let us, now, consider static force equilibrium (ignoring body forces) of the plate element in the x-direction. Summing up all the force components, we get

$$-\left(N_{xx} - \frac{\partial N_{xx}}{\partial x}\frac{dx}{2}\right)\cos\left(\alpha - \frac{\partial \alpha}{\partial x}\frac{dx}{2}\right)dy + \left(N_{xx} + \frac{\partial N_{xx}}{\partial x}\frac{dx}{2}\right)\cos\left(\alpha + \frac{\partial \alpha}{\partial x}\frac{dx}{2}\right)dy$$
$$-\left(N_{xy} - \frac{\partial N_{xy}}{\partial y}\frac{dy}{2}\right)\cos\alpha\, dx + \left(N_{xy} + \frac{\partial N_{xy}}{\partial y}\frac{dy}{2}\right)\cos\alpha\, dx = 0 \qquad (7.3)$$

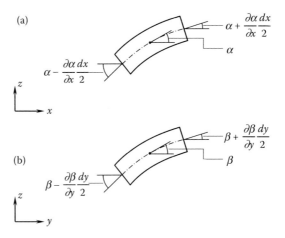

FIGURE 7.5 Deformed configurations of the differential plate element. (a) Middle surface slope in the *xz*-plane. (b) Middle surface slope in the *yz*-plane.

Elemental variation in the slope angle being small, we can write

$$\cos\left(\alpha - \frac{\partial \alpha}{\partial x}\frac{dx}{2}\right) \approx \cos\left(\alpha + \frac{\partial \alpha}{\partial x}\frac{dx}{2}\right) \approx \cos\alpha \qquad (7.4)$$

Then, Equation 7.3 reduces to

$$\frac{\partial N_{xx}}{\partial x} + \frac{\partial N_{xy}}{\partial y} = 0 \qquad (7.5)$$

Similarly, considering static force equilibrium in the *y*-direction, we get

$$\frac{\partial N_{yy}}{\partial y} + \frac{\partial N_{xy}}{\partial x} = 0 \qquad (7.6)$$

Next, we consider static force equilibrium in the *z*-direction. In this case, the transverse loads also come into consideration. Then, summing up all the force components, we get

$$-\left(N_{xx} - \frac{\partial N_{xx}}{\partial x}\frac{dx}{2}\right)\sin\left(\alpha - \frac{\partial \alpha}{\partial x}\frac{dx}{2}\right)dy + \left(N_{xx} + \frac{\partial N_{xx}}{\partial x}\frac{dx}{2}\right)\sin\left(\alpha + \frac{\partial \alpha}{\partial x}\frac{dx}{2}\right)dy$$

$$-\left(N_{yy} - \frac{\partial N_{yy}}{\partial y}\frac{dy}{2}\right)\sin\left(\beta - \frac{\partial \beta}{\partial y}\frac{dy}{2}\right)dx + \left(N_{yy} + \frac{\partial N_{yy}}{\partial y}\frac{dy}{2}\right)\sin\left(\beta + \frac{\partial \beta}{\partial y}\frac{dy}{2}\right)dx$$

$$-\left(N_{xy} - \frac{\partial N_{xy}}{\partial y}\frac{dy}{2}\right)\sin\left(\alpha - \frac{\partial \alpha}{\partial y}\frac{dy}{2}\right)dx + \left(N_{xy} + \frac{\partial N_{xy}}{\partial y}\frac{dy}{2}\right)\sin\left(\alpha + \frac{\partial \alpha}{\partial y}\frac{dy}{2}\right)dx$$

$$-\left(N_{xy} - \frac{\partial N_{xy}}{\partial x}\frac{dx}{2}\right)\sin\left(\beta - \frac{\partial \beta}{\partial x}\frac{dx}{2}\right)dy + \left(N_{xy} + \frac{\partial N_{xy}}{\partial x}\frac{dx}{2}\right)\sin\left(\beta + \frac{\partial \beta}{\partial x}\frac{dx}{2}\right)dy$$

$$-\left(V_{xz} - \frac{\partial V_{xz}}{\partial x}\frac{dx}{2}\right)\cos\left(\alpha - \frac{\partial \alpha}{\partial x}\frac{dx}{2}\right)dy + \left(V_{xz} + \frac{\partial V_{xz}}{\partial x}\frac{dx}{2}\right)\cos\left(\alpha + \frac{\partial \alpha}{\partial x}\frac{dx}{2}\right)dy$$

$$-\left(V_{yz} - \frac{\partial V_{yz}}{\partial y}\frac{dy}{2}\right)\cos\left(\beta - \frac{\partial \beta}{\partial y}\frac{dy}{2}\right)dx + \left(V_{yz} + \frac{\partial V_{yz}}{\partial y}\frac{dy}{2}\right)\cos\left(\beta + \frac{\partial \beta}{\partial y}\frac{dy}{2}\right)dx$$

$$+ q\,dx\,dy = 0 \qquad (7.7)$$

Analytical Solutions for Laminated Plates

For small angles, say less than approximately 10°, sine and cosine of the angle are approximately equal to the angle itself and unity, respectively. Also, such an angle can be approximately equated to the tangent of the angle. Then,

$$\sin\left(\alpha + \frac{\partial \alpha}{\partial x}\frac{dx}{2}\right) \approx \alpha + \frac{\partial \alpha}{\partial x}\frac{dx}{2} \quad \text{and} \quad \cos\left(\alpha + \frac{\partial \alpha}{\partial x}\frac{dx}{2}\right) \approx 1 \quad (7.8)$$

$$\alpha \approx \frac{\partial w_0}{\partial x} \quad \text{and} \quad \frac{\partial \alpha}{\partial x} \approx \frac{\partial^2 w_0}{\partial x^2} \quad (7.9)$$

and so on.

Then, Equation 7.7 reduces to

$$-\left(N_{xx} - \frac{\partial N_{xx}}{\partial x}\frac{dx}{2}\right)\left(\frac{\partial w_0}{\partial x} - \frac{\partial^2 w_0}{\partial x^2}\frac{dx}{2}\right)dy + \left(N_{xx} + \frac{\partial N_{xx}}{\partial x}\frac{dx}{2}\right)\left(\frac{\partial w_0}{\partial x} + \frac{\partial^2 w_0}{\partial x^2}\frac{dx}{2}\right)dy$$

$$-\left(N_{yy} - \frac{\partial N_{yy}}{\partial y}\frac{dy}{2}\right)\left(\frac{\partial w_0}{\partial y} - \frac{\partial^2 w_0}{\partial y^2}\frac{dy}{2}\right)dx + \left(N_{yy} + \frac{\partial N_{yy}}{\partial y}\frac{dy}{2}\right)\left(\frac{\partial w_0}{\partial y} + \frac{\partial^2 w_0}{\partial y^2}\frac{dy}{2}\right)dx$$

$$-\left(N_{xy} - \frac{\partial N_{xy}}{\partial y}\frac{dy}{2}\right)\left(\frac{\partial w_0}{\partial x} - \frac{\partial^2 w_0}{\partial x \partial y}\frac{dy}{2}\right)dx + \left(N_{xy} + \frac{\partial N_{xy}}{\partial y}\frac{dy}{2}\right)\left(\frac{\partial w_0}{\partial x} + \frac{\partial^2 w_0}{\partial x \partial y}\frac{dy}{2}\right)dx$$

$$-\left(N_{xy} - \frac{\partial N_{xy}}{\partial x}\frac{dx}{2}\right)\left(\frac{\partial w_0}{\partial y} - \frac{\partial^2 w_0}{\partial x \partial y}\frac{dx}{2}\right)dy + \left(N_{xy} + \frac{\partial N_{xy}}{\partial x}\frac{dx}{2}\right)\left(\frac{\partial w_0}{\partial y} + \frac{\partial^2 w_0}{\partial x \partial y}\frac{dx}{2}\right)dy$$

$$-\left(V_{xz} - \frac{\partial V_{xz}}{\partial x}\frac{dx}{2}\right)dy + \left(V_{xz} + \frac{\partial V_{xz}}{\partial x}\frac{dx}{2}\right)dy - \left(V_{yz} - \frac{\partial V_{yz}}{\partial y}\frac{dy}{2}\right)dx + \left(V_{yz} + \frac{\partial V_{yz}}{\partial y}\frac{dy}{2}\right)dx$$

$$+ q\,dx\,dy = 0 \quad (7.10)$$

After simple algebraic manipulation, we obtain

$$N_{xx}\frac{\partial^2 w_0}{\partial x^2} + N_{yy}\frac{\partial^2 w_0}{\partial y^2} + 2N_{xy}\frac{\partial^2 w_0}{\partial x \partial y} + \frac{\partial V_{xz}}{\partial x} + \frac{\partial V_{yz}}{\partial y}$$

$$+ \left(\frac{\partial N_{xx}}{\partial x} + \frac{\partial N_{xy}}{\partial y}\right)\frac{\partial w_0}{\partial x} + \left(\frac{\partial N_{xy}}{\partial x} + \frac{\partial N_{yy}}{\partial y}\right)\frac{\partial w_0}{\partial y} + q(x,y) = 0 \quad (7.11)$$

Utilizing Equations 7.5 and 7.6, we get from Equation 7.11, the following:

$$N_{xx}\frac{\partial^2 w_0}{\partial x^2} + N_{yy}\frac{\partial^2 w_0}{\partial y^2} + 2N_{xy}\frac{\partial^2 w_0}{\partial x \partial y} + \frac{\partial V_{xz}}{\partial x} + \frac{\partial V_{yz}}{\partial y} + q(x,y) = 0 \quad (7.12)$$

Let us now consider moment equilibrium in the *xz*-plane, as follows:

$$-\left(M_{xx} - \frac{\partial M_{xx}}{\partial x}\frac{dx}{2}\right)dy + \left(M_{xx} + \frac{\partial M_{xx}}{\partial x}\frac{dx}{2}\right)dy - \left(M_{xy} - \frac{\partial M_{xy}}{\partial y}\frac{dy}{2}\right)dx$$

$$+ \left(M_{xy} + \frac{\partial M_{xy}}{\partial y}\frac{dy}{2}\right)dx - \left(V_{xz} - \frac{\partial V_{xz}}{\partial x}\frac{dx}{2}\right)dy\frac{dx}{2}$$

$$- \left(V_{xz} + \frac{\partial V_{xz}}{\partial x}\frac{dx}{2}\right)dy\frac{dx}{2} = 0 \quad (7.13)$$

which can be readily simplified to

$$V_{xz} = \frac{\partial M_{xx}}{\partial x} + \frac{\partial M_{xy}}{\partial y} \quad (7.14)$$

Similarly, considering moment equilibrium in the yz-plane, it can be shown that

$$V_{yz} = \frac{\partial M_{yy}}{\partial y} + \frac{\partial M_{xy}}{\partial x} \quad (7.15)$$

Substituting Equations 7.14 and 7.15 in Equation 7.12, we get

$$N_{xx}\frac{\partial^2 w_0}{\partial x^2} + N_{yy}\frac{\partial^2 w_0}{\partial y^2} + 2N_{xy}\frac{\partial^2 w_0}{\partial x \partial y} + \frac{\partial^2 M_{xx}}{\partial x^2} + \frac{\partial^2 M_{yy}}{\partial y^2} + 2\frac{\partial^2 M_{xy}}{\partial x \partial y} + q(x,y) = 0 \quad (7.16)$$

which is the third equilibrium equation. For ready reference, let us write below all the three moderately large-deflection equilibrium equations, in terms of stress resultants, for plate bending

$$\frac{\partial N_{xx}}{\partial x} + \frac{\partial N_{xy}}{\partial y} = 0 \quad (7.17)$$

$$\frac{\partial N_{yy}}{\partial y} + \frac{\partial N_{xy}}{\partial x} = 0 \quad (7.18)$$

$$N_{xx}\frac{\partial^2 w_0}{\partial x^2} + N_{yy}\frac{\partial^2 w_0}{\partial y^2} + 2N_{xy}\frac{\partial^2 w_0}{\partial x \partial y} + \frac{\partial^2 M_{xx}}{\partial x^2} + \frac{\partial^2 M_{yy}}{\partial y^2} + 2\frac{\partial^2 M_{xy}}{\partial x \partial y} + q(x,y) = 0 \quad (7.19)$$

Note that we have not put any restriction in respect of the material. Thus, the above equations are valid for isotropic as well as anisotropic plates. We, however, have assumed that the angles α and β are relatively small and Equations 7.8 and 7.9 are valid. For the cases of plate bending, where out-of-plane deflections and the associated slope angles α and β are so small that the components of the in-plane force resultants in the z-direction can be neglected. In such cases, the above procedure can be repeated to show that the equilibrium equations for a thin plate are as follows:

$$\frac{\partial N_{xx}}{\partial x} + \frac{\partial N_{xy}}{\partial y} = 0 \quad (7.20)$$

$$\frac{\partial N_{yy}}{\partial y} + \frac{\partial N_{xy}}{\partial x} = 0 \quad (7.21)$$

$$\frac{\partial^2 M_{xx}}{\partial x^2} + \frac{\partial^2 M_{yy}}{\partial y^2} + 2\frac{\partial^2 M_{xy}}{\partial x \partial y} + q(x,y) = 0 \quad (7.22)$$

Analytical Solutions for Laminated Plates

Next, in order to derive the equilibrium equations in terms of the material stiffness parameters, let us go back to the laminate constitutive equations as per CLPT (Equations 5.42 in Chapter 5). Let us write these equations in the explicit form as follows:

$$\begin{Bmatrix} N_{xx} \\ N_{yy} \\ N_{xy} \\ M_{xx} \\ M_{yy} \\ M_{xy} \end{Bmatrix} = \begin{bmatrix} A_{11} & A_{12} & A_{16} & B_{11} & B_{12} & B_{16} \\ A_{12} & A_{22} & A_{26} & B_{12} & B_{22} & B_{26} \\ A_{16} & A_{26} & A_{66} & B_{16} & B_{26} & B_{66} \\ B_{11} & B_{12} & B_{16} & D_{11} & D_{12} & D_{16} \\ B_{12} & B_{22} & B_{26} & D_{12} & D_{22} & D_{26} \\ B_{16} & B_{26} & B_{66} & D_{16} & D_{26} & D_{66} \end{bmatrix} \begin{Bmatrix} \varepsilon_{xx}^0 \\ \varepsilon_{yy}^0 \\ \gamma_{xy}^0 \\ \kappa_{xx} \\ \kappa_{yy} \\ \kappa_{xy} \end{Bmatrix} \qquad (7.23)$$

The middle surface strains and curvatures are given by Equations 5.15 and 5.16, respectively. Then, using Equation 7.23, Equations 7.20 through 7.22 can be written as

$$A_{11}\frac{\partial^2 u_0}{\partial x^2} + 2A_{16}\frac{\partial^2 u_0}{\partial x \partial y} + A_{66}\frac{\partial^2 u_0}{\partial y^2} + A_{16}\frac{\partial^2 v_0}{\partial x^2} + (A_{12}+A_{66})\frac{\partial^2 v_0}{\partial x \partial y} + A_{26}\frac{\partial^2 v_0}{\partial y^2}$$
$$- B_{11}\frac{\partial^3 w_0}{\partial x^3} - 3B_{16}\frac{\partial^3 w_0}{\partial x^2 \partial y} - (B_{12}+2B_{66})\frac{\partial^3 w_0}{\partial x \partial y^2} - B_{26}\frac{\partial^3 w_0}{\partial y^3} = 0 \qquad (7.24)$$

$$A_{16}\frac{\partial^2 u_0}{\partial x^2} + (A_{12}+A_{66})\frac{\partial^2 u_0}{\partial x \partial y} + A_{26}\frac{\partial^2 u_0}{\partial y^2} + A_{66}\frac{\partial^2 v_0}{\partial x^2} + 2A_{26}\frac{\partial^2 v_0}{\partial x \partial y} + A_{22}\frac{\partial^2 v_0}{\partial y^2}$$
$$- B_{16}\frac{\partial^3 w_0}{\partial x^3} - (B_{12}+2B_{66})\frac{\partial^3 w_0}{\partial x^2 \partial y} - 3B_{26}\frac{\partial^3 w_0}{\partial x \partial y^2} - B_{22}\frac{\partial^3 w_0}{\partial y^3} = 0 \qquad (7.25)$$

$$D_{11}\frac{\partial^4 w_0}{\partial x^4} + 4D_{16}\frac{\partial^4 w_0}{\partial x^3 \partial y} + 2(D_{12}+2D_{66})\frac{\partial^4 w_0}{\partial x^2 \partial y^2} + 4D_{26}\frac{\partial^4 w_0}{\partial x \partial y^3}$$
$$+ D_{22}\frac{\partial^4 w_0}{\partial y^4} - B_{11}\frac{\partial^3 u_0}{\partial x^3} - 3B_{16}\frac{\partial^3 u_0}{\partial x^2 \partial y} - (B_{12}+2B_{66})\frac{\partial^3 u_0}{\partial x \partial y^2} - B_{26}\frac{\partial^3 u_0}{\partial y^3}$$
$$- B_{16}\frac{\partial^3 v_0}{\partial x^3} - (B_{12}+2B_{66})\frac{\partial^3 v_0}{\partial x^2 \partial y} - 3B_{26}\frac{\partial^3 v_0}{\partial x \partial y^2} - B_{22}\frac{\partial^3 v_0}{\partial y^3} = q \qquad (7.26)$$

Equations 7.24 through 7.26 are the governing static equilibrium equations in terms of middle surface displacements for a thin laminated plate. These equations can be greatly simplified for specially orthotropic laminates. (A specially orthotropic laminate is one, in which the plies are specially orthotropic and symmetric about the middle surface.) Of course, most laminates in real life are not so simple in their ply sequence. However, for the sake of simplicity and with a view to only demonstrating the analytical solution procedures, we shall limit our discussions primarily to specially orthotropic laminated plates and other laminated plates with simple ply sequences.

7.3.2.2 Buckling Equations for Laminated Plates

Let us consider a flat plate under uniformly distributed in-plane compressive and shear loads (Figure 7.6). At sufficiently small loads, the plate undergoes in-plane

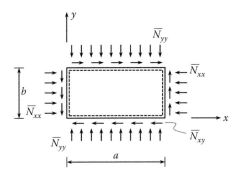

FIGURE 7.6 Rectangular plate under in-plane compressive and shear loads.

deformation in the form of compression, extension, and shear. In a laminated plate, depending on its ply sequence, coupling effect may be present and the plate may undergo bending and twisting under in-plane compressive load. However, the plate still remains stable in an approximately flat equilibrium configuration, referred to as the membrane prebuckled configuration. As the in-plane compressive load is gradually increased, at certain load, the plate becomes unstable and with small disturbance, excessive out-of-plane displacements result. This sudden occurrence of excessive out-of-plane displacements is called buckling and the load at which buckling occurs is the buckling load. The characteristic shape of the buckled plate is called the buckling mode.

Mathematically, buckling is an eigenvalue problem, where the eigenvalues are the buckling loads. The lowest buckling load is of critical importance in design and analysis and it is often referred to as the critical buckling load.

The buckling differential equations can be derived by utilizing the principle of minimum potential energy. The derivation procedure is not discussed here (see References 4, 19, and 20 for details); instead, we present the final expressions of the buckling differential equations, as follows:

$$\frac{\partial \delta N_{xx}}{\partial x} + \frac{\partial \delta N_{xy}}{\partial y} = 0 \tag{7.27}$$

$$\frac{\partial \delta N_{yy}}{\partial y} + \frac{\partial \delta N_{xy}}{\partial x} = 0 \tag{7.28}$$

$$\frac{\partial^2 \delta M_{xx}}{\partial x^2} + 2\frac{\partial^2 \delta M_{xy}}{\partial x \partial y} + \frac{\partial^2 \delta M_{yy}}{\partial y^2} = \bar{N}_{xx}\frac{\partial^2 \delta w_0}{\partial x^2} + 2\bar{N}_{xy}\frac{\partial^2 \delta w_0}{\partial x \partial y} + \bar{N}_{yy}\frac{\partial^2 \delta w_0}{\partial y^2} \tag{7.29}$$

where

$\delta N_{xx}, \delta N_{yy}, \delta N_{xy}$ Variations of the induced force resultants from their prebuckled equilibrium state

$\delta M_{xx}, \delta M_{yy}, \delta M_{xy}$ Variations of the induced moment resultants from their prebuckled equilibrium state

$\bar{N}_{xx}, \bar{N}_{yy}, \bar{N}_{xy}$ Applied compressive in-plane normal and shear force resultants on the plate edges

δw_0 Variation of the out-of-plane deflection of the middle surface of the plate

Equations 7.27 through 7.29 are the buckling equations in terms of variations in the induced stress resultants. With a view to bringing in the variations in displacements, let us now write the laminate constitutive equations as per CLPT (Equations 5.42 in Chapter 5) in terms of variations of force and moment resultants and middle surface strains and curvatures

$$\begin{Bmatrix} \delta N_{xx} \\ \delta N_{yy} \\ \delta N_{xy} \\ \delta M_{xx} \\ \delta M_{yy} \\ \delta M_{xy} \end{Bmatrix} = \begin{bmatrix} A_{11} & A_{12} & A_{16} & B_{11} & B_{12} & B_{16} \\ A_{12} & A_{22} & A_{26} & B_{12} & B_{22} & B_{26} \\ A_{16} & A_{26} & A_{66} & B_{16} & B_{26} & B_{66} \\ B_{11} & B_{12} & B_{16} & D_{11} & D_{12} & D_{16} \\ B_{12} & B_{22} & B_{26} & D_{12} & D_{22} & D_{26} \\ B_{16} & B_{26} & B_{66} & D_{16} & D_{26} & D_{66} \end{bmatrix} \begin{Bmatrix} \delta \varepsilon_{xx}^0 \\ \delta \varepsilon_{yy}^0 \\ \delta \gamma_{xy}^0 \\ \delta \kappa_{xx} \\ \delta \kappa_{yy} \\ \delta \kappa_{xy} \end{Bmatrix} \quad (7.30)$$

Utilizing Equation 7.30, buckling equations can be written as

$$A_{11}\frac{\partial^2 \delta u_0}{\partial x^2} + 2A_{16}\frac{\partial^2 \delta u_0}{\partial x \partial y} + A_{66}\frac{\partial^2 \delta u_0}{\partial y^2} + A_{16}\frac{\partial^2 \delta v_0}{\partial x^2} + (A_{12} + A_{66})\frac{\partial^2 \delta v_0}{\partial x \partial y}$$
$$+ A_{26}\frac{\partial^2 \delta v_0}{\partial y^2} - B_{11}\frac{\partial^3 \delta w_0}{\partial x^3} - 3B_{16}\frac{\partial^3 \delta w_0}{\partial x^2 \partial y} - (B_{12} + 2B_{66})\frac{\partial^3 \delta w_0}{\partial x \partial y^2}$$
$$- B_{26}\frac{\partial^3 \delta w_0}{\partial y^3} = 0 \quad (7.31)$$

$$A_{16}\frac{\partial^2 \delta u_0}{\partial x^2} + (A_{12} + A_{66})\frac{\partial^2 \delta u_0}{\partial x \partial y} + A_{26}\frac{\partial^2 \delta u_0}{\partial y^2} + A_{66}\frac{\partial^2 \delta v_0}{\partial x^2} + 2A_{26}\frac{\partial^2 \delta v_0}{\partial x \partial y}$$
$$+ A_{22}\frac{\partial^2 \delta v_0}{\partial y^2} - B_{16}\frac{\partial^3 \delta w_0}{\partial x^3} - (B_{12} + 2B_{66})\frac{\partial^3 \delta w_0}{\partial x^2 \partial y} - 3B_{26}\frac{\partial^3 \delta w_0}{\partial x \partial y^2}$$
$$- B_{22}\frac{\partial^3 \delta w_0}{\partial y^3} = 0 \quad (7.32)$$

$$D_{11}\frac{\partial^4 \delta w_0}{\partial x^4} + 2(D_{12} + 2D_{66})\frac{\partial^4 \delta w_0}{\partial x^2 \partial y^2} + 4D_{16}\frac{\partial^4 \delta w_0}{\partial x^3 \partial y} + 4D_{26}\frac{\partial^4 \delta w_0}{\partial x \partial y^3}$$
$$+ D_{22}\frac{\partial^4 \delta w_0}{\partial y^4} - B_{11}\frac{\partial^3 \delta u_0}{\partial x^3} - 3B_{16}\frac{\partial^3 \delta u_0}{\partial x^2 \partial y} - (B_{12} + 2B_{66})\frac{\partial^3 \delta u_0}{\partial x \partial y^2}$$
$$- B_{26}\frac{\partial^3 \delta u_0}{\partial y^3} - B_{16}\frac{\partial^3 \delta v_0}{\partial x^3} - (B_{12} + 2B_{66})\frac{\partial^3 \delta v_0}{\partial x^2 \partial y} - 3B_{26}\frac{\partial^3 \delta v_0}{\partial x \partial y^2}$$
$$- B_{22}\frac{\partial^3 \delta v_0}{\partial y^3} = \bar{N}_{xx}\frac{\partial^2 \delta w_0}{\partial x^2} + 2\bar{N}_{xy}\frac{\partial^2 \delta w_0}{\partial x \partial y} + \bar{N}_{yy}\frac{\partial^2 \delta w_0}{\partial y^2} \quad (7.33)$$

A word of explanation of the above equations is in order. The equations—whether in terms of variations in force and moment resultants or in terms of variations in displacements—have striking similarities with the moderately large-deflection equilibrium equations of plate bending. However, important differences exist and we must recognize that buckling equations are *not* equilibrium equations. In-plane force resultants $\bar{N}_{xx}$, $\bar{N}_{yy}$, and $\bar{N}_{xy}$ are the applied force resultants on the edges. The induced stress resultants as well as the in-plane and out-of-plane displacements are prefixed with the symbol of variation δ to insist that these are in fact variations in the respective parameters from their prebuckled equilibrium state. We shall refer to the variations in induced force/moment resultants and displacements as buckling force/moment resultants and buckling displacements. However, for the sake of simplicity, the symbol δ is dropped in the remainder of this book. Thus, the buckling equations would appear as follows:

In terms of buckling stress resultants:

$$\frac{\partial N_{xx}}{\partial x} + \frac{\partial N_{xy}}{\partial y} = 0 \tag{7.34}$$

$$\frac{\partial N_{yy}}{\partial y} + \frac{\partial N_{xy}}{\partial x} = 0 \tag{7.35}$$

$$\frac{\partial^2 M_{xx}}{\partial x^2} + 2\frac{\partial^2 M_{xy}}{\partial x \partial y} + \frac{\partial^2 M_{yy}}{\partial y^2} = \bar{N}_{xx}\frac{\partial^2 w_0}{\partial x^2} + 2\bar{N}_{xy}\frac{\partial^2 w_0}{\partial x \partial y} + \bar{N}_{yy}\frac{\partial^2 w_0}{\partial y^2} \tag{7.36}$$

In terms of buckling displacements:

$$A_{11}\frac{\partial^2 u_0}{\partial x^2} + 2A_{16}\frac{\partial^2 u_0}{\partial x \partial y} + A_{66}\frac{\partial^2 u_0}{\partial y^2} + A_{16}\frac{\partial^2 v_0}{\partial x^2} + (A_{12}+A_{66})\frac{\partial^2 v_0}{\partial x \partial y} + A_{26}\frac{\partial^2 v_0}{\partial y^2}$$
$$- B_{11}\frac{\partial^3 w_0}{\partial x^3} - 3B_{16}\frac{\partial^3 w_0}{\partial x^2 \partial y} - (B_{12}+2B_{66})\frac{\partial^3 w_0}{\partial x \partial y^2} - B_{26}\frac{\partial^3 w_0}{\partial y^3} = 0 \tag{7.37}$$

$$A_{16}\frac{\partial^2 u_0}{\partial x^2} + (A_{12}+A_{66})\frac{\partial^2 u_0}{\partial x \partial y} + A_{26}\frac{\partial^2 u_0}{\partial y^2} + A_{66}\frac{\partial^2 v_0}{\partial x^2} + 2A_{26}\frac{\partial^2 v_0}{\partial x \partial y}$$
$$+ A_{22}\frac{\partial^2 v_0}{\partial y^2} - B_{16}\frac{\partial^3 w_0}{\partial x^3} - (B_{12}+2B_{66})\frac{\partial^3 w_0}{\partial x^2 \partial y} - 3B_{26}\frac{\partial^3 w_0}{\partial x \partial y^2} - B_{22}\frac{\partial^3 w_0}{\partial y^3} = 0 \tag{7.38}$$

$$-B_{11}\frac{\partial^3 u_0}{\partial x^3} - 3B_{16}\frac{\partial^3 u_0}{\partial x^2 \partial y} - (B_{12}+2B_{66})\frac{\partial^3 u_0}{\partial x \partial y^2} - B_{26}\frac{\partial^3 u_0}{\partial y^3} - B_{16}\frac{\partial^3 v_0}{\partial x^3}$$
$$- (B_{12}+2B_{66})\frac{\partial^3 v_0}{\partial x^2 \partial y} - 3B_{26}\frac{\partial^3 v_0}{\partial x \partial y^2} - B_{22}\frac{\partial^3 v_0}{\partial y^3} + D_{11}\frac{\partial^4 w_0}{\partial x^4}$$
$$+ 4D_{16}\frac{\partial^4 w_0}{\partial x^3 \partial y} + 2(D_{12}+2D_{66})\frac{\partial^4 w_0}{\partial x^2 \partial y^2} + 4D_{26}\frac{\partial^4 w_0}{\partial x \partial y^3}$$
$$+ D_{22}\frac{\partial^4 w_0}{\partial y^4} = \bar{N}_{xx}\frac{\partial^2 w_0}{\partial x^2} + 2\bar{N}_{xy}\frac{\partial^2 w_0}{\partial x \partial y} + \bar{N}_{yy}\frac{\partial^2 w_0}{\partial y^2} \tag{7.39}$$

Note: Energy methods are highly useful in buckling problems. Energy-based buckling criterion and solution process are given in Section 7.3.4.3.

7.3.2.3 Vibration Equations for Laminated Plates

We provided a brief introduction to the concept of vibration in Chapter 6. As we know, any oscillatory motion of a body is called vibration. Oscillatory motion of a plate about its static equilibrium state is plate vibration. It can be either free or forced. When a plate vibrates due to an initial excitation at time $t = 0$ and it continues to vibrate under the action of forces inherent to itself and no external forces after time zero, it is called free vibration. The frequencies at which free vibration occurs are the natural frequencies of the plate. The natural frequencies and the mode shapes are critical parameters from structural design and analysis as well as functional point of view. Our objective, in this section, is to find the natural frequencies and corresponding mode shapes of laminated plates. Different combinations of ply sequences and boundary conditions result in a number of different types of plate vibration problems. With a view to only demonstrating the solution procedure, we shall limit our discussion to a few select ply sequences only under simply supported edges. Similarly, we shall not go to the details of derivations (see Reference 21) and rather limit ourselves to studying the features of the governing vibration equation.

The equation of motion of a vibrating plate can be derived by adopting the dynamic equilibrium approach, the variational approach, or the integral equation approach. We derived the plate-bending equations in Section 7.3.2.1 by adopting the classical static equilibrium approach. Under static force equilibrium, $\sum F_x = 0$, $\sum F_y = 0$, $\sum F_z = 0$. In the case of dynamic equilibrium, for transverse vibration of the plate, we have to equate the net force to the mass multiplied by its acceleration. Also, for free vibration, $q(x, y) = 0$. Neglecting rotary inertia, then, the equations of motions, in terms of force and moment resultants, for a plate in transverse vibration are obtained from Equations 7.20 through 7.22 as

$$\frac{\partial N_{xx}}{\partial x} + \frac{\partial N_{xy}}{\partial y} = 0 \tag{7.40}$$

$$\frac{\partial N_{yy}}{\partial y} + \frac{\partial N_{xy}}{\partial x} = 0 \tag{7.41}$$

$$\frac{\partial^2 M_{xx}}{\partial x^2} + \frac{\partial^2 M_{yy}}{\partial y^2} + 2\frac{\partial^2 M_{xy}}{\partial x \partial y} = \rho \frac{\partial^2 w_0}{\partial t^2} \tag{7.42}$$

in which, ρ is the area density of the laminated plate, that is, mass per unit area of the plate. (Note that mass densities for different laminae in the laminate are different and they are integrated across the plate thickness to obtain the area density.) The force and moment resultants are related to the laminate stiffness matrices and middle surface displacements by orthotropic constitutive relations given by Equation 7.23. Then, the equations of motion can be written in terms of the middle surface displacements as follows:

$$A_{11}\frac{\partial^2 u_0}{\partial x^2} + 2A_{16}\frac{\partial^2 u_0}{\partial x \partial y} + A_{66}\frac{\partial^2 u_0}{\partial y^2} + A_{16}\frac{\partial^2 v_0}{\partial x^2} + (A_{12} + A_{66})\frac{\partial^2 v_0}{\partial x \partial y} + A_{26}\frac{\partial^2 v_0}{\partial y^2}$$

$$- B_{11}\frac{\partial^3 w_0}{\partial x^3} - 3B_{16}\frac{\partial^3 w_0}{\partial x^2 \partial y} - (B_{12} + 2B_{66})\frac{\partial^3 w_0}{\partial x \partial y^2} - B_{26}\frac{\partial^3 w_0}{\partial y^3} = 0 \tag{7.43}$$

$$A_{16}\frac{\partial^2 u_0}{\partial x^2} + (A_{12}+A_{66})\frac{\partial^2 u_0}{\partial x \partial y} + A_{26}\frac{\partial^2 u_0}{\partial y^2} + A_{66}\frac{\partial^2 v_0}{\partial x^2} + 2A_{26}\frac{\partial^2 v_0}{\partial x \partial y} + A_{22}\frac{\partial^2 v_0}{\partial y^2}$$
$$- B_{16}\frac{\partial^3 w_0}{\partial x^3} - (B_{12}+2B_{66})\frac{\partial^3 w_0}{\partial x^2 \partial y} - 3B_{26}\frac{\partial^3 w_0}{\partial x \partial y^2} - B_{22}\frac{\partial^3 w_0}{\partial y^3} = 0 \qquad (7.44)$$

$$B_{11}\frac{\partial^3 u_0}{\partial x^3} + 3B_{16}\frac{\partial^3 u_0}{\partial x^2 \partial y} + (B_{12}+2B_{66})\frac{\partial^3 u_0}{\partial x \partial y^2} + B_{26}\frac{\partial^3 u_0}{\partial y^3} + B_{16}\frac{\partial^3 v_0}{\partial x^3}$$
$$+ (B_{12}+2B_{66})\frac{\partial^3 v_0}{\partial x^2 \partial y} + 3B_{26}\frac{\partial^3 v_0}{\partial x \partial y^2} + B_{22}\frac{\partial^3 v_0}{\partial y^3} - D_{11}\frac{\partial^4 w_0}{\partial x^4} - D_{22}\frac{\partial^4 w_0}{\partial y^4}$$
$$- 2(D_{12}+2D_{66})\frac{\partial^4 w_0}{\partial x^2 \partial y^2} - 4D_{16}\frac{\partial^4 w_0}{\partial x^3 \partial y} - 4D_{26}\frac{\partial^4 w_0}{\partial x \partial y^3} = \rho \frac{\partial^2 w_0}{\partial t^2} \qquad (7.45)$$

The plate vibration problem is similar to plate buckling problem. In a buckling problem, as we know, the governing equation involves variations in middle surface displacements and force/moment resultants from the prebuckled equilibrium state. In a similar way, in a vibration problem, the middle surface displacements and force/moment resultants, in fact, are variations in the respective parameters from the static equilibrium state of the plate. (For the sake of simplicity, we have not used the variational symbol δ.) Note that the variations in displacements during vibrations are functions of spatial coordinates as well as time, that is, $u_0 = u_0(x, y, t)$, $v_0 = v_0(x, y, t)$, and $w_0 = w_0(x, y, t)$. Similarly, the variations in force and moment resultants during vibrations are functions of time in addition to the spatial coordinates, that is, $N_{xx} = N_{xx}(x, y, t)$, etc. Also, note that the boundary conditions in plate vibration problems are the same as in plate buckling.

7.3.3 Boundary Conditions in a Laminated Plate

The middle surface displacements (or their derivatives) and the loads along the edges, which are constrained by the physical conditions of the plate along the edges, constitute the boundary conditions. Two types of boundary conditions exist. Boundary conditions specified in terms of the middle surface displacements (or their derivatives) are called geometric or kinematic boundary conditions. Transverse as well as in-plane displacements and slope of the middle surface are the geometric boundary conditions. The displacements may be specified in one or more of the three coordinate directions—x, y, and z. Slope is specified in the direction normal to the edge. Then, for an edge at $x = a$, which is normal to the x-direction, the possible geometric boundary conditions are

$$u_0(a, y) = \bar{u}_0 \qquad (7.46)$$

$$v_0(a, y) = \bar{v}_0 \qquad (7.47)$$

$$w_0(a, y) = \bar{w}_0 \qquad (7.48)$$

$$\frac{\partial w_0(a, y)}{\partial x} = \frac{\partial \bar{w}_0}{\partial x} \qquad (7.49)$$

in which, the terms with an over bar are the specified values.

Analytical Solutions for Laminated Plates

On the other hand, the boundary conditions specified in terms of the loads on the edges are called static or natural boundary conditions. In-plane and transverse force resultants and moment resultants are the static boundary conditions. On an edge, as we know there are five force/moment resultants, out of which, two in-plane force resultants and one bending moment resultant can be independently specified. Remaining two stress resultants—one twisting moment resultant and one shear force resultant are mutually dependent on each other and they have to be specified together. Thus, on an edge, four static boundary conditions are possible. For example, for an edge at $y = b$, which is normal to the y-direction, the possible static boundary conditions are

$$N_{yy}(x,b) = \bar{N}_{yy} \tag{7.50}$$

$$N_{yx}(x,b) = \bar{N}_{yx} \tag{7.51}$$

$$M_{yy}(x,b) = \bar{M}_{yy} \tag{7.52}$$

$$V_{yz}(x,b) + \frac{\partial M_{yx}(x,b)}{\partial x} = \bar{K}_{yz} \tag{7.53}$$

in which, as in the case of geometric boundary conditions, the terms with an over bar are the specified values.

It can be shown that on any edge of the rectangular plate, variation of the twisting moment is statically equivalent to a shear force resultant. In Equation 7.53, $\bar{K}_{yz}$ is the equivalent *Kirchhoff force* defined as the sum of applied shear force resultant $V_{yz}(x, b)$ and a statically equivalent variation of the twisting moment $\partial M_{yx}(x, b)/\partial x$.

The specified values of the boundary conditions, whether geometric or natural, can be either zero or nonzero. Boundary conditions specified as equal to zero are referred to as homogeneous and those with nonzero vales as inhomogeneous.

A plate, whether isotropic or anisotropic, can be free, simply supported, or clamped along its edges. A free edge of a plate is an edge that is free from any external loading. For laminated plates, owing to their anisotropic nature, the boundary conditions are a little more complex. We shall elaborate below the simply supported and clamped boundary conditions.

7.3.3.1 Simply Supported Boundary Condition

A simply supported boundary condition on an edge is one in which the transverse displacement and bending moment on the edge are zero. Thus, on a simply supported edge $x = a$, $w_0(a, y) = M_{xx}(a, y) = 0$. However, in a thin laminated plate, the definition of simply supported boundary condition is a little more involved and we must specify a combination of four geometric and natural conditions. The resulting four types of simply supported boundary conditions are as follows:

For edge, $x = a$:

$$\text{S1:} \quad w_0 = 0 \quad M_{xx} = 0 \quad u_0 = \bar{u}_0 \quad v_0 = \bar{v}_0 \tag{7.54}$$

$$\text{S2:} \quad w_0 = 0 \quad M_{xx} = 0 \quad N_{xx} = \bar{N}_{xx} \quad v_0 = \bar{v}_0 \tag{7.55}$$

$$\text{S3:} \quad w_0 = 0 \quad M_{xx} = 0 \quad u_0 = \bar{u}_0 \quad N_{xy} = \bar{N}_{xy} \tag{7.56}$$

$$S3: \quad w_0 = 0 \quad M_{xx} = 0 \quad N_{xx} = \bar{N}_{xx} \quad N_{xy} = \bar{N}_{xy} \tag{7.57}$$

Similar is the case with the other three edges.

7.3.3.2 Clamped Boundary Condition

A clamped or fixed boundary condition on an edge is one in which the transverse displacement and slope on the edge are zero. Thus, on a clamped edge $x = a$, $w_0(a, y) = \partial w_0(a, y)/\partial x = 0$. Like the simply supported boundary conditions, here too, we must specify a combination of four geometric and natural conditions. The resulting four types of clamped boundary conditions are as follows:

For edge, $x = a$:

$$C1: \quad w_0 = 0 \quad \frac{\partial w_0}{\partial x} = 0 \quad u_0 = \bar{u}_0 \quad v_0 = \bar{v}_0 \tag{7.58}$$

$$S2: \quad w_0 = 0 \quad \frac{\partial w_0}{\partial x} = 0 \quad N_{xx} = \bar{N}_{xx} \quad v_0 = \bar{v}_0 \tag{7.59}$$

$$S3: \quad w_0 = 0 \quad \frac{\partial w_0}{\partial x} = 0 \quad u_0 = \bar{u}_0 \quad N_{xy} = \bar{N}_{xy} \tag{7.60}$$

$$S3: \quad w_0 = 0 \quad \frac{\partial w_0}{\partial x} = 0 \quad N_{xx} = \bar{N}_{xx} \quad N_{xy} = \bar{N}_{xy} \tag{7.61}$$

In a similar way, clamped boundary conditions can be specified on the other three edges.

7.3.4 Solution Methods

Physical phenomena in any field are bound by laws of physics and virtually any phenomenon can be mathematically expressed in terms of algebraic, differential, or integral equations. Most often, differential equations are used to describe a physical problem. Three types of physical problems exist:

- Boundary value problems
- Initial value problems
- Eigenvalue problems

If the dependent variable or its derivatives are specified on the boundary, the problem is called a boundary value problem. Initial value problems are time-dependent problems, where the dependent variable or its derivatives are specified at time $t = 0$. The third type of problems are the eigenvalue problems, which is mathematically expressed as

$$A(u) + \lambda B(u) = 0 \tag{7.62}$$

in which A and B are differential operators, u is the eigenvector, and λ is the eigenvalue.

It can be seen in the subsequent sections that plate bending is a boundary value problem whereas plate buckling and plate vibration problems are eigenvalue problems.

Analytical Solutions for Laminated Plates

Solution methods available for plate problems (for that matter, any elasticity and other field problems) can be broadly classified as follows:

- Analytical methods
  - Exact methods, for example, Navier method, Levy method, etc.
  - Approximate methods, Ritz method, Galerkin method, etc.
- Numerical methods
  - Finite element method
  - Finite difference method
  - Numerical integration

An analytical method is one that gives us a solution to the governing differential equations at any point in the domain. The solution can be either exact or approximate. An exact solution is one that satisfies the governing equations at every point in the domain and the boundary conditions as well as the initial conditions. An approximate solution is one that satisfies the governing equations and boundary and initial conditions in an approximate way. It can be either closed form or an infinite series. An infinite series solution, in actual practice, involves a finite number of terms, and in that sense it is approximate. Numerical methods are also, in general, approximate methods that are highly useful in most practical applications, where exact analytical solutions are difficult or impossible to obtain.

We shall discuss in this chapter solutions of some of the laminated plate bending, buckling, and free vibration problems using the Navier, Levy, and Ritz methods.

7.3.4.1 Navier Method

The Navier method, which involves separation of variables, is an exact analytical method. In this method, a double trigonometric series expansion is chosen for the dependent variable. The choice of the trigonometric function is such that it satisfies all the boundary conditions including geometric as well as natural boundary conditions.

In the case of plate-bending problems, the chosen solutions for the dependent variables, viz. the in-plane and out-of-plane displacements of the middle surface, are typically of the following form:

$$u_0(x,y) = \sum_{n=1}^{\infty}\sum_{m=1}^{\infty} U_{mn} \sin\frac{m\pi x}{a} \cos\frac{n\pi y}{b} \qquad (7.63)$$

$$v_0(x,y) = \sum_{n=1}^{\infty}\sum_{m=1}^{\infty} V_{mn} \cos\frac{m\pi x}{a} \sin\frac{n\pi y}{b} \qquad (7.64)$$

$$w_0(x,y) = \sum_{n=1}^{\infty}\sum_{m=1}^{\infty} W_{mn} \sin\frac{m\pi x}{a} \sin\frac{n\pi y}{b} \qquad (7.65)$$

The in-plane and the out-of-plane displacements are coupled in an anisotropic laminated plate. As a consequence, we need all the three chosen functions mentioned above. However, in symmetric laminates, as we will see subsequently, owing to the absence of any coupling, the third function alone is sufficient.

In a plate-bending problem, our objective is to find primarily the out-of-plane displacements and stresses. In general, owing to the presence of coupling, the in-plane displacements are also required. Thus, we need to find the coefficients U_{mn}, V_{mn}, and W_{mn}. The general procedure is to substitute the chosen displacement functions in the governing differential equations and obtain a set of equations, in which the trigonometric functions are separated from the rest that involve the laminate stiffness parameters, the coefficients to be determined and the parameters m and n. The final result is a set of three algebraic equations from which the required displacement coefficients can be readily obtained. For each pair of the values of m and n, the contributions toward u_0, v_0, and w_0 can be computed, which can be summed up over a finite number of terms for obtaining the total deflection.

Note that for separation of variables to be feasible, the load $q(x, y)$ is also expanded in terms of a double Fourier sine series, as follows:

$$q(x,y) = \sum_{n=1}^{\infty}\sum_{m=1}^{\infty} Q_{mn} \sin\frac{m\pi x}{a} \sin\frac{n\pi y}{b} \tag{7.66}$$

For buckling problems, the chosen solutions for the buckling displacements are of the following form:

$$u_0(x,y) = U_{mn} \sin\frac{m\pi x}{a} \cos\frac{n\pi y}{b} \tag{7.67}$$

$$v_0(x,y) = V_{mn} \cos\frac{m\pi x}{a} \sin\frac{n\pi y}{b} \tag{7.68}$$

$$w_0(x,y) = W_{mn} \sin\frac{m\pi x}{a} \sin\frac{n\pi y}{b} \tag{7.69}$$

in which, each pair of m and n are the numbers of half waves in the deformed shape and the coefficients U_{mn}, V_{mn}, and W_{mn} are the amplitudes of displacements. Note that summation sign is not present in the above functions. In an eigenvalue problem, our aim is to find the eigenvalues (buckling loads) and the corresponding mode shape numbers (m and n), but not the amplitudes. The general procedure is to substitute the chosen functions in the governing buckling equations, which finally result in a set of algebraic equations, which, as we will see in the subsequent sections, is an eigenvalue problem in the form $[A + \lambda B]\{u\} = 0$. The amplitudes of displacements constitute the eigenvector $\{u\}$, which remain undetermined, whereas the eigenvalue λ are determined by equating determinant of the square matrix $[A + \lambda B]$ to zero. A number of eigenvalues, corresponding to different values of m and n, can be obtained, out of which, the minimum one is the critical buckling load.

For free vibration problems, the chosen solutions for the displacements during vibration are of the following form:

$$u_0(x,y,t) = U_{mn} \sin\frac{m\pi x}{a} \cos\frac{n\pi y}{b} T(t) \tag{7.70}$$

$$v_0(x,y,t) = V_{mn} \cos\frac{m\pi x}{a} \sin\frac{n\pi y}{b} T(t) \tag{7.71}$$

$$w_0(x,y,t) = W_{mn} \sin\frac{m\pi x}{a} \sin\frac{n\pi y}{b} T(t) \tag{7.72}$$

in which, the time-dependent function is generally taken as

$$T(t) = A\cos \omega t + B\sin \omega t \tag{7.73}$$

Vibration problem is also an eigenvalue problem and it is solved in a way similar to the buckling problem. The eigenvalues are the natural frequencies out of which the lowest one is the fundamental frequency.

As mentioned earlier, the Navier method is based on separation of variables. This is possible when:

- Laminate stacking sequence is such that $A_{16} = A_{26} = B_{16} = B_{26} = D_{16} = D_{26} = 0$, that is, the laminate is specially orthotropic, antisymmetric cross-ply, or antisymmetric angle-ply laminates
- All four edges are simply supported
- Applied edge shear force resultant is not present, that is, $\bar{N}_{xy} = 0$

As a result, applicability of this method is limited to cases when the above conditions are met.

7.3.4.2 Levy Method

When two opposite edges are simply supported and the other two edges are simply supported, free, or fixed, the Levy method can be employed. In this method for bending problems, single Fourier series expansions for plate displacements are assumed, as follows:

$$u_0(x,y) = \sum_{m=1}^{\infty} U_m(y)\sin\frac{m\pi x}{a} \tag{7.74}$$

$$v_0(x,y) = \sum_{m=1}^{\infty} V_m(y)\cos\frac{m\pi x}{a} \tag{7.75}$$

$$w_0(x,y) = \sum_{m=1}^{\infty} W_m(y)\sin\frac{m\pi x}{a} \tag{7.76}$$

in which, the plate is simply supported on edges $x = 0, a$. The load $q(x, y)$ is also expanded in a similar way, as follows:

$$q(x,y) = \sum_{m=1}^{\infty} Q_m(y)\sin\frac{m\pi x}{a} \tag{7.77}$$

The general procedure is to substitute the chosen functions in the governing equations, which results in separation of the sine function and finally a set of differential equations in y is obtained. The solution of these equations depends on the boundary conditions on edges $y = 0, b$.

For buckling problems, the chosen functions are

$$u_0(x,y) = U_m(y)\sin\frac{m\pi x}{a} \tag{7.78}$$

$$v_0(x,y) = V_m(y)\cos\frac{m\pi x}{a} \qquad (7.79)$$

$$w_0(x,y) = W_m(y)\sin\frac{m\pi x}{a} \qquad (7.80)$$

whereas, in vibration problems, the chosen functions can be

$$u_0(x,y,t) = U_m(y)\sin\frac{m\pi x}{a}T(t) \qquad (7.81)$$

$$v_0(x,y,t) = V_m(y)\cos\frac{m\pi x}{a}T(t) \qquad (7.82)$$

$$w_0(x,y,t) = W_m(y)\sin\frac{m\pi x}{a}T(t) \qquad (7.83)$$

and

$$T(t) = A\cos\omega t + B\sin\omega t \qquad (7.84)$$

7.3.4.3 Ritz Method

The Navier and the Levy solutions are available in a very few cases of laminate stacking sequence and boundary conditions. The Ritz method, however, is a general method applicable to a wide variety of problems. It is a direct energy-based approximate method for boundary value problems. Being a direct energy method, it does not use the governing differential equation. Instead, it employs possible deflection approximations to directly express approximate components of the total potential energy. Then, the principle of minimum potential energy is applied to obtain the solution.

7.3.4.3.1 Ritz Method for Plate Bending

In the case of bending of a rectangular plate, as per the Ritz method, the middle surface displacements of the plate can be approximated as

$$u_0(x,y) = \sum_{m=1}^{M}\sum_{n=1}^{N} U_{mn} f_{mn}^{(u)}(x,y) \qquad (7.85)$$

$$v_0(x,y) = \sum_{m=1}^{M}\sum_{n=1}^{N} V_{mn} f_{mn}^{(v)}(x,y) \qquad (7.86)$$

$$w_0(x,y) = \sum_{m=1}^{M}\sum_{n=1}^{N} W_{mn} f_{mn}^{(w)}(x,y) \qquad (7.87)$$

where $f_{mn}^{(u)}(x,y)$, $f_{mn}^{(v)}(x,y)$, and $f_{mn}^{(w)}(x,y)$ are some functions that essentially satisfy the geometric boundary conditions, that is, boundary conditions on the middle surface displacements and slope, but not necessarily the natural boundary conditions, that is, boundary conditions on moment and shear force. For a rectangular plate, the approximation functions $f_{mn}^{(u)}(x,y)$, $f_{mn}^{(v)}(x,y)$, and $f_{mn}^{(w)}(x,y)$ can be expressed as

$$f_{mn}^{(u)}(x,y) = X_m^{(u)}(x)Y_n^{(u)}(y) \tag{7.88}$$

$$f_{mn}^{(v)}(x,y) = X_m^{(v)}(x)Y_n^{(v)}(y) \tag{7.89}$$

$$f_{mn}^{(w)}(x,y) = X_m^{(w)}(x)Y_n^{(w)}(y) \tag{7.90}$$

Clearly, the choice of the approximation functions depends on the boundary conditions of the plate. For plate-bending problems, we need to determine the coefficients U_{mn}, V_{mn}, and W_{mn}, which is done by employing the minimum potential energy principle. The total potential energy has two parts—potential energy of internal forces or strain energy and potential energy of external forces or negative work done by the external forces,

$$\Pi = U + V \tag{7.91}$$

where U is the strain energy and V is the negative work done by the applied loads. Note that the strain energy is stored in the plate, whereas the work done by the external forces is expended and hence the potential energy of external forces is negative.

Now, the strain energy of the plate is given by

$$U = \frac{1}{2}\int_{-h/2}^{h/2}\int_0^b\int_0^a (\sigma_{xx}\varepsilon_{xx} + \sigma_{yy}\varepsilon_{yy} + \sigma_{zz}\varepsilon_{zz} + \tau_{xy}\gamma_{xy} + \tau_{yz}\gamma_{yz} + \tau_{zx}\gamma_{zx})\,dx\,dy\,dz \tag{7.92}$$

In CLT, the transverse strains are zero, that is, $\varepsilon_{zz} = \gamma_{yz} = \gamma_{zx} = 0$; then, expressing the remaining strains in terms of middle surface strains and curvatures as given by Equation 5.14, it can be shown, after carrying out integration across the thickness, that

$$U = \frac{1}{2}\int_0^b\int_0^a \left(N_{xx}\varepsilon_{xx}^0 + N_{yy}\varepsilon_{yy}^0 + N_{xy}\gamma_{xy}^0 + M_{xx}\kappa_{xx} + M_{yy}\kappa_{yy} + M_{xy}\kappa_{xy}\right)dx\,dy \tag{7.93}$$

In the case of pure plate bending, the middle surface in-plane strains are zero, that is, $\varepsilon_{xx}^0 = \varepsilon_{yy}^0 = \gamma_{xy}^0 = 0$. Then, utilizing Equations 7.23 and 5.16, we obtain the expression for strain energy of the plate as

$$U = \frac{1}{2}\int_0^b\int_0^a \left[D_{11}\left(\frac{\partial^2 w_0}{\partial x^2}\right)^2 + 4D_{16}\left(\frac{\partial^2 w_0}{\partial x^2}\right)\left(\frac{\partial^2 w_0}{\partial x \partial y}\right) + 2D_{12}\left(\frac{\partial^2 w_0}{\partial x^2}\right)\left(\frac{\partial^2 w_0}{\partial y^2}\right) \right.$$

$$\left. + 4D_{66}\left(\frac{\partial^2 w_0}{\partial x \partial y}\right)^2 + 4D_{26}\left(\frac{\partial^2 w_0}{\partial x \partial y}\right)\left(\frac{\partial^2 w_0}{\partial y^2}\right) + D_{22}\left(\frac{\partial^2 w_0}{\partial y^2}\right)^2\right]dx\,dy \tag{7.94}$$

The potential energy of external forces or the negative work done by the applied loads is given by

$$V = -\int_0^b\int_0^a q(x,y)w_0\,dx\,dy \tag{7.95}$$

The minimum potential energy principle states that the total potential energy of the plate is the minimum at equilibrium. Thus, the first derivatives of the total potential energy w.r.t. the coefficients are equated to zero, that is,

$$\frac{\partial \Pi}{\partial U_{mn}} = 0 \qquad (7.96)$$

$$\frac{\partial \Pi}{\partial V_{mn}} = 0 \qquad (7.97)$$

$$\frac{\partial \Pi}{\partial W_{mn}} = 0 \qquad (7.98)$$

which results in a set of three algebraic equations, solving which, the coefficients can be found. Note that, for symmetric laminates, in-plane displacements are not coupled with the out-of-plane displacements and the third equation alone, that is, Equation 7.98, is sufficient for determination of plate deflections.

7.3.4.3.2 Ritz Method for Plate Buckling

We shall first present a statement of buckling criterion based on energy. Toward this, let us consider a flat plate under in-plane compressive and shear loads (Figure 7.6, in Section 7.3.2.2). At sufficiently small loads, the plate remains in stable equilibrium. As we gradually increase the applied loads, at a certain load level, the equilibrium state of the plate changes from stable to unstable and bifurcation takes place. Let us consider an initial equilibrium configuration and let Π denote the total potential energy of the plate at this initial equilibrium configuration. Let us consider a state of the plate in a close neighborhood of the initial equilibrium configuration and let Π' denote the corresponding total potential energy such that $\Delta\Pi \equiv \Pi' - \Pi$ represents the increment in the total potential energy of the plate from an initial equilibrium configuration to a neighboring equilibrium configuration. Now, stability of the initial equilibrium will be decided based on the following conditions:

Stable, if

$$\Delta\Pi > 0 \qquad (7.99)$$

Unstable, if

$$\Delta\Pi < 0 \qquad (7.100)$$

Neutral, if

$$\Delta\Pi = 0 \qquad (7.101)$$

Thus, bifurcation of the initial equilibrium takes place when

$$\Delta\Pi = 0 \qquad (7.102)$$

which is the energy-based buckling criterion for a plate.

At the stable equilibrium states, the plate undergoes in-plane deformation in the form of compression, extension and shear, that is, it undergoes only membrane deformations

Analytical Solutions for Laminated Plates

and the strain energy stored in the plate is purely membrane. In general, however, the total potential energy has two parts—strain energy and the work done by the external loads. The strain energy, in turn, has two parts—membrane strain energy and the bending strain energy. (Note that bending strain energy includes strain energy of bending and twisting both.) Thus, the increment of total potential energy in our buckling criterion can be expressed as

$$\Delta \Pi = \Delta U_m + \Delta U_b + \Delta V \tag{7.103}$$

where ΔU_m is the increment of the membrane strain energy of the plate, ΔU_b the increment of the bending strain energy of the plate, and ΔV the increment of the work done by the external loads.

Now, during buckling, the increment of the membrane strain energy is given by

$$\Delta U_m = \int_0^b \int_0^a \left(N_{xx} \varepsilon_{xx}^0 + N_{yy} \varepsilon_{yy}^0 + N_{xy} \gamma_{xy}^0 \right) dx\,dy \tag{7.104}$$

Note that there is no 1/2 in the right-hand side of the equation above as the in-plane stress resultants are already acting in the plate when buckling takes place. The nonlinear strains are given by Equations 2.61, Chapter 2. For small strain but moderate rotations (say, $\approx 10°$), we can ignore all the terms of order 2 except the rotations of the transverse normals, that is, $(\partial w_0/\partial x)^2$, $(\partial w_0/\partial y)^2$, and $(\partial w_0/\partial x)(\partial w_0/\partial y)$. Then, Equation 7.104 can be expanded as follows:

$$\Delta U_m = \int_0^b \int_0^a \left[N_{xx} \frac{\partial u_0}{\partial x} + N_{yy} \frac{\partial v_0}{\partial y} + N_{xy} \left(\frac{\partial u_0}{\partial y} + \frac{\partial v_0}{\partial x} \right) \right] dx\,dy$$

$$+ \frac{1}{2} \int_0^b \int_0^a \left[N_{xx} \left(\frac{\partial w_0}{\partial x} \right)^2 + N_{yy} \left(\frac{\partial w_0}{\partial y} \right)^2 + 2 N_{xy} \left(\frac{\partial w_0}{\partial x} \right) \left(\frac{\partial w_0}{\partial y} \right) \right] dx\,dy \tag{7.105}$$

The first integral in the above equation is further manipulated by integration by parts, keeping the second unchanged. Then, the increment of the membrane strain energy becomes

$$\Delta U_m = \int_0^b \left([N_{xx} u_0]_0^a + [N_{xy} v_0]_0^a \right) dy + \int_0^a \left([N_{yy} v_0]_0^b + [N_{xy} u_0]_0^b \right) dy$$

$$- \int_0^b \int_0^a u_0 \left(\frac{\partial N_{xx}}{\partial x} + \frac{\partial N_{xy}}{\partial y} \right) dx\,dy - \int_0^b \int_0^a v_0 \left(\frac{\partial N_{xy}}{\partial y} + \frac{\partial N_{yy}}{\partial x} \right) dx\,dy$$

$$+ \frac{1}{2} \int_0^b \int_0^a \left[N_{xx} \left(\frac{\partial w_0}{\partial x} \right)^2 + N_{yy} \left(\frac{\partial w_0}{\partial y} \right)^2 + 2 N_{xy} \left(\frac{\partial w_0}{\partial x} \right) \left(\frac{\partial w_0}{\partial y} \right) \right] dx\,dy \tag{7.106}$$

Using Equations 7.5 and 7.6, we see that the third and fourth terms in Equation 7.106 vanish, whereas a closer look shows that the first and second terms are nothing but the work done by the in-plane external forces. As we know, the potential energy of external forces is the negative work done by the external forces. Then, Equation 7.106 can be written as

$$\Delta U_m = -\Delta V + \frac{1}{2} \int_0^b \int_0^a \left[N_{xx} \left(\frac{\partial w_0}{\partial x} \right)^2 + N_{yy} \left(\frac{\partial w_0}{\partial y} \right)^2 + 2 N_{xy} \left(\frac{\partial w_0}{\partial x} \right) \left(\frac{\partial w_0}{\partial y} \right) \right] dx\, dy \quad (7.107)$$

The next component in the potential energy increment is the increment of bending strain energy, ΔU_b, which is given by Equation 7.94. Of course, as we are dealing with the increment of bending strain energy, the displacements involved are from the initial equilibrium configuration. At this point, let us note that the applied in-plane force resultants are uniformly distributed and the following hold good:

$$(N_{xx})_{x=0} = (N_{xx})_{x=a} = \bar{N}_{xx} \quad (7.108)$$

$$(N_{yy})_{y=0} = (N_{yy})_{y=b} = \bar{N}_{yy} \quad (7.109)$$

$$(N_{xy})_{x=0} = (N_{xy})_{x=a} = (N_{xy})_{y=0} = (N_{xy})_{y=b} = \bar{N}_{xy} \quad (7.110)$$

Now, in a buckling problem, our aim is to find the applied loads at the bifurcation point. Then, we can replace N_{xx}, N_{yy}, and N_{xy} with $\bar{N}_{xx}$, $\bar{N}_{yy}$, and $\bar{N}_{xy}$, respectively. Finally, using Equations 7.94, 7.103, and 7.107, the expression for the increment of total potential energy can be written as

$$\Delta \Pi = \frac{1}{2} \int_0^b \int_0^a \left[D_{11} \left(\frac{\partial^2 w_0}{\partial x^2} \right)^2 + 4 D_{16} \left(\frac{\partial^2 w_0}{\partial x^2} \right) \left(\frac{\partial^2 w_0}{\partial x \partial y} \right) + 2 D_{12} \left(\frac{\partial^2 w_0}{\partial x^2} \right) \left(\frac{\partial^2 w_0}{\partial y^2} \right) \right.$$
$$\left. + 4 D_{66} \left(\frac{\partial^2 w_0}{\partial x \partial y} \right)^2 + 4 D_{26} \left(\frac{\partial^2 w_0}{\partial x \partial y} \right) \left(\frac{\partial^2 w_0}{\partial y^2} \right) + D_{22} \left(\frac{\partial^2 w_0}{\partial y^2} \right)^2 \right] dx\, dy$$
$$+ \frac{1}{2} \int_0^b \int_0^a \left[\bar{N}_{xx} \left(\frac{\partial w_0}{\partial x} \right)^2 + \bar{N}_{yy} \left(\frac{\partial w_0}{\partial y} \right)^2 + 2 \bar{N}_{xy} \left(\frac{\partial w_0}{\partial x} \right) \left(\frac{\partial w_0}{\partial y} \right) \right] dx\, dy \quad (7.111)$$

Minimization of the increment of total potential energy would require the following:

$$\frac{\partial \Delta \Pi}{\partial U_{mn}} = 0 \quad (7.112)$$

$$\frac{\partial \Delta \Pi}{\partial V_{mn}} = 0 \quad (7.113)$$

$$\frac{\partial \Delta \Pi}{\partial W_{mn}} = 0 \quad (7.114)$$

which results in a set of algebraic equations that can be expressed in matrix form as $[A + \lambda B]\{u\} = 0$. The determinant of the square matrix is then equated to zero to obtain the eigenvalues. Vibration problem is similar to buckling problem; it is also an eigenvalue problem. Note that the eigenvector is the vector of amplitudes of the

Analytical Solutions for Laminated Plates

displacements, whereas depending on whether it is a buckling problem or vibration problem, the eigenvalues are the buckling loads or natural frequencies.

7.3.4.3.3 Useful Integration Identities

Before proceeding to the next section, however, let us note the following integration identities that are used at various stages in the solution process:

$$\int_0^l \sin\frac{i\pi x}{l} \sin\frac{j\pi x}{l} dx = 0 \quad \text{for } i \neq j$$

$$= \frac{l}{2} \quad \text{for } i = j \tag{7.115}$$

$$\int_0^l \cos\frac{i\pi x}{l} \cos\frac{j\pi x}{l} dx = 0 \quad \text{for } i \neq j$$

$$= \frac{l}{2} \quad \text{for } i = j \tag{7.116}$$

$$\int_0^l \cos\frac{i\pi x}{l} \sin\frac{j\pi x}{l} dx = 0 \quad \text{for } i+j \text{ is even}$$

$$= \frac{2jl}{\pi(j^2 - i^2)} \quad \text{for } i+j \text{ is odd} \tag{7.117}$$

7.4 SOLUTIONS FOR BENDING OF LAMINATED PLATES

7.4.1 Specially Orthotropic Plate with All Edges Simply Supported: Navier Method for Bending

7.4.1.1 Deflection of Middle Surface

For a specially orthotropic plate, $[B] = 0$ and $A_{16} = A_{26} = D_{16} = D_{26} = 0$ and substituting the same in Equations 7.24 through 7.26 the governing differential equations are obtained as

$$A_{11}\frac{\partial^2 u_0}{\partial x^2} + A_{66}\frac{\partial^2 u_0}{\partial y^2} + (A_{12} + A_{66})\frac{\partial^2 v_0}{\partial x \partial y} = 0 \tag{7.118}$$

$$(A_{12} + A_{66})\frac{\partial^2 u_0}{\partial x \partial y} + A_{66}\frac{\partial^2 v_0}{\partial x^2} + A_{22}\frac{\partial^2 v_0}{\partial y^2} = 0 \tag{7.119}$$

$$D_{11}\frac{\partial^4 w_0}{\partial x^4} + 2(D_{12} + 2D_{66})\frac{\partial^4 w_0}{\partial x^2 \partial y^2} + D_{22}\frac{\partial^4 w_0}{\partial y^4} = q \tag{7.120}$$

As $[B] = 0$, we can see from the constitutive relation (Equation 7.23) that the in-plane strains of the middle surface are not coupled with the curvatures. As a result, the boundary conditions do not involve any coupling between the in-plane displacements u_0 and v_0 and the out-of-plane deflection w_0 and the boundary conditions are far simpler than the ones described by Equations 7.54 through 7.57. Then, the boundary conditions for a specially orthotropic rectangular plate, simply supported on all the edges (Figure 7.7) are as follows:

$$(w_0)_{x=0} = 0 \tag{7.121}$$

$$(w_0)_{x=a} = 0 \tag{7.122}$$

$$(w_0)_{y=0} = 0 \tag{7.123}$$

$$(w_0)_{y=b} = 0 \tag{7.124}$$

$$(M_{xx})_{x=0} = -\left(D_{11}\frac{\partial^2 w_0}{\partial x^2} + D_{12}\frac{\partial^2 w_0}{\partial y^2} \right)_{x=0} = 0 \tag{7.125}$$

$$(M_{xx})_{x=a} = -\left(D_{11}\frac{\partial^2 w_0}{\partial x^2} + D_{12}\frac{\partial^2 w_0}{\partial y^2} \right)_{x=a} = 0 \tag{7.126}$$

$$(M_{yy})_{y=0} = -\left(D_{12}\frac{\partial^2 w_0}{\partial x^2} + D_{22}\frac{\partial^2 w_0}{\partial y^2} \right)_{y=0} = 0 \tag{7.127}$$

$$(M_{yy})_{y=b} = -\left(D_{12}\frac{\partial^2 w_0}{\partial x^2} + D_{22}\frac{\partial^2 w_0}{\partial y^2} \right)_{y=b} = 0 \tag{7.128}$$

As we said before, the governing equations for a specially orthotropic laminate do not involve any coupling between the in-plane displacements u_0 and v_0 and the out-of-plane deflection w_0. As a result, Equation 7.120 alone is sufficient for determination of out-of-plane deflection. The solution procedure, as suggested by Navier, involves assuming a double Fourier sine series expansion for the middle surface deflection, as follows:

$$w_0(x,y) = \sum_{n=1}^{\infty}\sum_{m=1}^{\infty} W_{mn} \sin\frac{m\pi x}{a} \sin\frac{n\pi y}{b} \tag{7.129}$$

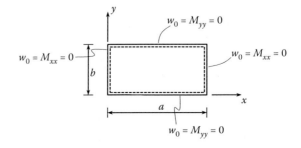

FIGURE 7.7 Rectangular plate simply supported on all four edges.

Analytical Solutions for Laminated Plates

Now, we need to find the coefficients W_{mn} so that the deflection $w_0(x, y)$ can be computed. Note that Equation 7.129 satisfies the boundary conditions given in Equations 7.121 through 7.128. The load $q(x, y)$ is also expanded in terms of a double Fourier sine series, as follows:

$$q(x,y) = \sum_{n=1}^{\infty}\sum_{m=1}^{\infty} Q_{mn} \sin\frac{m\pi x}{a} \sin\frac{n\pi y}{b} \qquad (7.130)$$

The coefficients Q_{mn} are obtained by multiplying both sides of the above equation by sine functions and double integrating, as follows:

$$\int_0^b \int_0^a q(x,y)\sin\frac{k\pi x}{a}\sin\frac{l\pi y}{b}\,dx\,dy$$

$$= \sum_{n=1}^{\infty}\sum_{m=1}^{\infty} \int_0^b \int_0^a Q_{mn}\sin\frac{m\pi x}{a}\sin\frac{k\pi x}{a}\sin\frac{n\pi y}{b}\sin\frac{l\pi y}{b}\,dx\,dy \qquad (7.131)$$

Now, using the integration identity given by Equation 7.115 in the expansion of the right-hand side of Equation 7.131, an expression for Q_{kl} is found. Since, k, l, m, n are arbitrary, we can replace k and l with m and n and get the following:

$$Q_{mn} = \frac{4}{ab}\int_0^b \int_0^a q(x,y)\sin\frac{m\pi x}{a}\sin\frac{n\pi y}{b}\,dx\,dy \qquad (7.132)$$

Then, substituting the chosen functions for the deflection and loads from Equations 7.129 and 7.130, respectively, in the governing equation, Equation 7.120, we get

$$\sum_{n=1}^{\infty}\sum_{m=1}^{\infty}\left[W_{mn}\left\{D_{11}\left(\frac{m\pi}{a}\right)^4 + 2(D_{12}+2D_{66})\left(\frac{m\pi}{a}\right)^2\left(\frac{n\pi}{b}\right)^2\right.\right.$$

$$\left.\left. + D_{22}\left(\frac{n\pi}{b}\right)^4\right\} - Q_{mn}\right]\sin\frac{m\pi x}{a}\sin\frac{n\pi y}{b} = 0 \qquad (7.133)$$

Since Equation 7.133 holds good for each point (x, y) in the domains $0 < x < a$ and $0 < y < b$, the terms inside the square brackets must vanish. Thus, the coefficients of the double Fourier series expansion of the plate middle surface displacement are obtained as

$$W_{mn} = \frac{Q_{mn}}{\dfrac{\pi^4}{a^4}\left[D_{11}m^4 + 2(D_{12}+2D_{66})m^2 n^2\left(\dfrac{a}{b}\right)^2 + D_{22}n^4\left(\dfrac{a}{b}\right)^4\right]} \qquad (7.134)$$

It may be noted that the coefficients Q_{mn} depends on the applied load. For example, for a uniformly distributed load $q(x, y) = q_0$, using Equation 7.132, it can be shown that

$$Q_{mn} = \frac{16 q_0}{mn\pi^2} \quad \text{for } m, n = 1, 3, 5, \ldots$$

$$= 0 \qquad \text{for } m, n = 2, 4, 6, \ldots \qquad (7.135)$$

Then, from Equation 7.134, W_{mn} is obtained and finally plate middle surface deflection is obtained from Equation 7.129, as follows:

$$w_0(x,y) = \sum_{n=1}^{\infty} \sum_{m=1}^{\infty} \frac{16 a^4 q_0 \sin\frac{m\pi x}{a} \sin\frac{n\pi y}{b}}{mn\pi^6 \left[D_{11} m^4 + 2(D_{12} + 2D_{66}) m^2 n^2 \left(\frac{a}{b}\right)^2 + D_{22} n^4 \left(\frac{a}{b}\right)^4 \right]} \quad (7.136)$$

$$\text{for } m, n = 1, 3, 5, \ldots$$
$$= 0 \quad \text{for } m, n = 2, 4, 6, \ldots$$

Let us use this expression to find the one-term solution for deflection at the center point of a specially orthotropic laminated square plate simply supported on all four edges. Then, substituting $x = y = a/2$, $b = a$, and $m = n = 1$, we get

$$w_0(a/2, b/2) = \frac{16 a^4 q_0}{\pi^6 \left[D_{11} + 2(D_{12} + 2D_{66}) + D_{22} \right]} \quad (7.137)$$

7.4.1.2 In-Plane Stresses

Plate middle surface displacement $w_0(x, y)$, obtained in the previous subsection is a function of x and y. Curvatures of the plate middle surface are obtained by differentiation of $w_0(x, y)$. Under pure bending, the middle surface in-plane strains are zero and thus, from Equation 5.24 in Chapter 5, the in-plane stresses at any point are obtained as follows:

$$\begin{Bmatrix} \sigma_{xx} \\ \sigma_{yy} \\ \tau_{xy} \end{Bmatrix}^{(k)} = z \begin{bmatrix} \bar{Q}_{11} & \bar{Q}_{12} & 0 \\ \bar{Q}_{12} & \bar{Q}_{22} & 0 \\ 0 & 0 & \bar{Q}_{66} \end{bmatrix}^{(k)} \begin{Bmatrix} \kappa_{xx} \\ \kappa_{yy} \\ \kappa_{xy} \end{Bmatrix} \quad (7.138)$$

that is,

$$\begin{Bmatrix} \sigma_{xx} \\ \sigma_{yy} \\ \tau_{xy} \end{Bmatrix}^{(k)} = -z \begin{bmatrix} \bar{Q}_{11} & \bar{Q}_{12} & 0 \\ \bar{Q}_{12} & \bar{Q}_{22} & 0 \\ 0 & 0 & \bar{Q}_{66} \end{bmatrix}^{(k)} \begin{Bmatrix} \frac{\partial^2 w_0}{\partial x^2} \\ \frac{\partial^2 w_0}{\partial y^2} \\ 2\frac{\partial^2 w_0}{\partial x \partial y} \end{Bmatrix} \quad (7.139)$$

where the superscript k refers to the kth ply corresponding to the point at which the stresses are being determined. Note that $\bar{Q}_{16} = \bar{Q}_{26} = 0$ for a specially orthotropic laminate. Then, using Equation 7.129 in Equation 7.139, we get the in-plane stresses for a specially orthotropic laminated plate, which we write in the explicit form, as follows:

$$\sigma_{xx}^{(k)}(x,y,z) = z \sum_{n=1}^{\infty} \sum_{m=1}^{\infty} W_{mn} \left[\bar{Q}_{11}^{(k)} \left(\frac{m\pi}{a}\right)^2 + \bar{Q}_{12}^{(k)} \left(\frac{n\pi}{b}\right)^2 \right] \sin\frac{m\pi x}{a} \sin\frac{n\pi y}{b} \quad (7.140)$$

$$\sigma_{yy}^{(k)}(x,y,z) = z \sum_{n=1}^{\infty} \sum_{m=1}^{\infty} W_{mn} \left[\bar{Q}_{12}^{(k)} \left(\frac{m\pi}{a}\right)^2 + \bar{Q}_{22}^{(k)} \left(\frac{n\pi}{b}\right)^2 \right] \sin\frac{m\pi x}{a} \sin\frac{n\pi y}{b} \quad (7.141)$$

Analytical Solutions for Laminated Plates

$$\tau_{xy}^{(k)}(x,y,z) = -2z \sum_{n=1}^{\infty} \sum_{m=1}^{\infty} W_{mn} \bar{Q}_{66}^{(k)} \left(\frac{m\pi}{a}\right)\left(\frac{n\pi}{b}\right) \cos\frac{m\pi x}{a} \cos\frac{n\pi y}{b} \qquad (7.142)$$

7.4.1.3 Interlaminar Stresses

Interlaminar stresses are identically zero as per CLT. However, they do exist especially near the free edges and their evaluation is critical in laminate design [22,23]. Interlaminar stresses in a plate can be determined by adopting a 3D elasticity-based approach. The 3D equilibrium equations, given by Equation 2.135, Chapter 2, take the following form for a static case with zero body forces:

$$\frac{\partial \sigma_{xx}}{\partial x} + \frac{\partial \tau_{xy}}{\partial y} + \frac{\partial \tau_{zx}}{\partial z} = 0 \qquad (7.143)$$

$$\frac{\partial \tau_{xy}}{\partial x} + \frac{\partial \sigma_{yy}}{\partial y} + \frac{\partial \tau_{yz}}{\partial z} = 0 \qquad (7.144)$$

$$\frac{\partial \tau_{zx}}{\partial x} + \frac{\partial \tau_{yz}}{\partial y} + \frac{\partial \sigma_{zz}}{\partial z} = 0 \qquad (7.145)$$

The interlaminar normal stress σ_{zz} and interlaminar shear stresses τ_{zx} and τ_{yz} in any ply are determined by integration of the above three partial differential equations in that ply. Thus,

$$\tau_{zx}^{(k)} = -\int_{z_{k-1}}^{z} \left(\frac{\partial \sigma_{xx}}{\partial x} + \frac{\partial \tau_{xy}}{\partial y}\right) dz + C_1^{(k)} \qquad (7.146)$$

$$\tau_{yz}^{(k)} = -\int_{z_{k-1}}^{z} \left(\frac{\partial \tau_{xy}}{\partial x} + \frac{\partial \sigma_{yy}}{\partial y}\right) dz + C_2^{(k)} \qquad (7.147)$$

$$\sigma_{zz}^{(k)} = -\int_{z_{k-1}}^{z} \left(\frac{\partial \tau_{zx}}{\partial x} + \frac{\partial \tau_{yz}}{\partial y}\right) dz + C_3^{(k)} \qquad (7.148)$$

Now, using Equations 7.140 through 7.142 in Equations 7.146 and 7.147, respectively, we get $\tau_{zx}^{(k)}$ and $\tau_{yz}^{(k)}$. $\tau_{zx}^{(k)}$ and $\tau_{yz}^{(k)}$, in turn, are substituted in Equation 7.148. Thus,

$$\tau_{zx}^{(k)} = -\left(\frac{z^2 - z_{k-1}^2}{2}\right) \sum_{n=1}^{\infty} \sum_{m=1}^{\infty} W_{mn} \left[\left(\frac{m\pi}{a}\right)^3 \bar{Q}_{11}^{(k)} + \left(\frac{m\pi}{a}\right)\left(\frac{n\pi}{b}\right)^2 \left(\bar{Q}_{12}^{(k)} + 2\bar{Q}_{66}^{(k)}\right)\right]$$

$$\times \cos\frac{m\pi x}{a} \sin\frac{n\pi y}{b} + C_1^{(k)} \qquad (7.149)$$

$$\tau_{yz}^{(k)} = -\left(\frac{z^2 - z_{k-1}^2}{2}\right) \sum_{n=1}^{\infty} \sum_{m=1}^{\infty} W_{mn} \left[\left(\frac{m\pi}{a}\right)^2 \left(\frac{n\pi}{b}\right)\left(\bar{Q}_{12}^{(k)} + 2\bar{Q}_{66}^{(k)}\right) + \left(\frac{n\pi}{b}\right)^3 \bar{Q}_{22}^{(k)}\right]$$

$$\times \sin\frac{m\pi x}{a} \cos\frac{n\pi y}{b} + C_2^{(k)} \qquad (7.150)$$

$$\sigma_{zz}^{(k)} = -\left(\frac{z^3 - 3zz_{k-1}^2 + 2z_{k-1}^3}{6}\right)\sum_{n=1}^{\infty}\sum_{m=1}^{\infty} W_{mn}\left[\left(\frac{m\pi}{a}\right)^4 \bar{Q}_{11}^{(k)} + \left(\frac{n\pi}{b}\right)^4 \bar{Q}_{22}^{(k)}\right.$$

$$\left. + 2\left(\frac{m\pi}{a}\right)^2\left(\frac{n\pi}{b}\right)^2\left(\bar{Q}_{12}^{(k)} + 2\bar{Q}_{66}^{(k)}\right)\right]\sin\frac{m\pi x}{a}\sin\frac{n\pi y}{b} + C_3^{(k)} \quad (7.151)$$

Equations 7.149 through 7.151 give us the interlaminar stresses at a point in the kth ply. The constants of integration $C_1^{(k)}$, $C_2^{(k)}$, and $C_3^{(k)}$ in the above equations are functions of x and y. For the bottommost ply, these are determined by using the boundary conditions $\tau_{zx}^{(k)} = \tau_{yz}^{(k)} = \sigma_{zz}^{(k)} = 0$ at $z = -h/2$, that is, bottom of the plate. For the remaining plies from second ply upwards, the following interface continuity conditions are utilized:

$$\tau_{zx}^{(k)}(x,y,z_k) = \tau_{zx}^{(k+1)}(x,y,z_k) \quad (7.152)$$

$$\tau_{yz}^{(k)}(x,y,z_k) = \tau_{yz}^{(k+1)}(x,y,z_k) \quad (7.153)$$

$$\sigma_{zz}^{(k)}(x,y,z_k) = \sigma_{zz}^{(k+1)}(x,y,z_k) \quad (7.154)$$

(See Figure 5.5 in Chapter 5 for coordinates of kth ply in a laminated plate.)

7.4.2 Specially Orthotropic Plate with Two Opposite Edges Simply Supported: Levy Method for Bending

7.4.2.1 Deflection of Middle Surface

Figure 7.8 shows a rectangular plate simply supported at two opposite edges. The other two edges can be any one of the support conditions—simply supported, free, or fixed. Let the plate be simply supported at $x = 0$ and $x = a$. Then, the known boundary conditions:

$$(w_0)_{x=0} = 0 \quad (7.155)$$

$$(w_0)_{x=a} = 0 \quad (7.156)$$

$$(M_{xx})_{x=0} = -\left(D_{11}\frac{\partial^2 w_0}{\partial x^2} + D_{12}\frac{\partial^2 w_0}{\partial y^2}\right)_{x=0} = 0 \quad (7.157)$$

$$(M_{xx})_{x=a} = -\left(D_{11}\frac{\partial^2 w_0}{\partial x^2} + D_{12}\frac{\partial^2 w_0}{\partial y^2}\right)_{x=a} = 0 \quad (7.158)$$

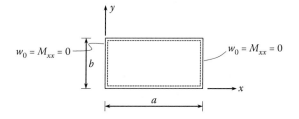

FIGURE 7.8 Rectangular plate simply supported on two opposite edges.

Analytical Solutions for Laminated Plates

For such a problem, Navier's double Fourier series solution procedure is not applicable. Levy proposed a single Fourier series expansion for transverse plate deflection, as follows:

$$w_0(x,y) = \sum_{m=1}^{\infty} W_m(y) \sin \frac{m\pi x}{a} \quad (7.159)$$

which satisfies the boundary conditions given in Equations 7.155 through 7.158. The load $q(x, y)$ is also expanded in a similar way, as follows:

$$q(x,y) = \sum_{m=1}^{\infty} Q_m(y) \sin \frac{m\pi x}{a} \quad (7.160)$$

Note that $W_m(y)$ and $Q_m(y)$ are functions of y. The functions Q_m are obtained by multiplying both sides of the above equation by a sine function and integrating, as follows:

$$\int_0^a q(x,y) \sin \frac{k\pi x}{a} dx = \sum_{m=1}^{\infty} \int_0^a Q_m(y) \sin \frac{m\pi x}{a} \sin \frac{k\pi x}{a} dx \quad (7.161)$$

Then, utilizing the integration identity given by Equation 7.115 in the expansion of right-hand side of Equation 7.161, an expression for $Q_k(y)$ can be obtained. Since k and m are arbitrary, we can replace k with m and obtain the following:

$$Q_m(y) = \frac{2}{a} \int_0^a q(x,y) \sin \frac{m\pi x}{a} dx \quad (7.162)$$

Then, substituting the deflection expansion and load expansion from Equations 7.159 and 7.160 in the governing equation, Equation 7.120, we get

$$\sum_{m=1}^{\infty} \left[D_{11} \left(\frac{m\pi}{a}\right)^4 W_m - 2(D_{12} + 2D_{66}) \left(\frac{m\pi}{a}\right)^2 \frac{d^2 W_m}{dy^2} + D_{22} \frac{d^4 W_m}{dy^4} - Q_m \right] \sin \frac{m\pi x}{a} = 0$$

(7.163)

Since Equation 7.163 holds good for all values of x in the domain $0 < x < a$, the terms inside the square brackets must vanish. Thus,

$$D_{11} \left(\frac{m\pi}{a}\right)^4 W_m - 2(D_{12} + 2D_{66}) \left(\frac{m\pi}{a}\right)^2 \frac{d^2 W_m}{dy^2} + D_{22} \frac{d^4 W_m}{dy^4} = Q_m \quad (7.164)$$

The solution of Equation 7.164, together with Equation 7.162, gives us $W_m(y)$, which when substituted in Equation 7.159 gives us the middle surface deflection of the plate.

Now, Equation 7.164 can be solved either numerically or analytically. The analytical solution involves two parts—homogeneous and particular solutions, that is,

$$W_m(y) = W_m^h(y) + W_m^p(y) \quad (7.165)$$

The auxiliary equation of the fourth-order ordinary differential equation (Equation 7.164) is

$$D_{11}\left(\frac{m\pi}{a}\right)^4 - 2(D_{12} + 2D_{66})\left(\frac{m\pi}{a}\right)^2 \lambda^2 + D_{22}\lambda^4 = 0 \quad (7.166)$$

and the homogeneous solution is

$$W_m^h(y) = Ce^{\lambda y} \quad (7.167)$$

where λ is the root of the auxiliary equation (Equation 7.166). The nature of the roots leads to different possible cases—roots are real and unequal, roots are real and equal and roots are complex. Our objective here is only to demonstrate the method of solution and we shall restrict our discussion only to the first case, that is, roots are real and unequal, as follows.

Let λ_1, λ_2, λ_3, and λ_4 be the four roots. Then,

$$\lambda_1 = \sqrt{\frac{1}{D_{22}}\left(\frac{m\pi}{a}\right)^2 \left[D_{12} + 2D_{66} + \sqrt{(D_{12} + 2D_{66})^2 - D_{11}D_{22}}\right]} \quad (7.168)$$

$$\lambda_2 = -\sqrt{\frac{1}{D_{22}}\left(\frac{m\pi}{a}\right)^2 \left[D_{12} + 2D_{66} + \sqrt{(D_{12} + 2D_{66})^2 - D_{11}D_{22}}\right]} \quad (7.169)$$

$$\lambda_3 = \sqrt{\frac{1}{D_{22}}\left(\frac{m\pi}{a}\right)^2 \left[D_{12} + 2D_{66} - \sqrt{(D_{12} + 2D_{66})^2 - D_{11}D_{22}}\right]} \quad (7.170)$$

$$\lambda_4 = -\sqrt{\frac{1}{D_{22}}\left(\frac{m\pi}{a}\right)^2 \left[D_{12} + 2D_{66} - \sqrt{(D_{12} + 2D_{66})^2 - D_{11}D_{22}}\right]} \quad (7.171)$$

and, noting that $\lambda_2 = -\lambda_1$ and $\lambda_4 = -\lambda_3$, the homogeneous part of the solution takes the following form:

$$W_m^h(y) = A_m \cosh \lambda_1 y + B_m \sinh \lambda_1 y + C_m \cosh \lambda_3 y + D_m \sinh \lambda_3 y \quad (7.172)$$

Then, from Equation 7.165,

$$W_m(y) = A_m \cosh \lambda_1 y + B_m \sinh \lambda_1 y + C_m \cosh \lambda_3 y + D_m \sinh \lambda_3 y + W_m^p(y) \quad (7.173)$$

And clubbing Equation 7.173 with Equation 7.159, the solution for transverse deflection of the plate is obtained as

$$w_0(x,y) = \sum_{m=1}^{\infty} \left[A_m \cosh \lambda_1 y + B_m \sinh \lambda_1 y + C_m \cosh \lambda_3 y + D_m \sinh \lambda_3 y + W_m^p(y)\right] \sin\frac{m\pi x}{a}$$

$$(7.174)$$

Analytical Solutions for Laminated Plates

In the above equation, we still have five unknowns. The constants A_m, B_m, C_m, and D_m are determined using the boundary conditions at $y = 0$ and $y = b$. (Recall, we said in the beginning of this section that the plate is simply supported on two opposite edges, whereas it can have arbitrary boundary conditions on the other two edges—simply supported, free, or fixed.)

The fifth unknown is the particular solution $W_m^p(y)$. When, Q_m is a constant (e.g., if $q(x, y) = q_0$, from Equation 7.162, $Q_m = 4q_0/m\pi$, $m = 1, 3, 5, \ldots$), the particular solution is obtained from Equation 7.164 as

$$W_m^p(y) = \frac{Q_m}{D_{11}}\left(\frac{a}{m\pi}\right)^4 \tag{7.175}$$

For the sake of demonstration only, let us consider a constant Q_m. Similarly, let us consider fixed boundary condition on edges $y = 0$ and $y = b$, that is,

$$(w_0)_{y=0} = 0 \tag{7.176}$$

$$(w_0)_{y=b} = 0 \tag{7.177}$$

$$\left(\frac{\partial w_o}{\partial y}\right)_{y=0} = 0 \tag{7.178}$$

$$\left(\frac{\partial w_o}{\partial y}\right)_{y=b} = 0 \tag{7.179}$$

The boundary conditions from Equations 7.176 through 7.179, when used in Equation 7.174, yield four equations, which can be expressed in the matrix form as follows:

$$\begin{bmatrix} 1 & 0 & 1 & 0 \\ \cosh\lambda_1 b & \sinh\lambda_1 b & \cosh\lambda_3 b & \sinh\lambda_3 b \\ 0 & \lambda_1 & 0 & \lambda_3 \\ \lambda_1 \sinh\lambda_1 b & \lambda_1 \cosh\lambda_1 b & \lambda_3 \sinh\lambda_1 b & \lambda_3 \cosh\lambda_1 b \end{bmatrix} \begin{bmatrix} A_m \\ B_m \\ C_m \\ D_m \end{bmatrix} = \frac{-1}{D_{11}(m\pi/a)^4} \begin{Bmatrix} Q_m \\ Q_m \\ 0 \\ 0 \end{Bmatrix} \tag{7.180}$$

A_m, B_m, C_m, and D_m are readily obtained from Equation 7.180, using which in Equation 7.174, we get the middle surface deflection of the plate.

7.4.2.2 In-Plane and Interlaminar Stresses

Equation 7.139, in the section on Navier's method for specially orthotropic plate, gives the expression for in-plane stresses in terms of middle surface deflection of the plate. In the case of Levy procedure, the middle surface deflection, obtained from Equations 7.174, 7.175 and 7.180, is substituted in Equation 7.139 for obtaining the in-plane stresses. On the other hand, Equations 7.146 through 7.148 give the expressions for interlaminar stresses in terms of in-plane stresses. Resulting expressions, however, are more conveniently utilized in a computer-aided computational environment.

7.4.3 Specially Orthotropic Plate with All Edges Simply Supported: Ritz Method for Bending

Closed form solution, for example, by Navier method, exists for all four edges simply supported specially orthotropic plates. However, for the sake of demonstration, we discuss the Ritz method in this section.

7.4.3.1 Deflection of Middle Surface

In a specially orthotropic laminate, as we know, the in-plane displacements are not coupled with the out-of-plane deflections and approximation function for the middle surface out-of-plane deflection only is sufficient. Let us choose an approximation function for the middle surface out-of-plane deflection of the plate as follows:

$$w_0(x,y) = \sum_{n=1}^{N}\sum_{m=1}^{M} W_{mn} \sin\frac{m\pi x}{a} \sin\frac{n\pi y}{b} \quad (7.181)$$

Further, let the applied transverse load be a uniformly distributed load q_0.

$$q(x,y) = q_0 \quad (7.182)$$

The total potential energy has two parts—strain energy and negative work done by the applied loads,

$$\Pi = U + V \quad (7.183)$$

where the strain energy U of the plate and negative work done by the external forces V are, respectively, given by Equations 7.94 and 7.95.

Further, for a specially orthotropic laminate, $[B] = 0$ and $A_{16} = A_{26} = D_{16} = D_{26} = 0$. Then, utilizing Equation 7.94, we obtain the expression for strain energy of the plate as

$$U = \frac{1}{2}\int_0^b\int_0^a \left[D_{11}\left(\frac{\partial^2 w_0}{\partial x^2}\right)^2 + D_{22}\left(\frac{\partial^2 w_0}{\partial y^2}\right)^2 + 2D_{12}\left(\frac{\partial^2 w_0}{\partial x^2}\right)\left(\frac{\partial^2 w_0}{\partial y^2}\right) + 4D_{66}\left(\frac{\partial^2 w_0}{\partial x \partial y}\right)^2 \right] dx\,dy$$

$$(7.184)$$

The negative work done by the external loads is

$$V = -\int_0^b\int_0^a q_0 w_0\, dx\, dy \quad (7.185)$$

Let us now determine the terms in Equation 7.184.

$$\int_0^b\int_0^a \left(\frac{\partial^2 w_0}{\partial x^2}\right)^2 dx\,dy = \int_0^b\int_0^a \left[-\sum_{n=1}^{N}\sum_{m=1}^{M}\left(\frac{m\pi}{a}\right)^2 W_{mn} \sin\frac{m\pi x}{a} \sin\frac{n\pi y}{b}\right]^2 dx\,dy \quad (7.186)$$

Analytical Solutions for Laminated Plates

We can utilize the integration identity, Equation 7.115, in the expansion of the summation in the right-hand side of Equation 7.186 and see that all terms other than those associated with W_{mn}^2 vanish, which leads to

$$\int_0^b \int_0^a \left(\frac{\partial^2 w_0}{\partial x^2}\right)^2 dx\,dy = \frac{ab}{4} \sum_{n=1}^{N} \sum_{m=1}^{M} \left(\frac{m\pi}{a}\right)^4 W_{mn}^2 \qquad (7.187)$$

Similarly,

$$\int_0^b \int_0^a \left(\frac{\partial^2 w_0}{\partial y^2}\right)^2 dx\,dy = \frac{ab}{4} \sum_{n=1}^{N} \sum_{m=1}^{M} \left(\frac{n\pi}{b}\right)^4 W_{mn}^2 \qquad (7.188)$$

Next,

$$\int_0^b \int_0^a \left(\frac{\partial^2 w_0}{\partial x^2}\right)\left(\frac{\partial^2 w_0}{\partial y^2}\right) dx\,dy = \frac{ab}{4} \sum_{n=1}^{N} \sum_{m=1}^{M} \left(\frac{m\pi}{a}\right)^2 \left(\frac{n\pi}{b}\right)^2 W_{mn}^2 \qquad (7.189)$$

and finally,

$$\int_0^b \int_0^a \left(\frac{\partial^2 w_0}{\partial x \partial y}\right)^2 dx\,dy = \frac{ab}{4} \sum_{n=1}^{N} \sum_{m=1}^{M} \left(\frac{m\pi}{a}\right)^2 \left(\frac{n\pi}{b}\right)^2 W_{mn}^2 \qquad (7.190)$$

Substituting Equations 7.187 through 7.190 in Equation 7.184, the strain energy expression is obtained as

$$U = \frac{ab}{8} \sum_{n=1}^{N} \sum_{m=1}^{M} W_{mn}^2 \left[D_{11}\left(\frac{m\pi}{a}\right)^4 + 2(D_{12} + 2D_{66})\left(\frac{m\pi}{a}\right)^2 \left(\frac{n\pi}{b}\right)^2 + D_{22}\left(\frac{n\pi}{b}\right)^4 \right] \qquad (7.191)$$

Differentiating U w.r.t. W_{mn}, we get

$$\frac{\partial U}{\partial W_{mn}} = \frac{ab}{4} W_{mn} \left[D_{11}\left(\frac{m\pi}{a}\right)^4 + D_{22}\left(\frac{n\pi}{b}\right)^4 + 2(D_{12} + 2D_{66})\left(\frac{m\pi}{a}\right)^2 \left(\frac{n\pi}{b}\right)^2 \right] \qquad (7.192)$$

On the other hand, the negative work done by external loads is further decomposed by substituting Equation 7.181 in Equation 7.185, as follows:

$$V = -q_0 \sum_{n=1}^{N} \sum_{m=1}^{M} \left(\frac{a}{m\pi}\right)\left(\frac{b}{n\pi}\right) W_{mn} (\cos m\pi - 1)(\cos n\pi - 1) \qquad (7.193)$$

The right-hand side of the equation above vanishes for all even values of m and n. Then,

$$V = -q_0 \sum_{n=1}^{N} \sum_{m=1}^{M} \left(\frac{4ab}{mn\pi^2}\right) W_{mn} \quad \text{for } m, n = 1, 3, 5, \ldots$$
$$= 0 \quad \text{for } m, n = 2, 4, 6, \ldots \qquad (7.194)$$

Differentiating V w.r.t. W_{mn}, we get

$$\frac{\partial V}{\partial W_{mn}} = -\frac{4abq_0}{mn\pi^2} \quad \text{for } m, n = 1, 3, 5, \ldots \qquad (7.195)$$
$$= 0 \quad \text{for } m, n = 2, 4, 6, \ldots$$

The minimum potential energy principle states that the total potential energy of the plate is the minimum at equilibrium. Thus, the first derivative of the total potential energy w.r.t. the coefficient W_{mn} is equated to zero, that is,

$$\frac{\partial \Pi}{\partial W_{mn}} = 0 \qquad (7.196)$$

Utilizing Equations 7.192 and 7.195 together with Equation 7.183, we get the following from Equation 7.196:

$$W_{mn} = \frac{16a^4 q_0}{mn\pi^6 \left[D_{11} m^4 + 2(D_{12} + 2D_{66}) m^2 n^2 \left(\frac{a}{b}\right)^2 + D_{22} m^4 \left(\frac{a}{b}\right)^4 \right]} \qquad (7.197)$$

$$\text{for } m, n = 1, 3, 5, \ldots$$
$$= 0 \quad \text{for } m, n = 2, 4, 6, \ldots$$

Substituting the expression for W_{mn} in Equation 7.181, we get the expression for plate middle surface deflection as

$$w_0(x, y) = \sum_{n=1}^{N} \sum_{m=1}^{M} \frac{16a^4 q_0 \sin\frac{m\pi x}{a} \sin\frac{n\pi y}{b}}{mn\pi^6 \left[D_{11} m^4 + 2(D_{12} + 2D_{66}) m^2 n^2 \left(\frac{a}{b}\right)^2 + D_{22} m^4 \left(\frac{a}{b}\right)^4 \right]} \qquad (7.198)$$

$$\text{for } m, n = 1, 3, 5, \ldots$$
$$= 0 \quad \text{for } m, n = 2, 4, 6, \ldots$$

Let us use this expression to find the one-term solution for deflection at the center point of a specially orthotropic laminated square plate simply supported on all four edges. Then, substituting $x = y = a/2$, $b = a$, and $m = n = 1$, we get

$$w_0(x, y) = \frac{16a^4 q_0}{\pi^6 \left[D_{11} + 2(D_{12} + 2D_{66}) + D_{22} \right]} \qquad (7.199)$$

We note that the solution is identical with the Navier's solution (refer Equation 7.137).

7.4.3.2 In-Plane and Interlaminar Stresses

Once the middle surface deflection is obtained, the procedure for stress calculation is the same as discussed in Sections 7.4.1.2 and 7.4.1.3. Thus, Equation 7.139, in the section on Navier method, gives the expression for in-plane stresses in terms of middle surface deflection of the plate. In the case of Ritz method, the middle surface deflection, obtained from Equation 7.198, is substituted in Equation 7.139 for obtaining the in-plane stresses. On the other hand, Equations 7.146 through 7.148 give the expressions for interlaminar stresses in terms of in-plane stresses.

7.4.3.3 Approximation Functions for General Boundary Conditions

We demonstrated above the Ritz method for obtaining solution of a specially orthotropic plate under simply supported boundary conditions. The Ritz method, as noted earlier, is applicable for plates with general boundary conditions. The procedure is the same for all boundary conditions; but, appropriate choice of approximation function is essential. The approximation function must satisfy the geometric boundary conditions, regardless of the values of the coefficients, but not necessarily the natural boundary conditions. Thus, the middle surface deflection $w_0(x, y)$, given by Equation 7.198, satisfies the geometric boundary conditions. For $w_0(x, y)$ to satisfy the natural boundary conditions, we have to take sufficient number of terms in the series.

General boundary conditions of a rectangular plate are combinations of the three conditions—free, simply supported, and clamped on the four edges $x = 0, a$ and $y = 0, b$. In specially orthotropic plates, owing to their middle surface symmetry, $[B] = 0$ and transverse deflections $w_0(x, y)$ are not coupled with the in-plane displacements $u_0(x, y)$ and $v_0(x, y)$. Thus, simply supported as well as clamped boundary conditions are not required to be further distinguished based on the nature of the in-plane boundary conditions on $u_0(x, y)$ and $v_0(x, y)$.

For simply supported boundary conditions on two opposite edges, a sine function can be used. For example, for edges simply supported on $x = 0$ and $x = a$ the approximation function can be

$$X_m^{(w)}(x) = \sin \frac{m\pi x}{a} \tag{7.200}$$

Alternately, a polynomial function can also be used, as follows:

$$X_m^{(w)}(x) = x^{m+1}(x-a)^2 \tag{7.201}$$

For clamped edges on $y = 0$ and $y = b$, a beam function is used, as follows:

$$Y_n^{(w)}(y) = \sin \lambda_n y - \sinh \lambda_n y + \alpha_n (\cosh \lambda_n y - \cos \lambda_n y) \tag{7.202}$$

where

$$\alpha_n = \frac{\sinh \lambda_n b - \sin \lambda_n b}{\cosh \lambda_n b - \cos \lambda_n b} \tag{7.203}$$

and λ_n are the roots of the following characteristic equation:

$$\cos \lambda_n b \cosh \lambda_n b - 1 = 0 \tag{7.204}$$

For mixed boundary conditions on opposite edges, for example, simply supported on $x = 0$ and free on $x = a$, the following can be used:

$$X_m^{(w)}(x) = \sinh \lambda_m a \sin \lambda_m x + \sin \lambda_m a \sinh \lambda_m x \tag{7.205}$$

where λ_n are the roots of the following characteristic equation:

$$\tan \lambda_m a - \tanh \lambda_m a = 0 \qquad (7.206)$$

In this way, the approximation functions can be constructed for different combinations of boundary conditions (see, for instance, Reference 4 for a more comprehensive discussion on boundary conditions).

7.4.4 Symmetric Angle-Ply Laminated Plate with All Edges Simply Supported: Ritz Method for Bending

In a symmetric angle-ply laminate, $[B] = 0$, that is, there is no extension–bending coupling. However, the extensional stiffness matrix $[A]$ and the bending stiffness matrix $[D]$ are fully populated. Transverse deflections $w_0(x, y)$ are not coupled with the in-plane displacements $u_0(x, y)$ and $v_0(x, y)$ and the same are governed by Equation 7.26, which reduces to

$$D_{11}\frac{\partial^4 w_0}{\partial x^4} + 4D_{16}\frac{\partial^4 w_0}{\partial x^3 \partial y} + 2(D_{12} + 2D_{66})\frac{\partial^4 w_0}{\partial x^2 \partial y^2} + 4D_{26}\frac{\partial^4 w_0}{\partial x \partial y^3} + D_{22}\frac{\partial^4 w_0}{\partial y^4} = q(x,y)$$

(7.207)

Simply supported boundary conditions as well as clamped boundary conditions cannot be further distinguished, based on the nature of the in-plane boundary conditions on u_0 and v_0, into $S1, S2, S3, S4$ and $C1, C2, C3, C4$, respectively. Thus, for a plate with all four edges simply supported, the boundary conditions are

$$(w_0)_{x=0} = 0 \qquad (7.208)$$

$$(w_0)_{x=a} = 0 \qquad (7.209)$$

$$(w_0)_{y=0} = 0 \qquad (7.210)$$

$$(w_0)_{y=b} = 0 \qquad (7.211)$$

$$(M_{xx})_{x=0} = -\left(D_{11}\frac{\partial^2 w_0}{\partial x^2} + D_{12}\frac{\partial^2 w_0}{\partial y^2} + 2D_{16}\frac{\partial^2 w_0}{\partial x \partial y}\right)_{x=0} = 0 \qquad (7.212)$$

$$(M_{xx})_{x=a} = -\left(D_{11}\frac{\partial^2 w_0}{\partial x^2} + D_{12}\frac{\partial^2 w_0}{\partial y^2} + 2D_{16}\frac{\partial^2 w_0}{\partial x \partial y}\right)_{x=a} = 0 \qquad (7.213)$$

$$(M_{yy})_{y=0} = -\left(D_{12}\frac{\partial^2 w_0}{\partial x^2} + D_{22}\frac{\partial^2 w_0}{\partial y^2} + 2D_{26}\frac{\partial^2 w_0}{\partial x \partial y}\right)_{y=0} = 0 \qquad (7.214)$$

$$(M_{yy})_{y=b} = -\left(D_{12}\frac{\partial^2 w_0}{\partial x^2} + D_{22}\frac{\partial^2 w_0}{\partial y^2} + 2D_{26}\frac{\partial^2 w_0}{\partial x \partial y}\right)_{y=b} = 0 \qquad (7.215)$$

Navier method, which we used for obtaining solutions of a specially orthotropic plate with all edges simply supported, is not suitable for angle-ply laminates due to the presence of the terms D_{16} and D_{26} in the governing differential equation (Equation 7.207). The problem, however, can be solved by other methods such as the Ritz method.

In the case of pure plate bending under transverse load, $N_{xx} = N_{yy} = N_{xy} = 0$. Utilizing Equation 7.94, we obtain the expression for strain energy of the plate as

$$U = \frac{1}{2}\int_0^b \int_0^a \left[D_{11}\left(\frac{\partial^2 w_0}{\partial x^2}\right)^2 + D_{22}\left(\frac{\partial^2 w_0}{\partial y^2}\right)^2 + 4D_{66}\left(\frac{\partial^2 w_0}{\partial x \partial y}\right)^2 + 2D_{12}\left(\frac{\partial^2 w_0}{\partial x^2}\right)\left(\frac{\partial^2 w_0}{\partial y^2}\right) \right.$$
$$\left. + 4D_{16}\left(\frac{\partial^2 w_0}{\partial x^2}\right)\left(\frac{\partial^2 w_0}{\partial x \partial y}\right) + 4D_{26}\left(\frac{\partial^2 w_0}{\partial y^2}\right)\left(\frac{\partial^2 w_0}{\partial x \partial y}\right) \right] dx\, dy \qquad (7.216)$$

From Equation 7.95, considering uniformly distributed load, that is, $q(x, y) = q_0$, the work done by the external loads is obtained as

$$V = -\int_0^b \int_0^a q_0 w_0 dx\, dy \qquad (7.217)$$

and the total potential energy is given by

$$\Pi = U + V \qquad (7.218)$$

As seen in Section 7.4.3, for simply supported boundary conditions, in the Ritz method, with deflection approximation function as given by Equation 7.181 can be adopted, that is,

$$w_0(x, y) = \sum_{m=1}^{M}\sum_{n=1}^{N} W_{mn} \sin\frac{m\pi x}{a} \sin\frac{n\pi y}{b} \qquad (7.219)$$

Substitution of Equation 7.219 in Equations 7.216 and 7.217 followed by minimization of the total potential energy w.r.t. the Ritz coefficients W_{mn} results in a system of $M \times N$ linear simultaneous equations, solving which we get the values of the coefficients W_{mn}, and finally, $w_0(x, y)$.

7.4.5 Antisymmetric Cross-Ply Laminated Plate with All Edges Simply Supported: Navier Method for Bending

In an antisymmetric cross-ply laminate, $A_{11} = A_{22}$, $D_{11} = D_{22}$, and $A_{16} = A_{26} = D_{16} = D_{26} = 0$. Also, $B_{ij} = 0$ except $B_{11} = -B_{22} \neq 0$, that is, extension–bending coupling exists. As a result, the three equilibrium differential equations of plate bending are coupled, as follows:

$$A_{11}\frac{\partial^2 u_0}{\partial x^2} + A_{66}\frac{\partial^2 u_0}{\partial y^2} + (A_{12} + A_{66})\frac{\partial^2 v_0}{\partial x \partial y} - B_{11}\frac{\partial^3 w_0}{\partial x^3} = 0 \qquad (7.220)$$

$$(A_{12} + A_{66})\frac{\partial^2 u_0}{\partial x \partial y} + A_{66}\frac{\partial^2 v_0}{\partial x^2} + A_{11}\frac{\partial^2 v_0}{\partial y^2} + B_{11}\frac{\partial^3 w_0}{\partial y^3} = 0 \qquad (7.221)$$

$$D_{11}\left(\frac{\partial^4 w_0}{\partial x^4} + \frac{\partial^4 w_0}{\partial y^4}\right) + 2(D_{12} + 2D_{66})\frac{\partial^4 w_0}{\partial x^2 \partial y^2} - B_{11}\left(\frac{\partial^3 u_0}{\partial x^3} - \frac{\partial^3 v_0}{\partial y^3}\right) = q \qquad (7.222)$$

It is seen from the above equations that the transverse deflections $w_0(x, y)$ are coupled with the in-plane displacements $u_0(x, y)$ and $v_0(x, y)$ and simply supported boundary conditions and clamped boundary conditions are required to be further categorized into $S1$, $S2$, $S3$, $S4$ and $C1$, $C2$, $C3$, $C4$, respectively.

Navier solution exists for antisymmetric cross-ply laminated rectangular plate with simply supported boundary conditions $S2$. These boundary conditions are given by

$$(w_0)_{x=0} = 0 \tag{7.223}$$

$$(w_0)_{x=a} = 0 \tag{7.224}$$

$$(w_0)_{y=0} = 0 \tag{7.225}$$

$$(w_0)_{y=b} = 0 \tag{7.226}$$

$$(M_{xx})_{x=0} = \left(B_{11} \frac{\partial u_0}{\partial x} - D_{11} \frac{\partial^2 w_0}{\partial x^2} - D_{12} \frac{\partial^2 w_0}{\partial y^2} \right)_{x=0} = 0 \tag{7.227}$$

$$(M_{xx})_{x=a} = \left(B_{11} \frac{\partial u_0}{\partial x} - D_{11} \frac{\partial^2 w_0}{\partial x^2} - D_{12} \frac{\partial^2 w_0}{\partial y^2} \right)_{x=a} = 0 \tag{7.228}$$

$$(M_{yy})_{y=0} = \left(B_{22} \frac{\partial v_0}{\partial y} - D_{12} \frac{\partial^2 w_0}{\partial x^2} - D_{22} \frac{\partial^2 w_0}{\partial y^2} \right)_{y=0} = 0 \tag{7.229}$$

$$(M_{yy})_{y=b} = \left(B_{22} \frac{\partial v_0}{\partial y} - D_{12} \frac{\partial^2 w_0}{\partial x^2} - D_{22} \frac{\partial^2 w_0}{\partial y^2} \right)_{y=b} = 0 \tag{7.230}$$

$$(v_0)_{x=0} = 0 \tag{7.231}$$

$$(v_0)_{x=a} = 0 \tag{7.232}$$

$$(u_0)_{y=0} = 0 \tag{7.233}$$

$$(u_0)_{y=b} = 0 \tag{7.234}$$

$$(N_{xx})_{x=0} = \left(A_{11} \frac{\partial u_0}{\partial x} + A_{12} \frac{\partial v_0}{\partial y} - B_{11} \frac{\partial^2 w_0}{\partial x^2} \right)_{x=0} = 0 \tag{7.235}$$

$$(N_{xx})_{x=a} = \left(A_{11} \frac{\partial u_0}{\partial x} + A_{12} \frac{\partial v_0}{\partial y} - B_{11} \frac{\partial^2 w_0}{\partial x^2} \right)_{x=a} = 0 \tag{7.236}$$

$$(N_{yy})_{y=0} = \left(A_{12} \frac{\partial u_0}{\partial x} + A_{11} \frac{\partial v_0}{\partial y} + B_{11} \frac{\partial^2 w_0}{\partial y^2} \right)_{y=0} = 0 \tag{7.237}$$

Analytical Solutions for Laminated Plates

$$(N_{yy})_{y=b} = \left(A_{12}\frac{\partial u_0}{\partial x} + A_{11}\frac{\partial v_0}{\partial y} + B_{11}\frac{\partial^2 w_0}{\partial y^2}\right)_{y=b} = 0 \qquad (7.238)$$

Assumed displacement functions in this case are

$$u_0(x,y) = \sum_{n=1}^{\infty}\sum_{m=1}^{\infty} U_{mn} \cos\frac{m\pi x}{a}\sin\frac{n\pi y}{b} \qquad (7.239)$$

$$v_0(x,y) = \sum_{n=1}^{\infty}\sum_{m=1}^{\infty} V_{mn} \sin\frac{m\pi x}{a}\cos\frac{n\pi y}{b} \qquad (7.240)$$

$$w_0(x,y) = \sum_{n=1}^{\infty}\sum_{m=1}^{\infty} W_{mn} \sin\frac{m\pi x}{a}\sin\frac{n\pi y}{b} \qquad (7.241)$$

Further, the transverse loading is expanded as

$$q(x,y) = \sum_{n=1}^{\infty}\sum_{m=1}^{\infty} Q_{mn} \sin\frac{m\pi x}{a}\sin\frac{n\pi y}{b} \qquad (7.242)$$

It can be seen that the boundary conditions (Equations 7.223 through 7.238) are satisfied by the assumed displacement functions and expansion of the transverse loading (Equations 7.239 through 7.242).

The coefficients Q_{mn} are given by Equation 7.132, as

$$Q_{mn} = \frac{4}{ab}\int_0^b\int_0^a q(x,y)\sin\frac{m\pi x}{a}\sin\frac{n\pi y}{b}\,dx\,dy \qquad (7.243)$$

The objective is to determine the coefficients U_{mn}, V_{mn}, and W_{mn} such that, using Equations 7.239 through 7.241, the plate displacements are obtained. Then, substituting the displacement functions in the governing equilibrium differential equations (Equations 7.220 through 7.222), we can obtain a system of algebraic equations, which can be written in the matrix form as follows:

$$\begin{bmatrix} F_{11} & F_{12} & F_{13} \\ F_{12} & F_{22} & F_{23} \\ F_{13} & F_{23} & F_{33} \end{bmatrix}\begin{Bmatrix} U_{mn} \\ V_{mn} \\ W_{mn} \end{Bmatrix} = \begin{Bmatrix} 0 \\ 0 \\ Q_{mn} \end{Bmatrix} \qquad (7.244)$$

in which, the terms in the square matrix are given by

$$F_{11} = A_{11}\left(\frac{m\pi}{a}\right)^2 + A_{66}\left(\frac{n\pi}{b}\right)^2 \qquad (7.245)$$

$$F_{12} = (A_{12} + A_{66})\left(\frac{m\pi}{a}\right)\left(\frac{n\pi}{b}\right) \qquad (7.246)$$

$$F_{13} = -B_{11}\left(\frac{m\pi}{a}\right)^3 \tag{7.247}$$

$$F_{22} = A_{11}\left(\frac{n\pi}{b}\right)^2 + A_{66}\left(\frac{m\pi}{a}\right)^2 \tag{7.248}$$

$$F_{23} = B_{11}\left(\frac{n\pi}{b}\right)^3 \tag{7.249}$$

$$F_{33} = D_{11}\left[\left(\frac{m\pi}{a}\right)^4 + \left(\frac{n\pi}{b}\right)^4\right] + 2(D_{12} + 2D_{66})\left(\frac{m\pi}{a}\right)^2\left(\frac{n\pi}{b}\right)^2 \tag{7.250}$$

Equation 7.244, together with Equations 7.245 through 7.250, give us the coefficients U_{mn}, V_{mn}, and W_{mn} in terms of Q_{mn}, laminate stiffness and plate dimensions, which, in turn, are used for determining the plate displacements.

7.4.6 Antisymmetric Angle-Ply Laminated Plate with All Edges Simply Supported: Navier Method for Bending

In an antisymmetric angle-ply laminate, $A_{16} = A_{26} = D_{16} = D_{26} = B_{11} = B_{12} = B_{22} = B_{66} = 0$. Extension–bending coupling of a different type exists. As a result, the three equilibrium differential equations of plate bending are coupled, as follows:

$$A_{11}\frac{\partial^2 u_0}{\partial x^2} + A_{66}\frac{\partial^2 u_0}{\partial y^2} + (A_{12} + A_{66})\frac{\partial^2 v_0}{\partial x \partial y} - 3B_{16}\frac{\partial^3 w_0}{\partial x^2 \partial y} - B_{26}\frac{\partial^3 w_0}{\partial y^3} = 0 \tag{7.251}$$

$$(A_{12} + A_{66})\frac{\partial^2 u_0}{\partial x \partial y} + A_{66}\frac{\partial^2 v_0}{\partial x^2} + A_{22}\frac{\partial^2 v_0}{\partial y^2} - B_{16}\frac{\partial^3 w_0}{\partial x^3} - 3B_{26}\frac{\partial^3 w_0}{\partial x \partial y^2} = 0 \tag{7.252}$$

$$D_{11}\frac{\partial^4 w_0}{\partial x^4} + 2(D_{12} + 2D_{66})\frac{\partial^4 w_0}{\partial x^2 \partial y^2} + D_{22}\frac{\partial^4 w_0}{\partial y^4} - B_{16}\left(3\frac{\partial^3 u_0}{\partial x^2 \partial y} + \frac{\partial^3 v_0}{\partial x^3}\right)$$
$$- B_{26}\left(\frac{\partial^3 u_0}{\partial y^3} + 3\frac{\partial^3 v_0}{\partial x \partial y^2}\right) = q \tag{7.253}$$

Solution exists for antisymmetric angle-ply laminated rectangular plate with simply supported boundary conditions $S3$. These boundary conditions are given by

$$(w_0)_{x=0} = 0 \tag{7.254}$$

$$(w_0)_{x=a} = 0 \tag{7.255}$$

$$(w_0)_{y=0} = 0 \tag{7.256}$$

$$(w_0)_{y=b} = 0 \tag{7.257}$$

Analytical Solutions for Laminated Plates

$$(M_{xx})_{x=0} = \left[B_{16}\left(\frac{\partial u_0}{\partial y} + \frac{\partial v_0}{\partial x}\right) - D_{11}\frac{\partial^2 w_0}{\partial x^2} - D_{12}\frac{\partial^2 w_0}{\partial y^2} \right]_{x=0} = 0 \quad (7.258)$$

$$(M_{xx})_{x=a} = \left[B_{16}\left(\frac{\partial u_0}{\partial y} + \frac{\partial v_0}{\partial x}\right) - D_{11}\frac{\partial^2 w_0}{\partial x^2} - D_{12}\frac{\partial^2 w_0}{\partial y^2} \right]_{x=a} = 0 \quad (7.259)$$

$$(M_{yy})_{y=0} = \left[B_{26}\left(\frac{\partial u_0}{\partial y} + \frac{\partial v_0}{\partial x}\right) - D_{12}\frac{\partial^2 w_0}{\partial x^2} - D_{22}\frac{\partial^2 w_0}{\partial y^2} \right]_{y=0} = 0 \quad (7.260)$$

$$(M_{yy})_{y=b} = \left[B_{26}\left(\frac{\partial u_0}{\partial y} + \frac{\partial v_0}{\partial x}\right) - D_{12}\frac{\partial^2 w_0}{\partial x^2} - D_{22}\frac{\partial^2 w_0}{\partial y^2} \right]_{y=b} = 0 \quad (7.261)$$

$$(u_0)_{x=0} = 0 \quad (7.262)$$

$$(u_0)_{x=a} = 0 \quad (7.263)$$

$$(v_0)_{y=0} = 0 \quad (7.264)$$

$$(v_0)_{y=b} = 0 \quad (7.265)$$

$$(N_{xy})_{x=0} = \left[A_{66}\left(\frac{\partial u_0}{\partial y} + \frac{\partial v_0}{\partial x}\right) - B_{16}\frac{\partial^2 w_0}{\partial x^2} - B_{26}\frac{\partial^2 w_0}{\partial y^2} \right]_{x=0} = 0 \quad (7.266)$$

$$(N_{xy})_{x=a} = \left[A_{66}\left(\frac{\partial u_0}{\partial y} + \frac{\partial v_0}{\partial x}\right) - B_{16}\frac{\partial^2 w_0}{\partial x^2} - B_{26}\frac{\partial^2 w_0}{\partial y^2} \right]_{x=a} = 0 \quad (7.267)$$

$$(N_{xy})_{y=0} = \left[A_{66}\left(\frac{\partial u_0}{\partial y} + \frac{\partial v_0}{\partial x}\right) - B_{16}\frac{\partial^2 w_0}{\partial x^2} - B_{26}\frac{\partial^2 w_0}{\partial y^2} \right]_{y=0} = 0 \quad (7.268)$$

$$(N_{xy})_{y=b} = \left[A_{66}\left(\frac{\partial u_0}{\partial y} + \frac{\partial v_0}{\partial x}\right) - B_{16}\frac{\partial^2 w_0}{\partial x^2} - B_{26}\frac{\partial^2 w_0}{\partial y^2} \right]_{y=b} = 0 \quad (7.269)$$

Assumed displacement functions in this case are

$$u_0(x,y) = \sum_{n=1}^{\infty}\sum_{m=1}^{\infty} U_{mn} \sin\frac{m\pi x}{a} \cos\frac{n\pi y}{b} \quad (7.270)$$

$$v_0(x,y) = \sum_{n=1}^{\infty}\sum_{m=1}^{\infty} V_{mn} \cos\frac{m\pi x}{a} \sin\frac{n\pi y}{b} \quad (7.271)$$

$$w_0(x,y) = \sum_{n=1}^{\infty}\sum_{m=1}^{\infty} W_{mn} \sin\frac{m\pi x}{a} \sin\frac{n\pi y}{b} \quad (7.272)$$

Further, as in the case of antisymmetric cross-ply laminate, the transverse loading is expanded in a similar way as

$$q(x,y) = \sum_{n=1}^{\infty}\sum_{m=1}^{\infty} Q_{mn} \sin\frac{m\pi x}{a} \sin\frac{n\pi y}{b} \quad (7.273)$$

It can be seen that the boundary conditions (Equations 7.254 through 7.269) are satisfied by the assumed displacement functions and expansion of the transverse loading (Equations 7.270 through 7.273).

As before, the coefficients Q_{mn} are given by Equation 7.232, as

$$Q_{mn} = \frac{4}{ab}\int_0^b\int_0^a q(x,y)\sin\frac{m\pi x}{a}\sin\frac{n\pi y}{b}\,dx\,dy \quad (7.274)$$

Then, substituting the displacement functions in the governing equilibrium differential equations (Equations 7.251 through 7.253), we can obtain a system of algebraic equations, which can be written in the matrix form as in Equation 7.244, in which, the terms in the square matrix are given by

$$F_{11} = A_{11}\left(\frac{m\pi}{a}\right)^2 + A_{66}\left(\frac{n\pi}{b}\right)^2 \quad (7.275)$$

$$F_{12} = (A_{12} + A_{66})\left(\frac{m\pi}{a}\right)\left(\frac{n\pi}{b}\right) \quad (7.276)$$

$$F_{13} = -3B_{16}\left(\frac{m\pi}{a}\right)^2\left(\frac{n\pi}{b}\right) - B_{26}\left(\frac{n\pi}{b}\right)^3 \quad (7.277)$$

$$F_{22} = A_{22}\left(\frac{n\pi}{b}\right)^2 + A_{66}\left(\frac{m\pi}{a}\right)^2 \quad (7.278)$$

$$F_{23} = -B_{16}\left(\frac{m\pi}{a}\right)^3 - 3B_{26}\left(\frac{m\pi}{a}\right)\left(\frac{n\pi}{b}\right)^2 \quad (7.279)$$

$$F_{33} = D_{11}\left(\frac{m\pi}{a}\right)^4 + D_{22}\left(\frac{n\pi}{b}\right)^4 + 2(D_{12} + 2D_{66})\left(\frac{m\pi}{a}\right)^2\left(\frac{n\pi}{b}\right)^2 \quad (7.280)$$

Equation 7.244, together with Equations 7.275 through 7.280, give us the coefficients U_{mn}, V_{mn}, and W_{mn} in terms of Q_{mn}, laminate stiffness and plate dimensions. These coefficients, in turn, are used for determining the plate displacements.

7.5 SOLUTIONS FOR BUCKLING OF LAMINATED PLATES

7.5.1 Specially Orthotropic Simply Supported Plate under In-Plane Uniaxial Compressive Loads: Navier Method for Buckling

For a specially orthotropic plate, $[B] = 0$ and $A_{16} = A_{26} = D_{16} = D_{26} = 0$. Also, for the case of uniaxial compressive loads in the x-direction, $\bar{N}_{yy} = \bar{N}_{xy} = 0$. Then, substituting these in Equations 7.37 through 7.39 the governing differential equations are obtained as

$$A_{11}\frac{\partial^2 u_0}{\partial x^2} + A_{66}\frac{\partial^2 u_0}{\partial y^2} + (A_{12} + A_{66})\frac{\partial^2 v_0}{\partial x \partial y} = 0 \quad (7.281)$$

$$(A_{12} + A_{66})\frac{\partial^2 u_0}{\partial x \partial y} + A_{66}\frac{\partial^2 v_0}{\partial x^2} + A_{22}\frac{\partial^2 v_0}{\partial y^2} = 0 \quad (7.282)$$

$$D_{11}\frac{\partial^4 w_0}{\partial x^4} + 2(D_{12} + 2D_{66})\frac{\partial^4 w_0}{\partial x^2 \partial y^2} + D_{22}\frac{\partial^4 w_0}{\partial y^4} = \bar{N}_{xx}\frac{\partial^2 w_0}{\partial x^2} \quad (7.283)$$

Note that in the above equations u_0, v_0, and w_0 are actually the variations in displacements, that is, δu_0, δv_0, and δw_0, respectively. As mentioned earlier, the symbol δ is dropped for the sake of simplicity in writing.

Next, note that in-plane buckling displacements u_0 and v_0 are not coupled with the out-of-plane buckling displacement w_0. As a result, the buckling boundary conditions for a specially orthotropic rectangular plate are rather simple, which, for the case of simply supported on all four edges, are

$$(w_0)_{x=0} = 0 \quad (7.284)$$

$$(w_0)_{x=a} = 0 \quad (7.285)$$

$$(w_0)_{y=0} = 0 \quad (7.286)$$

$$(w_0)_{y=b} = 0 \quad (7.287)$$

$$(M_{xx})_{x=0} = -\left(D_{11}\frac{\partial^2 w_0}{\partial x^2} + D_{12}\frac{\partial^2 w_0}{\partial y^2}\right)_{x=0} = 0 \quad (7.288)$$

$$(M_{xx})_{x=a} = -\left(D_{11}\frac{\partial^2 w_0}{\partial x^2} + D_{12}\frac{\partial^2 w_0}{\partial y^2}\right)_{x=a} = 0 \quad (7.289)$$

$$(M_{yy})_{y=0} = -\left(D_{12}\frac{\partial^2 w_0}{\partial x^2} + D_{22}\frac{\partial^2 w_0}{\partial y^2}\right)_{y=0} = 0 \quad (7.290)$$

$$(M_{yy})_{y=b} = -\left(D_{12}\frac{\partial^2 w_0}{\partial x^2} + D_{22}\frac{\partial^2 w_0}{\partial y^2}\right)_{y=b} = 0 \quad (7.291)$$

Navier solution exists for such a case, and a double Fourier series expansion is assumed for the variation in out-of-plane buckling displacement as follows:

$$w_0(x,y) = W_{mn}\sin\frac{m\pi x}{a}\sin\frac{n\pi y}{b} \quad (7.292)$$

Note that, unlike plate bending, summation of displacement components for various values of m and n is not required in the case of plate buckling. In fact, in Equation 7.292, each combination of m and n corresponds to one mode shape, where m and n are the numbers of half buckle waves in the x- and y-directions, respectively. Substituting Equation 7.292 in Equation 7.283, we get

$$\bar{N}_{xx} = \pi^2 \left[D_{11}\left(\frac{m}{a}\right)^2 + 2(D_{12} + 2D_{66})\left(\frac{n}{b}\right)^2 + D_{22}\left(\frac{n}{b}\right)^4\left(\frac{a}{m}\right)^2 \right] \quad (7.293)$$

7.5.2 Specially Orthotropic Simply Supported Plate under In-Plane Uniaxial Compressive Loads: Ritz Method for Buckling

The governing buckling differential equations and the buckling boundary conditions for a specially orthotropic laminated plate with all edges simply supported were presented in the previous section. Also, as seen in the previous section, Navier solution is available for this class of problems. However, for the sake of demonstration of the procedure and comparison of results, the Ritz method is discussed in this section.

In the case of specially orthotropic laminated plate under in-plane normal compressive force resultant $\bar{N}_{xx}$, the total potential energy functional can be obtained from Equation 7.111 as follows:

$$\Pi = \frac{1}{2}\int_0^b \int_0^a \left[D_{11}\left(\frac{\partial^2 w_0}{\partial x^2}\right)^2 + D_{22}\left(\frac{\partial^2 w_0}{\partial y^2}\right)^2 + 4D_{66}\left(\frac{\partial^2 w_0}{\partial x \partial y}\right)^2 \right.$$
$$\left. + 2D_{12}\left(\frac{\partial^2 w_0}{\partial x^2}\right)\left(\frac{\partial^2 w_0}{\partial y^2}\right) - \bar{N}_{xx}\left(\frac{\partial w_0}{\partial x}\right)^2 \right] dx\, dy \quad (7.294)$$

The Ritz approximation function for buckling displacement can be chosen as

$$w_0(x,y) = \sum_{m=1}^{M}\sum_{n=1}^{N} W_{mn} \sin\frac{m\pi x}{a} \sin\frac{n\pi y}{b} \quad (7.295)$$

The above equation satisfies the geometric boundary conditions as well as the natural boundary conditions given in the previous section. (Note that the Ritz method requires the approximation function to satisfy the geometric boundary conditions but not necessarily the natural boundary conditions. Note further that closed-form solution exists in the present case of specially orthotropic plate; we are discussing the approximate Ritz method only for the sake of demonstration!)

Substituting the approximation function for buckling displacement from Equation 7.295 in the total potential energy functional in Equation 7.294 and utilizing the integration identities in Equations 7.115 through 7.116, we can carry out direct integration and show that the integration terms in Equation 7.294 are as follows:

$$\int_0^b \int_0^a \left(\frac{\partial^2 w_0}{\partial x^2}\right)^2 dx\, dy = \frac{ab}{4}\sum_{n=1}^{N}\sum_{m=1}^{M}\left(\frac{m\pi}{a}\right)^4 W_{mn}^2 \quad (7.296)$$

Analytical Solutions for Laminated Plates

$$\int_0^b \int_0^a \left(\frac{\partial^2 w_0}{\partial y^2}\right)^2 dx\,dy = \frac{ab}{4} \sum_{n=1}^{N} \sum_{m=1}^{M} \left(\frac{n\pi}{b}\right)^4 W_{mn}^2 \qquad (7.297)$$

$$\int_0^b \int_0^a \left(\frac{\partial^2 w_0}{\partial x^2}\right)\left(\frac{\partial^2 w_0}{\partial y^2}\right) dx\,dy = \frac{ab}{4} \sum_{n=1}^{N} \sum_{m=1}^{M} \left(\frac{m\pi}{a}\right)^2 \left(\frac{n\pi}{b}\right)^2 W_{mn}^2 \qquad (7.298)$$

$$\int_0^b \int_0^a \left(\frac{\partial^2 w_0}{\partial x \partial y}\right)^2 dx\,dy = \frac{ab}{4} \sum_{n=1}^{N} \sum_{m=1}^{M} \left(\frac{m\pi}{a}\right)^2 \left(\frac{n\pi}{b}\right)^2 W_{mn}^2 \qquad (7.299)$$

$$\int_0^b \int_0^a \left(\frac{\partial w_0}{\partial x}\right)^2 dx\,dy = \frac{ab}{4} \sum_{n=1}^{N} \sum_{m=1}^{M} \left(\frac{m\pi}{a}\right)^2 W_{mn}^2 \qquad (7.300)$$

Then, substituting Equations 7.296 through 7.300 in Equation 7.294, the potential energy functional is obtained as

$$\Pi = \frac{ab}{8} \sum_{n=1}^{N} \sum_{m=1}^{M} W_{mn}^2 \left[D_{11}\left(\frac{m\pi}{a}\right)^4 + D_{22}\left(\frac{n\pi}{b}\right)^4 + 2(D_{12} + 2D_{66})\left(\frac{m\pi}{a}\right)^2 \left(\frac{n\pi}{b}\right)^2 - \bar{N}_{xx}\left(\frac{m\pi}{a}\right)^2 \right]$$

(7.301)

For buckling,

$$\frac{\partial \Pi}{\partial W_{mn}} = 0 \qquad (7.302)$$

It results in $M \times N$ algebraic equations, each corresponding to a unique pair of m and n. Each of these equations gives us a value of $\bar{N}_{xx}$ of which the minimum one is the critical buckling load, in general. We note here that, $\bar{N}_{xx}$ is given by

$$\bar{N}_{xx} = \pi^2 \left[D_{11}\left(\frac{m}{a}\right)^2 + 2(D_{12} + 2D_{66})\left(\frac{n}{b}\right)^2 + D_{22}\left(\frac{n}{b}\right)^4 \left(\frac{a}{m}\right)^2 \right] \qquad (7.303)$$

It is seen that the buckling load as per the Ritz method given by Equation 7.303 is identical with that as per the Navier method given by Equation 7.293. In other words, for a specially orthotropic laminated plate simply supported on all four edges, the Ritz method gives an exact solution.

7.5.3 Symmetric Angle-Ply Laminated Simply Supported Plate under In-Plane Uniaxial Compressive Loads: Ritz Method for Buckling

In a symmetric angle-ply laminate, $[B] = 0$, that is, there is no extension–bending coupling. However, the extensional stiffness matrix $[A]$ and the bending stiffness matrix $[D]$ are fully populated. Also, for the case of uniaxial compressive loads in the

x-direction, $\bar{N}_{yy} = \bar{N}_{xy} = 0$. Then, substituting these in Equations 7.37 through 7.39, the governing differential equations are obtained as

$$A_{11}\frac{\partial^2 u_0}{\partial x^2} + 2A_{16}\frac{\partial^2 u_0}{\partial x \partial y} + A_{66}\frac{\partial^2 u_0}{\partial y^2} + A_{16}\frac{\partial^2 v_0}{\partial x^2} + (A_{12} + A_{66})\frac{\partial^2 v_0}{\partial x \partial y} + A_{26}\frac{\partial^2 v_0}{\partial y^2} = 0$$
(7.304)

$$A_{16}\frac{\partial^2 u_0}{\partial x^2} + (A_{12} + A_{66})\frac{\partial^2 u_0}{\partial x \partial y} + A_{26}\frac{\partial^2 u_0}{\partial y^2} + A_{66}\frac{\partial^2 v_0}{\partial x^2} + 2A_{26}\frac{\partial^2 v_0}{\partial x \partial y} + A_{22}\frac{\partial^2 v_0}{\partial y^2} = 0$$
(7.305)

$$D_{11}\frac{\partial^4 w_0}{\partial x^4} + 2(D_{12} + 2D_{66})\frac{\partial^4 w_0}{\partial x^2 \partial y^2} + D_{22}\frac{\partial^4 w_0}{\partial y^4} + 4D_{16}\frac{\partial^4 w_0}{\partial x^3 \partial y} + 4D_{26}\frac{\partial^4 w_0}{\partial x \partial y^3} = \bar{N}_{xx}\frac{\partial^2 w_0}{\partial x^2}$$
(7.306)

Note that in-plane buckling displacements u_0 and v_0 are not coupled with the out-of-plane buckling displacement w_0. As a result, the buckling boundary conditions for a symmetric angle-ply laminated plate are rather simple, and simply supported boundary conditions as well as clamped boundary conditions cannot be further distinguished, based on the nature of the in-plane boundary conditions on u_0 and v_0, into S1, S2, S3, S4 and C1, C2, C3, C4, respectively. Thus, for a plate with all four edges simply supported, the boundary conditions are

$$(w_0)_{x=0} = 0 \qquad (7.307)$$

$$(w_0)_{x=a} = 0 \qquad (7.308)$$

$$(w_0)_{y=0} = 0 \qquad (7.309)$$

$$(w_0)_{y=b} = 0 \qquad (7.310)$$

$$(M_{xx})_{x=0} = -\left(D_{11}\frac{\partial^2 w_0}{\partial x^2} + D_{12}\frac{\partial^2 w_0}{\partial y^2} + 2D_{16}\frac{\partial^2 w_0}{\partial x \partial y}\right)_{x=0} = 0 \qquad (7.311)$$

$$(M_{xx})_{x=a} = -\left(D_{11}\frac{\partial^2 w_0}{\partial x^2} + D_{12}\frac{\partial^2 w_0}{\partial y^2} + 2D_{16}\frac{\partial^2 w_0}{\partial x \partial y}\right)_{x=a} = 0 \qquad (7.312)$$

$$(M_{yy})_{y=0} = -\left(D_{12}\frac{\partial^2 w_0}{\partial x^2} + D_{22}\frac{\partial^2 w_0}{\partial y^2} + 2D_{26}\frac{\partial^2 w_0}{\partial x \partial y}\right)_{y=0} = 0 \qquad (7.313)$$

$$(M_{yy})_{y=b} = -\left(D_{12}\frac{\partial^2 w_0}{\partial x^2} + D_{22}\frac{\partial^2 w_0}{\partial y^2} + 2D_{26}\frac{\partial^2 w_0}{\partial x \partial y}\right)_{y=b} = 0 \qquad (7.314)$$

Owing to the presence of the terms D_{16} and D_{26} in the governing differential equation (Equation 7.306), closed-form solutions such as Navier method are not available

Analytical Solutions for Laminated Plates 379

in the case of symmetric angle-ply laminates. The problem, however, can be solved by other methods such as the Ritz method or the Galerkin method.

For a symmetric angle-ply laminated plate under in-plane normal compressive force resultant $\bar{N}_{xx}$, the total potential energy functional is obtained from Equation 7.111 as follows:

$$\Pi = \frac{1}{2}\int_0^b \int_0^a \left[D_{11}\left(\frac{\partial^2 w_0}{\partial x^2}\right)^2 + D_{22}\left(\frac{\partial^2 w_0}{\partial y^2}\right)^2 + 4D_{66}\left(\frac{\partial^2 w_0}{\partial x \partial y}\right)^2 + 2D_{12}\left(\frac{\partial^2 w_0}{\partial x^2}\right)\left(\frac{\partial^2 w_0}{\partial y^2}\right) \right.$$
$$\left. + 4D_{16}\left(\frac{\partial^2 w_0}{\partial x^2}\right)\left(\frac{\partial^2 w_0}{\partial x \partial y}\right) + 4D_{26}\left(\frac{\partial^2 w_0}{\partial y^2}\right)\left(\frac{\partial^2 w_0}{\partial x \partial y}\right) - \bar{N}_{xx}\left(\frac{\partial w_0}{\partial x}\right)^2 \right] dx\, dy$$

(7.315)

The Ritz approximation function for buckling displacement, for a plate with all four edges simply supported, can be chosen as

$$w_0(x,y) = \sum_{m=1}^{M}\sum_{n=1}^{N} W_{mn} \sin\frac{m\pi x}{a}\sin\frac{n\pi y}{b} \tag{7.316}$$

Note that the above equation satisfies the geometric boundary conditions but not the natural boundary conditions. (Note further that Ritz method requires the approximation function to satisfy the geometric boundary conditions but not necessarily the natural boundary conditions.)

Equation 7.316 is then substituted in Equation 7.315 and direct integration is carried out. However, owing to the presence of the stiffness parameters D_{16} and D_{26} in the expression for the potential energy functional, integration identity given by Equation 7.117 is also required. Subsequent procedure is rather complicated and computation in a computerized environment is needed, in which employing minimum potential energy principle, we can obtain $M \times N$ algebraic equations, each corresponding to a unique pair of m and n. Each of these equations gives us a value of $\bar{N}_{xx}$, of which the minimum one is the critical buckling load.

7.5.4 Antisymmetric Cross-Ply Laminated Simply Supported Plate under In-Plane Uniaxial Compressive Loads: Navier Method for Buckling

In an antisymmetric cross-ply laminate, $A_{11} = A_{22}$, $D_{11} = D_{22}$, and $A_{16} = A_{26} = D_{16} = D_{26} = 0$. Also, $B_{ij} = 0$ except $B_{11} = -B_{22} \neq 0$, that is, extension–bending coupling exists. As a result, the three differential equations of plate buckling are coupled; they can be obtained from Equations 7.37 through 7.39, as follows:

$$A_{11}\frac{\partial^2 u_0}{\partial x^2} + A_{66}\frac{\partial^2 u_0}{\partial y^2} + (A_{12}+A_{66})\frac{\partial^2 v_0}{\partial x \partial y} - B_{11}\frac{\partial^3 w_0}{\partial x^3} = 0 \tag{7.317}$$

$$(A_{12}+A_{66})\frac{\partial^2 u_0}{\partial x \partial y} + A_{66}\frac{\partial^2 v_0}{\partial x^2} + A_{11}\frac{\partial^2 v_0}{\partial y^2} + B_{11}\frac{\partial^3 w_0}{\partial y^3} = 0 \tag{7.318}$$

$$D_{11}\left(\frac{\partial^4 w_0}{\partial x^4} + \frac{\partial^4 w_0}{\partial y^4}\right) + 2(D_{12}+2D_{66})\frac{\partial^4 w_0}{\partial x^2 \partial y^2} - B_{11}\left(\frac{\partial^3 u_0}{\partial x^3} - \frac{\partial^3 v_0}{\partial y^3}\right) = \bar{N}_{xx}\frac{\partial^2 w_0}{\partial x^2} \tag{7.319}$$

Transverse deflections $w_0(x, y)$ are coupled with the in-plane displacements $u_0(x, y)$ and $v_0(x, y)$ and simply supported boundary conditions and clamped boundary conditions are required to be further categorized into S1, S2, S3, S4 and C1, C2, C3, C4, respectively.

Navier solution exists for antisymmetric cross-ply laminated rectangular plate with simply supported boundary conditions S2. These boundary conditions are given by

$$(w_0)_{x=0} = 0 \tag{7.320}$$

$$(w_0)_{x=a} = 0 \tag{7.321}$$

$$(w_0)_{y=0} = 0 \tag{7.322}$$

$$(w_0)_{y=b} = 0 \tag{7.323}$$

$$(M_{xx})_{x=0} = \left(B_{11} \frac{\partial u_0}{\partial x} - D_{11} \frac{\partial^2 w_0}{\partial x^2} - D_{12} \frac{\partial^2 w_0}{\partial y^2} \right)_{x=0} = 0 \tag{7.324}$$

$$(M_{xx})_{x=a} = \left(B_{11} \frac{\partial u_0}{\partial x} - D_{11} \frac{\partial^2 w_0}{\partial x^2} - D_{12} \frac{\partial^2 w_0}{\partial y^2} \right)_{x=a} = 0 \tag{7.325}$$

$$(M_{yy})_{y=0} = \left(B_{22} \frac{\partial v_0}{\partial y} - D_{12} \frac{\partial^2 w_0}{\partial x^2} - D_{22} \frac{\partial^2 w_0}{\partial y^2} \right)_{y=0} = 0 \tag{7.326}$$

$$(M_{yy})_{y=b} = \left(B_{22} \frac{\partial v_0}{\partial y} - D_{12} \frac{\partial^2 w_0}{\partial x^2} - D_{22} \frac{\partial^2 w_0}{\partial y^2} \right)_{y=b} = 0 \tag{7.327}$$

$$(v_0)_{x=0} = 0 \tag{7.328}$$

$$(v_0)_{x=a} = 0 \tag{7.329}$$

$$(u_0)_{y=0} = 0 \tag{7.330}$$

$$(u_0)_{y=b} = 0 \tag{7.331}$$

$$(N_{xx})_{x=0} = \left(A_{11} \frac{\partial u_0}{\partial x} + A_{12} \frac{\partial v_0}{\partial y} - B_{11} \frac{\partial^2 w_0}{\partial x^2} \right)_{x=0} = 0 \tag{7.332}$$

$$(N_{xx})_{x=a} = \left(A_{11} \frac{\partial u_0}{\partial x} + A_{12} \frac{\partial v_0}{\partial y} - B_{11} \frac{\partial^2 w_0}{\partial x^2} \right)_{x=a} = 0 \tag{7.333}$$

$$(N_{yy})_{y=0} = \left(A_{12} \frac{\partial u_0}{\partial x} + A_{11} \frac{\partial v_0}{\partial y} + B_{11} \frac{\partial^2 w_0}{\partial y^2} \right)_{y=0} = 0 \tag{7.334}$$

$$(N_{yy})_{y=b} = \left(A_{12} \frac{\partial u_0}{\partial x} + A_{11} \frac{\partial v_0}{\partial y} + B_{11} \frac{\partial^2 w_0}{\partial y^2} \right)_{y=b} = 0 \tag{7.335}$$

Assumed buckling displacement functions in this case are

$$u_0(x,y) = \sum_{n=1}^{\infty}\sum_{m=1}^{\infty} U_{mn} \cos\frac{m\pi x}{a} \sin\frac{n\pi y}{b} \quad (7.336)$$

$$v_0(x,y) = \sum_{n=1}^{\infty}\sum_{m=1}^{\infty} V_{mn} \sin\frac{m\pi x}{a} \cos\frac{n\pi y}{b} \quad (7.337)$$

$$w_0(x,y) = \sum_{n=1}^{\infty}\sum_{m=1}^{\infty} W_{mn} \sin\frac{m\pi x}{a} \sin\frac{n\pi y}{b} \quad (7.338)$$

It can be seen that both the geometric as well as natural boundary conditions (Equations 7.320 through 7.335) are satisfied by the assumed buckling displacement functions (Equations 7.336 through 7.338).

Substituting the buckling displacement functions in the governing buckling differential equations (Equations 7.317 through 7.319), we can obtain a system of algebraic equations, which can be written in the matrix form as follows:

$$\begin{bmatrix} F_{11} & F_{12} & F_{13} \\ F_{12} & F_{22} & F_{23} \\ F_{13} & F_{23} & F_{33} - \bar{N}_{xx}\left(\frac{m\pi}{a}\right)^2 \end{bmatrix} \begin{Bmatrix} U_{mn} \\ V_{mn} \\ W_{mn} \end{Bmatrix} = \begin{Bmatrix} 0 \\ 0 \\ 0 \end{Bmatrix} \quad (7.339)$$

in which, the terms in the square matrix are given by

$$F_{11} = A_{11}\left(\frac{m\pi}{a}\right)^2 + A_{66}\left(\frac{n\pi}{b}\right)^2 \quad (7.340)$$

$$F_{12} = (A_{12} + A_{66})\left(\frac{m\pi}{a}\right)\left(\frac{n\pi}{b}\right) \quad (7.341)$$

$$F_{13} = -B_{11}\left(\frac{m\pi}{a}\right)^3 \quad (7.342)$$

$$F_{22} = A_{11}\left(\frac{n\pi}{b}\right)^2 + A_{66}\left(\frac{m\pi}{a}\right)^2 \quad (7.343)$$

$$F_{23} = B_{11}\left(\frac{n\pi}{b}\right)^3 \quad (7.344)$$

$$F_{33} = D_{11}\left[\left(\frac{m\pi}{a}\right)^4 + \left(\frac{n\pi}{b}\right)^4\right] + 2(D_{12} + 2D_{66})\left(\frac{m\pi}{a}\right)^2\left(\frac{n\pi}{b}\right)^2 \quad (7.345)$$

For nontrivial solutions, the determinant of the square matrix in Equation 7.339 above is equated to zero, which gives the buckling loads as

$$\bar{N}_{xx} = \left(\frac{a}{m\pi}\right)^2 \left(F_{33} + \frac{2F_{12}F_{23}F_{13} - F_{22}F_{13}^2 - F_{11}F_{23}^2}{F_{11}F_{22} - F_{12}^2}\right) \quad (7.346)$$

7.5.5 Antisymmetric Angle-Ply Laminated Simply Supported Plate under In-Plane Uniaxial Compressive Load: Navier Method for Buckling

In an antisymmetric angle-ply laminate, $A_{16} = A_{26} = D_{16} = D_{26} = B_{11} = B_{12} = B_{22} = B_{66} = 0$. B_{16} and B_{26} are nonzero and extension–bending coupling exists. As a result, the three differential equations of buckling are coupled, which for a plate under uniaxial compressive force resultant $\bar{N}_{xx}$ can be obtained from Equations 7.37 through 7.39, as follows:

$$A_{11}\frac{\partial^2 u_0}{\partial x^2} + A_{66}\frac{\partial^2 u_0}{\partial y^2} + (A_{12} + A_{66})\frac{\partial^2 v_0}{\partial x \partial y} - 3B_{16}\frac{\partial^3 w_0}{\partial x^2 \partial y} - B_{26}\frac{\partial^3 w_0}{\partial y^3} = 0 \quad (7.347)$$

$$(A_{12} + A_{66})\frac{\partial^2 u_0}{\partial x \partial y} + A_{66}\frac{\partial^2 v_0}{\partial x^2} + A_{22}\frac{\partial^2 v_0}{\partial y^2} - B_{16}\frac{\partial^3 w_0}{\partial x^3} - 3B_{26}\frac{\partial^3 w_0}{\partial x \partial y^2} = 0 \quad (7.348)$$

$$-3B_{16}\frac{\partial^3 u_0}{\partial x^2 \partial y} - B_{26}\frac{\partial^3 u_0}{\partial y^3} - B_{16}\frac{\partial^3 v_0}{\partial x^3} - 3B_{26}\frac{\partial^3 v_0}{\partial x \partial y^2} + D_{11}\frac{\partial^4 w_0}{\partial x^4}$$
$$+ 2(D_{12} + 2D_{66})\frac{\partial^4 w_0}{\partial x^2 \partial y^2} + D_{22}\frac{\partial^4 w_0}{\partial y^4} = \bar{N}_{xx}\frac{\partial^2 w_0}{\partial x^2} \quad (7.349)$$

As in the case of antisymmetric cross-ply laminates, in this case, too, the transverse deflections $w_0(x, y)$ are coupled with the in-plane displacements $u_0(x, y)$ and $v_0(x, y)$ and simply supported boundary conditions and clamped boundary conditions are required to be further categorized into S1, S2, S3, S4 and C1, C2, C3, C4, respectively.

Since extension–shear and bending–twisting couplings are not there, that is, $A_{16} = A_{26} = D_{16} = D_{26} = 0$, Navier solution exists for antisymmetric angle-ply laminated rectangular plate with simply supported boundary conditions. We discuss below the solution for simply supported S3 boundary conditions. These boundary conditions are given by

$$(w_0)_{x=0} = 0 \quad (7.350)$$

$$(w_0)_{x=a} = 0 \quad (7.351)$$

$$(w_0)_{y=0} = 0 \quad (7.352)$$

$$(w_0)_{y=b} = 0 \quad (7.353)$$

$$(M_{xx})_{x=0} = \left[B_{16}\left(\frac{\partial u_0}{\partial y} + \frac{\partial v_0}{\partial x}\right) - D_{11}\frac{\partial^2 w_0}{\partial x^2} - D_{12}\frac{\partial^2 w_0}{\partial y^2}\right]_{x=0} = 0 \quad (7.354)$$

Analytical Solutions for Laminated Plates

$$(M_{xx})_{x=a} = \left[B_{16}\left(\frac{\partial u_0}{\partial y} + \frac{\partial v_0}{\partial x}\right) - D_{11}\frac{\partial^2 w_0}{\partial x^2} - D_{12}\frac{\partial^2 w_0}{\partial y^2} \right]_{x=a} = 0 \quad (7.355)$$

$$(M_{yy})_{y=0} = \left[B_{26}\left(\frac{\partial u_0}{\partial y} + \frac{\partial v_0}{\partial x}\right) - D_{12}\frac{\partial^2 w_0}{\partial x^2} - D_{22}\frac{\partial^2 w_0}{\partial y^2} \right]_{y=0} = 0 \quad (7.356)$$

$$(M_{yy})_{y=b} = \left[B_{26}\left(\frac{\partial u_0}{\partial y} + \frac{\partial v_0}{\partial x}\right) - D_{12}\frac{\partial^2 w_0}{\partial x^2} - D_{22}\frac{\partial^2 w_0}{\partial y^2} \right]_{y=b} = 0 \quad (7.357)$$

$$(u_0)_{x=0} = 0 \quad (7.358)$$

$$(u_0)_{x=a} = 0 \quad (7.359)$$

$$(v_0)_{y=0} = 0 \quad (7.360)$$

$$(v_0)_{y=b} = 0 \quad (7.361)$$

$$(N_{xy})_{x=0} = \left[A_{66}\left(\frac{\partial u_0}{\partial y} + \frac{\partial v_0}{\partial x}\right) - B_{16}\frac{\partial^2 w_0}{\partial x^2} - B_{26}\frac{\partial^2 w_0}{\partial y^2} \right]_{x=0} = 0 \quad (7.362)$$

$$(N_{xy})_{x=a} = \left[A_{66}\left(\frac{\partial u_0}{\partial y} + \frac{\partial v_0}{\partial x}\right) - B_{16}\frac{\partial^2 w_0}{\partial x^2} - B_{26}\frac{\partial^2 w_0}{\partial y^2} \right]_{x=a} = 0 \quad (7.363)$$

$$(N_{xy})_{y=0} = \left[A_{66}\left(\frac{\partial u_0}{\partial y} + \frac{\partial v_0}{\partial x}\right) - B_{16}\frac{\partial^2 w_0}{\partial x^2} - B_{26}\frac{\partial^2 w_0}{\partial y^2} \right]_{y=0} = 0 \quad (7.364)$$

$$(N_{xy})_{y=b} = \left[A_{66}\left(\frac{\partial u_0}{\partial y} + \frac{\partial v_0}{\partial x}\right) - B_{16}\frac{\partial^2 w_0}{\partial x^2} - B_{26}\frac{\partial^2 w_0}{\partial y^2} \right]_{y=b} = 0 \quad (7.365)$$

We choose buckling displacement functions that satisfy the geometric boundary conditions. The following are the buckling displacement functions:

$$u_0(x,y) = \sum_{n=1}^{\infty}\sum_{m=1}^{\infty} U_{mn} \sin\frac{m\pi x}{a} \cos\frac{n\pi y}{b} \quad (7.366)$$

$$v_0(x,y) = \sum_{n=1}^{\infty}\sum_{m=1}^{\infty} V_{mn} \cos\frac{m\pi x}{a} \sin\frac{n\pi y}{b} \quad (7.367)$$

$$w_0(x,y) = \sum_{n=1}^{\infty}\sum_{m=1}^{\infty} W_{mn} \sin\frac{m\pi x}{a} \sin\frac{n\pi y}{b} \quad (7.368)$$

It can be seen that both the geometric as well as natural boundary conditions (Equations 7.350 through 7.365) are satisfied by the assumed buckling displacement functions (Equations 7.366 through 7.368).

Substituting the buckling displacement functions in the governing buckling differential equations (Equations 7.347 through 7.349), we can obtain a system of algebraic equations, which can be written in the matrix form as follows:

$$\begin{bmatrix} F_{11} & F_{12} & F_{13} \\ F_{12} & F_{22} & F_{23} \\ F_{13} & F_{23} & F_{33} - \bar{N}_{xx}\left(\dfrac{m\pi}{a}\right)^2 \end{bmatrix} \begin{Bmatrix} U_{mn} \\ V_{mn} \\ W_{mn} \end{Bmatrix} = \begin{Bmatrix} 0 \\ 0 \\ 0 \end{Bmatrix} \quad (7.369)$$

in which, the terms in the square matrix are given by

$$F_{11} = A_{11}\left(\dfrac{m\pi}{a}\right)^2 + A_{66}\left(\dfrac{n\pi}{b}\right)^2 \quad (7.370)$$

$$F_{12} = (A_{12} + A_{66})\left(\dfrac{m\pi}{a}\right)\left(\dfrac{n\pi}{b}\right) \quad (7.371)$$

$$F_{13} = -3B_{16}\left(\dfrac{m\pi}{a}\right)^2\left(\dfrac{n\pi}{b}\right) - B_{26}\left(\dfrac{n\pi}{b}\right)^3 \quad (7.372)$$

$$F_{22} = A_{22}\left(\dfrac{n\pi}{b}\right)^2 + A_{66}\left(\dfrac{m\pi}{a}\right)^2 \quad (7.373)$$

$$F_{23} = -B_{16}\left(\dfrac{m\pi}{a}\right)^3 - 3B_{26}\left(\dfrac{m\pi}{a}\right)\left(\dfrac{n\pi}{b}\right)^2 \quad (7.374)$$

$$F_{33} = D_{11}\left(\dfrac{m\pi}{a}\right)^4 + 2(D_{12} + 2D_{66})\left(\dfrac{m\pi}{a}\right)^2\left(\dfrac{n\pi}{b}\right)^2 + D_{22}\left(\dfrac{n\pi}{b}\right)^4 \quad (7.375)$$

For nontrivial solutions, the determinant of the square matrix in Equation 7.369 is equated to zero, which gives the buckling loads as

$$\bar{N}_{xx} = \left(\dfrac{a}{m\pi}\right)^2 \left(F_{33} + \dfrac{2F_{12}F_{23}F_{13} - F_{22}F_{13}^2 - F_{11}F_{23}^2}{F_{11}F_{22} - F_{12}^2} \right) \quad (7.376)$$

7.6 SOLUTIONS FOR VIBRATION OF LAMINATED PLATES

7.6.1 Specially Orthotropic Simply Supported Plate: Navier Method for Free Vibration

For a specially orthotropic plate, $[B] = 0$ and $A_{16} = A_{26} = D_{16} = D_{26} = 0$. Substituting these in Equations 7.43 through 7.45, the governing differential equations for free transverse vibration of a specially orthotropic plate are obtained as

Analytical Solutions for Laminated Plates

$$A_{11}\frac{\partial^2 u_0}{\partial x^2} + A_{66}\frac{\partial^2 u_0}{\partial y^2} + (A_{12} + A_{66})\frac{\partial^2 v_0}{\partial x \partial y} = 0 \tag{7.377}$$

$$(A_{12} + A_{66})\frac{\partial^2 u_0}{\partial x \partial y} + A_{66}\frac{\partial^2 v_0}{\partial x^2} + A_{22}\frac{\partial^2 v_0}{\partial y^2} = 0 \tag{7.378}$$

$$D_{11}\frac{\partial^4 w_0}{\partial x^4} + 2(D_{12} + 2D_{66})\frac{\partial^4 w_0}{\partial x^2 \partial y^2} + D_{22}\frac{\partial^4 w_0}{\partial y^4} + \rho\frac{\partial^2 w_0}{\partial t^2} = 0 \tag{7.379}$$

As seen, the in-plane displacements during vibration u_0 and v_0 are not coupled with the out-of-plane displacement during vibration w_0. (Note that displacements during vibration, as in the case of buckling, are actually variations in displacements.) As a result, the vibration boundary conditions for a specially orthotropic rectangular plate are rather simple, which are as follows:

$$(w_0)_{x=0} = 0 \tag{7.380}$$

$$(w_0)_{x=a} = 0 \tag{7.381}$$

$$(w_0)_{y=0} = 0 \tag{7.382}$$

$$(w_0)_{y=b} = 0 \tag{7.383}$$

$$(M_{xx})_{x=0} = -\left(D_{11}\frac{\partial^2 w_0}{\partial x^2} + D_{12}\frac{\partial^2 w_0}{\partial y^2}\right)_{x=0} = 0 \tag{7.384}$$

$$(M_{xx})_{x=a} = -\left(D_{11}\frac{\partial^2 w_0}{\partial x^2} + D_{12}\frac{\partial^2 w_0}{\partial y^2}\right)_{x=a} = 0 \tag{7.385}$$

$$(M_{yy})_{y=0} = -\left(D_{12}\frac{\partial^2 w_0}{\partial x^2} + D_{22}\frac{\partial^2 w_0}{\partial y^2}\right)_{y=0} = 0 \tag{7.386}$$

$$(M_{yy})_{y=b} = -\left(D_{12}\frac{\partial^2 w_0}{\partial x^2} + D_{22}\frac{\partial^2 w_0}{\partial y^2}\right)_{y=b} = 0 \tag{7.387}$$

As the in-plane displacements during vibration are uncoupled with the out-of-plane displacements during vibration, the solution for a specially orthotropic plate involves consideration of only the out-of-plane displacement during vibration. The solution can be obtained by adopting a method of separation, as follows:

$$w_0(x,y,t) = W(x,y)T(t) \tag{7.388}$$

and

$$W(x,y) = W_{mn} \sin\frac{m\pi x}{a} \sin\frac{n\pi y}{b} \tag{7.389}$$

$$T(t) = A\cos\omega t + B\sin\omega t \qquad (7.390)$$

Note that Equation 7.388 together with Equations 7.389 and 7.390, satisfy both the geometric as well as the natural boundary conditions. Note further that these equations also satisfy the governing equations. Then, substituting these equations in Equation 7.379, we get

$$\omega_{mn}^2 = \frac{\pi^4}{\rho}\left[D_{11}\left(\frac{m}{a}\right)^4 + 2(D_{12}+2D_{66})\left(\frac{m}{a}\right)^2\left(\frac{n}{b}\right)^2 + D_{22}\left(\frac{n}{b}\right)^4\right] \qquad (7.391)$$

where the suffix mn is added to indicate that, for each mode shape corresponding to each pair of m and n, there exists a natural frequency. The lowest frequency is called the fundamental natural frequency.

7.6.2 Specially Orthotropic Simply Supported Plate: Ritz Solution for Free Vibration

The governing free vibration differential equations and the boundary conditions for a specially orthotropic laminated plate with all edges simply supported were presented in the previous section. Also, as seen in the previous section, Navier solution is available for this class of problems. However, for the sake of demonstration of the procedure and comparison of results, the Ritz method is discussed in this section.

The total potential energy functional is given by

$$\Pi = \frac{1}{2}\int_0^b\int_0^a\left[D_{11}\left(\frac{\partial^2 w_0}{\partial x^2}\right)^2 + D_{22}\left(\frac{\partial^2 w_0}{\partial y^2}\right)^2 + 4D_{66}\left(\frac{\partial^2 w_0}{\partial x\partial y}\right)^2 \right.$$
$$\left. + 2D_{12}\left(\frac{\partial^2 w_0}{\partial x^2}\right)\left(\frac{\partial^2 w_0}{\partial y^2}\right) - \rho\left(\frac{\partial w_0}{\partial t}\right)^2\right]dx\,dy \qquad (7.392)$$

The spatial coordinates and time in the transverse displacement during vibration are separated by considering a solution as follows:

$$w_0(x,y,t) = W(x,y)T(t) \qquad (7.393)$$

and

$$W(x,y) = \sum_{n=1}^{\infty}\sum_{m=1}^{\infty} W_{mn}\sin\frac{m\pi x}{a}\sin\frac{n\pi y}{b} \qquad (7.394)$$

$$T(t) = e^{\omega t} \qquad (7.395)$$

The above equations satisfy the geometric boundary conditions as well as the natural boundary conditions given in the previous section. Substituting the approximation function for vibration displacement from Equation 7.393 together with Equations 7.394 and 7.395 in the total potential energy functional in Equation 7.392

Analytical Solutions for Laminated Plates

and utilizing the integration identities in Equations 7.115 and 7.116, we can carry out direct integration and show that the integration terms in Equation 7.392 are as follows:

$$\int_0^b \int_0^a \left(\frac{\partial^2 w_0}{\partial x^2}\right)^2 dx\,dy = \frac{ab}{4}T^2(t)\sum_{n=1}^{N}\sum_{m=1}^{M}\left(\frac{m\pi}{a}\right)^4 W_{mn}^2 \qquad (7.396)$$

$$\int_0^b \int_0^a \left(\frac{\partial^2 w_0}{\partial y^2}\right)^2 dx\,dy = \frac{ab}{4}T^2(t)\sum_{n=1}^{N}\sum_{m=1}^{M}\left(\frac{n\pi}{b}\right)^4 W_{mn}^2 \qquad (7.397)$$

$$\int_0^b \int_0^a \left(\frac{\partial^2 w_0}{\partial x^2}\right)\left(\frac{\partial^2 w_0}{\partial y^2}\right) dx\,dy = \frac{ab}{4}T^2(t)\sum_{n=1}^{N}\sum_{m=1}^{M}\left(\frac{m\pi}{a}\right)^2\left(\frac{n\pi}{b}\right)^2 W_{mn}^2 \qquad (7.398)$$

$$\int_0^b \int_0^a \left(\frac{\partial^2 w_0}{\partial x \partial y}\right)^2 dx\,dy = \frac{ab}{4}T^2(t)\sum_{n=1}^{N}\sum_{m=1}^{M}\left(\frac{m\pi}{a}\right)^2\left(\frac{n\pi}{b}\right)^2 W_{mn}^2 \qquad (7.399)$$

$$\int_0^b \int_0^a \left(\frac{\partial w_0}{\partial t}\right)^2 dx\,dy = \frac{ab}{4}\omega^2 T^2(t)\sum_{n=1}^{N}\sum_{m=1}^{M} W_{mn}^2 \qquad (7.400)$$

Then, substituting Equations 7.396 through 7.400 in Equation 7.392, the potential energy functional is obtained as

$$\Pi = \frac{ab}{8}T^2(t)\sum_{n=1}^{N}\sum_{m=1}^{M} W_{mn}^2\left[D_{11}\left(\frac{m\pi}{a}\right)^4 + D_{22}\left(\frac{n\pi}{b}\right)^4 + 2(D_{12}+2D_{66})\left(\frac{m\pi}{a}\right)^2\left(\frac{n\pi}{b}\right)^2 - \rho\omega_{mn}^2\right]$$

(7.401)

For dynamic equilibrium during vibration,

$$\frac{\partial \Pi}{\partial W_{mn}} = 0 \qquad (7.402)$$

It results in $M \times N$ algebraic equations, each corresponding to a unique pair of m and n. Each of these equations gives us a value of ω_{mn}, of which the minimum one is the natural frequency. We note that here, ω_{mn} is given by

$$\omega_{mn}^2 = \frac{\pi^4}{\rho}\left[D_{11}\left(\frac{m}{a}\right)^4 + 2(D_{12}+2D_{66})\left(\frac{m}{a}\right)^2\left(\frac{n}{b}\right)^2 + D_{22}\left(\frac{n}{b}\right)^4\right] \qquad (7.403)$$

which is identical with Equation 7.391. In other words, Ritz solution is an exact solution as Navier solution for transverse free vibration of specially orthotropic simply supported plate.

7.6.3 Symmetric Angle-Ply Laminated Plate with All Four Edges Simply Supported: Ritz Method for Free Vibration

In a symmetric angle-ply laminate, $[B] = 0$, that is, there is no extension–bending coupling. However, the extensional stiffness matrix $[A]$ and the bending stiffness matrix $[D]$ are fully populated. Then, governing equations for free vibration (Equations 7.43 through 7.45) reduce to

$$A_{11}\frac{\partial^2 u_0}{\partial x^2} + 2A_{16}\frac{\partial^2 u_0}{\partial x \partial y} + A_{66}\frac{\partial^2 u_0}{\partial y^2} + A_{16}\frac{\partial^2 v_0}{\partial x^2} + (A_{12}+A_{66})\frac{\partial^2 v_0}{\partial x \partial y} + A_{26}\frac{\partial^2 v_0}{\partial y^2} = 0 \tag{7.404}$$

$$A_{16}\frac{\partial^2 u_0}{\partial x^2} + (A_{12}+A_{66})\frac{\partial^2 u_0}{\partial x \partial y} + A_{26}\frac{\partial^2 u_0}{\partial y^2} + A_{66}\frac{\partial^2 v_0}{\partial x^2} + 2A_{26}\frac{\partial^2 v_0}{\partial x \partial y} + A_{22}\frac{\partial^2 v_0}{\partial y^2} = 0 \tag{7.405}$$

$$D_{11}\frac{\partial^4 w_0}{\partial x^4} + D_{22}\frac{\partial^4 w_0}{\partial y^4} + 2(D_{12}+2D_{66})\frac{\partial^4 w_0}{\partial x^2 \partial y^2} + 4D_{16}\frac{\partial^4 w_0}{\partial x^3 \partial y} + 4D_{26}\frac{\partial^4 w_0}{\partial x \partial y^3} + \rho \frac{\partial^2 w_0}{\partial t^2} = 0 \tag{7.406}$$

Simply supported boundary conditions as well as clamped boundary conditions cannot be further distinguished, based on the nature of the in-plane boundary conditions on u_0 and v_0, into S1, S2, S3, S4 and C1, C2, C3, C4, respectively. Thus, for a plate with all four edges simply supported, the boundary conditions are

$$(w_0)_{x=0} = 0 \tag{7.407}$$

$$(w_0)_{x=a} = 0 \tag{7.408}$$

$$(w_0)_{y=0} = 0 \tag{7.409}$$

$$(w_0)_{y=b} = 0 \tag{7.410}$$

$$(M_{xx})_{x=0} = -\left(D_{11}\frac{\partial^2 w_0}{\partial x^2} + D_{12}\frac{\partial^2 w_0}{\partial y^2} + 2D_{16}\frac{\partial^2 w_0}{\partial x \partial y}\right)_{x=0} = 0 \tag{7.411}$$

$$(M_{xx})_{x=a} = -\left(D_{11}\frac{\partial^2 w_0}{\partial x^2} + D_{12}\frac{\partial^2 w_0}{\partial y^2} + 2D_{16}\frac{\partial^2 w_0}{\partial x \partial y}\right)_{x=a} = 0 \tag{7.412}$$

$$(M_{yy})_{y=0} = -\left(D_{12}\frac{\partial^2 w_0}{\partial x^2} + D_{22}\frac{\partial^2 w_0}{\partial y^2} + 2D_{26}\frac{\partial^2 w_0}{\partial x \partial y}\right)_{y=0} = 0 \tag{7.413}$$

$$(M_{yy})_{y=b} = -\left(D_{12}\frac{\partial^2 w_0}{\partial x^2} + D_{22}\frac{\partial^2 w_0}{\partial y^2} + 2D_{26}\frac{\partial^2 w_0}{\partial x \partial y}\right)_{y=b} = 0 \tag{7.414}$$

From the governing equations as well as the boundary conditions of a symmetric angle-ply laminated plate, we see that the variations in in-plane displacements are uncoupled with the variations in out-of-plane displacements. As a result, the solution involves consideration of only the variation in out-of-plane displacement. However, owing to the presence of the terms D_{16} and D_{26}, Navier type closed-form solutions do not exist. Similar to the cases in bending and buckling, the problem, however, can be solved by other methods such as the Ritz method.

Now, the total potential energy functional is given by

$$\Pi = \frac{1}{2} \int_0^b \int_0^a \left[D_{11}\left(\frac{\partial^2 w_0}{\partial x^2}\right)^2 + 4D_{16}\left(\frac{\partial^2 w_0}{\partial x^2}\right)\left(\frac{\partial^2 w_0}{\partial x \partial y}\right) + 2D_{12}\left(\frac{\partial^2 w_0}{\partial x^2}\right)\left(\frac{\partial^2 w_0}{\partial y^2}\right) \right.$$
$$\left. + 4D_{66}\left(\frac{\partial^2 w_0}{\partial x \partial y}\right)^2 + 4D_{26}\left(\frac{\partial^2 w_0}{\partial x \partial y}\right)\left(\frac{\partial^2 w_0}{\partial y^2}\right) + D_{22}\left(\frac{\partial^2 w_0}{\partial y^2}\right)^2 - \rho\left(\frac{\partial w_0}{\partial t}\right)^2 \right] dx\, dy \quad (7.415)$$

The spatial coordinates and time in the transverse displacement during vibration are separated by considering a solution as follows:

$$w_0(x, y, t) = W(x, y) T(t) \quad (7.416)$$

and

$$W(x, y) = \sum_{n=1}^{\infty} \sum_{m=1}^{\infty} W_{mn} \sin\frac{m\pi x}{a} \sin\frac{n\pi y}{b} \quad (7.417)$$

$$T(t) = e^{\omega t} \quad (7.418)$$

The above equation satisfies the geometric boundary conditions (Equations 7.407 through 7.410) but not the natural boundary conditions (Equations 7.411 through 7.414). (Note that the Ritz method requires the approximation function to satisfy the geometric boundary conditions but not necessarily the natural boundary conditions.)

Equation 7.416, together with Equations 7.417 and 7.418, is then substituted in Equation 7.415 and direct integration is carried out. However, owing to the presence of the stiffness parameters D_{16} and D_{26} in the expression for the potential energy functional, integration identity given by Equation 7.117 is also required to obtain the terms associated with D_{16} and D_{26}. Subsequent procedure is rather complicated and computation in a computerized environment is needed, in which employing minimum potential energy principle, we can obtain $M \times N$ algebraic equations, each corresponding to a unique pair of m and n. Each of these equations gives us a value of ω_{mn}, of which the minimum one is the natural frequency.

7.6.4 Antisymmetric Cross-Ply Laminated Simply Supported Plate: Navier Method for Free Vibration

In an antisymmetric cross-ply laminate, $A_{11} = A_{22}$, $D_{11} = D_{22}$, and $A_{16} = A_{26} = D_{16} = D_{26} = 0$. Also, $B_{ij} = 0$ except $B_{11} = -B_{22} \neq 0$, that is, extension–bending coupling exists. The three differential equations (Equations 7.43 through 7.45) take the following forms for free transverse vibration of plate:

$$A_{11}\frac{\partial^2 u_0}{\partial x^2} + A_{66}\frac{\partial^2 u_0}{\partial y^2} + (A_{12} + A_{66})\frac{\partial^2 v_0}{\partial x \partial y} - B_{11}\frac{\partial^3 w_0}{\partial x^3} = 0 \quad (7.419)$$

$$(A_{12} + A_{66})\frac{\partial^2 u_0}{\partial x \partial y} + A_{66}\frac{\partial^2 v_0}{\partial x^2} + A_{11}\frac{\partial^2 v_0}{\partial y^2} + B_{11}\frac{\partial^3 w_0}{\partial y^3} = 0 \quad (7.420)$$

$$B_{11}\left(\frac{\partial^3 u_0}{\partial x^3} - \frac{\partial^3 v_0}{\partial y^3}\right) - D_{11}\left(\frac{\partial^4 w_0}{\partial x^4} + \frac{\partial^4 w_0}{\partial y^4}\right) - 2(D_{12} + 2D_{66})\frac{\partial^4 w_0}{\partial x^2 \partial y^2} = \rho\frac{\partial^2 w_0}{\partial t^2} \quad (7.421)$$

Transverse deflections $w_0(x, y)$ are coupled with the in-plane displacements $u_0(x, y)$ and $v_0(x, y)$ and simply supported boundary conditions and clamped boundary conditions are required to be further categorized into S1, S2, S3, S4 and C1, C2, C3, C4, respectively.

Navier solution exists for antisymmetric cross-ply laminated rectangular plate with simply supported boundary conditions S2. These boundary conditions are given by

$$(w_0)_{x=0} = 0 \quad (7.422)$$

$$(w_0)_{x=a} = 0 \quad (7.423)$$

$$(w_0)_{y=0} = 0 \quad (7.424)$$

$$(w_0)_{y=b} = 0 \quad (7.425)$$

$$(M_{xx})_{x=0} = \left(B_{11}\frac{\partial u_0}{\partial x} - D_{11}\frac{\partial^2 w_0}{\partial x^2} - D_{12}\frac{\partial^2 w_0}{\partial y^2}\right)_{x=0} = 0 \quad (7.426)$$

$$(M_{xx})_{x=a} = \left(B_{11}\frac{\partial u_0}{\partial x} - D_{11}\frac{\partial^2 w_0}{\partial x^2} - D_{12}\frac{\partial^2 w_0}{\partial y^2}\right)_{x=a} = 0 \quad (7.427)$$

$$(M_{yy})_{y=0} = \left(B_{22}\frac{\partial v_0}{\partial y} - D_{12}\frac{\partial^2 w_0}{\partial x^2} - D_{22}\frac{\partial^2 w_0}{\partial y^2}\right)_{y=0} = 0 \quad (7.428)$$

$$(M_{yy})_{y=b} = \left(B_{22}\frac{\partial v_0}{\partial y} - D_{12}\frac{\partial^2 w_0}{\partial x^2} - D_{22}\frac{\partial^2 w_0}{\partial y^2}\right)_{y=b} = 0 \quad (7.429)$$

$$(v_0)_{x=0} = 0 \quad (7.430)$$

$$(v_0)_{x=a} = 0 \quad (7.431)$$

$$(u_0)_{y=0} = 0 \quad (7.432)$$

$$(u_0)_{y=b} = 0 \quad (7.433)$$

$$(N_{xx})_{x=0} = \left(A_{11}\frac{\partial u_0}{\partial x} + A_{12}\frac{\partial v_0}{\partial y} - B_{11}\frac{\partial^2 w_0}{\partial x^2}\right)_{x=0} = 0 \quad (7.434)$$

$$(N_{xx})_{x=a} = \left(A_{11}\frac{\partial u_0}{\partial x} + A_{12}\frac{\partial v_0}{\partial y} - B_{11}\frac{\partial^2 w_0}{\partial x^2}\right)_{x=a} = 0 \quad (7.435)$$

$$(N_{yy})_{y=0} = \left(A_{12}\frac{\partial u_0}{\partial x} + A_{11}\frac{\partial v_0}{\partial y} + B_{11}\frac{\partial^2 w_0}{\partial y^2}\right)_{y=0} = 0 \quad (7.436)$$

$$(N_{yy})_{y=b} = \left(A_{12}\frac{\partial u_0}{\partial x} + A_{11}\frac{\partial v_0}{\partial y} + B_{11}\frac{\partial^2 w_0}{\partial y^2}\right)_{y=b} = 0 \quad (7.437)$$

The solution can be obtained by adopting a method of separation, in which the displacements are assumed as follows:

$$u_0(x,y,t) = U(x,y)T(t) \quad (7.438)$$

$$v_0(x,y,t) = V(x,y)T(t) \quad (7.439)$$

$$w_0(x,y,t) = W(x,y)T(t) \quad (7.440)$$

and

$$U(x,y) = U_{mn}\cos\frac{m\pi x}{a}\sin\frac{n\pi y}{b} \quad (7.441)$$

$$V(x,y) = V_{mn}\sin\frac{m\pi x}{a}\cos\frac{n\pi y}{b} \quad (7.442)$$

$$W(x,y) = W_{mn}\sin\frac{m\pi x}{a}\sin\frac{n\pi y}{b} \quad (7.443)$$

$$T(t) = A\cos\omega t + B\sin\omega t \quad (7.444)$$

It can be seen that both the geometric as well as natural boundary conditions (Equations 7.422 through 7.437) are satisfied by the assumed vibration displacement functions (Equations 7.438 through 7.444). Further, these equations also satisfy the governing equations. Then, substituting these equations in Equations 7.419 through 7.421, we get a system of algebraic equations, which can be written in the matrix form as follows:

$$\begin{bmatrix} F_{11} & F_{12} & F_{13} \\ F_{12} & F_{22} & F_{23} \\ F_{13} & F_{23} & F_{33} - \rho\omega_{mn}^2 \end{bmatrix} \begin{Bmatrix} U_{mn} \\ V_{mn} \\ W_{mn} \end{Bmatrix} = \begin{Bmatrix} 0 \\ 0 \\ 0 \end{Bmatrix} \quad (7.445)$$

in which, the terms in the square matrix are given by

$$F_{11} = A_{11}\left(\frac{m\pi}{a}\right)^2 + A_{66}\left(\frac{n\pi}{b}\right)^2 \quad (7.446)$$

$$F_{12} = (A_{12} + A_{66})\left(\frac{m\pi}{a}\right)\left(\frac{n\pi}{b}\right) \tag{7.447}$$

$$F_{13} = -B_{11}\left(\frac{m\pi}{a}\right)^3 \tag{7.448}$$

$$F_{22} = A_{11}\left(\frac{n\pi}{b}\right)^2 + A_{66}\left(\frac{m\pi}{a}\right)^2 \tag{7.449}$$

$$F_{23} = B_{11}\left(\frac{n\pi}{b}\right)^3 \tag{7.450}$$

$$F_{33} = D_{11}\left[\left(\frac{m\pi}{a}\right)^4 + \left(\frac{n\pi}{b}\right)^4\right] + 2(D_{12} + 2D_{66})\left(\frac{m\pi}{a}\right)^2\left(\frac{n\pi}{b}\right)^2 \tag{7.451}$$

For nontrivial solutions, the determinant of the square matrix in Equation 7.445 is equated to zero, which gives

$$\omega_{mn}^2 = \frac{1}{\rho}\left(F_{33} + \frac{2F_{12}F_{23}F_{13} - F_{22}F_{13}^2 - F_{11}F_{23}^2}{F_{11}F_{22} - F_{12}^2}\right) \tag{7.452}$$

As stated previously, the suffix mn indicates that, for each mode shape corresponding to each pair of m and n, there exists a natural frequency; the fundamental natural frequency is the minimum of the discrete frequencies obtained from various combinations of m and n. (Note that fundamental frequency need not necessarily correspond to $m = n = 1$.)

7.6.5 Antisymmetric Angle-Ply Laminated Simply Supported Plate: Navier Method for Free Vibration

In an antisymmetric angle-ply laminate, $A_{16} = A_{26} = D_{16} = D_{26} = B_{11} = B_{12} = B_{22} = B_{66} = 0$. B_{16} and B_{26} are nonzero and extension–bending coupling exists. As a result, the three differential equations of free vibration are coupled and they are obtained from Equations 7.43 through 7.45 as follows:

$$A_{11}\frac{\partial^2 u_0}{\partial x^2} + A_{66}\frac{\partial^2 u_0}{\partial y^2} + (A_{12} + A_{66})\frac{\partial^2 v_0}{\partial x \partial y} - 3B_{16}\frac{\partial^3 w_0}{\partial x^2 \partial y} - B_{26}\frac{\partial^3 w_0}{\partial y^3} = 0 \tag{7.453}$$

$$(A_{12} + A_{66})\frac{\partial^2 u_0}{\partial x \partial y} + A_{66}\frac{\partial^2 v_0}{\partial x^2} + A_{22}\frac{\partial^2 v_0}{\partial y^2} - B_{16}\frac{\partial^3 w_0}{\partial x^3} - 3B_{26}\frac{\partial^3 w_0}{\partial x \partial y^2} = 0 \tag{7.454}$$

$$3B_{16}\frac{\partial^3 u_0}{\partial x^2 \partial y} + B_{26}\frac{\partial^3 u_0}{\partial y^3} + B_{16}\frac{\partial^3 v_0}{\partial x^3} + 3B_{26}\frac{\partial^3 v_0}{\partial x \partial y^2} - D_{11}\frac{\partial^4 w_0}{\partial x^4}$$

$$- D_{22}\frac{\partial^4 w_0}{\partial y^4} - 2(D_{12} + 2D_{66})\frac{\partial^4 w_0}{\partial x^2 \partial y^2} = \rho\frac{\partial^2 w_0}{\partial t^2} \tag{7.455}$$

Analytical Solutions for Laminated Plates

As in the case of antisymmetric cross-ply laminates, in this case, too, the transverse deflections $w_0(x, y)$ are coupled with the in-plane displacements $u_0(x, y)$ and $v_0(x, y)$ and simply supported boundary conditions and clamped boundary conditions are required to be further categorized into $S1$, $S2$, $S3$, $S4$ and $C1$, $C2$, $C3$, $C4$, respectively.

Since extension–shear and bending–twisting couplings are not there, that is, $A_{16} = A_{26} = D_{16} = D_{26} = 0$, Navier solution exists for antisymmetric angle-ply laminated rectangular plate with simply supported boundary conditions. We discuss below the solution for simply supported $S3$ boundary conditions. These boundary conditions are given by

$$(w_0)_{x=0} = 0 \tag{7.456}$$

$$(w_0)_{x=a} = 0 \tag{7.457}$$

$$(w_0)_{y=0} = 0 \tag{7.458}$$

$$(w_0)_{y=b} = 0 \tag{7.459}$$

$$(M_{xx})_{x=0} = \left[B_{16} \left(\frac{\partial u_0}{\partial y} + \frac{\partial v_0}{\partial x} \right) - D_{11} \frac{\partial^2 w_0}{\partial x^2} - D_{12} \frac{\partial^2 w_0}{\partial y^2} \right]_{x=0} = 0 \tag{7.460}$$

$$(M_{xx})_{x=a} = \left[B_{16} \left(\frac{\partial u_0}{\partial y} + \frac{\partial v_0}{\partial x} \right) - D_{11} \frac{\partial^2 w_0}{\partial x^2} - D_{12} \frac{\partial^2 w_0}{\partial y^2} \right]_{x=a} = 0 \tag{7.461}$$

$$(M_{yy})_{y=0} = \left[B_{26} \left(\frac{\partial u_0}{\partial y} + \frac{\partial v_0}{\partial x} \right) - D_{12} \frac{\partial^2 w_0}{\partial x^2} - D_{22} \frac{\partial^2 w_0}{\partial y^2} \right]_{y=0} = 0 \tag{7.462}$$

$$(M_{yy})_{y=b} = \left[B_{26} \left(\frac{\partial u_0}{\partial y} + \frac{\partial v_0}{\partial x} \right) - D_{12} \frac{\partial^2 w_0}{\partial x^2} - D_{22} \frac{\partial^2 w_0}{\partial y^2} \right]_{y=b} = 0 \tag{7.463}$$

$$(u_0)_{x=0} = 0 \tag{7.464}$$

$$(u_0)_{x=a} = 0 \tag{7.465}$$

$$(v_0)_{y=0} = 0 \tag{7.466}$$

$$(v_0)_{y=b} = 0 \tag{7.467}$$

$$(N_{xy})_{x=0} = \left[A_{66} \left(\frac{\partial u_0}{\partial y} + \frac{\partial v_0}{\partial x} \right) - B_{16} \frac{\partial^2 w_0}{\partial x^2} - B_{26} \frac{\partial^2 w_0}{\partial y^2} \right]_{x=0} = 0 \tag{7.468}$$

$$(N_{xy})_{x=a} = \left[A_{66} \left(\frac{\partial u_0}{\partial y} + \frac{\partial v_0}{\partial x} \right) - B_{16} \frac{\partial^2 w_0}{\partial x^2} - B_{26} \frac{\partial^2 w_0}{\partial y^2} \right]_{x=a} = 0 \tag{7.469}$$

$$(N_{xy})_{y=0} = \left[A_{66}\left(\frac{\partial u_0}{\partial y} + \frac{\partial v_0}{\partial x}\right) - B_{16}\frac{\partial^2 w_0}{\partial x^2} - B_{26}\frac{\partial^2 w_0}{\partial y^2}\right]_{y=0} = 0 \quad (7.470)$$

$$(N_{xy})_{y=b} = \left[A_{66}\left(\frac{\partial u_0}{\partial y} + \frac{\partial v_0}{\partial x}\right) - B_{16}\frac{\partial^2 w_0}{\partial x^2} - B_{26}\frac{\partial^2 w_0}{\partial y^2}\right]_{y=b} = 0 \quad (7.471)$$

The solution can be obtained by adopting a method of separation, in which the displacements are assumed as follows:

$$u_0(x,y,t) = U(x,y)T(t) \quad (7.472)$$

$$v_0(x,y,t) = V(x,y)T(t) \quad (7.473)$$

$$w_0(x,y,t) = W(x,y)T(t) \quad (7.474)$$

and

$$U(x,y) = U_{mn} \sin\frac{m\pi x}{a} \cos\frac{n\pi y}{b} \quad (7.475)$$

$$V(x,y) = V_{mn} \cos\frac{m\pi x}{a} \sin\frac{n\pi y}{b} \quad (7.476)$$

$$W(x,y) = W_{mn} \sin\frac{m\pi x}{a} \sin\frac{n\pi y}{b} \quad (7.477)$$

$$T(t) = A\cos\omega t + B\sin\omega t \quad (7.478)$$

It can be seen that both the geometric as well as natural boundary conditions (Equations 7.456 through 7.471) are satisfied by the assumed vibration displacement functions (Equations 7.472 through 7.478).

Substituting the assumed vibration displacement functions in the governing vibration differential equations (Equations 7.453 through 7.455), we can obtain a system of algebraic equations, which can be written in the matrix form as follows:

$$\begin{bmatrix} F_{11} & F_{12} & F_{13} \\ F_{12} & F_{22} & F_{23} \\ F_{13} & F_{23} & F_{33} - \rho\omega_{mn}^2 \end{bmatrix} \begin{Bmatrix} U_{mn} \\ V_{mn} \\ W_{mn} \end{Bmatrix} = \begin{Bmatrix} 0 \\ 0 \\ 0 \end{Bmatrix} \quad (7.479)$$

in which, the terms in the square matrix are given by

$$F_{11} = A_{11}\left(\frac{m\pi}{a}\right)^2 + A_{66}\left(\frac{n\pi}{b}\right)^2 \quad (7.480)$$

$$F_{12} = (A_{12} + A_{66})\left(\frac{m\pi}{a}\right)\left(\frac{n\pi}{b}\right) \quad (7.481)$$

$$F_{13} = -3B_{16}\left(\frac{m\pi}{a}\right)^2\left(\frac{n\pi}{b}\right) - B_{26}\left(\frac{n\pi}{b}\right)^3 \quad (7.482)$$

$$F_{22} = A_{22}\left(\frac{n\pi}{b}\right)^2 + A_{66}\left(\frac{m\pi}{a}\right)^2 \quad (7.483)$$

$$F_{23} = -B_{16}\left(\frac{m\pi}{a}\right)^3 - 3B_{26}\left(\frac{m\pi}{a}\right)\left(\frac{n\pi}{b}\right)^2 \quad (7.484)$$

$$F_{33} = D_{11}\left(\frac{m\pi}{a}\right)^4 + 2(D_{12} + 2D_{66})\left(\frac{m\pi}{a}\right)^2\left(\frac{n\pi}{b}\right)^2 + D_{22}\left(\frac{n\pi}{b}\right)^4 \quad (7.485)$$

For nontrivial solutions, the determinant of the square matrix in Equation 7.479 above is equated to zero, which gives us the natural frequencies as follows:

$$\omega_{mn}^2 = \frac{1}{\rho}\left(F_{33} + \frac{2F_{12}F_{23}F_{13} - F_{22}F_{13}^2 - F_{11}F_{23}^2}{F_{11}F_{22} - F_{12}^2}\right) \quad (7.486)$$

7.7 SUMMARY

In this chapter, we reviewed the analytical solutions for three broad classes of plate problems—bending, buckling, and vibration. The basic principles of solid mechanics and macromechanics of laminated composites are extended and governing equations are developed for bending, buckling, and vibration of laminated composite rectangular plates. Some of the key points to be noted are

- Analytical treatment of a laminated plate is based on the basic assumptions and restrictions of CLPT.
- Plate bending is an equilibrium problem and the governing equations are derived by considering static equilibrium of a differential plate element.
- The governing equations for buckling of a laminated composite plate are expressed in terms of buckling displacements, that is, variations in displacements w.r.t. the prebuckled equilibrium configuration; it is an eigenvalue problem.
- Free vibration of a composite plate is also an eigenvalue problem; the governing equations in this case can be derived by considering dynamic equilibrium of a differential plate element.
- The boundary conditions in a laminated composite plate are more complex due to the anisotropic nature of the material system.
- Various analytical solution methods are available. The Navier method and the Levy method are two of the exact methods; but they have limited applicability. The Ritz method is an approximate energy-based method and it has wide applicability.

EXERCISE PROBLEMS

7.1 Consider a rectangular plate simply supported on all the four edges under a uniformly distributed load of $4 \times 10^{-3}\,\text{N/mm}^2$. The size of the plate is 400 mm × 300 mm × 3 mm and its ply sequence is [0°/90°/0°] where each ply is of equal thickness. Determine at the center of the plate (i) middle surface deflection, (ii) in-plane stresses on the bottom face, and (iii) interlaminar stresses at the interface between the bottom and middle plies.

Use Navier method and restrict your calculations to the first term, that is, $m = n = 1$. Assume the following material data:

$$E_1 = 150 \text{ GPa}, E_2 = 8 \text{ GPa}, G_{12} = 4 \text{ GPa, and } \nu_{12} = 0.2$$

7.2 Write a MATLAB/C/C++ code for determining the following:

  i. Middle surface deflection at any point, $w_0(x, y)$
  ii. In-plane stresses at any point, $\sigma_{xx}(x, y, z)$, $\sigma_{yy}(x, y, z)$, and $\tau_{xy}(x, y, z)$
  iii. Interlaminar stresses at any point, $\tau_{xz}(x, y, z)$, $\tau_{yz}(x, y, z)$, and $\sigma_{zz}(x, y, z)$

Use the following as input data:

  i. Plate dimensions, a and b
  ii. Material data, E_1, E_2, G_{12}, and ν_{12}
  iii. Ply sequence giving details of ply orientation, θ and ply thickness, t
  iv. Uniformly distributed load, q_0

7.3 Use the code developed in Exercise 7.2 and solve the problem in Exercise 7.1. Compare the results with $m = n = 1$ with those obtained in Exercise 7.1. Update the results with higher values of m and n till convergence is achieved.

7.4 Consider the plate in Exercise 7.1 and determine, using the code developed in Exercise 7.2, the middle surface deflections in the plate in Exercise 7.1 at regular intervals. Draw "w_0 versus x" along $y = 150$ mm and "w_0 versus y" along $x = 200$ mm plots.

7.5 Equation 7.180 gives us the constants A_m, B_m, C_m, and D_m in the Levy method for bending of a specially orthotropic plate simply supported on edges $x = 0$ and $x = a$ and fixed on edges $y = 0$ and $y = b$. Derive the equation if the plate is simply supported on all the edges.

Hint: Substitute $(w_0)_{y=0} = (w_0)_{y=b} = (M_{yy})_{y=0} = (M_{yy})_{y=b}$ in Equation 7.174.

7.6 Write a MATLAB/C/C++ code in line with the one in Exercise 7.2 for Levy method.

7.7 Solve the problem in Exercise 7.1 by Levy method.

Hint: Consider edges $x = 0$ and $x = a$ as simply supported and $y = 0$ and $y = b$ as arbitrary. Consider the edges $y = 0$ and $y = b$ as simply supported for determining the constants A_m, B_m, C_m, and D_m.

7.8 Write a MATLAB/C/C++ code to determine middle surface deflection of a symmetric angle-ply laminated plate in bending by the Ritz method. Use input data as in Exercise 7.2.

Hint: Substitute Equation 7.219 in Equations 7.216 and 7.217. Differentiate $\Pi = U + V$ w.r.t. W_{mn} and obtain a system of $M \times N$ linear simultaneous equations by setting

$$\frac{\partial \Pi}{\partial W_{mn}} = 0$$

7.9 Consider a rectangular plate simply supported on all the four edges under a uniformly distributed load of 4×10^{-3} N/mm². The size of the plate is 400 mm × 300 mm × 4 mm and its ply sequence is [0°/45°/−45°/0°] where each ply is of equal thickness. Determine the middle surface deflection at the center of the plate. Use Navier method and restrict your calculations to the first term, that is, $m = n = 1$. Assume the following material data

$$E_1 = 150 \text{ GPa}, E_2 = 8 \text{ GPa}, G_{12} = 4 \text{ GPa, and } \nu_{12} = 0.2.$$

7.10 Write a MATLAB/C/C++ code, in line with the one in Exercise 7.2, for middle surface deflection of an antisymmetric cross-ply laminated plate simply supported on all edges. Solve the problem in Exercise 7.9 for $m = n = 1$ and compare the results with Exercise 7.9. Update the results with higher values of m and n till convergence is achieved.

7.11 Consider the specially orthotropic plate in Exercise 7.1. Determine its critical buckling load by Navier method.

7.12 Consider the antisymmetric laminate in Exercise 7.9. Determine its critical buckling load by Navier method.

7.13 Consider an antisymmetric angle-ply rectangular laminated plate simply supported on all the four edges. The size of the plate is 600 mm × 500 mm × 8 mm and its ply sequence is [0°/30°/60°/90°/90°/−60°/−30°/0°] where each ply is of equal thickness. Determine its critical buckling load $\bar{N}_{xx}$ using Navier method. Assume the following material data:

$$E_1 = 150 \text{ GPa}, E_2 = 8 \text{ GPa}, G_{12} = 4 \text{ GPa, and } \nu_{12} = 0.2.$$

Hint: Manual computation will be tedious; write a MATLAB/C/C++ code for determining the [A], [B], and [D] matrices.

7.14 Consider the problem in Exercise 7.13. Determine the critical buckling load $\bar{N}_{yy}$ using Navier load.

Hint: Rewrite the ply sequence by rotating the coordinate axis by 90°.

7.15 Consider a specially orthotropic laminated plate simply supported on all the four edges. The size of the plate is 600 mm × 500 mm × 8 mm and its ply sequence is $[0°/90°/0°/90°]_s$ where each ply is of equal thickness. Determine its natural frequency corresponding to $m = n = 1$ using Navier method. Assume the following material data:

$$E_1 = 150 \text{ GPa}, E_2 = 8 \text{ GPa}, G_{12} = 4 \text{ GPa}, \nu_{12} = 0.2, \text{ and } \rho = 1.52 \text{ g/cm}^3.$$

7.16 Write a MATLAB/C/C++ code for computing the natural frequencies of a symmetric angle-ply laminated plate by Ritz method. Use the following as input data:

  i. Plate dimensions, a and b
  ii. Material data, E_1, E_2, G_{12}, ν_{12}, and ρ
  iii. Ply sequence giving details of ply orientation, θ and ply thickness, t

Hint: Substitute Equations 7.416 through 7.418 in Equation 7.415 and obtain a system of $M \times N$ algebraic equations by setting

$$\frac{\partial \Pi}{\partial W_{mn}} = 0$$

7.17 Consider the antisymmetric angle-ply laminated plate in Exercise 7.13. Determine its natural frequency corresponding to $m = n = 1$ using Navier method. Assume density of composite $\rho = 1.52$ g/cm³.

REFERENCES AND SUGGESTED READING

1. S. P. Timoshenko and S. Woinnwsky-Krieger, *Theory of Plates and Shells*, second edition, McGraw-Hill, International Edition, 1959.
2. E. Ventsel and T. Krauthammer, *Thin Plates and Shells—Theory, Analysis and Applications*, Marcel Dekker, New York, 2001.
3. J. M. Whitney, *Structural Analysis of Laminated Anisotropic Plates*, Technomic Publishing Company, Lancaster, 1987.
4. J. N. Reddy, *Mechanics of Laminated Composite Plates and Shells*, CRC Press, Boca Raton, FL, 2004.
5. R. M. Jones, *Mechanics of Composite Materials*, Taylor & Francis, New York, 1998.
6. V. V. Vasiliev, (Robert M. Jones, English Edition Editor), *Mechanics of Composite Structures*, Taylor & Francis, Washington, DC, 1993.
7. L. P. Kollar and G. S. Springer, *Mechanics of Composite Structures*, Cambridge University Press, Cambridge, 2003.
8. M. E. Tuttle, *Structural Analysis of Polymeric Composite Materials*, Marcel Dekker, New York, 2004.
9. B. D. Agarwal, L. J. Broutman, and K. Chandrashekhara, *Analysis and Performance of Fiber Composites*, John Wiley & Sons, New York, 2006.
10. J. T. S. Wang, On the solution of plates of composite materials, *Journal of Composite Materials*, 3(3), 1969, 590–592.
11. J. E. Ashton and M. E. Waddoups, Analysis of anisotropic plates, *Journal of Composite Materials*, 3(1), 1969, 148–165.
12. J. E. Ashton, Analysis of anisotropic plates II, *Journal of Composite Materials*, 3(3), 1969, 470–479.
13. J. M. Whitney and A. W. Leissa, Analysis of heterogeneous anisotropic plates, *Journal of Applied Mechanics*, 36(2), 1969, 261–266.
14. J. M. Whitney, Bending-extensional coupling in laminated plates under transverse loading, *Journal of Composite Materials*, 3(1), 1969, 20–28.
15. J. E. Ashton, Anisotropic plate analysis—Boundary conditions, *Journal of Composite Materials*, 4(2), 1970, 162–171.
16. J. E. Ashton, Approximate solutions for unsymmetrically laminated plates, *Journal of Composite Materials*, 3(1), 1969, 189–191.
17. R. M. Jones, Buckling and vibration of rectangular unsymmetrically laminated cross-ply plates, *AIAA Journal*, 11(12), 1973, 1626–1632.
18. J. N. Reddy and A. A. Khdeir, Buckling and vibration of laminated composite plates using various plate theories, *AIAA Journal*, 27(12), 1989, 1808–1817.
19. R. M. Jones, *Buckling of Bars, Plates, and Shells*, Bull Ridge Publishing, Blacksburg, Virginia, 2006.
20. S. P. Timoshenko and J. M. Gere, *Theory of Elastic Stability*, second edition, McGraw-Hill Book Company, Singapore, 1963.
21. S. S. Rao, *Vibration of Continuous Systems*, John Wiley & Sons, Hoboken, NJ, 2007.
22. R. Byron Pipes and N. J. Pagano, Interlaminar stresses in composite laminates under uniform axial extension, *Journal of Composite Materials*, 4(4), 1970, 538–546.
23. N. J. Pagano and R. Byron Pipes, The influence of stacking sequence on laminate strength, *Journal of Composite Materials*, 5, 1971, 50–57.

8

Finite Element Method

8.1 CHAPTER ROAD MAP

The solutions to beam and plate problems by analytical methods are addressed in Chapters 6 and 7. The applicability of these methods is limited to basic structural elements with simple geometry under simple loading. Real-life structures, however, are complex; they generally comprise several basic structural elements such as beams, plates, shells, etc. and are subjected to complex loads. As a result, in a product design environment, analytical methods are not sufficient and the use of numerical methods is nearly essential.

The most popular numerical method is the finite element method (FEM), and this chapter is devoted to present the fundamental aspects involved in it. The subject of FEM is too vast to be covered in one chapter. Also, a detailed discussion on it is beyond the scope of this book. The objective of this chapter is to acquaint the reader with the basic concepts and address the general procedure of FEM, especially from the point of view of stress analysis of structures. With this in mind, the basic concepts in the FEM are discussed first. Next, the general procedure is presented. Owing to its versatility, interdisciplinary nature, and wide application, it is possible that the subject would appear too complex. In this respect, an understanding of the basic procedure is highly helpful. Finally, some simple elements are developed to demonstrate how the method can be used in structural stress analysis problems.

Familiarity with the concepts in solid mechanics and mechanics of laminated composite material (Chapters 2, 4, and 5) is a prerequisite for effective assimilation of the contents of this chapter.

8.2 PRINCIPAL NOMENCLATURE

A number of symbols are used in this chapter. A conscious effort is made to be consistent in using these symbols and they are described where they occur. In this section, the principal notations are listed for ready reference.

$[A]$	Extensional stiffness matrix
A	Area of cross section
$\{a\}$	Vector of polynomial coefficients
$a_1, a_2, \ldots$	Coefficients of a polynomial function
$[B]$	Strain–displacement matrix—a matrix obtained by partial differentiation of the matrix of shape functions
	Also, extension-bending coupling stiffness matrix
$[C]$	Elastic stiffness matrix of a material
$[D]$	Bending stiffness matrix
E	Total number of elements
	Also, Young's modulus

f	A field variable
$[g]$	A matrix of the polynomial variable(s)
h	Thickness of a plate
$[K], [K^{(e)}]$	Global and element stiffness matrices (stress analysis)
$[K_G], [K_G^{(e)}]$	Global and element geometric stiffness matrices (buckling analysis)
l, b	Length and breadth, respectively
$\{M\}$	Vector of moment resultants
M	Total number of degrees of freedom in the structure
m	Total number of polynomial coefficients in a polynomial function
$[N]$	Matrix of shape functions
$\{N\}$	Vector of force resultants
$N_1, N_2, \ldots$	Shape functions
n	Degree of a polynomial function
$\{u\}$	Vector of displacements at a point
u, v, w	Displacements at a point
$\{P\}, \{P^{(e)}\}$	Vector of nodal forces for the whole body and element, respectively
W	Work done on a body by external forces
$[X]$	A square matrix of the nodal coordinates
x, y, z	Element coordinates
$\{\mathcal{B}\}$	Vector of body forces at a point
$\mathcal{B}_x, \mathcal{B}_y, \mathcal{B}_z$	Body forces at a point
$\{\varepsilon\}$	Vector of strains
$\varepsilon_{xx}, \varepsilon_{yy}, \varepsilon_{zz}$	Normal strains
$\{\mathcal{F}\}$	Vector of surface forces at a point
$\mathcal{F}_x, \mathcal{F}_y, \mathcal{F}_z$	Surface forces at a point
$\gamma_{xy}, \gamma_{yz}, \gamma_{zx}$	Shear strains
ν	Poisson's ratio
$\{\phi_i\}$	Vector of nodal values for the structure
$\{\phi_i^{(e)}\}$	Vector of nodal values for element e
$\phi_1, \phi_2, \ldots$	Nodal values for the structure
$\phi_1^{(e)}, \phi_2^{(e)}, \ldots$	Nodal values for element e
$[\lambda^{(e)}]$	Transformation matrix for element e
π	Strain energy of a body
π_p	Potential energy of a body
$\{\sigma\}$	Vector of stresses
$\sigma_{xx}, \sigma_{yy}, \sigma_{zz}$	Normal stresses
$\gamma_{xy}, \gamma_{yz}, \gamma_{zx}$	Shear stresses
$\theta_x, \theta_y, \theta_z$	Rotations

8.3 INTRODUCTION

The FEM has been extensively dealt with in the literature; in addition to numerous articles, a large number of standard texts giving detailed presentation of the method are available [1–8]. It is a versatile tool that can be applied to problems in virtually any field. Its application in composites has also been addressed in standard texts [9–12]. There are a number of commercially available general-purpose finite element softwares, for example, ANSYS, ABAQUS, NASTRAN, and NISA. These softwares have highly user-friendly interfaces, very large computing capability, and appealing display features for preprocessing as well as postprocessing. There is a general tendency among many students and practicing engineers to use these packages and produce output without bothering to understand what FEM or the software does! However, the correctness of the output depends on several factors, including input material data, type of elements, boundary conditions, etc. Here lies the necessity to know the basics of FEM for proper utilization of any of these packages.

The analysis of a physical problem involves two broad aspects:

- Mathematical formulation of the problem
- Solution to the mathematical model

In general, any physical problem can be described in terms of mathematical expressions—differential equations, integral equations, and algebraic equations, of which differential equations are the most common. These equations are valid within certain domains under certain prescribed conditions, depending on which a physical problem is categorized as either initial value problem or boundary value problem. In an initial value problem, the dependent variable or its derivatives are prescribed initially, that is, at time $t_0 = 0$, whereas in a boundary value problem, the dependent variable and its derivatives are prescribed on the boundary of the domain.

Depending on the nature, three broad classes of problems are encountered (see, for instance, References 4 and 5; these are

- Equilibrium or steady-state problems
- Transient or propagation problems
- Eigenvalue problems

Equilibrium problems are time-independent problems, in which we determine the static or quasi-static response of the body to the applied generic loads. Transient problems are time-dependent; in these problems, the applied loads vary with time and we are interested in finding the response of the body under such time-varying loads. On the other hand, eigenvalue problems are a sort of combination of both of the above. In these problems, the time dimension is involved indirectly and we are interested in determining values of certain parameters that correspond to some critical state of the body.

The formulation of mathematical models for real-life problems and their analytical solutions are virtually impossible in most cases and invariably numerical methods—most commonly FEM—are used for making estimates of various parameters. In composites, owing to their anisotropic nature, analysis is more complex and FEM is extensively used. Typical examples of applications of FEM in the design and development of composite products are indicated in Table 8.1.

TABLE 8.1

Typical Applications of FEM in Composites

Nature of Problems	Examples
Equilibrium problems	Analysis for • Displacements • Stress distribution • Temperature distribution
Transient problems	Analysis for • Crack propagation • Response to impact and fatigue loads • Ablation
Eigenvalue problems	Analysis for • Buckling loads and their mode shapes • Natural frequencies and their mode shapes

8.4 BASIC CONCEPTS IN FINITE ELEMENT METHOD

The FEM is built on some unique concepts; some of the essential concepts necessary to understand the basic finite element procedure are reviewed in this section.

8.4.1 Elements and Nodes

In the FEM, the domain or the body is divided into a finite number of subdomains that share common edges and are interconnected at certain points. These subdomains are called the elements and the interconnecting points are called the nodes (Figure 8.1). The nodes are placed at the corners of the elements. In some elements, nodes are also placed on the edges between the corner nodes. The concept of elements and nodes is the basis of the FEM. The focus is initially limited to the mathematical formulations that govern the individual elements. Element formulations are generic in nature and various types of elements have been developed. The body or the domain is considered as an assemblage of elements and thus virtually any kind of body can be handled by this method. As mentioned above, governing equations are developed for individual elements and the body or domain is simulated as an assemblage of elements. The connection between adjacent elements is through the common nodes. External loads are applied at the nodes and load transfer between elements is through the nodes.

A physical problem is described in terms of a number of field variables—known and unknown. For example, displacement, stress, pressure, temperature, etc. are commonly encountered field variables. It is required that the field variables involved in a problem are identified first. Of the several field variables, it is necessary to identify a basic variable around which the element equations are developed. The primary aim is to first find the nodal values of the basic variable. The nodal values are then utilized for estimating the basic field variable and other relevant field variables everywhere in the body.

For stress analysis problems in solid mechanics, the basic variable is displacement. Nodal displacements are determined first and approximating functions (see Section 8.4.3) are used for estimating the displacement distribution. Strain–displacement relations and constitutive relations are then used for estimating strains and stresses, respectively.

For heat conduction problems, temperature is the basic variable. Nodal temperatures are determined first. Temperature distribution, thermal stresses, strains, etc. are then estimated.

At any point of the body, the basic variable can possess certain degrees of freedom. For example, in a general case, displacement can have up to six degrees of freedom at a point—three translations along three axes and three rotations about these axes.

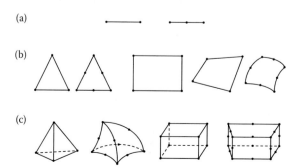

FIGURE 8.1 Typical elements. (a) One-dimensional (1D) elements. (b) Two-dimensional (2D) elements. (c) Three-dimensional (3D) elements.

In the FEM, the body is divided into a finite number of elements with a finite number of nodes. The basic variable has certain known degrees of freedom at each node (nodal degrees of freedom). Thus, the problem is reduced from the determination of the basic variable at theoretically infinite locations to a finite number of nodal values.

8.4.2 Discretization

The first step in a finite element analysis is discretization of the body into elements. It is also called as mesh generation. In the process of discretization, the body, which can be of irregular geometric shape, is modeled with finite elements of regular shapes. The accuracy of modeling depends on how closely the original body is simulated by the assemblage of elements. A number of factors related to discretization influence the accuracy of the final output of a finite element analysis; type of elements, size and number of elements, location of nodes, etc. are some of the key factors.

The elements can be 1D, 2D, or 3D (Figure 8.1). The choice of type of elements depends primarily on three parameters—geometry of the body, number of spatial coordinates required to describe the system, and desired accuracy/details of output data. For example, a uniaxially loaded bar of constant cross-sectional area and material properties varying along the axis can be described conveniently by its axial coordinates alone; in this case, an obvious choice of elements is 1D elements and general information on deformation, stress, and strain can be obtained. If, however, local stress distribution across the area of cross section near the load application point is required, 3D elements are needed and possibly a local analysis near the joint will serve the purpose. Similarly, a thin-shell structure, for example, a pressure vessel under biaxial membrane state of stress, can be modeled with 2D elements. But if the skin thickness is high and it is subjected to bending, leading to complex stress distribution across the thickness, 3D elements will be preferred. Figure 8.2 shows typical examples of discretization using 1D, 2D, and 3D elements.

An important aspect of discretization is to take advantage of the symmetry of the system. The overall computational time and effort can be greatly reduced by modeling a part of the body. As an example, consider a rectangular plate with a central hole under tension (Figure 8.3). Note that the plate together with the loading is symmetric w.r.t. the axial and lateral center lines and modeling of only one-fourth of the plate is sufficient.

The next important aspect in discretization is the size and number of elements. Elements of smaller sizes are needed in areas of rapidly changing geometry, load, and material properties so that the body is appropriately represented by the mesh of elements. Smaller elements are also required in areas, such as cutout, joint, etc., where

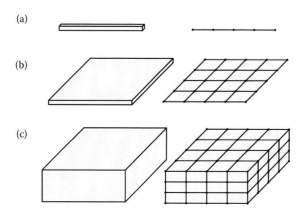

FIGURE 8.2 Discretization. (a) One-dimensional elements. (b) Two-dimensional elements. (c) Three-dimensional elements.

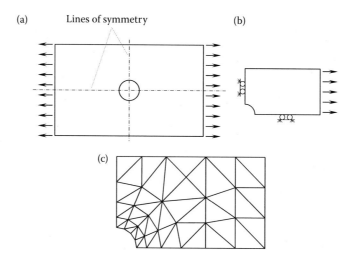

FIGURE 8.3 Discretization of a rectangular plate with a central hole. (a) Total plate. (b) One quadrant of the plate. (c) Mesh with varying sizes of elements.

stress concentration is expected (Figure 8.3c). Smaller size of elements implies larger number of elements, larger number of equations to be solved, and longer computational time and capacity. Thus, the choice of element size must be based on proper judgment.

Another important aspect is the numbering scheme for elements and nodes. Each element in an FEM is given a unique number. Similarly, each node is given a unique number. A proper numbering scheme is critical as it influences the bandwidth of the global characteristic matrix, which in turn influences the efficiency of the solution process (see, for instance, References 5 and 7).

8.4.3 Approximating Function and Shape Function

The distribution of a field variable within the body as well as within the elements is not known. Sometimes, the distribution may be known but the exact distribution may be too complicated for mathematical manipulations. Thus, some approximations are introduced. An approximating function is a function used for representing the distribution of a field variable within an element in terms of the element coordinates and other constants. The most commonly used approximating functions are polynomial functions, although trigonometric functions are also sometimes used. For 1D, 2D, and 3D elements, the generalized forms of a polynomial function are

1D elements:

$$f(x) = a_1 + a_2 x + a_3 x^2 + \cdots + a_m x^n \tag{8.1}$$

2D elements:

$$f(x,y) = a_1 + a_2 x + a_3 y + a_4 x^2 + a_5 y^2 + \cdots + a_{m-1} x^n + a_m y^n \tag{8.2}$$

3D elements:

$$f(x,y,z) = a_1 + a_2 x + a_3 y + a_4 z + a_5 x^2 + a_6 y^2 + a_7 z^2 + \cdots + a_{m-2} x^n + a_{m-1} y^n + a_m z^n \tag{8.3}$$

Finite Element Method

in which

- f — A field variable
- x, y, z — Element coordinates
- $a_1, a_2, \ldots$ — Coefficients of the polynomial, known as generalized coordinates
- n — Degree of the polynomial
- m — Total number of polynomial coefficients

The approximating function can be conveniently expressed in matrix form as a multiple of two matrices—one, a square matrix of element coordinates and two, a vector of coefficients of the polynomial. For example, let us consider a 1D element (Figure 8.4). The approximating function is

$$f(x) = a_1 + a_2 x \tag{8.4}$$

which can be written as

$$f(x) = [g(x)]\{a\} \tag{8.5}$$

where $[g(x)]$ is a matrix (size: 1×2) of the polynomial variable and $\{a\}$ is a vector (size: 2×1) of polynomial coefficients. They are given by

$$[g(x)] = [1 \quad x] \tag{8.6}$$

$$\{a\} = \begin{Bmatrix} a_1 \\ a_2 \end{Bmatrix} \tag{8.7}$$

The idea in the FEM is to determine the field variables from the nodal values. The nodal values in this case are $\phi_1^{(e)}$ and $\phi_2^{(e)}$, where $\phi_1^{(e)} = f(0)$ and $\phi_2^{(e)} = f(l)$. The vector of nodal values for element e, $\{\phi_i^{(e)}\}$ is given by

$$\{\phi_i^{(e)}\} \equiv \begin{Bmatrix} \phi_1^{(e)} \\ \phi_2^{(e)} \end{Bmatrix} = \begin{Bmatrix} f(0) \\ f(l) \end{Bmatrix} \tag{8.8}$$

Then, using Equations 8.5 and 8.8, the vector of nodal displacements $\{\phi_i^{(e)}\}$ and the vector of polynomial coefficients $\{a\}$ can be related as

$$\{\phi_i^{(e)}\} = [X]\{a\} \quad \text{or} \quad \{a\} = [X]^{-1}\{\phi_i^{(e)}\} \tag{8.9}$$

where $[X]$ is a square matrix of the nodal coordinates and it is given by

$$[X] = \begin{bmatrix} 1 & 0 \\ 1 & l \end{bmatrix} \tag{8.10}$$

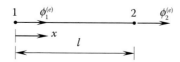

FIGURE 8.4 1D element.

Then, substituting Equation 8.9 in Equation 8.5, we get

$$f(x) = [g(x)][X]^{-1}\{\phi_i^{(e)}\} \qquad (8.11)$$

or

$$f(x) = [N]\{\phi_i^{(e)}\} \qquad (8.12)$$

where

$$[N] = [g(x)][X]^{-1} = \begin{bmatrix} \dfrac{l-x}{l} & \dfrac{x}{l} \end{bmatrix} \qquad (8.13)$$

is the matrix of shape functions. Thus, the shape functions are the elements of a square matrix that relates the field variable to the nodal values of the field variable. There is one shape function for each node. The numerical value of each shape function is "one" at the corresponding node, "zero" at other nodes, and between "one" and "zero" at other locations of the element.

Note: The approximating function is also known as interpolation function, interpolation model, etc. In structural mechanics problem, it is commonly called the displacement model. The term "interpolation function" is also used in place of shape function. However, the distinction between the two concepts must be kept in mind. To avoid confusion, we shall stick to the terms "approximating function" and "shape function."

As mentioned earlier, the most commonly used approximating functions are polynomial functions. Selection of the approximating function is important and it is guided primarily by three factors.

First, the total number of polynomial coefficients should be equal to the total number of nodal degrees of freedom. The number of degrees of freedom per node multiplied by the number of nodes in an element gives the total number of degrees of freedom for the element and the polynomial function should be so chosen as to have the total number of polynomial coefficients equal to this number.

Second, the approximating function should satisfy convergence requirements. The FEM is an approximate method and its accuracy can be improved by several means, as follows [4,5,7]:

- p-method—by increasing the order of the polynomial function
- r-method—by changing the location of nodes along the element edge without changing the number of elements
- h-method—by increasing the number of elements, that is, by reducing the size of elements

There are three basic convergence requirements that should be satisfied by the approximating function so that with successive steps the results approach the exact solution. These requirements are

- The approximating function must be continuous within the element.
- The approximating function must appropriately represent the basic field variable and its partial derivatives up to the highest order appearing in the functional I.

- The basic field variable and its partial derivatives up to one order less than the highest-order derivative appearing in the functional I must be continuous at the boundaries of the element.

Note: The functional I is the functional used in the variational formulation.

Third, the chosen approximating function should be such that the distribution of a field variable obtained by using an approximating function should be independent of the local coordinate system.

8.4.4 Element Characteristic Matrices and Vectors

Some of the commonly encountered problems in structural mechanics are as follows:

- Equilibrium or steady-state problems
  - Static stress analysis
  - Quasi-static thermal analysis
- Transient or propagation problems
  - Transient vibration analysis
- Eigenvalue problems
  - Buckling analysis
  - Free vibration analysis

Finite element formulations are most conveniently expressed in matrix forms, which for the above classes of problems can be expressed in matrix forms as follows (see, for instance, Reference 5 for details):

Stress analysis:

$$[K]\{\phi\} = \{P\} \tag{8.14}$$

Thermal analysis:

$$([K_1] + [K_2])\{T\} + [K_3]\{\dot{T}\} = \{P\} \tag{8.15}$$

Transient vibration analysis:

$$[K]\{\phi\} + [C]\{\dot{\phi}\} + [M]\{\ddot{\phi}\} = \{P\} \tag{8.16}$$

Buckling analysis:

$$([K] - \lambda[K_G])\{\phi\} = \{0\} \tag{8.17}$$

Free vibration analysis:

$$([K] - \lambda[M])\{\phi\} = \{0\} \tag{8.18}$$

In the above equations, a number of matrices have been introduced. The matrices within square brackets are the characteristic matrices and the vectors within curly brackets are the characteristic vectors. These characteristic matrices and vectors are known by different names in different classes of problems. For example, in structural mechanics problems, the characteristic matrix is called the stiffness matrix; it relates the nodal displacements to the nodal forces. In thermal analysis problems, the characteristic matrix is called the conductivity matrix and it relates the nodal temperatures to the nodal fluxes.

Note that the above equations are at the global level for the entire body or domain. In the FEM, the equations are first developed for the individual elements, that is, the element characteristic matrices and vectors are developed. We shall use the superscript (e) to denote the element characteristic matrices and vectors. Then, the matrices and vectors in these equations are

Stress analysis:
- $K, K^{(e)}$ Global and element stiffness matrices
- $\phi, \phi^{(e)}$ Nodal displacements
- $P, P^{(e)}$ Nodal forces

Thermal analysis:
- $K_1, K_2, K_3, K_1^{(e)}, K_2^{(e)}, K_3^{(e)}$ Global and element conductivity matrices
- $T, \dot{T}, T^{(e)}, \dot{T}^{(e)}$ Nodal temperatures and their derivatives
- $P, P^{(e)}$ Nodal heat fluxes

Transient vibration analysis:
- $K, K^{(e)}$ Global and element stiffness matrices
- $M, M^{(e)}$ Global and element mass matrices
- $C, C^{(e)}$ Global and element damping matrices
- $\phi, \dot{\phi}, \ddot{\phi}, \phi^{(e)}, \dot{\phi}^{(e)}, \ddot{\phi}^{(e)}$ Nodal displacements and their time derivatives
- $P, P^{(e)}$ Nodal forces

Buckling analysis:
- $K, K^{(e)}$ Global and element stiffness matrices
- $K_G, K_G^{(e)}$ Global and element geometric stiffness matrices
- $\phi, \phi^{(e)}$ Nodal displacements

Free vibration analysis:
- $K, K^{(e)}$ Global and element stiffness matrices
- $M, M^{(e)}$ Global and element mass matrices
- $\phi, \ddot{\phi}, \phi^{(e)}, \ddot{\phi}^{(e)}$ Nodal displacements and their time derivatives

8.4.5 Derivation of Element Characteristic Matrices

There are primarily three approaches for the derivation of the characteristic matrices and vectors:

- Direct approach
- Variational approach
  - Rayleigh–Ritz method
- Weighted residual approach
  - Galerkin method
  - Least squares method

The direct approach is based on direct physical reasoning and principles of science and engineering. Thus, it is useful in getting an insight into the FEM. This method can be applied for solving simple problems; however, for complex problems, it is not suitable and one has to rely on either variational or weighted residual approach.

The variational approach is based on the fact that many problems in science and engineering can be expressed in the variational form. In the variational form, a functional I is defined in terms of the field variables and their derivatives. The solution, which can be either exact or approximate, is the one that minimizes the functional.

In the weighted residual methods, the finite element equations are derived from the governing equations directly without reference to any functional. The Galerkin method and the least squares method are two common methods that belong to this approach.

8.4.6 Finite Element Equations by the Variational Approach

The element characteristic matrix and vector play a central role in the FEM. Extensive derivation of these matrices and vectors by different methods for different types of physical problems is beyond the scope of this book. Instead, we shall consider the case of static analysis by the variational approach based on the principle of minimum potential energy that states that the equilibrium configuration of a body, out of all the possible configurations that satisfy compatibility and kinematic boundary conditions, makes the potential energy the minimum.

The potential energy of an elastic body is given by

$$\pi_p = \pi - W \tag{8.19}$$

where

- π_p Potential energy of the body
- π Strain energy of the body
- W Work done on the body by the external forces

The strain energy of an initially strain-free body is given by the following volume integral:

$$\pi = \frac{1}{2} \iiint_V \begin{Bmatrix} \varepsilon_{xx} \\ \varepsilon_{yy} \\ \varepsilon_{zz} \\ \gamma_{yz} \\ \gamma_{zx} \\ \gamma_{xy} \end{Bmatrix}^T \begin{Bmatrix} \sigma_{xx} \\ \sigma_{yy} \\ \sigma_{zz} \\ \tau_{yz} \\ \tau_{zx} \\ \tau_{xy} \end{Bmatrix} dV \tag{8.20}$$

or

$$\pi = \frac{1}{2} \iiint_V \{\varepsilon\}^T \{\sigma\} dV \tag{8.21}$$

where $\{\varepsilon\}$ and $\{\sigma\}$ are the vectors of strains and stresses. (Refer to Chapter 2 for a detailed discussion on the terms $\{\varepsilon\}$ and $\{\sigma\}$.)

Note: The symbols $\iiint_V$ and $\iint_S$ are used for volume integral and surface integral, respectively.

Introducing the constitutive relation (refer Equation 2.157, Chapter 2) into Equation 8.21, we get

$$\pi = \frac{1}{2} \iiint_V \{\varepsilon\}^T [C] \{\varepsilon\} dV \tag{8.22}$$

where $[C]$ is the elastic stiffness matrix of the material. In the presence of initial strains $\{\varepsilon_0\}$, Equation 8.22 takes the following form:

$$\pi = \frac{1}{2} \iiint_V \{\varepsilon\}^T [C] \{\varepsilon\} dV - \iiint_V \{\varepsilon\}^T [C] \{\varepsilon_0\} dV \tag{8.23}$$

or

$$\pi = \frac{1}{2} \iiint_V \{\varepsilon\}^T [C] \{\varepsilon - 2\varepsilon_0\} dV \tag{8.24}$$

On the other hand, the works done on the body by the body forces, surface forces, and external concentrated forces are as follows:

$$W = \iiint_V \begin{Bmatrix} u \\ v \\ w \end{Bmatrix}^T \begin{Bmatrix} \mathcal{B}_x \\ \mathcal{B}_y \\ \mathcal{B}_z \end{Bmatrix} dV + \iint_S \begin{Bmatrix} u \\ v \\ w \end{Bmatrix}^T \begin{Bmatrix} \mathcal{F}_x \\ \mathcal{F}_y \\ \mathcal{F}_z \end{Bmatrix} dS + W_c \tag{8.25}$$

or

$$W = \iiint_V \{u\}^T \{\mathcal{B}\} dV + \iint_S \{u\}^T \{\mathcal{F}\} dS + W_c \tag{8.26}$$

where

$\{\mathcal{B}\} = \begin{Bmatrix} \mathcal{B}_x \\ \mathcal{B}_y \\ \mathcal{B}_z \end{Bmatrix}$ Vector of body forces at a point

$\{\mathcal{F}\} = \begin{Bmatrix} \mathcal{F}_x \\ \mathcal{F}_y \\ \mathcal{F}_z \end{Bmatrix}$ Vector of surface forces at a point

$\{u\} = \begin{Bmatrix} u \\ v \\ w \end{Bmatrix}$ Vector of displacements at a point

W_c Work done by the external concentrated forces

Then, combining Equations 8.19, 8.24, and 8.26, we get the expression for the potential energy of the body as follows:

$$\pi_p = \frac{1}{2} \iiint_V \{\varepsilon\}^T [C] \{\varepsilon - 2\varepsilon_0\} dV - \iiint_V \{u\}^T \{\mathcal{B}\} dV - \iint_S \{u\}^T \{\mathcal{F}\} dS - W_c \tag{8.27}$$

Note that V denotes the volume of the body and S denotes the surface over which surface forces (tractions) are prescribed. The above expression gives the potential energy of the whole body. Now, we need an expression for the potential energy of an element. Using Equation 8.27, we can express the potential energy of an element without the contribution of the external concentrated forces as follows:

$$\pi_p^{(e)} = \frac{1}{2} \iiint_{V^{(e)}} \{\varepsilon\}^T [C] \{\varepsilon - 2\varepsilon_0\} dV - \iiint_{V^{(e)}} \{u\}^T \{\mathcal{B}\} dV - \iint_{S^{(e)}} \{u\}^T \{\mathcal{F}\} dS \tag{8.28}$$

where $V^{(e)}$ denotes the volume of the element and $S^{(e)}$ denotes the surface of the element over which surface forces are prescribed. Note that Equation 8.28 does not include the contribution of work done by concentrated forces.

As we noted in Section 8.4.1, the basic variable in the solid mechanics problem is displacement. Now, we bring in the strain–displacement relation (refer Equations 2.44 through 2.46 and 2.52 through 2.54, Chapter 2) and express the strain vector as follows:

$$\{\varepsilon\} = \begin{Bmatrix} \varepsilon_{xx} \\ \varepsilon_{yy} \\ \varepsilon_{zz} \\ \gamma_{yz} \\ \gamma_{zx} \\ \gamma_{xy} \end{Bmatrix} = \begin{Bmatrix} \dfrac{\partial u}{\partial x} \\ \dfrac{\partial v}{\partial y} \\ \dfrac{\partial w}{\partial z} \\ \dfrac{\partial v}{\partial x} + \dfrac{\partial u}{\partial y} \\ \dfrac{\partial w}{\partial y} + \dfrac{\partial v}{\partial z} \\ \dfrac{\partial u}{\partial z} + \dfrac{\partial w}{\partial x} \end{Bmatrix} = \begin{bmatrix} \dfrac{\partial}{\partial x} & 0 & 0 \\ 0 & \dfrac{\partial}{\partial y} & 0 \\ 0 & 0 & \dfrac{\partial}{\partial z} \\ \dfrac{\partial}{\partial y} & \dfrac{\partial}{\partial x} & 0 \\ 0 & \dfrac{\partial}{\partial z} & \dfrac{\partial}{\partial y} \\ \dfrac{\partial}{\partial z} & 0 & \dfrac{\partial}{\partial x} \end{bmatrix} \begin{Bmatrix} u(x,y,z) \\ v(x,y,z) \\ w(x,y,z) \end{Bmatrix} \quad (8.29)$$

As we discussed in Section 8.4.3, the matrix of shape functions relates the field variable inside an element to the nodal values of the field variable, as follows:

$$\begin{Bmatrix} u(x,y,z) \\ v(x,y,z) \\ w(x,y,z) \end{Bmatrix} = \begin{bmatrix} N_{11} & N_{12} & \cdots & N_{1n} \\ N_{21} & N_{22} & \cdots & N_{2n} \\ N_{31} & N_{32} & \cdots & N_{3n} \end{bmatrix} \begin{Bmatrix} \phi_1^{(e)} \\ \phi_2^{(e)} \\ \vdots \\ \phi_n^{(e)} \end{Bmatrix} \quad (8.30)$$

or

$$\{u\} = [N]\{\phi_i^{(e)}\} \quad (8.31)$$

Note that n is the number of degrees of freedom per element. Substituting the above in Equation 8.29, we get

$$\begin{Bmatrix} \varepsilon_{xx} \\ \varepsilon_{yy} \\ \varepsilon_{zz} \\ \gamma_{yz} \\ \gamma_{zx} \\ \gamma_{xy} \end{Bmatrix} = \begin{bmatrix} \dfrac{\partial}{\partial x} & 0 & 0 \\ 0 & \dfrac{\partial}{\partial y} & 0 \\ 0 & 0 & \dfrac{\partial}{\partial z} \\ \dfrac{\partial}{\partial y} & \dfrac{\partial}{\partial x} & 0 \\ 0 & \dfrac{\partial}{\partial z} & \dfrac{\partial}{\partial y} \\ \dfrac{\partial}{\partial z} & 0 & \dfrac{\partial}{\partial x} \end{bmatrix} \begin{bmatrix} N_{11} & N_{12} & \cdots & N_{1n} \\ N_{21} & N_{22} & \cdots & N_{2n} \\ N_{31} & N_{32} & \cdots & N_{3n} \end{bmatrix} \begin{Bmatrix} \phi_1^{(e)} \\ \phi_2^{(e)} \\ \vdots \\ \phi_n^{(e)} \end{Bmatrix} \quad (8.32)$$

or

$$\{\varepsilon\} = [B]\{\phi_i^{(e)}\} \tag{8.33}$$

where

$$[B] = \begin{bmatrix} \dfrac{\partial}{\partial x} & 0 & 0 \\ 0 & \dfrac{\partial}{\partial y} & 0 \\ 0 & 0 & \dfrac{\partial}{\partial z} \\ \dfrac{\partial}{\partial y} & \dfrac{\partial}{\partial x} & 0 \\ 0 & \dfrac{\partial}{\partial z} & \dfrac{\partial}{\partial y} \\ \dfrac{\partial}{\partial z} & 0 & \dfrac{\partial}{\partial x} \end{bmatrix} \begin{bmatrix} N_{11} & N_{12} & \cdots & N_{1n} \\ N_{21} & N_{22} & \cdots & N_{2n} \\ N_{31} & N_{32} & \cdots & N_{3n} \end{bmatrix} \tag{8.34}$$

Then, we substitute Equations 8.31 and 8.33 in Equation 8.28 and obtain the expression for the potential energy of the element without the contribution of external concentrated forces, as follows:

$$\pi_p^{(e)} = \frac{1}{2} \iiint_{V^{(e)}} \{\phi_i^{(e)}\}^T [B]^T [C][B]\{\phi_i^{(e)}\} dV - \iiint_{V^{(e)}} \{\phi_i^{(e)}\}^T [B]^T [C]\{\varepsilon_0\} dV$$

$$- \iiint_{V^{(e)}} \{\phi_i^{(e)}\}^T [N]^T \{\mathcal{B}\} dV - \iint_{S^{(e)}} \{\phi_i^{(e)}\}^T [N]^T \{\mathcal{F}\} dS \tag{8.35}$$

The nodal displacements are independent of the element coordinates and they can be brought outside the integration. Then, Equation 8.35 can be reframed as follows:

$$\pi_p^{(e)} = \frac{1}{2} \{\phi_i^{(e)}\}^T \left[\iiint_{V^{(e)}} [B]^T [C][B] dV \right] \{\phi_i^{(e)}\}$$

$$- \{\phi_i^{(e)}\}^T \left\{ \iiint_{V^{(e)}} [B]^T [C]\{\varepsilon_0\} dV \right\}$$

$$- \{\phi_i^{(e)}\}^T \left\{ \iiint_{V^{(e)}} [N]^T \{\mathcal{B}\} dV \right\}$$

$$- \{\phi_i^{(e)}\}^T \left\{ \iint_{S^{(e)}} [N]^T \{\mathcal{F}\} dS \right\} \tag{8.36}$$

The total potential energy of the body can be obtained by summing up the potential energies of all the elements, with due consideration of the contribution of the external concentrated forces. Thus,

$$\pi_p = \sum_{e=1}^{E} \pi^{(e)} - W_c \tag{8.37}$$

Note that E is the total number of elements. The concentrated forces are applied at the nodes and the second term W_c in Equation 8.37 is given by

$$W_c = \begin{Bmatrix} \phi_1 \\ \phi_2 \\ \vdots \\ \vdots \\ \phi_M \end{Bmatrix}^T \begin{Bmatrix} P_{c1} \\ P_{c2} \\ \vdots \\ \vdots \\ P_{cM} \end{Bmatrix} = \{\phi_i\}^T \{P_c\} \tag{8.38}$$

where $\{\phi_i\}$ is the vector of nodal displacements and $\{P_c\}$ is the vector of externally applied nodal concentrated forces. Note the difference between $\{\phi_i\}$ and $\{\phi_i^{(e)}\}$. Also, note that M is the total number of nodal degrees of freedom in the body. The first term in Equation 8.37 is derived after certain modifications to the sizes of the matrices and vectors involved therein. The sizes of various matrices and vectors in Equation 8.36 are given in Table 8.2.

Note that n is the total number of degrees of freedom per element. The square matrix of size $n \times n$ and the vectors of size $n \times 1$ (Sl. No. 7–11, Table 8.2) are expanded by appropriately inserting zeros to size $M \times M$ and $M \times 1$, respectively. In other words, these matrix and vectors are expanded to the "structure size." These expanded matrices and vectors are summed up or assembled. (More about the process of expansion and assembly

TABLE 8.2
Matrices and Vectors in Element Potential Energy Expression (Equation 8.35)

Sl. No.	Matrix/Vector	Size
1	$[B]$	$6 \times n$
2	$[C]$	6×6
3	$\{\varepsilon_0\}$	6×1
4	$\{\mathcal{B}\}$	3×1
5	$\{\mathcal{F}\}$	3×1
6	$[N]$	$3 \times n$
7	$\left[\iiint_{V^{(e)}} [B]^T [C][B] dV \right]$	$n \times n$
8	$\left\{ \iiint_{V^{(e)}} [B]^T [C]\{\varepsilon_0\} dV \right\}$	$n \times 1$
9	$\left\{ \iiint_{V^{(e)}} [N]^T \{\mathcal{B}\} dV \right\}$	$n \times 1$
10	$\left\{ \iint_{S^{(e)}} [N]^T \{\mathcal{F}\} dS \right\}$	$n \times 1$
11	$\{\phi_i^{(e)}\}$	$n \times 1$

is given in Section 8.4.8.) Then, using Equations 8.36 and 8.38 in Equation 8.37, we get the following:

$$\pi_p = \frac{1}{2}\{\phi_i\}^T \sum_{e=1}^{E}\left[\iiint_{V^{(e)}}[B]^T[C][B]dV\right]\{\phi_i\} - \{\phi_i\}^T \sum_{e=1}^{E}\left\{\iiint_{V^{(e)}}[B]^T[C]\{\varepsilon_0\}dV\right\}$$

$$-\{\phi_i\}^T \sum_{e=1}^{E}\left\{\iiint_{V^{(e)}}[N]^T\{\mathcal{B}\}dV\right\} - \{\phi_i\}^T \sum_{e=1}^{E}\left\{\iint_{S^{(e)}}[N]^T\{\mathcal{F}\}dS\right\} - \{\phi_i\}^T\{P_c\}$$

(8.39)

The matrix and vectors involved in the global potential energy expression are given in Table 8.3.

Now, we apply the principle of minimum potential energy, which gives us

$$\frac{\partial \pi_p}{\partial \phi_1} = \frac{\partial \pi_p}{\partial \phi_2} = \cdots = \frac{\partial \pi_p}{\partial \phi_M} = 0 \qquad (8.40)$$

There are total M numbers of equations in Equation 8.40. Each of these equations corresponds to one row of the following vector equation:

$$\left\{\frac{\partial \pi_p}{\partial \phi_i}\right\} = \{0\} \qquad (8.41)$$

Upon carrying out the partial differentiation, Equation 8.41 along with Equation 8.39 leads to the following:

$$\sum_{e=1}^{E}\left[\iiint_{V^{(e)}}[B]^T[C][B]dV\right]\{\phi_i\} = \sum_{e=1}^{E}\left\{\iiint_{V^{(e)}}[B]^T[C]\{\varepsilon_0\}dV\right\}$$

$$+\sum_{e=1}^{E}\left\{\iiint_{V^{(e)}}[N]^T\{\mathcal{B}\}dV\right\} + \sum_{e=1}^{E}\left\{\iint_{S^{(e)}}[N]^T\{\mathcal{F}\}dS\right\} + \{P_c\} \qquad (8.42)$$

TABLE 8.3
Matrices and Vectors in Structure Potential Energy Expression (Equation 8.39)

Sl. No.	Matrix/Vector	Size
1	$\sum_{e=1}^{E}\left[\iiint_{V^{(e)}}[B]^T[C][B]dV\right]$	→ $M \times M$
2	$\sum_{e=1}^{E}\left\{\iiint_{V^{(e)}}[B]^T[C]\{\varepsilon_0\}dV\right\}$	→ $M \times 1$
3	$\sum_{e=1}^{E}\left\{\iiint_{V^{(e)}}[N]^T\{\mathcal{B}\}dV\right\}$	→ $M \times 1$
4	$\sum_{e=1}^{E}\left\{\iint_{S^{(e)}}[N]^T\{\mathcal{F}\}dS\right\}$	→ $M \times 1$
5	$\{P_c\}$	→ $M \times 1$
6	$\{\phi_i\}$	→ $M \times 1$

or

$$\sum_{e=1}^{E}[K^{(e)}]\{\phi_i\} = \sum_{e=1}^{E}\{P_i^{(e)}\} + \sum_{e=1}^{E}\{P_b^{(e)}\} + \sum_{e=1}^{E}\{P_s^{(e)}\} + \{P_c\} \qquad (8.43)$$

or

$$[K]\{\phi_i\} = \{P\} \qquad (8.44)$$

where various matrices and vectors are given as follows:
Global stiffness matrix (size $= M \times M$):

$$[K] = \sum_{e=1}^{E}[K^{(e)}] \qquad (8.45)$$

Global vector of nodal displacements (size $= M \times 1$):

$$\{\phi_i\} = \begin{Bmatrix} \phi_1 \\ \phi_2 \\ \vdots \\ \vdots \\ \phi_M \end{Bmatrix} \qquad (8.46)$$

Global vector of nodal loads (size $= M \times 1$):

$$\{P\} = \sum_{e=1}^{E}\{P_i^{(e)}\} + \sum_{e=1}^{E}\{P_b^{(e)}\} + \sum_{e=1}^{E}\{P_s^{(e)}\} + \{P_c\} \qquad (8.47)$$

Element stiffness matrix (size $= n \times n$):

$$[K^{(e)}] = \left[\iiint_{V^{(e)}} [B]^T[C][B]dV \right] \qquad (8.48)$$

Element load vector due to initial strains (size $= n \times 1$):

$$\{P_i^{(e)}\} = \left\{ \iiint_{V^{(e)}} [B]^T[C]\{\varepsilon_0\}dV \right\} \qquad (8.49)$$

Element load vector due to body forces (size $= n \times 1$):

$$\{P_b^{(e)}\} = \left\{ \iiint_{V^{(e)}} [N]^T\{\mathcal{B}\}dV \right\} \qquad (8.50)$$

Element load vector due to surface forces (size = $n \times 1$):

$$\{P_s^{(e)}\} = \left\{ \iint_{S^{(e)}} [N]^T \{\mathcal{F}\} dS \right\} \tag{8.51}$$

Vector of concentrated forces (size = $M \times 1$):

$$\{P_c\} = \begin{Bmatrix} P_{c1} \\ P_{c2} \\ \vdots \\ \vdots \\ P_{cM} \end{Bmatrix} \tag{8.52}$$

Equations 8.42 through 8.44 are the general forms of finite element equation for stress analysis problems in solid mechanics.

8.4.7 Coordinate Transformation

In the finite element modeling, often local coordinates are used in addition to the global coordinates (Figure 8.5). For convenience, the element characteristic matrices are developed in the local coordinate systems. Once the individual element characteristic matrices have been developed, the global equations are developed by assembling the element characteristic matrices. However, before proceeding to assembly, it is imperative to transform the element characteristic matrices to a common coordinate system, which typically is the global coordinate system.

Coordinate transformation is conveniently done by utilizing the transformation matrix $[\boldsymbol{\lambda}^{(e)}]$. Let us consider a stress analysis problem. Then, the transformation matrix $[\boldsymbol{\lambda}^{(e)}]$ relates the vector of nodal displacements in the local coordinates to that in the global coordinates as follows:

$$\{\phi_i^{(e)}\}_{local} = [\boldsymbol{\lambda}^{(e)}]\{\phi_i^{(e)}\}_{global} \tag{8.53}$$

The transformation matrix relates the load vector in the local coordinates to that in the global coordinates in the same way, that is,

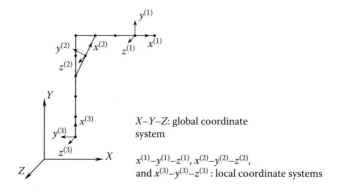

X–Y–Z: global coordinate system

$x^{(1)}$–$y^{(1)}$–$z^{(1)}$, $x^{(2)}$–$y^{(2)}$–$z^{(2)}$, and $x^{(3)}$–$y^{(3)}$–$z^{(3)}$: local coordinate systems

FIGURE 8.5 Global and local coordinate systems.

Finite Element Method

$${\{\boldsymbol{P}^{(e)}\}}_{local} = [\boldsymbol{\lambda}^{(e)}]\{\boldsymbol{P}^{(e)}\}_{global} \tag{8.54}$$

Now, as we know, work done is given by the sum of multiples of nodal displacements with corresponding nodal forces. Also, work done is a scalar quantity and it does not depend on the coordinate system. Then,

$$\{\boldsymbol{\phi}_i^{(e)}\}_{local}^T \{\boldsymbol{P}^{(e)}\}_{local} = \{\boldsymbol{\phi}_i^{(e)}\}_{global}^T \{\boldsymbol{P}^{(e)}\}_{global} \tag{8.55}$$

Using Equation 8.53 in Equation 8.55, we get

$$\{\boldsymbol{\phi}_i^{(e)}\}_{global}^T [\boldsymbol{\lambda}^{(e)}]^T \{\boldsymbol{P}^{(e)}\}_{local} = \{\boldsymbol{\phi}_i^{(e)}\}_{global}^T \{\boldsymbol{P}^{(e)}\}_{global} \tag{8.56}$$

that is,

$$[\boldsymbol{\lambda}^{(e)}]^T \{\boldsymbol{P}^{(e)}\}_{local} = \{\boldsymbol{P}^{(e)}\}_{global} \tag{8.57}$$

which gives us

$$[\boldsymbol{\lambda}^{(e)}]^T [\boldsymbol{K}^{(e)}]_{local} \{\boldsymbol{\phi}_i^{(e)}\}_{local} = [\boldsymbol{K}^{(e)}]_{global} \{\boldsymbol{\phi}_i^{(e)}\}_{global} \tag{8.58}$$

We substitute Equation 8.53 in Equation 8.58 and obtain the following:

$$[\boldsymbol{\lambda}^{(e)}]^T [\boldsymbol{K}^{(e)}]_{local} [\boldsymbol{\lambda}^{(e)}] \{\boldsymbol{\phi}_i^{(e)}\}_{global} = [\boldsymbol{K}^{(e)}]_{global} \{\boldsymbol{\phi}_i^{(e)}\}_{global} \tag{8.59}$$

It then leads to

$$[\boldsymbol{K}^{(e)}]_{global} = [\boldsymbol{\lambda}^{(e)}]^T [\boldsymbol{K}^{(e)}]_{local} [\boldsymbol{\lambda}^{(e)}] \tag{8.60}$$

Equation 8.60 gives us element stiffness matrix in the global coordinate system from that in the local coordinate system, where

$[\boldsymbol{K}^{(e)}]_{global}$ Element stiffness matrix of element e in the global coordinates
$[\boldsymbol{K}^{(e)}]_{local}$ Element stiffness matrix of element e in the local coordinates
$[\boldsymbol{\lambda}^{(e)}]$ Transformation matrix for element e

Note that the transformation matrix consists of the direction cosines.

8.4.8 Assembly

The coordinate transformation discussed above is done for expressing all the element characteristic matrices and vectors in one common coordinate (global coordinate) system. The element matrices and vectors are then assembled for developing the system equations. The procedure of assembly is the same for all types of elements. The underlying principle is based on the requirement that at a common node (i.e., a node shared by more than one element), the nodal value must be the same for all the corresponding elements. We shall describe the procedure with the help of a structural mechanics

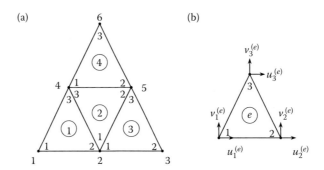

FIGURE 8.6 Assembly of element matrices. (a) Cross section of a triangular prismatic bar discretized with four triangular elements. (b) Definition of the element.

example. The cross section of a prismatic bar is discretized using triangular elements as shown in Figure 8.6a. Element definition is given in Figure 8.6b. Each element has three nodes, each node has two degrees of freedom, and each element has total six degrees of freedom. Thus, the element stiffness matrix is of size 6×6, as indicated below:

$$[K^{(e)}] = \begin{bmatrix} K_{11}^{(e)} & K_{12}^{(e)} & K_{13}^{(e)} & K_{14}^{(e)} & K_{15}^{(e)} & K_{16}^{(e)} \\ K_{12}^{(e)} & K_{22}^{(e)} & K_{23}^{(e)} & K_{24}^{(e)} & K_{25}^{(e)} & K_{26}^{(e)} \\ K_{13}^{(e)} & K_{23}^{(e)} & K_{33}^{(e)} & K_{34}^{(e)} & K_{35}^{(e)} & K_{36}^{(e)} \\ K_{14}^{(e)} & K_{24}^{(e)} & K_{34}^{(e)} & K_{44}^{(e)} & K_{45}^{(e)} & K_{46}^{(e)} \\ K_{15}^{(e)} & K_{25}^{(e)} & K_{35}^{(e)} & K_{45}^{(e)} & K_{55}^{(e)} & K_{56}^{(e)} \\ K_{16}^{(e)} & K_{26}^{(e)} & K_{36}^{(e)} & K_{46}^{(e)} & K_{56}^{(e)} & K_{66}^{(e)} \end{bmatrix} \quad (8.61)$$

There are a total of six nodes in the assembly and 12 degrees of freedom. Thus, the size of the global stiffness matrix is 12×12. For the purpose of assembly, the element stiffness matrix of each element is expanded to a matrix of size 12×12. Toward this, first local node numbers are associated with the respective global node numbers. Then, element stiffness matrix components are placed corresponding to the local degrees of freedom associated with the local nodes. All the vacant places are then filled with zeros (Tables 8.4 through 8.7). Thus, we have four expanded element stiffness matrices, which are algebraically added to obtain the global stiffness matrix. In a general case,

$$[\tilde{K}] = \sum_{e=1}^{E} [\tilde{K}^{(e)}] \quad (8.62)$$

where
- $[\tilde{K}]$ Global stiffness matrix of size $M \times M$
- $[\tilde{K}^{(e)}]$ Expanded element stiffness matrix of size $M \times M$
- E Total number of elements
- M Total number of degrees of freedom in the structure

In a similar way, the global vector of nodal displacements and load vectors are also assembled and we get the following:

$$\{\tilde{\phi}\} = \sum_{i=1}^{E} \{\tilde{\phi}_i^{(e)}\} \quad (8.63)$$

TABLE 8.4
Expanded Element Stiffness Matrix for Element No. 1 (Refer Section 8.4.8)

Global Node Number				1				2			3	4		5	6		
	Local Node Number			1				2			–	3		–	–		
		Global DoF		1	2	3	4	5	6	7	8	9	10	11	12		
			Local DoF	1	2	3	4	–	–	5	6	–	–	–	–
1	1	1	1	$K_{11}^{(1)}$	$K_{12}^{(1)}$	$K_{13}^{(1)}$	$K_{14}^{(1)}$	0	0	$K_{15}^{(1)}$	$K_{16}^{(1)}$	0	0	0	0
		2	2	$K_{12}^{(1)}$	$K_{22}^{(1)}$	$K_{23}^{(1)}$	$K_{24}^{(1)}$	0	0	$K_{25}^{(1)}$	$K_{26}^{(1)}$	0	0	0	0
2	2	3	3	$K_{13}^{(1)}$	$K_{23}^{(1)}$	$K_{33}^{(1)}$	$K_{34}^{(1)}$	0	0	$K_{35}^{(1)}$	$K_{36}^{(1)}$	0	0	0	0
		4	4	$K_{14}^{(1)}$	$K_{24}^{(1)}$	$K_{34}^{(1)}$	$K_{44}^{(1)}$	0	0	$K_{45}^{(1)}$	$K_{46}^{(1)}$	0	0	0	0
3	–	5	–	0	0	0	0	0	0	0	0	0	0	0	0
		6	–	0	0	0	0	0	0	0	0	0	0	0	0
4	3	7	5	$K_{15}^{(1)}$	$K_{25}^{(1)}$	$K_{35}^{(1)}$	$K_{45}^{(1)}$	0	0	$K_{55}^{(1)}$	$K_{56}^{(1)}$	0	0	0	0
		8	6	$K_{16}^{(1)}$	$K_{26}^{(1)}$	$K_{36}^{(1)}$	$K_{46}^{(1)}$	0	0	$K_{56}^{(1)}$	$K_{66}^{(1)}$	0	0	0	0
5	–	9	–	0	0	0	0	0	0	0	0	0	0	0	0
		10	–	0	0	0	0	0	0	0	0	0	0	0	0
6	–	11	–	0	0	0	0	0	0	0	0	0	0	0	0
		12	–	0	0	0	0	0	0	0	0	0	0	0	0

$$\{\tilde{P}\} = \sum_{i}^{E} \{\tilde{P}_i^{(e)}\} \tag{8.64}$$

where

$\{\tilde{\phi}\}$	Global vector of nodal displacements of size $M \times 1$
$\{\tilde{\phi}_i^{(e)}\}$	Expanded element vector of nodal displacements of size $M \times 1$
$\{\tilde{P}\}$	Global load vector of size $M \times 1$
$\{\tilde{P}_i^{(e)}\}$	Expanded element load vector of size $M \times 1$

TABLE 8.5
Expanded Element Stiffness Matrix for Element No. 2 (Refer Section 8.4.8)

Global Node Number				1		2				3	4		5		6	
	Local Node Number			–		1			–		3		2		–	
		Global DoF		1	2	3	4	5	6	7	8	9	10	11	12	
			Local DoF	–	–	1	2	–	–	5	6	3	4	–	–
1	–	1	–	0	0	0	0	0	0	0	0	0	0	0	0
		2	–	0	0	0	0	0	0	0	0	0	0	0	0
2	1	3	1	0	0	$K_{11}^{(2)}$	$K_{12}^{(2)}$	0	0	$K_{15}^{(2)}$	$K_{16}^{(2)}$	$K_{13}^{(2)}$	$K_{14}^{(2)}$	0	0
		4	2	0	0	$K_{12}^{(2)}$	$K_{22}^{(2)}$	0	0	$K_{25}^{(2)}$	$K_{26}^{(2)}$	$K_{23}^{(2)}$	$K_{24}^{(2)}$	0	0
3	–	5	–	0	0	0	0	0	0	0	0	0	0	0	0
		6	–	0	0	0	0	0	0	0	0	0	0	0	0
4	3	7	5	0	0	$K_{15}^{(2)}$	$K_{25}^{(2)}$	0	0	$K_{55}^{(2)}$	$K_{56}^{(2)}$	$K_{35}^{(2)}$	$K_{45}^{(2)}$	0	0
		8	6	0	0	$K_{16}^{(2)}$	$K_{26}^{(2)}$	0	0	$K_{56}^{(2)}$	$K_{66}^{(2)}$	$K_{36}^{(2)}$	$K_{46}^{(2)}$	0	0
5	2	9	3	0	0	$K_{13}^{(2)}$	$K_{23}^{(2)}$	0	0	$K_{35}^{(2)}$	$K_{36}^{(2)}$	$K_{33}^{(2)}$	$K_{34}^{(2)}$	0	0
		10	4	0	0	$K_{14}^{(2)}$	$K_{24}^{(2)}$	0	0	$K_{45}^{(2)}$	$K_{46}^{(2)}$	$K_{34}^{(2)}$	$K_{44}^{(2)}$	0	0
6	–	11	–	0	0	0	0	0	0	0	0	0	0	0	0
		12	–	0	0	0	0	0	0	0	0	0	0	0	0

TABLE 8.6
Expanded Element Stiffness Matrix for Element No. 3 (Refer Section 8.4.8)

Global Node Number				1		2				3		4		5		6	
	Local Node Number			–		1				2		–		3		–	
		Global DoF		1	2	3	4	5	6	7	8	9	10	11	12
			Local DoF	–	–	1	2	3	4	–	–	5	6	–	–
1	–	1	–	0	0	0	0	0	0	0	0	0	0	0	0
		2	–	0	0	0	0	0	0	0	0	0	0	0	0
2	1	3	1	0	0	$K^{(3)}_{11}$	$K^{(3)}_{12}$	$K^{(3)}_{13}$	$K^{(3)}_{14}$	0	0	$K^{(3)}_{15}$	$K^{(3)}_{16}$	0	0
		4	2	0	0	$K^{(3)}_{12}$	$K^{(3)}_{22}$	$K^{(3)}_{23}$	$K^{(3)}_{24}$	0	0	$K^{(3)}_{25}$	$K^{(3)}_{26}$	0	0
3	2	5	3	0	0	$K^{(3)}_{13}$	$K^{(3)}_{23}$	$K^{(3)}_{33}$	$K^{(3)}_{34}$	0	0	$K^{(3)}_{35}$	$K^{(3)}_{36}$	0	0
		6	4	0	0	$K^{(3)}_{14}$	$K^{(3)}_{24}$	$K^{(3)}_{34}$	$K^{(3)}_{44}$	0	0	$K^{(3)}_{45}$	$K^{(3)}_{46}$	0	0
4	–	7	–	0	0	0	0	0	0	0	0	0	0	0	0
		8	–	0	0	0	0	0	0	0	0	0	0	0	0
5	3	9	5	0	0	$K^{(3)}_{15}$	$K^{(3)}_{25}$	$K^{(3)}_{35}$	$K^{(3)}_{45}$	0	0	$K^{(3)}_{55}$	$K^{(3)}_{56}$	0	0
		10	6	0	0	$K^{(3)}_{16}$	$K^{(3)}_{26}$	$K^{(3)}_{36}$	$K^{(3)}_{46}$	0	0	$K^{(3)}_{56}$	$K^{(3)}_{66}$	0	0
6	–	11	–	0	0	0	0	0	0	0	0	0	0	0	0
		12	–	0	0	0	0	0	0	0	0	0	0	0	0

TABLE 8.7
Expanded Element Stiffness Matrix for Element No. 4 (Refer Section 8.4.8)

Global Node Number				1		2		3				4		5		6	
	Local Node Number			–		–		–				1		2		3	
		Global DoF		1	2	3	4	5	6	7	8	9	10	11	12
			Local DoF	–	–	–	–	–	–	1	2	3	4	5	6
1	–	1	–	0	0	0	0	0	0	0	0	0	0	0	0
		2	–	0	0	0	0	0	0	0	0	0	0	0	0
2	–	3	–	0	0	0	0	0	0	0	0	0	0	0	0
		4	–	0	0	0	0	0	0	0	0	0	0	0	0
3	–	5	–	0	0	0	0	0	0	0	0	0	0	0	0
		6	–	0	0	0	0	0	0	0	0	0	0	0	0
4	1	7	1	0	0	0	0	0	0	$K^{(4)}_{11}$	$K^{(4)}_{12}$	$K^{(4)}_{13}$	$K^{(4)}_{14}$	$K^{(4)}_{15}$	$K^{(4)}_{16}$
		8	2	0	0	0	0	0	0	$K^{(4)}_{12}$	$K^{(4)}_{22}$	$K^{(4)}_{23}$	$K^{(4)}_{24}$	$K^{(4)}_{25}$	$K^{(4)}_{26}$
5	2	9	3	0	0	0	0	0	0	$K^{(4)}_{13}$	$K^{(4)}_{23}$	$K^{(4)}_{33}$	$K^{(4)}_{34}$	$K^{(4)}_{35}$	$K^{(4)}_{36}$
		10	4	0	0	0	0	0	0	$K^{(4)}_{14}$	$K^{(4)}_{24}$	$K^{(4)}_{34}$	$K^{(4)}_{44}$	$K^{(4)}_{45}$	$K^{(4)}_{46}$
6	3	11	5	0	0	0	0	0	0	$K^{(4)}_{15}$	$K^{(4)}_{25}$	$K^{(4)}_{35}$	$K^{(4)}_{45}$	$K^{(4)}_{55}$	$K^{(4)}_{56}$
		12	6	0	0	0	0	0	0	$K^{(4)}_{16}$	$K^{(4)}_{26}$	$K^{(4)}_{36}$	$K^{(4)}_{46}$	$K^{(4)}_{56}$	$K^{(4)}_{66}$

The global stiffness matrix developed above leads us to the general system equations of the following form:

$$[\tilde{K}]\{\tilde{\phi}_i\} = \{\tilde{P}\} \tag{8.65}$$

We need to solve these equations to determine the values of $\{\tilde{\phi}_i\}$. However, $[\tilde{K}]$ is a singular matrix as we have not yet imposed the boundary conditions. In solid mechanics problems, it implies that the structure undergoes rigid body movement if no boundary condition is imposed. Thus, boundary conditions are required to be applied so that the equilibrium state is achieved. As we know, there are two types of boundary conditions—geometric or forced or essential boundary conditions, and free or natural boundary conditions. It can be shown that imposition of geometric boundary conditions implicitly means satisfaction of free boundary conditions; thus, we need to impose only the geometric boundary conditions.

Several methods are available for imposing the boundary conditions. We shall briefly examine one such method here.

Equation 8.65 can be reframed by partitioning the square matrix and vectors as follows:

$$\begin{bmatrix} K & K_1 \\ K_2 & K_3 \end{bmatrix} \begin{Bmatrix} \phi_i \\ \phi_1 \end{Bmatrix} = \begin{Bmatrix} P_1 \\ P_2 \end{Bmatrix} \quad (8.66)$$

that is,

$$[K]\{\phi_i\} + [K_1]\{\phi_1\} = \{P_1\} \quad (8.67)$$

and

$$[K_2]\{\phi_i\} + [K_3]\{\phi_1\} = \{P_2\} \quad (8.68)$$

where
- $\{\phi_i\}$ Vector of unknown nodal displacements
- $\{\phi_1\}$ Vector of known nodal displacements (boundary conditions)
- $\{P_1\}$ Vector of known nodal loads (applied forces)
- $\{P_2\}$ Vector of unknown nodal loads (reactions)

Equation 8.67 gives us

$$[K]\{\phi_i\} = \{P\} \quad (8.69)$$

which is the final system equation to be solved. Here,
- $[K]$ Modified stiffness matrix
- $\{\phi_i\}$ Modified vector of nodal displacements (to be determined)
- $\{P\}$ Modified vector of applied nodal loads given by

$$\{P\} = \{P_1\} - [K_1]\{\phi_1\}$$

The nodal reactions are obtained from Equation 8.68.

Note that we have considered an equilibrium problem of structural mechanics for demonstrating the procedure of assembly and imposition of boundary conditions. In the case of other types of problems, in principle, the procedure remains the same with suitable changes.

8.4.9 Solution Methods

The finite element formulations lead to the development of a large number of simultaneous algebraic equations that are required to be solved for the determination of the unknown nodal values. Several solution methods suitable for different types of problems have been developed. Some of the commonly used methods are as follows:

- Equilibrium problems
  - Gaussian elimination method
  - Choleski method
- Eigenvalue problems
  - Transformation methods, for example, Jacobi method
  - Iterative methods, for example, power method and Rayleigh–Ritz subspace method
- Transient problems
  - Runge–Kutta method, Adams–Moulton method, etc.

Equilibrium problems are of the following general form:

$$[A]\{x\} = \{b\} \tag{8.70}$$

where $[A]$ is a square matrix and $\{x\}$ and $\{b\}$ are vectors of which $\{x\}$ is to be determined.

In the Gaussian elimination method, the system of equations are transformed into an equivalent form

$$[A']\{x\} = \{b'\} \tag{8.71}$$

where $[A']$ is an upper triangular matrix (i.e., a square matrix with nonzero elements in the diagonal and above and all zeros below the diagonal). So,

$$[A'] = \begin{bmatrix} a'_{11} & a'_{12} & a'_{13} & \cdots & a'_{1n} \\ 0 & a'_{22} & a'_{23} & \cdots & a'_{2n} \\ 0 & 0 & a'_{33} & \cdots & a'_{3n} \\ \vdots & \vdots & \vdots & \ddots & \vdots \\ 0 & 0 & 0 & 0 & a'_{nn} \end{bmatrix} \tag{8.72}$$

Then the system of equations can be easily solved by back-substitution.

In the Choleski method, the square matrix is expressed as a multiple of a lower triangular matrix and a unit upper triangular matrix, that is, the system equations are written as

$$[L][U]\{x\} = \{b\} \tag{8.73}$$

where

$$[L] = \begin{bmatrix} l_{11} & 0 & 0 & \cdots & 0 \\ l_{12} & l_{22} & 0 & \cdots & 0 \\ l_{13} & l_{23} & l_{33} & \cdots & 0 \\ \vdots & \vdots & \vdots & \ddots & \vdots \\ l_{1n} & l_{2n} & l_{3n} & \cdots & l_{nn} \end{bmatrix} \tag{8.74}$$

and,

$$[U] = \begin{bmatrix} 1 & u_{12} & u_{13} & \cdots & u_{1n} \\ 0 & 1 & u_{23} & \cdots & u_{2n} \\ 0 & 0 & 1 & \cdots & u_{3n} \\ \vdots & \vdots & \vdots & \ddots & \vdots \\ 0 & 0 & 0 & 0 & 1 \end{bmatrix} \quad (8.75)$$

For solving, $[U]\{x\}$ is replaced by a vector $\{z\}$, that is,

$$[U]\{x\} = \{z\} \quad (8.76)$$

which implies

$$\{L\}\{z\} = \{b\} \quad (8.77)$$

First, $[L]$ being a lower triangular matrix, $\{z\}$ is readily solved and then, $\{x\}$ is obtained by back-substitution.

Eigenvalue problems are of the following general form:

$$([A] - \lambda[B])\{x\} = \{0\} \quad (8.78)$$

Here, $[A]$ and $[B]$ are square matrices, λ is the eigenvalue, and $\{x\}$ is the eigenvector. In buckling problems, $[A]$ is the stiffness matrix, $[B]$ is the geometric stiffness matrix, λ is the buckling factor, and $\{x\}$ is the mode shape of buckling displacements. On the other hand, in free vibration problems, $[A]$ is the stiffness matrix, $[B]$ is the mass matrix, λ is the square of natural frequency, and $\{x\}$ is the mode shape of the vibrating body. A meaningful solution to Equation 8.78 can be obtained when $\{x\} \neq \{0\}$ for which the determinant of the square matrix has to be equated to zero. Thus, the eigenvalues are determined from the following:

$$\|[A] - \lambda[B]\| = 0 \quad (8.79)$$

In the expanded form,

$$\begin{vmatrix} a_{11} - \lambda b_{11} & a_{12} - b_{12} & \cdots & a_{1n} - b_{1n} \\ a_{21} - b_{21} & a_{22} - \lambda b_{22} & \cdots & a_{2n} - b_{2n} \\ \vdots & \vdots & \ddots & \vdots \\ a_{n1} - b_{n1} & a_{n2} - b_{n2} & \cdots & a_{nn} - \lambda b_{nn} \end{vmatrix} = 0 \quad (8.80)$$

It results in an algebraic equation, called the characteristic equation, of nth degree for λ. The n roots of this equation are the eigenvalues. For each eigenvalue, the ratios of the components of the eigenvector can be determined. Note that the eigenvector corresponding to an eigenvalue is not unique and it gives only a shape.

There are two types of methods for solving eigenvalue problems—transformation methods and iterative methods. Transformation methods are useful for finding all the eigenvalues and eigenvectors. The Jacobi is a commonly used transformation method that is based on principles of linear algebra. On the other hand, iterative methods are useful for finding limited number of eigenvalues and eigenvectors. For example, the power method allows one to determine the largest eigenvalue and the corresponding eigenvectors, whereas the Rayleigh–Ritz method is used for finding the lowest eigenvalue and the corresponding eigenvectors.

The transient or propagation problems are time-dependent and they are expressed as a set of simultaneous linear differential equations, as follows:

$$[A]\{x\} + [B]\{\dot{x}\} + [C]\{\ddot{x}\} = \{P\} \qquad (8.81)$$

In transient vibration problems, $[A]$ is the stiffness matrix, $[B]$ is the damping matrix, and $[C]$ is the mass matrix. The vector $\{P\}$ is the vector of spatial and time-dependent external loads. Numerical integration methods such as Runge–Kutta method are used for obtaining solution of these problems.

8.5 BASIC FINITE ELEMENT PROCEDURE

The FEM is a highly versatile method useful for solving a very wide variety of problems. It is a vast subject that incorporates concepts from various fields such as mathematics, solid mechanics, fluid mechanics, heat transfer, and so on. The basic principle, however, is simple and it works in a very orderly manner.

In the FEM, the domain or the body is divided into a finite number of elements. Certain basic field variables, for example, displacement or temperature, is identified for the physical problem. The actual variation of the field variable within the body is unknown and its determination is the first objective. However, instead of determining the continuous variation, the values of the field variable at the nodes are determined and the actual variation is approximated within each element by some simple approximating function. Also, the loads and boundary conditions are applied at the nodes. For each element, an element characteristic matrix is obtained by using certain variational principle or equilibrium condition. Now, the focus is shifted from individual elements to the entire body and the field equations are written for the body. Toward this, the element characteristic matrices are assembled to form global field equations. The nodal values of the field variable are obtained by solving these field equations from which other desired information is extracted.

It can be seen from the above that the basic procedure of the FEM has broadly five steps as follows:

Step 1: Discretization or mesh generation—This is the first step in the analysis of a structure or a domain by using FEM.
Step 2: Element formulation
- Identification of variables
- Selection of approximating function
- Formation of element characteristic matrix

  The next step is element formulation, in which the finite element equations are developed for each element leading to the generation of element characteristic matrix. Toward this, first, the field variables are identified. For stress analysis problems, displacement is the basic field variable around which element equations are developed. Second, the approximating function (mostly polynomial function) is chosen. Third, the element characteristic matrix and load vector for each element are derived.
Step 3: Assembly
- Transformation of element characteristic matrices
- Incorporation of boundary conditions
- Formation of global characteristic matrix

Once the element equations are developed, the focus is shifted to the entire body or domain and the element characteristic matrices are appropriately assembled. Element characteristic matrices, developed in the local coordinate systems, are transformed to the global coordinate system before carrying out assembly. Then, boundary conditions are imposed and the final global equations are developed.

Step 4: Solution—The final global equations are solved as discussed in Section 8.4.9. The solution process is a straightforward step in linear problems, where it is a one-step process. However, in nonlinear problems, the loads are applied in several substeps and solutions are obtained for each of these substeps. For each substep, the configuration of the body or domain is modified as per the output of the solution of the previous substep and the process is continued till the final level of loads.

Step 5: Generation of output—The final step in the finite element procedure is the generation of output. Solution in respect of the basic field variable, for example, displacements in stress analysis problems, is utilized to generate other useful data such as strains, stresses, etc.

8.6 DEVELOPMENT OF ELEMENTS

With the knowledge of basic concepts in the FEM, we now consider the development of finite elements in some limited number of specific cases. We shall restrict our discussion to the stress analysis of the following:

- One-dimensional
  - Bar element
  - Torsion element
  - Planar beam element
  - General beam element
- Two-dimensional
  - Rectangular membrane element
  - Rectangular bending plate element
  - Rectangular general plate element
  - Rectangular general plate element with layered composites

8.6.1 One-Dimensional Elements

8.6.1.1 Bar Element

A bar element is a 1D element with only one degree of freedom at each node. Thus, it allows axial extension or compression but no lateral translation or rotation. Figure 8.7 describes the bar element. Only one coordinate axis is needed to describe the element and the element coordinate system is defined by the *x*-axis along the axial direction

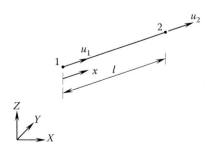

FIGURE 8.7 Bar element.

of the element and its origin at node 1. The length of the element is l and the cross-sectional area is A. The total number of degrees of freedom in the element is two and we consider a polynomial function with two polynomial coefficients as follows:

$$u(x) = a_1 + a_2 x \tag{8.82}$$

In line with the finite element procedure, we express the displacement in terms of shape functions and nodal displacements as follows:

$$\{u(x)\} = [N]\{\phi_i^{(e)}\} \tag{8.83}$$

Note that sizes of the displacement vector $\{u(x)\}$, shape function matrix $[N]$, and nodal displacement vector $\{\phi_i^{(e)}\}$ are $\{u(x)\} \to 1 \times 1$, $[N] \to 1 \times 2$, and $\{\phi_i^{(e)}\} \to 2 \times 1$. Now, with a view to deriving the expression of the shape function matrix, we write the displacement vector as follows:

$$\{u(x)\} = \begin{bmatrix} 1 & x \end{bmatrix} \begin{Bmatrix} a_1 \\ a_2 \end{Bmatrix} \tag{8.84}$$

The nodal values are $\phi_1^{(e)}$ and $\phi_2^{(e)}$, that is, $\phi_1^{(e)} = u(0)$ and $\phi_2^{(e)} = u(l)$. The vector of nodal values $\{\phi_i^{(e)}\}$ is given by

$$\{\phi_i^{(e)}\} \equiv \begin{Bmatrix} \phi_1^{(e)} \\ \phi_2^{(e)} \end{Bmatrix} = \begin{bmatrix} 1 & 0 \\ 1 & l \end{bmatrix} \begin{Bmatrix} a_1 \\ a_2 \end{Bmatrix} \tag{8.85}$$

or

$$\begin{Bmatrix} a_1 \\ a_2 \end{Bmatrix} = \begin{bmatrix} 1 & 0 \\ 1 & l \end{bmatrix}^{-1} \{\phi_i^{(e)}\} \tag{8.86}$$

Substituting Equation 8.86 in Equation 8.84, we get

$$\{u(x)\} = \begin{bmatrix} 1 & x \end{bmatrix} \begin{bmatrix} 1 & 0 \\ 1 & l \end{bmatrix}^{-1} \{\phi_i^{(e)}\} \tag{8.87}$$

Comparing Equation 8.83 with Equation 8.87, we obtain the expression for the matrix of shape functions as

$$[N] = \begin{bmatrix} 1 & x \end{bmatrix} \begin{bmatrix} 1 & 0 \\ 1 & l \end{bmatrix}^{-1} = \begin{bmatrix} \dfrac{l-x}{l} & \dfrac{x}{l} \end{bmatrix} \tag{8.88}$$

From Equation 8.48, we know

$$[K^{(e)}] = \left[\iiint_{V^{(e)}} [B]^T [C] [B] dV \right] \tag{8.89}$$

Finite Element Method

For 1D problem, the strain vector is given by

$$\varepsilon_{xx} = \frac{\partial u(x)}{\partial x} \quad \text{or} \quad \{\varepsilon\} = \left[\frac{\partial}{\partial x}\right]\{u(x)\} \tag{8.90}$$

which, together with Equation 8.33, gives us

$$\{\varepsilon\} = \left[\frac{\partial}{\partial x}\right][N]\{\phi_i^{(e)}\} = [B]\{\phi_i^{(e)}\} \tag{8.91}$$

Thus,

$$[B] = \left[\frac{\partial}{\partial x}\right][N] = \left[\frac{\partial}{\partial x}\right]\left[\frac{l-x}{l} \quad \frac{x}{l}\right] = \left[-\frac{1}{l} \quad \frac{1}{l}\right] \tag{8.92}$$

For 1D problem with isotropic material, the material stiffness matrix is given by $[C] = [E]$, where E is the Young's modulus. Then, the element stiffness matrix is obtained as

$$[K^{(e)}] = A \int_0^l dx \begin{Bmatrix} -\dfrac{1}{l} \\ \dfrac{1}{l} \end{Bmatrix} [E] \begin{bmatrix} -\dfrac{1}{l} & \dfrac{1}{l} \end{bmatrix} \tag{8.93}$$

that is,

$$[K^{(e)}] = \frac{AE}{l}\begin{bmatrix} 1 & -1 \\ -1 & 1 \end{bmatrix} \tag{8.94}$$

Note that the total number of degrees of freedom for the element is two and the size of the element stiffness matrix is (2×2). Further, the element stiffness matrix developed above is in the local coordinate system. It has to be transformed to the global coordinate system, as discussed in Section 8.4.7, before carrying out assembly.

EXAMPLE 8.1

Find the stress at mid-height of the axially loaded bar shown in Figure 8.8a. Take Young's modulus of the bar material as $E = 70$ GPa.

Solution

The bar is discretized using three elements as shown in Figure 8.8b.
Element stiffness matrices are calculated as follows (Equation 8.94):

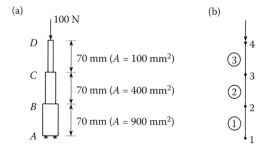

FIGURE 8.8 Axially loaded stepped bar (Example 8.1). (a) Details of the bar. (b) Discretization of the bar.

$$[K^{(1)}] = \frac{900 \times 70{,}000}{70}\begin{bmatrix} 1 & -1 \\ -1 & 1 \end{bmatrix} = \begin{bmatrix} 9 & -9 \\ -9 & 9 \end{bmatrix} \times 10^5$$

$$[K^{(2)}] = \frac{400 \times 70{,}000}{70}\begin{bmatrix} 1 & -1 \\ -1 & 1 \end{bmatrix} = \begin{bmatrix} 4 & -4 \\ -4 & 4 \end{bmatrix} \times 10^5$$

$$[K^{(2)}] = \frac{100 \times 70{,}000}{70}\begin{bmatrix} 1 & -1 \\ -1 & 1 \end{bmatrix} = \begin{bmatrix} 1 & -1 \\ -1 & 1 \end{bmatrix} \times 10^5$$

The assembly of the element stiffness matrices, before imposition of boundary conditions, is done as follows:

$$[\tilde{K}] = \begin{bmatrix} 9 & -9 & 0 & 0 \\ -9 & 9+4 & -4 & 0 \\ 0 & -4 & 4+1 & -1 \\ 0 & 0 & -1 & 1 \end{bmatrix} \times 10^5 = \begin{bmatrix} 9 & -9 & 0 & 0 \\ -9 & 13 & -4 & 0 \\ 0 & -4 & 5 & -1 \\ 0 & 0 & -1 & 1 \end{bmatrix} \times 10^5$$

Similarly, the nodal displacement vector and the nodal load vector, before the imposition of boundary conditions, are

$$\{\tilde{\phi}_i\} = \begin{Bmatrix} \phi_1 \\ \phi_2 \\ \phi_3 \\ \phi_4 \end{Bmatrix}$$

and

$$\{\tilde{P}\} = \begin{Bmatrix} R_1 \\ 0 \\ 0 \\ -100 \end{Bmatrix}$$

The equilibrium equations before the imposition of boundary conditions are given by

$$\begin{bmatrix} 9 & -9 & 0 & 0 \\ -9 & 13 & -4 & 0 \\ 0 & -4 & 5 & -1 \\ 0 & 0 & -1 & 1 \end{bmatrix} \begin{Bmatrix} 0 \\ \phi_2 \\ \phi_3 \\ \phi_4 \end{Bmatrix} \times 10^5 = \begin{Bmatrix} R_1 \\ 0 \\ 0 \\ -100 \end{Bmatrix}$$

The known boundary condition is $\phi_1 = 0$. Using Equations 8.66 through 8.68, the above equation is reframed as follows:

$$\begin{bmatrix} \begin{bmatrix} 13 & -4 & 0 \\ -4 & 5 & -1 \\ 0 & -1 & 1 \end{bmatrix} & \begin{bmatrix} -9 \\ 0 \\ 0 \end{bmatrix} \\ [-9 \quad 0 \quad 0] & [9] \end{bmatrix} \begin{Bmatrix} \begin{Bmatrix} \phi_2 \\ \phi_3 \\ \phi_4 \end{Bmatrix} \\ \{0\} \end{Bmatrix} \times 10^5 = \begin{Bmatrix} \begin{Bmatrix} 0 \\ 0 \\ -100 \end{Bmatrix} \\ \{R_1\} \end{Bmatrix}$$

Finite Element Method

We see that the final global stiffness matrix, nodal displacement vector, and load vector are

$$[K] = \begin{bmatrix} 13 & -4 & 0 \\ -4 & 5 & -1 \\ 0 & -1 & 1 \end{bmatrix} \times 10^5$$

$$\{\phi_i\} = \begin{Bmatrix} \phi_2 \\ \phi_3 \\ \phi_4 \end{Bmatrix}$$

$$\{P\} = \begin{Bmatrix} 0 \\ 0 \\ -100 \end{Bmatrix}$$

Thus, the final equilibrium equations are given by

$$[K] = \begin{bmatrix} 13 & -4 & 0 \\ -4 & 5 & -1 \\ 0 & -1 & 1 \end{bmatrix} \begin{Bmatrix} \phi_2 \\ \phi_3 \\ \phi_4 \end{Bmatrix} \times 10^5 = \begin{Bmatrix} 0 \\ 0 \\ -100 \end{Bmatrix}$$

and

$$[-9 \quad 0 \quad 0] \begin{Bmatrix} \phi_2 \\ \phi_3 \\ \phi_4 \end{Bmatrix} \times 10^5 = \{R_1\}$$

Solving the above equations, we get the nodal displacements and support reaction as follows:

$$\begin{Bmatrix} \phi_2 \\ \phi_3 \\ \phi_4 \end{Bmatrix} = -\begin{Bmatrix} 1.11 \\ 3.61 \\ 13.61 \end{Bmatrix} \times 10^{-4} \text{ mm}$$

and

$$R_1 = -100 \text{ N}$$

For Element 2, matrix $[B]$ is given by (Equation 8.92)

$$[B] = \begin{bmatrix} -\dfrac{1}{l} & \dfrac{1}{l} \end{bmatrix} = \begin{bmatrix} -\dfrac{1}{70} & \dfrac{1}{70} \end{bmatrix}$$

And, the strain vector is given by

$$\{\varepsilon\} = [B]\{\phi_i^{(2)}\} = -\begin{bmatrix} -\dfrac{1}{70} & \dfrac{1}{70} \end{bmatrix} \begin{Bmatrix} 1.11 \\ 3.61 \end{Bmatrix} \times 10^{-4} = -3.57 \times 10^{-6}$$

The corresponding stress is given by

$$\{\sigma\} = [C]\{\varepsilon\} = -(70 \times 10^3) \times (3.57 \times 10^{-6}) = -0.25\,\text{MPa}$$

8.6.1.2 Torsion Element

A torsion element (Figure 8.9) is a 1D element with only one degree of freedom at each node. Thus, it allows rotation about its own axis but no axial extension or compression or lateral translation. The element coordinate system is defined by the x-axis along the axial direction of the element and its origin at node 1. The length of the element is l and the cross-sectional area is A. The total number of degrees of freedom in the element is two and we consider a polynomial function with two polynomial coefficients as follows:

$$\theta_x(x) = a_1 + a_2 x \tag{8.95}$$

The torsional displacement (angle of twist) is expressed in terms of shape functions and nodal displacements as follows:

$$\{\theta_x(x)\} = [N]\{\phi_i^{(e)}\} \tag{8.96}$$

Then, adopting the same procedure as that used in the case of bar element, we can show that the matrix of shape functions is

$$[N] = \begin{bmatrix} \dfrac{l-x}{l} & \dfrac{x}{l} \end{bmatrix} \tag{8.97}$$

The strain–displacement relation, however, is marginally different. Assuming a circular cross section, the strain vector at a distance r from the axis can be expressed as

$$\{\varepsilon\} = r\left[\dfrac{\partial}{\partial x}\right]\{\theta_x(x)\} \tag{8.98}$$

which gives us

$$\{\varepsilon\} = r\left[\dfrac{\partial}{\partial x}\right][N]\{\phi_i^{(e)}\} = [B]\{\phi_i^{(e)}\} \tag{8.99}$$

Thus,

$$[B] = r\left[\dfrac{\partial}{\partial x}\right][N] = r\left[\dfrac{\partial}{\partial x}\right]\begin{bmatrix} \dfrac{l-x}{l} & \dfrac{x}{l} \end{bmatrix} = \begin{bmatrix} -\dfrac{r}{l} & \dfrac{r}{l} \end{bmatrix} \tag{8.100}$$

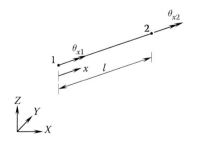

FIGURE 8.9 Torsion element.

For 1D torsional problem with isotropic material, the material stiffness matrix is given by $[C] = [G]$, where G is the shear modulus. Now, the element stiffness matrix is given by Equation 8.48, as follows:

$$[K^{(e)}] = \left[\iiint_{V^{(e)}} [B]^T [C][B] dV \right] \qquad (8.101)$$

which leads to

$$[K^{(e)}] = \int_0^l dx \iint_A r^2 dA \begin{Bmatrix} -\dfrac{1}{l} \\ \dfrac{1}{l} \end{Bmatrix} [G] \begin{bmatrix} -\dfrac{1}{l} & \dfrac{1}{l} \end{bmatrix} \qquad (8.102)$$

that is,

$$[K^{(e)}] = \dfrac{GJ}{l} \begin{bmatrix} 1 & -1 \\ -1 & 1 \end{bmatrix} \qquad (8.103)$$

Note: $J = \iint_A r^2 dA$ is the polar moment of inertia.

8.6.1.3 Planar Beam Element

In this section, we consider a planar beam element that allows lateral translation and rotation. Thus, it can resist bending moment in one principal plane containing the longitudinal axis. It has two nodes and two degrees of freedom at each node. Figure 8.10 describes the beam element. As in the cases of the bar element and torsion element, here too only one coordinate axis is needed to describe the element and the element coordinate system is defined by the x-axis along the axial direction of the element and its origin at node 1. The length of the element is l and the cross-sectional area is A. The bending plane is taken as the xy-plane. The element nodal displacement vector is as follows:

$$\{\phi_i^{(e)}\} \equiv \begin{Bmatrix} \phi_1^{(e)} \\ \phi_2^{(e)} \\ \phi_3^{(e)} \\ \phi_4^{(e)} \end{Bmatrix} = \begin{Bmatrix} v_1 \\ \theta_{z1} \\ v_2 \\ \theta_{z2} \end{Bmatrix} = \begin{Bmatrix} v(0) \\ \theta_z(0) \\ v(l) \\ \theta_z(l) \end{Bmatrix} \qquad (8.104)$$

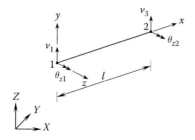

FIGURE 8.10 Planar beam element.

The basic variable is identified as $v(x)$. Note that the rotation $\theta_z(x)$ is not independent and it can be expressed in terms of the other independent displacement as follows:

$$\theta_z(x) = \frac{\partial v(x)}{\partial x} \quad (8.105)$$

The total number of degrees of freedom is four, that is, $n = 4$ and we consider a polynomial function with four polynomial coefficients, as follows:

$$v(x) = a_1 + a_2 x + a_3 x^2 + a_4 x^3 \quad (8.106)$$

The displacement, in terms of shape functions and nodal displacements, is given by

$$\{v(x)\} = [N]\{\phi_i^{(e)}\} \quad (8.107)$$

Note that sizes of the displacement vector, shape function matrix, and nodal displacement vector are $\{v(x)\} \to 1 \times 1$, $[N] \to 1 \times 4$, and $\{\phi_i^{(e)}\} \to 4 \times 1$. Now, with a view to deriving the expression of the shape function matrix, we write the displacement vector as follows:

$$\{v(x)\} = \begin{bmatrix} 1 & x & x^2 & x^3 \end{bmatrix} \begin{Bmatrix} a_1 \\ a_2 \\ a_3 \\ a_4 \end{Bmatrix} \quad (8.108)$$

The nodal values are $\phi_1^{(e)} = v(0)$, $\phi_2^{(e)} = \theta_z(0)$, $\phi_3^{(e)} = v(l)$, and $\phi_4^{(e)} = \theta_z(l)$. Then, using Equations 8.104, 8.105, and 8.108, the vector of nodal values $\{\phi_i^{(e)}\}$ is expressed as

$$\{\phi_i^{(e)}\} = \begin{Bmatrix} v_1 \\ \theta_{z1} \\ v_2 \\ \theta_{z2} \end{Bmatrix} = \begin{bmatrix} 1 & 0 & 0 & 0 \\ 0 & 1 & 0 & 0 \\ 1 & l & l^2 & l^3 \\ 0 & 1 & 2l & 3l^2 \end{bmatrix} \begin{Bmatrix} a_1 \\ a_2 \\ a_3 \\ a_4 \end{Bmatrix} \quad (8.109)$$

or

$$\begin{Bmatrix} a_1 \\ a_2 \\ a_3 \\ a_4 \end{Bmatrix} = \begin{bmatrix} 1 & 0 & 0 & 0 \\ 0 & 1 & 0 & 0 \\ 1 & l & l^2 & l^3 \\ 0 & 1 & 2l & 3l^2 \end{bmatrix}^{-1} \{\phi_i^{(e)}\} \quad (8.110)$$

Substituting Equation 8.110 in Equation 8.108, we get

$$\{v(x)\} = \begin{bmatrix} 1 & x & x^2 & x^3 \end{bmatrix} \begin{bmatrix} 1 & 0 & 0 & 0 \\ 0 & 1 & 0 & 0 \\ 1 & l & l^2 & l^3 \\ 0 & 1 & 2l & 3l^2 \end{bmatrix}^{-1} \{\phi_i^{(e)}\} \quad (8.111)$$

Comparing Equation 8.107 with Equation 8.111, we obtain the expression for the matrix of shape functions as

$$[N] = [1 \quad x \quad x^2 \quad x^3] \begin{bmatrix} 1 & 0 & 0 & 0 \\ 0 & 1 & 0 & 0 \\ 1 & l & l^2 & l^3 \\ 0 & 1 & 2l & 3l^2 \end{bmatrix}^{-1} \quad (8.112)$$

After carrying out the algebra involved in the above expression, we can show that the shape functions are given by

$$[N] = \frac{1}{l^3}[(2x^3 - 3lx^2 + l^3) \quad (lx^3 - 2l^2x^2 + l^3x) \quad (-2x^3 + 3lx^2) \quad (lx^3 - l^2x^2)] \quad (8.113)$$

From basic beam theory, the strain–displacement relation can be expressed as

$$\varepsilon_{xx} = y \frac{\partial^2 v(x)}{\partial x^2} \quad (8.114)$$

or in the matrix form

$$\{\varepsilon\} = y \left[\frac{\partial^2}{\partial x^2}\right] \{v(x)\} \quad (8.115)$$

which gives us

$$\{\varepsilon\} = y \left[\frac{\partial^2}{\partial x^2}\right] [N]\{\phi_i^{(e)}\} = [B]\{\phi_i^{(e)}\} \quad (8.116)$$

Thus

$$[B] = y \left[\frac{\partial^2}{\partial x^2}\right] [N] \quad (8.117)$$

Using Equation 8.113 in Equation 8.117, we get

$$[B] = \frac{y}{l^3}[(12x - 6l) \quad (6lx - 4l^2) \quad (-12x + 6l) \quad (6lx - 2l^2)] \quad (8.118)$$

For 1D bending problem with isotropic material, the material stiffness matrix is given by $[C] = [E]$, where E is the Young's modulus. Now, the element stiffness matrix is given by Equation 8.48, as follows:

$$[K^{(e)}] = \left[\iiint_{V^{(e)}} [B]^T[C][B]dV\right] \quad (8.119)$$

which leads to

$$[K^{(e)}] = \iint_A y^2 \, dA \int_0^l \frac{1}{l^6} \begin{Bmatrix} 12x-6l \\ 6lx-4l^2 \\ -12x+6l \\ 6lx-2l^2 \end{Bmatrix} [E] \begin{Bmatrix} 12x-6l \\ 6lx-4l^2 \\ -12x+6l \\ 6lx-2l^2 \end{Bmatrix}^T dx \quad (8.120)$$

Upon carrying out the algebraic manipulation involved in the above expression and noting that $I_{zz} = \iint_A y^2 \, dA$ is the area moment of inertia about the z-axis, we can show that

$$[K^{(e)}] = \frac{EI_{zz}}{l^3} \begin{bmatrix} 12 & 6l & -12 & 6l \\ 6l & 4l^2 & -6l & 2l^2 \\ -12 & -6l & 12 & -6l \\ 6l & 2l^2 & -6l & 4l^2 \end{bmatrix} \quad (8.121)$$

8.6.1.4 General Beam Element

In this section, we consider a general beam element (Figure 8.11) that allows axial extension or compression, lateral translation, and rotation. Thus, it can resist axial force, bending moments in the two principal planes containing the longitudinal axis, and twisting moment about its longitudinal axis. It has two nodes and six degrees of freedom at each node. As in the case of the bar element, here too only one coordinate axis is needed to describe the element and the element coordinate system is defined by the x-axis along the axial direction of the element and its origin at node 1. The length of the element is l and the cross-sectional area is A. The element nodal displacement vector is as follows:

$$\{\phi_i^{(e)}\} \equiv \begin{Bmatrix} \phi_1^{(e)} \\ \phi_2^{(e)} \\ \phi_3^{(e)} \\ \phi_4^{(e)} \\ \phi_5^{(e)} \\ \phi_6^{(e)} \\ \phi_7^{(e)} \\ \phi_8^{(e)} \\ \phi_9^{(e)} \\ \phi_{10}^{(e)} \\ \phi_{11}^{(e)} \\ \phi_{12}^{(e)} \end{Bmatrix} = \begin{Bmatrix} u_1 \\ v_1 \\ w_1 \\ \theta_{x1} \\ \theta_{y1} \\ \theta_{z1} \\ u_2 \\ v_2 \\ w_2 \\ \theta_{x2} \\ \theta_{y2} \\ \theta_{z2} \end{Bmatrix} = \begin{Bmatrix} u(0) \\ v(0) \\ w(0) \\ \theta_x(0) \\ \theta_y(0) \\ \theta_z(0) \\ u(l) \\ v(l) \\ w(l) \\ \theta_x(l) \\ \theta_y(l) \\ \theta_z(l) \end{Bmatrix} \quad (8.122)$$

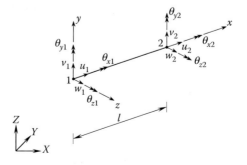

FIGURE 8.11 Beam element.

Finite Element Method

The basic variables are identified as $u(x)$, $v(x)$, $w(x)$, and $\theta_x(x)$. Note that there are two more rotations, $\theta_y(x)$ and $\theta_z(x)$, but they are not independent and can be expressed in terms of the other independent displacements as follows:

$$\theta_y(x) = \frac{\partial w(x)}{\partial x} \quad \text{and} \quad \theta_z(x) = \frac{\partial v(x)}{\partial x} \tag{8.123}$$

The total number of degrees of freedom is 12, that is, $n = 12$, and we consider the following polynomial functions with 12 polynomial coefficients, as follows:

$$\begin{Bmatrix} u(x) \\ v(x) \\ w(x) \\ \theta_x(x) \end{Bmatrix} = \begin{Bmatrix} a_1 + a_5 x \\ a_2 + a_6 x + a_9 x^2 + a_{11} x^3 \\ a_3 + a_7 x + a_{10} x^2 + a_{12} x^3 \\ a_4 + a_8 x \end{Bmatrix} \tag{8.124}$$

Note that there are four independent displacements: $u(x)$, $v(x)$, $w(x)$, and $\theta_x(x)$. The polynomial functions for these displacements are the same as discussed in the previous three cases and the element stiffness matrix can actually be derived by appropriate superposition. For ready reference, let us write the element stiffness matrices for bar, torsion, and planar beam elements below:

$$\left[K_a^{(e)} \right] = \frac{AE}{l} \begin{bmatrix} 1 & -1 \\ -1 & 1 \end{bmatrix} \tag{8.125}$$

$$\left[K_t^{(e)} \right] = \frac{GJ}{l} \begin{bmatrix} 1 & -1 \\ -1 & 1 \end{bmatrix} \tag{8.126}$$

$$\left[K_{xy}^{(e)} \right] = \frac{EI_{zz}}{l^3} \begin{bmatrix} 12 & 6l & -12 & 6l \\ 6l & 4l^2 & -6l & 2l^2 \\ -12 & -6l & 12 & -6l \\ 6l & 2l^2 & -6l & 4l^2 \end{bmatrix} \tag{8.127}$$

$$\left[K_{xz}^{(e)} \right] = \frac{EI_{yy}}{l^3} \begin{bmatrix} 12 & 6l & -12 & 6l \\ 6l & 4l^2 & -6l & 2l^2 \\ -12 & -6l & 12 & -6l \\ 6l & 2l^2 & -6l & 4l^2 \end{bmatrix} \tag{8.128}$$

where

- $\left[K_a^{(e)} \right]$ Element stiffness matrix for bar element (with elements $K_{a11}^{(e)}$, $K_{a12}^{(e)}$, and $K_{a22}^{(e)}$)
- $\left[K_t^{(e)} \right]$ Element stiffness matrix for torsion element (with elements $K_{t11}^{(e)}$, $K_{t12}^{(e)}$, and $K_{t22}^{(e)}$)
- $\left[K_{xy}^{(e)} \right]$ Element stiffness matrix for planar bending element in the xy-plane (with elements $K_{xy11}^{(e)}$, $K_{xy12}^{(e)}$, ..., $K_{xy44}^{(e)}$)
- $\left[K_{xz}^{(e)} \right]$ Element stiffness matrix for planar bending element in the xz-plane (with elements $K_{xz11}^{(e)}$, $K_{xz12}^{(e)}$, ..., $K_{xz44}^{(e)}$)

TABLE 8.8
Stiffness Matrix of a General Beam Element

	$\phi_1^{(e)}$	$\phi_2^{(e)}$	$\phi_3^{(e)}$	$\phi_4^{(e)}$	$\phi_5^{(e)}$	$\phi_6^{(e)}$	$\phi_7^{(e)}$	$\phi_8^{(e)}$	$\phi_9^{(e)}$	$\phi_{10}^{(e)}$	$\phi_{11}^{(e)}$	$\phi_{12}^{(e)}$
$\phi_1^{(e)}$	$K_{a11}^{(e)}$	0	0	0	0	0	$K_{a12}^{(e)}$	0	0	0	0	0
$\phi_2^{(e)}$	0	$K_{xy11}^{(e)}$	0	0	0	$K_{xy12}^{(e)}$	0	$K_{xy13}^{(e)}$	0	0	0	$K_{xy14}^{(e)}$
$\phi_3^{(e)}$	0	0	$K_{xz11}^{(e)}$	0	$K_{xz12}^{(e)}$	0	0	0	$K_{xz13}^{(e)}$	0	$K_{xz14}^{(e)}$	0
$\phi_4^{(e)}$	0	0	0	$K_{t11}^{(e)}$	0	0	0	0	0	$K_{t12}^{(e)}$	0	0
$\phi_5^{(e)}$	0	0	$K_{xz12}^{(e)}$	0	$K_{xz22}^{(e)}$	0	0	0	$K_{xz23}^{(e)}$	0	$K_{xz24}^{(e)}$	0
$\phi_6^{(e)}$	0	$K_{xy12}^{(e)}$	0	0	0	$K_{xy22}^{(e)}$	0	$K_{xy23}^{(e)}$	0	0	0	$K_{xy24}^{(e)}$
$\phi_7^{(e)}$	$K_{a12}^{(e)}$	0	0	0	0	0	$K_{a22}^{(e)}$	0	0	0	0	0
$\phi_8^{(e)}$	0	$K_{xy13}^{(e)}$	0	0	0	$K_{xy23}^{(e)}$	0	$K_{xy33}^{(e)}$	0	0	0	$K_{xy34}^{(e)}$
$\phi_9^{(e)}$	0	0	$K_{xz13}^{(e)}$	0	$K_{xz23}^{(e)}$	0	0	0	$K_{xz33}^{(e)}$	0	$K_{xz34}^{(e)}$	0
$\phi_{10}^{(e)}$	0	0	0	$K_{t12}^{(e)}$	0	0	0	0	0	$K_{t22}^{(e)}$	0	0
$\phi_{11}^{(e)}$	0	0	$K_{xz14}^{(e)}$	0	$K_{xz24}^{(e)}$	0	0	0	$K_{xz34}^{(e)}$	0	$K_{xz44}^{(e)}$	0
$\phi_{12}^{(e)}$	0	$K_{xy14}^{(e)}$	0	0	0	$K_{xy24}^{(e)}$	0	$K_{xy34}^{(e)}$	0	0	0	$K_{xy44}^{(e)}$

The element stiffness matrix for bar element $\left[K_a^{(e)}\right]$ is associated with the axial displacements $\phi_1^{(e)} = u_1$ and $\phi_7^{(e)} = u_2$, which correspond to the first and seventh rows in the element displacement vector (Equation 8.122). Thus, in the superimposed matrix, the elements of $\left[K_a^{(e)}\right]$ are placed in the first and seventh rows and columns. In a similar way, the elements of $\left[K_t^{(e)}\right]$, $\left[K_{xy}^{(e)}\right]$, and $\left[K_{xz}^{(e)}\right]$ are placed, and the final element stiffness matrix for a general beam element is obtained as given in Table 8.8.

8.6.2 Two-Dimensional Elements

8.6.2.1 Rectangular Membrane Element

A rectangular membrane element (Figure 8.12) is a 2D element. It is considered to lie in the xy-plane. The element coordinate system is defined by x- and y-axes along the longitudinal and lateral directions, respectively, and its origin at node 1. The length and breadth of the element are l and b, respectively, and the thickness is h. There are two degrees of freedom at each node and the element allows in-plane extension or compression but no transverse translation or rotation.

The basic variables are identified as $u = u(x,y)$ and $v = v(x,y)$. The total number of degrees of freedom in the element is eight and we consider polynomial functions with eight polynomial coefficients as follows:

$$\begin{Bmatrix} u(x,y) \\ v(x,y) \end{Bmatrix} = \begin{Bmatrix} a_1 + a_3 x + a_5 y + a_7 xy \\ a_2 + a_4 x + a_6 y + a_8 xy \end{Bmatrix} \tag{8.129}$$

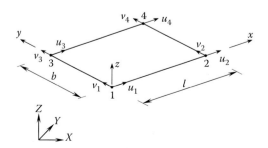

FIGURE 8.12 Rectangular membrane element.

or

$$\begin{Bmatrix} u(x,y) \\ v(x,y) \end{Bmatrix} = [g(x,y)]\{a\} \quad (8.130)$$

where $[g(x,y)]$ is a matrix (size: 2×8) of polynomial variables and $\{a\}$ is a vector (size: 8×1) of polynomial coefficients that are given by

$$[g(x,y)] = \begin{bmatrix} 1 & 0 & x & 0 & y & 0 & xy & 0 \\ 0 & 1 & 0 & x & 0 & y & 0 & xy \end{bmatrix} \quad (8.131)$$

and

$$\{a\} = \begin{Bmatrix} a_1 \\ a_2 \\ \vdots \\ a_8 \end{Bmatrix} \quad (8.132)$$

The nodal displacement vector for the element is as follows:

$$\{\phi_i^{(e)}\} \equiv \begin{Bmatrix} \phi_1^{(e)} \\ \phi_2^{(e)} \\ \phi_3^{(e)} \\ \phi_4^{(e)} \\ \phi_5^{(e)} \\ \phi_6^{(e)} \\ \phi_7^{(e)} \\ \phi_8^{(e)} \end{Bmatrix} = \begin{Bmatrix} u_1 \\ v_1 \\ u_2 \\ v_2 \\ u_3 \\ v_3 \\ u_4 \\ v_4 \end{Bmatrix} = \begin{Bmatrix} u(0,0) \\ v(0,0) \\ u(l,0) \\ v(l,0) \\ u(0,b) \\ v(0,b) \\ u(l,b) \\ v(l,b) \end{Bmatrix} \quad (8.133)$$

Then, using Equations 8.130 and 8.133, the vector of nodal displacements $\{\phi_i^{(e)}\}$ and the vector of polynomial coefficients $\{a\}$ can be related as

$$\{\phi_i^{(e)}\} = [X]\{a\} \quad \text{or} \quad \{a\} = [X]^{-1}\{\phi_i^{(e)}\} \quad (8.134)$$

where

$$[X] = \begin{bmatrix} 1 & 0 & 0 & 0 & 0 & 0 & 0 & 0 \\ 0 & 1 & 0 & 0 & 0 & 0 & 0 & 0 \\ 1 & 0 & l & 0 & 0 & 0 & 0 & 0 \\ 0 & 1 & 0 & l & 0 & 0 & 0 & 0 \\ 1 & 0 & 0 & 0 & b & 0 & 0 & 0 \\ 0 & 1 & 0 & 0 & 0 & b & 0 & 0 \\ 1 & 0 & l & 0 & b & 0 & lb & 0 \\ 0 & 1 & 0 & l & 0 & b & 0 & lb \end{bmatrix} \quad (8.135)$$

Substituting Equation 8.134 in Equation 8.130, we get

$$\begin{Bmatrix} u(x,y) \\ v(x,y) \end{Bmatrix} = [g(x,y)][X]^{-1}\{\phi_i^{(e)}\} \quad (8.136)$$

The displacements, in terms of shape functions and nodal displacements, are given by

$$\begin{Bmatrix} u(x,y) \\ v(x,y) \end{Bmatrix} = [N]\{\phi_i^{(e)}\} \qquad (8.137)$$

Comparing Equation 8.136 with Equation 8.137, we can express the matrix of shape functions as

$$[N] = [g(x,y)][X]^{-1} \qquad (8.138)$$

Note that the size of the shape function matrix $[N]$ is 2×8.

From basic plate theory, the strain–displacement relation can be expressed as

$$\{\varepsilon\} \equiv \begin{Bmatrix} \varepsilon_{xx} \\ \varepsilon_{yy} \\ \gamma_{xy} \end{Bmatrix} = \begin{Bmatrix} \dfrac{\partial u}{\partial x} \\ \dfrac{\partial v}{\partial y} \\ \dfrac{\partial u}{\partial y} + \dfrac{\partial v}{\partial x} \end{Bmatrix} \qquad (8.139)$$

which can also be written as

$$\{\varepsilon\} = \begin{bmatrix} \dfrac{\partial}{\partial x} & 0 \\ 0 & \dfrac{\partial}{\partial y} \\ \dfrac{\partial}{\partial y} & \dfrac{\partial}{\partial x} \end{bmatrix} \begin{Bmatrix} u(x,y) \\ v(x,y) \end{Bmatrix} \qquad (8.140)$$

which gives us

$$\{\varepsilon\} = \begin{bmatrix} \dfrac{\partial}{\partial x} & 0 \\ 0 & \dfrac{\partial}{\partial y} \\ \dfrac{\partial}{\partial y} & \dfrac{\partial}{\partial x} \end{bmatrix} [N]\{\phi_i^{(e)}\} = [B]\{\phi_i^{(e)}\} \qquad (8.141)$$

where

$$[B] = \begin{bmatrix} \dfrac{\partial}{\partial x} & 0 \\ 0 & \dfrac{\partial}{\partial y} \\ \dfrac{\partial}{\partial y} & \dfrac{\partial}{\partial x} \end{bmatrix} [N] \qquad (8.142)$$

Using Equation 8.138 in Equation 8.142, we get

$$[B] = \begin{bmatrix} \dfrac{\partial}{\partial x} & 0 \\ 0 & \dfrac{\partial}{\partial y} \\ \dfrac{\partial}{\partial y} & \dfrac{\partial}{\partial x} \end{bmatrix} [g(x,y)][X]^{-1} \quad (8.143)$$

After carrying out the partial differentiation on [$g(x,y)$], we can show that

$$[B] = [G(x,y)][X]^{-1} \quad (8.144)$$

where

$$[G(x,y)] = \begin{bmatrix} 0 & 0 & 1 & 0 & 0 & 0 & y & 0 \\ 0 & 0 & 0 & 0 & 0 & 1 & 0 & x \\ 0 & 0 & 0 & 1 & 1 & 0 & x & y \end{bmatrix} \quad (8.145)$$

For 2D membrane problem with isotropic material, the material stiffness matrix is given by

$$[C] = \dfrac{E}{1-\nu^2} \begin{bmatrix} 1 & \nu & 0 \\ \nu & 1 & 0 \\ 0 & 0 & \dfrac{1-\nu}{2} \end{bmatrix} \quad (8.146)$$

where E and ν are the Young's modulus and Poisson's ratio, respectively. Now, the element stiffness matrix is given by Equation 8.48, as follows:

$$[K^{(e)}] = \left[\iiint\limits_{V^{(e)}} [B]^T [C][B] dV \right] \quad (8.147)$$

or

$$[K^{(e)}] = h \int_0^b \int_0^l [B]^T [C][B] \, dx \, dy \quad (8.148)$$

Note that in the above expression, [C] is a known matrix of size 3 × 3. [B] is given by Equation 8.144 in terms of element coordinates (x,y) and element length and breadth (l,b). Clearly, upon integration, we get [$K^{(e)}$] in terms of the element dimensions (l,b,h) and material properties (E,ν). Note further that the size of [B] is 3 × 8 and that of [$K^{(e)}$] is 8 × 8.

8.6.2.2 Rectangular Bending Plate Element

A rectangular bending plate element (Figure 8.13) is a 2D element. It is considered to lie in the xy-plane. The element coordinate system is defined by x- and y-axes along the longitudinal and lateral directions, respectively, and its origin at node 1. The length and breadth of the element are l and b, respectively, and the thickness is h. There are three degrees of freedom at each node and the element allows transverse (out-of-plane) translation and rotations about the x- and y-axes but no in-plane extension or compression and rotation about the z-axis.

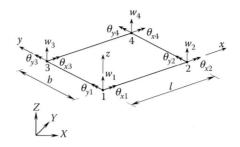

FIGURE 8.13 Rectangular bending plate element.

The basic variable is identified as $w(x,y)$. Note that the rotations $\theta_x(x,y)$ and $\theta_y(x,y)$ are not independent and they can be expressed in terms of the independent displacement as follows:

$$\theta_x(x,y) = \frac{\partial w(x,y)}{\partial y} \quad \text{and} \quad \theta_y(x,y) = \frac{\partial w(x,y)}{\partial x} \quad (8.149)$$

The total number of degrees of freedom in the element is 12 and we consider a polynomial function with 12 polynomial coefficients as follows:

$$w(x,y) = a_1 + a_2 x + a_3 y + a_4 x^2 + a_5 xy + a_6 y^2 + a_7 x^3 + a_8 x^2 y + a_9 xy^2 \\ + a_{10} y^3 + a_{11} x^3 y + a_{12} xy^3 \quad (8.150)$$

or

$$\{w(x,y)\} = [g(x,y)]\{a\} \quad (8.151)$$

where $[g(x,y)]$ is a matrix (size: 1×12) of polynomial variables and $\{a\}$ is a vector (size: 12×1) of polynomial coefficients that are given by

$$[g(x,y)] = [1 \quad x \quad y \quad x^2 \quad xy \quad y^2 \quad x^3 \quad x^2 y \quad xy^2 \quad y^3 \quad x^3 y \quad xy^3] \quad (8.152)$$

and

$$\{a\} = \begin{Bmatrix} a_1 \\ a_2 \\ \vdots \\ a_{12} \end{Bmatrix} \quad (8.153)$$

The nodal displacement vector for the element is as follows:

$$\{\phi_i^{(e)}\} \equiv \begin{Bmatrix} \phi_1^{(e)} \\ \phi_2^{(e)} \\ \phi_3^{(e)} \\ \phi_4^{(e)} \\ \phi_5^{(e)} \\ \phi_6^{(e)} \\ \phi_7^{(e)} \\ \phi_8^{(e)} \\ \phi_9^{(e)} \\ \phi_{10}^{(e)} \\ \phi_{11}^{(e)} \\ \phi_{12}^{(e)} \end{Bmatrix} = \begin{Bmatrix} w_1 \\ \theta_{x1} \\ \theta_{y1} \\ w_2 \\ \theta_{x2} \\ \theta_{y2} \\ w_3 \\ \theta_{x3} \\ \theta_{y3} \\ w_4 \\ \theta_{x4} \\ \theta_{y4} \end{Bmatrix} = \begin{Bmatrix} w(0,0) \\ \theta_x(0,0) \\ \theta_y(0,0) \\ w(l,0) \\ \theta_x(l,0) \\ \theta_y(l,0) \\ w(0,b) \\ \theta_x(0,b) \\ \theta_y(0,b) \\ w(l,b) \\ \theta_x(l,b) \\ \theta_y(l,b) \end{Bmatrix} \quad (8.154)$$

Using Equations 8.149 and 8.150 in Equation 8.154, the vector of nodal displacements $\{\phi_i^{(e)}\}$ and the vector of polynomial coefficients $\{a\}$ can be related as

$$\{\phi_i^{(e)}\} = [X]\{a\} \quad \text{or} \quad \{a\} = [X]^{-1}\{\phi_i^{(e)}\} \tag{8.155}$$

where

$$[X] = \begin{bmatrix} 1 & 0 & 0 & 0 & 0 & 0 & 0 & 0 & 0 & 0 & 0 & 0 \\ 0 & 0 & 1 & 0 & 0 & 0 & 0 & 0 & 0 & 0 & 0 & 0 \\ 0 & 1 & 0 & 0 & 0 & 0 & 0 & 0 & 0 & 0 & 0 & 0 \\ 1 & l & 0 & l^2 & 0 & 0 & l^3 & 0 & 0 & 0 & 0 & 0 \\ 0 & 0 & 1 & 0 & l & 0 & 0 & l^2 & 0 & 0 & l^3 & 0 \\ 0 & 1 & 0 & 2l & 0 & 0 & 3l^2 & 0 & 0 & 0 & 0 & 0 \\ 1 & 0 & b & 0 & 0 & b^2 & 0 & 0 & 0 & b^3 & 0 & 0 \\ 0 & 0 & 1 & 0 & 0 & 2b & 0 & 0 & 0 & 3b^2 & 0 & 0 \\ 0 & 1 & 0 & 0 & b & 0 & 0 & 0 & b^2 & 0 & 0 & b^3 \\ 1 & l & b & l^2 & lb & b^2 & l^3 & l^2b & lb^2 & b^3 & l^3b & lb^3 \\ 0 & 0 & 1 & 0 & l & 2b & 0 & l^2 & 2lb & 3b^2 & l^3 & 3lb^2 \\ 0 & 1 & 0 & 2l & b & 0 & 3l^2 & 2lb & b^2 & 0 & 3l^2b & b^3 \end{bmatrix} \tag{8.156}$$

Substituting Equation 8.155 in Equation 8.151, we get

$$\{w(x,y)\} = [g(x,y)][X]^{-1}\{\phi_i^{(e)}\} \tag{8.157}$$

The displacements, in terms of shape functions and nodal displacements, are also given by

$$\{w(x,y)\} = [N]\{\phi_i^{(e)}\} \tag{8.158}$$

Comparing Equation 8.158 with Equation 8.157, we obtain the matrix of shape functions as

$$[N] = [g(x,y)][X]^{-1} \tag{8.159}$$

Note that the size of the shape function matrix $[N]$ is 1×12.

From basic plate-bending theory, the strain–displacement relation can be expressed as

$$\{\varepsilon\} \equiv \begin{Bmatrix} \varepsilon_{xx} \\ \varepsilon_{yy} \\ \gamma_{xy} \end{Bmatrix} = -z \begin{Bmatrix} \dfrac{\partial^2 w}{\partial x^2} \\ \dfrac{\partial^2 w}{\partial y^2} \\ 2\dfrac{\partial^2 w}{\partial x \partial y} \end{Bmatrix} \tag{8.160}$$

which can also be written as

$$\{\varepsilon\} = -z \begin{Bmatrix} \dfrac{\partial^2}{\partial x^2} \\ \dfrac{\partial^2}{\partial y^2} \\ 2\dfrac{\partial^2}{\partial x \partial y} \end{Bmatrix} \{w(x,y)\} \tag{8.161}$$

which, in turn, gives us

$$\{\varepsilon\} = -z \begin{Bmatrix} \dfrac{\partial^2}{\partial x^2} \\ \dfrac{\partial^2}{\partial y^2} \\ 2\dfrac{\partial^2}{\partial x \partial y} \end{Bmatrix} [N]\{\phi_i^{(e)}\} = -z[B]\{\phi_i^{(e)}\} \tag{8.162}$$

Thus,

$$[B] = \begin{Bmatrix} \dfrac{\partial^2}{\partial x^2} \\ \dfrac{\partial^2}{\partial y^2} \\ 2\dfrac{\partial^2}{\partial x \partial y} \end{Bmatrix} [N] = \begin{Bmatrix} \dfrac{\partial^2}{\partial x^2} \\ \dfrac{\partial^2}{\partial y^2} \\ 2\dfrac{\partial^2}{\partial x \partial y} \end{Bmatrix} [g(x,y)][X]^{-1} \tag{8.163}$$

After carrying out the partial differentiation on $[g(x,y)]$, we can show that

$$[B] = [G(x,y)][X]^{-1} \tag{8.164}$$

where

$$[G(x,y)] = \begin{bmatrix} 0 & 0 & 0 & 2 & 0 & 0 & 6x & 2y & 0 & 0 & 6xy & 0 \\ 0 & 0 & 0 & 0 & 0 & 2 & 0 & 0 & 2x & 6y & 0 & 6xy \\ 0 & 0 & 0 & 0 & 2 & 0 & 0 & 4x & 4y & 0 & 6x^2 & 6y^2 \end{bmatrix} \tag{8.165}$$

For 2D plate problem with isotropic material, the material stiffness matrix is given by

$$[C] = \dfrac{E}{1-\nu^2} \begin{bmatrix} 1 & \nu & 0 \\ \nu & 1 & 0 \\ 0 & 0 & \dfrac{1-\nu}{2} \end{bmatrix} \tag{8.166}$$

where E and ν are the Young's modulus and Poisson's ratio, respectively. Now, the standard expression for the element stiffness matrix is given by Equation 8.48, as follows:

$$[K^{(e)}] = \left[\iiint_{V^{(e)}} [B]^T [C][B] dV \right] \quad (8.167)$$

Note that here we have to replace $[B]$ with $-z[B]$ (Equation 8.162). Then, we can write

$$[K^{(e)}] = \int_{-h/2}^{h/2} z^2 \, dz \int_0^b \int_0^l [B]^T [C][B] \, dx \, dy \quad (8.168)$$

or

$$[K^{(e)}] = \frac{h^3}{12} \int_0^b \int_0^l [B]^T [C][B] \, dx \, dy \quad (8.169)$$

Note that in the above expression, $[C]$ is a known matrix of size 3×3. $[B]$ is given by Equation 8.164 together with Equations 8.156 and 8.165 in terms of element coordinates (x,y) and element length and breadth (l,b). Clearly, upon integration, we get $[K^{(e)}]$ in terms of the element dimensions (l,b,h) and material properties (E,ν). Note further that the size of $[B]$ is 3×12 and that of $[K^{(e)}]$ is 12×12.

8.6.2.3 Rectangular General Plate Element

A rectangular general plate element (Figure 8.14) is a 2D element. It is considered to lie in the xy-plane. The element coordinate system is defined by x- and y-axes along the longitudinal and lateral directions, respectively, and its origin at node 1. The length and breadth of the element are l and b, respectively, and the thickness is h. There are five degrees of freedom at each node—three translations and two rotations. This element can be considered as a combination of the previous two plate elements, viz. rectangular membrane plate element and rectangular bending plate element. The basic variables are identified as $u(x,y)$, $v(x,y)$, and $w(x,y)$. As we know, the rotations $\theta_x(x,y)$ and $\theta_y(x,y)$ are

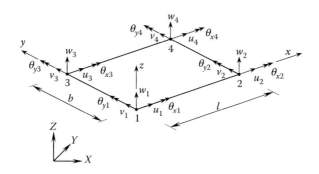

FIGURE 8.14 Rectangular general plate element.

not independent and they can be expressed in terms of the independent displacement as follows:

$$\theta_x(x,y) = \frac{\partial w(x,y)}{\partial y} \quad \text{and} \quad \theta_y(x,y) = \frac{\partial w(x,y)}{\partial x} \tag{8.170}$$

The total number of degrees of freedom in the element is 20 and we consider polynomial functions with 20 polynomial coefficients as follows:

$$\begin{Bmatrix} u(x,y) \\ v(x,y) \\ w(x,y) \end{Bmatrix} = \begin{Bmatrix} a_1 + a_4 x + a_7 y + a_{10} x^2 + a_{13} xy + a_{16} y^2 \\ a_2 + a_5 x + a_8 y + a_{11} x^2 + a_{14} xy + a_{17} y^2 \\ a_3 + a_6 x + a_9 y + a_{12} x^2 + a_{15} xy + a_{18} y^2 + a_{19} x^2 y + a_{20} xy^2 \end{Bmatrix} \tag{8.171}$$

or

$$\begin{Bmatrix} u(x,y) \\ v(x,y) \\ w(x,y) \end{Bmatrix} = [g(x,y)]\{a\} \tag{8.172}$$

where $[g(x,y)]$ is a matrix (size: 3×20) of polynomial variables and $\{a\}$ is a vector (size: 20×1) of polynomial coefficients that are given by

$$[g(x,y)] = \begin{bmatrix} 1 & 0 & 0 & x & 0 & 0 & y & \cdots & 0 \\ 0 & 1 & 0 & 0 & x & 0 & 0 & \cdots & 0 \\ 0 & 0 & 1 & 0 & 0 & x & 0 & \cdots & xy^2 \end{bmatrix}_{(3 \times 20)} \tag{8.173}$$

and

$$\{a\} = \begin{Bmatrix} a_1 \\ a_2 \\ \vdots \\ a_{20} \end{Bmatrix}_{(20 \times 1)} \tag{8.174}$$

The nodal displacement vector for the element is as follows:

$$\{\phi_i^{(e)}\} \equiv \begin{Bmatrix} \phi_1^{(e)} \\ \phi_2^{(e)} \\ \phi_3^{(e)} \\ \phi_4^{(e)} \\ \phi_5^{(e)} \\ \phi_6^{(e)} \\ \vdots \\ \phi_{20}^{(e)} \end{Bmatrix}_{(20 \times 1)} = \begin{Bmatrix} u_1 \\ v_1 \\ w_1 \\ \theta_{x1} \\ \theta_{y1} \\ u_2 \\ \vdots \\ \theta_{y4} \end{Bmatrix}_{(20 \times 1)} = \begin{Bmatrix} u(0,0) \\ v(0,0) \\ w(0,0) \\ \theta_x(0,0) \\ \theta_y(0,0) \\ u_2(l,0) \\ \vdots \\ \theta_y(l,b) \end{Bmatrix}_{(20 \times 1)} \tag{8.175}$$

Using Equations 8.170 and 8.171 in Equation 8.175, the vector of nodal displacements $\{\phi_i^{(e)}\}$ and the vector of polynomial coefficients $\{a\}$ can be related as

$$\{\phi_i^{(e)}\} = [X]\{a\} \quad \text{or} \quad \{a\} = [X]^{-1}\{\phi_i^{(e)}\} \tag{8.176}$$

where

$$[X] = \begin{bmatrix} 1 & 0 & 0 & 0 & 0 & 0 & \cdots & 0 \\ 0 & 1 & 0 & 0 & 0 & 0 & \cdots & 0 \\ 0 & 0 & 1 & 0 & 0 & 0 & \cdots & 0 \\ 0 & 0 & 0 & 0 & 0 & 0 & \cdots & 0 \\ 0 & 0 & 0 & 0 & 0 & 1 & \cdots & 0 \\ 1 & 0 & 0 & l & 0 & 0 & \cdots & 0 \\ \vdots & \vdots & \vdots & \vdots & \vdots & \vdots & \ddots & \vdots \\ 0 & 0 & 0 & 0 & 0 & 1 & \cdots & b^2 \end{bmatrix}_{(20 \times 20)} \tag{8.177}$$

and $\{a\}$ is a vector of size 20×1 as indicated by Equation 8.174.

Substituting Equation 8.176 in Equation 8.172, we get

$$\begin{Bmatrix} u(x,y) \\ v(x,y) \\ w(x,y) \end{Bmatrix} = [g(x,y)][X]^{-1}\{\phi_i^{(e)}\} \tag{8.178}$$

The displacements, in terms of shape functions and nodal displacements, are also given by

$$\begin{Bmatrix} u(x,y) \\ v(x,y) \\ w(x,y) \end{Bmatrix} = [N]\{\phi_i^{(e)}\} \tag{8.179}$$

Comparing Equation 8.178 with Equation 8.179, we obtain the matrix of shape functions as

$$[N] = [g(x,y)][X]^{-1} \tag{8.180}$$

Note that the size of the shape function matrix $[N]$ is 3×20.

For a plate under membrane and bending displacement, the strain–displacement relation can be expressed as

$$\{\varepsilon\} \equiv \begin{Bmatrix} \varepsilon_{xx} \\ \varepsilon_{yy} \\ \gamma_{xy} \end{Bmatrix} = \{\varepsilon^0\} + z\{\kappa\} \tag{8.181}$$

where the in-plane strains and curvatures, respectively, are given by

$$\{\varepsilon^0\} \equiv \begin{Bmatrix} \varepsilon_{xx}^0 \\ \varepsilon_y^0 \\ \gamma_{xy}^0 \end{Bmatrix} = \begin{Bmatrix} \dfrac{\partial u}{\partial x} \\ \dfrac{\partial v}{\partial y} \\ \dfrac{\partial u}{\partial y} + \dfrac{\partial v}{\partial x} \end{Bmatrix} \quad (8.182)$$

$$\{\kappa\} \equiv \begin{Bmatrix} \kappa_{xx} \\ \kappa_{yy} \\ \kappa_{xy} \end{Bmatrix} = -\begin{Bmatrix} \dfrac{\partial^2 w}{\partial x^2} \\ \dfrac{\partial^2 w}{\partial y^2} \\ 2\dfrac{\partial^2 w}{\partial x \partial y} \end{Bmatrix} \quad (8.183)$$

Then, the strain–displacement relations can be written as

$$\{\varepsilon\} = \{\varepsilon^0\} + z\{\kappa\} = [\partial(z)] \begin{Bmatrix} u(x,y) \\ v(x,y) \\ w(x,y) \end{Bmatrix} \quad (8.184)$$

where $[\partial(z)]$ is a matrix of partial differential operators and it is given by

$$[\partial(z)] = \begin{bmatrix} \dfrac{\partial}{\partial x} & 0 & -z\dfrac{\partial^2}{\partial x^2} \\ 0 & \dfrac{\partial}{\partial y} & -z\dfrac{\partial^2}{\partial y^2} \\ \dfrac{\partial}{\partial y} & \dfrac{\partial}{\partial x} & -2z\dfrac{\partial^2}{\partial x \partial y} \end{bmatrix} \quad (8.185)$$

Note that $[\partial(z)]$ is a function of z. Equation 8.184, in turn, gives us

$$\{\varepsilon\} = [\partial(z)][N]\{\phi_i^{(e)}\} = [B]\{\phi_i^{(e)}\} \quad (8.186)$$

Thus,

$$[B] = [\partial(z)][N] = [\partial(z)][g(x,y)][X]^{-1} \quad (8.187)$$

After carrying out the partial differentiation on $[g(x,y)]$, we can show that

$$[B] = [G(x,y,z)][X]^{-1} \quad (8.188)$$

where

$$[G(x,y,z)] = \begin{bmatrix} 0 & 0 & 0 & 1 & 0 & 0 & \cdots & 0 \\ 0 & 0 & 0 & 0 & 0 & 0 & \cdots & -2xz \\ 0 & 0 & 0 & 0 & 1 & 0 & \cdots & -4yz \end{bmatrix}_{(3 \times 20)} \quad (8.189)$$

For 2D plate problem with isotropic material, the material stiffness matrix is given by

$$[C] = \frac{E}{1-\nu^2} \begin{bmatrix} 1 & \nu & 0 \\ \nu & 1 & 0 \\ 0 & 0 & \frac{1-\nu}{2} \end{bmatrix} \quad (8.190)$$

where E and ν are the Young's modulus and Poisson's ratio, respectively. Now, the standard expression for the element stiffness matrix is given by

$$[K^{(e)}] = \left[\iiint_{V^{(e)}} [B]^T [C][B] dV \right] \quad (8.191)$$

Note that $[B]$ is a function of x, y, and z. Then, we can write

$$[K^{(e)}] = \int_{-h/2}^{h/2} \int_0^b \int_0^l [B]^T [C][B] \, dx \, dy \, dz \quad (8.192)$$

Note that in the above expression, $[C]$ is a known matrix of size 3×3. $[B]$ is given by Equation 8.188 together with Equations 8.177 and 8.189 in terms of element coordinates (x,y,z) and element length and breadth (l,b). Clearly, upon integration, we get $[K^{(e)}]$ in terms of the element dimensions (l,b,h) and material properties (E,ν). Note further that the size of $[B]$ is 3×20 and that of $[K^{(e)}]$ is 20×20.

8.6.2.4 Rectangular General Plate Element with Laminated Composites

So far, we have considered the development of elements with isotropic material. In this section, we discuss the development of the element stiffness matrix of a general plate element with layered composites. Element description is the same as in the case of rectangular general plate element. While the treatment of the strain–displacement relation needs some special attention, other details such as number of nodes, degrees of freedom, approximation functions, shape functions, etc. are similar to those presented in Section 8.6.2.3. However, for the sake of completeness and ready reference, we mention some of the key expressions.

The basic field variables are expressed as follows:

$$\begin{Bmatrix} u(x,y) \\ v(x,y) \\ w(x,y) \end{Bmatrix} = [g(x,y)]\{a\} \quad (8.193)$$

where $[g(x,y)]$ is of size 3×20, given by Equation 8.173, and $\{a\}$ is vector of size 20×1, given by Equation 8.174.

The nodal displacement vector $\{\phi_i^{(e)}\}$ is of size 20×1 and it is related to the vector $\{a\}$ as follows:

$$\{\phi_i^{(e)}\} = [X]\{a\} \quad \text{or} \quad \{a\} = [X]^{-1}\{\phi_i^{(e)}\} \tag{8.194}$$

where $[X]$ is given by Equation 8.177.

Substituting Equation 8.194 in Equation 8.193, we get

$$\begin{Bmatrix} u(x,y) \\ v(x,y) \\ w(x,y) \end{Bmatrix} = [g(x,y)][X]^{-1}\{\phi_i^{(e)}\} \tag{8.195}$$

The strain–displacement relation needs some attention; it can be expressed as

$$\begin{Bmatrix} \varepsilon^0 \\ \kappa \end{Bmatrix} = \begin{Bmatrix} \dfrac{\partial u}{\partial x} \\ \dfrac{\partial v}{\partial y} \\ \dfrac{\partial u}{\partial y} + \dfrac{\partial v}{\partial x} \\ -\dfrac{\partial^2 w}{\partial x^2} \\ -\dfrac{\partial^2 w}{\partial y^2} \\ -2\dfrac{\partial^2 w}{\partial x \partial y} \end{Bmatrix} = [\partial] \begin{Bmatrix} u(x,y) \\ v(x,y) \\ w(x,y) \end{Bmatrix} \tag{8.196}$$

where $[\partial]$ is a matrix of partial differential operators and it is given by

$$[\partial] = \begin{bmatrix} \dfrac{\partial}{\partial x} & 0 & 0 \\ 0 & \dfrac{\partial}{\partial y} & 0 \\ \dfrac{\partial}{\partial y} & \dfrac{\partial}{\partial x} & 0 \\ 0 & 0 & -\dfrac{\partial^2}{\partial x^2} \\ 0 & 0 & -\dfrac{\partial^2}{\partial y^2} \\ 0 & 0 & -2\dfrac{\partial^2}{\partial x \partial y} \end{bmatrix} \tag{8.197}$$

Substituting Equation 8.195 in Equation 8.196, we find

$$\begin{Bmatrix} \varepsilon^0 \\ \kappa \end{Bmatrix} = [\partial][g(x,y)][X]^{-1}\{\phi_i^{(e)}\} \tag{8.198}$$

After carrying out the partial differentiation on $[g(x,y)]$, we can show that

$$\begin{Bmatrix} \varepsilon^0 \\ \kappa \end{Bmatrix} = [G(x,y)][X]^{-1}\{\phi_i^{(e)}\} \tag{8.199}$$

where $G(x,y)$ is given by

$$[G(x,y)] = \begin{bmatrix} 0 & 0 & 0 & 1 & 0 & 0 & 0 & 0 & 2x & 0 & 0 & y & 0 & 0 & 0 & 0 & 0 & 0 & 0 \\ 0 & 0 & 0 & 0 & 0 & 0 & 1 & 0 & 0 & 0 & 0 & 0 & x & 0 & 0 & 2y & 0 & 0 & 0 \\ 0 & 0 & 0 & 0 & 1 & 0 & 1 & 0 & 0 & 0 & 2x & 0 & x & y & 0 & 2y & 0 & 0 & 0 \\ 0 & 0 & 0 & 0 & 0 & 0 & 0 & 0 & 0 & 0 & 0 & -2 & 0 & 0 & 0 & 0 & 0 & 0 & -2y & 0 \\ 0 & 0 & 0 & 0 & 0 & 0 & 0 & 0 & 0 & 0 & 0 & 0 & 0 & 0 & 0 & 0 & -2 & 0 & -2x \\ 0 & 0 & 0 & 0 & 0 & 0 & 0 & 0 & 0 & 0 & 0 & 0 & 0 & -2 & 0 & 0 & 0 & -4x & -4y \end{bmatrix}$$

(8.200)

Now, we need to bring in some considerations from mechanics of laminated composites. The constitutive relation for laminated composites is given by (refer Equation 5.43, Chapter 5):

$$\begin{Bmatrix} N \\ M \end{Bmatrix} = \begin{bmatrix} A & B \\ B & D \end{bmatrix} \begin{Bmatrix} \varepsilon^0 \\ \kappa \end{Bmatrix} \tag{8.201}$$

where
  $\{N\}$ Vector of force resultants (note that in this section, N is not used as the matrix of shape functions)
  $\{M\}$ Vector of moment resultants
  $[A]$ Extensional stiffness matrix
  $[B]$ Extension-bending coupling stiffness matrix (note that in this section, B is not used as the strain–displacement matrix)
  $[D]$ Bending stiffness matrix

Let us now consider the strain energy of the element due to applied loads, which is given by

$$\pi^{(e)} = \frac{1}{2} \iiint_V \{\varepsilon\}^T \{\sigma\} dV \tag{8.202}$$

The vector $\{\varepsilon\}$ consists of two parts—mid-plane strains $\{\varepsilon^0\}$ and curvature $\{\kappa\}$. Then, we write the strain energy expression as

$$\pi^{(e)} = \frac{1}{2} \int_A \int_{-h/2}^{h/2} \{\{\varepsilon^0\} + z\{\kappa\}\}^T \{\sigma\} dz\, dA \tag{8.203}$$

Note that the stress resultants are given by

$$\{N\} = \int_{-h/2}^{h/2} \{\sigma\}dz \quad \text{and} \quad \{M\} = \int_{-h/2}^{h/2} z\{\sigma\}dz \quad (8.204)$$

Then, with some matrix operations, it can be shown that the strain energy expression reduces to

$$\pi^{(e)} = \frac{1}{2}\int_A \begin{Bmatrix} \varepsilon^0 \\ \kappa \end{Bmatrix}^T \begin{Bmatrix} N \\ M \end{Bmatrix} dA \quad (8.205)$$

Substituting Equation 8.201 in the above expression, we get

$$\pi^{(e)} = \frac{1}{2}\int_A \begin{Bmatrix} \varepsilon^0 \\ \kappa \end{Bmatrix}^T \begin{bmatrix} A & B \\ B & D \end{bmatrix} \begin{Bmatrix} \varepsilon^0 \\ \kappa \end{Bmatrix} dA \quad (8.206)$$

which, with the help of Equation 8.199, becomes

$$\pi^{(e)} = \frac{1}{2}\{\phi_i^{(e)}\}^T [X]^{-1^T} \int_A [G(x,y)]^T \begin{bmatrix} A & B \\ B & D \end{bmatrix} [G(x,y)]dA[X]^{-1}\{\phi_i^{(e)}\} \quad (8.207)$$

The total strain energy of the body due to applied loads is obtained as

$$\pi = \frac{1}{2}\{\phi\}^T \left[\sum_{e=1}^{E}[X]^{-1^T} \int_A [G(x,y)]^T \begin{bmatrix} A & B \\ B & D \end{bmatrix} [G(x,y)]dA[X]^{-1}\right]\{\phi\} \quad (8.208)$$

The strain energy expression in the above equation represents the first component of the total potential energy of the body given by Equation 8.35. To derive the element stiffness matrix, we can substitute this and minimize the potential energy. Alternatively, simply by comparison, we can conclude that expression inside the large square brackets is the stiffness matrix of the body and the element stiffness matrix is

$$[K^{(e)}] = [X]^{-1^T} \int_A [G(x,y)]^T \begin{bmatrix} A & B \\ B & D \end{bmatrix} [G(x,y)]dA[X]^{-1} \quad (8.209)$$

Note that in the above expression, [A], [B], and [D] are known matrices (each of size: 3×3) that depend on the orthotropic material properties and ply sequence. [G(x,y)] (size: 6×20) is given by Equation 8.200 in terms of element coordinates (x,y). [X] has a size of 20×20 and it involves element length and breadth (l,b). Clearly, upon integration, we get [$K^{(e)}$] in terms of the element dimensions (l,b) and material properties A, B, and D. Note further that the size of [$K^{(e)}$] is 20×20.

8.7 SUMMARY

In this chapter, we provided a brief discussion on the FEM. It is a vast subject and a versatile tool useful for solving virtually any kind of problem in any field; we had a look at the essential concepts that are needed to understand the general procedure. The concept of elements is what makes this method unique and versatile. The body or domain is discretized into an assemblage of elements, finite element equations are developed for the individual elements and global equations are obtained by assembly of the elements and imposition of boundary conditions. On solving the global equations and further postprocessing, we get the desired results. The formulation of element equations is the central concept, for which primarily three approaches can be adopted. In this chapter, we used the variational approach and demonstrated the development of elements in some of the common 1D and 2D cases including laminated composites.

EXERCISE PROBLEMS

8.1 A rectangular plate with two holes under uniform uniaxial tension is shown in Figure 8.15. Discretize the plate using triangular elements. Take advantage of symmetry wherever applicable.

8.2 Consider a bar element as shown in Figure 8.16. Determine the matrix of shape functions for the following polynomial function:

$$u(x) = a_1 + a_2 x + a_3 x^2$$

8.3 Consider an aluminum rod (Figure 8.17a) subjected to an axial tensile force of 12 kN. Discretize the rod using four 1D elements and determine the extension and stresses. Assume $E = 70$ GPa.

8.4 Solve the problem in Exercise 8.3, if the cross section gradually changes (Figure 8.17b). Use four constant cross section 1D elements for discretization.

8.5 Write a code in MATLAB/C/C++ for the problems in Exercises 8.3 and 8.4. Discretize with more number of elements and compare the results.

8.6 Consider the rod in Exercise 8.3. If the rod is free at both the ends (Figure 8.17c) and it is pulled at either end by equal tensile force of 25 kN, determine the axial displacements and stresses. Discretize the rod using four 1D elements of constant cross-sectional diameter.

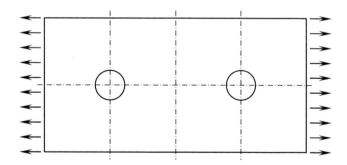

FIGURE 8.15 Rectangular plate with two holes (Exercise 8.1).

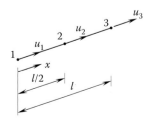

FIGURE 8.16 Bar element with three nodes (Exercise 8.2).

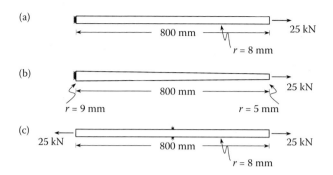

FIGURE 8.17 Rod under uniaxial tensile force (Exercises 8.3 through 8.6). (a) Constant cross section rod with one end free. (b) Varying cross section rod with one end free. (c) Constant cross section rod with both ends free.

Hint: The half-way cross section, that is, the cross section at a distance of 200 mm from any end, remains stationery.

8.7 Consider a steel rod of length 300 mm and diameter 6 mm fixed at one end. If the torque is 2 kN.mm, determine the shear stress at the fixed end. Discretize the rod using three 1D elements. Assume $G = 77$ GPa.

Hint: Use Equation 8.99 for shear strain.

8.8 Consider the structure shown in Figure 8.18. The structure is discretized with three planar beam elements as shown. Determine the element stiffness matrices, and carry out transformation and assembly operations to find the global stiffness matrix. If a point load of 0.5 kN is applied, determine the nodal displacements and stresses. The cross section of the structural member is circular with a radius of 25 mm. Assume $E = 200$ GPa and $\nu = 0.3$.

Hint: Use a mathematical interactive tool such as MATLAB for carrying out the matrix operations.

8.9 Write a code in MATLAB/C/C++ to find the element stiffness matrix of a general beam element (see Section 8.6.1.4) and associated general displacements, strains, and stresses.

8.10 Consider a cantilever beam with geometrical details as given in Figure 8.17a. A point load of 30 kN, oriented at 45° to all the three axes, is applied at the free end. Discretize the beam using four 1D general elements and determine the following:

a. Size of the global stiffness matrix after incorporating the boundary conditions
b. Global displacement vector after incorporating the boundary conditions
c. Global load vector after incorporating the boundary conditions

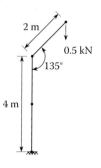

FIGURE 8.18 Planar structure with one-dimensional elements (Exercise 8.8).

8.11 Use the code developed in Exercise 8.9 and solve the problem in Exercise 8.10 for displacements, strains, and stresses.

8.12 Consider a rectangular membrane element of size 10 mm × 8 mm. Determine (i) strain–displacement matrix $[B]$, (ii) material stiffness matrix $[C]$, and (iii) element stiffness matrix $[K^{(e)}]$. What are the sizes of these matrices? Assume $t = 0.5$ mm, $E = 200$ GPa, and $\nu = 0.3$.

Hint: Use an interactive mathematical tool such as MATLAB for doing matrix operations.

8.13 Solve the problem in Exercise 8.12 for a rectangular bending plate element.

8.14 In Equations 8.173, 8.177, and 8.189, the matrices $[g(x,y)]$, $[X]$, and $[G(x,y,z)]$ are given partially. Write these matrices in their full form.

Hint: The sizes of the matrices are indicated in the corresponding equations.

8.15 Solve the problem in Exercise 8.12 for a rectangular general plate element.

8.16 Write a code in MATLAB/C/C++ solving a general plate element problem. The output of the code should be

  a. Strain–displacement matrix $[B]$
  b. Material stiffness matrix $[C]$
  c. Element stiffness matrix $[K^{(e)}]$
  d. Global stiffness matrix after incorporation of boundary conditions
  e. Nodal displacement vector
  f. Load vector

  Consider the following as input data

  a. Geometrical details of the plate—length, breadth, and thickness
  b. Geometrical details of the plate element l, b, and h (depending on the number of elements along the length and breadth)
  c. Material properties E and ν
  d. Boundary conditions
  e. Applied loads

8.17 Consider a rectangular plate of dimensions 500 mm × 400 mm × 6 mm subjected to a uniformly distributed load of 5×10^{-4} N/mm². The plate is fixed on all the four edges. Discretize the plate using general plate elements of size 50 mm × 50 mm and determine the (i) nodal displacements and (ii) nodal stresses using the code developed in Exercise 8.12. Assume the following material properties: $E = 70$ GPa and $\nu = 0.3$.

Hint: Convert the uniformly distributed load into an equivalent set of nodal loads.

8.18 If the material of the plate in Exercise 8.15 is replaced with carbon/epoxy (stacking sequence: $[0°/90°/0°]_s$, each ply being 0.5 mm in thickness), determine the element stiffness matrix.

Hint: Use a mathematical tool such as MATLAB for matrix operations including determination of the laminate stiffness matrices $[A]$, $[B]$, and $[D]$.

REFERENCES AND SUGGESTED READING

1. O. C. Zienkiewicz and R. L. Taylor, *The Finite Element Method, Volume 1, Basic Formulation and Linear Problems*, McGraw-Hill, London, 1989.
2. T. J. R. Hughes, *The Finite Element Method—Linear Static and Dynamic Finite Element Analysis*, Dover Publications, New York, 2000.

3. K. J. Bathe, *Numerical Procedures in the Finite Element Method*, Prentice Hall, Englewood Cliffs, NJ, 1995.
4. J. N. Reddy, *An Introduction to the Finite Element Method*, second edition, Tata McGraw-Hill, New York, 2003.
5. S. S. Rao, *The Finite Element Method in Engineering*, Elsevier, Burlington, 2005.
6. R. D. Cook, D. S. Malkus, and M. E. Plesha, *Concepts and Applications of Finite Element Analysis*, John Wiley & Sons, New York, 2000.
7. C. S. Krishnamoorthy, *Finite Element Analysis—Theory and Programming*, Tata McGraw-Hill, New Delhi, 1997.
8. D. W. Pepper and J. C. Heinrich, *The Finite Element Method—Basic Concepts and Applications*, Taylor & Francis, Boca Raton, FL, 2006.
9. E. J. Barbero, *Finite Element Analysis of Composite Materials*, CRC Press, Taylor & Francis, Boca Raton, FL, 2008.
10. E. J. Barbero, *Finite Element Analysis of Composite Materials Using ANSYS®*, second edition, CRC Press, Taylor & Francis, Boca Raton, FL, 2014.
11. E. J. Barbero, *Finite Element Analysis of Composite Materials Using ABAQUS™*, CRC Press, Taylor & Francis, Boca Raton, FL, 2013.
12. S. V. Hoa and W. Feng, *Hybrid Finite Element Method for Stress Analysis of Laminated Composites*, Kluwer Academic Publishers, Boston, MA, 1998.

Part II

Materials, Manufacturing, Testing, and Design

9

Reinforcements and Matrices for Polymer Matrix Composites

9.1 CHAPTER ROAD MAP

An introduction to composite materials is given in Chapter 1, which is followed by presentation of topics related to mechanics and analysis in Chapters 2 through 8. These topics are primarily computational in nature. In addition to these, the product development engineer needs to be familiar with a number of other topics that are associated primarily with the composite shop floor. One such aspect is in respect of the materials that a composite product is made up of. This chapter is devoted to present, from the view point of a product development engineer, the common materials used as reinforcements and matrix in PMCs.

As we know, reinforcements and matrix are the two primary constituents. The continuous phase, that is, the matrix is a polymer in PMCs, and it is addressed first. Different classes of polymers are briefly discussed and characteristic properties and applications of some of the commonly used thermosetting resins are addressed. For the sake of completeness, thermoplastics and rubbers are briefly touched upon. Reinforcements are the discrete constituents, and characteristic properties, manufacturing principle, and applications of some of the common fibers are addressed in this chapter. Reinforcements are available in various physical forms and familiarity with them is important especially from a manufacturing point of view. A discussion on the physical forms of reinforcements is given in the end.

For a reader whose objective of study is primarily to familiarize with composites manufacturing and testing, Chapters 2 through 8 can be skipped and this chapter, that is, Chapter 9 can be taken up immediately after going through the introductory discussion in Chapter 1.

9.2 POLYMERS

In a PMC material, a polymer such as epoxy or polyester is used as the matrix material that is reinforced with very fine diameter fibers such as carbon, glass, etc.

A polymer is a natural or synthetic compound of usually high molecular weight consisting of many repeating units of smaller molecules (monomers) that can be linked in linear, branched, or cross-linked form. A linear polymer is one in which the monomers form long chains without any branches or cross-links. It is the simplest in form but it is also the poorest in terms of strength and stiffness properties. A branched polymer has a 2D molecular structure with branches connected to the main linear chains. A cross-linked polymer is one in which the adjacent linear chains are connected by covalent bonds in a highly complex 3D form. (A covalent bond is a very strong attraction between the atoms.)

Polymers are classified as follows:

- Thermosets
- Thermoplastics
  - Noncrystalline
  - Crystalline
- Rubbers

9.2.1 Thermosets

Thermosets are polymers that chemically react under suitable environment, such as high temperature, to a permanently solid and infusible state. In such a solid state, the molecular structure is highly complex and 3D; thus, the cross-linking process is irreversible and softening of the solid polymer upon heating is not possible. Both thermosets and thermoplastics are used for making composites. However, thermosets are more popular as matrix material. Common thermosetting resins (matrix materials are more commonly known as resins in the shop floor) are epoxy, phenolic, polyester, vinyl ester, etc.

9.2.2 Thermoplastics

Thermoplastics are those polymers that can be repeatedly softened by application of heat and hardened by cooling. They are largely either linear or branched polymers and their change in form upon heating is more physical than chemical. Thermoplastics have high melt viscosity; as a result, wetting of the reinforcements by these polymers is generally difficult. However, thermoplastics are recyclable and advanced processing techniques are being developed for composites with thermoplastic matrix. Characteristics of thermoplastics are compared with those of thermosets in Table 9.1 [1].

Common thermoplastics used in composites are nylon, polyethylene, and polysulfone.

9.2.3 Rubber

Both natural as well as synthetic rubbers are available in the market. Natural rubber is obtained from a sticky milky colloidal fluid produced by various rubber plants, of which *Hevea brasiliensis* is the most common commercially exploited rubber plant [2,3]. Other rubber plants include *Castilla elastica*, *Ficus elastica*, etc. Incisions are made into the bark of the rubber plants and the whitish fluid flows down from the incision marks, which is collected and refined to obtain rubber. Refined natural rubber is composed primarily of a chemical called polyisoprene.

TABLE 9.1

Comparison of Thermoplastics with Thermosets

	Thermosets	Thermoplastics
On heating	Decompose on heating beyond certain limit	Soften on heating
Mechanical loading	Exhibit low strains at failure	Exhibit low strains at failure
Cross-linking	Long cure cycle	Short cure cycle
Processability	Tacky, thus difficult to handle	Nontacky, thus easy to handle
	Generally low temperature of processing, thus easy to process	Generally high temperature of processing, thus difficult to process
	Good solvent resistance	Excellent solvent resistance
Storage	Definite shelf life	Indefinite shelf life

Rubber is also made synthetically and a wide range of synthetic rubbers such as silicone rubber, Viton rubber, EPDM, nitrile rubber, polyurethane, etc. dominate the market today.

Natural rubber has high elongation at failure, low Young's modulus, and very effective waterproofness.

Rubber is extensively used in many household and industrial applications. Tire and tubes are the single largest products that consume almost 60% of total rubber production worldwide. In the other general goods sector, rubbers are used for making hoses, belts, adhesives, mats, gloves, carpet, toys, gaskets, and many others.

Rubbers, however, are rarely used as matrix material for making a composite product.

9.3 COMMON THERMOSETS FOR PMCs

Thermosets are the common resins used in PMCs especially in continuous fiber-reinforced composites. There are a number of these resins and numerous possibilities exist for selecting a resin together with associated curatives, modifiers, and additives [4,5]. A detailed discussion is far beyond the scope of this book; rather a brief introductory note is given in the following sections on three common thermosetting resin systems—epoxy, polyester, and phenolic resins.

9.3.1 Epoxy Resins

Epoxy resins are a class of thermosetting polymers that have found extensive applications in industrial, aerospace, and other high-end sectors. Typically, in an epoxy system, there are two components—a base epoxy resin and an epoxy curative. In addition, a modifier is also added to obtain specific desired property. Each of these three components imparts specific physical and mechanical characteristics to the resin system during processing as well as a processed material. By proper choice of the three components, it is possible to tailor the resin system and unique combinations of physical and mechanical properties are achieved. As a result, epoxies have established themselves as a highly versatile material system and they are routinely used as adhesive, molding compounds, and matrices in PMCs with continuous as well as short fiber reinforcements.

9.3.1.1 Base Epoxy Resin

The base resin in an epoxy system is chemically an organic compound containing epoxide molecules. An epoxide molecule has an epoxide group in its molecular structure. As shown in Figure 9.1, there are one oxygen and two carbon atoms in an epoxide group. The epoxide groups are a common feature in all types of epoxide molecules, but the other details attached to the epoxide groups vary from one type of epoxide molecule to another. A wide variety of epoxy resins are commercially available, which can be broadly categorized as follows:

- Glycidyl epoxies
  - Glycidyl ether
  - Glycidyl ester
  - Glycidyl amine
- Nonglycidyl epoxies
  - Aliphatic
  - Cycloaliphatic

The most common commercial epoxy till today is diglycidyl ether of bisphenol-A (DGEBA). Other important epoxy resins include novolac epoxy and brominated epoxy.

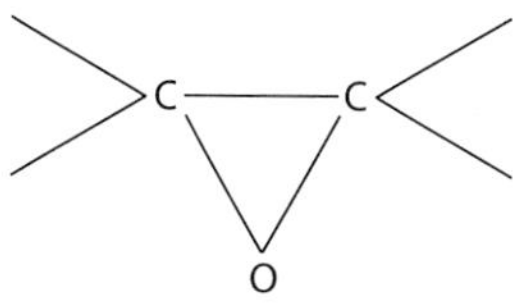

FIGURE 9.1 Epoxide group.

9.3.1.1.1 Curatives

In the initial "green" or uncured state, the epoxide molecules do not normally react with each other at room temperature. Curatives (commonly called as hardeners) are added to the base epoxy resin and the process of curing starts. During the process of curing or thermosetting reaction, the small base molecules and the curatives chemically react with each other into a complex 3D network. In this process of cross-linking, large molecules are formed and the resin system hardens into a rigid mass. Depending upon the type of curatives, the curing process may demand application of high temperature and pressure. Also, curing duration may vary from a few minutes to several hours depending upon the curatives.

A wide variety of hardeners—primarily amines, derivatives of amine, and anhydrides—are commercially available, of which amines are the most common. Typical examples of amine hardeners include diethylene triamine (DETA), triethylene triamine (TETA), tetraethylene pentamine (TEPA), etc.

9.3.1.1.2 Modifiers

In an epoxy formulation, modifiers are added to the base epoxy resin and the hardener to impart specific physical and mechanical properties to the uncured as well as cured resin. Different types of modifiers include flame retardants, fillers, pigments and dyes, diluents, rubbers, etc. These are available in different physical forms such as liquid, powder, flakes, fibers, etc. Some of the common types of modifiers are listed in Table 9.2 [5–7].

9.3.1.1.3 Properties

Mechanical and physical properties of cured epoxy resin system are useful from two angles—first, to compare and select the appropriate resin system for a particular usage, and second, in micromechanical analysis of composites. Properties of cured epoxy systems depend upon the type of base resin, hardener, modifiers, and their mixing proportions. There are many commercially available resins and hardeners and, virtually innumerable possibilities exist. However, based on the required properties, the choice of the resin system can be quickly narrowed down. Thus, the resin, hardeners, and modifiers are chosen based on the application, composite manufacturing method, and the end properties required. In micromechanics, individual parameters, such as density, modulus, etc., of reinforcements and cured resin are used as input data for determining the composite laminate properties (refer Chapter 3).

TABLE 9.2

Common Categories of Modifiers in Epoxy Resins

Modifier Categories	Properties Imparted to the Resin System
Flame retardants	Reduction in flammability of epoxy composites
Fillers	Shrinkage reduction, electrical and thermal conductivity, density reduction, increase in viscosity, etc.
Pigments and dyes	Specific colors to the finished product
Diluents	Viscosity reduction for increased ease of processing
Rubbers	Toughness, fatigue and fracture resistance, flexibility, etc.

An epoxy system is associated with a number of physical and mechanical properties. Some of these correspond to the base uncured resin and the hardeners, and the rest to the cured epoxy. Representative properties of uncured and cured epoxy resins are given in Table 9.3. It may be noted that the properties of epoxies given in the table are indicative [5–10]; depending on the formulation and processing, epoxies with a wide range of properties can be synthesized and manufacturer's data sheet should be consulted for the design of a product.

For the composites engineer, on the shop floor, characteristics of the uncured resin system are of primary concern. From this angle, physical form, viscosity, pot life, cure characteristics, shelf life, storage condition, etc. need to be considered. Commonly, epoxy resins are available as liquids, powders, flakes, semisolid paste, etc. Resins in a solid or semisolid form are inconvenient to process as a laminating system. The process of mixing the resin and hardener to prepare the resin system varies and the details can be obtained from the manufacturer. Generally, resins and hardeners in solid/semisolid form are heated to a higher temperature and the molten materials are mixed at the specified ratio. Also, at times, the resin mix is kept at elevated temperature so as to maintain the melt viscosity within limit. Highly viscous resins are not suitable for composites processing as it leads to nonuniform/inadequate wetting of the reinforcements.

Pot life is the length of time from the instant the hardener is mixed with resin until the time the resin mix is usable for its intended use. It may vary from a few minutes to a couple of hours. Too short a pot life is inconvenient for manual composites processing such as hand lay-up as well as for automated processing such as filament winding. However, too long a pot life is also not desirable as it may lead to problems like resin dripping during processing.

Cure characteristics of an epoxy system depend on the chemistry of the resin and hardeners, their mixing ratio, and the cure environment in terms of temperature and pressure. Cure time may vary from a few minutes to several hours. Also, cure can take place at room temperature or at elevated temperature. Elevated temperature curing may be done in a stepwise fashion, in which the temperature is increased from room temperature to the final cure temperature with one or two holds at intermediate temperatures.

Another important parameter worth mentioning is the glass transition temperature (T_g) of cured epoxy resin. It is the temperature below which a polymer retains its structural rigidity. Above this temperature, the individual molecular segments tend to move relative to each other, and the resin system becomes rubbery with drastic reduction in

TABLE 9.3
Representative Properties of Epoxy Resins

Parameter	Value
Liquid resin	
Viscosity at 25°C (cps)	500–7000
Cast resin	
Density (g/cm^3)	1.1–1.3
Shore D hardness	50–80
Glass transition temperature (°C)	50–160
Tensile strength (MPa)	30–90
Tensile modulus (GPa)	1.2–3.7
Failure elongation (%)	1.4–2.4

Note: Data provided in the table are indicative; for precise and authentic information on properties of specific brand, manufacturers' current data sheets can be consulted.

the modulus. The glass transition temperature of a cured resin system depends upon its molecular structure and it is related to the cure temperature. In general, the T_g of a high-temperature curing system is higher than that for a room temperature curing system.

9.3.1.2 Applications

Aerospace composite applications, in general, demand high strength and stiffness, high toughness, high temperature resistance, and good dimensional stability. Epoxy resins provide unique combinations of these properties which is rarely possible with other resins. Thus, the aerospace composites market is dominated by epoxy resins. Aerospace applications, however, form only a small fraction of the total usage of epoxy resins. Table 9.4 lists major applications of epoxy resins.

9.3.2 Polyester Resins

Polyester resins are unsaturated thermosetting resins made by dissolving polyester oligomers in a solvent.

9.3.2.1 Polyester Oligomer

Polyester oligomers are produced by condensation reaction of difunctional acids and difunctional alcohols. Commonly used acids are phthalic acid and maleic acid, whereas glycols such as propylene glycols are used as alcohols. The oligomer, whose structure can vary widely, determines the mechanical, thermal, and chemical properties of the resin system. In unsaturated polyester resins (commonly known as polyester resin), the acids used for making the oligomer are partly saturated and partly unsaturated.

TABLE 9.4
Applications of Epoxy Resins

Usage	Description
Protective coatings	• The coatings industry consumes the largest share of epoxy resins produced worldwide
	• Used in industrial and automotive applications to provide a hard, durable, and rustproof surface
	• Used as
	– Powder coatings on domestic appliances for corrosion protection
	– Fusion-bonded powder coatings on steel pipes and fittings for corrosion protection
	– Primers on metal surfaces to improve adhesion of paints
	– Food-grade coatings on metal cans and containers, etc.
Tooling	• Used in industrial applications for making composite tooling, patterns, molds, and castings
Bonding	• Used as structural adhesives in a very wide range of applications, including common household appliances to highly specialized space vehicles
Construction industry	• Used as coatings, flooring, etc.
Composites application	• Used in composites industry especially for high-end applications as laminating resin for making various carbon/epoxy, glass/epoxy, and Kevlar/epoxy composite parts
	• Typical usages include:
	– Aircraft wings, fuselage
	– Rocket motor casing, airframe, etc.
Electrical and electronics applications	• Used in motors, generators, transformers, inductors, insulators, switchgears, bushings, etc.
	• Used for insulation and encapsulation of electrical parts from short circuits, dust, and moisture
	• Used in printed circuit boards, integrated circuits, transistors, etc.

The saturated acid is the major part in the composition, and based on the type of the saturated acid used in the oligomer, different classes of polyester resins are made. Common classes of polyester resins are orthophthalic, isophthalic, terephthalic, dicyclopentadiene (DCPD), chlorendic, and bisphenol-A fumarate.

Orthophthalic polyester resins are made from phthalic anhydride, maleic anhydride, and propylene glycol. These resins, also called ortho resins or general-purpose (GP) resins, possess good structural properties. These are among the cheapest of all polyester resins and thus, GP resins are the most popular polyester resins used in commercial applications.

Isophthalic polyester resins are made from isophthalic acid. These resins are more expensive than and superior to the orthophthalic resins in terms of strength and stiffness, chemical resistance, and thermal properties.

Terephthalic polyester resins are produced from terephthalic acid. In terms of physical, mechanical, and chemical properties, terephthalic resins are nearly similar to the isophthalic resins, whereas, their thermal characteristics are marginally better.

DCPD, chlorendic, and bisphenol-A fumarate polyester resins are specialty polyester resins. DCPD polyester resins are produced using DCPD. These resins are cheap, and characterized by low shrinkage, very good surface finish, rapid cure, and UV resistance. They have largely replaced the ortho resins in marine applications.

Chlorendic resins are made from HET (hexachlorocyclopentadiene) acid, an unsaturated acid such as fumaric acid and glycol. These resins are very rigid and possess excellent chemical resistance, thermal stability (high heat deflection temperature [HDT]), and fire retardancy.

Bisphenol A fumarate resins are made from bisphenol A, propylene oxide, and fumaric acid. Bisphenol A and propylene oxide are reacted to form glycol, which in turn is reacted with fumaric acid to produce bisphenol A polyester. Like the chlorendic resins, these resins are also characterized by rigidity, excellent chemical resistance, and thermal stability.

9.3.2.1.1 Solvent

The most common solvent used for dissolving the polyester oligomers is styrene. It acts as a cross-linking agent and reduces the resin viscosity as well. During curing, the styrene molecules react among themselves and polyester oligomers forming a 3D network, and the resin solidifies.

9.3.2.1.2 Additives

In a polyester resin, the polyester oligomer and the solvent are the primary components. In addition to these, a number of additives, in small proportions, are added with specific purposes. Initiators and promoters are the primary additives. The initiator initiates the chemical reaction for cure. A commonly used initiator is methyl ethyl ketone peroxide (MEKP). MEKP and other initiators decompose very slowly and they cannot cure polyester resin completely at room temperature. Thus, a promoter such as cobalt naphthenate (CoNap) is added to the resin for rapidly decomposing the initiator. The initiator and promoter are mixed with the resin by the fabricator during manufacture of the final part.

Polyester resin during storage cures or "gels" at room temperature, although at a much slower rate. Inhibitors and retarders are added by the resin manufacturer for increasing the resin storage life. The difference between the two is that while the inhibitors delay the start of cure, the retarders slow down the rate of cure.

9.3.2.1.3 Properties

Representative mechanical and physical properties of different classes of unreinforced polyester resins are listed in Table 9.5. This property data, generated using various

TABLE 9.5
Properties of Unreinforced Polyester Resins

Type of Resin	Liquid Resin		Cast Resin						
	Specific Gravity	Viscosity	Barcol Hardness	Heat Deflection Temperature	Tensile Strength	Tensile Modulus	Failure Strain	Flexural Strength	Flexural Modulus
	–	(cps)	(BHN)	(°C)	(MPa)	(GPa)	(%)	(MPa)	(GPa)
Orthophthalic	1.01–1.15	200–400	35–48	80–110	55–75	3.4–4.0	2.1–3.5	80–140	3.4–4.4
Isophthalic	1.06–1.14	300–700	39–43	85–107	75–95	3.4–4.1	2.4–4.2	125–160	3.8–4.3
Terephthalic	1.04–1.14	400–900	45–47	80–145	65–90	3.0–4.0	2.7–4.0	114–150	3.0–4.0
Dicyclopentadiene	1.08–1.14	–	44–47	76–110	40–85	3.3–3.7	1.3–3.7	60–160	3.5–4.3
Chlorendic	1.12–1.15	250–550	40–45	140–280	20–50	3.4–3.5	1.6–2.4	110–120	3.8–5.2
Bisphenol-A fumarate	1.08–1.14	–	35–40	124–140	40–70	2.8–3.0	1.4–2.6	110–120	3.0–3.4
Vinyl ester	1.04–1.15	350–750	34–52	100–210	70–86	3.4–3.8	1.5–6.0	110–140	3.4–4.0

Note: Data provided above are indicative; for more precise and authentic information on specific grade, manufacturers' data sheets can be consulted.

sources [11–15], are intended for making an overall idea; however, the final choice of the resin for specific application as well as design calculations should be based on the data sheet available from the resin manufacturer for a particular resin formulation and associated additives.

Mechanical properties of the cured polyester resin depend upon a number of parameters that include type of the base acids and alcohols making the oligomer, solvent and its proportion, additives, etc. In general, isophthalic resins have the highest mechanical properties. Orthophthalic resins have adequate mechanical and thermal properties; they are inferior to the isophthalic resins, but cheap and, thus popular in general purpose application. Chlorendic and bisphenol A fumarate resins are very rigid and highly stable in thermal and chemical environments.

9.3.2.2 Applications

While epoxy resins dominate the aerospace composites market, polyester resins are the most widely used resin systems in overall market for resins. More than 80% of total polyester resins are used for making FRP products, of which GFRP are the most common. Fiber-reinforced polyesters are used in a very wide variety of applications in different industrial sectors that include automotive, buildings and construction, chemical, consumer goods, electrical, energy, marine, recreational and sporting goods, etc. Polyester resins are also used without any reinforcement; typical casting applications include gel coats, decorative products, etc. Table 9.6 lists some of the major applications of polyester resins.

9.3.3 Vinyl Ester Resins

Vinyl ester resins have certain commonalities with polyester resins. Similar to the polyester resins, these are also unsaturated thermosetting resins made by dissolving an oligomer in a solvent and cured by mixing similar catalysts and accelerators. However, the vinyl ester oligomer chemistry is different from that of polyester. It is made by reacting an unsaturated carboxylic acid with an epoxy.

Using different types of epoxy base, different types of vinyl ester resins are made; these are GP vinyl ester (based on bisphenol A epoxy), fire retardant vinyl ester (based on brominated epoxy), novolac vinyl ester (based on novolac epoxy), etc.

Vinyl ester resins are versatile and their properties and applications vary widely. Owing to the presence of epoxy, these resins, in general, have better chemical resistance

TABLE 9.6
Applications of Polyester Resins

Sector	Applications
Automotive	Car accessories and body parts such as door, bumper, dashboard, etc., seating, bus body, hoods for goods vehicles
Building and construction	Building panels, corrugated sheets, doors, modular house
	Bathtubs, bathroom fixtures, cultured marble
	Swimming pools
Chemical	Corrosion resistant tanks, pipes, industrial vessels, sewer lines, waste water treatment equipment, pollution control equipment
Electrical and electronics	Circuit boards, insulators, switch gears, appliance covers
Energy	Wind mill blades, platform for oil exploration activities
Marine	Boats, yachts, dinghies, gel coats
Miscellaneous	Decorative items, buttons, recreational and sporting goods such as bowling balls, helmets

coupled with higher failure strain and mechanical properties. GP vinyl ester resins have excellent mechanical properties, good heat resistance, and chemical resistance. (Typical representative properties of vinyl ester resins are given along with different types of polyester resins in Table 9.5.) Novolac vinyl ester resins have higher heat resistance than the GP resins.

Applications of vinyl ester resins are many, of which, use as corrosion-resistant reinforced plastics is predominant. Typical applications of these resins are tanks, pipes, flooring, lining, electrical equipment, etc. Bisphenol A and chlorendic polyester resins are good competitors for vinyl ester resins; however, owing to the improved mechanical properties the latter are gaining popularity.

9.3.4 Phenolic Resins

Phenolic resins are commonly produced by condensation reaction of phenol and formaldehyde in the presence of a catalyst [16,17]. A variety of these resins can be made by adjusting the ratio of phenol to formaldehyde, the reaction temperature, or the catalyst. Out of these, two types of phenolic resins are common—resole and novolac.

Resole phenolic resins are made by using an alkaline catalyst and excess formaldehyde. Initially, the reaction is controlled so as to achieve a noncross-linked, low-molecular-weight resins, which finally cure at elevated temperature in a single-stage process and a three-dimensionally cross-linked solid infusible polymer is obtained.

Novolac phenolic resins are made by using an acidic catalyst, in which the polymerization process is a two-stage process. In the first stage, a low-molecular-weight resin is produced, which does not cure without a hardener. In the second stage, in the presence of a hardener, 3D cross-linking takes place at elevated temperature.

Phenolics are an old resin system; but, their use in continuous fiber-reinforced advanced composites is somewhat limited due to low failure strain/mechanical properties, brittle nature. Also, curing process is somewhat complicated as special care by means of vacuum bagging in autoclave curing may be needed to remove volatile contents and water that are produced during curing reactions. These resins, however, have very attractive thermal characteristics in terms of dimensional stability and retention of mechanical properties at high temperatures. Phenolic resins are used in applications where fire safety and high-temperature capabilities are of critical importance. As reinforced plastics phenolic resins are commonly used as ablative liners and insulating barriers to structural parts that are subjected to moderate to extremely high temperatures. Filament-wound phenolic resin-based pipes are used in fire safety critical applications such as mining, tunneling, offshore oil exploration, etc. Phenolic resin-based sheet molding compounds (SMCs) are used in aircrafts and aerospace vehicles as specialized parts such as fins, lugs, etc. Phenolic resin-based engineering plastics are also used in many commercial applications that include electrical devices such as switchgear, circuit breakers, connectors, etc.; consumer appliances such as handles, knobs, etc.; and automotive parts such as disk brake piston, solenoids, etc.

9.4 REINFORCEMENTS

Reinforcements are the primary load-bearing component in an advanced PMC material. Different materials in fiber forms are used as reinforcements, of which the following are significant:

- Glass fibers
- Carbon fibers
- Aramid fibers
- Boron fibers
- Natural fibers
- Whiskers and ceramic fibers

Materials in their bulk form contain flaws, which affect their strength and stiffness properties. In the fiber form, internal flaws are absent and the net load-bearing area as a fraction of the gross cross-sectional area is very high. Further, fibers have high degree of molecular and crystallographic alignment. As a result, mechanical properties of materials in their fiber forms are higher than in the bulk forms. Materials used for fiber reinforcements are of lower density as compared to other conventional structural materials such as steel, aluminum, etc. and, thus specific strength and stiffness (strength and stiffness divided by density) are very high for fibrous reinforcements. Representative properties of common fibers and conventional structural materials are given in Table 9.7 [18–39]).

TABLE 9.7
Representative Properties of Common Fibers and Conventional Structural Materials

Name of Fiber	Specific Gravity	Fiber Diameter (μm)	Coefficient of Thermal Expansion (m/m/°C)	Tensile Strength (MPa)	Tensile Modulus (GPa)	Elongation at Break (%)	Specific Tensile Modulus (GPa/g/cm^3)	Specific Tensile Strength (MPa/g/cm^3)
E-glass	2.58	5–20	52×10^{-6}	3450	76	4.8	30	1.3
S-glass	2.48	5–10	5.6×10^{-6}	4600	88	5.7	36	1.9
C-glass	2.50	–	6.7×10^{-6}	3170	69	4.8	28	1.3
IM carbon	1.8	8	-0.6×10^{-6}	3900	260	1.4	110	2.2
HM carbon	1.9	6	-0.7×10^{-6}	4500	400	1.2	220	2.5
UHM	2.0	8	-1.4×10^{-6}	3000	710	1.0	335	1.7
Kevlar 29	1.44	12	-4.0×10^{-6}	2900	70	4.0	48.6	2.0
Kevlar 49	1.45	12	-4.9×10^{-6}	3000	120	2.8	82.8	2.1
Steel	7.80	–	1.3×10^{-6}	1000	205	6	26.3	0.1
Aluminum	2.78	–	2.2×10^{-6}	450	70	10	25.2	0.2

Note: Data provided above are indicative; for precise information on specific brand, manufacturers' data sheets can be consulted.

9.5 COMMON REINFORCEMENTS FOR PMCs

9.5.1 Glass Fibers

Glass fibers are the most widely used commercial reinforcements for PMCs. These are of high tensile strength but relatively low modulus fibers.

9.5.1.1 Types of Glass Fibers

Different types of glass fibers are commercially available. Traditionally, these are known by certain letter designations such as E-glass, S-glass, D-glass, C-glass, etc. All these types can, however, be categorized into two broad categories—general purpose (GP) fibers and specialty fibers. Representative compositions of different types of glass fibers are tabulated in Table 9.8 [19–22]. Various oxides are present in the chemical compositions of glass fibers, of which silica is the main constituent. Other major constituents are aluminum oxide, calcium oxide, boron oxide, and magnesium oxide. Several other oxides are also present in small quantities; however, in certain special cases, sodium oxide and zirconium oxide constitute good proportions of the overall composition.

9.5.1.1.1 GP Glass Fibers

E-glass fibers are the GP glass fibers and constitute over 90% of the total glass fiber production. E-glass fibers (E denoting electrical) were originally developed for electrical applications, where low electrical conductivity is a requirement. Essential constituents in an E-glass fiber composition are SiO_2, Al_2O_3, and CaO. In addition to these, B_2O_3 may or may not be present. Thus, based on the presence or otherwise of boron in the chemical composition, two primary subtypes of E-glass fibers are commercially available—boron-free and boron-containing E-glass fibers.

Chemical compositions of commercially available boron-free and boron-containing E-glass fibers differ substantially. Boron-containing E-glass fibers have lower dielectric constant than the boron-free type and, they are used in electrical applications such as electronic circuit boards, etc. On the other hand, boron-containing and boron-free types are both good from the point of mechanical properties. Further, they are available at low prices and, thus, both types of E-glass fibers are extensively used for making GFRP products for a wide range of commercial applications.

TABLE 9.8

Chemical Composition of Different Types of Glass Fibers (in Weight %)

	SiO_2	Al_2O_3	CaO	B_2O_3	MgO	Others (Na_2O, K_2O, TiO_2, ZnO, ZrO_2, Fe_2O_3, etc.)
E-glass with boron	54	14	18	8	4	2
E-glass without boron	59	14	21	–	4	2
A-glass	68	3	8	3	2	16 ($Na_2O = 14$, Rest = 2)
C-glass	65	4	13	5	3	10 ($Na_2O = 8$, Rest = 2)
D-glass	74	–	–	23	–	3
ECR-glass	58	12	21	–	2	7 (ZnO = 3, $TiO_2 = 2$)
AR-glass	65	2	6	4	–	23 ($ZrO_2 = 2$)
R-glass	60	23	9	–	7	1
S-glass	65	25	–	–	10	–

Note: Data given above are indicative; for precise information on specific brand, manufacturers' data sheets can be consulted.

9.5.1.1.2 Specialty Glass Fibers

Glass fibers other than the E-glass fibers belong to the specialty types. Each of these types has certain specific characteristic property associated with it.

S-glass fibers (S denoting strength) are premium glass fibers that provide the best of mechanical, thermal, and chemical properties of all commercial glass fibers. A variant of these fibers is S-2 glass fibers. Both S-glass as well as S-2 glass fibers are commonly used in aerospace and military applications. D-glass fibers (D denoting dielectric constant) are of borosilicate composition containing high percentage of B_2O_3 in addition to SiO_2. Owing to the high proportion of boron oxide, dielectric constants of different D-glass fibers are much lower as compared to the E-glass fibers. Thus, these fibers are favored in high performance electrical applications. Corrosion resistant glass fibers have been developed; these include C-glass, ECR-glass, and AR-glass. ECR-glass fibers are suitable for use in acidic environment, whereas, AR-glass fibers, owing to the presence of zirconium oxide in their composition, provide good alkali resistance and are used in fiber reinforced concrete applications.

9.5.1.2 Production of Glass Fiber

Glass fiber production is a three-stage process: raw materials handling, glass melting, and fiber forming. Typical production cycle is schematically shown in Figure 9.2. A number of raw materials are used for the production of glass fibers. Typical raw materials include silica sand or "glass making sand" (for silica), china clay (for alumina), limestone (for calcium oxide), boric acid (for boric oxide), dolomite (for magnesia), etc. Handling of these raw materials is done either in batch mode or in continuous mode. Proper care is needed to drive away impurities from the raw materials. Individual materials are weighed as per the desired product recipe, mixed thoroughly, and introduced into a furnace. The furnace typically consists of three sections. The first section is the melting unit that receives and melts the mix of raw materials at high temperature. The temperature of glass melting depends upon the composition and it can be around 1500°C. Homogenization of the melt takes place in the first section and then the melt goes to the second section where cooling, refining, and further homogenization take place. In the final section, the temperature is further reduced to a level of glass fiber formation.

A platinum–rhodium alloy tank called bushing is used in the fiber-forming stage. It has numerous tiny nozzles of 1–2 mm diameter. The bushing is electrically heated so as

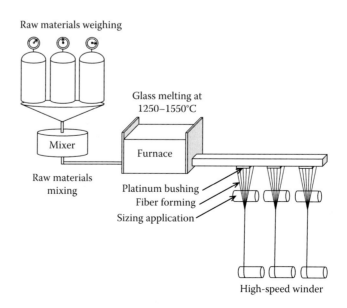

FIGURE 9.2 Schematic representation of glass fiber production (also see Reference 20).

to control the drawing temperature of the molten glass that flows, usually under gravity, through these nozzles and fibers are formed. These are very rapidly cooled, collected, and pulled at a high speed by a rotating drum. In this process, the fibers are solidified and elongated, and fiber diameter reduces to about 5–20 μm.

Application of sizing: Glass fibers are abrasive in nature and they need protective measure for avoiding fiber breakage during collection on the rotating drum, packaging, and also during subsequent weaving operations. For this, within milliseconds of solidifying, glass fibers are coated with certain sizing solutions by an applicator roller. The sizing helps in improving adhesion within the individual filaments as well as with the resin system.

9.5.1.3 Forms of Glass Fiber Reinforcements

Glass fibers are available in various forms; these are explained in Table 9.9.

9.5.1.4 Properties of Glass Fibers

Properties of glass fibers are dependent on the type of the glass fiber. Typical properties of three types, viz. E-glass, S-glass, and C-glass representing GP, strength, and chemical resistance, are shown in Table 9.10 [18–23]. In general, glass fibers possess very high tensile strength, good heat and fire resistance, resistance to chemical and biological degradation, moisture resistance, and good thermal properties due to low CTE. S-glass fibers have the highest mechanical properties, minimum specific gravity, and the highest temperature durability as well.

9.5.1.5 Applications of Glass Fibers

Glass fibers are extensively used in GFRP products for a wide range of applications. E-glass fibers are the most common of all types of glass fibers and are used in almost

TABLE 9.9
Forms of Glass Fiber Reinforcements

Forms	Description
Continuous strands	These are bundles of 204 filaments that are gathered after application of sizing. The strands are wound onto rotating mandrels and packed as spools of glass fibers
Chopped strands	These are cut pieces of strands of about 25 mm in length, normally used for making molding compounds
CSM	It is a nonwoven 2D mat, in which chopped strands of 25–50 mm are evenly spread and bound by using suitable binders such as PVA. CSMs are commercially available in continuous rolls and different surface densities. These are normally used for making low-strength products
Roving	These are bundles of continuous strands in the form of bands. These are commonly used for making high-quality products such as tanks, pressure vessels, etc.
Yarn	These are an assembly of continuous processed strands with slight twist that can be used in weaving operations
Woven roving	It is a coarse fabric made by weaving number of rovings, in which the rovings are woven in both weft and warp directions. These are usually used in making flat laminated products by hand lay-up or press molding
Woven yarn	It is a relatively fine fabric made by weaving yarns. Sizings are applied onto the yarns for ease of weaving and, often the sizing is removed by passing the fabric through a hot air chamber. It is resized for making it compatible with specific resin system. These are commercially available in different thicknesses such as 5 mil, 7 mil, 13 mil, etc.
Surface mat	It is a very fine mat of strands that are used for better surface finish

TABLE 9.10
Typical Properties of Glass Fibers

Parameter	Unit	E-Glass	S-Glass	C-Glass
Specific gravity	–	2.58	2.48	2.50
Softening point	(°C)	845	1055	730
Coefficient of thermal expansion	(m/m/°C)	5.2×10^{-6}	5.6×10^{-6}	6.7×10^{-6}
Tensile modulus	(GPa)	76	88	69
Tensile strength	(MPa)	3450	4600	3170
Elongation at break	(%)	4.8	5.7	4.8
Specific tensile modulus	(GPa/g/cm³)	30	36	28
Specific tensile strength	(GPa/g/cm³)	1.34	1.85	1.27

Note: Data given above are indicative; for precise information on specific brand, manufacturers' data sheets can be consulted.

all the sectors in the glass fiber market. Common usage of E-glass fibers includes car body parts, wind mill blades, consumer goods such as chairs, tables, desert air cooler body, storage tanks, pipes, different types of boats, etc. Other specific applications include radomes (D-glass), storage tanks and pipes in highly corrosive environment (C-glass), reinforced concrete (AR-glass), military applications (S-glass), aerospace vehicle parts (high silica glass), etc.

9.5.2 Carbon Fibers

Carbon fibers are very thin fibers of 5–15 μm diameter and composed primarily of carbon (more than 92% by weight).

9.5.2.1 Types of Carbon Fiber

As indicated in Table 9.11, carbon fibers can be classified in different ways [24]. Carbon exists in crystalline, quasicrystalline, and amorphous forms. In the amorphous form, the carbon fibers are isotropic in nature and low in modulus and strength. General purpose (GP) carbon fibers are of this category. High performance carbon fibers are crystalline, anisotropic in nature and they exhibit high modulus and strength. Activated carbon fibers contain numerous surface micropores.

A more commonly used method of classification of carbon fibers is to use tensile modulus and strength as the basis. Modulus and strength of carbon fibers vary in a rather wide range. Theoretical modulus of single-crystal graphite is 1000 GPa and carbon fibers of extremely high modulus (nearly 900 GPa) have been produced. Carbon fibers of elastic modulus in the range of 500–900 GPa are generally referred to as ultra-high modulus carbon fibers. Ultra-high modulus, high modulus, and intermediate modulus carbon fibers are anisotropic in structure and modulus is directly proportional to the degree of anisotropy. These fibers exhibit very high tensile strength as well. These are high performance carbon fibers. Low modulus fibers are also of low strength and these fibers are isotropic in structure.

Carbon fibers are manufactured using primarily three different types of precursor materials—polyacrylonitrile (PAN), pitch, and rayon. From this angle, we have three types of carbon fibers—PAN based, pitch based, and rayon based. Pitch precursors used can be both isotropic as well as anisotropic (mesophase pitch). PAN-based and mesophase pitch-based carbon fibers are highly anisotropic in structure and they are of high modulus and strength. Isotropic pitch-based and rayon-based carbon fibers exhibit relatively lower modulus and strength.

TABLE 9.11
Types of Carbon Fibers

Basis of Classification	Types	Description
Based on structure and general properties	General purpose	• Isotropic due to amorphous form • Low in strength and modulus
	High performance	• Anisotropic due to highly crystalline form • High in strength and modulus
	Activated carbon	• Characterized by the presence of numerous surface micropores
Based on tensile strength and modulus	Ultra-high modulus	• Tensile modulus ≈500–900 GPa • Tensile strength ≈2100–3900 GPa
	High modulus	• Tensile modulus ≈300–500 GPa • Tensile strength ≈2800–5500 GPa
	Intermediate modulus	• Tensile modulus ≈100–300 GPa • Tensile strength ≈1400–7000 GPa
	Low modulus	• Tensile modulus ≈30–100 GPa • Tensile strength ≈700–1000 GPa
Based on precursor material	PAN based	• Precursor material is PAN
	Pitch based	• Precursor material is pitch
	Rayon based	• Precursor material is rayon
Final heat treatment temperature	High-heat-treatment (Type I)	• Heat treatment temperature ≈1500–2800°C • High modulus type
	Intermediate-heat-treatment (Type II)	• Heat treatment temperature ≈1000–1500°C • High strength type
	Low-heat-treatment (Type III)	• Heat treatment temperature ≈300–1000°C • Low modulus and low strength

During manufacture of carbon fibers, the fibers are subjected to heat treatment at different stages. Based on the final heat treatment temperature, carbon fibers can be classified into high-heat-treatment carbon fibers (Type I), intermediate-heat-treatment carbon fibers (Type II), and low-heat-treatment carbon fibers (Type III). These three types are associated with high modulus, high strength, and low modulus/strength, respectively.

9.5.2.2 Production of Carbon Fiber

Carbon fibers can be produced from almost any material that contains carbon; however, only three precursor materials are used commercially. These are PAN, pitch, and rayon. Although these source materials are different from each other, the basic stages in the manufacture of carbon fibers remain the same [24–26]. These stages are spinning, stabilization, carbonization, and sizing. For achieving higher modulus, an additional stage of graphitization is optionally added after carbonization.

9.5.2.2.1 PAN-Based Carbon Fibers

PAN is the most widely used source material for carbon fiber. The primary stages involved in the manufacture of PAN-based carbon fibers are as follows (Figure 9.3):

- Polymerization
- Spinning
- Stabilization
- Carbonization
- Graphitization
- Surface treatment and sizing

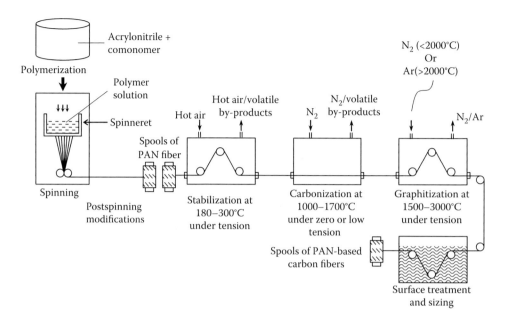

FIGURE 9.3 Schematic representation of manufacture of PAN-based carbon fiber (also see References 25 and 27).

The first stage in the manufacture of carbon fiber is a process of polymerization, during which a monomer called acrylonitrile (a colorless or yellowish liquid chemical compound, $CH_2 = CH-N$) is reacted with an appropriate comonomer. Polymerization methods used are solution polymerization and solvent-water suspension polymerization. The resultant product is the PAN polymer $(CH_2 = CH-N)_n$. It is a white solid/semisolid material with melting temperature of about 350°C. However, it has a much lower glass transition temperature of about 80°C and it degrades prior to melting.

PAN is converted into PAN fibers by adopting different textile spinning methods such as wet spinning, dry spinning, melt-assisted spinning, etc. The process of spinning involves forcing a PAN polymer solution to pass through a large number of tiny holes in a steel plate, called a spinneret. The PAN fibers typically have a diameter of 10–20 μm and density of 1.17 g/cm³. These fibers have a structure of oriented large molecules.

The next stage in the carbon fiber manufacture is stabilization, during which the PAN fibers are stretched and heated to about 180–300°C in an oxidizing environment. In the process of stabilization, the molecules get further oriented and cross-linked. As a result, the stabilized fibers do not decompose during pyrolysis at higher temperatures.

Stabilization is followed by carbonization. It involves heating the fibers up to a temperature of about 1000–1700°C in an inert atmosphere. The resultant fibers are rich in carbon content (up to about 95%) and lighter. Further, the fibers shrink in diameter, but not in the longitudinal direction. In this process, the molecules get further oriented and improved mechanical properties are resulted.

Carbonization is optionally followed by graphitization for producing very high modulus carbon fibers. In this case, the fibers are stretched and heated up to about 1500–2800°C in an inert atmosphere.

The final stage in the carbon fiber manufacture is surface treatment and application of a suitable sizing. The sizing helps minimize fiber damage during spooling, handling, and subsequent composite processing operations such as prepregging, winding, etc.

9.5.2.2.2 Carbon Fiber from Pitch

Pitch is a thermoplastic material of complex mixture of aromatic hydrocarbons. It is made typically from certain by-products of petroleum and coal-processing industry.

Typical sources for making bulk pitch for carbon fiber manufacturing are petroleum asphalt, coal tar, and PVC. These raw materials contain certain solid impurities; these are removed for making proper pitch and subsequent quality carbon fiber. Two types of pitches are used for making carbon fiber—isotropic pitch and anisotropic (mesophase pitch). Refined pitch is modified, in respect of molecular weight and chemical composition, by different methods and isotropic pitch is obtained. Anisotropic pitch is obtained by different methods involving heat treatment above 300°C of isotropic pitch. It contains a liquid crystal phase in the form of droplets called mesophase.

Stages of pitch-based carbon fiber manufacture are similar to those for PAN-based carbon fiber. However, the spinning methods and other process parameters such as heat treatment temperature, rate of heating, tension, etc. are different. Most commonly used method for spinning of pitch is melt spinning. Spun fibers are typically stabilized at 250–350°C in an oxidizing environment. Stabilization is followed by carbonization at 700–2000°C in an inert atmosphere. Carbonization chamber provides for gradual increase in temperature during which volatile products are released. Carbonization is optionally followed by graphitization in an inert atmosphere at 2000–3000°C.

9.5.2.2.3 Carbon Fiber from Rayon

Rayon fibers of different grades are used for making carbon fibers. Manufacturing cycle of rayon-based carbon fibers is very similar to PAN-based carbon fibers. Stabilization of rayon fibers is a low-temperature decomposition process at temperature below 400°C. During the initial heat treatment up to 125°C, adsorbed water is removed. Heat treatment environment can be either inert or reactive. An oil bath is used for removal of tarry products. Little or no tension is applied to the rayon fibers during stabilization as tensioning is not effective in rayon fibers at low temperatures. Carbonization of stabilized fibers is carried out at 1000–1500°C in an inert atmosphere. Stretching is found to be effective during carbonization in improving mechanical properties. For further improvement in mechanical properties, graphitization is carried out at 2500°C under appropriate tension.

9.5.2.3 Forms of Carbon Fiber Reinforcements

Continuous carbon fibers are available in 1D, 2D, and 3D forms. Tows and yarns are 1D, fabrics are 2D, and preforms are 3D in form. In addition to these, carbon fibers are also available as discontinuous reinforcements (staple fibers). Some of the important forms of carbon fibers are briefly explained in Table 9.12.

9.5.2.4 Properties of Carbon Fibers

Carbon fibers possess excellent mechanical and physical properties that make them unique in the field of aerospace composites. Some of the highly useful properties of carbon fibers and carbon fiber composites are as follows:

- High tensile modulus and strength
- Low density
- High specific tensile modulus and strength
- Excellent creep and fatigue characteristics
- Excellent corrosion resistance
- Excellent thermal stability
- Low CTE
- High electrical and thermal conductivity

Carbon fibers, however, possess certain disadvantages as well—notably, low strain to failure, low compressive strength as compared to tensile strength, and high degree of anisotropy.

TABLE 9.12
Forms of Carbon Fiber Reinforcements

Forms	Description
Tow	It is a bundle of continuous untwisted carbon filaments. Tows are available in different specifications such as 3k, 6k, 12k, 24k, etc., where 3k means 3000 filaments per tow, and so on
Yarn	A carbon fiber yarn is a bundle of continuous carbon filaments with slight twist. The twist given to the yarn makes it suitable for further processing such as weaving
Chopped carbon fibers	These are cut pieces of carbon fibers, typically 6 mm or less in length. These are normally used for making molding compounds
Carbon fabrics	Carbon fabrics of different types, commercially available in continuous rolls, are 2D in shape. However, from the point of view of reinforcement direction, they can be unidirectional (UD), bidirectional (BD), or multidirectional. Further, these fabrics may be either nonwoven, woven, or knitted
UD carbon fabrics	These are fabrics, in which nearly all the fibers are only in one direction. Both woven as well as nonwoven UD fabrics are available.
	In a nonwoven UD fabric, the uniaxial fibers are held in position by using a suitable binder
	On the other hand, in a woven fabric, in addition to the primary uniaxial carbon fibers, secondary fibers are also used and woven to form a stable fabric. The secondary fibers are used in very little proportion as compared to the primary fibers, and thus the fabric remains practically UD
BD carbon fabrics	BD carbon fabrics are made by weaving carbon tows as well as carbon yarns in the warp and weft directions. Different weaves are possible, of which three basic patterns are plain, twill, and satin
Multidirectional carbon fabric	Multidirectional fabrics are typically tridirectional (fibers at 0° and ±60°) or quadridimensional (fibers at 0°, 90°, and ±45°). These fabrics possess equal strength and stiffness properties in the directions of the fibers
Knitted carbon fabric	A knitted fabric is made up of flexible yarns that are formed into loops. A looped yarn is intertangled with the previous looped yarn and the process is continued to obtain the desired knitted fabric. Knitted fabrics are available in different knitting patterns; these are highly flexible and drapeable, and are available as flat as well as tubular fabric forms
Carbon preforms	These are near net shape rigid structures of carbon fibers. Typically, the fibers are placed in three or more directions. Unlike the fabrics, the thickness dimension in a preform is comparable with the other two dimensions, and reinforcements are put in the thickness direction as well

A number of commercial carbon fibers from different manufacturers are available; typical properties [29–32] of these fibers are presented in Figures 9.4 and 9.5 and Table 9.13. Note that the information provided in these figures and table is indicative and is meant for making an overall idea about carbon fiber properties. For information on specific grade of carbon fiber required during product design, data sheet from manufacturer should be consulted. Carbon fiber properties vary widely and are greatly dependent on the microstructure of the fibers, precursor material used, and processing parameters adopted during manufacture. In general, mesophase pitch-based carbon fibers exhibit the highest tensile modulus. Orientation of the microstructure is improved by stretching and heat treatment during manufacture; the result of such structural changes is higher tensile modulus, electrical conductivity, and thermal conductivity. However, an increase in tensile modulus is generally associated with a decrease in failure strain.

9.5.2.5 Applications of Carbon Fibers

High performance carbon fibers, owing to their very high mechanical properties, are extensively used for making CFRP products for a wide range of applications in the aerospace industry. In addition to the aerospace sector, CFRP products have been gaining

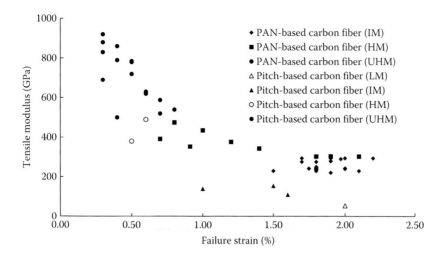

FIGURE 9.4 Tensile modulus versus tensile failure strain of carbon fibers. (Data from data sheets of commercial brands from various carbon fiber manufacturers: www.toraycfa.com; www.ngfworld.com; www.hexcel.com; www.tohotenax.com.)

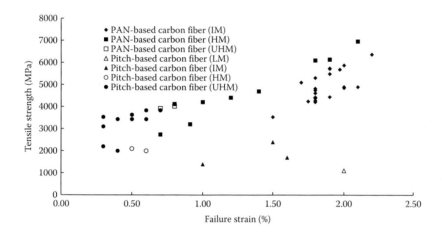

FIGURE 9.5 Tensile modulus versus tensile strength of carbon fibers. (Data from data sheets of commercial brands from various carbon fiber manufacturers: www.toraycfa.com; www.ngfworld.com; www.hexcel.com; www.tohotenax.com.)

TABLE 9.13
Typical Properties of Carbon Fibers

Parameter	Unit	PAN-Based Carbon Fiber			Pitch-Based Carbon Fiber			
		IM	HM	UHM	LM	IM	HM	UHM
Density	(g/cm^3)	1.73–1.93	1.76–1.80	1.77–1.78	1.65	1.70–1.85	2.00	2.00–2.19
Filament diameter	(μm)	5.0–7.1	4.4–7.0	5.0–7.0	10	10	–	7.0–10.0
CTE (×–10^{-6})	(m/m/°C)	0.38–1.10	0.64–0.70	–	1.4	0.8–1.0	–	1.4–1.5
Tensile modulus	(GPa)	221–296	303–475	540–588	54	110–155	380–490	500–920
Tensile strength	(MPa)	3530–6370	2740–6964	3920–4020	1100	1400–2400	2000–2100	2000–3830
Elongation at break	(%)	0.7–2.2	1.8–2.1	0.91–1.75	2.0	1.0–1.6	0.4–0.6	0.3–0.7
Specific tensile modulus	(GPa/g/cm^3)	124.4–166.3	170.2–265.4	305.1–330.3	32.7	54.7–83.8	245.0	247.6–420.1
Sp. tensile strength	(GPa/g/cm^3)	1.5–3.5	2.4–3.9	1.8–2.4	0.7	1.0–1.3	1.0	1.0–1.8

Source: www.toraycfa.com; www.ngfworld.com; www.hexcel.com; www.tohotenax.com; L. L. Clements, Organic fibers, *Handbook of Composites* (S. T. Peters, ed.), second edition, Chapman & Hall, London, 1998, pp. 202–241.

Note: Data given above are indicative; for precise information on specific brand, manufacturers' data sheets can be consulted.

increasing acceptance in other sectors such as automobile, sports, marine, biomedical, etc. High performance at low weight is a critical requirement in the aerospace sector, for which CFRP products are highly suitable. Typical applications of CFRP in the aerospace sectors are primary and secondary structures of civilian as well as military aircrafts, motor case, nozzle, airframe structures of space vehicles, satellite parts, etc. In the automobile sector, carbon fibers are used for making parts such as automobile bodies, gears, bearings, cams, etc. Another major consumer of carbon fibers is the sporting goods sector; common sporting goods made by carbon fibers are golf club, bicycle frame, tennis racket, etc.

9.5.3 Aramid Fibers

Aramids are a type of generic organic material made from polyamides. Broadly, they belong to the nylon family; however, they are different from the common nylon. Common nylon is an aliphatic polyamide containing amide–carboxylic bonds. Aramids, too, contain amide–carboxylic bonds; but they have an aromatic ring structure. Aramid fibers are a class of high performance fibers that exhibit good chemical and thermal stability, high toughness, and exceptional tensile strength and modulus. The high performance characteristics of aramid fibers are attributed to the aromatic ring structure.

9.5.3.1 Types of Aramid Fibers

Aramids can be divided into two types—para-aramids and meta-aramids [33]. The terms para and meta refer to the position of the carboxylic and amine groups in the monomer ring. Para-aramid has a symmetric structure with the bonds at 180° to each other. On the other hand, in the meta-aramids, the bonds are at 120°. A number of commercial brands are available under these two types of aramids. Some of the common aramid fibers are Kevlar® and Nomex® from Dupont, Teijinconex® and Technora® from Teijin, and Twaron® from Akzo. Kevlar is one of the oldest aramid fibers, and it, together with its different varieties, is also the most popular aramid fiber. Common Kevlar fibers are Kevlar, Kevlar-29, Kevlar-49, Kevlar-68, Kevlar-129, and Kevlar-149.

9.5.3.2 Production of Aramid Fibers

There are broadly two stages in the production of aramid fibers—polymer preparation and fiber preparation. The base polymer for para-aramid is poly-*p*-phenylene terephthalamide (PPD-T). PPD-T is prepared by polycondensation of *p*-phenylene diamine (PPD) and terephthaloyl chloride (TCl) in an amine solvent at low temperature. Para-aramid fiber is produced by extruding and spinning an anisotropic solution of PPD-T in concentrated sulfuric acid. There are two spinning methods available: wet spinning process and dry-jet spinning process. The basic process involves spinning the crystalline solution at high temperature in a spinneret followed by coagulating it in a quench bath. In this process, the hot polymer solution gets oriented and continuous fiber is formed, and the coagulated fiber is further washed, dried, heat-treated, and finally, collected on a rotating bobbin for packaging. In the wet-spinning process, the spinneret is kept right inside the coagulation liquid. This process is associated with certain drawbacks, which are overcome in the dry-jet spinning process. In the dry-jet spinning process, the spinneret is kept at a distance above the coagulating liquid. The air gap between the spinneret nozzle and the coagulating liquid helps achieve higher degree of structural alignment and resultant higher mechanical fiber properties. Figure 9.6 shows schematic representation of the dry-jet spinning process.

The base polymer for meta-aramid is MPD-I (poly-*m*-phenylene isophthalamide). MPD-I is prepared by polycondensation of *m*-phenylene diamine and isophthaloyl

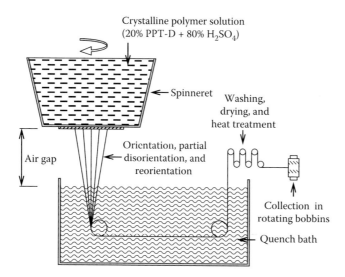

FIGURE 9.6 Schematic representation of dry-jet spinning process for aramid fiber manufacture (also see Reference 37).

chloride in an amine solvent at low temperature. The MPD-I solution is directly spun in a spinneret and extruded through a coagulation bath.

9.5.3.3 Forms of Aramid Fibers

Aramid fibers are commercially available as continuous as well as discontinuous reinforcements. Continuous forms of aramid fibers include yarns, rovings, and fabrics. On the other hand, different discontinuous forms of aramid fibers are staple fibers, spun yarns, and pulp.

A yarn of aramid fibers is a collection of a number of continuous filaments, and it is spun directly during the manufacture of the fibers. A roving, on the other hand, is a collection of a number of yarns. The number of filaments in a yarn may vary from a few numbers to a few thousands, and it depends upon the type of aramid fiber and its specified denier. (Denier, in textile terminology, is the linear density of the yarn or roving in g per 9 km.) Aramid yarns are relatively flexible and nonbrittle; they can be woven in different weave patterns such as plain, basket, satin, and crowfoot, and fabrics of different surface densities and surface fineness are commercially available. Fabrics with woven rovings are also available; however, these are relatively coarse.

Aramid staple fibers are commercially available; these are typically 6 mm and above in length, and used for making spun yarns. Staple fibers are also used in felts and other applications. A unique discontinuous form of aramid fibers is aramid pulp. It is a highly fibrillated form with 2–4 mm fibers.

9.5.3.4 Properties of Aramid Fibers

Representative properties of a number of aramid fibers are given in Table 9.14. Aramid fibers are lighter than both glass fibers as well as carbon fibers. In general, aramid fibers are superior to most glass fibers in terms of tensile modulus/strength as well as specific tensile modulus/strength, and comparable to some grades of carbon fibers. As stated earlier, aramid fibers are nonbrittle and possess higher failure strain as compared to carbon fibers. A unique property of para-aramid fibers is negative CTE, which is useful in designing zero-thermal-expansion composite laminate. Another characteristic property of aramid fibers and their composites is toughness. Aramid fibers, however, are not good in compression and bending. Further, moisture absorption is also a

TABLE 9.14
Typical Properties of Aramid Fibers

	Unit	Kevlar-29	Kevlar-49	Kevlar-129	Kevlar-149	Nomex	Twaron	Technora
Manufacturer	–	DuPont	DuPont	DuPont	DuPont	DuPont	Teijin	Teijin
Type	–	Para-aramid	Para-aramid	Para-aramid	Para-aramid	Para-aramid	Para-aramid	Para-aramid
Density	(g/cm^3)	1.44	1.44	1.45	1.47	1.38	1.44	1.39
Filament diameter	(μm)	12	12	12	12	16	12	12
Longitudinal CTE ($\times 10^{-6}$)	(m/m/°C)	−4.0	−4.9	–	–	15	−3.5	−6.2
Tensile modulus	(GPa)	70	112	95	150	17	120	70
Tensile strength	(MPa)	2920	3000	3800	2800	600	2700	3400
Elongation at break	(%)	4.0	2.8	3.3	1.5	2.2	2.0	4.5
Specific tensile modulus	(GPa/g/cm^3)	48.6	82.8	68.3	97.3	12.3	83.3	50.4
Specific tensile strength	(GPa/g/cm^3)	2.0	2.1	2.3	1.6	0.4	1.9	2.4

Source: *Comprehensive Composite Materials, Vol. 1, Fiber Reinforcements and General Theory of Composites*, H. H. Yang, Aramid fibers, pp. 199–229, Copyright 2000, from Elsevier; www.dupont.com; www.teijinaramid.com.

Note: Data provided in the table are indicative; for more authentic information, current data sheets from respective manufacturers can be consulted.

concern with aramid fibers. In general, useful properties of aramid fibers and aramid fiber composites can be listed as follows:

- High tensile modulus and strength
- Low density
- High specific tensile modulus and strength
- Excellent toughness characteristics
- Good fire resistance (can be ignited but fire does not continue once the source is removed)
- Negative longitudinal CTE

9.5.3.5 Applications of Aramid Fibers

Aramid fibers are extensively used in both composites and noncomposites applications, in diverse industrial sectors including aerospace, sports, automotive, and marine. Use of aramid fibers depends upon the specific type and fiber form. Meta-aramid fibers such as Nomex, owing to their excellent thermal resistance, good textile characteristics but poor mechanical properties, are used in protective clothing, reinforced belts and hoses, industrial coated fabrics, etc. Para-aramid fibers such as Kevlar and its various subtypes, Twaron, etc. possess excellent mechanical properties. Continuous filament yarns and rovings of para-aramids are used in composite pressure vessels, rocket motor casings, sporting goods, rope, and cable, etc. Fabrics are used in facings of sandwich constructions in aircrafts and helicopters, boat hulls, etc. Staple fibers are used in automotive applications such as brake and clutch linings, gaskets, etc.

In addition to glass, carbon, and aramid fibers, there are other reinforcing fibers used in PMCs; notable among these are boron and extended chain polyethylene fiber. Ceramic fibers and whiskers are commonly used in MMCs and CMCs. A unique class of reinforcing fibers is natural fibers, and it has become a subject of active research now.

9.5.4 Boron Fibers

Boron fibers are high performance fibers with tensile strength of 3100–4200 MPa and tensile modulus of 360 GPa. These are made by chemical vapor deposition of a gaseous mixture of hydrogen and boron trichloride on fine diameter tungsten filaments at high temperature of 1200°C. While the tungsten core is fine with a diameter of 13 μm, boron filaments are rather fat with an overall diameter ranging from 100–140 μm; thus, these fibers do not bend easily. With a relatively low density of 2.6 g/cm^3, boron fibers compare well with other high performance fibers in terms of specific tensile strength and modulus. Another advantage of boron fibers is their high compressive properties, and they retain their mechanical properties at high temperature of up to 800°C. These are used in boron/epoxy composites for aircraft applications, certain sports goods, and MMCs.

9.5.5 Extended Chain Polyethylene Fibers

Extended chain polyethylene (PE) fibers are a type of high performance organic fibers with some extraordinary properties, though not as popular as glass, carbon, or aramids. These fibers are produced by solid-state extrusion and gel spinning methods, and they possess a combination of favorable and unfavorable properties. Notable among the favorable properties are very low density (0.97 g/cm^3), high tensile modulus (130 GPa) and strength (2700 GPa), outstanding fatigue and impact resistance, excellent environmental resistance, etc. However, they are poor in creep, compression, and transverse directional properties. Further, these fibers have a very low service temperature, and untreated extended chain PE fibers do not have good compatibility with resin matrix. Typical applications of these fibers include bulletproof vests, military helmets, ropes, cables, nets, surgical gloves, sports goods, etc.

9.5.6 Ceramic Fibers and Whiskers

Ceramic fibers with oxide as well as nonoxide compositions are used in composites. Typical examples of oxide fibers are alumina (Al_2O_3), alumina–silica (Al_2O_3–SiO_2), and zirconite (Zr-Al_2O_3), whereas silicon carbide (SiC) is a typical nonoxide fiber. Ceramic fibers are available in continuous as well as discontinuous forms, and due to their high-temperature stability, they are commonly used as reinforcements for MMCs and CMCs in high-temperature applications.

Whiskers are single-crystal fibers with nearly zero defects. They are very short in length, but their aspect ratios are very high. Whiskers, owing to their nearly perfect crystal alignment and defect-free structure, possess high mechanical properties compared to their bulk forms.

9.5.7 Natural Fibers

Interest in natural fibers as an alternate class of fibers for composites has been growing and considerable amount of work has been done especially in the recent past (see, for instance, References 40–42 and the bibliography contained therein). There are three broad sources of natural fibers: plants, animals, and minerals. Based on the specific part of a plant that is used, plant fibers are categorized into bast fiber, seed fiber, leaf fiber, fruit fiber, and stalk fiber. In a similar way, animal fibers can also be categorized into wool or hair, silk fiber, and avian fiber. Examples of various plant and animal fibers are given in Table 9.15.

Natural fibers possess a few favorable characteristics. In general, they are light, biocompatible, and environment friendly. However, mechanical properties of natural fibers are generally poor and nonuniform. Poor compatibility with matrix system and moisture absorption are two other drawbacks. Present applications of natural fibers are in automotive industry (car dash board), construction industry (doors, windows), etc.

TABLE 9.15
Examples of Natural Fibers

Type	Type	Example
Plant fibers	Bast fiber	Flax, jute, kenaf, hemp, ramie
	Seed fiber	Cotton
	Leaf fiber	Banana, sisal
	Fruit fiber	Coconut fiber (coir)
	Stalk fiber	Bamboo, straws of rice, barley, wheat
Animal fibers	Wool or hair	Sheep wool, goat hair, horse hair, etc.
	Silk fiber	Silk fiber
	Avian fibers	Feathers

9.6 PHYSICAL FORMS OF REINFORCEMENTS

Filaments are the very basic form of reinforcements in advanced PMCs. They are characterized by their very fine diameter and very high aspect ratio (l/d ratio) and they possess very high structural properties. Individual filaments, however, are not directly useful in the composite processing shop floor. Reinforcements are available in different physical forms. Reinforcements in the forms of continuous fibers, short fibers, fabrics, and 3D preforms are used in PMCs processing. In this section, from the point of view of composite processing, we will have a brief discussion on these physical forms [20,43].

9.6.1 Continuous and Short Fibers

Continuous fibers are 1D reinforcements and they are available as strands, tows, rovings, and yarns.

Strand is a term associated normally with glass fibers. A strand is a bundle of untwisted filaments. Similarly, tow, a term associated with carbon fibers, refers to a bundle of untwisted carbon filaments. Tow size is specified in terms of number of filaments in one tow, for example, 12k tow would mean 12,000 filaments per tow.

A roving is a collection of untwisted strands or tows. Rovings are directly used in the manufacture of composite products by processes such as filament winding and pultrusion.

Yarns, on the other hand, are collections of twisted strands or tows. Twisting helps improve handleability in processes such as weaving.

Short fibers are obtained by chopping continuous fiber strands into lengths of about 25 mm. These are used for making molding compounds and CSMs. Chopped strands are also used in spray-up process.

9.6.2 Fabrics and Mats

Fabrics and mats are 2D reinforcements. They are used in wet lay-up process. A major use of fabrics is in making prepregs.

Fabrics are primarily of two types—woven fabrics and nonwoven fabrics.

Woven fabrics are made by weaving yarns, tows, or rovings on a loom. The fibers are placed in the warp and weft directions, and interlaced in different ways to make different weave styles. Warp is the 0° direction, that is, parallel to the length direction of the roll, whereas, weft, also known as fill, is the 90° direction. Relative amounts of fibers in the warp and weft directions depend on the type of weave style, and in general, equal or nearly equal amounts fibers are placed in both the directions. These fabrics are called as balanced fabrics. In certain cases, most of the fibers are placed in the warp direction

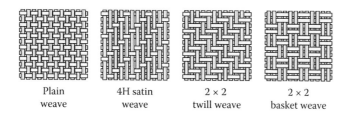

FIGURE 9.7 Common 2D weave styles.

only and these fibers are held in position by very fine threads in the weft directions. These fabrics, known as unidirectional fabrics, possess exceptionally high strength and stiffness properties in the warp direction but very low properties in the weft direction.

Woven fabrics are available in different surface densities, thicknesses, and finish qualities. Fabrics made from twisted yarns have finer finish than fabrics made from untwisted rovings. Selection of fabric depends on a number of factors such as structural strength, wettability, drapeability, composite ply thickness, etc. These characteristics are influenced by the fabric's weave style. Different weave styles are employed in fabric making. Common weave styles are plain weave, satin weave, twill weave, basket weave, etc. and their subtypes such as $4H$ satin, $8H$ satin, 2×2 twill, 2×1 twill, etc. Figure 9.7 shows some common weave styles. Plain weave has the maximum number of interlaces per unit area and it is very stable with maximum resistance to in-plane shear movement. But, plain weave fabrics are not very drapeable, and thus, are not suitable for lay-up on highly contoured surfaces. Owing to too many interlaces, strength of these fabrics is the least as compared to other weaves. Also, wettability of these fabrics during impregnation is poor. Basket weaves are marginally stronger than the plain weave. Satin weaves have minimum interlaces per unit area. They are strong, highly drapeable, and suitable for lay-up on highly contoured surfaces. On the other hand, from the point of view of wettability, twill weave fabrics, possibly, are the most convenient. For making a nonwoven or noncrimp fabric, the yarns or rovings are placed in the desired direction and then stitched using polyester threads. Unlike the woven fabrics, the yarns and rovings do not bend over each other and they remain straight; thus, nonwoven fabrics are stronger than the woven fabrics. Fibers can be placed in almost any direction as per final requirement, and it is possible to make multi-ply fabrics with fiber in more than one direction. Multi-ply fabrics are quite thick, and fabrication time can be drastically reduced. However, high ply thickness also causes problem of poor wettability of the fabrics.

In addition to the fabrics, mats such as CSM, surface mat, etc. are also regularly used in composite fabrication. CSMs are made by binding randomly distributed chopped glass strands by a suitable binder such as poly vinyl alcohol (PVA). In surface mats, very fine strands are used. Owing to the random nature of the reinforcements, the mats are not suitable for high performance structural applications and they are used in nonstructural commercial applications.

9.6.3 Preforms

Preforms are a type of reinforcement form, in which the fibers are arranged, typically, in three dimensions. In the composite processing stage using preforms, resin is injected into the pores in the preform. Resin transfer molding and its variants are common composite processing methods that use preforms. There are different methods to make preforms. Braiding is a common method for making continuous fiber preforms. Other methods include stitching, in which fabrics, mats or combinations thereof are stitched. For short fiber preforms, spray guns that spray chopped strands and a binder are used.

9.6.4 Molding Compounds

Molding compounds are a form of ready-to-use material used in molding operations. They are made by using reinforcements in different forms, resin, fillers, plasticizers, and other ingredients. There are different types of molding compounds—SMC, bulk molding compound (BMC), and injection molding compound.

9.6.4.1 Sheet Molding Compounds

SMC is a material in the form of sheet that contains uniformly distributed short fibers held by uncured thermosetting resins. Normally, chopped strands of glass and polyester or vinylester are used as the reinforcements as matrix, respectively. Typically, SMC consists of about 30% (by weight) chopped glass fiber. Continuous glass strands are also used in some cases. The process of manufacture of SMC involves making a paste of resin with some ingredients followed by application of the paste onto chopped glass strands. The ingredients used for making the paste have their own specific purposes and some of them are optional (Table 9.16). Specified quantity of the paste is applied on a carrier film. As the carrier film moves and comes under a chopper, chopped glass strands fall on the resin paste on the carrier film. Another carrier film with resin paste is then introduced such that the chopped glass strands get sandwiched between the two layers of resin paste on the carrier films. Doctor blades are used for adjusting the thickness of paste. The material is compacted by compaction rollers for achieving complete wetting of the glass fibers. The process of SMC making is schematically illustrated in Figure 9.8.

9.6.4.2 Bulk Molding Compounds

BMC, also known as dough molding compound (DMC), is a ready-to-mold material in the bulk form. It is made by mixing the resin paste with the fibers that are generally 6–12 mm in length. Fiber volume fraction in BMC is low, typically 15%–20%,

TABLE 9.16

Typical Ingredients for Making SMC

Ingredients	Description
Unsaturated polyester resin	It is the base resin in the paste and it binds the chopped glass fibers on curing
Styrene	It is a comonomer and solvent. It acts as a cross-linking agent and it reduces the resin viscosity
Catalyst	It initiates the cross-linking process in the resin. Normally, 0.3%–1.5% (by wt.) of catalyst is added
Fillers	Fillers constitute a major weight fraction of about 40% of the final part. Calcium carbonate, hydrated alumina, etc. are common fillers. They reduce shrinkage in the final part during curing, enhance dimensional stability and appearance of the molded part, and reduce the overall cost of the part. Certain fillers impart specific characteristics. For example, hydrated alumina provides flame retardancy
Thickener	It starts a chemical thickening process that converts the resin mix into a handleable paste. Common thickeners include oxides and hydroxides of calcium and magnesium
Mold release agent	It is added in small fraction for easy release of the molded part from the mold
Inhibitor	It is added for preventing premature cross-linking of the resin

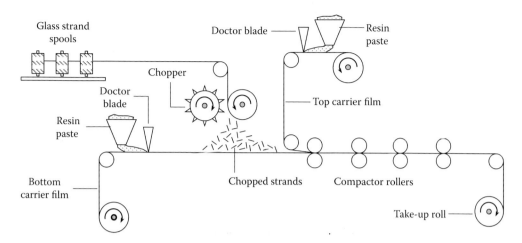

FIGURE 9.8 Schematic illustration of SMC manufacturing process. (Adapted with permission from S. K. Mazumdar, *Composites Manufacturing—Materials, Product and Process Engineering*, CRC Press, Boca Raton, FL, 2002.)

and as a result, mechanical properties of BMC composites are lower. Commonly used resins are unsaturated polyester, vinyl ester, and phenolics. Chopped glass strands are common reinforcements, however, in some special cases, prepregs are also cut in the form of flakes and used as a molding compound.

9.6.4.3 Injection Molding Compounds

Injection molding compounds are molding compounds formulated specifically for injection molding process. Injection molding is very common in thermoplastic composites manufacturing process. Epoxy, polyester, and vinylester are used for making thermoset injection molding compounds. Fibers used are very short, typically 1.5 mm in length.

9.6.5 Prepregs

Prepregs are a form of reinforcement, in which the fibers or fabrics are preimpregnated and stored for use at a later date. In a prepreg, the reinforcements are wetted with the resin and the resin is advanced to a state called as "B-stage." In this condition, the prepregs can be stored for a limited period at low temperature and can be used on a later date. The main advantage of using prepregs is that the impregnation of the reinforcements is done during raw material manufacture itself, and no impregnation is required during actual component manufacture. Thus, fiber volume fraction can be more accurately controlled, fibers can be more precisely oriented, and overall quality of final product is improved.

Prepregs are available in different forms. Preimpregnated roving prepregs, also known as towpregs, are unidirectional prepregs and are frequently used in filament winding applications. Unidirectional prepregs are also available in the form of tapes with fibers placed in the 0° direction. Woven fabrics are preimpregnated for making 2D flat prepregs. They are routinely used in lay-up of flat-to-medium contoured-to-highly contoured parts.

Prepregs are manufactured either by solvent impregnation or by hot melt process. Figure 9.9 schematically shows the manufacturing process of prepregs by solvent impregnation. The reinforcements in the form of fabrics or rovings are drawn through a resin bath containing low viscosity resin. The impregnated reinforcements are pulled at a constant speed through a series of rollers and an oven with different chambers maintained at different temperatures. The positions of the compactor rollers are so

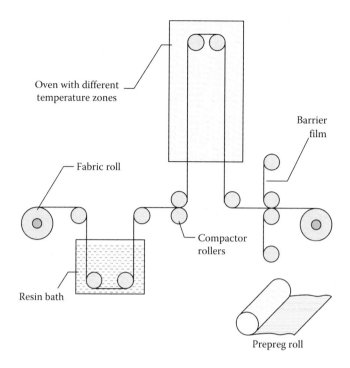

FIGURE 9.9 Schematic illustration of prepreg manufacture.

adjusted as to achieve the desired resin content in the prepreg. Advancement of the resin takes place inside the oven, and the parameters such as speed, temperatures at different chambers, etc. are set such that the resin comes to B-stage. At this state, the resin does not flow and the product is obtained as a preimpregnated fabric, which is slightly tacky. For ease of handling and storage, polythene films are introduced and the prepregs are rolled on rollers for storage. The prepregs have limited shelf life and stored at low temperature in a cold storage equipment.

9.7 SUMMARY

In this chapter, we presented an introduction to the constituents of PMCs. The key points can be summarized as follows:

- The matrix of a PMC material is a polymer, which is a natural or synthetic compound of usually high molecular weight consisting of many repeating units of smaller molecules (monomers).
- Polymers are of three types—thermosets, thermoplastics, and rubbers, of which thermosets are of primary interest in this book.
- Epoxies, polyesters, vinyl ester, and phenolics are some of the most common thermosetting resins used in PMCs industry.
- Epoxy resins are a class of thermosetting polymers consisting of primarily two components—a base epoxy system and an epoxy curative. In the composites field, they have found extensive applications in industrial, aerospace, and other high-end applications.
- Polyester resins are unsaturated thermosetting resins made by dissolving polyester oligomers in a solvent. They are the most commonly used resins in PMCs for commercial applications.
- Vinyl ester resins are similar to polyester resins but with an epoxy base. Applications of vinyl ester resins are many, of which, use as corrosion-resistant reinforced plastics is predominant.

- Phenolic resins are a class of resins made by condensation reaction of phenol and formaldehyde in the presence of a catalyst. Among several other applications, fire safety and high-temperature applications of phenolics are of critical importance.
- Various materials in the fiber form are used as reinforcements. Of these, glass, carbon, aramid, boron, extended chain polyethylene, ceramic fibers, and natural fibers are worthy of mention.
- Glass fibers are high tensile strength but relatively low modulus fibers used for making a wide range of GFRP products primarily for commercial applications. Depending on chemical compositions, there are several types of glass fibers such as E-glass, S-glass, etc. that are available in different physical forms such as roving, mats, etc.
- Carbon fibers are composed primarily of carbon and they are produced from three main precursor materials—PAN, pitch, and rayon. They have exceptional mechanical and thermal properties and are used in aerospace, defence, and high-end commercial applications.
- Aramids are a class of organic fibers with exceptional mechanical properties. Kevlar is one of the oldest and most commonly used aramid fibers.

EXERCISE PROBLEMS

9.1 What are thermosets and thermoplastics? How are they different from each other?

9.2 Write a comparative note on the properties and applications of epoxies, polyester, and phenolics.

9.3 What are the different types of glass fibers? Give a brief note on their chemical compositions.

9.4 How are glass fibers produced? Give a brief description.

9.5 Write a note on the different types of carbon fibers and their classification.

9.6 Write a note on production of carbon fibers.

9.7 Discuss and make a comparative note on the properties of various major classes of reinforcing fibers.

9.8 Make a comparative note on the applications of different types of reinforcing materials.

9.9 What are the different physical forms that reinforcements are available in? Discuss and make a comparative note of these forms in glass, carbon fibers, and aramid fibers.

9.10 What are prepregs and molding compounds? Write a brief note giving the similarities and differences between the two ready-to-use material forms.

9.11 Write a brief description on production prepregs and molding compounds.

REFERENCES AND SUGGESTED READING

1. P. K. Mallick, *Fiber-Reinforced Composites: Materials, Manufacturing and Design*, third edition, CRC Press, Boca Raton, FL, 2013.
2. J. A. Brydson, *Rubbery Materials and Their Compounds*, Elsevier Applied Science, London, 1988.
3. K. Nagdi, *Rubber as an Engineering Material: Guideline for Users*, Hanser Publishers, Munich, 1993.
4. T. D. Juska and P. M. Puckett, Matrix resins and fiber/matrix adhesion, *Composites Engineering Handbook* (P. K. Mallick, ed.), Marcel Dekker Inc., New York, 1997, pp. 101–165.
5. I. K. Varma and V. B. Gupta, Thermosetting resin-properties, *Comprehensive Composite Materials, Volume 2, Polymer Matrix Composites* (A. Kelly and C. Zweben, editors-in-chief; R. Talreja and J.-A. E. Manson, vol. ed.), Elsevier, Oxford, 2000, pp. 1–56.
6. M. A. Boyle, C. J. Martin, and J. D. Neuner, Epoxy resins, *ASM Handbook, Vol. 21, Composites* (D. B. Miracle and S. L. Donaldson, *vol. chairs*), ASM International, Ohio, 2001, pp. 78–89.

7. L. S. Penn and H. Wang, Epoxy resins, *Handbook of Composites* (S. T. Peters, ed.), second edition, Chapman & Hall, London, 1998, pp. 48–73.
8. C. A. May, Epoxy resins, *Engineering Materials Handbook, Volume 1, Composites* (T. J. Reinhart, technical chairman), ASM International, Ohio, 1987, pp. 66–77.
9. J. S. Puglisi and M. A. Chaudhari, Epoxies (EP), *Engineering Materials Handbook, Vol. 2, Engineering Plastics* (J. N. Epel, *consultant*), ASM International, Ohio, 1997, pp. 240–241.
10. M. Y. Shelley, Epoxy resins, *Polymer Data Handbook* (J. E. Mark, ed.), Oxford University Press, New York, 1999, pp. 90–96.
11. T. Pepper, Polyester resins, *ASM Handbook, Vol. 21, Composites* (D. B. Miracle and S. L. Donaldson, vol. chairs), ASM International, Ohio, 2001, pp. 90–96.
12. C. D. Dudgeon, Polyester resins, *Engineering Materials Handbook, Vol. 1, Composites* (T. J. Reinhart, technical chairman), ASM International, Ohio, 1987, pp. 90–96.
13. C. D. Dudgeon, Unsaturated polyesters, *Engineering Materials Handbook, Vol. 2, Engineering Plastics* (J. N. Epel, *consultant*), ASM International, Ohio, 1997, pp. 246–251.
14. F. A. Cassis and R. C. Talbot, Polyester and vinyl ester resins, *Handbook of Composites* (S. T. Peters, ed.), second edition, Chapman & Hall, London, 1998, pp. 34–47.
15. T. P. O'Hearn, Vinyl esters, *Engineering Materials Handbook, Vol. 2, Engineering Plastics* (J. N. Epel, *consultant*), ASM International, Ohio, 1997, pp. 272–275.
16. S. P. Qureshi and P. Resins, *ASM Handbook, Vol. 21, Composites* (D. B. Miracle and S. L. Donaldson, vol. chairs), ASM International, Ohio, 2001, pp. 120–125.
17. H. J. Harrington, Phenolics, *Engineering Materials Handbook, Vol. 2, Engineering Plastics* (J. N. Epel, *consultant*), ASM International, Ohio, 1997, pp. 242–245.
18. F. T. Wallenberger, Introduction to reinforcing fibers, *ASM Handbook, Vol. 21, Composites* (D. B. Miracle and S. L. Donaldson, vol. chairs), ASM International, Ohio, 2001, pp. 23–26.
19. F. T. Wallenberger, J. C. Watson, and H. Li, Glass fibers, *ASM Handbook, Vol. 21, Composites* (D. B. Miracle and S. L. Donaldson, vol. chairs), ASM International, Ohio, 2001, pp. 27–34.
20. D. J. Vaughan, Fiberglass reinforcement, *Handbook of Composites* (S. T. Peters, ed.), second edition, Chapman & Hall, London, 1998, pp. 131–155.
21. D. W. Dwight, Glass fiber reinforcements, *Comprehensive Composite Materials, Vol. 1, Fiber Reinforcements and General Theory of Composites* (A. Kelly and C. Zweben, editors-in-chief; T.-W. Chou, vol. eds.), Elsevier, Oxford, 2000, pp. 231–261.
22. D. M. Miller, Glass fibers, *Engineering Materials Handbook, Vol. 1, Composites* (T. J. Reinhart, technical chairman), ASM International, Ohio, 1987, pp. 45–48.
23. G. D. Sims and W. R. Broughton, Glass fiber reinforced plastics-properties, *Comprehensive Composite Materials, Vol. 2, Polymer Matrix Composites* (A. Kelly and C. Zweben, editors-in-chief; R. Talreja and J.-A. E. Manson, vol. eds.), Elsevier, Oxford, 2000, pp. 151–198.
24. D. D. L. Chung, *Carbon Fiber Composites*, Butterworth-Heinemann, Boston, 1994.
25. P. J. Walsh, Carbon fibers, *ASM Handbook, Vol. 21, Composites* (D. B. Miracle and S. L. Donaldson, vol. chairs), ASM International, Ohio, 2001, pp. 35–40.
26. K. Lafdi and M. A. Wright, Carbon fibers, *Handbook of Composites* (S. T. Peters, ed.), second edition, Chapman & Hall, London, 1998, pp. 169–201.
27. R. J. Diefendorf, Carbon/graphite fibers, *Engineering Materials Handbook, Vol. 1, Composites* (T. J. Reinhart, technical chairman), ASM International, Ohio, 1987, pp. 49–53.
28. P. A. Smith, Carbon fiber reinforced plastics-properties, *Comprehensive Composite Materials, Vol. 2, Polymer Matrix Composites* (A. Kelly and C. Zweben, editors-in-chief; R. Talreja and J.-A. E. Manson, vol. eds.), Elsevier, Oxford, 2000, pp. 107–150.
29. www.toraycfa.com
30. www.ngfworld.com
31. www.hexcel.com
32. www.tohotenax.com
33. L. L. Clements, Organic fibers, *Handbook of Composites* (S. T. Peters, ed.), second edition, Chapman & Hall, London, 1998, pp. 202–241.
34. M. Jassal and S. Ghosh, Aramid fibers—An overview, *Indian Journal of Fiber and Textile Research*, 27, 2002, 290–306.
35. K. K. Chang, Aramid fibers, *ASM Handbook, Vol. 21, Composites* (D. B. Miracle and S. L. Donaldson, vol. chairs), ASM International, Ohio, 2001, pp. 41–45.
36. M. W. Wardle, Aramid fiber reinforced plastics-properties, *Comprehensive Composite Materials, Vol. 2, Polymer Matrix Composites* (A. Kelly and C. Zweben, editors-in-chief; R. Talreja and J.-A. E. Manson, vol. eds.), Elsevier, Oxford, 2000, pp. 199–229.
37. H. H. Yang, Aramid fibers, *Comprehensive Composite Materials, Vol. 1, Fiber Reinforcements and General Theory of Composites* (A. Kelly and C. Zweben, editors-in-chief; T.-W. Chou, vol. eds.), Elsevier, Oxford, 2000, pp. 199–229.
38. www.dupont.com
39. www.teijinaramid.com

40. F. T. Wallenberger and N. E. Weston (eds.), *Natural Fibers, Plastics and Composites*, Kluwer Academic Publishers, Boston, 2004.
41. H. Ku, H. Wang, N. Pattarachaiyakoop, and M. Trada, A review on the tensile properties of natural fiber reinforced polymer composites, *Composites: Part B*, 42, 2011, 856–873.
42. Y. Xie, C. A. S. Hill, Z. Xiao, H. Militz, and C. Mai, Silane coupling agents used for natural fiber/polymer composites: A review, *Composites: Part A*, 41, 2010, 806–819.
43. S. K. Mazumdar, *Composites Manufacturing—Materials, Product and Process Engineering*, CRC Press, Boca Raton, FL, 2002.

10
Manufacturing Methods for Polymer Matrix Composites

10.1 CHAPTER ROAD MAP

After providing an introduction to composites in Chapter 1 followed by a detailed presentation on mechanics and analysis of composites in Chapters 2 through 8, common reinforcements, their physical forms, and matrix materials were discussed in Chapter 9. The objective of this chapter is to find an answer to the next logical question: how are these reinforcements and matrix materials converted into useful composite products? There are several PMC manufacturing methods that are employed for making a composite product. These methods involve certain basic processing steps; these essential processing steps are introduced first. PMC manufacturing methods can be categorized into three groups and some of the common manufacturing methods representing these three groups are chosen for discussion. The selection of a manufacturing method is an important decision in the development of a product; various aspects of the manufacturing method selection are presented. There are other critical aspects in composites manufacturing; two such areas, viz. process modeling and composites machining, are discussed at the end of this chapter. Our discussion will be limited to thermoset resin matrix composites.

As stated in the roadmap to Chapter 9, for a reader whose objective of study is primarily to familiarize with composites manufacturing and testing, Chapters 2 through 8 can be skipped. In that sense, Chapters 1 and 9 should be covered before proceeding to this chapter.

10.2 INTRODUCTION

Composites technology is process intensive. For anyone in the field of composites, good knowledge of manufacturing processes is essential to know how the reinforcements and matrix materials are converted efficiently into a useful product.

As reflected by numerous research articles, composites manufacturing is an extensively researched vast subject. Several full volumes of text have been devoted to discuss the various aspects of composites manufacturing in detail [1–3]. There are also several texts in which composites manufacturing is presented in lesser detail to complement discussions on other topics of overall composites technology [4–6]. In this chapter, the manufacturing of composites is treated as an important link in the chain of various topics that are involved in the final goal, that is, development of a composite product.

10.3 BASIC PROCESSING STEPS

All composites manufacturing methods involve four basic steps [1]. These are

- Impregnation
- Lay-up
- Consolidation
- Solidification

These steps are essential in all composites manufacturing methods. However, the methodology of implementation and the exact order in which these steps are implemented vary depending upon the manufacturing method and physical form of the reinforcement and matrix.

10.3.1 Impregnation

Impregnation or wetting is the step in which the fibers are wetted with the resin. The objective of this step is to ensure that each filament is wetted with resin all around. The matrix and the fiber–matrix interface play an important role in the load transfer mechanism in a composite material; resin-starved areas imply discontinuity in the load transfer path, and thus, proper wetting of fibers is essential for the manufacture of sound composites parts. Several process parameters influence proper impregnation of the fibers; some of the key parameters are viscosity of resin, its surface tension, and capillary action.

Impregnation is done in different ways in different composites manufacturing methods. In all wet lay-up and wet winding processes, impregnation is done during the actual fabrication process. For example, in wet filament winding, fibers are wetted in a resin bath prior to depositing them on the mandrel. Similarly, in wet hand lay-up, fabric plies are placed on a mold and wetted with resin using an application brush. On the other hand, in the dry lay-up and dry winding processes, impregnation of the reinforcements is done during prepreg/towpreg manufacturing.

10.3.2 Lay-Up

Lay-up is the step in which the reinforcements are placed as per the designed orientation. The structural performance of a composite part depends greatly on the ply sequence. The methods of placing the reinforcements vary from one process to another. For example, in filament winding, winding programs are used to control relative motions of the mandrel and carriage unit. The impregnated fibers are deposited by the carriage unit on the mandrel along a predesignated path. In a hand lay-up process, specified numbers of fabric plies are placed at specific fiber orientations so as to obtain the desired laminate thickness. In contrast to the above two and other such manufacturing methods, in which reinforcements are deposited in layers, there are manufacturing methods in which short reinforcements are placed at random orientation, resulting in globally isotropic composite parts. There are other types of manufacturing methods, such as RTM, that use preforms. In a preform, fibers are placed in a specified pattern, typically in three dimensions.

10.3.3 Consolidation

Consolidation is the process of removing excess resin and entrapped air during composite processing. The objective is to create intimate contact between laminae, and thereby sound composite parts. In hand lay-up processes, excess resin is squeezed out by the application of a roller during lay-up. Further, during curing, typically, pressure is applied for consolidation. Consolidation takes place simultaneously during lay-up in processes like filament winding, in which winding tension in the rovings generate sufficient consolidation pressure during winding, and no external consolidation pressure is necessary.

10.3.4 Solidification

Solidification is the final step that gives the composite part a physical solid shape. It essentially involves the cure of the resin matrix, during which cross-linking takes place and the resin solidifies. Typically, a positive pressure, vacuum, and temperature are applied. The duration of the curing process and requirement of pressure, vacuum, and temperature depend upon the cure kinetics of the resin system. For example, certain resins emit by-products during curing; application of vacuum is necessary for the removal of these by-products, and proper consolidation. Similarly, certain other resins cure at room temperature, and so on. The process of curing is discussed in more detail in Section 10.6.

Note: Given above is a brief discussion on the four essential steps involved in any composites manufacturing method. It may, however, be noted again that these steps or processes need not necessarily take place sequentially; in many cases, more than one processes go on simultaneously.

10.4 COMPOSITES MANUFACTURING PROCESSES

A number of manufacturing processes are available for making composite parts. These processes can be grouped as follows:

- Open mold processes
  - Wet lay-up
  - Prepreg lay-up
  - Spray-up
  - Rosette lay-up
- Closed mold processes
  - Compression molding
  - RTM and its variants
- Continuous molding processes
  - Pultrusion
  - Filament winding
  - Tape winding
  - Fiber placement

Note: The classification of composites manufacturing methods given above is mainly for the convenience of discussion. Also, the list of processing techniques is not exhaustive.

10.4.1 Open Mold Processes

Open mold processes, also referred to as contact molding processes, are one of the most common composites processing techniques that involve placing either dry fabrics or prepregs on an open mold [7–8]. A major advantage of these processes is that the cost of tooling and equipment and machinery is generally low. They are, however, labor intensive; thus, the final quality of the component is highly dependent on individual skill. These processes are suitable for medium- to large-sized parts in limited numbers. Lay-up processes can be either wet lay-up or prepreg lay-up depending upon whether dry fabrics or prepregs are used. Spray-up and rosette lay-up are special lay-up processes as discussed below.

10.4.1.1 Wet Lay-Up

10.4.1.1.1 *Basic Process*

Wet lay-up process is schematically shown in Figure 10.1. The basic processing steps are as follows:

- A release agent such as wax polish or polyvinyl alcohol (PVA) is applied on the mold surface. The release agent is necessary for easy extraction of the component after curing.
- First, a gel coat is applied on the mold surface. The gel coat is a high-viscosity resin that gives better surface finish. Low-surface-density reinforcing material such as a surface mat is put on the gel coat.
- Dry cut pieces (generally referred to as developments) of reinforcement fabrics are then laid-up as per the desired ply sequence.
- Each layer of fabric is wetted with liquid resin. Usually, a brush is used for applying resin on the dry fabric. Other methods of applying resin are also in vogue. For example, the resin can be poured and then spread using a spatula on the dry fabric. Alternatively, the fabric can be wetted separately and then laid-up on the mold. Clearly, other such variations can be thought of, which depends upon the ingenuity of the processing engineer.
- Excess resin and air voids are squeezed out by using a roller.
- The composite is then cured, usually at ambient environment.
- The cured part is extracted, and finishing operation such as parting, trimming, removal of high spots, etc. are carried out.

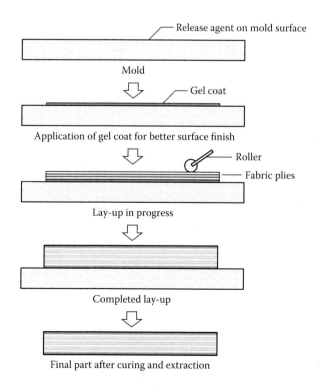

FIGURE 10.1 Schematic illustration of wet lay-up process.

10.4.1.1.2 Tooling and Capital Equipment

In most cases of wet lay-up, room-temperature curing at atmospheric pressure is adopted. As a result, demand on tooling as well as capital equipment is very low. Steel, wood, glass fiber–reinforced plastics (GFRP), etc. are common mold-making materials. Both male mold and female mold are used. Male mold gives better inside finish, and is typically used for making parts such as bath tub, etc. On the other hand, female mold gives better outer finish, and is typically used for parts such as yacht hull, etc. For large components, molds made from GFRP are also used. These composite molds are made by laying up glass fabrics on a pattern that resembles the part. The pattern, in turn, is made by adopting a process called loft template technique. Steel or wooden templates, machined as per the contour at regular cross section of the part, are placed parallel to each other and the gaps are filled with plaster. Excess plaster material is removed taking reference from the templates and the pattern is realized. Figure 10.2 gives a schematic representation of the wet lay-up process using composite mold and plaster pattern.

10.4.1.1.3 Basic Raw Materials

Commonly used reinforcements in wet lay-up processes are glass, carbon, and Kevlar fabrics. These are either bidirectional or unidirectional woven fabrics. Specially manufactured multilayered fabrics, contour-woven socks, etc. are also used. Another special type of reinforcing material is a unidirectional sheet that is made by winding rovings on a cylindrical drum. The circumferentially wound rovings are cut along meridian and a unidirectional sheet is obtained for immediate consumption. Surface mat and other low surface density materials are used for reinforcing gel coats. Common resins are polyester, epoxy, and vinyl ester. These are typically low-viscosity resins, whereas gel coats are specially formulated high-viscosity resins that give good surface finish.

10.4.1.1.4 Advantages and Disadvantages

The advantages of the wet lay-up process include the following:

- It is a simple and versatile process useful for making a wide range of products using a wide range of reinforcements.

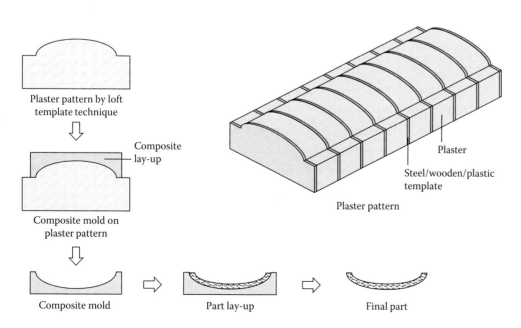

FIGURE 10.2 Wet lay-up using composite mold and plaster pattern.

- Initial capital investment is very low as the tooling can be very simple. Also, expensive equipment and machinery are generally not essential.

The disadvantages of this process include the following:

- It is a labor-intensive process.
- Part quality is highly dependent on individual skill and thus not consistent from part to part.
- It is suitable for medium- to large-sized components in limited numbers only.
- The process is somewhat messy and, due to its open mold nature, it involves health concern.

10.4.1.2 Prepreg Lay-Up

10.4.1.2.1 Basic Process

Prepreg lay-up is similar to wet lay-up in respect of the basic lay-up process. Both are open mold manual processes. However, differences do exist. First, in the case of wet lay-up, impregnation of the fabric is done during component processing. On the contrary, in the case of prepreg lay-up, impregnation is done at the raw material preparation stage itself, and the ready-to-use impregnated and B-staged fabrics, that is, the prepregs, are kept at subzero temperature in a cold storage equipment for future use. Another key difference is in respect of the curing process. In most wet lay-up processes, curing is done at room temperature and no pressure or vacuum is applied. On the other hand, prepreg lay-up processes generally involve curing in an autoclave at high temperature under external pressure and vacuum. Thus, prepreg lay-up processes are also known as autoclave process or vacuum bagging process.

The basic processing steps are as follows [8–10]:

- The prepreg rolls are taken out of the cold storage unit and brought to room temperature.
- The prepreg rolls are unrolled and the prepreg sheets are laid on a cutting table in a clean room under controlled temperature and humidity. Different sizes of prepreg cut pieces, that is, developments, in different numbers, as per manufacturing drawings, are required to make a component. For large complex components, the shape, size, and number of the developments are determined using specialized manufacturing software so as to minimize wastage. Prepreg sheet cutting is either manual or mechanized. Templates are used to aid in manual cutting. In the case of mechanized cutting, cutting equipments are used for speed and accuracy. These machines can cut several layers of prepreg sheets stacked one above the other.
- The mold is cleaned and the release agent is applied on the mold surface.
- Prepreg sheets are provided with a backing film either on one side or both the sides for easy handling. After cutting the prepreg sheets, these backing films are removed. It is important to ensure complete removal of the backing films as otherwise delaminations would be formed in the final part.
- The developments are then laid-up on the mold as per the manufacturing ply sequence. After laying up each layer, a roller is used for removing the entrapped air between the prepreg sheets. The process is continued until all the developments as per the process design are laid-up.
- The green composite is then vacuum-bagged and cured in an autoclave or an oven. After curing and cooling, the vacuum bag is removed and the part is taken out. The process of vacuum bagging and autoclave curing needs little more attention and is elaborated in the section on curing.

10.4.1.2.2 Tooling and Capital Equipment

Prepreg lay-up requires open molds that can be made out of steel, wood, or composites. Mold design and material are greatly influenced by the scheme of curing. Autoclave is a key capital equipment in the prepreg lay-up process. It is a pressure chamber with provisions for the application of heat and vacuum. The pressurization medium is either air or an inert gas such as nitrogen or carbon dioxide; an inert gas is preferable as air involves the risk of fire accidents. The temperature, pressure, and vacuum are computer controlled.

10.4.1.2.3 Basic Raw Materials

Carbon/epoxy prepregs are the most common material used for making structural parts in the aerospace industry. Other prepregs used are glass/epoxy and Kevlar/epoxy. For thermal insulation and ablative applications, carbon/phenolic and glass/phenolic prepregs are used.

10.4.1.2.4 Advantages and Disadvantages

The major advantages of prepreg lay-up process are

- As compared to the wet lay-up process, it is a neat process.
- Complex parts can be made by this process.
- High fiber volume fractions (more than 60%) can be achieved. Thus, parts made using prepregs are structurally very strong and stiff.
- Tooling cost is generally low, and the process is suitable for making prototype parts.

The disadvantages of prepreg lay-up process include the following:

- It is a highly labor-intensive process.
- Autoclave curing process is expensive due to high initial capital investment as well as high cost of autoclave consumables. Thus, parts made by this process are expensive.

10.4.1.3 Spray-Up

10.4.1.3.1 Basic Process

Spray-up is similar to the wet lay-up process except the method of depositing the reinforcements. In the wet lay-up process, fabric plies are laid-up and wetted one by one, whereas in the spray lay-up, a spray gun is used for depositing a mixture of chopped reinforcements mixed with resin on the mold. The basic steps are as follows:

- The mold is cleaned, and wax polish or other suitable release agent is applied.
- A gel coat is applied on the mold surface and allowed to harden. The gel coat gives a highly polished surface finish to the part.
- A barrier coat is applied on the hardened gel coat. It prevents fibers from printing through the gel coat.
- A spray gun is used to spray a mixture of chopped fibers and resin on the mold surface. Continuous rovings are fed into the gun; the gun chops the rovings to a predetermined size. A tank containing the resin mix is connected to the spray gun. The gun mixes the resin mix with the chopped fibers, and the mixture of chopped fibers and resin is sprayed.
- Once the desired thickness is built-up, rollers and brushes are used for the removal of entrapped air and further wetting, if needed.

- The composite part is then cured followed by extraction and finishing operations of the final part. Curing is usually done at ambient environment; however, oven curing at elevated temperatures is also adopted for accelerating the cure process.

Spray-up is sometimes used for making certain specialized parts, wherein after building up the first skin, corrugated material, or foam, is placed to act as core material. A second skin is then laid-up by spray lay-up. Sandwich structures made in this way are structurally stiffer than usual spray lay-up components, but more expensive.

10.4.1.3.2 Tooling and Capital Equipment

Tooling requirement for spray lay-up is similar to the wet lay-up process. Simple male or female open molds, made out of metal, wood, or FRP, are used.

A spray gun having features to receive continuous rovings or strands and cut the rovings to specified length is an essential equipment. The spray gun can receive either catalyst-mixed or hardener-mixed polyester resin from a tank or mix the catalyst or hardener with the resin inside the gun itself. Further, the spraying process can be mechanized for which robotic spray guns are used.

10.4.1.3.3 Basic Raw Materials

The most common raw materials are E-glass rovings and polyester resin. In specific cases, other reinforcements such as carbon and Kevlar in forms like continuous strand mat, fabric, and other core materials are also used.

10.4.1.3.4 Advantages and Disadvantages

The main advantages of this process are

- Spray lay-up is a very economical process that is suitable for making parts of almost any size—small, medium, or large.
- The tooling requirement is very simple.
- Complex parts can be made easily by this method.
- It can be easily automated.

The main disadvantages are

- Parts made by spray lay-up do not have very high structural properties.
- The quality of the product is highly dependent on the skill of the operator, and thus, parameters like fiber volume fraction and part thickness are neither uniform nor repeatable.
- The process is not suitable for good dimensional accuracy. Also, good surface finish is possible only on one side.

10.4.1.4 Rosette Lay-Up

Rosette lay-up is a unique composite lay-up technique. It can actually be considered as an open mold process as well as closed mold process. What differentiates this process from other conventional processes is the 3D geometry of the plies. In most other common layered composite construction, the plies are, in general, parallel or nearly parallel to the predominant surface of the component. For example, the plies are 2D in a flat laminate, cylindrical in a circular cylindrical shell, conical in a conical shell, and so on. The thickness of the part increases as the plies are laid-up one above the other. On the contrary, in a rosette construction, the plies are laid-up across the thickness.

The common applications of rosette lay-up are in the ablative liners for rocket nozzles. In a nozzle liner made by rosette lay-up, the edges of the plies are exposed to the

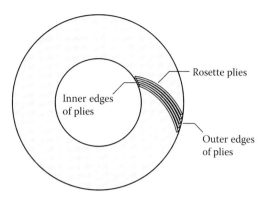

FIGURE 10.3 Cross-sectional representation of rosette lay-up of a cylindrical part.

hot and high-velocity gases of propellant burning. As only the edges are exposed, catastrophic ply erosion is avoided and controlled consumption of the ablative composites takes place, and the structural parts are protected. Figure 10.3 shows a typical rosette construction. The 3D orientation of the plies is important from the point of view of efficient functional performance of the part. It may also be noted that, owing to geometrical constraints, arbitrary shapes of rosette plies are not desirable. Note that if care is not taken while designing the shape of the plies, gaps will be created in the outer periphery of the part. The 3D orientation of the rosette ply at any point is defined in terms of three angles—arc angle, helix angle, and surface angle. These three angles are mathematically related to each other, and the rosette surface is determined based on such mathematical considerations. The mold for rosette lay-up consists of a block with the rosette surface on which the first ply is laid-up and the lay-up process is continued till the part is filled with the predecided numbers of plies.

10.4.2 Closed Mold Processes

There are several closed mold processes that are regularly employed for making composite parts. Although grouped together, these processes vary greatly from each other in respect of working principle, types of parts made, rate of production, tooling and machinery required, and basic raw materials used. We may identify these characteristics as we discuss some of the common closed mold processes.

10.4.2.1 Compression Molding Process

Compression molding is a popular composites manufacturing process suitable for making small- to medium-sized parts in large volumes as well as small numbers. In this process, typically, a female mold and a matching male mold are used [11–13]. In view of the matching molds, this is also known as matched-die-mold process. Compression molding is based on the principle of consolidating the charge (charge is the material placed between the two halves of the molds) by external pressure exerted by a hydraulic press followed by curing the part in the press itself. Variants are also in vogue, in which case, the consolidation pressure may be applied by the tool itself and curing is done in an oven.

10.4.2.1.1 Basic Process

The basic processing steps involved in compression molding are as follows (see Figure 10.4 for schematic representation):

- The top and bottom halves of the mold are cleaned and assembled on to the top and bottom platens of the hydraulic press. A suitable release agent such as wax polish is applied. (Typically, a hydraulic press is used for the application of

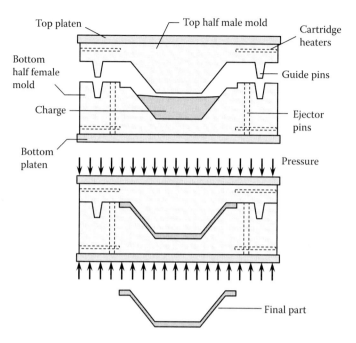

FIGURE 10.4 Schematic representation of compression molding. (Adapted with permission from S. K. Mazumdar, *Composites Manufacturing—Materials, Product and Process Engineering*, CRC Press, Boca Raton, FL, 2002.)

consolidation pressure. In this respect, a variant of this process is a completely tooling based compression molding process, in which the consolidation pressure is applied by the tool itself, and thus no platens are assembled to the molds.)
- If required, the mold is preheated to about 140°C, where the preheating temperature is decided based on the resin cure kinetics.
- Based on the mass of the final component, the quantity of charge is determined. The charge is then placed on the bottom female mold.
- The mold is closed by bringing the top half mold down. The press applies the required pressure to close the molds. The charge flows inside the cavity between the top and bottom halves of the mold and gets consolidated.
- Curing of the component takes place at high temperature for which either platen heating or direct heating of the molds by cartridge heaters or both are resorted to.
- After curing, the top platen is moved up and the component is demolded from the bottom half mold. Ejector pins are used for demolding.

Note: For large components, intermittent compaction is required, and only partial quantity of charge is placed at one time.

10.4.2.1.2 Tooling and Capital Equipment

Tooling involves a combination of a male mold and a female matched mold that are normally made out of steel. The molds are designed with provisions for outlet for the entrapped air and excess resin, stopper blocks for controlling the thickness of the part, guide pins for alignment, ejector pins for demolding the part, required numbers of cartridge heaters for mold heating, and anchor points for handling and assembly with the platens of the press.

A hydraulic press of required tonnage is the typical capital equipment used in compression molding. The upward and downward movements of the platens are guided so as to maintain parallelism between the platens. Curing is done in the press itself, and thus, provisions are made for heating of the platens that in turn heat up the molds.

10.4.2.1.3 Basic Raw Materials

The commonly used raw materials in compression molding processes are molding compounds such as sheet molding compounds, bulk molding compound, and thick molding compounds. Certain specific applications use prepreg flakes as the charge for which carbon/phenolic prepreg sheets are cut to 25 mm × 25 mm flakes. Compression molding is also used for making laminated parts with good surface finish on both the sides. In such a case, prepreg lay-up or wet lay-up is carried out using materials like carbon/epoxy.

10.4.2.1.4 Advantages and Disadvantages

The key advantages of the compression molding process are

- The process can be very fast and thus suitable for high volume production such as automotive applications.
- Components made by compression molding have good surface finish on both the sides.
- Good dimensional accuracy and repetitive quality can be achieved.
- Suitable for making near net shape components.
- Structural composite components with stiffening ribs can be made. Also, local stiffening for cut-outs can be easily provided. These composites involve detailed ply design that include ply drop-offs for achieving varying rib thickness/width.

Compression molding process also suffers from a few limitations; these are

- Generally, high-performance structural components cannot be made by compression molding of SMC or BMC. However, this limitation can be overcome by using prepreg or fabric lay-up.
- Initial investment is generally high.

10.4.2.2 Resin Transfer Molding Process

The RTM process involves pumping a resin mix into a porous preform that is kept inside the male and female half molds [14,15]. Curing of the resin takes place in the mold and a near net shape composite part is obtained. The process is an efficient one, and is suitable for making structural components of small to medium sizes in low to medium volume production quantities.

10.4.2.2.1 Basic Process

The RTM process is schematically shown in Figure 10.5. The processing steps involved are as follows:

- A release agent is applied on the top and bottom molds, and the preform is placed appropriately on the bottom mold. Cores and inserts, as required, may be incorporated in the preform.
- The top molds are closed by placing the top mold and the two are clamped.
- The molds are heated to the specified temperature.
- The resin mix is prepared, kept in the resin tank, and then pumped into the pores in the preform through several inlets. Resin injection into the porous preform takes place at high temperature under pressure. Optionally, the process of resin flow is assisted by providing vacuum.
- Once the mold is completely filled with resin and the preform is thoroughly wetted by the resin, the ports are closed and the inside pressure is increased for further consolidation.

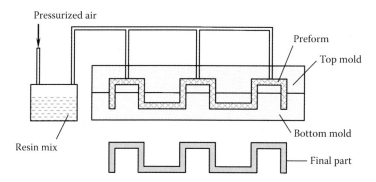

FIGURE 10.5 Schematic representation of the RTM process.

- Curing takes place inside the mold, after which the part is demolded and finishing operations are carried out.

10.4.2.2.2 Tooling and Capital Equipment

The tooling for RTM is usually made out of steel or aluminum. It involves top and bottom molds that have to be sufficiently strong and stiff for taking the pressure of resin transfer. However, compared to the compression molding process, working pressures in an RTM process are lower, and thus, the molds are also lighter and less stiff. In addition to the tooling, a resin dispensing equipment with a system for resin distribution is required.

10.4.2.2.3 Basic Raw Materials

Different types of preforms as well as fabrics are used as reinforcements in the RTM process. Normally, glass, carbon, and aramid fibers are used for making preforms, out of which E-glass is the most common. Polyester and epoxies are the typical resins used in this process. For aerospace applications, carbon/epoxy systems are very common. Other resins include vinylester and phenolics. A requirement for the resin for RTM is low viscosity so as to achieve proper flow of the resin and complete wetting of the preform. For high-viscosity resins, suitable modifications are required in the equipment for injecting the resin. Certain resin conditioning is also carried out in the case of such high-viscosity resins. Fillers such as alumina trihydrate, calcium carbonate, micro balloons, etc. are added for imparting specific characteristics and for economy.

10.4.2.2.4 Advantages and Disadvantages

The advantages of the RTM process include the following:

- Near net shape parts with good dimensional tolerances can be made by RTM. Final machining and other finishing operations are minimum.
- Parts made by this route have good surface finish on all the sides.
- RTM is suitable for making structural parts as the reinforcements can be accurately placed as per design requirement. It gives high fiber volume fractions up to 65%.
- RTM offers good joining and assembly features in the final product. Metallic inserts can be easily accommodated that are used subsequently in assembly operations.
- As compared to other closed mold processes such as compression molding and injection molding, in general, tooling cost is lower in RTM, and thus, initial investment is low. As a result, this process is suitable for prototype making.
- Automation is possible in RTM, and thus, high volume production is possible.

Manufacturing Methods for Polymer Matrix Composites

There are certain disadvantages of RTM as well; these are

- Tooling design is generally complex.
- Several trial runs may be required to establish process parameters so as to achieve proper flow of rein and void-free and dry fiber-free components.

There are several other thermoset composites manufacturing processes that are similar to the RTM. These processes can be considered as variants of RTM; they work on similar principles with marginal differences. Some of the RTM variants are vacuum-assisted resin transfer molding (VARTM), reaction injection molding (RIM), structural reaction injection molding (SRIM), reinforced reaction injection molding (RRIM), Seemann composite resin infusion molding process (SCRIMP), etc. VARTM is a process in which the resin is drawn by vacuum into the dry fabrics laid-up on the bottom mold. For this, a vacuum tight cavity is created by placing a top cover on the bottom mold. SCRIMP is a patented process with a specialized resin distribution system. RIM, SRIM, and RRIM are processes that involve mixing of resins at high velocity and injection of the resin mix into the preform.

10.4.3 Continuous Molding Processes

Continuous molding processes include pultrusion, winding processes, fiber placement, etc. Winding processes are those in which continuous reinforcement in the form of either roving or tape is deposited on a rotating mandrel. Two distinct winding processes are in practice—filament winding and tape winding.

10.4.3.1 Pultrusion

In the pultrusion process, fibers are pulled through a resin bath and a heated die to make the part [16–19]. The die gives the cross-sectional shape, which is either solid or hollow, to the parts. The parts are continuous in nature and of constant cross section.

10.4.3.1.1 Basic Process

The pultrusion process is shown schematically in Figure 10.6. The processing steps involved are as follows:

- The pultrusion setup is prepared for the specific run. It includes positioning the required numbers of spools on the creel stand, assembling the die, and setting up the guide rollers, resin bath, puller mechanism, and cutting saw. The number of rovings, which is normally 100 or above, is decided based on the cross-sectional area of the part and desired fiber volume fraction.

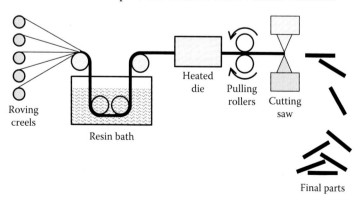

FIGURE 10.6 Schematic illustration of the pultrusion process. (Adapted with permission from S. K. Mazumdar, *Composites Manufacturing—Materials, Product and Process Engineering*, CRC Press, Boca Raton, FL, 2002.)

- The fiber rovings from the spools on the creel stand are drawn, guided by a system of ceramic eyelets and rollers, and pulled through a resin bath.
- The resin-impregnated fibers are then pulled at constant speed through a heated die. The temperature of the die and the pulling speed depend on the cure characteristics of the resin. The resin cures inside the die and the cured composite takes the inside cross-sectional shape of the die as its own cross-sectional shape.
- The composite is gripped in the pulling rollers and pulled through a sufficient distance so as to allow it to cool after which it is cut to the required length by a cutting saw.
- Finishing operations are carried out on the final part.

10.4.3.1.2 Tooling and Capital Equipment

Steel dies of constant cross section along the length are used in pultrusion. Usually, a taper is provided at the entrance for smooth entry of the fiber yarns. The dies are chrome plated so as to reduce abrasion.

The pultrusion setup consists of a creel stand for holding the roving spools. Dry reinforcements are usually fragile. Some reinforcements such as glass and carbon are abrasive. Thus, the incoming rovings are required to be guided properly for which usually ceramic eyelets are used. A resin bath, which is either open or enclosed, is used for impregnating the reinforcements. In an alternate method, resin is directly injected under pressure into the die, and the reinforcements get impregnated inside the die. The pulling mechanism typically consists of two sets of gripping rollers with rubber pads rotating in opposite directions. The rubber pads are tailor-made for the pultruded part geometry.

10.4.3.1.3 Basic Raw Materials

Glass, carbon, and aramid fibers are typical reinforcements used in the pultrusion process, out of which E-glass fibers are the most common. Owing to the continuous nature of the pultruded parts, reinforcements used in pultrusion are primarily continuous. Further, pultruded parts are mostly unidirectional, and thus, are made by using rovings. However, in certain cases, where bidirectional and multidirectional structural properties are required, continuous fabrics and mats are also used.

Resin used in pultrusion is of low viscosity, long pot life, and fast reactivity. A commonly used resin is unsaturated polyester. Fillers are added to the resin for economy as well as to impart specific characteristics. Other resins used in specific cases of structural requirements are epoxy, phenolic, and vinylester.

10.4.3.1.4 Advantages and Disadvantages

Pultrusion is an excellent process for the mass production of commercial composite parts. It has several advantages that make it popular. Some of these advantages are

- It is an automated continuous process with high production rate; thus, it is suitable for high-volume applications.
- It is a cost-effective process; processing cost is low due to its continuous nature, raw materials used are generally cheap and scrap is minimum.

The pultrusion process, however, suffers from a few drawbacks as indicated below:

- Parts produced are of constant cross section along the length. Parts with varying cross section and complex geometry cannot be made by this method.
- It is not possible to make thin-walled parts by pultrusion.

- Reinforcements are primarily in the longitudinal direction only; tailoring of the reinforcements is not generally possible.

10.4.3.2 Tape Winding

Tape winding is a winding process in which continuous prepreg tapes are wound on an axisymmetric mandrel—conical, contoured, or cylindrical. Tape-wound components are usually used as ablative liners in rocket nozzle applications. It can be implemented as either parallel winding or angled tape winding. In either case, the edges of the fabric are exposed to the inner and outer surfaces. The difference lies in the orientation of the prepreg tapes. In parallel winding, the tape forms a near cylindrical shape in one full round, and in a longitudinal cross section, the tape center line is parallel to the axis of the component. On the other hand, in angled tape winding, the tape forms a near conical shape in one full round, and the tape center line, in a longitudinal cross section, is at an angle to the center line of the component. A schematic representation of the tape winding process is given in Figure 10.7.

The basic processing steps involved in tape winding are as follows:

- The prepreg sheets are cut to form a tape of desired width and the tapes are rolled on spools. The width is determined based on the part thickness required, half-cone angle of the mandrel meridian, and the prepreg thickness (Figure 10.8). It is given by $T/\sin \theta$.
- The mandrel is loaded on the tape-winding machine and a spool is loaded on the carriage unit of the machine. The mandrel is wax polished.
- The tape is drawn through a system of guide rollers and tensioning mechanism, the backing films on the tape are removed and the tape is stuck on the smaller end of the mandrel.
- The machine is then started and the winding program moves the carriage unit in a controlled manner as the mandrel is rotated (Figure 10.8). The carriage unit motion is given by $\Delta = t/\sin \phi$.
- A roller is used for applying consolidation pressure that removes air gaps.
- After completion of the winding, the part is cured in an autoclave or hydroclave, parted/machined, and extracted.

The tooling requirement for tape winding is simple. Generally, metallic mandrels made from steel are used for making tape-wound components.

A tape-winding machine is a lathe with a carriage unit that has features to hold the tape spool. The carriage unit, further, has a tensioning mechanism, a system of rollers to guide the prepreg tape, and a roller that deposits the tape on the mandrel. Usually, two axes of motion—mandrel rotation and translation of the carriage unit parallel to

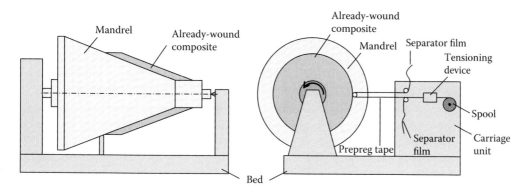

FIGURE 10.7 Schematic representation of the tape winding process.

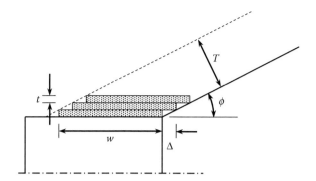

FIGURE 10.8 Tape winding process parameters.

the machine axis—are sufficient for parallel winding. For angled tape winding, the axis of the tape depositing roller has to be at an angle to the machine axis, and accordingly, the carriage unit has to have an additional feature of orienting itself.

The common raw materials used for making tape-wound components include phenolic prepregs of glass and carbon fabrics. The prepregs are provided with backing films on both the sides, and stored at subzero temperatures.

Tape winding is a very convenient process for making axisymmetric parts for ablative and thermal applications. The process, however, has somewhat limited applications.

10.4.3.3 Fiber Placement

Fiber placement is a method with a high degree of automation, which makes it possible to make simple to complex parts with accurate placement of reinforcement. In this method, either prepreg tows or prepreg slit tapes are drawn from spools and laid under roller compaction pressure by a fiber depositing head along a predesignated path on a rotating mandrel or a stationary mold [20,21]. The lay-up process is a continuous cut-restart process, in which the individual tows in the bundle of tows are cut at different intervals along the fiber path; the process is restarted and continued till the entire surface is covered and subsequent layers are deposited. It is an efficient process, suitable for complex parts, and has been successfully used in making aircraft and aerospace vehicle parts such as fuselage sections, engine parts, payload adapters, etc.

10.4.3.3.1 Basic Process

The fiber placement process is somewhat similar to filament winding (filament winding is discussed in the next section). The basic processing steps involved in it can be identified as follows:

- The mandrel is loaded on the fiber placement machine. Alternatively, when a stationary mold is used, it is positioned on the bed in the fiber placement setup.
- The towpreg spools are loaded on the creel stand and a bundle of tows are drawn and fed to the fiber depositing head. The tows are kept in requisite tension by a tensioning device.
- Referencing of the mandrel w.r.t. the machine is done, that is, the mandrel geometry is put in the machine coordinates.
- The fiber depositing head pulls the bundle of tows and places it on the work surface of the mandrel. A local heater is switched on for controlled heating of the tows and they are steered along the programmed path on the mandrel. The fiber depositing head has got several degrees of freedom and its relative motions ensure that the tows are accurately placed in the proper orientation. The roller pressure helps maintain appropriate thickness and width of the towpreg band and necessary compaction.

- Individual tows are cut or added at different points so as to maintain the required width along the path.
- After completion of each course, the fiber depositing head is brought back near the starting point and the next course is started. The programming is done in such a way as to ensure that the tow band is placed just next to the previous band without any overlap or gap.
- The process is continued until the entire work surface is covered and subsequent layers are placed to achieve the designed thickness.
- The composite is then cured in an oven or an autoclave and the final component is obtained after extraction and machining, if required.

10.4.3.3.2 Tooling and Capital Equipment

Depending on part configuration, both male as well as female mandrels and molds are used in fiber placement. Structural steel is commonly used for making mandrels and molds. The work surface should be firm enough not to deflect under action of the compaction roller. The tool design must cater for provision of features for referencing.

The central feature in a fiber placement machine is the fiber depositing head and the associated electronic system for controlling its motions. As noted earlier, it has a number of simultaneously controllable axes of motions that are programmed for accurate steering of the tows on the work surface. It has (i) a roller that finally deposits the tows under pressure, (ii) a cutting device that cuts the individual tows, (iii) a heating device for local heating of the tows at the point of delivery, (iv) tow restart rollers, and (v) a band collimator. Other parts of a typical fiber placement machine are a creel stand for holding the spools of prepreg tows, a headstock, and a tailstock.

10.4.3.3.3 Basic Raw Materials

Prepreg tows or slit tapes of glass, carbon, aramid fibers, etc. can be used in a fiber placement setup. Typically, the width of a flattened tow or slit tape is between 3 and 6 mm. Slit tapes are made by slitting wider prepregs in a slitter to the desired width and spooled on a core. Backing films are introduced during spooling and the same is removed during despooling and fiber placement process.

Tackiness of the material is a key parameter in fiber placement. When the tows are despooled and fed to the fiber-depositing head, they should not stick to each other and to the guide rollers and compaction roller. However, after compaction under the compaction roller and heating, the tows should stick to the underlying surface. Thus, low or nil tack is desirable at room temperature and high tack at elevated temperature.

10.4.3.3.4 Advantages and Disadvantages

The primary advantages of fiber placement can be identified as follows:

- Complex part configurations can be realized with accurate fiber orientation by fiber placement. Material can be laid-up on both convex as well as concave surfaces.
- It can be automated to a great extent and part quality is highly reliable.
- It is an efficient process and wastage of material is minimum.

There are certain drawbacks as well; these are

- Fiber placement is possible with only prepregs; thus, the availability of appropriate towpreg or slit prepreg tapes can be a critical factor.
- It is a capital-intensive process. It essentially needs a fiber placement machine and frequently an autoclave. As a result, it may turn out to be expensive.
- Skilled manpower is essential.

An introductory presentation on some of the common/representative manufacturing processes is given above. Note that the discussion is generic in nature. Needless to say, product-specific modifications with more details would be necessary in the shop floor. Also, each of the processes presented above has much more details to be addressed; a detailed discussion is not intended here. The processes discussed above belong to all the three broad categories—open mold processes, closed mold processes, and continuous molding processes. Filament winding, a continuous molding process, is an important one and it is dealt in some detail in the next section.

10.5 FILAMENT WINDING

Filament winding is a common continuous molding process used for making a wide range of products with tubular configuration. Historically, this process is reported to have been used by the Egyptians way back in 1370 BC [22]; modern composite filament winding is of relatively recent origin and it has been in use since the mid-1940s. Products with axisymmetric tubular configurations such as pressure vessels, rocket motor casings, pipes, storage tanks, launch tubes, etc. are regularly manufactured by filament winding. Another class of highly efficient filament-wound structures are the grid-stiffened structures [23–25]. With the advancement in the areas of computing and simulation, nonaxisymmetric and other geometrically complex configurations are also possible now with filament winding; rotor blades, elbows, T-junction, driveshafts, bushings, etc. are some of the typical examples of such products [1].

10.5.1 Filament Winding Fundamentals and the Basic Process

The filament winding process essentially involves a mandrel around which rovings are deposited along certain predesignated paths. Winding programs are used to ensure uniform deposition of fibers on the mandrel surface so as to obtain the designed thickness and ply orientation. The green composite on the mandrel is cured and the component is extracted by removing the mandrel. The four basic steps of composites manufacture (refer Section 10.3) are applicable to filament winding as well and it is interesting to see how these steps are incorporated.

10.5.1.1 Impregnation

Impregnation is done commonly in three ways and thus the rovings used are of three types—wet, wet rerolled, and dry [26,27]. In the first case, dry rovings are drawn under tension from the spools on a creel stand, made to pass through a resin bath for thorough wetting, and deposited on a rotating mandrel. In this case, wetting of dry tows takes place in the resin bath during winding and the process is normally referred to as wet winding. Wet winding is relatively simple and cost-effective. It ensures complete wetting of the fibers and the resultant cured composite is generally void-free. However, it is a somewhat messy process and uniform fiber volume fraction is difficult to achieve. In the second and third cases, wet rerolled and prepreg tows (towpregs), respectively, are wound on the mandrel. Wet rerolled rovings are dry rovings that are wetted using controlled volume of resin and respooled before using them in winding. They are either used immediately after respooling or stored in a freezer for future use. On the other hand, towpregs are preimpregnated rovings, in which the cross-linking in the resin is advanced to the B-stage. Winding with wet rerolled rovings and towpregs does not involve wetting of the fibers during winding operation and the process is generally known as dry winding. It is a neat shop floor process that results in uniform fiber volume fraction in the cured component. However, it is relatively more expensive. Also, in some cases, the tows may not spread properly and the resultant composite material can have defects like insufficient resin content and the presence of voids especially in thick shells such as the areas near the pole openings of a pressure vessel.

Note: The term "dry winding" is also used to refer to winding with dry rovings followed by impregnation by resin and curing; this method is however not common.

10.5.1.2 Lay-Up

The wetted rovings are deposited on the rotating mandrel by a fiber-depositing unit called pay-out-eye. The pay-out-eye is attached to a carriage unit that has different axes of motion. The mandrel rotation and the motions of the carriage unit along with the pay-out-eye are controlled by a winding program, and the relative motions of the carriage unit and the mandrel result in the desired path along which the fibers are deposited. The angle of winding (to be defined in Section 10.5.2) along the fiber path on the mandrel surface is the ply orientation angle. At this point, a distinctive feature of filament winding may be noted. In a wet lay-up or prepreg lay-up process, the ply orientation angle is the same at any point in a ply; whereas in filament winding, depending on the type of winding and meridional contour of the component, the angle of winding may or may not remain constant in a ply. For example, the angle of winding in a hoop ply is the same at any point in the ply; but in a helical ply, the angle of winding can vary. Thus, in a strict sense, in a filament-wound structure with helical plies, the ply sequence can vary from one cross section to another.

10.5.1.3 Consolidation

During filament winding, certain tension is applied to the rovings so as to keep them straight between the pay-out-eye and the fiber touch-down-point on the mandrel. Also, when the rovings are pulled through the resin bath and guided through the system of rollers, certain amount of tension is created. The roving tension has a component that acts in the inward normal direction, which generates a consolidation pressure. Thus, consolidation takes place during winding itself. The consolidation pressure is directly proportional to the roving tension and inversely to the local radius of curvature in a plane containing the fiber path. It can be represented as

$$p = \frac{kT}{r} \tag{10.1}$$

where p is the consolidation pressure, T is the winding tension in the rovings, r is the local radius of curvature in the plane containing the fiber path, and k is a constant involving number of rovings, bandwidth, and number of circuits. Note that for hoop winding on a circular cylindrical mandrel, r is nothing but the radius of the cylinder; the consolidation pressure is the maximum for a given mandrel and roving tension. On the other hand, for the same mandrel, on account of large r, a low angle helical winding is associated with low consolidation pressure; for example, at an angle of winding of $0°$, r becomes infinite, making the consolidation pressure zero. This is why a low angle helical winding is likely to result in inferior laminate quality.

10.5.1.4 Solidification

Curing of a filament-wound part can be done in different ways:

- Curing at room temperature
- Curing at elevated temperature in an oven
- Curing at elevated temperature under pressure and vacuum in an autoclave
- Inline curing using UV radiation
- Inline curing using electron beam

The requirement of temperature application is dependent on the resin system. Room-temperature curing can be done simply by keeping the mandrel with the wound composite either on the winding machine or on support stands. For wet wound systems, the mandrel can be kept in rotation to avoid resin flow under gravity. Most high-performance epoxy resin systems need the application of high temperature. Consolidation pressure is generated during the winding process by the winding tension in the rovings; additional curing pressure is generally not required and thus an autoclave is not necessary. In some cases, for example, towpreg winding with close cross-overs, insufficient resin flow may result in the creation of undesired air voids and consolidation pressure during winding is supplemented with positive pressure and vacuum during curing in an autoclave. Pressure and vacuum application in an autoclave in the cure of a filament-wound component, however, is associated with the risk of creating kink in the plies by fiber buckling [3,26]. UV radiation and electron beam are used for simultaneous winding and layer-by-layer curing of filament-wound products. Such inline curing is particularly useful in the cure of large components with large thickness.

10.5.1.5 Basic Processing Steps

Filament-wound product realization involves a number of basic processing steps; these steps are as follows:

- The mandrel is provided with a release agent such as wax polish or PVA solution or release film, and loaded on the winding machine.
- Resin mix is prepared and poured in the resin bath.
- Spools of fiber rovings are loaded on the creel stand, and the rovings are drawn from the spools to the mandrel through the system of tensioning device, resin bath, doctor blades, guide rollers, and comb. The guide rollers and comb are adjusted so as to avoid fiber fudging.
- The rovings are attached to the mandrel at certain designated location. The fiber attachment can be done by bonding or similar other means. Generally, the initial portion of the fibers may not follow the designed path and they may not spread properly; appropriate winding program can ensure to keep such portion outside the actual component.
- Tensioning device is started so as to create appropriate tension in the fiber rovings.
- Doctor blade setting is adjusted for controlled pick-up of resin by the tows. In certain cases of resin system, the resin bath is heated in a controlled manner for maintaining the resin viscosity within a desired range so as to ensure uniform resin pick-up and complete wetting of the tows.
- The winding program is loaded in the computer system of the winding machine, and the rotation of the mandrel and motions of the carriage unit are started. It is important to carry out appropriate referencing so as to move the carriage unit to the starting point before starting the winding program. The mandrel rotation draws the rovings on to itself and the relative motions of the carriage unit and the mandrel rotation deposit the rovings along the predefined path on the mandrel as per the winding program.
- After the completion of one circuit (one full round around the mandrel is called as one circuit) or a group of circuits, the winding program ensures that the next fiber band is placed just near the already-placed fiber band with prespecified marginal overlaps, and the process is continued till the programmed numbers of circuits are completed. The completion of the programmed numbers of circuits in one ply ensures coverage of the complete mandrel surface area.

The process of winding is further continued till the designed numbers of plies are deposited and desired part thickness is obtained.
- During the process of winding, excess resin from the already-wound surface are scrapped with a soft scrapper blade.
- In certain cases, depending on design requirements, the filament-wound plies are intermingled with fabric lay-ups. Rollers are used for the consolidation of these laid-up plies.
- The part is then cured in an oven. In certain cases, especially with very large components, curing is carried out on the machine itself by enclosing the component with insulating walls and heating the air inside the enclosed space.
- The mandrel is extracted from the cured part.

10.5.2 Computational Aspects of Filament Winding

The computational aspects of filament winding are somewhat involved as compared to some other composites manufacturing processes such as wet lay-up. These aspects must be addressed with care and approximations should be avoided for realizing quality products. Details of fiber path computations, stability of fiber path, and winding program generation are beyond the scope of this book; interested reader can consult, for instance, References 28–31 among others. Winding programs can be generated for specific part geometry using mathematical tools and programming languages such as MATLAB, C, C++, etc. and the output can be integrated with the computer numerical control (CNC) system of the winding machine. Alternatively, commercially available packages such as CADFILL® and CADWIND® can be utilized. However, familiarity with the basics of these computational aspects is essential.

As noted earlier, filament winding can be done for axisymmetric as well as nonaxisymmetric parts of various configurations. Winding can be performed using end domes or without. The end domes provide useful features for the fibers to take reversal at the poles. For small components, if the end domes are not parts of the final component, they can be parted by machining. However, for large components without a closed end, with a view to reducing wastage, end domes are not used during winding, and winding is carried out using pins for fiber reversal. However, we shall restrict our discussion to cylindrical part with end domes only.

10.5.2.1 Geodesic and Nongeodesic Windings

During filament winding, fibers are kept under tension. A key manufacturing criterion for the feasibility of winding is stability of the fiber path. If the fiber path is not determined with proper consideration for stability, the fiber roving will not stay put on the desired path and slip away to some other undesired shape. From the point of view of stability of the fiber path, two types of winding cases can be found: geodesic winding and nongeodesic winding. The physical significance of a geodesic path is that it is the shortest path for a given angular difference between the two points; thus, it is stable by nature. Figure 10.9 shows two geodesic paths between two points on a cylindrical surface. Note that a cylindrical surface can be developed into a rectangular shape. The shortest path between two points on a flat surface is a straight line and thus a geodesic path on the development of a cylindrical surface is also a straight line. Note further that there is only one geodesic path between two points for a given angular difference—Geodesic-I and Geodesic-II in Figure 10.9 correspond to an angular difference of θ and $2\pi + \theta$, respectively. For an axisymmetric component, geodesic path is given by *Clairut's rule* that leads to

$$r \sin \alpha = R_0 \tag{10.2}$$

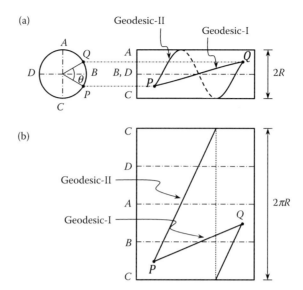

FIGURE 10.9 Geodesic paths on a cylindrical surface. (a) Details in the end and front views. (b) Details on the development of the cylindrical surface.

where r, α, and R_0 are radius at any point on the surface of the component, angle of winding at the point, and radius at the pole where the fibers take reversal. It is easy to see that for geodesic winding, the pole openings are equal.

On the other hand, any path that is not geodesic is nongeodesic and theoretically infinite numbers of nongeodesic paths can be constructed between two points. Nongeodesic winding is inherently unstable, and it needs the special consideration of friction characteristics of the mandrel surface and the wet composite material. Nongeodesic winding is unavoidable in some products such as rocket motor casings. In such a case, the feasibility of winding is ensured as long as

$$\mu_r \leq \mu_a \tag{10.3}$$

In Equation 10.3, μ_r is the friction factor required between the fibers being deposited and the already-wound substrate or mandrel surface so as to avoid slippage of fibers from the designed path. Note that it is an analytical parameter that is obtained by computation based on the geometry of the mandrel, fiber path, and winding tension. On the other hand, μ_a is the available friction factor at the corresponding location and it is the physical parameter that depends on the material characteristics such as surface roughness, resin viscosity, etc.

10.5.2.2 Helical, Hoop, and Polar Windings

Depending on geometrical considerations, filament winding paths can be of three different types:

- Helical winding
- Hoop or circumferential winding
- Polar winding

Helical winding is where the fiber path is basically a helix (Figure 10.10a). It is characterized by its angle of winding or helical angle or helix angle. The angle of winding at any point on the fiber path is defined as the angle between the tangent to the fiber path and the tangent to the meridional contour at that point (Figure 10.11). Theoretically, it can vary between 0° and 90°; however, too small or too large angles of

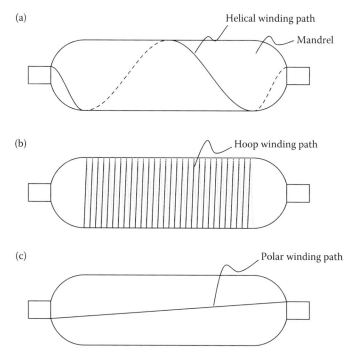

FIGURE 10.10 Schematic representation of filament winding paths. (a) Helical winding. (b) Hoop winding. (c) Polar winding.

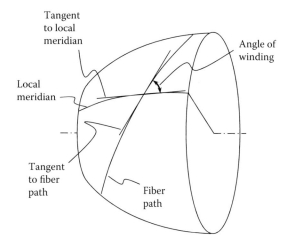

FIGURE 10.11 Angle of winding.

winding in the cylindrical portion are uncommon. The angle of winding is 90° at the poles where the fibers take reversal; for geodesic path, it gradually decreases from the poles toward the central cylindrical portion.

Hoop winding is a special case of helical winding (Figure 10.10b) in which the angle of winding is 90°. (As the fiber band has finite width, the angle of winding is actually less than 90°. It is easy to note that larger the fiber bandwidth the smaller is the actual angle of winding.) Polar winding, on the other hand, is winding in a plane passing through a point at one pole and the opposite point at the other pole (Figure 10.10c).

10.5.2.3 Programming Basics

As mentioned earlier, fibers are deposited along the predefined path by simultaneous controlled movements of the various axes of the carriage unit and mandrel rotation. During the winding of a ply, axis movements are controlled by a winding program. The basic principle of the winding program generation is based on geometrical

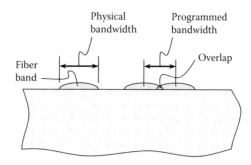

FIGURE 10.12 Programmed bandwidth and physical bandwidth.

considerations: (i) the line representing the free-standing fiber between the exit point on the pat-out-eye and the fiber-touch-down point on the mandrel is a three-dimensionally straight line and (ii) the line referred above is tangent to the pat-out-eye surface and the mandrel surface at any instant during the winding operation.

10.5.2.3.1 Programming for Hoop Winding

Hoop winding program generation is a rather straightforward affair that can be done with certain experimental input data. For a hoop ply of uniform thickness, a single-block winding program is sufficient in which mandrel rotation and carriage translation are required to be specified. The mandrel rotation in a hoop winding program is given by

$$C = \frac{360l}{b} \tag{10.4}$$

where C is the mandrel rotation, l is the translation of the carriage parallel to the mandrel axis, which is nothing but the length of the hoop ply to be wound, and b is the programmed bandwidth. The term "programmed bandwidth" should be differentiated from the term "physical bandwidth." As shown in Figure 10.12, when wound around a mandrel, the fibers under tension spread to some width, which is the physical bandwidth. The actual value of the physical bandwidth is likely to vary along the fiber path within some range depending on various local conditions related primarily to the fibers, matrix, substrate, winding tension, mandrel geometry, and pay-out-eye geometry. Thus, winding is done in such a way as to place the adjacent fiber bands with some overlaps. The distance between the center lines of adjacent fiber bands is the programmed bandwidth. Note that the programmed bandwidth is constant and is marginally narrower than the physical bandwidth so as to ensure that physically there is no gap between adjacent fiber bands.

The programmed bandwidth is chosen based on the number of rovings and required ply thickness. While it is possible to theoretically estimate the ply thickness, a more reliable information can be obtained by experimentation (refer Example 10.1).

EXAMPLE 10.1

Determine the programmed bandwidth for four rovings of 12 k carbon fibers so as to obtain an average ply thickness of 0.6 mm. Assume the carbon filament diameter and fiber volume fraction as 7 μm and 0.6, respectively.

Solution

The cross-sectional area of filaments in one roving is estimated as

$$CSA_{fib} = \frac{12,000 \times \pi \times 0.007^2}{4} = 0.4618 \, \text{mm}^2$$

The cross-sectional area of composite ply for four rovings is then given by

$$CSA_{com} = \frac{4 \times 0.4618}{0.6} = 3.0788 \text{ mm}^2$$

The programmed bandwidth required for an average ply thickness of 0.6 mm is obtained as

$$b = \frac{3.0788}{0.6} = 5.13 \text{ mm}$$

EXAMPLE 10.2

Solve the problem in Example 10.1, if the ply thickness is experimentally found to be 0.4 mm for two rovings at a programmed bandwidth of 5 mm. What is the estimated fiber volume fraction?

Solution

The required programmed bandwidth is arrived at as follows:

$$b = 5 \times \frac{4}{2} \times \frac{0.4}{0.6} = 6.67 \text{ mm}$$

The estimated fiber volume fraction is

$$V_f = 0.6 \times \frac{5.13}{6.67} = 0.46$$

10.5.2.3.2 Programming for Helical Winding

As noted earlier, the angle of winding may continuously vary along the helical fiber path. Also, the radial distance of the meridional contour may vary. To take these variations, typically, the fiber path is divided into a number of small segments and linear approximation can be made within each segment. Thus, a helical winding program contains a number of blocks, and each block contains information on incremental movement of each axis of motion. A convenient way to generate winding program is to consider an envelope around the mandrel on which the pay-out-eye is programmed to move. 3D lines, representing the free-standing fiber between the fiber exit point on the pay-out-eye and fiber-touch-down point on the mandrel surface, can be considered at the end of each segment of the fiber path such that these lines are tangent to the fiber path. The 3D coordinates, together with certain corrections for pay-out-eye dimensions, of the points of intersection of the 3D lines with the pay-out-eye envelope can be conveniently used for determining the incremental translations of the carriage unit and mandrel rotation. Winding quality is further improved by incorporating incremental pay-out-eye rotations.

There are filament winding terminologies/concepts associated with helical winding. First among these is the *number of starts*. Generally, helical winding is done in such a way as to progress coverage of the mandrel surface from multiple start points (Figure 10.13) in any cross section such that if the number of starts is N, after the completion of the first N circuits, $(N+1)$th circuit is placed just next to the first circuit, $(N+2)$th circuit is placed just next to the second circuit, and so on. The number of starts divides the circumference into N equal segments. If the number of circuits for each circumferential

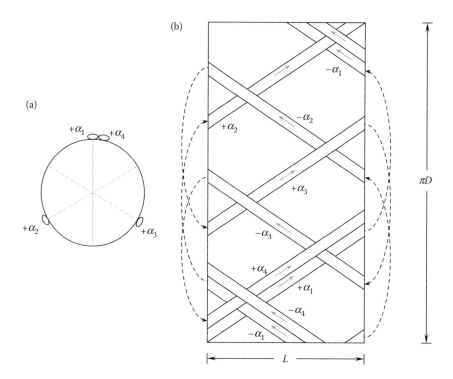

FIGURE 10.13 Schematic representation of multiple starts in helical winding (3-start helical winding on a cylinder of length L and diameter D). (a) End view. (b) Development of the cylinder. Note: The arrows indicate the directions in which the fiber bands are deposited during winding.

segment is n, the total number of circuits in one ply is Nn. Then, the programmed bandwidth in a helical ply is given by

$$b = \frac{2\pi r \cos \alpha}{Nn} \qquad (10.5)$$

where r is the local radial distance of the meridional contour and α is the local angle of winding. In a design environment, depending on the ply thickness requirement, the required programmed bandwidth is computed and the number of starts and number of circuits are chosen.

The number of starts cannot be arbitrarily chosen; its choice is greatly influenced by total mandrel rotation in one circuit. The total mandrel rotation in a circuit is typically adjusted by providing *dwells* (a dwell is an incremental mandrel rotation keeping other axes at rest) at the poles such that the revised total mandrel rotation $\sum C$ corresponds to a suitable number of starts.

10.5.3 Basic Raw Materials

The commonly used reinforcements in filament winding are continuous rovings of glass, carbon, and aramid. Polyester, epoxy, and vinylester are the common resins. Glass and polyester are normally used in low-cost applications. Aerospace applications demand high structural performance, in which case carbon/epoxy and Kevlar/epoxy filament-wound parts are common. The choice of reinforcements is largely dependent on structural requirements and cost. The resin system in a wet winding setup has to meet certain criteria that are unique to filament winding [26]. A low-viscosity resin system in the range of 2000 cps or lower is desirable in wet winding for proper impregnation of the filaments. Pot life should be long so that the frequency of changing resin in the resin bath can be reduced. Also, from a safety point of view, resin toxicity should be low.

10.5.4 Tooling and Capital Equipment

Tooling in filament winding includes primarily a mandrel with a central shaft for holding it on the winding machine. In addition to it, specific associated tools and fixtures are required for holding add-on parts, supporting, handling, extraction, etc. While the associated tools and fixtures are generally fabricated using structural steel based on specific design requirement, the mandrel design has some generic requirements, as follows:

- *Stiffness*: The mandrel must be sufficiently stiff so that its deflection under self-weight, associated fixtures, and added composite is within the acceptable limit. For small products, a typical steel tubular construction with or without end domes is generally sufficient. For large products, however, mandrel deflection is an important design consideration where typically a central shaft or a framed structure is provided.
- *Hard outer surface*: The outer winding surface of the mandrel must be hard so as to ensure that the rovings under winding tension do not bite into the mandrel material. It should also be smooth so that it aids the extraction of the mandrel.
- *Weight*: Mandrel weight can be a major design consideration for large products; the total weight of the mandrel with associated fixtures and added composite material has to be within the capacity of the winding machine. A lighter mandrel is easy to handle.
- *Handling*: Features must be provided on the mandrel for handling it along with the green composite with belts, tackles, lifting beams, etc. It is an issue that needs special attention in large products with nonuniform weight distribution lengthwise. Typical handling needs are in mandrel assembly, loading on the winding machine, unloading, movement to curing facility, extraction, etc.
- *Thermal expansion*: Thermal expansion of the mandrel material during curing has to be kept in mind while designing the mandrel.
- *Mandrel extraction*: An essential requirement of the mandrel is that it has to be extracted from the part. For cylindrical or conical products, the end domes can be parted by machining, and the mandrel can be extracted by pulling. However, for pressure vessels with smaller pole openings than the central cross section, extraction by pulling the mandrel is not possible. In such cases, the mandrel is designed with (i) collapsible steel segments, (ii) breakable rigid foam, (iii) breakable plaster material, (iv) soluble sand, and (v) a combination of collapsible, breakable, and soluble elements. It can also be designed as an inflatable mandrel and lost mandrel. In some cases, a mandrel is designed incorporating some parts of the final component; these parts become a part of the component. Mandrel extraction puts some contradictory requirements and design can be rather tricky from this point of view.
- *Dimensional accuracy*: Dimensional accuracy of the final component is directly dependent on the mandrel design. In this regard, thermal expansion and shrinkage of the tooling material(s) during curing need special attention. Further, proper design considerations must go into the configurations of the associated fixtures holding add-on parts during winding.
- *Cost*: Cost considerations can play a major role in some mandrel designs. It may be of interest to note that sometimes the repetitive costs of materials consumed in mandrel preparation can be comparable with the one time cost of the mandrel.

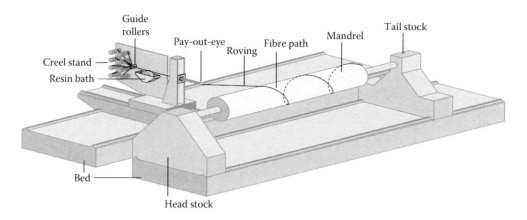

FIGURE 10.14 Schematic representation of filament winding setup.

The capital equipment required in filament winding includes essentially a winding machine, an oven, and a machining facility. While horizontal filament winding machines are the most common, vertical winding machines are also available. Vertical winding machines provide some advantages as the center of gravity (c.g.) of the mandrel is in line with the machine axis and mandrel deflection is avoided. However, wet winding on a vertical filament winding machine is not advisable as resin flow during winding may cause serious quality problems. There are different types of horizontal winding machines. Figure 10.14 gives a schematic representation of a typical filament winding setup. Filament winding machine is similar to a lathe with two centers to hold the mandrel. The carriage unit has an attached creel stand on which the spools are positioned. Sometimes the creel stand is kept separately as a standalone unit. The carriage unit also supports a system of guide rollers, a tensioning device, a resin bath with controlled heating mechanism, a comb, and a pay-out-eye. The pay-out-eye, in the form of an eye let, concave-shaped roller, etc., is an essential feature that finally deposits the fiber on the mandrel. The fiber path on the mandrel is resulted by the relative motions of the mandrel and the carriage unit. The minimum requirement of motions for helical winding are mandrel rotation and translation of the pay-out-eye along the axis of the machine. However, for better winding performance, more motions as indicated below are necessary. While all of these are not required for helical winding of axisymmetric parts, winding of nonaxisymmetric parts need more flexibility. These motions are

- Rotation of the mandrel about the axis of the machine
- Translation of the pay-out-eye (or translation of the carriage unit) parallel to the machine axis
- Translation of the pay-out-eye (or translation of the carriage unit) normal to the machine axis in the horizontal plane
- Translation of the pay-out-eye normal to the machine axis in the vertical plane
- Rotation of the pay-out-eye about its own horizontal axis normal to the machine axis
- Rotation of the pay-out-eye about its own vertical axis

Modern filament-winding machines are integrated with sophisticated CNC system that controls the movements of the axes mentioned above. These axes can be controlled simultaneously by winding programs to deposit the fiber rovings along the predefined paths on the mandrel in proper orientation so as to achieve desired fiber spread and designed winding pattern and ply thickness.

10.5.5 Advantages and Disadvantages

Filament winding has been in use for several decades. As a composite processing technology, it is rather synonymous with certain classes of applications, such as pressure vessels, pipes, etc. There are a number of advantages that make filament winding the first choice for these classes of products; these are

- Components of any axisymmetric shell of revolution can be wound with maximum cost and structural efficiency. Thus, pipes, pressure vessels, tanks, shafts, etc. of any size can be made.
- In the recent past, remarkable developments have been made in the front of computer programming, and nonaxisymmetric parts, such as elbow, tee, wind mill blade, etc. are also made easily by filament winding.
- The mandrel rotation and carriage unit motions are computer controlled, and accordingly, the fiber orientation angles are precisely controlled.
- The process is highly automated.
- Fiber tension during winding results in sufficient consolidation pressure, and no additional consolidation pressure is required during curing.

Filament winding also suffers from a few drawbacks; these are

- It is suitable for closed components with convex shape. Winding cannot be done on a concave surface. Whenever an open structure is made by filament winding, machining is required for parting, for example, leaf spring.
- While geometrically any helical angle is feasible, from a practical point of view, too small an angle is not suitable. Also, helical angles depend on relative dimensions of cylinder radius and pole opening radius, etc., and thus, there are restrictions on the angle of orientation of the fibers.
- Fiber volume fraction is generally limited to about 60%. Owing to the presence of cross-overs, it can actually be much lower in some cases.

10.6 CURING

We have seen in Section 10.3 that any composites manufacturing process essentially involves four basic steps—impregnation, lay-up, consolidation, and solidification. These are implemented in different ways in different processes. Of these four basic steps, solidification or curing is largely dependent on the resin system and it plays a crucial role in the final quality of the part. In the following subsections, important aspects of curing are addressed in brief (see, for instance, References 3 and 9 among others for more details).

10.6.1 Tools and Equipment

Curing is done in different ways:

- At room temperature with or without vacuum
- At elevated temperature with or without vacuum
- At elevated temperature with or without vacuum under positive pressure
- Inline curing by UV radiation or E-beam

For curing at elevated temperature without pressure, an oven is used. Air (or an inert gas like nitrogen) inside the oven is heated and the heat is contained by providing insulating lining on the oven walls. Air/gas temperature as well as component temperatures are monitored during the curing process for which provisions are incorporated for thermocouples.

Elevated temperature curing under positive pressure is done in an autoclave, a hydroclave, or a hydraulic press.

An autoclave is a pressure vessel with provisions for heat input to the circulating air/gas inside the chamber. A vacuum system for applying negative pressure, temperature and pressure monitoring system, and a control system for controlling the operating parameters are the other essential features in an autoclave. Air, nitrogen, or carbon dioxide is used as the pressurization medium. Air is cheap but it has fire hazard. Among the inert gases, nitrogen is more common. During curing, the air/gas is injected into the chamber for pressurizing it and heated by electrical means or gas firing. The air/gas is circulated inside the chamber for uniform distribution of temperature. A vacuum bag is invariably used so as to make the isostatically applied pressure effective. Autoclave curing is widely used for realizing high-quality aerospace products. But the process involves a number of in-process consumables such as vacuum bag, bleeder, breather, sealant putty, etc., which makes it relatively expensive. Hydroclave is similar to an autoclave where the pressurization medium is water; it is operated at relatively high pressure and commonly used for cure of phenolic nozzle liners.

10.6.2 Vacuum Bagging

Vacuum is applied during curing for the removal of entrapped air, gases, and volatile product. It is an essential element in autoclave curing where the applied pressure is isostatic in nature. It is also done in the curing of a part in an oven. The green composite part is vacuum bagged before loading it inside the autoclave for curing. Figure 10.15 schematically shows the process of vacuum bagging; typical steps involved in it are described below:

- A perforated release film is applied on the green composite part. It helps the composite part to avoid getting stuck to the bagging material. The perforations allow the entrapped air, excess resins, and volatile products to escape during curing. Optionally, a peel ply is applied on the composite part before putting the release film. The peel ply creates a good surface that can be adhesively bonded to other mating surface at a later stage.

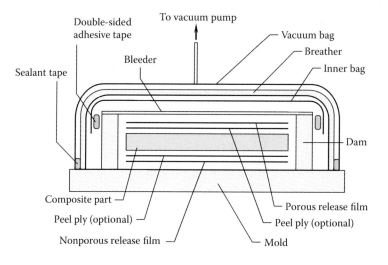

FIGURE 10.15 Schematic representation of vacuum bagging of prepreg lay-up. (Adapted with permission from S. K. Mazumdar, *Composites Manufacturing—Materials, Product and Process Engineering*, CRC Press, Boca Raton, FL, 2002.)

- A porous fabric, called bleeder is applied, on the top of the release film. It absorbs moisture and excess resin coming from the prepregs.
- A nonporous film, called barrier film, is then applied on the bleeder.
- The breather, a porous fabric similar to the bleeder, is applied on the barrier. It ensures uniform application of pressure. It also allows moisture, air, and volatiles to escape.
- Finally, the vacuum bag is put. Sealant tapes are used for creating airtight joint between the bagging material and the mold. If porous material is used for making the mold, the complete mold may be enclosed inside the bag. Vacuum is created inside the bag by connecting the bag to a vacuum pump with a hose and nozzle.
- The vacuum bagged composite part is then pushed inside the autoclave and cured.

10.6.3 Curing of Epoxy Composites

Autoclave curing involves the application of heat, pressure, and vacuum on the composite part. The cure cycle primarily depends on the type of the resin system. However, the size and shape of the composite part and the mold do affect the final cure cycle. Also, the size of the autoclave can influence the cure cycle. In general, during the ramp, that is, heating phase, the part temperature lags behind the air or gas temperature. It is easy to visualize that for a thicker component, the temperature deep inside the component will be at a lower level than the skin temperature, which will be more or less equal to the chamber temperature. A typical autoclave cure cycle involving two ramps, two constant temperature holds, and a cooling phase is shown in Figure 10.16. As seen in the figure, broadly three zones can be identified in the cure cycle: (i) first ramp followed by first hold, (ii) second ramp followed by second hold, and (iii) cooling phase. In a typical epoxy composite curing operation, heat is provided as input and air and component temperatures together with vacuum and pressure levels are monitored. The points of application of vacuum and pressure during the cure cycle are crucial from the point of realization of quality product without delaminations, voids, and other defects.

During the first ramp and hold, for an addition curing thermoset resin such as epoxy, the semisolid resin melts on heating, loses its viscosity drastically, and flows. Vacuum is applied right in the beginning that helps excess resin, entrapped air, and volatiles to escape. In this phase, as reflected by a sharp decline in the viscosity, the resin remains in a fluid state and the application of pressure may cause voids to get entrapped without

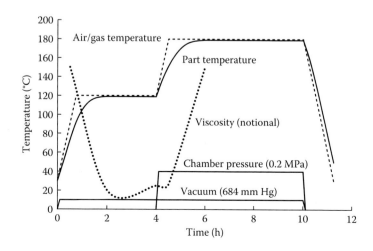

FIGURE 10.16 Typical autoclave cure cycle.

any path for escape. From this point of view, pressure is not required during the first ramp and hold. However, without any pressure, the hydrostatic resin pressure can be very low, resulting in the formation of voids [3]. Also, without any pressure on the composites, heat transfer is not very effective. Thus, from this viewpoint, pressure application is needed even during the first ramp and hold. Thus, during this phase, either pressure is not applied or applied to a less than full level.

The second ramp and the hold are the actual curing phase during which initially the resin viscosity drops marginally and then drastically increases, the resin cross-links and gels to its solid state. Full chamber pressure is applied for consolidation; as a result, voids get eliminated and intimate contact takes place between the composite plies.

The other important aspects in the cure cycle are in respect of rate of heating and cooling and durations of the holds. Residual thermal stresses are detrimental to the health of the component and it is believed that slow heating and cooling rates generally result in reduced residual rates. High ramp rates can also result in thermal gradient across the thickness especially in thick laminates leading to nonuniform cure and improper consolidation.

10.6.4 Curing of Phenolic Composites

Curing of addition curing resin system such as epoxy is rather simple compared to that of condensation curing resin system. Phenolic resin is a typical example of condensation curing resin system. Unlike addition curing resins, condensation curing resins such as phenolics and polyimides emit by-products during the cross-linking process. These by-products that include water, volatiles, and solvents have to be removed. Volatile management during curing of a phenolic composite part is an extremely important task; without proper removal of the volatiles, the resultant part can be highly porous with voids and delaminations. In a press molding operation, high platen pressures (higher than volatile pressures) (i) reduce the volatile evolution rate and (ii) ensure removal of the volatiles completely. In an autoclave curing, generally, multiple ramps and holds are introduced in the cure cycle while maintaining vacuum for the removal of volatiles. Full pressure is applied after all the by-products have been removed. Another method adopted in autoclave curing is to carry out intermediate hot-debulking. In this method, the part is laid-up partially, debulked under vacuum at high temperature, cooled down, and partial lay-up is resumed. The process is continued till all the plies are laid-up and final curing is done. This process is however very time consuming.

10.7 MANUFACTURING PROCESS SELECTION

We had seen in Chapter 9 that there are a number of reinforcement and matrix materials that can be used for making composite products. The reinforcements are available in various physical forms and there are various manufacturing processes that can be adopted for converting the reinforcements and matrix into a useful product. An important issue here is the selection of the manufacturing process. Several aspects associated with the product development have to be considered for making a judicious choice. The major aspects associated with manufacturing process selection are identified below:

- Related to product description
  - Configuration of the product
  - Size of the product
- Related to product requirements
  - Structural property requirement
  - Surface finish

TABLE 10.1
Selection of Manufacturing Methods—Open Mold Processes

	Wet Lay-Up	Prepreg Lay-Up	Spray-Up	Rosette Lay-Up
Configuration of product	Flat and low to medium curvature panels	Flat and low to medium curvature panels	Flat and low to medium curvature panels	Axisymmetric parts, low to medium curvature panels
Size of product	Any size	Any size	Any size	Small to medium
Structural property	Medium to high	High	Low	Low
Surface finish	Good on one side	Good on one side	Good on one side	Poor
Reliability and repeatability	Low to medium	Medium to high	Low to medium	Medium
Suitability for high volume production	Low	Low	Medium	Low
Tooling requirements	Low to medium	Low to medium	Low to medium	Low to medium
Need for skilled manpower	High	High	High	High
Feasibility of automation	Low	Low to medium	Medium	Low
Cycle time	Short to medium	Medium	Short	Long
Cost	Low	High	Low to medium	Medium to high

- Reliability and repeatability
- Production requirement
■ Related to process requirements
- Tooling requirements
- Need and availability of skilled manpower
■ Related to process characteristics
- Automation
- Cycle time
- Cost

An indicative assessment of the suitability of various composites manufacturing processes w.r.t. each of the above aspects is made in Tables 10.1 through 10.3 (also see Reference 3). Note that some of the processes can be employed in virtually any possible case of a particular aspect. For example, wet lay-up can be used for laying up parts of virtually any configuration; however, it is certainly not suitable for all part configurations. Here, our emphasis is to check the general suitability of the manufacturing processes in various cases.

TABLE 10.2
Selection of Manufacturing Methods—Closed Mold Processes

	Compression Molding	Resin Transfer Molding
Configuration of product	Near net shape solid (nonhollow) parts, flat and low curvature panels	Any configuration
Size of product	Small to medium	Any size
Structural property	Low to high	High
Surface finish	Good on all sides	Good on all sides
Reliability and repeatability	High	Medium to high
Suitability for high volume production	Medium to high	Medium to high
Tooling requirements	Medium to high	Medium to high
Need for skilled manpower	High	High
Feasibility of automation	Low to medium	Low
Cycle time	Short	Short
Cost	Low to medium	Low to medium

TABLE 10.3

Selection of Manufacturing Methods—Continuous Molding Processes

	Pultrusion	Filament Winding	Tape Winding	Fiber Placement
Configuration of product	Long sections of constant cross section	Axisymmetric (also, nonaxisymmetric) outwardly convex thin shells	Axisymmetric (also, nonaxisymmetric) outwardly convex thick shells	Any configuration (shell/panel)
Size of product	Any size	Any size	Any size	Any size
Structural property	High	High	Low	High
Surface finish	Good	Good on inner surface	Poor	Good on inner surface
Reliability and repeatability	Medium to high	Medium to high	Medium to high	High
Suitability for high volume production	High	Medium to high	Medium	Medium to high
Tooling requirements	Medium	Medium to high	Medium	Medium to high
Need for skilled manpower	High	High	High	High
Feasibility of automation	High	High	High	High
Cycle time	Short	Medium to long	Medium to long	Medium to long
Cost	Low to medium	Medium to high	Medium to high	Medium to high

10.7.1 Configuration of the Product

An as-molded composite part can be of various configurations: (i) solid or hollow, (ii) thin- or thick-walled, (iii) flat or curved, (iv) axisymmetric or nonaxisymmetric, (v) channels with open or closed cross sections, etc. We had discussions on 10 key composite processing methods under three groups in Sections 10.4 and 10.5. As is seen there and noted in Tables 10.1 through 10.3, some of these methods are suitable for practically any part configuration, whereas some others have specific capability.

Among the open mold processes, the first three processes, viz. wet lay-up, prepreg lay-up, and spray-up, can be used for making flat panels and panels with low to medium curvatures. Prepreg lay-up and spray-up can also be employed in complex configurations involving large curvatures (small radii of curvature). Rosette lay-up is suitable for moderately thick axisymmetric parts and panels with low to medium curvatures.

Compression molding is suitable for making near net shape solid parts and panels with low curvatures. Complex configurations are also possible with appropriate molds. With RTM, too, near net shape parts with complex configurations can be made.

Continuous molding processes, in general, are most suitable for regular configurations. Constant cross section long sections are most conveniently made by pultrusion. Filament winding as well as tape winding are generally suitable for axisymmetric parts. Nonaxisymmetric parts can also be made by using complex winding programs. However, for filament winding to be feasible, the part profile along a fiber path must be outwardly convex. On the other hand, as compared to filament winding, fiber placement provides more flexibility in the part configurations.

10.7.2 Size of the Product

Monolithic composite parts can be made in a wide range of sizes. Wet lay-up, prepreg lay-up, and spray-up processes can be used for making parts of virtually any size except very small sizes. Rosette lay-up is suitable for axisymmetric parts and panels of small to medium sizes.

Compression molding and RTM processes are suitable for making parts of small to medium sizes. The size of a compression molded part is directly influenced by the size of the mold and the capacity of the press. In the case of RTM, on the other hand, tooling plays a critical role and with properly designed tooling, large components can also be manufactured.

Pultrusion is most suitable for making long sections with small uniform cross-sectional areas. Filament winding, tape winding, and fiber placement processes are suitable for making components of small to large sizes. In these cases, mandrel design and capacity of winding machine directly control the size of the component. These are not suitable for very small parts. On the other hand, with proper mandrel design and adequate machine capability, very large components can be made by filament winding and fiber placement processes.

10.7.3 Structural Property Requirement

The structural properties of a composite part are greatly influenced by the reinforcements, their physical forms, and stacking sequence. In addition to these factors, the manufacturing method adopted to make a part also has a very significant role in imparting structural properties. Filament winding, pultrusion, RTM, etc. can be employed to make highly directional composite parts. On the other hand, compression-molded randomly oriented components and open-molded CSM products are poor in structural properties.

Wet lay-up and prepreg lay-up processes are suitable, in general, to meet requirements of medium to high and high structural properties; autoclave cured prepreg laid-up composite parts are excellent in applications requiring high structural properties. Spray-up and rosette lay-up, however, do not impart high structural characteristics. In fact, rosette lay-up is primarily used for making parts for ablative applications.

Compression molding and RTM can be used for making high strength or high stiffness parts. Compression molding is also used for making components for use in nonstructural applications.

Continuous molding processes, other than tape winding, are suitable for making parts with high structural properties. However, structural characteristics vary greatly. Pultrusion imparts high strength and stiffness in the axial direction. Filament winding is suitable for making shell structures with high membrane properties. The membrane strength and stiffness of a filament-wound shell depend greatly on the fiber orientation. In filament winding, the very nature of the winding process imposes certain constraints on the fiber orientation capabilities; thus, structural characteristics are influenced. Fiber placement, on the other hand, is more flexible.

10.7.4 Surface Finish

Surface finish is a desired characteristic in many composite product applications. Machining can improve surface finish, but it is not desirable from the point of view of requirement of structural properties.

Wet lay-up, prepreg lay-up, and spray-up processes impart excellent surface finish on the side of the part in contact with the mold during processing, but the other side is rather poor in surface finish. Surface machining on the outer side can improve its finish, but it greatly reduces the strength and stiffness characteristics and sometimes causes surface delaminations. Rosette lay-up has its specific ablative application and the as-molded part is usually machined on the outer as well as the inner surfaces to obtain the desired configuration and surface finish.

Owing to the closed mold nature, compression molding and RTM are very efficient in imparting good surface finish on all the sides of the composite as-molded parts.

Pultruded sections are generally good in surface finish. Finish is good only on the inner surface of a filament-wound part; it is rather poor on the outer side. Hoop-wound outer surface is relatively better in finish than helical-wound outer surface. On the other hand, tape-wound parts usually need machining so as to achieve the desired surface finish.

10.7.5 Reliability and Repeatability

The reliability of a composite part and the repeatability of its quality depend significantly on the manufacturing method adopted to make it.

In general, open mold processes do not fare well in this aspect. Human skill plays a significant role in these processes; it is especially true in wet lay-up, wherein dry fabric layers are cut and wetted with resin manually. Lay-up is also done manually and curing is done at room temperature, in an oven, or in an autoclave. Defects like voids, delaminations, disoriented fibers, nonuniform fiber volume fraction, etc. result in poor reliability and repeatability. In prepreg lay-up, preimpregnated plies with controlled resin content are used and some of the operations can be mechanized; curing is done mostly in an autoclave and reliability and repeatability are better. Spray-up is also highly dependent on operator skill and nonuniform fiber volume fraction; varying part thickness and random nature of chopped fibers make reliability and repeatability poor. Prepreg ply developments are laid-up either in an open mold or in a closed mold. As compared to wet lay-up processes, part quality is more reliable and repeatable.

Compression molding and RTM yield reliable parts with repeatable quality. Closed mold processes are especially suitable for maintaining dimensional repeatability of as-molded parts.

Continuous molding processes can be placed as above average in rank in respect of reliability and repeatability. These processes can be automated to a large extent and helps improve reliability and repeatability. However, various process parameters affect the quality of the final product in each of the processes in this class. For example, in filament winding, a close eye must be kept on the process variables such as resin viscosity, resin temperature, winding tension, doctor blade setting, number of spools, etc. In tape winding, tape tension, roller pressure, Chang's index, etc. are important parameters. In other words, in-process quality control is critical for ensuring reliability of a part and repeatability of its quality.

10.7.6 Production Requirement

Manufacturing processes are not uniformly suitable for different production requirements. In general, open mold processes are suitable for making only limited numbers of products. Common characteristics, viz. labor-intensive nature, low scope for automation, relatively long cycle time, and low repeatability, associated with these processes make them unsuitable for large volume production. Prepreg lay-up can be partly automated and they fare relatively better than wet lay-up in meeting production volume requirements. Spray-up can also be automated to a large extent and it can be made suitable for high production volume. On the other hand, rosette lay-up is suitable for low to medium production volumes; also, parts made by rosette lay-up are used in specific applications with typically limited requirements.

Closed mold processes, viz. compression molding and RTM, are typically suitable for meeting high volume production requirements of small parts. For example, compression molding is regularly employed for the production of automotive parts using molding compounds. These methods are also employed for making medium to large size parts involving complex ply sequence or specially designed preforms. In some cases, extensive machining of cured parts is also required. In such cases, production volumes are usually limited.

Pultrusion is a quick process and is suitable for high volume production. Filament-wound parts of simple configurations such as pipes and tubes can be made in large volumes. However, in some other cases of specialized applications, owing to complex configurations and complex tooling, production volumes can be low to medium. Generally, tape winding is also employed for making parts for specialized applications, and production volumes are low or, at best, medium.

10.7.7 Tooling Requirements

Tooling requirement is a common feature for any composite processing technique. The nature of the tooling depends on the processing technique and the configuration of the part.

Tooling requirements for open mold processes, except rosette lay-up, are generally low or, at best, medium. Typically, simple male and female open molds, made out of steel, wood, and FRP, are used. Owing to the nature of open molding process and room-temperature curing, a simple mold with machined surface matching with the profile of the part is sufficient in wet lay-up and no intricate tool design is involved. In prepreg lay-up, vacuum bagging, and high-temperature curing processes increase tooling requirements a little more as compared to wet lay-up. Spray-up is similar to wet lay-up in regard of tooling needs. In the case of rosette lay-up, owing to its very nature of laying up plies across the laminate thickness, two halves with matching profiles are needed for making axisymmetric parts and tooling requirements are little high. For open panels, however, simple tools are sufficient even in rosette lay-up.

As compared to the open mold processes, closed mold processes, viz. compression molding and RTM, demand more attention to tool design and fabrication. The molds in these processes need to meet a number of processing parameters. Additionally, in the case of compression molding, the molds have to meet high strength and stiffness requirements. Also, these processes are employed for making parts with controlled dimensions and in some cases rather highly complex molds are required.

Tooling requirements in continuous molding processes can be considered as medium or medium to high. The complexity of the tooling also depends on the part to be made. In general, a die is used in pultrusion and a mandrel in filament winding, tape winding, and fiber placement. The mandrel has to meet a number of processing requirements. The extraction of the mandrel after curing is a critical processing requirement, and in some cases, for example, in a pressure vessel with small pole openings, the mandrel design may become rather complex.

10.7.8 Automation and Skilled Manpower Needs

The choice of manufacturing process can be greatly influenced by need and availability of skilled manpower. Obviously, a choice of a processing technique requiring a large pool of skilled manpower is not a good one if skilled manpower is in short supply.

Wet lay-up and prepreg lay-up processes are highly labor-intensive and the scope for automation is relatively low. In spray-up process, it is possible to implement automation and thereby the need for skilled manpower can be reduced. In respect of skilled manpower requirement, rosette lay-up process is similar to prepreg lay-up and it is skilled manpower-intensive.

Compression molding and RTM are relatively less-skilled manpower-intensive. Automation is possible to some extent, which reduces the requirement of skilled manpower. In some special cases, however, complex parts are made in limited numbers, where skilled manpower is needed in large numbers.

Continuous molding processes can be greatly automated and demand for skilled manpower can be reduced.

10.7.9 Cycle Time

The cycle time of manufacturing a composite part is dependent on the processing technique to a great extent. Wet lay-up involves generally room-temperature curing and cycle time of part fabrication is short. Vacuum bagging and autoclave curing operations lead to relatively longer cycle time in prepreg lay-up. In this regard, spray-up is similar to wet lay-up and cycle time is typically short. On the other hand, rosette lay-up, due to its unique nature of ply orientation, is a relatively long operation.

Compression molded and resin transfer molded parts are generally of short cycle time. In these cases, automation is possible and short curing resins can be used leading to short cycle time. However, in special cases of compression molding as well as RTM, parts involving very detailed ply sequencing or complex preforms are made, where cycle times are relatively longer.

Among the continuous molding processes, pultruded parts have generally short cycle time. Filament winding, tape winding, and fiber placement processes take relatively longer time. In some cases, parts requiring complex tooling and detailed ply sequencing and multistage winding are made; cycle times of manufacture can be rather long in such cases.

10.7.10 Cost

Cost is a critical factor in choosing the manufacturing process in most applications except possibly in some limited cases of strategic importance. Even in such cases, given other parameters equal in ranking, a cheaper processing technique is always preferable. The major cost elements in the development of a composite product are costs of raw materials, capital equipments, tooling, in-process consumables, processing (including manpower and machine running costs), and design and analysis.

In general, wet lay-up is a cheap process that involves simple tooling, inexpensive raw materials, no capital equipment, and minimum in-process consumables. Prepreg lay-up, on the other hand, is an expensive process primarily on account of expensive initial investment in capital equipment, high processing cost, and expensive in-process consumables. Spray-up involves inexpensive raw materials, simple molds, and processing tools; as a result, it is generally an inexpensive process. Rosette lay-up, due to its unique nature and possible complex molds, expensive raw material, and need for autoclave curing, can be relatively more expensive than the other open mold processes.

Initial investment in compression molding can be high and it depends on the capacity of the press. Tooling cost varies from low to high depending on part configuration. Similarly, material cost also varies depending on its type. In general, for high volume production, the cost of a compression molded part tends to be low to medium and for parts with limited volume requirements and special specifications, it tends to be medium to high. In the case of RTM, initial investment involves primarily the cost of the resin dispensing system. Costs of preforms and tooling are two other major cost elements. By and large, the cost of a resin transfer molded part tends to be low to medium in a setup for high volume production. On the other hand, cost tends to go up for parts with limited volume requirements and complex specifications.

Initial investments are generally high for continuous molding processes. By and large, these processes need highly sophisticated machines and equipments for making quality components. Tooling costs are dependent on size and complexity of the tooling and in some cases tool preparation becomes a major recurring cost element.

10.8 OTHER TOPICS IN COMPOSITES MANUFACTURING

10.8.1 Process Modeling

There are many process parameters that influence the manufacturing process and the final quality of a composite product. The possible process parameters associated with common manufacturing methods are identified and listed in Table 10.4. These parameters can be broadly termed as input parameters; by and large, we can either control them or make a choice from among possible alternatives during manufacturing. For example, we can choose an acceptable range of resin viscosity during wet lay-up.

TABLE 10.4

Typical Process Parameters Associated with Common Manufacturing Methods

Manufacturing Method	Manufacturing Parameters
Wet lay-up	• Resin condition—viscosity of resin, quantity of resin per unit area of fabric • Fabric condition—thickness of fabric, style of fabric • Lay-up parameters—type of brush and roller
Prepreg lay-up	• Prepreg condition—Chang's index, tackiness, resin content • Lay-up parameters—type of roller, roller pressure • Curing parameters—temperature, pressure, vacuum, autoclave characteristics
Spray-up	• Resin condition—viscosity of resin • Fiber condition—lengths of cut pieces, tex • Spraying parameters—spray pressure, nozzle type, etc.
Rosette lay-up	• Prepreg condition—Chang's index, tackiness, resin content • Lay-up parameters—type of roller, roller pressure • Curing parameters—temperature, pressure, vacuum, autoclave characteristics
Compression molding	• Charge condition—type of charge (molding compound or prepreg flakes), prepreg condition • Molding parameters—quantity of charge per lot, load, curing temperature
Resin transfer molding	• Resin condition—resin viscosity, resin temperature • Preform condition—type of reinforcement, fiber orientation • Resin injection parameters—resin injection pressure, gate location
Pultrusion	• Resin condition—resin viscosity, resin temperature, cure characteristics of resin • Roving/tow condition—tex of rovings • Pultrusion parameters—number of spools, guide roller setting, pulling tension, die temperature
Filament winding (wet winding)	• Resin condition—resin viscosity, resin temperature • Roving/tow condition—tex of rovings • Winding parameters—doctor blade setting, number of spools, bandwidth, number of starts
Tape winding	• Prepreg condition—Chang's index, tackiness, resin content • Winding parameters—tape tension, roller pressure, pitch • Curing parameters—temperature, pressure, vacuum, autoclave characteristics
Fiber placement	• Towpreg/tape condition—tackiness • Fiber placement parameters—fiber tension, roller pressure • Curing parameters—temperature, oven characteristics

Similarly, during prepreg lay-up, advancement of resin in the prepreg, as represented by Chang's index, can be controlled to be within acceptable limits, and so on. There is also another set of parameters associated with the manufactured part or the laminate. They include laminate density, fiber volume fraction, degree of cure, surface finish, voids, etc. These laminate parameters are basically the output parameters and represent the quality of the composite part. Some of the output parameters such as voids represent defects and it is desirable to reduce them; whereas some others such as fiber volume fraction, degree of cure, etc. should be high, in general.

The process parameters have significant effect on the resultant values of the laminate parameters and it is reasonable to conclude that an optimum choice of these parameters is essential to manufacture a quality product. For making a choice of the process parameters, two broad approaches are in vogue. The first approach is basically a trial-and-error approach that is guided by experience and intuition of the process engineer and shop floor technicians. In general, it is time-consuming, expensive, and inefficient. On the other hand, the second approach is precise and more efficient; it is based on process modeling [32,33].

A process model is a mathematical representation of a manufacturing method in which the resultant laminate or part parameters are expressed in terms of the processing variables associated with it. It is developed based on physical laws, appropriate initial and boundary conditions, experimental observations, and suitable assumptions. Four broad interrelated areas can be identified in process modeling in composites manufacturing. These are

- Thermochemical aspects
- Resin flow characteristics
- Residual stresses and strains
- Voids

First, the thermochemical aspects include basically temperature distribution in the part during processing that has effects on the viscosity of the resin and the degree of cure. Temperature distribution depends on material properties and the curing environment. The curing conditions, including room temperature, heat input, oven/autoclave temperature, pressure, and vacuum, form the initial conditions and boundary conditions in the mathematical representation of the thermochemical aspects in process modeling.

Second, uniform wetting of the fibers with the resin is essential in any manufacturing process and resin flow characteristics are of critical importance in process modeling. Nonuniform wetting may lead to resin-starved areas as well as low fiber volume fraction. Resin viscosity, temperature profile, pressure and vacuum during curing, vacuum port location, resin injection gate location, etc. are the controlling variables that are considered in simulating the resin flow characteristics. The variables that are applicable in a specific model depend on the type of the manufacturing method. For example, resin flow in RTM would depend on resin injection gate location; whereas in filament winding, doctor blade setting controls resin flow to a great extent. On the other hand, resin viscosity is an applicable parameter in both the cases.

Third, residual stresses and strains are developed in a composite part during curing. Dissimilar thermal coefficients of fibers and resin result in thermal strains in the plies at the end of cooling. In this respect, ply sequence and material properties play an important role; unsymmetric ply sequence with dissimilar materials may cause high residual stresses and strains, which in turn result in distortion as well as interlaminar cracks. An improper cure cycle involving sudden increase and decrease in temperature can lead to uneven cure and result in increased residual stresses and strains. Residual stresses and strains and associated distortions are undesirable and they should be reduced to the minimum. In this regard, process models simulating cure characteristics can be used efficiently.

The fourth major area in process modeling is in respect of voids that are created during part manufacturing. Voids are undesirable as they decrease part quality. They are generated at random during lay-up as entrapped air or gas bubbles and they change in size and shape during curing due to (i) changes in pressure and temperature, (ii) volatile generation, and (iii) differential thermal deformation. Models have been developed taking into account some of these aspects and the same can be used to design cure cycles so as to minimize the presence of voids in the resultant part.

10.8.2 Machining of Composites

Composites manufacturing is different from conventional metallic manufacturing processes in several aspects. One such aspect is the fact that composites manufacturing involves the addition of material during processing, whereas manufacturing processes in metals generally involve removal of material. Also, large parts can be made with composites resulting in reduced part count. Thus, the requirement of machining in composites is much less and sometimes it is undesirable to machine a composite part. However, it cannot be avoided totally and a certain amount of machining is invariably needed in the manufacture of almost any composite product. Machining of composites needs special care and an introductory discussion is given in the following subsections. For more information, the interested reader can consult, for instance, References 1, 34–37 and the bibliography given therein.

10.8.2.1 Requirements of Composites Machining

Machining is required in various circumstances in composites manufacturing; the common requirements are

- Edge machining
- Surface machining
- Machining for parts making
- Providing holes and other features

10.8.2.1.1 Edge Machining

Edges of as-molded flat panels and shells are generally not clean and they require machining to remove the boundary strips having disoriented fibers/fabric, resin globules, missing plies, etc. Machining along the edges is also required to achieve proper dimensions. For example, while designing a mold for making a flat laminate by either wet lay-up or prepreg lay-up, typically allowances are kept along the four edges. The strips along the edges are removed by employing standard cutting operation using one of the several either hand-held or mechanized tools such as hand saw, circular saw, band saw, etc. Similarly, in a filament-wound cylindrical component, the circular strips at the two ends are removed by either parting or facing. In some cases, for example, compression molded parts, edge machining may involve simply trimming by grinding or filing.

10.8.2.1.2 Surface Machining

Surface machining is required (i) to maintain laminate thickness within specified tolerances, (ii) to achieve desired surface finish, and (iii) to prepare matching profiles for adhesive bonding. In the open mold processes and continuous molding processes, it is nearly impossible to maintain laminate thickness within tight tolerances and good surface finish is achieved only on one side. When ply edges are exposed on both sides of a laminate—as in the case of rosette lay-up and tape winding—surface machining is required on both the sides, if good finish is desired. Also, bonded joints are incorporated in many applications, where matching surfaces need machining.

Surface machining of axisymmetric components is done generally by turning. Straight cylindrical turning, taper turning, and profile turning are all employed depending on the part meridional geometry. Grinding and sanding are frequently done for better finish on flat as well as contoured surfaces. On the other hand, milling may be needed to achieve controlled thickness of flat panels.

10.8.2.1.3 Machining for Part Making

In many cases of composites manufacturing, a large panel or a long shell is made first, and subsequently, smaller parts are obtained from the parent panel or shell by machining. A typical example is the case of test specimens that are machined out from a parent laminate by either sawing or cutting. Similarly, for the realization of small- to medium-sized shells, often, a long shell is made by filament winding and parts of desired lengths are obtained by carrying out parting operation.

10.8.2.1.4 Providing Holes and Other Features

Holes are required in composite panels and shells for incorporating bolted and riveted joints, and to provide features for integrating with adapters, nozzles, etc. Drilling is routinely done to provide holes in composite laminates. Often, reaming is done for better hole quality. Holes can also be provided with countersunking. In many composites manufacturing processes, it is difficult to provide special features such as slots, steps, etc. in the as-molded composite parts and machining operations like milling and turning become essential.

10.8.2.2 Critical Aspects of Composites Machining

The underlying principles of machining in composites are basically the same as in metals. Thus, metal-machining operations like cutting, sawing, turning, milling, grinding, drilling, reaming, etc. are also used in composites. However, owing to the anisotropic and nonhomogeneous nature of composites, machining in composites is highly challenging and is associated with several critical aspects. Some of the common problems faced during composites machining are

- Tool life
- Dust removal
- Damage to work piece

10.8.2.2.1 Tool Life

Advanced composites such as carbon/epoxy are highly abrasive in nature and wear and tear of cutting tools is a major concern. High-speed steel tools, coated with tungsten carbide or diamond, have longer life. High-speed steel tools can be used for glass fiber–reinforced composites. Carbide and polycrystalline diamond (PCD) tool inserts are superior in terms of tool life and machining quality. These tools, although more expensive, are commonly used in carbon fiber and Kevlar fiber–reinforced composites.

10.8.2.2.2 Dust Removal

Unlike metal machining, where the chips are formed in ribbons, composite machining generates powders that cause health concern. Carbon dust can result in short circuit and damage the CNC unit of the lathe. Note also that coolant and lubricants are generally not used in composites machining, and unless proper care is taken, the machining dust can heavily pollute the air in the machine shop. An efficient dust extraction

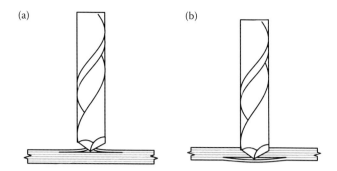

FIGURE 10.17 Delamination of laminate during drilling. (a) Peel-up delamination at entry. (b) Push-out delamination at exit. (Adapted with permission from S. Abrate, Machining of composite materials, *Composites Engineering Handbook* (P. K. Mallick, ed.), Marcel Dekker, New York, 1997, pp. 777–810.)

system involving vacuum suction hoses is essential for removal of the powders as they are formed.

10.8.2.2.3 Damage to Work Piece

Unless proper care is taken during machining, it leads to low-quality machined surface and also damages the work piece. Common damages associated with composites machining are delamination, fiber pull-out, uncut fibers, fuzzing, and matrix degradation.

The most common machining operation in composites is drilling holes. As shown in Figure 10.17, delamination during drilling can occur either by peel-up of the top plies at entry or by push-out of the bottom plies at exit [34,35]. At the entry, the tool's mechanical action has a tendency to peel up the top plies. High thermal stresses can also cause delamination in the top plies. On the other hand, the tool acts as a punch at the exit and causes delamination in the bottom plies. Appropriate backing plate and reduced feed rate at exit help minimize this type of delamination. It is important that appropriate drill bit with proper machining parameters is used for drilling of composites. Machining parameters depend on several factors, including laminate material, laminate thickness, laminate ply sequence, hole diameter, cutting speed, feed rate, and tool material. For unidirectional as well as multidirectional carbon/epoxy laminates, carbide and PCD drill bits are commonly used. Typical cutting speeds are in the range of 40–60 m/min at feed rates of 0.02–0.08 mm/rev.

Machining operations like turning and sawing also cause delaminations and other damages. Cutting force normal to the laminate during the parting operation tends to push the uncut plies down, especially when a worn-out tool is used and causes edge delaminations. In general, a combination of high cutting speed (mm/min) and low feed rate (in mm/rev or mm/tooth) minimizes the possibility of delamination and results in smooth machined edge. Very high cutting speed may lead to local heating and deterioration of resin, which clogs the teeth of the saw during sawing.

Aramid fibers are ductile in nature and low in compressive strength. During machining of aramid fiber composites, the fibers tend to recede within the matrix and they do not get sheared off. Also, owing to their ductile nature, aramid fibers absorb a great deal of energy leading to low-quality machined surface characterized by uncut fibers, fiber kinks, and uneven surface. The machining approach to avoid such issues should be to stress the fibers in tension so that they can be cut by shear action [35].

10.9 SUMMARY

Manufacturing is a critical aspect in the composite product development cycle. In this respect, the following aspects have been reviewed in this chapter: (i) basic concepts,

(ii) common manufacturing processes, (iii) manufacturing process selection, (iv) process modeling, and (v) composites machining. The following major points have been noted:

- Composites processing involves essentially four basic steps—impregnation, lay-up, consolidation, and solidification.
- The methodology of implementation and the exact order, in which the basic steps are implemented, vary depending upon the manufacturing process and physical form of reinforcement and matrix.
- Composites manufacturing processes can be grouped in three broad categories—open mold processes, closed mold processes, and continuous molding processes.
- Filament winding is an important continuous molding process especially suitable for making tubular products such as pressure vessels, pipes, storage tanks, etc. Like in other composites manufacturing processes, the basic processing steps are applicable here too. Additionally, the computational aspects involving programming are of critical significance.
- The manufacturing process selection depends on a number of factors such as configuration of the product, tooling requirement, cost, etc. These factors can be related to the product description, product requirements, process description, and process requirements, and some of them are interrelated. The manufacturing methods can be qualitatively ranked in each of these aspects for making a suitable selection.
- There are also other critical topics in composites manufacturing; two such topics of importance are process modeling and composites machining.

EXERCISE PROBLEMS

10.1 What are the essential processing steps in any composites manufacturing method? Write a brief note giving details of their significance in the quality of the final product.

10.2 How are the four basic processing steps incorporated in (i) prepreg lay-up, (ii) resin transfer molding, (iii) pultrusion, and (iv) filament winding (wet winding) processes?

10.3 An E-glass/epoxy laminate of size 500 mm × 500 mm is designed with eight bidirectional balanced fabric plies. Determine the quantity of resin and hardener to be mixed. Assume the following:

Resin to hardener mix ratio = 100:20
Surface density of fabric = 430 g/m^2
Desired fiber volume fraction = 0.6
Density of E-glass fiber = 2.54 g/cm^3
Density of resin mix = 1.1 g/cm^3

Hint: Refer Chapter 3 to determine weight fractions corresponding to the desired fiber volume fraction.

10.4 Write a note giving details of advantages and disadvantages of the spray-up process.

10.5 Consider a circular cylindrical part of inner and outer diameters of 150 mm and 230 mm, respectively. The part is laid-up in a rosette construction. The ply developments are rectangular and they are laid-up in such a way that the inner and outer edges are parallel to the axis of the cylindrical part. Now, the arc angle at a point in a ply is defined as the angle between the tangent lines as shown in Figure 10.18. If the arc angle is 90° on the inner circumference, determine its value on the outer circumference. How many plies are

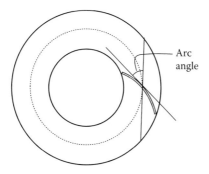

FIGURE 10.18 Definition of arc angle (Exercise 10.5).

required to lay-up the part completely? What will happen if the arc angle is changed to 75° on the inner circumference? Assume uniform ply thickness.

Hint: Given the information as above, the arc angle is given by

$$\alpha = \sin^{-1}\left(\frac{Nt}{2\pi r}\right)$$

where N, r, and t are the number of plies, radius at the concerned point, and ply thickness, respectively.

10.6 Consider a tape-wound conical part of dimensions as shown in Figure 10.19. Determine the prepreg tape width and the pitch in the winding program if the winding is done as parallel winding. Assume uniform ply thickness of 0.5 mm. Consider machining allowance of 2 mm on both faces.

10.7 Compare filament winding and pultrusion in respect of (i) how the basic processing steps are incorporated, (ii) typical structural characteristic of the end products, and (iii) common applications.

10.8 Identify two typical products made by each of the 10 composites manufacturing methods discussed in this chapter. Discuss why these manufacturing methods are suitable for making the corresponding identified products.

10.9 It is found experimentally that three spools of carbon rovings at a programmed bandwidth of 6 mm results in a ply thickness of 0.4 mm. Determine the number of mandrel rotations required to make a unidirectional ply of width 2400 mm and thickness 0.6 mm by hoop winding using two spools.

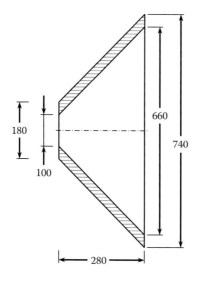

FIGURE 10.19 Tape-wound conical part (Exercise 10.6). All dimensions are in mm.

10.10 A cylindrical mandrel of diameter 240 mm and length 600 mm with end domes is to be wound with geodesic helical plies at an angle of winding of 30° (at the cylindrical portion). Given an experimental data that three spools of carbon rovings at a programmed bandwidth of 6 mm results in a ply thickness of 0.4 mm, determine the number of circuits required for one complete coverage of the mandrel with a minimum thickness of 0.6 mm. What is the actual ply thickness estimated? Assume four spools are used in the winding operation.

Hint: The number of circuits is an integer. Also, each helical ply actually consists of two plies—one at 30° and the other at −30°.

10.11 Consider a cylindrical mandrel with hemispherical end domes. Consider a geodesic helical fiber path. Show that the mandrel rotation corresponding to the fiber path in one end dome is 90°, irrespective of the diameters at the cylinder and pole openings.

Hint: The geodesic helical path in the end dome of a hemispherical shape is a great circle; so it would be a straight line in the side view.

10.12 Consider a cylindrical mandrel of radius R and length L. Also, consider a helical path on it with an angle of winding α. If the end domes are hemispherical in shape with equal pole openings, show that the total mandrel rotation θ for one complete circuit is given by

$$\theta = \frac{2L \tan \alpha}{R} + \pi$$

Hint: The geodesic path on the development of a cylinder is a straight line at angle α to the lengthwise edge.

10.13 Consider a cylindrical mandrel of diameter 400 mm and length 800 mm with two hemispherical end domes. If the pole opening diameters are 100 mm at either end, what is the geodesic helical angle of winding at the cylindrical portion? Determine the total mandrel rotation involved in one complete circuit.

10.14 Consider the cylindrical mandrel in the exercise above. Given an experimental data that three spools of carbon rovings at a programmed bandwidth of 6 mm results in a ply thickness of 0.4 mm, determine the number of circuits required for one complete coverage of the mandrel with a minimum thickness of 0.6 mm and the total quantity of fiber needed for one complete coverage. Assume, tex of the fiber = 800 g/km.

10.15 Write a brief note on the importance of real-time decision regarding the point of application of pressure in an autoclave curing of epoxy composite product.

10.16 Write a note on the manufacturing process selection.

10.17 List down various process parameter that can be included in the development of a process model for (i) prepreg lay-up, (ii) compression molding, (iii) pultrusion, and (iv) filament winding.

10.18 Write a brief note on the need for machining of composites. What are the different types of machining involved in PMCs?

REFERENCES AND SUGGESTED READING

1. S. K. Mazumdar, *Composites Manufacturing—Materials, Product and Process Engineering*, CRC Press, Boca Raton, FL, 2002.

2. T. G. Gutowski (ed.), *Advanced Composites Manufacturing*, John Wiley & Sons, New York, NY, 1997.
3. F. C. Campbell, *Manufacturing Processes for Advanced Composites*, Elsevier, Amsterdam, the Netherlands, 2004.
4. P. K. Mallick, *Fiber-Reinforced Composites: Materials, Manufacturing and Design*, third edition, CRC Press, Boca Raton, FL, 2013.
5. E. J. Barbero, *Introduction to Composite Materials Design*, second edition, CRC Press, Boca Raton, FL, 2015.
6. F. C. Campbell, *Structural Composite Materials*, ASM International, Ohio, 2010.
7. D. Cripps, T. J. Searle, and J. Summerscales, Open mold techniques for thermoset composites, *Comprehensive Composite Materials, Volume 2, Polymer Matrix Composites* (A. Kelly and C. Zweben, editors-in-chief; R. Talreja and J.-A. E. Manson, Vol. eds.), Elsevier, Amsterdam, the Netherlands, 2000, pp. 737–762.
8. D. R. Sidwell, Hand lay-up and bag molding, *Handbook of Composites* (S. T. Peters, ed.), second edition, Chapman & Hall, London, 1998, pp. 352–377.
9. G. Dillon, P. Mallon, and M. Monaghan, The autoclave processing of composites, *Advanced Composites Manufacturing* (T. G. Gutowski, ed.), John Wiley & Sons, New York, NY, 1997, pp. 207–258.
10. J. C. Seferis, R. W. Hillermeier, and F. U. Buehler, Prepregging and autoclaving of thermoset composites, *Comprehensive Composite Materials, Volume 2, Polymer Matrix Composites* (A. Kelly and C. Zweben, editors-in-chief; R. Talreja and J.-A. E. Manson, Vol. eds.), Elsevier, Amsterdam, the Netherlands, 2000, pp. 701–736.
11. C. W. Peterson, G. Ehnert, R. Liebold, K. Horsting, and R. Kuhfusz, Compression molding, *ASM Handbook, Volume 21, Composites* (D. B. Miracle and S. L. Donaldson, Vol. Chairs), ASM International, Ohio, 2001, pp. 516–535.
12. J. M. Castro and R. M. Griffith, Press molding processes, *Composite Materials Handbook* (P. K. Mallick, ed.), Marcel Dekker, New York, 1997, pp. 481–513.
13. E. Haque and Burr (Bud) L. Leach, Matched metal compression molding of polymer composites, *Handbook of Composites* (S. T. Peters, ed.), second edition, Chapman & Hall, London, 1998, pp. 379–396.
14. L. Fong and S. G. Advani, Resin transfer molding, *Handbook of Composites* (S. T. Peters, ed.), second edition, Chapman & Hall, London, 1998, pp. 433–455.
15. C. D. Rudd, Resin transfer molding and structural reaction injection molding, *ASM Handbook, Vol 21, Composites* (D. B. Miracle and S. L. Donaldson, Vol. Chairs), ASM International, Ohio, 2001, pp. 492–500.
16. B. A. Wilson, Pultrusion, *Handbook of Composites* (S. T. Peters, ed.), second edition, Chapman & Hall, London, 1998, pp. 488–524.
17. J. E. Sumerak, The pultrusion process for continuous automated manufacture of engineered composite profiles, *Composite Materials Handbook* (P. K. Mallick, ed.), Marcel Dekker, New York, 1997, pp. 549–577.
18. J. E. Sumerak, Pultrusion, *ASM Handbook, Volume 21, Composites* (D. B. Miracle and S. L. Donaldson, Vol. Chairs), ASM International, Ohio, 2001, pp. 550–564.
19. J. P. Fanucci, S. Nolet, and S. McCarthy, Pultrusion of composites, *Advanced Composites Manufacturing* (T. G. Gutowski, ed.), John Wiley & Sons, New York, NY, 1997, pp. 259–295.
20. M. N. Grimshaw, Automated tape laying, *ASM Handbook, Volume 21, Composites* (D. B. Miracle and S. L. Donaldson, Vol. Chairs), ASM International, Ohio, 2001, pp. 480–485.
21. D. O. Evans, Fiber placement, *ASM Handbook, Volume 21, Composites* (D. B. Miracle and S. L. Donaldson, Vol. Chairs), ASM International, Ohio, 2001, pp. 477–479.
22. Yu. M. Tarnopol'skii, S. T. Peters, and A. I. Beli, Filament winding, *Handbook of Composites* (S. T. Peters, ed.), second edition, Chapman & Hall, London, 1998, pp. 456–475.
23. V. V. Vasiliev, V. A. Barynin, and A. F. Rasin, Anisogrid lattice structures—Survey of development and application, *Composite Structures*, 54(2–3), 2001, 361–370.
24. V. V. Vasiliev and A. F. Razin, Anisogrid composite lattice structures for spacecraft and aircraft application, *Composite Structures*, 76(1–2), 2006, 182–189.
25. S. T. Peters, Filament winding—Introduction and overview, *Composite Filament Winding* (S. T. Peters, ed.), ASM International, Ohio, 2011, pp. 1–5.
26. S. T. Peters and Yu. M. Tarnopol'skii, Filament winding, *Composites Engineering Handbook* (P. K. Mallick, ed.), Marcel Dekker, New York, 1997, pp. 515–548.
27. S. T. Peters, Filament winding, *ASM Handbook, Volume 21, Composites* (D. B. Miracle and S. L. Donaldson, Vol. Chairs), ASM International, Ohio, 2001, pp. 536–549.
28. S. Koussios, Integral design for filament winding materials, *Winding Patterns, and Roving Dimensions for Optimal Pressure Vessels, Composite Filament Winding* (S. T. Peters, ed.), ASM International, Ohio, 2011, pp. 19–34.
29. V. A. Bunakov and V. D. Protasov, Composite pressure vessels, *Handbook of Composites* (A. Kelly and Yu. N. Rabotnov, eds.), *Volume 2, Structures and Design* (C. T. Herakovich and Yu. M. Tarnopol'skii, ed.), North Holland, Amsterdam-New York-Oxford-Tokyo, 1989, pp. 464–530.

30. A. P. Priestley, Programming techniques, computer-aided manufacturing, and simulation software, *Composite Filament Winding* (S. T. Peters, ed.), ASM International, Ohio, 2011, pp. 35–47.
31. J. E. Green, Automated filament winding systems, *Composite Filament Winding* (S. T. Peters, ed.), ASM International, Ohio, 2011, pp. 7–18.
32. S. G. Advani and E. M. Sozer, *Process Modeling in Composites Manufacturing*, Marcel Dekker, New York, 2003.
33. S. G. Advani, Process modeling, *ASM Handbook, Volume 21, Composites* (D. B. Miracle and Steven L. Donaldson, Vol. Chairs), ASM International, Ohio, 2001, pp. 423–433.
34. S. Abrate, Machining of composite materials, *Composites Engineering Handbook* (P. K. Mallick, ed.), Marcel Dekker, New York, 1997, pp. 777–810.
35. K. E. Kohkonen and N. Potdar, Composite machining, *Handbook of Composites* (S. T. Peters, ed.), second edition, Chapman & Hall, London, 1998, pp. 596–609.
36. M. J. Paleen and J. J. Kilwin, Hole drilling in polymer-matrix composites, *ASM Handbook, Volume 21, Composites* (D. B. Miracle and S. L. Donaldson, Vol. Chairs), ASM International, Ohio, 2001, pp. 646–650.
37. L. F. Kuberski, Machining, trimming and routing of polymer-matrix composites, *ASM Handbook, Volume 21, Composites* (D. B. Miracle and S. L. Donaldson, Vol. Chairs), ASM International, Ohio, 2001, pp. 616–619.

11

Testing of Composites and Their Constituents

11.1 CHAPTER ROAD MAP

The primary objective of this book is to take the reader through the complete cycle of development of a composite product. In this journey, testing is a crucial part and this chapter is devoted to discuss issues involved in the testing of composites and their constituents. First, an introductory remark including test objectives, standards for testing, and general philosophy of testing is made. Mechanical and nonmechanical parameters at different levels are identified and their test methodologies are presented. Different levels of tests discussed here include (i) tests on the constituent raw materials—reinforcement and matrix, (ii) tests at the lamina/laminate level, (iii) tests at the element level, and (iv) tests at the component level—subscale and full-scale tests. It will be seen that composites testing is rather involved and relatively complex due to the anisotropic nature of the materials and unique manufacturing philosophy. Thus, our objective in this chapter is primarily to acquaint the reader with the concept and process of testing of composite materials.

Familiarity with the introductory concepts of composites, characteristic parameters of composites at the micro and macro levels, and materials and manufacturing methodologies is essential for the effective assimilation of the topics covered here. Thus, Chapters 1, 3, 4, 9, and 10 should be covered before proceeding to this chapter.

11.2 INTRODUCTION

Testing is an inseparable part in any composite product development program. We will see in the next section in the discussion on test objectives that tests are also required as research and development work for technological growth. Composite materials and products are built starting with the constituent raw materials and we will see in this chapter that tests are done at different levels. We also know that composites are anisotropic in nature; compared to isotropic materials, the number of material parameters, which are required from design and analysis as well as from quality control and quality assurance points of view, is rather large. As a result, the concept and process of testing are more involved in composites. Extensive work has been done in the field of composites testing. Test methods discussed here are not exhaustive; for carrying out a specific test and reporting the results, one should refer to the corresponding applicable standards, of which ASTM standards are listed in the reference. For more information on testing principles, methodology, specifications, standards, and material behavior, the interested reader may consult available texts (see, for instance, References 1–5).

11.2.1 Test Objectives

The simple and obvious objective of carrying out a test is to generate some useful data in respect of various mechanical and nonmechanical characteristics of the material or structure. The data obtained from a test can be utilized in different ways. Based on the intended utility of the test data, for the sake of convenience of discussion and for efficient planning of a test program, the test objectives can be broadly categorized as follows (also see References 6 and 7):

- Design and analysis
  - Material selection
  - Design calculation
  - Performance prediction
- Quality control and quality assurance
  - Acceptance of materials
  - Stage clearance during processing
  - Final acceptance of a product
- Research and development
  - New materials development

The first and one of the most common test objectives is to generate data for use in the design and analysis of a product. Material selection and preliminary design calculation are often done using material data readily available in the literature. (Note, however, that material data available in the literature are also obtained from tests carried out by someone else!) The end characteristics, whether mechanical or nonmechanical, of a composite material depend on the constituents and the process adopted. Thus, for a final detail design and performance prediction, it is essential to use material data obtained from a material characterization exercise using the same constituents in a similar processing environment.

The second major category of test objectives is in the area of quality control and quality assurance. Sometimes, in a product development program, it may be possible to use readily available material data in design and analysis, and a detailed material characterization process can be skipped. However, we must know the materials that are actually used in the manufacture of the product and a limited testing process may still be required from quality control and quality assurance points of view. Tests are carried out for the acceptance of a particular batch of material for use in making a product. Tests are also regularly carried out for clearing a partially processed part to the subsequent processing operations as well as for acceptance of the final product.

The third major objective of carrying out tests is generation of data for materials research. The ultimate goal is to achieve improvement in existing materials and invention of new and more efficient material systems.

Note that the above three categories of test objectives are not mutually exclusive and there exist overlaps. For example, we can take the case of tests for the determination of mechanical properties of the constituents, say tensile strength of fibers. Most often, as a part of quality control steps, fibers are tested for tensile strength to accept a particular batch of fibers. Fiber strength is generally not used directly in the design calculation of a composite product and tensile testing of fibers is not generally done for generating any design input. However, adopting a micromechanics approach in the preliminary design phase, fiber tensile strength can be used to determine the lamina tensile strength, which in turn can be used in the design of the product. Other similar cases can be found. Clearly, the above categorization of test objectives is a simplification made for the sake of convenience of discussion.

It must also be noted that the intensity of the test program would depend on the test objectives. In general, tests carried out with the objective of design data generation are more elaborate and detailed in nature. The tests carried out for quality control are generally simple and quick. On the other hand, in the third category, the tests can be rather diverse in nature.

11.2.2 Building Block Approach

The overall process of development and production that includes design and analysis, testing, initial development, and final series production of a composite product can be viewed from a building block approach involving a number of levels in the form of a pyramid [1,8,9]. At the bottommost level are the constituent materials, viz. fibers and resin, which has the widest base. The next level is the lamina, which is based on the constituents. The process is continued till the final product, which is at the topmost level, is reached. Each level is built on the knowledge of the previous lower level. Each level has fewer elements but more complexities compared to the previous lower level. Tests can be associated to each of these levels and a number of levels of tests can be found. For the sake of convenience of discussion, we can broadly categorize the tests into the following levels and sublevels:

- Coupon-level tests
    - Constituent-level tests—tests on fibers and resin
    - Lamina-level tests
    - Laminate-level tests
- Structural element-level tests
- Component-level tests
    - Subscale component-level tests
    - Full-scale component-level tests

As we know, the basic building block in the design of a composite product is a lamina. Placing a number of laminae in a desired sequence, we get a laminate, which is the primary form of any composite product. However, a composite product is not manufactured by using readily available laminae; rather it is made by using readily available fibers and resin. Thus, in a complete testing program for a composite product, the starting point is testing of the constituents, viz. fibers and resin. Statistically significant numbers of samples from each batch are tested for various mechanical and nonmechanical parameters in respect of both fibers and resin. Generally, tests at this level are simple but more focused. At the initial stage, these tests may be carried out for generating material specifications, whereas at the subsequent stages, often, these tests form a part of the quality control and quality assurance procedure.

Tests in the next level are on the mechanical and nonmechanical properties of the lamina. Data from these tests are commonly used as inputs to the analytical and numerical tools in design calculations for configuring sizes, shapes, and ply sequence of various design elements and their performance prediction.

The common design philosophy is to use the lamina properties, obtained from the lamina-level tests, as the primary design input data and carry out design calculations by adopting a macromechanics-based approach. An alternate design philosophy is also in vogue, in which, laminate properties are directly used as a primary design input. In such a case, laminate-level test coupons with plies resembling the component ply sequence are tested for various mechanical and nonmechanical parameters. In addition to this, laminate-level test data are also used as a check for reliability of the mathematical models used in the design calculations.

TABLE 11.1
Tests in Polymer Matrix Composites

Main Class of Tests	Subclass of Tests	Mechanical or Nonmechanical	Purpose	Example
Tests for constituent-level properties	Tests on reinforcement	Both	Quality assurance	Tests for density, breaking strength, etc.
	Tests on matrix	Both	Quality assurance	Tests for viscosity, tensile strength, etc.
Tests for lamina/laminate-level properties	–	Both	Design data generation, quality assurance	Tests for fiber volume fraction, tensile properties, etc.
Tests for element-level properties	–	Mechanical	Design data generation	Tests for joint strength, discontinuities, etc.
Tests at component level	Subscale	Mechanical	Design data generation	Nonstandard structural elements, subscale pressure vessel test, subscale wing box test, etc.
	Full-scale	Mechanical/functional	Acceptance and qualification	Pressure vessel test, aircraft fatigue test, etc.

The laminates in a composite product may have holes, cutouts, bonded interfaces, and other forms of discontinuities. Tests conducted for the generation of data for the design of such structural elements or quality control in such areas can be put in this level of testing. While standard coupons are available in some cases such as lamina/laminate tensile strength with a hole, interface adhesive strength, etc., specially designed subscale test components are also devised for testing in some cases.

Component-level tests are case-specific tests that are conducted for the determination of the general failure behavior of the product. In the subscale component testing, either the whole product made to a subscale or a selected critical portion is tested. Often, the critical portion selected is a major joint in the product. On the other hand, in the full-scale testing, individual component, subcomponent, or the final product is tested under simulated test condition. In general, limited numbers of full-scale tests are carried out for determining the actual available margin that accounts for both design and manufacturing uncertainties. Full-scale tests may be conducted at a lower level of critical load case on each product as an acceptance procedure. Full-scale tests are time and cost intensive, and the scope and extent of such tests are dependent on the type of the product and they are often organization specific.

Tests conducted at different levels in the PMC industry are enumerated in Table 11.1.

11.2.3 Test Standards

Standard test methods are available for most of the mechanical and nonmechanical parameters of fibers, resin, prepregs, laminae, and laminates. Typically, a standard covers the following aspects of a test:

- Scope of the test
- Test specimen and sampling

- Apparatus, equipment, and machinery
- Physical principle
- Test procedure
- Calculation and reporting of results

The primary utility of standards is the efficiency and reliability achieved by using them in a test program. Standards are prepared and constantly modified with time. Specifications and standards are an essential part in a composite product development program, in general and testing, in particular. A number of organizations are available worldwide, working for the preparation and improvement of standards. Some of the common standards in the field of composites are listed below (see, for instance, References 2 and 3 for more information on standards and specifications).

- American Society for Mechanical Engineers (ASME)
- ASTM International, formerly known as American Society for Testing and Materials (ASTM)
- British Standards from BSI Group (BS)
- German Institute for Standardization (DIN)
- Indian Standards Institute (ISI)
- International Organisation for Standardization (ISO)
- SAE International, formerly known as Society of Automotive Engineers (SAE)
- United States Defence Standard (MIL)

In this introductory discussion, we had a brief look at three broad aspects of composites testing—test objectives, building block approach, and test standards. The concepts of various mechanical and nonmechanical tests are discussed in the remaining sections in this chapter. However, it should be kept in mind that there are other topics of general importance. For example, laminate and specimen preparation, test article preparation, statistical variability, data acquisition and data interpretation, etc. can play a significant role in the success of a test program [10,11].

11.3 TESTS ON REINFORCEMENT

Reinforcements are used in different physical forms—filament, yarn, fabric, mat, etc. The characterization of reinforcements is required primarily from quality assurance and R&D points of view and rarely for the generation of design data. Various mechanical and nonmechanical parameters of reinforcements are tested and the results are compared with specified values for acceptance or otherwise.

11.3.1 Nonmechanical Tests on Reinforcement

The test methods for the following nonmechanical parameters of reinforcements are discussed in this section:

- Density
- Moisture content
- Filament diameter
- Tex
- Fabric construction
- Areal density of fabric

Table 11.2 gives a summary of the test methods for these nonmechanical parameters.

TABLE 11.2
Standard Test Methods for Nonmechanical Parameters of Reinforcements

Parameter	Description	Principle of Testing	Applicable ASTM Standard
Density	Mass per unit volume	Method of displacement	ASTM D792
		Archimedes principle	ASTM D3800
		Density-gradient column	ASTM D1505
Moisture content	Mass of moisture per unit mass of reinforcement	Moisture removal by oven drying and weighing	ASTM D123
Filament diameter	Diameter of individual filament	Planimetering, indirect method based on density and linear density	ASTM D3379
Tex	Mass per unit length	Direct measurement and vibroscope procedure	ASTM D1577
Fabric construction	Filament count and weave	Direct counting	ASTM D3775 (filament count)
Fabric areal density	Mass per unit area	Direct measurement	ASTM D3776

11.3.1.1 Density of Fiber

The density of high-modulus continuous as well as discontinuous fibers can be determined as per ASTM D3800, by using either the buoyancy method (based on Archimedes principle) or the sink–float method [12].

In the buoyancy method, the test specimen is a suitable size sample of cut fibers of minimum 0.5 g. The sample is weighed in air and then in a liquid with lower density. The difference in weights in air and the liquid is the buoyancy force, that is, the weight of the displaced liquid. From the knowledge of the weight of the displaced liquid, its mass can be obtained and this mass divided by the density of the liquid gives the volume of the displaced liquid. The volume of the sample is the same as that of the displaced liquid. Then, the density of the fiber is obtained by dividing the mass of the fiber sample by its volume. Thus,

$$\rho_f = \frac{W_1}{(W_1 - W_2/\rho_l)} \quad (11.1)$$

where
- W_1 Weight of the fiber sample in air (g)
- W_2 Weight of the fiber sample in the test liquid (g)
- ρ_f Density of fiber (g/cm³)
- ρ_l Density of test liquid (g/cm³)

In the sink–float method, the test specimen is a cut fiber of approximately 50 mm in length with a loose overhand knot. The sample is allowed to sink in a mixture of two miscible liquids, where one of the liquids is lighter than the fiber and the other heavier. The liquids should be such that the sample gets thoroughly wetted by the liquid mixture. The heavier liquid is then added in increments, and after each addition, the contents are gently mixed without disturbing the sample. The process is continued till the sample is suspended at an intermediate point. Lighter liquid is also added, if necessary, for achieving the equilibrium point. At such an equilibrium point, the density of the sample is equal to the density of the liquid mix. A hydrometer is used to obtain the density of the liquid mix, which, in turn, gives the density of the fiber.

11.3.1.2 Moisture Content

Typically, a test method for the determination of moisture content of rovings, yarns, or fabrics uses a sample, which is a small piece of fabric or a small roll of roving weighing approximately 2 g. The procedure involves drying and weighing a glass weighing bottle. The sample is placed in the glass weighing bottle, weighed at room temperature, and then heated at 105–110°C for moisture removal. The sample with the container is then weighed again and moisture content of the sample is calculated using the two weights as follows:

$$\text{M.C.} = \left(\frac{W_2 - W_3}{W_3 - W_1}\right) \times 100 \tag{11.2}$$

where
- W_1 Weight of the empty glass weighing bottle (g)
- W_2 Weight of the glass weighing bottle with the sample before moisture removal (g)
- W_3 Weight of the glass weighing bottle with the sample after oven drying (g)
- M.C.   Moisture content (%)

11.3.1.3 Filament Diameter

The diameter of an individual filament can be determined by direct planimeter measurement of the filament area of cross section on a photomicrograph using a magnification factor of 2000–3000. An alternate indirect method gives the filament area of cross section as the ratio of tex (in g/km) to density (in g/mm²), with proper care for the conversion of units, of the fiber. The linear density and density of fibers are determined from a bundle of fibers; thus, the indirect method gives only an average filament diameter.

EXAMPLE 11.1

Determine the filament diameter for carbon fiber tows with the following given parameters:

Density of fiber: 1.78 g/cm³, tex: 800, and number of filaments per tow: 12 k.

Solution

The average cross-sectional area of a filament is obtained as

$$A_f = \frac{800 \times 10^{-6}/12000}{1.78 \times 10^{-6}} = 0.0375 \, \text{mm}^2$$

Then, the average filament diameter is given by

$$d_f = \sqrt{\frac{0.0375 \times 4}{\pi}} = 0.218 \, \text{mm}$$

11.3.1.4 Tex

Tex is a commonly used unit for linear density, that is, mass per unit length, of either an individual filament or a bundle in the form of strand, yarn, etc. It stands for mass per 1000 m. Two methods are prescribed by ASTM D1577—first, by direct measurement of mass of known length of an individual fiber or a bundle of fibers and second, by the vibroscope procedure [13].

In the direct method, tex is readily obtained by dividing the mass (in mg) by the length (in mm) and multiplying the ratio by 1000. This method is not suitable for individual filaments of lengths less than 30 mm.

In the other method, a vibroscope is used for finding the fundamental resonant frequency of an individual fiber. From the fundamental resonant frequency of transverse vibration, under known conditions of length of the individual fiber and applied tension, the linear density can be calculated. This method is suitable, especially for staple fibers with low linear density.

11.3.1.5 Fabric Construction

Two primary parameters that define fabric construction are the filament count and weave of fabric. These parameters are crucial in the sense that fabric characteristics such as handleability, drapeability, stability, resultant fabric/ply thickness, transition of fiber properties to fabric properties, etc. are directly influenced by fabric construction. ASTM D3775 gives a standard procedure for the measurement of the filament count of fabrics; it involves the direct counting of filaments under appropriate magnification [14]. The common weave styles are plain weave, satin weave, twill weave, basket weave, etc.; there are several subtypes such as $4H$ satin, $8H$ satin, 2×2 twill, 2×1 twill, etc. A discussion on weave styles is given in Chapter 9.

11.3.1.6 Areal Density of Fabric

Fabric areal density or mass per unit area can be obtained from the tests prescribed by ASTM D3776 [15]. The procedure involves direct measurement of the mass of the specimen of specified dimensions. Areal density is readily obtained by dividing the net mass of the fabric by the product of the length and width of the specimen.

11.3.2 Mechanical Tests on Reinforcement

The test methods for the following mechanical parameters of reinforcements are discussed:

- Tensile strength and modulus of a single filament
- Tensile strength and modulus of a tow
- Breaking strength of fabric

Table 11.3 gives a summary of the test methods for these mechanical parameters.

11.3.2.1 Tensile Properties by Single-Filament Tensile Testing

Tensile strength and modulus of fibers can be obtained by adopting the single-filament tensile testing method as per ASTM D3379 [16]. The test specimen is a single filament of sufficient length, which is separated with utmost care from a dry strand. The filament is mounted on a slotted cardboard tab as shown in Figure 11.1. The tabbed specimen is

TABLE 11.3

Standard Test Methods for Mechanical Parameters of Reinforcements

Sl. No.	Parameter	Principle of Testing	Applicable ASTM Standard(s)
1	Tensile strength and modulus of a single filament	Loading till failure	ASTM D3379
2	Tensile strength and modulus of a bundle of filaments	Loading till failure	ASTM D4018
3	Breaking strength of a fabric	Loading till failure	ASTM D7018/D5034/D5035

Testing of Composites and Their Constituents

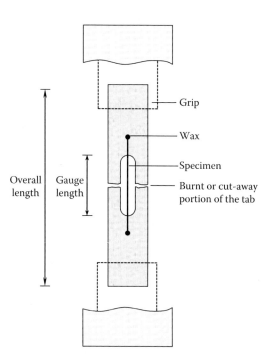

FIGURE 11.1 Single-filament tensile testing. (Adapted from ASTM D3379-75, *Standard Test Method for Tensile Strength and Young's Modulus for High-Modulus Single-Filament Materials*, ASTM International, 1989.)

loaded in a tensile testing machine and gripped inside the jaws. The cardboard tab is then burnt away very carefully in the midregion on both sides and the unsupported specimen is pulled at constant cross-head speed till failure. The applied load and corresponding elongation are continuously recorded. A large number of specimens are tested so that statistically meaningful results can be obtained.

Tensile strength can be readily obtained by dividing the failure load by the average filament cross-sectional area, that is,

$$T_f = \frac{F}{A} \tag{11.3}$$

where
- T_f Tensile strength of fiber (N/mm^2)
- F Failure load (N)
- A Average filament area of cross section (mm^2)

The prescribed technique for the determination of filament area of cross section as per ASTM D3379 is by planimetering the cross section on a photomicrograph using a magnification factor of 2000–3000. An alternate technique for area calculation is an indirect one, in which the area is obtained by dividing the linear density (in g/mm) by density (in g/mm^2) of the fiber.

Tensile modulus is determined from the load-elongation data. A large number of specimens with different gauge lengths are tested, apparent compliance (elongation per unit load) is calculated, and gauge length–compliance curve, which is typically linear, is plotted. The intercept of the curve, that is, compliance at zero gauge length is the system compliance. The elongation in the specimen is obtained from the cross-head movement, and thus, system compliance has to be subtracted from the apparent compliance to determine the true compliance. Finally, modulus is obtained by dividing the specimen gauge length (in mm) by the product of true compliance (in mm/N) and filament cross-sectional area (in mm²), as follows:

$$E_f = \frac{L}{CA} \quad (11.4)$$

where

E_f Tensile modulus of fiber (N/mm²)
L Gauge length of the specimen (mm)
C True compliance (mm/N)
A Average filament area of cross section (mm²)

11.3.2.2 Tensile Properties by Tow Tensile Testing

The job of selecting individual filament and mounting it on the tabs is a difficult one and as a result, this method is not very accurate, especially for the determination of modulus. Alternatively, tensile strength and modulus of fibers can also be obtained by adopting the tow tensile testing method prescribed by ASTM D4018 [17]. In this method, in addition to the tensile strength and modulus, density and linear density or mass per unit length of fibers are also obtained as secondary output. The tensile test specimen is a bundle of fibers, impregnated with a suitable resin and consolidated. The resin should be compatible with the fiber and any size on it. Impregnation of the fibers with resin followed by consolidation imparts easy handleability to the specimens and allows the individual filaments in the specimen to be loaded uniformly.

Tabs may or may not be used in the specimens. Tabbed specimens have a gauge length of 150 mm between tabs, whereas untabbed specimens have sufficient length so that a gauge length of 150 mm is available between grips.

The tensile test involves pulling the specimen of resin impregnated and consolidated fibers in a calibrated tensile testing machine till failure. For finding the fiber modulus, a calibrated extensiometer is attached to the specimen.

The tensile strength and modulus of the fiber are calculated as follows:

$$\left(\sigma_{1f}^T\right)_{ult} = \left(\frac{\rho_f}{\rho_f^{(l)}}\right) P \quad (11.5)$$

$$E_{1f} = \left(\frac{\rho_f}{\rho_f^{(l)}}\right)\left(\frac{P_u - P_l}{\varepsilon_u - \varepsilon_l}\right) \times 10^{-3} \quad (11.6)$$

where

$\left(\sigma_{1f}^T\right)_{ult}$ Longitudinal tensile strength of fiber (MPa)
ρ_f Density of fiber (g/cm³)
$\rho_f^{(l)}$ Linear density of fiber (g/m)
P Maximum tensile load (N)
E Longitudinal Young's modulus of fiber (GPa)
P_u Tensile load at upper strain limit (N)
P_l Tensile load at lower strain limit (N)
ε_u Upper strain limit
ε_l Lower strain limit

Testing of Composites and Their Constituents

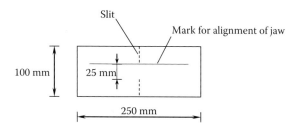

FIGURE 11.2 Schematic representation of grab test specimen. (Adapted from ASTM Standard D5034-09, *Standard Test Method for Breaking Strength and Elongation of Textile Fabrics (Grab Test)*, ASTM International, 2013.)

11.3.2.3 Breaking Strength of Fabric

The breaking strength of a fabric is the tensile failure load per unit width, specified as N per 25 mm. ASTM D5035 gives a strip method for testing breaking strength and elongation of fabrics using specimens of widths 25 mm and 50 mm and minimum length 150 mm [18]. Specimens are usually cut along warp and fill directions, as required. The breaking strength is readily calculated from the load at failure.

ASTM D5034 gives an alternate method for fabric breaking strength and elongation by the grab test [19]. In this case, the specimen is a fabric piece of width 100 mm and length 200–250 mm. In the specimen, two side slits are provided midway between the two ends except the central 25 mm (Figure 11.2).

11.4 TESTS ON MATRIX

11.4.1 Nonmechanical Tests on Matrix

The density of cast resin and the viscosity of liquid resin are two basic nonmechanical parameters of matrix materials that are highly useful in the design and processing of PMCs. Glass transition temperature (T_g) is another critical parameter, which influences in-process thermal cycles and service temperature. In addition to these, other parameters such as volatile content, etc. are of great significance in specific resin systems. In the following sections, we discuss the test methods for density, viscosity, and T_g.

11.4.1.1 Density

Density and specific gravity or relative density of cast resin can be determined either by the liquid displacement method (ASTM D792) or the density-gradient method (ASTM D1505) [20,21].

In the liquid displacement method, the test specimen is a solid piece of cast resin with smoothened edges. Its weight should normally be not more than 50 g and volume not less than 1 cm³. The method, like the buoyancy method for the density measurement of fibers, is based on Archimedes principle. The sample is weighed in air and its mass in air is recorded. Then the sample is weighed in water. A sinker, if required, and a fine wire for hanging the sample are also used. The weights of the partially immersed wire and fully immersed sinker, when used, are suitably adjusted in the calculation process. The difference in weights in air and water is the weight of the displaced water. The density of water being known ($\approx$ 1 g/cm³), the volume of the displaced water, which is equal to the volume of the sample, is readily calculated. Thus,

$$\rho_m = \frac{W_1}{W_1 - W_2} \qquad (11.7)$$

where

W_1 Weight of the cast resin sample in air (g)
W_2 Weight of the cast resin sample fully immersed in water (g)
ρ_m Density of cast resin (g/cm³)

EXAMPLE 11.2

In a density determination test for cast epoxy resin, the following data are recorded:

Weight of the cast resin sample in air: 30 g
Weight of the sample with a sinker fully immersed and wire partially immersed in water: 75 g
Weight of the sinker fully immersed and wire partially immersed in water: 70.2 g

Solution

The weight of the sample in water is given by

$$W_2 = 75 - 70.2 = 4.8\,\text{g}$$

The density of the cast resin sample is obtained as

$$\rho_m = \frac{30}{30 - 4.8} = 1.19\,\text{g/cm}^3$$

In the density-gradient method, the specimen is cut from the cast resin to a convenient shape and size, taking care to ensure that the center of volume of the cut piece is clearly identifiable. A density-gradient column containing a solution with linearly varying density from bottom to top is used. The solution is made by mixing two suitable liquids and maintained at a precise temperature. (A list of suitable liquids is given by ASTM D1505.) A standard glass float of known density is gently dropped and its equilibrium position w.r.t. an arbitrary reference plane is recorded. The glass float is removed without disturbing the density gradient and equilibrium positions of another standard glass float with different density and the specimen are recorded in turn. The density of the specimen is then determined by linear interpolation, as follows:

$$\rho_m = (\rho_2 - \rho_1)\left(\frac{z_m - z_1}{z_2 - z_1}\right) + \rho_1 \qquad (11.8)$$

where

ρ_m, ρ_1, ρ_2 Densities of the cast resin sample and the two standard glass floats, respectively (g/cm³)
z_m, z_1, z_2 Distances, w.r.t. an arbitrary datum, of the cast resin sample and the two standard glass floats, respectively (mm)

11.4.1.2 Viscosity

The viscosity of a liquid resin has direct influence on proper impregnation of reinforcement during manufacturing. It is a critical manufacturing parameter that is regularly evaluated as a part of acceptance and quality control process. Several types of viscosity measurement instruments are available. Of these three notable ones are rotational-type viscometers, U-tube viscometers, and flow-cup viscometers.

Classical Brookfield viscometers are of the rotational type and they are commonly used in the composites laboratory and shop floor. ASTM D2393 gives a standard method of viscosity measurement of resins [22]. It is based on the principle that the torque required to turn an object in a viscous fluid indicates the viscosity of the fluid. Typically, a spindle is dipped in the fluid and torque is applied on to the spindle through a calibrated spring. The viscosity of the fluid is indicated by the spring deflection.

Ostwald's U-tube viscometer is based on capillary principle. It consists of a U-shaped glass tube with two vertical arms with precise dimensions and features. The fluid is sucked through one arm and then allowed to flow down. The other arm has two marks at two levels and the time taken by the fluid to pass through the two marks indicates the viscosity of the fluid.

The flow-cup viscometer consists of a cup with an orifice. The fluid is allowed to flow through the orifice and the time taken by a known volume of the fluid indicates its kinematic viscosity.

11.4.1.3 Glass Transition Temperature

When a cured polymer matrix is heated and the temperature is gradually increased, at a certain temperature, the state of the material changes from glassy to rubbery. (A glassy state is one in which the individual molecular segments have only vibrational motion without any relative motion, whereas in a rubbery state, the individual molecular segments can move relative to each other.) This change in state is called glass transition and the temperature at which it occurs is called the glass transition temperature, T_g. The modulus of the matrix reduces drastically by several orders as the temperature is increased beyond this level. It is also associated with change in heat capacity and CTE. The significance of the glass transition temperature is that it indicates the maximum service temperature for the resin matrix.

T_g depends on the molecular structure of the material and it is influenced by the cure temperature. T_g of a resin matrix cured at high temperature is higher than that of the same material cured at a low temperature.

There are several methods available for the determination of T_g of a matrix material:

- Differential scanning calorimetry (DSC)
- Thermomechanical analysis (TMA)
- Dynamic mechanical analysis (DMA)

In the following paragraphs, we briefly discuss the DSC method.

11.4.1.3.1 T_g by Differential Scanning Calorimetry

The heat capacity of a cast resin matrix changes during glass transition and the glass transition temperature can be obtained from the heat flow versus temperature curve. T_g is associated with a shift in the curve, which may be pronounced in some cases and mild in others. A DSC method for the determination of T_g of a material is given by ASTM E1356 [23,24]. A specimen of suitable mass, generally between 5 and 20 g, and a comparable quantity of a reference material of similar heat capacity are heated in the test chamber at 10°C/min. Often, an empty pan is taken as the reference. The test chamber is capable of providing uniform controlled heating or cooling of the specimen and the reference either to a constant temperature or to a constant rate. It is provided with temperature sensors to indicate specimen and reference temperatures and differential sensors to indicate heat flow difference between the specimen and reference. Initially, the differential heat flow w.r.t. temperature remains constant. Heat capacity (i.e., the quantity of heat required to increase the temperature by 1°C) of a polymer is higher above its glass transition, and as a result, the heat flow increases across the glass transition region. Heat flow versus temperature curve is plotted and key temperatures

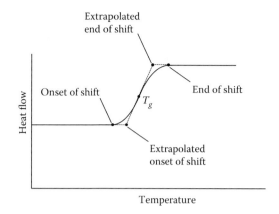

FIGURE 11.3 Schematic representation of a typical heat flow versus temperature curve obtained from a DSC run. (Adapted from ASTM Standard E1356-08, *Standard Test Method for Assignment of the Glass Transition Temperatures by Differential Scanning Calorimetry*, ASTM International, 2014.)

are identified; typically, T_g is assigned to the midtemperature between the extrapolated onset of shift and extrapolated end of shift (Figure 11.3).

Note: Glass transition takes place over a temperature zone, and glass transition temperature, which lies in this temperature zone, depends on not only the material but also the method adopted and the test environment. Thus, in a strict sense, the test methods discussed above actually *assign* T_g to the material, and while reporting T_g, it is necessary to mention the method used in the test.

11.4.2 Mechanical Tests on Matrix

The matrix plays a supporting role to the reinforcements in giving resultant mechanical characteristics such as strength and stiffness to a PMC material. Generally, the mechanical properties of the matrix material by themselves are not of much significance in the structural design of a composite part; however, these properties are necessary for specification and quality control purposes. Commonly tested mechanical properties of cast resins are tensile, compressive, and shear properties. In addition, other characteristics such as hardness, impact resistance, etc. may also be required in specific cases. Here, we briefly discuss the determination of tensile, compressive, and shear properties of cast resins.

11.4.2.1 Tensile Properties

The commonly used test specimen for the determination of tensile properties of neat cast resin is a dog-bone-shaped flat piece of material. ASTM D638 recommends a number of specimen types, of which, the most commonly used specimen is 165 mm in overall length, 19 mm in overall width, and 3.2 mm in thickness [25]. The gauge section length and width are 50 and 13 mm, respectively. The test specimen is loaded in tension by grabbing it in the grips of the testing machine. Test is carried out at constant cross-head movement and tensile strength is calculated by dividing the maximum force by the average original cross-sectional area of the specimen in the gauge section. Extensometers are used and force–extension curves are automatically and continuously recorded. The initial straight portion of the curve is extended and modulus is calculated by dividing the difference in stresses corresponding to any two points on the line by the difference in strains corresponding to the same points.

11.4.2.2 Compressive Properties

ASTM D695 gives a standard test method for the evaluation of compressive properties of neat cast resin [26]. The commonly used specimen is a short cylinder or prism, typical sizes being 12.7 mm (diameter) × 25 mm (height) for a cylinder and 12.7 mm (side) × 12.7 mm (side) × 25 mm (height) for a prism. A compression tool, consisting of two hardened, ground and flat plates, is used for holding the test specimen. Care must be taken to ensure that the end faces of the specimen are parallel to the faces of the compression tool plates. Further, the axis of the compression tool must be aligned with the center line of the plunger of the testing machine. The specimen is tested by applying compressive force under constant cross-head rate of 1.3 mm/min till the yield point is reached. After the yield point, the cross-head speed is increased to 5–6 mm/min and loading is continued till failure. The compressive strength is calculated by dividing the maximum compressive force by the original minimum cross-sectional area of the specimen. When stress–strain data are required, a compressometer is attached to the specimen for recording and plotting a force–displacement curve. The modulus of elasticity is calculated from a tangent drawn to the initial linear portion of this curve.

11.4.2.3 Shear Properties

The shear properties of cast resins can be determined either by the V-notched beam method or the torsion cylinder method. The V-notched beam (ASTM D5379) is typically used for the evaluation of shear properties at the lamina/laminate level [27]; it is discussed in detail subsequently. ASTM E143 describes a method for the determination of the shear modulus of structural materials using either a solid cylinder or a hollow tube [28]. For cast resin materials, a solid cylinder usually in the dog-bone shape is molded and the specimen is loaded under external torque that causes uniform twist in the specimen. A T–δ curve is plotted, where deviation δ is given by

$$\delta = L\left(\theta - \frac{T}{K}\right) \quad (11.9)$$

where
- L Gauge length (mm)
- T Applied torque (N-mm)
- θ Angle of twist per unit length (radians/mm)
- K A constant chosen so that the term inside the brackets in Equation 11.9 is a constant within the proportionality limit

Then, shear modulus is calculated as follows:

$$G = \frac{\Delta T}{J \Delta \theta} \quad (11.10)$$

where
- ΔT Increment in applied torque (N-mm)
- $\Delta \theta$ Increment in angle of twist per unit length (radians/mm)

11.5 TESTS FOR LAMINA/LAMINATE PROPERTIES

A laminate, which is the basic structural element in a laminated composite product, is designed as a stack of a number of laminae at different orientations. The mechanical properties of lamina such as longitudinal tensile strength and modulus, transverse tensile strength and modulus, etc. directly go into the design calculations. These mechanical properties are dependent on specific nonmechanical properties of the laminae.

Thus, laminae mechanical as well as nonmechanical parameters are essential inputs in the design and analysis of a composite product.

11.5.1 Nonmechanical Tests on Laminae

There are a number of nonmechanical, that is, chemical and physical, properties that characterize a composite material [29]. Table 11.4 gives a summary of these properties together with their very brief testing details. Here, we discuss the test methods for the following nonmechanical parameters of cured composites:

- Density
- Constituent content—fiber volume fraction and resin content
- Void content
- Glass transition temperature

11.5.1.1 Density of Composites

The principles and methodology of density measurement for a cured composite material are similar to those for cast resin (refer Section 11.4.1.1). The common methods are the liquid displacement method given in ASTM D792 and the density-gradient method given in ASTM D1505 [20,21].

11.5.1.2 Constituent Content

The fiber content, resin content, and void content of a composite material can be determined by a process of matrix removal given in ASTM D3171 [30,31]. Depending on the reinforcements, matrix removal can be done by either the matrix burning method or the matrix digestion method. The matrix burning method involves burning off the matrix material at high temperatures. At these temperatures, carbon and aramid fibers get oxidized, whereas glass fibers remain stable. Thus, the matrix burning method is suitable for glass fiber composites, but not for carbon and aramid fiber composites. On the other hand, in the acids used in the matrix digestion method, glass fibers get dissolved, whereas carbon and aramid fibers are largely inert to these acids. Thus, the matrix digestion method is used for carbon and aramid fiber composites, but not for glass fiber composites.

The specimen is a properly cut piece of suitable size; it should be representative of the material being tested. For the determination of volume fractions, the density is required and the same specimen is first used for density measurement.

TABLE 11.4
Standard Test Methods for Nonmechanical Parameters of Composites

Sl. No.	Parameter	Principle of Testing	Applicable ASTM Standard(s)
1	Density	Method of displacement	ASTM D792
		Archimedes principle	ASTM D3800
		Density-gradient column	ASTM D1505
2	Constituent content—fiber volume fraction and resin content	Matrix removal by matrix digestion and matrix burning	ASTM D3171
3	Void content	Theoretical calculation involving experimental and theoretical densities	ASTM D2734
4	Glass transition temperature	Differential scanning calorimetry	ASTM E1356

Testing of Composites and Their Constituents

In the matrix burning method, the specimen is dried in an air-circulating oven, and then placed in a desiccated preweighed crucible and weighed. The specimen is then heated to approximately 565°C by keeping the crucible with the specimen in a preheated electric muffle furnace. At this temperature, the matrix gets burnt off, while the glass fibers remain unaffected. The crucible with the specimen is then placed in a desiccator and cooled to room temperature and weighed.

In the matrix digestion method, a suitable acid is used for dissolving the matrix. Various acids/solutions used are as follows:

- Nitric acid
- Sulfuric acid/hydrogen peroxide
- Ethylene glycol/potassium hydroxide
- Sodium hydroxide
- Hydrochloric acid

The specimen is placed in a beaker containing an acid as indicated above and heated till the matrix gets digested fully. Care has to be taken to ensure that overdigestion and fiber loss do not occur. The contents are then cooled, filtered using a preweighed glass filter, dried, and weighed.

Both the methods work on the common processing step of removal of matrix from the composite material sample and the constituent contents are calculated as follows:

Fiber mass fraction

$$W_f = \frac{M_3 - M_1}{M_2 - M_1} \tag{11.11}$$

Fiber volume fraction

$$V_f = \left(\frac{M_3 - M_1}{M_2 - M_1}\right)\frac{\rho_c}{\rho_f} \tag{11.12}$$

Matrix mass fraction

$$W_m = \frac{M_2 - M_3}{M_2 - M_1} \tag{11.13}$$

Matrix volume fraction

$$V_m = \left(\frac{M_2 - M_3}{M_2 - M_1}\right)\frac{\rho_c}{\rho_m} \tag{11.14}$$

Void volume fraction

$$V_v = \frac{(M_2 - M_1)\rho_f \rho_m - (M_3 - M_1)\rho_c \rho_m - (M_2 - M_3)\rho_c \rho_f}{(M_2 - M_1)\rho_f \rho_m} \tag{11.15}$$

where
- M_1 Mass of the empty crucible/filter (g)
- M_2 Mass of the crucible/filter with the specimen before matrix removal (g)
- M_3 Mass of the crucible/filter with the specimen after matrix removal (g)
- ρ_f Density of fiber (g/cm³)

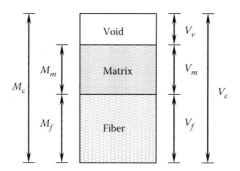

FIGURE 11.4 Idealized composite element.

ρ_m Density of matrix (g/cm³)
ρ_c Density of composite (g/cm³)

11.5.1.3 Void Content

The void content of a composite material can be determined by using the method given by ASTM D2734 [32]. Owing to the presence of voids, the actual density of a composite material can be expected to be less than the theoretical density. We can obtain a relation between the theoretical and measured composite densities and void volume fraction by considering an idealized composite element as shown in Figure 11.4.

By definition, the theoretical and measured densities of composites are given as follows:

Theoretical composite density

$$\rho_{ct} = \frac{M_c}{v_c - v_v} \quad (11.16)$$

Measured composite density

$$\rho_{cm} = \frac{M_c}{v_c} \quad (11.17)$$

The difference between the reciprocals of the two forms of composite density gives us

$$\frac{1}{\rho_{cm}} - \frac{1}{\rho_{ct}} = \frac{v_v}{M_c} \quad (11.18)$$

Noting that $M_c = v_c \rho_{cm}$, it is readily found that void volume fraction is given by

$$V_v \equiv \frac{v_v}{v_c} = 1 - \frac{\rho_{cm}}{\rho_{ct}} \quad (11.19)$$

where

ρ_{ct} Theoretical density of the composite material (g/cm³)
ρ_{cm} Actual measured density of the composite material (g/cm³)

The theoretical density of the composite material is calculated by using the densities of the fiber and matrix and their mass fractions, as follows:

$$\rho_{ct} = 1 / \left(\frac{W_f}{\rho_f} + \frac{W_m}{\rho_m} \right) \quad (11.20)$$

or

$$\rho_{ct} = \rho_f V_f + \rho_m V_m \qquad (11.21)$$

11.5.1.4 Glass Transition Temperature

The principles and methodology of T_g measurement for a cured composite material are similar to those for cast resin (refer Section 11.4.1.3).

11.5.2 Tests for Mechanical Properties of a Lamina

Lamina mechanical properties are critical parameters for the design and analysis of a composite product. These parameters can be grouped as follows:

- Tensile properties
- Compressive properties
- Shear properties
- Flexural properties

Out of the four groups mentioned above, the first three are fundamental in nature, whereas the fourth is a mixture of the first three. Table 11.5 summarizes the common test methods for the evaluation of mechanical properties of a lamina [33].

TABLE 11.5
Standard Test Methods for Mechanical Parameters of a Lamina

Sl. No.	Parameter	Principle of Testing	Common Standard(s)
1	Tensile properties Tensile moduli Tensile strengths Ultimate tensile strains Major Poisson's ratio in tension in 1–2 plane	Tensile loading of test specimen till failure Modulus is obtained from the slope of the stress–strain plot Strength and ultimate strain are obtained from the stress and strain at failure Poisson's ratio is obtained as the ratio of axial and lateral strains	ASTM D3039, BS 2782, ISO 527, and CRAG
2	Compressive properties Compressive moduli Compressive strengths Ultimate compressive strains Major Poisson's ratio in compression in 1–2 plane	Compressive loading of test specimen till failure Modulus is obtained from the slope of the stress–strain plot Strength and ultimate strain are obtained from the stress and strain at failure Poisson's ratio is obtained as the ratio of axial and lateral strains	ASTM D3410, D695, D6641, and D5467
3	Shear properties In-plane shear modulus in the 1–2 plane Interlaminar shear moduli In-plane shear strength in the 1–2 plane Interlaminar shear strengths	Tensile, compressive, or bending loading of test specimen, in pure or predominantly shear stress, till failure Moduli are obtained from the slope of the stress–strain plot Strength and ultimate strain are obtained from the stress and strain at failure	ASTM D3518, D5379, D4255, and D2344
4	Flexural properties Flexural modulus Flexural strength	Bending loading of test specimen, in pure or predominantly bending stress, till failure Moduli are obtained from the slope of the stress–strain plot Strength is obtained from the stress at failure	ASTM D790 and D6272

11.5.2.1 Tension Testing

The objective of tension testing is to determine (i) tensile modulus, (ii) tensile strength, (iii) ultimate tensile strain, and (iv) Poisson's ratio [34]. These are primary mechanical properties that are required in the design of most structures. At the lamina level, tensile properties are evaluated commonly in the 0° and 90° directions for both unidirectional as well as bidirectional composites.

11.5.2.1.1 Test Specimen and Specimen Preparation

The basic concept of tension testing is to pull a suitable test specimen and deduce the required material parameters from the specimen geometry and loading data. It is necessary to ensure that the specimen fails in the gauge section under uniaxial stress state, and to achieve this, a number of standard specimens have been proposed. For isotropic materials and material with low orthotropy, dog-bone-shaped or dumb-bell-shaped specimens, that is, specimens with tapering width given by ASTM D 638, can be used [25]. Load application can be made with the help of a pin through a hole at either end of the specimen. In such a specimen, highly localized stresses may develop around the holes and shear, and bearing or tearing failure may occur (Figure 11.5). In the case of specimens with tapering width, the high orthotropy of unidirectional composites may result in shear failure of the specimen near the grips parallel to the fiber direction, effectively reducing the specimen to a straight-edged one. Thus, it is more common to use straight-edged specimens with some gripping arrangements at the two ends.

A number of standard specimens and test procedures are in vogue; they include, among others, ASTM D3039, BS2782, ISO 527, and CRAG. Of these, ASTM D 3039 [35] is probably the most common and widely accepted. The standard specimen as per ASTM (Figure 11.6) is of simple rectangular configuration. Depending on the type of

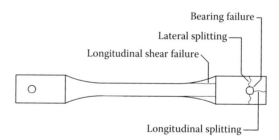

FIGURE 11.5 Dog-bone-shaped tensile test specimen with end holes and possible unwanted modes of failure.

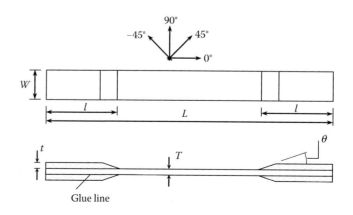

FIGURE 11.6 Tension test specimen. (Adapted from ASTM Standard D3039/D3039M-08, *Standard Test Method for Tensile Properties of Polymer Matrix Composite Materials*, ASTM International, 2008.)

TABLE 11.6
Standard Tensile Test Specimen as Recommended by ASTM D3039

Type of Material	Unidirectional (0°)	Unidirectional (90°)	Balanced and Symmetric	Random and Discontinuous
Overall length, L (mm)	250	175	250	250
Width, W (mm)	15	25	25	25
Thickness, T (mm)	1.0	2.0	2.5	2.5
Tab length, l (mm)	56	25	a	a
Tab thickness, t (mm)	1.5	1.5	a	a
Tab bevel angle, θ (°)	7–90	90	a	a

[a] Emery cloths are used as tabs.

material, end tabs may or may not be provided. The recommended dimensions of the specimen are described in Table 11.6. Note that specimen dimensions vary depending on material types, laminate fabrication method used, etc. For example, the specimen thickness would depend on the actual thickness of individual plies.

During testing, load is introduced to the specimen by means of a shearing action between the grips and the specimen, wherein the normal gripping force times the friction coefficient at the specimen and grip surfaces is the tensile force on the specimen. End tabs are used for efficient load transfer from the clamping grips to the specimen. First, they protect the outer fibers of the specimen from the high direct clamping forces. Second, they provide higher friction surfaces and thereby reduce the clamping forces. [±45°], [0/90°], or bidirectional fabric–reinforced E-glass/epoxy composite is commonly used as the tab material. Owing to the abrupt change in thickness, a tabbed specimen is associated with high stress concentration around the tip of the tab and unwanted failure may occur at the tabbed portion. A bevel angle in the tab is recommended by ASTM D3039 for unidirectional composites at 0°. Emery cloth may be used as tab for composites with fabric or mat reinforcements.

Typically, the laminate for a unidirectional composite specimen is prepared by filament winding around a flat-sided mandrel. The required laminate thickness can be directly achieved, and for proper control of thickness, flat metallic plates with spacers are used. Alternatively, unidirectional plies can be made by filament winding and the required numbers of plies are laid-up in a mold. For fabric and mat composites, too, commonly a mold is used for laying up the fabrics/mats and wetting.

Four continuous strips of tab material are bonded using suitable epoxy-based adhesive to the laminate. Proper surface preparation before bonding is essential. Also, it is necessary to use a suitable jig for positive positioning of the tab strips on the laminate. Finally, the specimens are machined out from the laminate. Proper alignment of the fibers is extremely important for obtaining correct tensile properties. A misalignment of fibers by 1° w.r.t. specimen axis can result in the reduction of the tensile strength of a unidirectional composite by about 30%.

11.5.2.1.2 Test Procedure and Data Reduction

It is important, prior to mounting the specimen on the machine, to accurately measure the cross-sectional area of the specimen in the gauge length. Usually, three readings are taken and the average is used for subsequent calculation of stress. The specimen is then mounted on the testing machine by placing the two ends in the grips of the machine. Either an extensometer or a strain gauge is used for recording the specimen elongation or strain. A biaxial extensometer is mounted on the gauge section of the specimen for recording axial and lateral displacements. Alternatively, 0°/90° strain gauges are bonded back to back on the specimen for directly recording the strains. The back-to-back bonding of two strain gauges helps in checking the presence of any bending in the specimen. When Poisson's ratio is not intended to be determined, the uniaxial extensometer or longitudinal strain gauge is sufficient.

Utmost care must be taken to align the specimen axis in the loading direction. Grips are tightened and the specimen is loaded either at constant strain rate or at constant cross-head movement. The standard strain rate is 0.01/min and the standard cross-head movement is 2 mm/min. All the specimens are loaded till failure and a sufficient number of data points are recorded so as to obtain an adequate representation of the stress–strain behavior of the material. The failure mode of each specimen must be studied carefully and noted; results with failure outside the gauge section should be rejected.

Tensile strength can be calculated from the ultimate load and the average cross-sectional area, as follows:

$$X^T = \frac{(P)_{ult}}{A} \quad (11.22)$$

where
- X^T Tensile strength of composite (MPa)
- $(P)_{ult}$ Ultimate tensile load (N)
- A Average cross-sectional area (mm²)

For calculation of the modulus, stress–strain curve is required. From each data point, stress is readily calculated by dividing the applied load by the average area of cross section, that is,

$$\sigma = \frac{P}{A} \quad (11.23)$$

where
- σ Axial tensile stress at any data point (MPa)
- P Tensile load at any data point (N)
- A Average cross-sectional area (mm²)

On the other hand, strain at each data point is calculated from the extensometer displacement as follows:

$$\varepsilon_a = \frac{\delta_a}{L_g} \quad \text{and} \quad \varepsilon_l = \frac{\delta_l}{L_g} \quad (11.24)$$

where
- ε_a Axial tensile strain at any data point
- ε_l Lateral tensile strain at the data point
- δ_a Axial extensometer displacement at the data point (mm)
- δ_l Lateral extensometer displacement at the data point (mm)
- L_g Extensometer gauge length (mm)

When strain gauges are used, strains are obtained directly from the strain gauge readings. The schematic representation of a typical stress–strain curve is given in Figure 11.7. Unidirectional composites exhibit largely linear behavior. However, at the start of the test, a certain amount of nonlinearity is usually observed. This nonlinearity is a local aberration that can be attributed to phenomena such as slippage and

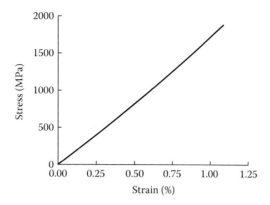

FIGURE 11.7 Schematic representation of typical tensile stress–strain curve.

initial settlement of the specimen, machine backlash, etc. In the modulus calculation process, this starting nonlinear portion of the curve is not considered. Modulus can be calculated as either tangent modulus or secant/chord modulus. The tangent modulus is calculated as the slope of the tangent to the initial part of the curve. It can also be calculated as the slope of the tangent to the curve at a specified strain level. The secant or chord modulus is calculated as the slope of the secant or chord between two points on the curve, typically at strain values of 0.0005 and 0.0025. Thus,

$$E^{(T)} = \frac{\Delta \sigma_a}{\Delta \varepsilon_a} \tag{11.25}$$

where
- $E^{(T)}$ Tensile modulus (GPa)
- $\Delta \sigma_a$ Stress difference on the tangent (MPa)
- $\Delta \varepsilon_a$ Corresponding axial strain difference on the tangent

and the Poisson's ratio is calculated as

$$\nu^{(T)} = -\frac{\Delta \varepsilon_l}{\Delta \varepsilon_a} \tag{11.26}$$

where
- $\nu^{(T)}$ Poisson's ratio in tension
- $\Delta \varepsilon_l$ Lateral strain difference corresponding to the longitudinal strain difference used in modulus calculation
- $\Delta \varepsilon_a$ Axial strain difference

Note: In the discussion above, we used the terminologies *axial* and *lateral* to mean along the lengthwise axis of the specimen and across, respectively. The specimen is used for the evaluation of tensile properties in both the longitudinal and transverse directions. The term "longitudinal" is used to describe the fiber direction in unidirectional composites and weft direction in fabrics. Similarly, the term "transverse" is used for across the fiber direction and warp direction.

11.5.2.1.3 NOL Ring Test

NOL ring is a simple specimen for the determination of unidirectional tensile strength. It is typically of diameter 150 mm, thickness 1.5 mm, and width 6–8 mm. The test fixture consists of either two or four steel segments. The segments are assembled to form a flat circular disk and the test specimen is placed around it. The segments are pulled in two opposite directions till the specimen breaks. This is a very simple and quick method, but the test process is generally associated with certain bending; the failure mode is not a true representation of tensile failure and it underestimates the strength of the material. Thus, these rings can be used for data comparison and in-process quality control purposes, but not for design data generation.

11.5.2.2 Compression Testing

The objective of compression testing is to determine (i) compressive modulus, (ii) compressive strength, (iii) ultimate compressive strain, and (iv) Poisson's ratio in compression.

In a compression test, a gradually increasing compressive force is applied on the test specimen till it fails and the response of the specimen versus force is recorded. The force versus strain (or force versus displacement) data, ultimate force at failure, and specimen geometry are utilized to calculate the required compressive properties. In a composite material, the response to compressive force and the final failure mechanisms are rather complex, and for accurate estimate of the material properties, it is necessary that the specimen experiences uniform compression and it fails in an acceptable failure mode. To ensure this, suitable test specimen and test fixtures are essential.

11.5.2.2.1 Failure Modes

There are two broad failure modes that are associated with a material under compressive loads—material failure and global buckling. The material failure of a composite material depends on the loading direction w.r.t. fiber direction (refer to Sections 3.5.2.2 and 3.5.2.4 in Chapter 3). Under longitudinal compression, that is, when the force is parallel to the fiber direction, the following failure modes are possible:

- Fiber microbuckling in extension
- Fiber microbuckling in shear
- Transverse tensile failure of matrix and fiber/matrix interface
- Shear failure by fiber kinking

Under transverse compression, that is, when the force is perpendicular to the fibers, the possible failure modes are as follows:

- Compression failure of matrix
- Shear failure of matrix
- Fiber crushing
- Fiber-to-matrix interface failure

For fabric and mat composites under compression, the failure modes are combinations of the above and more complex.

Failure modes in a test specimen due to longitudinal compression are axial splitting, shear failure, and kink zone failure. On the other hand, transverse compression can result in shear failure of the specimen along an inclined plane. These failure modes (Figure 11.8a–e) are generally accepted as true representation of compression failure of a composite material.

The other failure mode, viz. global buckling of the specimen (Figure 11.8f), is unacceptable and it has to be avoided, for which adequate restraining of the specimen inside the test fixture is required. Another way to minimize the possibility of global buckling is to reduce the gauge length (unsupported length) of the specimen. Over-restraining and too short specimen gauge length should be avoided as the former may result in an

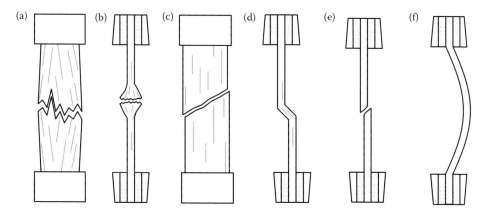

FIGURE 11.8 Typical failure modes in a compression test specimen. (a) Widthwise axial splitting. (b) Thickness-wise axial splitting. (c) Widthwise shear failure. (d) Kinking zone formation. (e) Thickness-wise shear failure. (f) Global buckling.

overestimate of the compressive properties and the latter may result in unwanted premature failure due to high stress concentration.

11.5.2.2.2 Test Specimen and Test Fixture

As shown in Figure 11.9, there are three mechanisms by which compressive force is applied on an axial compression test specimen [36]:

- By shear loading
- By direct loading at the end
- By a combination of shear and direct loading

ASTM D3410, D695, and D6641 present three methods that utilize, respectively, the above force introduction mechanisms [26,37,38]. In addition to these, beam bending is also employed for load transfer in a sandwich beam compression test specimen. ASTM D5467 presents a compression test method by sandwich beam bending [39]. There are several other compression test methods that have been proposed by various standard organizations and research institutes. Of these, the method suggested by ASTM D3410, probably, is the most popular; given below is a brief description of this method.

The standard specimen as per ASTM D3410 is a simple flat strip of constant rectangular cross section with or without tabs (Figure 11.10). The underlying idea in deciding on the specimen width, thickness, and gauge length is to prevent Euler buckling in the

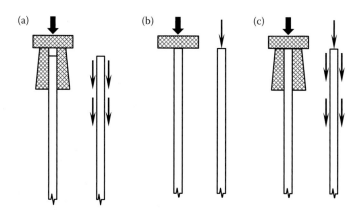

FIGURE 11.9 Load transfer mechanisms in a compression test. (a) Shear loading. (b) End loading. (c) Combination of shear loading and end loading.

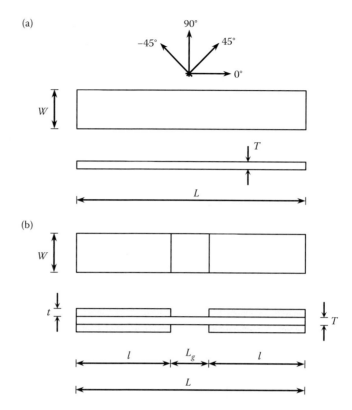

FIGURE 11.10 Compression test specimen. (a) Specimen without tabs. (b) Specimen with tabs. (Adapted from ASTM standard D3410/D3410M-03 [37].)

specimen gauge length. Thus, these dimensions depend on the expected compressive strength and expected compressive modulus; more specifically, the lower the expected compressive strength and expected compressive modulus, the greater the required specimen thickness.

Another requirement in choosing the gauge length is to ensure stress decay for uniform uniaxial compression in the gauge section. In a shear-loaded compression specimen, the gauge length required to ensure uniform uniaxial compressive stress increases with increasing specimen thickness. The required gauge length also depends on the ratio of longitudinal modulus to shear modulus. Obviously, the choice of gauge length is a trade-off between buckling-free short length and uniform-compression long length. ASTM D3410 recommends, based on an assumption of pinned-end column buckling, the specimen dimensions as given in Table 11.7.

TABLE 11.7
Standard Compression Test Specimen as Recommended by ASTM D3410

Type of Material	Unidirectional (0°)	Unidirectional (90°)	Specially Orthotropic
Overall length, L (mm)	140–155	140–155	140–155
Gauge length, L_g (mm)	10–25	10–25	10–25
Width, W (mm)	10	25	25
Thickness, T (mm)	1.00–10.91	1.00–10.91	1.00–10.91
Tab length, l (mm)	65	65	65
Tab thickness, t (mm)	1.5	1.5	1.5

Note: The recommended specimen thickness is the minimum thickness that depends on the expected compression strength, expected longitudinal modulus, and gauge length.

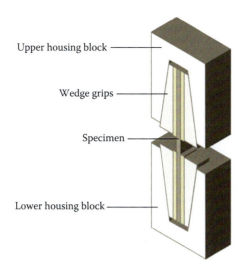

FIGURE 11.11 Simplified sectional schematic representation of shear loading compression test fixture (IITRI compression test fixture adopted by ASTM D3410.)

The shear loading test fixture was originally developed by Illinois Institute of Technology Research Institute (IITRI). It is a little complex and bulky (Figure 11.11); however, it is quite versatile as it can accommodate a wide range of specimen dimensions. It consists of an upper housing block and a lower housing block. Each of these blocks houses a pair of wedge grips, inside which the specimen is held. Alignment rods and ball bearings are used for aligning the two blocks so as to eliminate eccentric loading of the specimen.

11.5.2.2.3 Test Procedure and Data Reduction

Each test specimen is inspected and critical specimen dimensions are recorded. Strain gauges are bonded on both the sides of the specimen in the gauge section. Next, the test fixture is cleaned and properly lubricated, and the specimen is aligned in the fixture. Finally, the test fixture is mounted on the testing machine, the test is initiated, and the load is gradually increased till the specimen fails. Care must be taken to accept only those results that correspond to acceptable specimen failure modes.

Compressive strength can be calculated from the ultimate force and the average cross-sectional area, as follows:

$$X^C = \frac{(P)_{ult}}{A} \tag{11.27}$$

where
- X^C Compressive strength of composite (MPa)
- $(P)_{ult}$ Ultimate compressive force (N)
- A Average cross-sectional area (mm²)

Stress–strain curve is used for the calculation of the modulus. At each data point, stress is readily calculated by dividing the applied load by the average area of cross section, that is,

$$\sigma = \frac{P}{A} \tag{11.28}$$

where

- σ Axial compressive stress at any data point (MPa)
- P Compressive force at any data point (N)
- A Average cross-sectional area at the gauge section (mm²)

Axial and lateral strains are directly read from the two back-to-back strain gauges:

$$\varepsilon_a = \frac{\varepsilon_{a1} + \varepsilon_{a2}}{2} \quad \text{and} \quad \varepsilon_l = \frac{\varepsilon_{l1} + \varepsilon_{l2}}{2} \tag{11.29}$$

where

- ε_a Average axial compressive strain at any data point
- $\varepsilon_{a1}, \varepsilon_{a2}$ Gauge 1 and gauge 2 axial compressive strains, respectively, at the same data point
- ε_l Average lateral compressive strain at the data point
- $\varepsilon_{l1}, \varepsilon_{l2}$ Gauge 1 and gauge 2 lateral compressive strains, respectively, at the same data point

The schematic representation of a typical stress–strain curve is given in Figure 11.12. Compressive modulus can be calculated as either a tangent modulus or a secant/chord modulus. The tangent modulus is calculated as the slope of the tangent to the initial part of the curve. It can also be calculated as the slope of the tangent to the curve at a specified strain level. The secant or chord modulus is calculated as the slope of the secant or chord between two points on the curve. Thus,

$$E^C = \frac{\Delta \sigma_a}{\Delta \varepsilon_a} \tag{11.30}$$

where

- E^C Compressive modulus (GPa)
- $\Delta \sigma_a$ Axial stress difference on the tangent (MPa)
- $\Delta \varepsilon_a$ Corresponding axial strain difference on the tangent

and the compressive Poisson's ratio is calculated as

$$\nu^{(C)} = -\frac{\Delta \varepsilon_l}{\Delta \varepsilon_a} \tag{11.31}$$

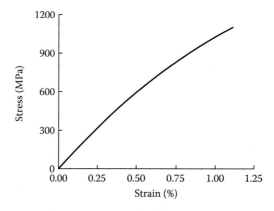

FIGURE 11.12 Schematic representation of a typical compressive stress–strain curve.

where
- $\nu^{(C)}$ Compressive Poisson's ratio
- $\Delta\varepsilon_l$ Lateral strain difference corresponding to the axial strain difference used in modulus calculation
- $\Delta\varepsilon_a$ Axial strain difference

Notes: Refer to the note at the end of Section 11.5.2.1 for a discussion on the usage of the terms *axial, lateral, longitudinal,* and *transverse*.

In this section, we provided a brief discussion on the IITRI compression test method that has been adopted by ASTM D3410. Other common compression test methods include the end loading method as per ASTM D695 and the combined loading compression method as per ASTM D6641.

11.5.2.3 Shear Testing

The objective of shear testing is to determine (i) shear modulus and (ii) shear strength. A number of test methods are available for the evaluation of shear properties of a composite material [40]. These methods can be broadly categorized into two categories—in-plane shear tests and interlaminar shear tests. (In the local 1–2 coordinate system, the in-plane shear stress, strength, and modulus are τ_{12}, S_{12}, and G_{12}, whereas interlaminar parameters are τ_{13}, τ_{23}, S_{13}, S_{23}, G_{13}, and G_{23}. Refer to Chapter 4 for definitions of in-plane and interlaminar shear parameters. Also, note that the terms S_{12}, S_{13}, and S_{23} are used here to denote shear strength and not compliance matrix elements.) In this section, we discuss the following common shear test methods:

- Uniaxial tension test of ±45° laminate
- V-notch beam shear test
- Rail shear test
- Short beam shear test

Out of the above four test methods, the first three are for in-plane shear properties, whereas the fourth is for interlaminar shear properties.

11.5.2.3.1 Uniaxial Tension Test of ±45° Laminate (ASTM D3518)

This method is used for the evaluation of in-plane compressive modulus and strength of composite materials [41]. The test specimen is a standard tension test specimen with symmetric ply sequence at ±45°.

The stress state in a ±45° symmetric laminate (Figure 11.13) under uniaxial tensile force in the x-direction is such that the local in-plane normal stresses σ_{11} and σ_{22} are

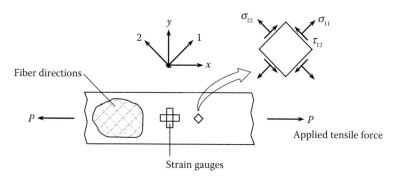

FIGURE 11.13 State of stress in a uniaxial tension test of ±45° specimen. (Adapted with permission from D. F. Adams, L. A. Carlsson, and R. B. Pipes, *Experimental Characterization of Advanced Composite Materials*, CRC Press, Boca Raton, FL, 2003.)

functions of uniaxial normal stress σ_{xx} and shear stress τ_{xy}, whereas the local in-plane shear stress τ_{12} is a function of only σ_{xx}. Thus,

$$\sigma_{11} = \frac{\sigma_{xx}}{2} + \tau_{xy} \qquad (11.32)$$

$$\sigma_{22} = \frac{\sigma_{xx}}{2} - \tau_{xy} \qquad (11.33)$$

$$\tau_{12} = \pm \frac{\sigma_{xx}}{2} \qquad (11.34)$$

where

$\sigma_{11}, \sigma_{22}, \tau_{12}$ In-plane normal and shear stresses in the local or material coordinates
σ_{xx}, τ_{xy} Applied in-plane normal and shear stresses

On the other hand, the in-plane normal and shear strains in the local coordinates are given by

$$\varepsilon_{11} = \frac{\varepsilon_{xx} + \varepsilon_{yy}}{2} \qquad (11.35)$$

$$\varepsilon_{22} = \frac{\varepsilon_{xx} + \varepsilon_{yy}}{2} \qquad (11.36)$$

$$\gamma_{12} = \varepsilon_{xx} - \varepsilon_{yy} \qquad (11.37)$$

where

$\varepsilon_{11}, \varepsilon_{22}, \gamma_{12}$ In-plane normal and shear strains in the local or material coordinates
$\varepsilon_{xx}, \varepsilon_{yy}$ Axial and lateral in-plane normal strains

Note that ε_{yy} is negative.

The standard tension test is performed on the specimen in a tensile testing machine and the shear strength is readily obtained from Equation 11.31. Thus,

$$S_{12} = \frac{(P)_{ult}}{2A} \qquad (11.38)$$

where $(P)_{ult}$ is the maximum tensile force on the specimen before ultimate failure. The ultimate shear strain is obtained from Equation 11.34, in which the normal strains correspond to the ultimate failure. However, identifying the failure point, owing to the typically nonlinear nature of shear stress–shear strain curve (Figure 11.14), is not a straightforward task. ASTM D3518 recommends that, if ultimate failure does not occur within 5% shear strain, P may be taken as the tensile force corresponding to 5% shear strain. The in-plane shear modulus is calculated as

$$G_{12} = \frac{\Delta \tau_{12}}{\Delta \gamma_{12}} \qquad (11.39)$$

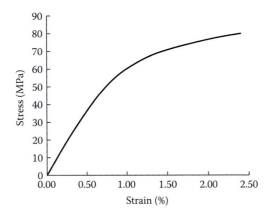

FIGURE 11.14 Schematic representation of typical shear stress–strain curve obtained from a uni-axial tension test of ±45° specimen.

where the in-plane shear stress difference $\Delta\tau_{12}$ and the corresponding in-plane shear strain difference $\Delta\gamma_{12}$ are taken from the initial linear portion of the stress–strain curve.

11.5.2.3.2 V-Notch Beam Shear Test (ASTM D 5379)

This shear test method [27] was originally proposed by Iosipescu and the same has been subsequently modified by Wyoming University and adapted by ASTM. This is used for the evaluation of in-plane as well as interlaminar shear properties of composite materials.

The schematics of the specimen and the test fixture are shown in Figure 11.15. The specimen is a beam of nominal length and height of 76 and 20 mm, respectively. The

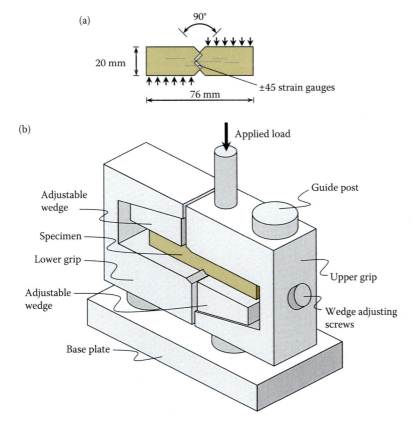

FIGURE 11.15 Schematic representation of V-notch beam shear test. (a) Specimen. (b) Test fixture. (Adapted from ASTM Standard D5379/D5379M-12, *Standard Test Method for Shear Properties of Composite Materials by the V-Notched Beam Method*, ASTM International, 2012.)

width is chosen as required. The beam is provided with two notches in the midspan as shown in the figure. Two two-element strain rosettes, one on either side of the specimen, are bonded with the elements being aligned in the ±45° directions. The fixture consists of an upper grip and a lower grip. The upper grip is attached to the cross-head of the testing machine. The specimen is inserted into the fixture taking care to align the V-notch in the line of load application. The upper grip is driven downward by the cross-head and the relative displacement between the two grips results in the development of a shear plane between the notches. The idealized free-body diagram, bending moment diagram, and shear force diagram (Figure 11.16) show the presence of pure shear stresses at the gauge section. Finite element analysis of the test specimen also shows that a state of uniform shear stress exists at the center of the specimen except around the roots of the notches.

The average shear stress and shear strain across the notched section are given by

$$\tau = \frac{P}{A} \tag{11.40}$$

$$\gamma = \varepsilon_{(45)} - \varepsilon_{(-45)} \tag{11.41}$$

where

τ, γ Shear stress and strain, respectively, at any data point
P Applied force at any data point
A Area of cross section between the notches
$\varepsilon_{(45)}, \varepsilon_{(-45)}$ Normal strains measured in the +45° and −45° directions, respectively, at any data point

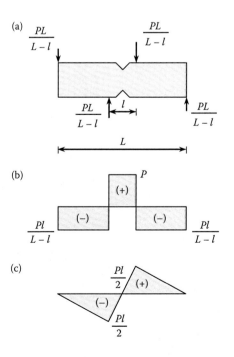

FIGURE 11.16 Idealization of V-notch beam shear test specimen under load. (a) Free body diagram. (b) Shear force diagram. (c) Bending moment diagram. (Adapted from ASTM Standard D5379/D5379M-12, *Standard Test Method for Shear Properties of Composite Materials by the V-Notched Beam Method*, ASTM International, 2012.)

Then, shear strength and modulus are obtained as

$$S = \frac{(P)_{ult}}{A} \tag{11.42}$$

$$G = \frac{\Delta \tau}{\Delta \gamma} \tag{11.43}$$

where
- S, G Shear strength and modulus, respectively
- $(P)_{ult}$ Ultimate force at failure
- A Area of cross section between the notches
- $\Delta \tau$ Difference in shear stresses between points on the shear stress–shear strain curve
- $\Delta \gamma$ Difference in shear strains between the same points on the shear stress–shear strain curve

Typically, the two points for modulus calculation are taken from the initial straight portion of the stress–strain curve. Further, note that we have not used any subscripts that indicate whether the shear properties are in-plane or interlaminar. The shear properties in the 1–2 plane are in-plane, those in the 1–3 and 2–3 planes are interlaminar. Thus, it is important to cut the specimen by suitable orientation of the fiber direction and ply plane w.r.t. the notch plane (Figure 11.17).

11.5.2.3.3 Rail Shear Test (ASTM D4255)

There are two rail shear test methods—two rail shear test method and three rail shear test method [42]. Both of these are used for the evaluation of in-plane shear properties.

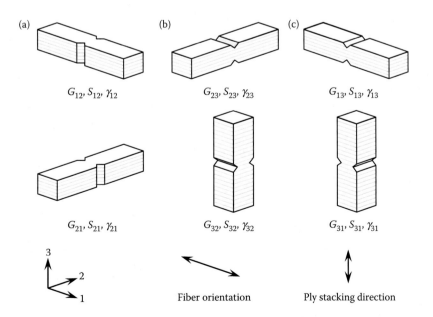

FIGURE 11.17 Fiber orientation in V-notch beam shear test specimen. (a) In-plane shear properties in 1–2 plane. (b) Interlaminar shear properties in 2–3 plane. (c) Interlaminar shear properties in 1–3 plane. (Note that the 1 direction is along the fiber direction and the 1–2 plane is parallel to the plies.) (Adapted from ASTM Standard D5379/D5379M-12, *Standard Test Method for Shear Properties of Composite Materials by the V-Notched Beam Method*, ASTM International, 2012.)

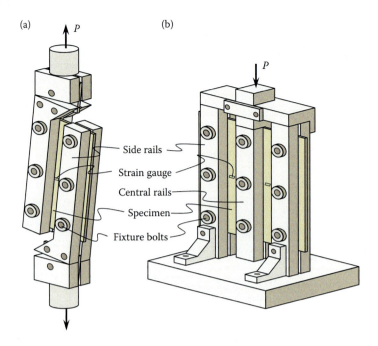

FIGURE 11.18 Schematic representation of rail shear test. (a) Two rail shear test fixture with specimen under tension. (b) Three rail shear test fixture with specimen under compression. (Adapted from ASTM Standard D4255/D4255M-01, *Standard Test Method for In-Plane Shear Properties of Polymer Matrix Composite Materials by the Rail Shear Method*, ASTM International, 2007.)

The schematics of the specimens and the test fixtures for the two rail and three rail shear tests are shown in Figure 11.18. In each case, the specimen is a flat laminate. The nominal size of the specimen is 152 mm × 76 mm for the two rail test, and 152 mm × 137 mm for the three rail test. The thickness of the specimen is chosen as appropriate.

In the two rail shear tests, the specimen is clamped between two pairs of clamping rails along the lengthwise edges by means of three bolts on each edge. The holes provided are clearance holes so as to ensure that load transfer from the rails to the specimen is by frictional force between them. Both tensile force as well as compressive force may be applied; however, the tensile version is more common. On the application of an axial force on the test fixture, an in-plane shear stress is induced in the specimen and strain in the specimen is continuously recorded. (A strain gauge is bonded either at 45° or at −45° for strain measurement.) Then, shear strength and modulus are obtained as

$$S = \frac{(P)_{ult}}{A} = \frac{(P)_{ult}}{lt} \tag{11.44}$$

$$G = \frac{\Delta \tau}{\Delta \gamma} = \frac{\Delta P}{2lt\Delta \varepsilon_{45}} \tag{11.45}$$

where
- S, G In-plane shear strength and modulus, respectively
- $(P)_{ult}$ Ultimate force at failure
- A Area of lengthwise cross section
- $\Delta \tau$ Difference in in-plane shear stresses between points on the shear stress–shear strain curve
- $\Delta \gamma$ Difference in in-plane shear strains between the same points on the shear stress–shear strain curve (note: $\gamma = 2\varepsilon_{45}$)

In the three rail shear test, the specimen is clamped between two pairs of clamping rails, which are attached to a base plate, along the lengthwise edges by means of three bolts on each edge. In addition, one more pair of rails is clamped to the specimen in the middle. The middle rails are guided through a slot in the top plate. Usually, a compressive force is applied on the middle rail that induces a pure in-plane shear stress in the specimen. The shear strength and modulus are obtained as

$$S = \frac{(P)_{ult}}{2A} = \frac{(P)_{ult}}{2lt} \quad (11.46)$$

$$G = \frac{\Delta \tau}{\Delta \gamma} = \frac{\Delta P}{4lt \Delta \varepsilon_{45}} \quad (11.47)$$

where the variables are as defined in case of two rail shear test.

11.5.2.3.4 Short Beam Shear Test (ASTM D2344)

The short beam shear test works on the three-point beam bending principles [43]. Typically, under three-point loading, a beam is supported at the two ends and loaded in the middle; both bending as well as shear stresses are generated in the beam. The bending stresses are the maximum at the top and bottom faces of the beam—compressive at the top face and tensile at the bottom under the loading point. For an elastic material, the bending stresses vary linearly through the thickness and, by definition, they change sign on the neutral plane. In other words, the bending stresses are zero on the neutral plane. On the other hand, the interlaminar shear stresses, which vary parabolically through the thickness, are zero at the top and bottom faces and maximum on the neutral plane. Thus, the neutral plane is in a state of pure shear.

The objective in a short beam shear test is to minimize the bending stresses by keeping the span-to-thickness ratio low such that the shear stresses reach their ultimate level before the bending tensile or compressive stresses, and shear failure takes place in the neutral plane. However, testing of the specimen is often associated with high stress concentration around the load application point and the supports, and much of the specimen is in a state of mixed stresses. As a result, failure of the specimen is not due to pure shear. In spite of such serious short comings, owing to its simplicity, this test is popular especially for quick QC check for material screening.

ASTM D2344 suggests two standard specimens for short beam shear test—flat beam and curved beam (Figure 11.19). The span-to-thickness ratio is restricted to 4:1. Other recommended dimensions are as follows:

Specimen thickness (minimum), $h = 2$ mm
Specimen length, $L = 6$ h
Specimen width, $b = 2$ h

The specimen is supported on 3-mm-diameter rollers and a central concentrated force is applied with the help of a 6-mm-diameter loading nose. Usually, load versus cross-head movement data and the final failure load are recorded. The interlaminar strength (rather the short beam strength) is calculated as

$$S_{13} = \frac{3(P)_{ult}}{4bh} \quad (11.48)$$

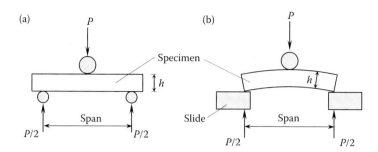

FIGURE 11.19 Short beam shear test. (a) Flat specimen. (b) Curved specimen. (Adapted from ASTM Standard D2344/D2344M-00, *Standard Test Method for Short-Beam Strength of Polymer Matrix Composite Materials and Their Laminates*, ASTM International, 2006.)

Note that strain gauges are not used, and failure strain and shear modulus are not evaluated.

11.5.2.4 Flexural Testing

The objective of flexural testing is to determine the flexural modulus and flexural strength. However, unlike tensile, compressive, and shear properties, flexural modulus and strength are not fundamental properties. Flexure or bending of a beam results in a stress state that is a mixture of tensile, compressive, and shear stresses. As a result, flexural modulus and strength cannot be directly used in design calculations. In spite of this, owing to the inherent simplicity of flexural test specimen and test procedure, these tests are frequently carried out especially for comparison of materials in quality control processes.

Two common flexural test methods are—three-point flexure test and four-point flexure test [44].

11.5.2.4.1 Three-Point Flexure Test (ASTM D790)

The specimen in a three-point flexure test is a flat beam of constant rectangular cross section [45]. The exact specimen dimensions are not critical as long as an appropriate support span-to-specimen thickness (L/h) ratio is chosen. The L/h ratio should be large enough to ensure bending failure of the specimen. The recommended support span-to-specimen thickness ratio depends on the ratio of the material tensile strength-to-interlaminar shear strength, and it ranges from 16:1 to 60:1. Typical L/h ratios for unidirectional 0° glass/epoxy composites and carbon/epoxy composites are 16:1 and 40:1, respectively. The nominal thickness is 2 mm for 0° composites and 4 mm for 90° composites. Similarly, a nominal width of 12.5 mm is common.

The specimen is mounted on a properly aligned test fixture in the testing machine and loaded with the help of a loading nose by moving the cross-head at a constant rate. The cross-head rate is so selected as to restrict the maximum strain rate on the outer surface to below 0.01 mm/mm. The common cross-head rate is 1–5 mm/min. The midspan deflection is measured by using a linear voltage differential transformer (LVDT) or an extensometer. Alternatively, the cross-head motion relative to the supports is used as the midspan deflection. Occasionally, a strain gauge is bonded on the specimen under the loading point to measure the strain directly.

In the three-point bending of a beam, the bending moment is the maximum at the midspan (Figure 11.20). The maximum bending stress, which occurs at the outer surface of the specimen, is given by

$$\sigma^{(b)} = \frac{3PL}{2bh^2} \tag{11.49}$$

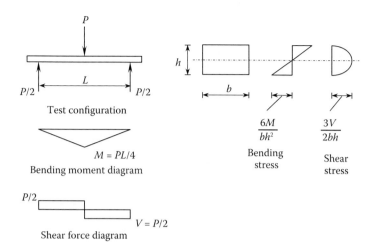

FIGURE 11.20 Three-point flexure test configuration and associated bending stress and shear stress diagrams.

where
- $\sigma^{(b)}$ Bending stress on the outer surface (MPa)
- P Applied force (N)
- L Span of the specimen (mm)
- b Width of the specimen (mm)
- h Thickness of the specimen (mm)

The bending strength is obtained by substituting P with the ultimate force $(P)_{ult}$. When deflection is measured, the bending strain is calculated as

$$\varepsilon^{(b)} = \frac{6\delta h}{L^2} \qquad (11.50)$$

where δ is the midspan deflection.

The flexural modulus is calculated from the initial linear portion of the bending stress–bending strain plot as follows:

$$E^{(b)} = \frac{L^3}{4bh^3}\left(\frac{\Delta P}{\Delta \delta}\right) \qquad (11.51)$$

where
- ΔP Difference in applied forces at two points in the initial linear portion of the bending stress–bending strain plot (N)
- $\Delta \delta$ Difference in midspan deflections at the corresponding points on the stress–strain plot (mm)
- L Span of the specimen (mm)
- b Width of the specimen (mm)
- h Thickness of the specimen (mm)

11.5.2.4.2 Four-Point Flexure Test (ASTM D6272)

The specimen in the four-point flexure test [46] is similar to that in the three-point bending specimen. The recommended support span-to-specimen thickness ratio is 16:1. The specimen is loaded by two loading nose cylinders. The load span can be either one-third of the support span or one-half (Figure 11.21).

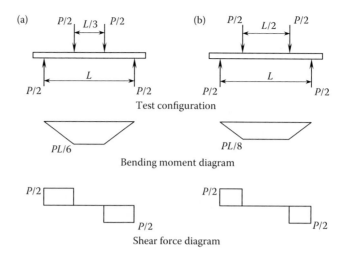

FIGURE 11.21 Four-point bending. (a) Load span is one-third of support span. (b) Load span is one half of support span.

In the four-point bending of a beam, the bending moment is the maximum and constant in the entire load span (Figure 11.21). The maximum bending stress, which occurs at the outer surface of the specimen, depends on the load span. When the load span is one-third of the support span, the maximum bending stress is given by

$$\sigma^{(b)} = \frac{PL}{bh^2} \tag{11.52}$$

and when the load span is one-half of the support span, the maximum bending stress is given by

$$\sigma^{(b)} = \frac{3PL}{4bh^2} \tag{11.53}$$

where the parameters on the right-hand side are as given before and shown in Figure 11.20.

When the load span is one-third of the support span, the maximum bending strain is calculated as

$$\varepsilon^{(b)} = 4.7 \frac{\delta h}{L^2} \tag{11.54}$$

and when the load span is one-half of the support span, the bending strain is calculated as

$$\varepsilon^{(b)} = 4.36 \frac{\delta h}{L^2} \tag{11.55}$$

where δ is the midspan deflection.

The flexural modulus is calculated from the initial linear portion of the bending stress–bending strain. When load span is one-third of the support span,

$$E^{(b)} = \frac{0.21 L^3}{bh^3} \left(\frac{\Delta P}{\Delta \delta} \right) \tag{11.56}$$

and when load span is one-half of the support span,

$$E^{(b)} = \frac{0.17L^3}{bh^3}\left(\frac{\Delta P}{\Delta \delta}\right) \quad (11.57)$$

11.5.2.5 Fracture Toughness Test

Laminated composite materials often contain zones where the laminae are not bonded. Such laminae separations are referred to as delaminations. Delaminations may develop either during processing of the laminate or during service life under load. The presence of delaminations adversely affects the performance of a laminate and it has been a widely studied subject. Fracture mechanics is the subject that deals with crack initiation and crack growth in a material.

As shown in Figure 11.22, there are three modes of crack propagation—Mode-I, Mode-II, and Mode-III [47]. Mode-I is the crack opening mode, in which the direction of crack growth is normal to the loading direction; the loading direction is normal to the plane of the crack and the separated parts move away from each other by a crack opening action. In Mode-II, crack growth is in line with the direction of the loads; the separated parts slide w.r.t. each other by a shearing action. Mode-III crack growth is by in-plane loads but the direction of loading is normal to the direction of crack growth; the separated parts slide away w.r.t. each other by a tearing action. In an isotropic material, the resistance to crack propagation is the minimum in the crack opening mode. Thus, Mode-I is the most common mode in an isotropic material such that, even if the loading direction is in the plane of the crack, the direction of crack propagation may reorient to Mode-I. On the other hand, in a laminated composite material, the matrix is the weak link and crack growth is invariably constrained between the laminae. As a result, all the three modes of crack propagation are seen in composites. In simple terms, the resistance to crack propagation is fracture toughness. There are several concepts that have been developed to characterize fracture toughness:

- Strain energy release rate
- Stress intensity factor
- J-integral
- Crack-tip opening displacement
- Strain energy density

Strain energy release rate, G, is a simple energy-based concept; it does not require complex mathematical treatment involving stress analysis at the crack tip and it is amenable to simple experimental evaluation. Consequently, out of all the above fracture toughness concepts, strain energy release rate is the most commonly used parameter for characterizing fracture toughness.

When a body containing a crack is subjected to external forces, a portion of the work done by the external forces is stored as strain energy and the rest drives the crack

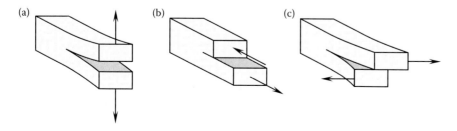

FIGURE 11.22 Fracture modes. (a) Mode-I (opening). (b) Mode-II (shearing). (c) Mode-III (tearing).

to grow. Strain energy release rate is a parameter that represents the energy available for crack growth. Fracture criterion based on strain energy release rate states that crack growth occurs when the energy available for crack growth is more than the work required to create a new crack area. Mathematically, strain energy release rate is defined as

$$G = \frac{d(W-U)}{dA} \quad (11.58)$$

where
- G Strain energy release rate (J/m²)
- W Work done by external forces (J)
- U Elastic strain energy stored in the body (J)
- dA Increment of new crack area (m²)

Now, denoting the energy required for the creation of new crack area by S, the condition for crack growth becomes

$$G \equiv \frac{d(W-U)}{dA} \geq \frac{dS}{dA} \quad (11.59)$$

At the threshold point, strain energy release rate is referred to as the critical strain energy release rate denoted by G_c, that is,

$$G_c = \frac{dS}{dA} \quad (11.60)$$

In other words, the strain energy release rate-based fracture criterion can be stated as

$$G \geq G_c \quad (11.61)$$

By adopting simple mechanics of materials approach, the strain energy release rate can be expressed as

$$G = \frac{P^2}{2} \frac{dC}{dA} \quad (11.62)$$

where
- P Applied external force (N)
- C Compliance (m/N) defined as $C = \delta/P$, δ being the displacement of the body under the loading point
- dA Increment of new crack area (m²)

G, in general, is dependent on crack size. However, at the critical point, it becomes independent of crack length and thus G_c is a material property. Note further that S is the energy required for creating new crack area; thus, dS/dA is resistance to crack growth. Thus, $G_c = dS/dA$ represents fracture toughness of the material.

A number of test methods have been developed for the evaluation of fracture toughness. Given in the following subsections is a brief description of two of these methods—double-cantilever beam (DCB) test for Mode-I and end-notched flexure (ENF) test for Mode-II.

11.5.2.5.1 DCB Test (ASTM D5528)

The DCB specimen (Figure 11.23) is employed to evaluate the Mode-I interlaminar fracture toughness, G_{Ic}, of continuous fiber–reinforced composite materials. The specimen is a simple beam of constant rectangular cross section with the following dimensions—length, $L = 125$ mm (minimum), width, $b = 20$–25 mm, and thickness, $h = 3$–5 mm. A precrack is created in the midplane of the specimen by inserting a nonadhesive layer such as a Teflon film. The precrack length, a_0, measured from the crack tip to the line of load application is typically 50 mm. Either piano hinges or end-blocks are used for applying tensile load as shown in Figure 11.23c and d. In a DCB test, load is applied typically at a cross-head rate of 0.5 mm/min, and applied load, P, crack opening displacement, δ, and crack length, a are recorded. While the crack opening displacement or the displacement of the loaded points is estimated from the cross-head motion, the crack length is monitored by using a traveling optical microscope and a cross-hair. At low loads, the displacement δ increases at constant initial crack length of a_0. On gradual increase in loads, at certain load level, the crack propagation starts. For the first 5 mm of crack growth, the loads associated with each 1 mm of crack growth are recorded and beyond this point loads at every 5 mm of crack growth are recorded. Generally, after 25 mm of crack growth, loading is stopped and the specimen is unloaded. Figure 11.24a shows a typical load–displacement plot.

Visual identification of the point of delamination initiation is often a tricky and operator-dependent task. Thus, to ensure repeatability, three methods have been suggested—onset of nonlinearity in the load–displacement curve, visual observation, and 5% offset intersection. The 5% offset intersection point is where the load–displacement curve is intersected by a line drawn from the origin and offset by 5% increase in compliance (δ/P).

Data from the load–displacement plot are used for the calculation of G_{Ic}. ASTM D5528 [48] gives three methods for calculation of G_{Ic}—modified beam theory (MBT), compliance calibration (CC), and modified compliance calibration (MCC). The MBT results in the most conservative estimate of G_{Ic}, according to which, the strain energy release rate is given by

$$G_I = \frac{3P\delta}{2ab} \qquad (11.63)$$

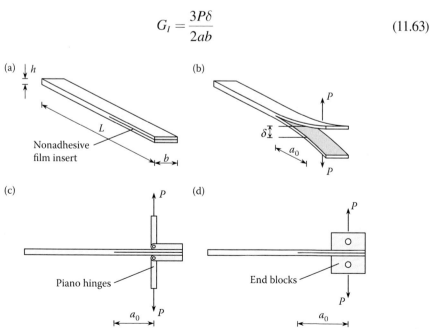

FIGURE 11.23 Double-cantilever beam test. (a) Specimen configuration. (b) Specimen when loaded. (c) Loading with piano hinges. (d) Loading with end-blocks. (Adapted from ASTM Standard 5528-01, *Standard Test Method for Mode I Interlaminar Fracture Toughness of Unidirectional Fiber-Reinforced Polymer Matrix Composites*, ASTM International, 2007.)

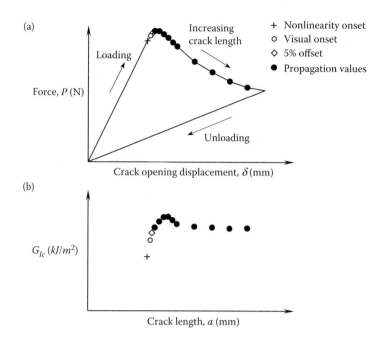

FIGURE 11.24 Schematic representation of DCB test data. (a) Load–displacement plot during crack growth. (b) R-curve.

where
- G_I Mode-I strain energy release rate (N)
- P Applied load (N)
- δ Crack opening displacement or displacement of loaded points (mm)
- a Crack length (mm)
- b Specimen width (mm)

The strain energy release rate as given by Equation 11.63 needs to be corrected for any possible rotation at the delamination front. Thus, the crack length a is increased by an additional increment Δ such that the critical strain energy release rate is given by

$$G_{Ic} = \frac{3P\delta}{2(a+\Delta)b} \qquad (11.64)$$

The value of Δ in Equation 11.64 is experimentally obtained by plotting $C^{1/3}$ ($C = \delta/P$) as a function of crack length a. The critical strain energy release rates are plotted as a function of crack length to prepare the resistance curve or the R-curve (Figure 11.24b). G_{Ic} at nonlinearity onset is typically the minimum and thus it is a conservative estimate of fracture toughness. The initiation of crack growth is normally associated with fiber bridging, which results in higher apparent G_{Ic}. As crack propagation continues, fiber bridging settles down and G_{Ic} becomes largely independent of crack length, a, and thus it is a material property, which in this context is the fracture toughness.

11.5.2.5.2 ENF Test

The ENF specimen (Figure 11.25) is used to evaluate Mode-II interlaminar fracture toughness, G_{IIc}, of continuous fiber–reinforced composite materials. The specimen is typically a 120-mm-long beam of constant rectangular cross section. Normal cross-sectional dimensions are—width, $b = 20$ mm, thickness, $2h = 3–5$ mm. The beam is tested in a three-point loading setup with a span of 100 mm. A precrack is created in

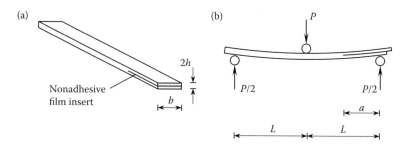

FIGURE 11.25 Schematic representation of ENF test. (a) Specimen configuration. (b) Specimen under load.

the midplane of the specimen by inserting a nonadhesive layer such as Teflon film. The precrack length, a_0, measured from the crack tip to the line of load application is typically 50 mm such that $a_0/L = 0.5$.

Under a three-point bend setup, shear loading is generated at the crack tip. Fracture test can be done with or without a precrack. When testing with a precrack, the specimen is loaded to create a stable crack at the end of the insert film and then the specimen is positioned in the test fixture to achieve an initial a_0/L ratio of 0.5. Typically, the specimen is loaded at a cross-head rate of 0.5–1.0 mm/min and center beam deflection and crack tip advancement with an LVDT and traveling microscope, respectively. The load–displacement plot, as in the case of the DCB test for Mode-I, allows one to identify the points corresponding to onset of nonlinearity, visual observation of crack initiation, 5% offset intersection, and propagation. Then, using beam theory principles, Mode-II fracture toughness is calculated as follows:

$$G_{IIc} = \frac{9P^2 C a^2}{2b(2L^3 + 3a^3)} \tag{11.65}$$

where C is the specimen compliance given by

$$C = \frac{2L^3 + 3a^3}{8E_{11} b h^3} \tag{11.66}$$

Combining Equations 11.65 and 11.66, and introducing a correction to the crack length, we get

$$G_{IIc} = \frac{9P^2 (a+\Delta)^2}{16 b^2 E_{11} h^3} \tag{11.67}$$

11.5.2.5.3 Other Fracture Test Methods

In addition to the DCB and the ENF test methods, there are a number of test methods for fracture toughness evaluation. Notable among these methods is the edge-cracked torsion (ECT) test for the evaluation of Mode-III fracture toughness. The ECT specimen is a flat rectangular panel with an edge crack in the middle plane. The specimen is loaded at one corner on the cracked edge and held at the three remaining corners. Under such a loading and supporting system, the panel undergoes twisting, and crack propagation takes place in Mode-III.

For mixed-mode fracture toughness evaluation, the mixed-mode bending (MMB) test is utilized. The MMB specimen is similar to those for DCB and ENF tests and the loading conditions are a combination of them. ASTM D6671 describes a standard test method based on this principle for the evaluation of mixed-mode fracture toughness for Mode-I and Mode-II [49].

11.5.2.6 Fatigue Testing

Often, the stresses or strains developed at a point in a mechanical structure are time dependent. Such variation occurs either due to variations in the nature and magnitudes of the applied loads or due to movement of the structure. A special case of time-dependent variations in stresses/strains is cyclic variation. When a material is subjected to cyclically repetitive stresses or strains, its strength and stiffness reduce significantly and premature failure takes place. The phenomenon of reduction in its strength and stiffness of a material, under cyclic loading, is referred to as fatigue.

Under fatigue loading, crack initiation takes place at a point of stress concentration at zones of imperfections. In metals, the initial imperfections can be simply a minute crack or some discontinuity. Under cyclic loading, initially, the crack propagates slowly; the crack size exceeds certain critical size and eventually fracture takes place, after a certain sufficient number of cycles of loading, at a load level that is lower than static failure loads. During the latter part, crack propagation is rather rapid and the final failure is nearly catastrophic without any prior sign.

On the other hand, in composite materials, there are many initial imperfection sites that include local broken fibers, matrix cracking, delamination, debond, void, etc. Crack propagation is coupled with fiber bridging and it takes place along the paths of minimum resistance—typically matrix and fiber–matrix interface. The damage mechanism is complex. However, final fatigue failure is rather gradual; it provides sufficient prior signal before failure and results in relatively longer fatigue life and higher fatigue loading.

Typical fatigue loading applied in a test is schematically shown in Figure 11.26. The cyclic stress mode can be sinusoidal, triangular, or of other appropriate shape about a mean stress [50]. The difference between the maximum and the minimum stresses is referred to as the range. Another characteristic parameter is the load ratio, R, defined as the ratio of the minimum stress to the maximum stress, that is, $R = S_{min}/S_{max}$. The mean stress can be either tensile or compressive. Also, depending on the mean stress and the range, three combinations of maximum and minimum stresses can be identified. Thus, we have three broad types of cyclic loading—tension–tension, compression–compression, and tension–compression. Tests have been devised for the determination

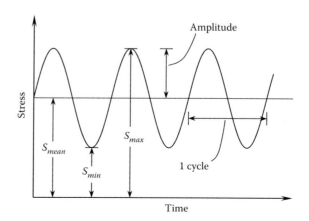

FIGURE 11.26 Schematic representation of a typical applied fatigue loading. (Adapted with permission from P. T. Curtis, Fatigue, *Mechanical Testing of Advanced Fiber Composites* (J. M. Hodgekinson, ed.), Woodhead Publishing, Cambridge, 2000, pp. 248–268.)

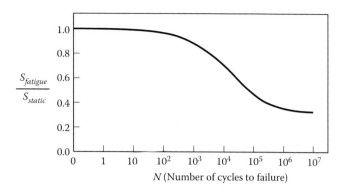

FIGURE 11.27 Schematic representation of S–N curve.

of fatigue characteristics in all these three modes. However, tension–tension fatigue testing is the most common.

Fatigue characteristics are usually expressed in terms of S–N curves, in which the number of cycles is plotted on a logarithmic scale on the x-axis and fatigue strength (or ratio of fatigue strength to static strength) is plotted on the y-axis (Figure 11.27). Note that the shape of the S–N curve depends on factors such as the material type, ply sequence, load ratio, etc. In some cases, it is linear; whereas in other cases, it is curvilinear. Clearly, fatigue strength is dependent on fatigue life or the number of cycles to failure. For larger applied fatigue stress, fatigue life is short and vice versa. Fatigue characteristics are also expressed in terms of failure strains as ε–N curves.

ASTM D3479 describes a tension–tension fatigue testing of PMC material [51]. The specimen is the same as in the static tensile test method given by ASTM D3039. The specimen is inserted and grabbed in the grips of the testing machine. The testing machine essentially has a movable head and a stationary head. The movable head can be moved w.r.t. the stationary head at controlled velocity so as to apply cyclic load on the specimen. Two approaches are in vogue—stress loading and strain loading. In the case of stress loading, the applied load is in terms of stress, and in the case of strain loading, it is strain. The maximum and minimum load levels, that is, S_{max} and S_{min} or ε_{max} and ε_{min} are selected and corresponding load ratio is calculated and reported. The frequency of cyclic load should be so chosen as to minimize localized heating of the specimen. In general, the frequency of the order of 10 Hz is chosen for specimens with predominantly unidirectional fibers.

11.5.3 Note on Tests for Laminate Properties

A single lamina is too thin to be of any practical use and the evaluation of lamina properties involves testing of specimens that are invariably made from laminates and not from laminae. For the evaluation of lamina mechanical properties, with the exception of in-plane shear properties by uniaxial tension test of ±45° specimen, laminates are made by orienting all the laminae in the same direction. Usually, the same laminates are also used for cutting specimens for nonmechanical properties such as density, fiber volume fraction, etc. In general, lamina mechanical and nonmechanical properties are sufficient for design calculations.

However, sometimes, mechanical properties are also evaluated at the laminate level. Laminates for specimen preparation are made with the desired ply sequence. Nonmechanical properties are often required at the laminate level from a quality control point of view; for example, to verify the fiber volume fraction in the cured component or degree of cure. In such cases, samples are cut either from the excess portions of the component or from representative laminates. The general principles of specimen preparation, test procedure, and data reduction remain the same as in the tests for lamina properties.

11.6 TESTS FOR ELEMENT-LEVEL PROPERTIES

The design of a composite structure often demands data that cannot be obtained from tests at lamina or laminate levels alone. Lamina- and laminate-level tests are relatively simple and they are carried out to determine specific material properties that correspond to specific failure modes. In the real design world, one comes across many cases of failure modes not addressed in simple coupon tests at lamina or laminate levels. Tests carried out to characterize material behavior in such cases are referred to as element-level tests [52]. For example, structural elements such as panels with cutouts, bolted joints, adhesively bonded joints, etc. can be designed with data that are obtained from tests at element level. Several failure modes may be present in an element-level test and, typically, the final failure mode is associated with the weakest path.

Standard structural elements are available for many cases. Several of these standard structural element tests are pretty similar to those at the coupon tests at lamina- or laminate-level. The primary difference lies in the fact that the element-level tests are associated with some form of discontinuities. For example, in a simple tension test as per ASTM D3039, one is concerned with the evaluation of tensile properties of a lamina or a flat laminate. On the other hand, in the case of a tension-loaded structural element testing, one would be concerned with the material behavior associated with, possibly, some holes, notches, ply drop-offs, etc. Clearly, the failure mode is more complex.

In the following sections, we discuss three types of element-level tests—tests for laminates with holes, bolted joints, and adhesively bonded joints.

11.6.1 Open-Hole Tests

Cutouts and holes in a laminate are common features in real structures. These are geometric discontinuities that are associated with stress concentration, that is, high stresses near the geometric discontinuities. Owing to stress concentration, the strength of a laminate is greatly reduced by the presence of such cutouts and holes; the reduction is generally more than the reduction in the effective area of cross section. It is common to express stress concentration in terms of a factor called stress concentration factor, K, which is defined as the ratio of the maximum stress to the nominal stress (Figure 11.28), that is,

$$K = \frac{\sigma_{max}}{\sigma_{nom}} \qquad (11.68)$$

where K is the stress concentration factor and σ_{max} and σ_{nom} are the maximum and nominal stresses, respectively. The strength of a notched plate, on a conservative basis, can be expressed in terms of the stress concentration factor as

$$\sigma_N = K\sigma_0 \qquad (11.69)$$

where σ_N and σ_0 are the notched and unnotched strengths, respectively. It is well known that K is 3 for isotropic materials and Equation 11.69 can be used for a conservative estimate of notched strength. On the other hand, in a composite material, the stress concentration factor, which is influenced by a number of factors, including the degree of anisotropy, laminate stacking sequence, fiber–matrix interface adhesion, matrix

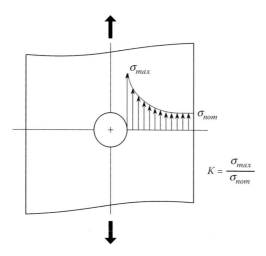

FIGURE 11.28 Stress concentration near a hole in a plate.

toughness, etc., is much higher—4 for glass/epoxy composites to as high as 9 for carbon/epoxy composites. Most composites, unlike isotropic materials, exhibit rather brittle failure behavior; the high stress concentrations do not get accommodated by local yielding and the final laminate failure is caused by a complex mechanism influenced by individual lamina stresses and strains. Also, laminate unnotched strength is dependent on hole diameter. As a result, the stress concentration factor by itself is not an appropriate means for estimating the notched strength of a laminated composite plate.

Several models, mostly empirical and semiempirical, have been developed for estimating notched strength. Two of the commonly used models are point stress criterion and average stress criterion. The point stress criterion states that failure takes place when the stress at a certain distance d_0 from the hole exceeds the unnotched strength. In the average stress criterion, on the other hand, the stresses are averaged over a certain distance and equated to the unnotched strength. The distance d_0 is determined experimentally and notched strength can be estimated for any arbitrary hole size. Frequently, these notched strengths are used by designers in the design of laminated composite structures having cutouts.

ASTM D5766 and D6484, respectively, provide two methods for the determination of open-hole tensile strength and open-hole compressive strengths of multidirectional composite laminates [53,54].

The specimen in the open-hole tensile strength test is a simple flat piece with a constant rectangular cross section and a central hole (Figure 11.29). Typically, the specimen is 150–300 mm in length, 36 mm in width, and 2–4 mm in thickness and the hole diameter is 6 mm. Shorter specimens are normally used with quasi-isotropic ply

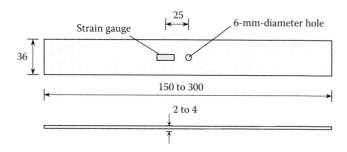

FIGURE 11.29 Open-hole tensile test specimen. (Adapted from ASTM Standard D5766/D5766M-11, *Standard Test Method for Open-Hole Tensile Strength of Polymer Matrix Composite Laminates*, ASTM International, 2011.)

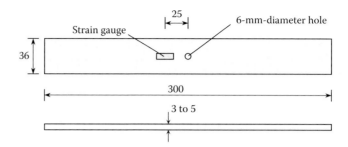

FIGURE 11.30 Open-hole compression test specimen. (Adapted from ASTM Standard D6484/D6484M-09, *Standard Test Method for Open-Hole Compressive Strength of Polymer Matrix Composite Laminates*, ASTM International, 2009.)

configurations. The laminate used for making specimens should be balanced and symmetric; typically, a ply sequence $[\pm 45_i/0_j/90_k]_{ns}$ is chosen for laminates with unidirectional laminae and $[45_i/0_j]_{ns}$ for bidirectional laminae. The specimens are tested in a uniaxial tension test setup as per ASTM D3039; force versus cross-head displacement and optionally, force versus strain/extensometer displacement are recorded. When strain data are required and extensometer is used, the hole should be located within the extensometer gauge length. The calculation of notched strength as well as stresses for stress–strain curve is calculated based on the gross cross-sectional area.

The specimen in the open-hole compressive strength test is also a flat piece with constant rectangular cross section and a central hole (Figure 11.30). Typically, the specimen is 300 mm in length, 36 mm in width, and 3–5 mm in thickness and the hole diameter is 6 mm. As in the case of open-hole tensile strength test, balanced and symmetric laminates are used for making the specimens; typically, a ply sequence $[\pm 45_i/0_j/90_k]_{ns}$ is chosen for laminates with unidirectional laminae and $[45_i/0_j]_{ns}$ for bidirectional laminae.

The specimen in the open-hole compressive strength test is pretty similar to that in an open-hole tensile strength test, but the fixturing is very different and much complex. The test fixture consists primarily of two short grips, two long grips, and two support plates (Figure 11.31). The specimen is held at one end in the long grip assembly and at the other end in the short grip assembly. The long grips, which cover the gauge section of the specimen, are provided with cutouts around the specimen hole so that damage propagation is not constrained and failure load is artificially not increased. The compressive force is introduced into the specimen by shear at the grips and the test is

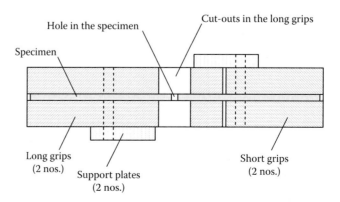

FIGURE 11.31 Schematic representation of open-hole compression test fixture in its longitudinal cross section. (Adapted from ASTM Standard D6484/D6484M-09, *Standard Test Method for Open-Hole Compressive Strength of Polymer Matrix Composite Laminates*, ASTM International, 2009.)

continued till failure of the specimen at the notched region. The calculation of strength is carried out based on gross cross-sectional area.

11.6.2 Bolted Joint

As we know, several failure modes may be present in an element-level test, which is particularly true in the case of a joint. Typical failure modes in the composite laminate in a bolted joint are

- Net tension/compression failure
- Bearing failure
- Shear-out failure
- Cleavage-tension failure

In addition to the laminate failure modes, a bolted joint failure can take place by bolt failure in different modes. Laminate ply sequence and various parameters that define the joint configuration, such as fastener type and size, laminate width and thickness, hole size and edge distance, are the primary factors that influence the occurrence of a particular failure mode. It is necessary to select the type of the joint test depending on the likely failure mode.

11.6.2.1 Bearing Strength

ASTM D5961 gives four procedures for the evaluation of bearing response of multidirectional composite laminates—Procedure A for double-shear in tensile loading, Procedure B for single-shear in tensile or compressive loading of a two-piece specimen, Procedure C for single-shear in tensile loading of a one-piece specimen, and Procedure D for double-shear in compressive loading [55]. Single-shear is associated with bending in addition to shear and it is comparatively more difficult to simulate in a test setup, but it is more representative of most real-life applications. Figure 11.32 shows a schematic representation of the single-shear two-piece test specimen with double-fastener joint as per Procedure B. A single-fastener joint specimen is also in vogue. In this case, eccentric loading results in high bending and lower joint strength, which is not generally representative of joints with multiple rows of fasteners. Thus, the single-fastener configuration is mostly used for fastener screening purposes, whereas the double-fastener configuration is used for both design data generation as well as fastener screening.

The test specimen as per Procedure B consists of two flat pieces of constant rectangular cross section fastened through one or two holes. The laminate stacking sequence

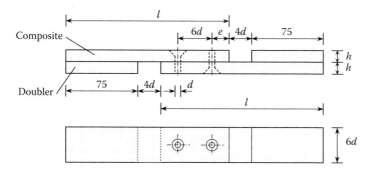

FIGURE 11.32 Schematic representation of single-shear test specimen with double-fastener joint for bearing strength. (Typical dimensions: $d = 6$ mm, $e = 18$ mm, $h = 4$ mm, and $l = 135$). (Adapted from ASTM Standard D5961/D5961M-10, *Standard Test Method for Bearing Response of Polymer Matrix Composite Laminates*, ASTM International, 2010.)

is balanced and symmetric—typically $[45_i/0_j/-45_i/90_k]_{ns}$ for laminates with unidirectional laminae and $[45_i/0_j]_{ns}$ for laminates with bidirectional laminae. A doubler, preferably of the same material, is bonded or frictionally held to the end of each piece so that the line of force action lies in the interface between the two pieces. The specimen is held either directly or with the help of a supporting fixture in grips of the testing machine and loaded either in tension or in compression. The longitudinal hole deformation is determined with the help of one or more extensometers. Both applied force and hole deformation are recorded, and the bearing stress–bearing strain curve is plotted. During testing, the specimen should fail in bearing of the composite laminate, and loading is continued till maximum force is reached. The bearing strength is obtained from the information on maximum applied force, hole diameter, and laminate thickness.

11.6.2.2 Bearing/By-Pass Strength

The load that is transferred around a hole divided by the laminate gross cross-sectional area is referred to as the by-pass strength. MIL-HDBK-17-1F prescribes testing of bearing/by-pass strength when the load transferred by a laminate around a hole is more than 20% of the total load per bolt [1]. A number of test specimens are available, which can be divided into three categories—passive, independent bolt load, and coupled bolt load/by-pass load. The recommended configuration is the independent bolt load configuration, in which the bolt load is independently measured so that the load transferred by the laminate around the hole can be directly calculated from the total load, and thereby, bearing/by-pass strength can be evaluated.

11.6.2.3 Shear-Out Strength

It is virtually impossible to have shear-out failure when the edge distance is three times the hole diameter (i.e., $e = 3d$) and the plies are not predominantly in the same direction. However, in certain design cases, shear-out strength data are required, which can be obtained using a bearing test specimen with lower edge distance-to-hole diameter ratio (e/d).

11.6.2.4 Fastener Pull-Through Strength

Fastener pull-through of a joint, in which two composite laminates are mechanically fastened, is the maximum normal force that can be applied before the two laminates come apart. Composites are typically weak in the transverse direction, and the pull-through strength of a joint involving composite laminates is rather critical from a design point of view. MIL-HDBK-17-1F suggests two procedures that involve two square composite plates, one of which is rotated by 45° w.r.t. the other. The plates are joined together by a central fastener. While the loading mechanisms are different in the two procedures, the resultant loading on the joint is the same and the fastener is loaded in tension. The load–deflection curve is plotted, the final failure load and failure mode are recorded.

11.6.3 Bonded Joint

In composite structures, bonded joints are made in three different routes—secondary bonding, cobonding, and cocuring. In secondary bonding, two precured parts are bonded with a thin adhesive layer. In the cobonding process, one of the two parts is precured, and the adhesive layer is applied on it and the other part is built-up by lay-up, followed by curing. On the other hand, in the cocuring process, both the parts are simultaneously cured, with or without any adhesive layer between them.

There are two aspects in the characterization of a bonded joint—characterization of the adhesive and characterization of the joint. Some of the adhesive characterization tests use only metallic adherends; these tests may be used for the evaluation of adhesive properties for design data generation, material screening, or research and development, but they are not useful for the evaluation of joint properties. On the other hand, bonded joint characterization tests are carried out with a view to validating the integrity of the overall joint.

11.6.3.1 Adhesive Characterization

The characterization of an adhesive primarily involves the evaluation of shear properties and tensile properties. ASTM D5656 gives a standard test method (Figure 11.33) for the evaluation of stress–strain behavior of adhesives in shear by tension loading [56]. The test specimen consists of two thick adherends. An adhesive layer of controlled thickness is provided between the adherends and cured under specified environment. Extensometers are mounted on each edge of the specimen such that the arms of the extensometer are equidistant on either side of the bond line and midway between the notches. The specimen is loaded till failure and load versus deflection data are recorded continuously. Typically, the load–deflection curve in shear is nonlinear, associated with a "knee" formation. Shear properties of the adhesive are calculated using this curve.

ASTM D2095 gives a standard method for the evaluation of tensile properties of adhesives by using a bar-and-rod specimen [57]; the two adherends are made using either similar or dissimilar materials. The test specimen consists of either two bars or two rods that are adhesively joined by a butt-joint. The specimen is tested in tensile load, and the tensile strength is calculated by dividing the breaking force by the bond area.

11.6.3.2 Bonded Joint Characterization

The validation of the structural integrity of a bonded joint would demand testing of a specimen that simulates the actual joint in respect of configuration, adherends, and adhesive. The adherends are made out of composite laminate and/or metal; the laminate ply sequence, joint fabrication process, and quality control procedures must also resemble those of the actual joint. It is difficult to develop a standard test specimen that is truly representative of an actual bonded joint in all aspects. Occasionally, in critical joints, nonstandard test specimens are designed and tested to generate data for the design of the joints. Nevertheless, there are standard specimens, which address certain types of joints.

ASTM D3528 gives a standard test method for the evaluation of adhesive shear strength in tension [58]. The adhesive bond surfaces are symmetrically loaded that simulates a peel-free condition. The specimen is useful for the evaluation of adhesive

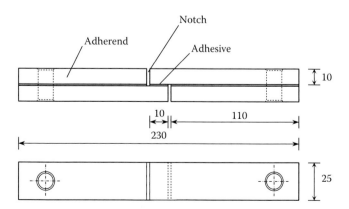

FIGURE 11.33 Schematic representation of test specimen for adhesive characterization in shear. (Adapted from ASTM Standard D5656-10, *Standard Test Method for Thick-Adherend Metal Lap-Shear Joints for Determination of the Stress–Strain Behavior of Adhesives in Shear by Tension Loading*, ASTM International, 2010.)

shear strength of a low-peel adhesively bonded joint for process control and specification purposes. However, it typically overestimates the strength of the actual joint. ASTM D3165 gives another standard test method that involves a single lap joint [59]. The use of doublers reduces peel stresses. This test specimen resembles many actual joint configurations and it is widely used in the industry.

11.7 TESTS AT COMPONENT LEVEL

11.7.1 Subscale Component Testing

A real-life structure consists of many components and subcomponents assembled by means of a number of joints of different types. Some of these components, subcomponents, and joints may be of complex configurations associated with complex internal load distribution, in which cases, lower-level tests are not sufficient for either design data generation or design validation and quality assurance. This is where nonstandard element and subscale component tests are required. These tests are performed with different objectives and at different stages of the overall development cycle of a product. Sometimes, in the early phase of the product development cycle, a subscale component is made and tested with a view to merely getting acquainted with the nitty-gritties of the relevant technologies and generating confidence.

Focus areas of the nonstandard element and subscale component tests are generally around joints and areas with complex shape and dissimilar materials, realized by complex fabrication process. A nonstandard test specimen must be designed for the critical load cases. It must simulate the actual component as closely as possible in terms of geometrical configuration, materials, processing parameters, applied loads, and loading environment. Of course, simplifications are made by omitting noncritical features, by replacing repetitive units by a single unit, by making geometrical simplification, and so on.

Another very important aspect in the nonstandard element and subscale testing is the interpretation and use of test results. Test results can be used as design input, but care must be taken of the effect of the simplifications made in the test specimen. In this respect, some of the issues that should be considered are as follows: effect of scaling down, effect of omitting noncritical features, effect of using a single unit in place of a number of repetitive units, etc.

11.7.2 Full-Scale Component Testing

Full-scale component tests are carried out to validate the design and manufacturing process and to assess the in-service performance of the final product. The final product is invariably an assembly of a number of components and subcomponents. During the development phase, these components and subcomponents typically undergo design and manufacturing iterations, and the individual components and subcomponents are qualified by full-scale testing at each stage. Thus, at each stage during the development cycle, the design, manufacturing process and the quality assurance procedure of the individual components and subcomponents get validated and the margins of safety get established. A part with appropriate margin of safety qualifies to the next phase; otherwise, design and processing modifications would be needed.

Full-scale tests are expensive and their costs increase as the levels go up. Fortunately, the building block approach helps reduce the requirement of full-scale tests. Results of lower-level tests together with findings of analysis—typically finite element analysis—are utilized to decide on the extent of full-scale tests at the higher levels.

11.8 SUMMARY

Test objectives, general test procedure, and test philosophy in composites are discussed in this chapter. It is an inseparable part of the development of a composite product and it covers a wide spectrum involving different levels. Salient points discussed in this chapter can be listed as follows:

- Tests are conducted to meet requirements in one or more of the following three areas:
  - Design and analysis
  - Quality control and quality assurance
  - Research and development
- A composite product is built by combining constituents. Thus, the building block approach to composites testing is highly useful. It involves testing at different levels and sublevels. At the bottommost level are the coupon-level tests for fibers and resins and at the topmost level are the full-scale component-level tests. Lower-level tests are simple and relatively inexpensive, but more in number. The complexities and costs of the tests increase as we proceed up the building block. At the top level, however, the tests are limited in number.
- There are a number of standards organizations that extensively provide standard test methods for most of the tests especially at the lower levels in the building block. For carrying out specific tests and reporting test results, standards should be referred to.
- Both mechanical and nonmechanical tests are carried out on fibers, resins, laminae, and laminates. Structural element- and component-level tests are primarily mechanical.
- Nonmechanical tests are carried out for the determination of density, moisture content, constituent content, glass transition temperature, etc. Coupon-level mechanical tests pertain primarily to evaluation of tensile, compressive, shear, and flexural properties.
- Structural element-level and subscale component-level tests are developed with a focus mainly on joint and other critical zones in the component.

EXERCISE PROBLEMS

11.1 Write a brief note on the building block approach to testing in composites. What are the tests that are typically required in a composite product development program?

11.2 Why do we need standards in composites testing? List down the common standards available for testing in composites.

11.3 Name the nonmechanical and mechanical parameters of reinforcements that are normally evaluated? Give a brief note covering test principles and applicable standards.

11.4 Determine the diameter of filaments in glass rovings, if the following data are given: density of fiber = 2.54 g/cm³, tex = 1200 g/km, and number of ends (strands) = 30.

11.5 In a tow tensile test for the determination of tensile strength and tensile modulus of 12k carbon fiber, the following data have been recorded:

   Maximum tensile load, $P = 2012$ N
   Density of fiber, $\rho_f = 1.78 \, \text{g/cm}^3$
   Linear density of fiber, $\rho_f^{(l)} = 0.8 \, \text{g/m}$
   Upper strain limit, $\varepsilon_u = 6 \times 10^{-3}$

Lower strain limit, $\varepsilon_l = 1 \times 10^{-3}$
Tensile load at upper strain limit, $P_u = 642$ N
Tensile load at lower strain limit, $P_l = 106$ N
Determine the longitudinal tensile strength and modulus of the fiber.

11.6 Name the nonmechanical and mechanical parameters of resin that are normally evaluated? Give a brief note covering test principles and applicable standards.

11.7 In a density determination test for cast polyester resin, the following data were recorded:

Weight of the cast resin sample in air = 40 g
Weight of the sample with a sinker fully immersed and wire partially immersed in water = 72.6 g
Weight of the sinker fully immersed and wire partially immersed in water = 70.5 g

Determine the density of the cast resin sample.

11.8 Write a brief note on the determination/assigning glass transition temperature by the DSC method.

11.9 In an acid digestion test for constituent contents of a carbon/epoxy sample, the following data were recorded:

Mass of empty filter = 55.31 g
Mass of filter with sample before matrix removal = 55.82 g
Mass of filter with sample after matrix removal = 55.66 g
If the experimentally determined densities of the fiber, matrix, and composite sample are 1.79 g/cm³, 1.15 g/cm³, and 1.53 g/cm³, respectively, determine the mass fractions and volume fractions of the fiber and matrix.

11.10 The following data are collected from a tensile test of a unidirectional composite specimen:

Applied Load (kN)	Extensometer Reading (mm)
2	0.020
4	0.039
6	0.057
8	0.075
10	0.093
12	0.112
14	0.131
16	0.150
18	0.169
20	0.188
22	0.206
24	0.225
26	0.243
28	0.262
30	0.282
32	0.300
34	0.318

The final failure took place at 34.24 kN. If the gauge length of the extensometer is 25 mm and the recorded average cross-sectional area of the specimen is 23.79 mm², determine the tensile strength and tensile modulus.

11.11 The following data have been reduced from the records of a compression test:

Applied Load (kN)	Average Axial Strain ($\times 10^{-6}$)
0	0
2	2153
4	3958
6	5599
8	7111
10	8541
12	9955
14	11,434
16	13,072
18	14,983
20	17,294
22	20,148
24	24,023

The final failure took place at 24.73 kN. If the average area of cross section is 24.79 mm^2, determine the (i) compressive strength, (ii) compressive modulus, and (iii) secant modulus.

11.12 Write a short note on load transfer mechanisms and failure modes in a compression test specimen.

11.13 Write a short note on various test methods used for the evaluation of shear properties.

11.14 In a uniaxial tension test of a specimen with ±45° plies, the failure load and average cross-sectional area are recorded as 5552 N and 49.4 mm^2, respectively. Determine the in-plane shear strength.

11.15 In a V-notch shear beam test, a short beam of length 76 mm, width 12 mm, and height 20 mm was tested. The beam was provided with two notches of 2 mm depth each. If the ultimate load at failure is 9.8 kN, determine the in-plane shear strength. Also, determine the in-plane shear modulus, if shear stress difference and shear strain difference between two points chosen on the experimental stress–strain curve are 4.8 MPa and 0.8×10^{-3}, respectively.

11.16 Discuss the principle of flexural testing. Comment on the utility of flexural testing in the design and development of a product.

11.17 What are the typical modes of failure in a composite bolted joint?

11.18 Discuss the significance of subscale and full-scale component testing in a product development program.

REFERENCES AND SUGGESTED READING

1. Composite Materials Handbook-17 (CMH-17), *Volume 1, Polymer Matrix Composites, Guidelines for Characterization of Structural Materials*, SAE International, Wichita State University, 2012.
2. F. T. Traceski, *Specifications and Standards for Plastics and Composites*, ASM International, Ohio, 1990.
3. M. H. Geier, *Quality Handbook for Composite Materials*, Chapman & Hall, London, 1994.
4. D. F. Adams, L. A. Carlsson, and R. B. Pipes, *Experimental Characterization of Advanced Composite Materials*, CRC Press, Boca Raton, FL, 2003.
5. J. M. Hodgekinson (ed.), *Mechanical Testing of Advanced Fiber Composites*, Woodhead Publishing, Cambridge, 2000.
6. D. W. Wilson and L. A. Carlsson, Mechanical property measurements, *Composite Materials Handbook* (P. K. Mallick, ed.), Marcel Dekker, New York, 1997, pp. 1067–1145.
7. S. Turner, General principles and perspectives, *Mechanical Testing of Advanced Fiber Composites* (J. M. Hodgekinson, ed.), Woodhead Publishing, Cambridge, 2000, pp. 4–35.

8. R. E. Fields, Overview of testing and certification, *ASM Handbook, Volume 21, Composites* (D. B. Miracle and S. L. Donaldson, Vol. Chairs), ASM International, Ohio, 2001, pp. 734–740.
9. Yu. M. Tarnopol'skii and V. L. Kulakov, Mechanical tests, *Handbook of Composites* (S. T. Peters, ed.), second edition, Chapman & Hall, London, 1998, pp. 778–793.
10. F. L. Matthews, Specimen preparation, *Mechanical Testing of Advanced Fiber Composites* (J. M. Hodgekinson, ed.), Woodhead Publishing, Cambridge, 2000, pp. 36–42.
11. L. C. Wolstenholme, Statistical modeling and testing of data variability, *Mechanical Testing of Advanced Fiber Composites* (J. M. Hodgekinson, ed.), Woodhead Publishing, Cambridge, 2000, pp. 314–339.
12. ASTM Standard D3800M-11, *Standard Test Method for Density of High Modulus Fibers*, ASTM International, 2011.
13. ASTM Standard D1577-07, *Standard Test Methods for Linear Density of Textile Fibers*, ASTM International, 2012.
14. ASTM Standard D3775-12, *Standard Test Method for Warp (End) and Filling (Pick) Count of Woven Fabrics*, ASTM International, 2012.
15. ASTM Standard D3776/D3776M-09, *Standard Test Methods for Mass per Unit Area (Weight) of Fabric*, ASTM International, 2013.
16. ASTM D3379-75, *Standard Test Method for Tensile Strength and Young's Modulus for High-Modulus Single-Filament Materials*, ASTM International, 1989.
17. ASTM Standard D4018-11, *Standard Test Methods for Properties of Continuous Filament Carbon and Graphite Fiber Tows*, ASTM International, 2011.
18. ASTM Standard D5035-11, *Standard Test Method for Breaking Strength and Elongation of Textile Fabrics (Strip Method)*, ASTM International, 2011.
19. ASTM Standard D5034-09, *Standard Test Method for Breaking Strength and Elongation of Textile Fabrics (Grab Test)*, ASTM International, 2013.
20. ASTM Standard D792-08, *Standard Test Methods for Density and Specific Gravity (Relative Density) of Plastics by Displacement*, ASTM International, 2008.
21. ASTM Standard D1505-10, *Standard Test Method for Density of Plastics by the Density Gradient Technique*, ASTM International, 2010.
22. ASTM Standard D2393-86, *Test Method for Viscosity of Epoxy Resins and Related Components*, ASTM International, 1986.
23. W. Grellmann and S. Seidler (ed.), *Polymer Testing*, Hanser, Munich, 2007.
24. ASTM Standard E1356-08, *Standard Test Method for Assignment of the Glass Transition Temperatures by Differential Scanning Calorimetry*, ASTM International, 2014.
25. ASTM Standard D638-10, *Standard Test Method for Tensile Properties of Plastics*, ASTM International, 2010.
26. ASTM Standard D695-10, *Standard Test Method for Compressive Properties of Rigid Plastics*, ASTM International, 2010.
27. ASTM Standard D5379/D5379M-12, *Standard Test Method for Shear Properties of Composite Materials by the V-Notched Beam Method*, ASTM International, 2012.
28. ASTM Standard E143-02, *Standard Test Method for Shear Modulus at Room Temperature*, ASTM International, 2008.
29. M. J. Parker, Test methods for physical properties, *Comprehensive Composite Materials, Volume 5, Test Methods, Nondestructive Evaluation, and Smart Materials* (A. Kelly and C. Zweben, editors-in-chief; L. Carlsson, R. L. Crane, and K. Uchino, vol. eds.), Elsevier, Oxford, 2000, pp. 183–226.
30. S. Bugaj, Constituent materials testing, *ASM Handbook, Volume 21, Composites* (D. B. Miracle and S. L. Donaldson, Vol. Chairs), ASM International, Ohio, 2001, pp. 749–758.
31. ASTM Standard D3171-11, *Standard Test Methods for Constituent Content of Composite Materials*, ASTM International, 2011.
32. ASTM Standard D2734-09, *Standard Test Methods for Void Content of Reinforced Plastics*, ASTM International, 2009.
33. D. F. Adams, Test methods for mechanical properties, *Comprehensive Composite Materials, Volume 5, Test Methods, Nondestructive Evaluation, and Smart Materials* (A. Kelly and C. Zweben, editors-in-chief; Leif Carlsson, Robert L. Crane, and Kenji Uchino, vol. eds.), Elsevier, Oxford, 2000, pp. 113–148.
34. E. W. Godwin, Tension, *Mechanical Testing of Advanced Fiber Composites* (J. M. Hodgekinson, ed.), Woodhead Publishing, Cambridge, 2000, pp. 43–74.
35. ASTM Standard D3039/D3039M-08, *Standard Test Method for Tensile Properties of Polymer Matrix Composite Materials*, ASTM International, 2008.
36. F. L. Matthews, Compression, *Mechanical Testing of Advanced Fiber Composites* (J. M. Hodgekinson, ed.), Woodhead Publishing, Cambridge, 2000, pp. 75–99.
37. ASTM Standard D3410/D3410M-03, *Standard Test Method for Compressive Properties of Polymer Matrix Composite Materials with Unsupported Gage Section by Shear Loading*, ASTM International, 2008.

38. ASTM Standard D6641/D6641M-09, *Standard Test Method for Compressive Properties of Polymer Matrix Composite Materials Using a Combined Loading Compression (CLC) Test Fixture*, ASTM International, 2009.
39. ASTM Standard D5467/D5467M-97, *Standard Test Method for Compressive Properties of Unidirectional Polymer Matrix Composite Materials Using a Sandwich Beam*, ASTM International, 2010.
40. W. R. Broughton, Shear, *Mechanical Testing of Advanced Fiber Composites* (J. M. Hodgekinson, ed.), Woodhead Publishing, Cambridge, 2000, pp. 100–123.
41. ASTM Standard D3518/D3518M-94, *Standard Test Method for In-Plane Shear Response of Polymer Matrix Composite Materials by Tensile Test of a ±45° Laminate*, ASTM International, 2007.
42. ASTM Standard D4255/D4255M-01, *Standard Test Method for In-Plane Shear Properties of Polymer Matrix Composite Materials by the Rail Shear Method*, ASTM International, 2007.
43. ASTM Standard D2344/D2344M-00, *Standard Test Method for Short-Beam Strength of Polymer Matrix Composite Materials and Their Laminates*, ASTM International, 2006.
44. J. M. Hodgkinson, Flexure, *Mechanical Testing of Advanced Fiber Composites* (J. M. Hodgkinson, ed.), Woodhead Publishing, Cambridge, 2000, pp. 124–142.
45. ASTM Standard D790-10, *Standard Test Methods for Flexural Properties of Unreinforced and Reinforced Plastics and Electrical Insulating Materials*, ASTM International, 2010.
46. ASTM Standard D6272-10, *Standard Test Method for Flexural Properties of Unreinforced and Reinforced Plastics and Electrical Insulating Materials by Four-Point Bending*, ASTM International, 2010.
47. P. Robinson and J. M. Hodgkinson, Interlaminar fracture toughness, *Mechanical Testing of Advanced Fiber Composites* (J. M. Hodgkinson, ed.), Woodhead Publishing, Cambridge, 2000, pp. 170–210.
48. ASTM Standard 5528-01, *Standard Test Method for Mode I Interlaminar Fracture Toughness of Unidirectional Fiber-Reinforced Polymer Matrix Composites*, ASTM International, 2007.
49. ASTM Standard D6671/D6671M-13, *Standard Test Method for Mixed Mode I-Mode II Interlaminar Fracture Toughness of Unidirectional Fiber Reinforced Polymer Matrix Composites*, ASTM International, 2013.
50. P. T. Curtis, Fatigue, *Mechanical Testing of Advanced Fiber Composites* (J. M. Hodgekinson, ed.), Woodhead Publishing, Cambridge, 2000, pp. 248–268.
51. ASTM Standard D3479/D3479M-96, *Standard Test Method for Tension-Tension Fatigue of Polymer Matrix Composite Materials*, ASTM International, 2007.
52. L. A. Gintert, Element and subcomponent testing, *ASM Handbook, Volume 21, Composites* (D. B. Miracle and S. L. Donaldson, Vol. Chairs), ASM International, Ohio, 2001, pp. 778–793.
53. ASTM Standard D5766/D5766M-11, *Standard Test Method for Open-Hole Tensile Strength of Polymer Matrix Composite Laminates*, ASTM International, 2011.
54. ASTM Standard D6484/D6484M-09, *Standard Test Method for Open-Hole Compressive Strength of Polymer Matrix Composite Laminates*, ASTM International, 2009.
55. ASTM Standard D5961/D5961M-10, *Standard Test Method for Bearing Response of Polymer Matrix Composite Laminates*, ASTM International, 2010.
56. ASTM Standard D5656-10, *Standard Test Method for Thick-Adherend Metal Lap-Shear Joints for Determination of the Stress–Strain Behavior of Adhesives in Shear by Tension Loading*, ASTM International, 2010.
57. ASTM Standard D2095-96, *Standard Test Method for Tensile Strength of Adhesive by Means of Bar and Rod Specimens*, ASTM International, 2008.
58. ASTM Standard D3528-96, *Standard Test Method for Strength Properties of Double Lap Shear Adhesive Joints by Tension Loading*, ASTM International, 2008.
59. ASTM Standard D3165-07, *Standard Test Method for Strength Properties of Adhesives in Shear by Tension Loading of Single-Lap-Joint Laminated Assemblies*, ASTM International, 2007.

12

Nondestructive Testing of Polymer Matrix Composites

12.1 CHAPTER ROAD MAP

It was noted in Chapter 11 that testing is an inseparable part of any composite product development program. Tests can be broadly classified into destructive and nondestructive, of which the destructive tests in general were discussed in Chapter 11. Our objective in this chapter is to familiarize the reader with the basic concepts in nondestructive testing in the field of PMCs. First, an introductory discussion is given on the topics on defects in PMCs and various available nondestructive techniques. Some of the common nondestructive techniques are chosen and a general discussion on each of these techniques, covering underlying concept, test equipment, advantages, and disadvantages, is presented.

The topics discussed in this chapter do not demand an in-depth understanding of the topics presented in the previous chapters. However, for effective assimilation of the concepts, it is suggested that the reader goes through the introductory topics of composites, basic mechanics of composites, constituent materials, composites manufacturing, and testing in Chapters 1, 3 through 5, and 9 through 11.

12.2 INTRODUCTION

The test methods, except some component-level acceptance tests, discussed in Chapter 11 are destructive in nature. In the coupon-level tests, the test sample gets either destroyed or consumed. In the component-level tests, depending on the nature and test objectives, the component may get destroyed; for example, in the burst test of a pressure vessel, the test article is tested until it bursts and is destroyed. If the test objective is only verification of acceptance, the test article is loaded below the ultimate loads and the component can be put to service. However, even in a test where the applied loads are below the ultimate loads, the component undergoes essentially deformations; microlevel damages may occur and, from this point of view, all mechanical tests are destructive. In contrast to destructive testing, nondestructive testing (NDT) or nondestructive evaluation/examination (NDE) neither destroys nor causes any damage to the part and utility of the part remains intact.

NDT is a critical phase in the life cycle of a composite product. It is essential from quality assurance and reliability points of view and various NDT techniques are routinely used in the composites industry. The objectives of NDT are manifold [1], which, for the sake of simplicity of discussion, can be categorized as follows:

- Quality assurance of manufactured composite product
- In-service quality monitoring

Quality is infused into a composite product in stages as it is manufactured. The critical stages are identified and NDT is carried out for checking the existence of any possible defects; the findings of NDT are analyzed and the realized parts are either cleared with/without suggested repair work or rejected. NDT in this respect is primarily process-oriented.

During service, a composite product undergoes degradation due to cyclic loading, continuous high stressing, impact, exposure to hostile environment, etc. As a result, an existing noncritical crack may grow beyond the safe limit and new defects may nucleate. A product in-service is usually inspected by NDT methods at regular intervals; based on the findings of inspection, possible repair works are undertaken and the residual life of the product is estimated.

12.2.1 Defects in Polymer Matrix Composites

PMCs are grossly different from conventional metals due to their anisotropic and heterogeneous nature, and unique processing techniques. Different types of defects are created in a composite material during processing as well as during service [2]. These defects are also grossly different from those in metals, and in general, relatively more in number. Typical defects in a PMC material are as follows:

- Defects in the fibers
    - Fiber breakage
    - Nonuniform fiber distribution
    - Fiber waviness
    - Improper fiber orientation
- Defects in the matrix
    - Foreign objects and contaminations
    - Porosity and voids
    - Matrix cracking—crazing and microcracks
    - Incomplete cure
- Defects in the fiber/matrix interface
    - Delaminations
    - Interlaminar cracks
- Other defects
    - Debonds

The presence of fiber breakage indicates loss of continuity in the load-carrying path. Nonuniform fiber distribution may lead to resin-rich or resin-starved regions. These defects result in stress concentrations leading to the creation of cracks. During service life, the cracks may grow and, beyond a certain limit, the failure of the part may occur. Fiber waviness and fiber misalignment adversely affect the strength and stiffness characteristics of the composite material.

Foreign objects in resin lead to stress concentration and creation of cracks. Contamination of resin, on the other hand, results in improper cure of the resin and poor fiber/matrix bond, leading to inefficient load transfer between fibers and matrix. Porosity in the resin may result due to improper curing and the pores can coalesce, leading to the formation of voids. Voids are also formed if entrapped air between plies during lay-up is not removed. Pores and voids result in stress concentration, and in some cases, delaminations that may be severely detrimental to the overall health of the component. Matrix cracks can occur either during service or during acceptance testing. Matrix crazing is the formation of superficial microcracks usually in the gel coat of a composite part. In the structural plies, microcracks are formed in the 90° plies. Matrix cracking leads to the reduction of stiffness of the composite material. On sustained loading, the cracks may grow beyond a certain critical size and spontaneous crack growth and eventual failure may result.

Delaminations are separations created between plies in a laminated composite material. The primary factor that causes delaminations is improper curing. Improper ply sequence can also result in delaminations. Delaminations and fiber/matrix debond are detrimental particularly under in-plane compressive loading.

12.2.2 NDT Techniques

NDT techniques are well developed for conventional metallic materials. A number of NDT techniques are available for the detection of defects in composite materials too; these are primarily extensions of those originally developed for metals. Visual inspection and coin tapping are often employed, both in metals as well as composites, to obtain first-hand information about the health of a component. For reliable and more specific details, almost invariably other methods of NDT are resorted to. Most of these methods are active in nature, that is, during testing, some form of energy is applied on the test specimen and its response is monitored for drawing conclusions about its health. The response of a PMC material to input energy is grossly different from what it is in metals. This difference is primarily on account of the unique characteristics of PMCs in respect of anisotropic and heterogeneous material construction, unique fracture mechanisms, and higher damping. What follows is that, although the basic principles of the NDT methods remain the same in metals and PMCs, suitable modifications are essential [3].

In this chapter, we discuss the following NDT methods used in the field of PMCs:

- Ultrasonic testing
- Radiographic testing
- Acoustic emission testing
- Infrared thermography
- Eddy current testing
- Shearography

The most common NDT methods in PMCs are ultrasonic testing and radiographic testing. These methods complement each other and are effective in the detection of a wide variety of defects. Acoustic emission (AE) testing and infrared thermography provide their own unique advantages and they are popular in many applications. Eddy current testing is based on the principle of electromagnetic induction; it is applicable to only conducting materials and its use is primarily limited to CFRPs. Shearography is an optical method that is routinely used in the rubber industry and has wide acceptance in the aerospace composites applications as well. Table 12.1 gives a summary of these methods.

NDT has been extensively researched in the past and continues to get attention now as well [4–11]. We have had a brief discussion on the introductory concepts such as defects in composites and various NDT techniques. We shall now proceed to see how some of the common NDT techniques work. A detailed review of these techniques is beyond the scope of this book. The interested reader may consult articles as indicated above and available standard texts (see, for instance, a list of reference books given in Reference 12.

TABLE 12.1
NDT Techniques for Polymer Matrix Composites

NDT Techniques	Principle	Detectable Types of Defects
Ultrasonic testing	Transmission of ultrasonic waves through a material is affected by defects and the signals received from the transmitted/reflected waves give information on the health of the material.	Fiber misalignment, foreign objects, porosity, matrix cracking, delamination, interlaminar cracks
Radiographic testing	The attenuation of x-rays and gamma rays traveling through a material is affected by material characteristics and defects are seen as differential shades in the captured image.	Fiber misalignment, foreign objects, porosity, debonds, delaminations, interlaminar cracks
Acoustic emission	The stress distribution in a loaded structure changes suddenly when damage mechanism processes occur. Damage nucleation and propagation are indicated by transient stress waves known as acoustic emission.	Fiber breakage, delaminations, interlaminar cracks, debonds
Infrared thermography	The surface temperature distribution and surface temperature decay rate are affected by thermal characteristics of a material. Variations in a thermographic image indicate the presence of defects.	Foreign objects and contaminations, delaminations, interlaminar cracks, impact damage
Eddy current testing	Eddy current induced in a conducting material by an alternating current in a probing coil is affected by material discontinuities and the same is reflected in changes in the impedance of the receiving coil.	Fiber breakage, fiber misalignment, foreign objects
Shearography	Optical interference	Foreign objects and contaminations, delaminations, interlaminar cracks, impact damage

12.3   ULTRASONIC TESTING

12.3.1   Basic Concept of Ultrasonic Testing

Ultrasonic testing involves sending ultrasonic waves through the specimen material and the signals received from the transmitted/reflected waves are analyzed for making conclusions regarding possible defects [13,14]. Ultrasonic waves, also known as stress waves, are mechanical (sound) waves that propagate in solids, liquids, and gases at frequencies above the normal human audible range of 20 Hz–20 kHz. The amplitude, velocity, and frequency of ultrasonic waves are related by

$$\lambda = \frac{c}{f} \tag{12.1}$$

where λ, c, and f are the amplitude, velocity, and frequency of the ultrasonic waves in the medium through which they are propagating. Ultrasonic waves with low frequency have higher penetrative power and more suitable for composite laminate of higher thickness. On the other hand, the high-frequency waves are more sensitive to defects. The test frequency is chosen depending on a number of factors such as part thickness, type of material, degree of anisotropy, type of defect, etc. For PMCs, it typically lies between 1 and 15 MHz, of which, lower frequencies are used in thick laminates and higher frequencies in thin laminates.

During propagation within a material, the ultrasonic waves are affected by the material characteristics, any discontinuity, material boundary, and the environment that surrounds the material. When the waves come across any material boundary, the waves partly get reflected at the boundary and the rest get transmitted. The relative proportions of reflected and transmitted waves depend on two parameters—angle of incidence of the waves and acoustic impedances of the materials on both sides of the boundary. The acoustic impedance of the material is given by [15]

$$Z = \rho c \tag{12.2}$$

where Z and ρ are the acoustic impedance and density of the material, respectively, and c is the velocity of the waves in the material. The fraction of reflected energy intensity, that is, energy reflected per unit area, to the incident energy intensity is given by

$$I_r = \left(\frac{Z_1 - Z_2}{Z_1 + Z_2}\right)^2 \tag{12.3}$$

where Z_1 is the acoustic impedance of the material in which the stress waves are propagating and Z_2 is the acoustic impedance of the material on which the waves are incident. Similarly, the fraction of transmitted energy intensity to the incident energy intensity is given by

$$I_t = 1 - I_r = \frac{4Z_1 Z_2}{(Z_1 + Z_2)^2} \tag{12.4}$$

When ultrasonic waves are applied on a material, the first boundary encountered by the waves is between the coupling medium and the material. As a result, some of the incident energy is reflected from the front wall and the rest is transmitted through the material. The transmitted waves are of lower energy, that is, they are attenuated; clearly, for dissimilar materials, the attenuation levels are higher. The couplant or the coupling medium is a material with comparable acoustic impedance such that efficient transmission of energy to the test material takes place. A defect in the material acts as a discontinuity in the path of the waves, which partly get reflected at the surface of the defect and the rest get transmitted through the test material. Finally, the transmitted waves meet the boundary at the back wall and a similar phenomenon of partial reflection and transmission occurs. The reflected and transmitted waves are captured as peaks in a display unit. The number of peaks, their location, and size are compared with those with a reference specimen, and conclusions are drawn w.r.t. type, size, and location of possible defects in the material.

12.3.2 Test Equipment

The ultrasonic test equipment typically consists of the following:

- Pulse generator
- Transmitter
- Transmitting transducer
- Receiving transducer
- Amplifier
- Display unit

High-voltage electrical pulses of controlled energy are produced by an electronic device referred to as the pulse generator and transmitted to the transducer. A transducer, in general, is a device that converts one form of energy to another. In the case of ultrasonic testing, it converts electrical energy to mechanical energy, in the form of ultrasound waves, and vice versa. Typically, it uses a piezoelectric crystal or ceramic material. A piezoelectric material is a material with polarized molecules. When an electric field is forced on such a material, the molecules align themselves in the electric field and the material undergoes dimensional changes. On the contrary, when mechanical forces are applied on the material causing dimensional changes in a piezoelectric material, an electric field is created. A suitable couplant with high acoustic impedance, such as oil, is used between the piezoelectric transducer and the laminate for efficient transmission of signal between them. (*Note:* Acoustic impedance of a material is the product of its density and velocity of sound in the material.) In addition to the piezoelectric transducers, one more type of transducers, viz. electromagnetic acoustic transducers (EMATs), are in use. A key advantageous feature of the EMATs is that no couplant is required with them.

The short high-voltage electrical pulses received by the transducer are converted to high-frequency ultrasonic waves, which propagate through the test material. The waves get reflected/transmitted at various boundaries and they are finally converted by the receiving transducer to electrical energy and amplified by the amplifier. The electrical signals are displayed by the display unit in the form of peaks with different heights.

12.3.3 Through-Transmission Technique

The ultrasonic through-transmission technique is schematically shown in Figure 12.1. In this method, the transmitting and receiving transducers are held on the front and the back surfaces of the laminate such that the transmitted signals are received and converted into electric signals by the receiving transducer, amplified, and finally displayed on the cathode ray tube (CRT) display unit. The characteristic display in the through-transmission technique shows two sharp peaks. The first, which is the larger of the two peaks, corresponds to the uninterrupted waves that propagate through the laminate. When the waves come across a defect, a partial loss of energy referred to as attenuation occurs. The attenuated waves are seen in the form of the smaller peak that appears second.

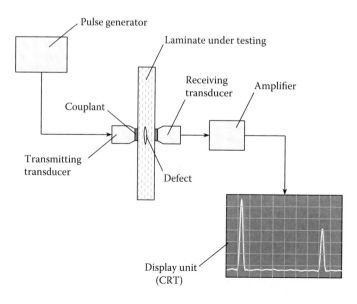

FIGURE 12.1 Schematic representation of ultrasonic through-transmission testing.

12.3.4 Pulse-Echo Technique

The ultrasonic pulse-echo technique is schematically shown in Figure 12.2. In this method, only one transducer is used that works in a transmitting mode as well as a receiving mode. The input waves are transmitted by the transducer in the transmitting mode and the reflected waves are received by it in the receiving mode. At a defect location, three peaks are seen in the CRT screen. The first peak is the largest and it corresponds to the waves reflected by the front surface of the laminate. When the waves come across any defect, attenuation occurs; parts of these attenuated waves are reflected from the internal defect surfaces and they appear as the second peak. Similarly, the waves that get reflected from the back surface of the laminate appear as the third peak.

12.3.5 Data Representation: A-Scan, B-Scan, and C-Scan

During ultrasonic testing of a composite part, it is probed by placing the transducer(s) at different locations at regular intervals on the surface and an ultrasonic map is created. There are three basic ways to represent these data—A-scan, B-scan, and C-scan.

12.3.5.1 A-Scan

An A-scan (Figure 12.3) is a graphical representation, at a point on the surface of a laminate, of the amount of ultrasonic energy received w.r.t. time. The horizontal and vertical axes are used for travel time and amplitudes of the ultrasonic pulses, respectively. It consists of a series of peaks, which, in a pulse-echo test, may correspond to the initial pulse, pulse reflected from a defect or pulse reflected from the back wall. The amplitude of a peak is compared with that of a transmitted/reflected pulse through/from a reference material and nature of a discontinuity, which can be a defect or laminate boundary, is estimated. The relative position of a peak on the time axis, on the other hand, is utilized to determine the depth of a defect in the laminate.

12.3.5.2 B-Scan

A B-scan (Figure 12.4) is a 2D graphical representation of defects in the laminate cross section. The horizontal axis of the scan is for the linear position of the transducer

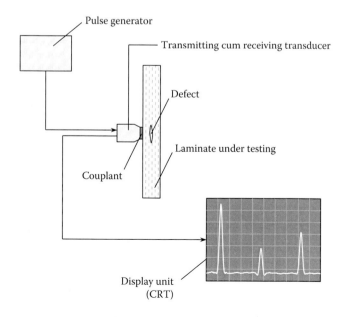

FIGURE 12.2 Schematic representation of pulse-echo technique.

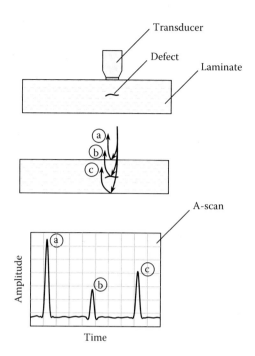

FIGURE 12.3 Schematic A-scan in a pulse-echo ultrasonic test. (a) Wave reflected from the front wall. (b) Wave reflected from the defect. (c) Wave reflected from the back wall.

and the vertical axis is for travel time. The transducer is moved linearly on the surface of the part and a number of very closely placed A-scans are recorded for the same defects. A trigger point is introduced in the A-scan data such that whenever the signal density crosses the trigger level, a point is created on the B-scan and as the transducer is moved, lines representing the defects are created.

12.3.5.3 C-Scan

A C-scan (Figure 12.5) is a 2D graphical representation of defects in the laminate plan view. The transducer is moved parallel to the laminate surface in a scan pattern so as to cover the complete laminate plane. Usually, a trigger point is introduced in the A-scans such that relative amplitude or travel time is recorded at regular intervals and planner representation of the defects is obtained.

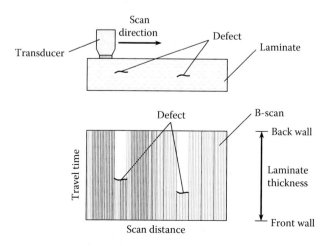

FIGURE 12.4 Schematic B-scan in a pulse-echo ultrasonic test.

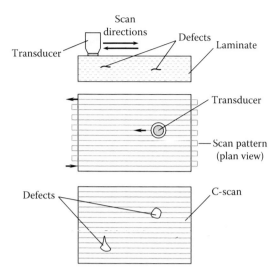

FIGURE 12.5 Schematic C-scan in an ultrasonic test.

12.3.6 Advantages and Disadvantages

Ultrasonic testing has a number of advantages that make it the most popular NDE technique. It is also associated with some disadvantages. These are enumerated below:

Advantages:

- Ultrasonic testing is a highly versatile technique; it is useful for detecting a wide range of defects, including both surface as well as interior defects.
- In addition to flaw detection, it can also be used for thickness measurement of a part.
- The process can be automated and the details of the defects can be obtained in various formats.
- In an ultrasonic testing, no elaborate part preparation is required.
- It is a very quick technique that provides instantaneous results.
- In the pulse-echo technique, access to only one side of the part is sufficient.
- Ultrasonic testing is a highly convenient process as it can be carried out using portable equipment.
- Ultrasonic testing does not involve any hazardous process or any hazardous material that can pose a risk to the safety of the operator or to the part.

Disadvantages:

- Ultrasonic testing is effective in detecting defects that are planner in general orientation; it is difficult to detect linear defects parallel to the applied beam by this method.
- The method is not suitable for composite laminates with transverse matrix cracks and fiber misalignment.
- It is difficult to carry out ultrasonic testing of components with rough surfaces.
- At least one surface of the part must be accessible so that it is possible to probe by holding the ultrasonic transducer on the surface. Also, too small parts and parts with complex configuration are difficult to scan. For complex part geometries, specially made probes may be needed.
- For efficient transfer of signal, a coupling medium is required between the part and the piezoelectric transducer. As a result, the process tends to become slow.

- Proper training of operators is essential for carrying out ultrasonic testing as well as interpretation of output data.
- It is essential to have reference standards for equipment calibration as well as characterization of defects.

12.4 RADIOGRAPHIC TESTING

12.4.1 Basic Concept of Radiographic Testing

During radiographic testing of a material, it is exposed to x-ray or gamma ray radiation that penetrates it and forms an image on a special type of film. The image is examined and interpreted to draw conclusions regarding the existence of possible defects and their characteristics [16,17].

From Equation 12.1, we know that frequency and wavelength are inversely related. Let us now consider the energy of the wave, which is directly related to frequency and inversely related to wavelength, as follows:

$$E = hf = \frac{hc}{\lambda} \tag{12.5}$$

where
- c Velocity of light
- h A constant called Planck's constant
- E Energy of the wave
- f Frequency of the wave
- λ Wavelength of the wave

X-rays and gamma rays are electromagnetic waves of extremely short wavelengths. Thus, these rays have very high frequency and energy (Table 12.2). As a matter of fact, in the electromagnetic spectrum, gamma rays top the list in an increasing order of wavelengths, followed by x-rays. (The electromagnetic spectrum consists of radio waves, microwaves, infrared radiations, optical waves, ultraviolet radiations, x-rays, and gamma rays.) Owing to their extremely short wavelengths and high energy levels, x-rays and gamma rays can penetrate through objects that are otherwise opaque to visual lights.

When x-rays or gamma rays are directed on a material, the energy of the radiation is partially attenuated by absorption and scattering of some of the photons within the material. The remaining photons travel through the material, the number of which is dependent on the energy of the photons, thickness, density, and atomic weight of the material. Mathematically, the intensity of the transmitted radiation is given by

$$I = I_0 e^{-\mu x} \tag{12.6}$$

TABLE 12.2
Characteristic Features of X-Rays and Gamma Rays

Parameters	X-Rays	Gamma Rays
Wavelength (m)	1×10^{-11} to 1×10^{-8}	$<1 \times 10^{-11}$
Frequency (Hz)	3×10^{16} to 3×10^{19}	$>3 \times 10^{19}$
Energy (J)	2×10^{-17} to 2×10^{-14}	$>2 \times 10^{-17}$

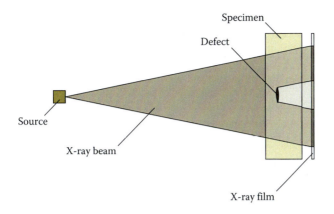

FIGURE 12.6 Schematic representation of radiographic testing.

where
- I Intensity of the transmitted radiation
- I_0 Intensity of the incident radiation
- $\mu = \mu(\lambda,\rho)$ Linear absorption coefficient of the material, which is a function of the wavelength of the rays and density of the material
- x Thickness of the material

The ratio of the linear absorption coefficient to the density of the material, that is, μ/ρ, is a material constant called the mass absorption or attenuation coefficient, which tends to increase with increasing atomic weight. The attenuation of the radiation increases exponentially with thickness and density. A defect in material represents a change in density and it results in a difference in attenuation. When a photographic film is exposed to the radiations and subsequently processed in a developing solution, the exposed areas become dark. The degree of darkening depends on the amount of exposure; minimum exposure (i.e., maximum attenuation) would result in minimum darkening and vice versa. Thus, a defect in the material can be detected by differences in the shades in the film (Figure 12.6).

12.4.2 Radiographic Test Setup

The radiographic test setup consists of basically three parts—the radiation source, the specimen on a support table, and the radiation-sensitive film on a holder. The source of radiation is the primary element in the setup and it is different for different types of radiation.

12.4.2.1 X-Ray Radiography

The source of x-rays is an x-ray tube (Figure 12.7), which is a vacuum tube with a cathode and an anode. The vacuum tube is a sealed container in which the cathode is an incandescent filament inside a focusing cup and the anode is a plate usually made of tungsten. A low voltage is applied to heat the filament that excites the electrons. A high voltage, referred to as the tube voltage, is applied between the cathode and anode; it causes the electrons to break free from the filament and accelerate toward the anode, which is the target. The focusing cup helps focus the stream of electrons (called the tube current) on a small area known as the focal point on the target. The collision of the high-velocity electrons with the target produces x-rays.

X-rays of different characteristics are needed for different materials and different thickness ranges. In general, radiations with higher penetrating power are required for thicker parts and denser materials. The low voltage applied to the filament is regulated

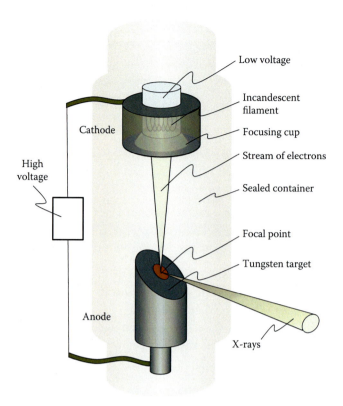

FIGURE 12.7 Schematic representation of x-ray tube. (Adapted from *Non-Destructive Testing of Fiber Reinforced Plastics Composites, Volume 1*, A. F. Blom and P. A. Gradin, Radiography, pp. 1–23, Copyright 1987, from Elsevier Applied Science.)

so as to control the intensity (i.e., number of photons passing through a unit normal area in a unit time) of the tube current. On the other hand, tube voltage is regulated for controlling both the energy and intensity of the radiation. The higher the tube voltage, the higher the speed of the electrons colliding with the target and the higher the energy and intensity of the radiation. The energy of the radiation is expressed in electron volts. The nominal energy level required for x-ray radiography of PMCs is lower than that for metals; typically, 4–8 MeV are common in PMCs.

12.4.2.2 Gamma Ray Radiography

Gamma rays are similar to x-rays in nature; the difference between the two lies in the sources of radiation. Unlike x-rays, gamma rays are produced by certain radioactive isotopes such as cobalt-60 and iridium-192. Unlike x-rays, gamma rays are emitted by these isotopes in two or three discrete wavelengths. Typically, a gamma ray source is a small pellet of the isotope shielded inside a portable stainless-steel device called camera. It has a cranking mechanism for making an exposure during radiographic testing.

Film radiography is the oldest and most commonly used radiographic technique. Filmless techniques are also in vogue; two of these techniques are real-time radiography and computed tomography.

12.4.3 Real-Time Radiography

Computer-assisted real-time radiography provides an opportunity for a quick inspection of a product, especially in an assembly line. A computerized system is connected to the radiographic test setup. The image obtained is digitized, processed, and stored using a computer. The data can be analyzed in real time; however, the real-time radiographs are inferior to film-based radiographs.

12.4.4 Computed Tomography

Computed tomography is a sophisticated version of radiographic testing that enables the technician to obtain a 3D representation of defects in the specimen [18]. The test setup provides for moving the radiation source along a circular path about a vertical axis around the specimen. Alternatively, the specimen is rotated about its vertical axis on a turntable. A collimator is used such that only a thin slice of the specimen normal to the vertical axis is exposed to radiation. At each angular position of the specimen w.r.t. the radiation source, the fan-shaped beam of radiation penetrates a slice of the specimen and an attenuation profile is recorded. A number of attenuation profiles at regular angular interval are recorded and the data are processed using a computer-based algorithm to reconstruct the cross-sectional profile of the specimen. The collimator is moved vertically and cross-sectional profiles of various slices at regular vertical intervals are obtained. A 3D image is then created by combining the cross-sectional profiles of all the slices. Figure 12.8 gives a schematic representation of the computed tomographic process.

12.4.5 Advantages and Disadvantages

Radiographic tests and ultrasonic tests are often employed as complementary to each other; data from one test can be used for the interpretation of data from the other and thereby a wide range of defects can be detected with maximum accuracy. Like ultrasonic testing, radiographic testing too has its own advantages and disadvantages. These are briefly enumerated below:

Advantages:

- Radiographic testing is useful for detecting a wide range of defects that include delaminations, debonds, voids, foreign material, etc.
- It can also be used, by taking tangential shots, for thickness measurement of a part.

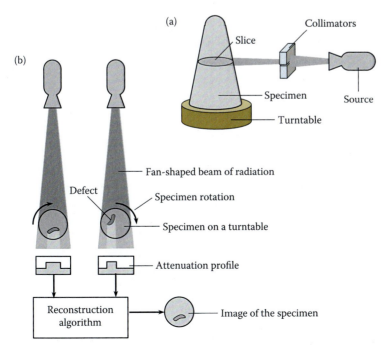

FIGURE 12.8 Schematic representation of computed tomography. (a) Exposure of a slice of the specimen. (b) Attenuation profiles in the slice at two angular locations.

- There is nearly no limitation on the type of materials that can be tested by radiographic techniques.
- Practically no part preparation is required in this method.
- Direct physical access to the part is not required.
- Permanent record of defects can be obtained.

Disadvantages:

- The location of defects in the thickness direction cannot be obtained in radiographic testing with shots normal to the laminate.
- Access to both the sides of the part is required.
- It is a slow process involving exposure of the part to the radiation followed by development of the films.
- Radiographic testing is an expensive process as it requires highly expensive equipment for the generation of x-rays or gamma rays. Similarly, a proper support table with suitable fixturing is necessary for holding and manipulating the part during inspection. Also, the civil works that house the testing facility are generally highly expensive.
- Radiographic testing is a highly hazardous process and adequate protective measures to the operators and nearby personnel are essential.
- A high level of training of operators is essential for carrying out radiographic testing as well as interpretation of output data.

12.5 ACOUSTIC EMISSION

12.5.1 Basic Concept of Acoustic Emission

12.5.1.1 Acoustic Emission

When a structure is loaded, it undergoes deformations and strain energy is stored in it. The stress distribution in a loaded structure changes suddenly when damage mechanism processes such as crack initiation and propagation, delamination, fiber breakage, debond, etc. occur. These processes are associated with a rapid release of strain energy. The released strain energy is partially consumed in crack propagation by new crack surface creation and plastic zone growth. Some of the released strain energy generates transient stress waves. AE is a term that refers to the generation of such transient stress waves [19–21].

AE waves are similar to ultrasonic waves in nature and they follow the same rules of reflection, refraction, attenuation, and other associated phenomena. They are of two types—burst emission and continuous emission. Burst emission corresponds to an individual event and it is a discrete signal of high amplitude that quickly dies down. Continuous emission, on the other hand, is characterized by sustained levels of amplitudes that correspond to rapidly occurring events.

12.5.1.2 AE Sources

As indicated earlier, the common AE sources in a PMC material are the various processes of damage such as matrix crack initiation and propagation, delamination, fiber breakage, debond, etc. Owing to the anisotropic nature, failure in a PMC material is a progressive phenomenon characterized by the initiation of matrix microcracking at a much lower level of load than the final failure load. Under gradually increasing loads, crack propagation associated with fiber bridging, delamination, debonding, and, finally, fiber fracture take place. Clearly, failure process in a PMC material is a complex phenomenon and the associated AE signals are a mixture of different types. In addition to the signals generated from the damaged zones, noises from unwanted mechanical, thermal, and electromagnetic sources clutter the raw data.

12.5.1.3 Kaiser Effect and Felicity Effect

The Kaiser effect is an important concept in the field of AE. It states that a material does not release AE at load levels lower than or equal to the previous maximum. To understand it in simpler words, let us consider a material subjected to a maximum load P_1. Let us unload and reload the material to a load P_2. Then, as per the Kaiser effect, acoustic activity will take place during reloading only if $P_2 > P_1$. The Kaiser effect was originally postulated for isotropic materials such as metals. This, however, does not hold good in composite materials. The breakdown of the Kaiser effect in composite materials is referred to as the Felicity effect. In this connection, a parameter called Felicity ratio is used and it is defined as the ratio of the load at which Felicity effect occurs, that is, the load, lower than the previous maximum, at which acoustic activity takes place during reloading, to the previously applied maximum load. The Kaiser effect and Felicity effect principles are often utilized in periodic inspection of structures for the detection of possible damage growth.

12.5.2 AE Test Setup

In an AE test, the test article is loaded gradually as per a test plan. Under the applied loads, as explained previously, AE waves are generated. The signals are detected and processed to locate and characterize possible defects in the test article. The test setup is schematically shown in Figure 12.9. There are basically two parts in the test setup—data acquisition unit and data analysis unit. The data acquisition part consists primarily of the AE sensors and the signal cables. A number of sensors in an array are attached directly to the structure for detecting the signals. The signals are preamplified, filtered, postamplified, and fed to the computer for processing, storage, and display.

12.5.3 Data Acquisition

The stress waves generated from a source propagate through the material and eventually reach the material surface, where they are detected by the AE sensors. Depending on the test objectives and test article characteristics, the sensors may be employed in single-channel, small multichannel array, or in elaborate multichannel array. In general,

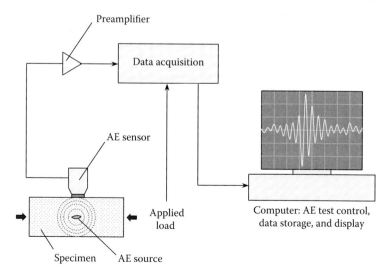

FIGURE 12.9 Schematic representation of acoustic emission test setup.

in an AE test, piezoelectric transducers are used as sensors. The sensors should be sensitive in the appropriate frequency range, which for PMC materials is usually 20–1000 kHz. Suitable gel-type couplants are used for mounting the sensors on the component surface such that the sensors can be removed after the test. For permanent bonding of sensors, epoxy-based adhesives are used.

12.5.4 Data Analysis

Data analysis is carried out for extracting useful information from the raw data. It involves primarily three steps—data processing, data interpretation, and evaluation of the structural integrity of the test article. Depending on test objectives, data analysis can be done as (i) real-time analysis during the test, (ii) post-test analysis, and (iii) both real-time and post-test analyses. Real-time data analysis is done when test load conditions are controlled based on the data analysis output. Usually, it involves plotting a number of graphs, in some predesignated format, of key parameters w.r.t. the applied loads. Post-test analysis is done when real-time analysis is not required. On the other hand, when real-time analysis is inconclusive and more detailed information is required, both real-time and post-test analyses are carried out.

The signals received from the sensors are generally of very low voltage and some amount of preamplification is required before any signal processing. A preamplifier of appropriate type is connected to the sensor output. It boosts the impedance of the signals to a level that is suitable for the long signal cables and other electronic units. Another important function of the preamplifier is to remove the unwanted noise, for which three different types of filters—high-pass filters, low-pass filters, and band-pass filters—are fitted to the preamplifier. The preamplified and boosted signals are then fed to the main amplifier unit, which gives further boost to the strength of the signals.

The main amplifier output can be treated by analog processing as well as digital processing. In analog processing, the burst events, which occur at random, are separated from the continuous events by employing a certain threshold. The continuous signals are usually processed by using root-mean-square (RMS) techniques. The RMS techniques are especially useful to detect the progressive failure of PMCs where the events occur rapidly in nearly continuous fashion.

The waveform of the AE signals is characterized by a number of parameters (Figure 12.10); notable among them are event count, ring-down count, amplitude, energy, event duration, and rise duration. The digital processing techniques revolve around the

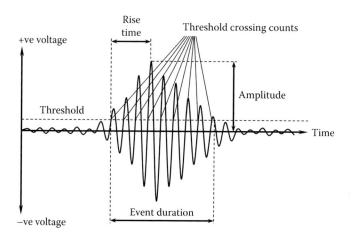

FIGURE 12.10 AE waveform parameters. (Adapted with permission from P. K. Mallick, Nondestructive tests, *Composites Engineering Handbook* (P. K. Mallick, ed.), Marcel Dekker Inc., New York, 1997, pp. 1147–1181.)

digital measurement of these parameters. For example, in the ring-down counting, the number of amplified pulses in excess of certain threshold amplitude is counted. Similarly, event counting involves counting of groups of pulses that exceed the threshold amplitude. Often, these counting techniques are employed, together with Felicity principle, in detecting damage initiation and damage growth in a composite structure under loads.

12.5.5 Advantages and Disadvantages

AE is a powerful method for the NDT of PMCs. Some of its main advantages and disadvantages are listed below:

Advantages:

- AE is a method for real-time and continuous health monitoring of a structure under load.
- It allows the detection of a growing defect much before its critical size is reached.
- It is a cost-effective and quick method; in an AE test setup, it is possible to deploy an array of limited number of sensors for global monitoring of the structure.
- It can be operated from a remote location in an unattended mode.

Disadvantages:

- AE testing demands the application of loads on the structure.
- It is not suitable for determination of damage size.
- AE testing and data analysis become very complicated due to the anisotropic nature of composites and high noise levels.

12.6 INFRARED THERMOGRAPHY

12.6.1 Basic Concept of IR Thermography

Infrared thermography, in the field of NDT, is a method in which a defect is detected by inspecting the specimen's surface temperature distribution and/or surface temperature decay rate [22]. A defect in the specimen, when compared to the surrounding material, possesses dissimilar thermal properties. As a result, it appears as a variation in a thermographic image.

Thermographic NDT methods are of two types—active or dynamic and passive or static. In active thermography, the specimen is heated either on its surface or in its interior locations for a limited period by means of some external means. As the heat dissipates into the interior of the specimen and the surrounding, the surface temperature distribution and temperature decay rates are monitored. An infrared camera is used in the bands of 3–6 μm for short wavelengths and 9–12 μm for long wavelengths. It converts the temperature distribution into a thermograph, in which a bright shade represents a hot spot and a dark shade a cold spot. Typically, a defect is seen either as a hot spot or as a cold spot.

In passive thermography, too, an infrared camera is used; however, in this case, the specimen is kept under normal conditions. Any matter above the absolute temperature emits electromagnetic radiations due to the motions of the atomic particles. Surface temperature distribution is monitored in a steady state and a thermographic image is made. For the determination of defects, it is compared with the thermographic image obtained from a defect-free specimen.

12.6.2 Types of Active Thermographic Methods

A number of active thermographic methods are in vogue; these methods can be further classified into optical methods and mechanical methods. In the optical methods, heat is applied by external excitation via optical means using different types of lamps, infrared heaters, or jets of hot air/water. Three commonly used optical active methods are—pulse thermography, lock-in thermography, and step heating thermography. Mechanical methods, on the other hand, involve heating of some interior locations, typically crack tips, by means of mechanical vibrations. Vibrothermography is a common mechanical active thermographic method.

12.6.2.1 Pulse Thermography

PT is a popular IR thermographic method in which the specimen is exposed to a short heat pulse from an external source. For nonconducting materials, such as PMCs, the pulse duration is a few seconds, whereas for conducting materials, it is generally less than a second. A schematic representation of the pulse thermography is given in Figure 12.11. The light wave from the source travels in air by radiation before it reaches the specimen surface. The infrared radiation partially gets reflected at the specimen surface and the rest travels by conduction as thermal wave inside the material. When encountered with a defect, some portion of the thermal wave gets reflected back to the surface and increases the local temperature. As result, the defect is seen as a hot spot in the thermographic image.

12.6.2.2 Lock-In Thermography

The reflected light wave from the specimen surface, in pulse thermography, may create noise, making it difficult to detect defects; this issue is overcome in the lock-in-thermographic method. In this method, instead of a single pulse, the intensity of the lamp is controlled and a continuous sinusoidal light wave is applied on the specimen. The thermal wave traveling through the material partially gets reflected from the defect and returns to the specimen surface. The reflected wave and the incoming wave interfere at the specimen surface and interference patterns are formed. An infrared camera

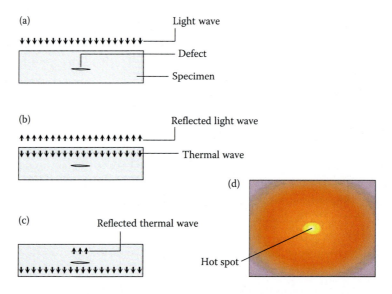

FIGURE 12.11 Pulse thermography. (a) Light wave traveling by radiation from source toward the specimen. (b) Partial reflection of light wave from the specimen surface and conduction of the rest as thermal wave. (c) Partial reflection of thermal wave from the defect surface. (d) Schematic representation of defect as a hotspot in the thermographic image.

monitors the interference patterns and the defects are detected by analyzing the phase and amplitude data.

12.6.2.3 Vibrothermography

Vibrothermography is based on the fact that when a specimen with flaws is subjected to external mechanical vibration at certain frequencies, the crack surfaces rub against each other and produce heat. Thus, the crack tips and surfaces are seen as hot spots in a thermographic image.

12.6.3 Advantages and Disadvantages

Infrared thermography is a unique NDT method and it offers several advantages over others. It can often be used effectively where other methods such as ultrasonic testing and radiography do not produce reliable results. Similarly, like other NDT techniques, it also has its own limitations. Some of its major advantages and disadvantages are as follows:

Advantages:

- Infrared thermography is a noncontact method and inspection can be done from a distance from the structure.
- It is fast method as a relatively large surface can be covered in a single shot.
- It provides a pictorial representation of the surface area and, as a result, data interpretation is quick.

Disadvantages:

- Infrared thermography is suitable mainly for surface and superficial defects only; its effectiveness reduces with increase in part thickness.
- The anisotropic nature of composites may produce different thermal properties in different directions, resulting in difficulty in data interpretation.
- Infrared camera may sometimes be expensive.

12.7 EDDY CURRENT TESTING

12.7.1 Basic Concept of Eddy Current Testing

Eddy current testing is based on the phenomenon of electromagnetic induction [23]. A conductor carrying electrical current is associated with a magnetic field around it. The magnetic flux is normal to the direction of the electric current. An alternating current in a conductor produces a magnetic field with corresponding varying strength and direction. Now, if a second conductor is placed close to the first, the varying magnetic field induces electric current in the second conductor. Further, if the first conductor is a coil and the second conductor is a flat object, the induced current follows a closed-loop circular path in the plane of the object (Figure 12.12). These induced currents, due to their resemblance with eddies, are called eddy currents.

Eddy current testing, as an NDT technique, involves inducing eddy currents in the specimen. The specimen is nothing but the second conductor referred above. A probe containing a coiled conductor carrying alternating current is used for excitation of the specimen. The probe is also used for receiving signals back from the specimen. Alternatively, a separate probe is used for receiving signals. The alternating current in the probing coil generates an oscillating magnetic field, referred to as the primary

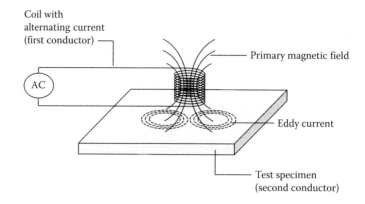

FIGURE 12.12 Schematic representation of eddy current testing setup.

magnetic field, which in turn, produces eddy current in the specimen. The eddy current produces a secondary magnetic field, which interacts with the primary magnetic field through mutual inductance.

The eddy current is affected by defects such as near-surface cracks. The variations such as changes in the amplitude and pattern of the eddy current and its associated magnetic field cause changes in the impedance of the receiving probe coil. An eddyscope instrument connected to the receiving coil plots the changes in impedance amplitude and phase angle, which are used for identifying defects in the material.

12.7.2 Advantages and Disadvantages

The advantages and disadvantages associated with eddy current testing can be enumerated as follows:

Advantages:

- Eddy current testing is a simple, reliable, and quick method for the NDT of conducting materials. By employing a number of channels in an array, a large specimen area can be covered in a single pass, which results in drastic reduction in inspection time and increased reliability.
- Part preparation required is a minimum.
- Access to both the sides of the part is not essential.
- It is a noncontact method, that is, the probe does not need to be in contact with the part.

Disadvantages:

- The applicability of eddy current testing is limited to only conductive materials; among PMCs, it is suitable mainly for carbon fiber–reinforced composites.
- The depth of penetration is limited and surface and near-surface defects can only be reliably detected by this technique; it is not suitable for detecting defects deep inside the material.
- It is not suitable for the detection of planner defects aligned in the directions of the eddy currents.
- The surface finish of the test specimen may interfere with the electromagnetic fields.
- Highly skilled trained manpower is essential for the interpretation of the data recorded.

12.8 SHEAROGRAPHY

Many optics-based techniques are in vogue for various engineering applications. Clearly, some source of light is used in these methods. Several of them have found use in the field of NDT of composites. Some of the optical NDT methods are moire interferometry, holographic interferometry, speckle photography, shearography, etc. All of these methods have their own unique flaw detection capabilities, advantages, and disadvantages. In this section, we discuss in brief the basic concepts and advantages and disadvantages of shearography.

12.8.1 Basic Concept of Shearography

The principle of optical interference is used in shearography [24–26]. The test specimen is illuminated by a divergent beam from a point laser source. The surface roughness of the specimen is required to be of the order of the wavelength of the incident light or more. The beam gets scattered from the surface; it is made to pass through a special shear lens and the image is recorded either by conventional wet photographic means or digitally. The sheared image and the direct image interfere with each other and form a speckle pattern. (A speckle pattern is formed by mutual interference of waves of same frequency.)

Now, if the specimen is loaded by some means such as acoustic or mechanical vibration, thermal loading, etc., a stress/strain distribution is created, which is affected by the presence of any defect. As a result, the speckle pattern from a loaded specimen also gets affected and defect detection is done by comparing the speckle patterns of the specimen in the unloaded and loaded states.

12.8.2 Advantages and Disadvantages

Shearography, as an NDT method, has its own advantages and disadvantages as enumerated as follows:

Advantages:

- Shearography is a noncontact method; inspection of the test specimen can be done without bringing any probe or test equipment in contact with it.
- It provides a full-field measurement of the test specimen.
- It provides a quick and reliable method of inspection and is suitable for on-field inspection.

Disadvantages:

- In shearography, the specimen is required to be loaded although to a lower level than operating conditions.
- Surface roughness should be comparable with the wavelength of light or more.
- Data interpretation is complex and high level of training and experience is needed for reliable flaw detection.

12.9 SUMMARY

A brief discussion on NDT of PMCs is given in this chapter. The key points that can be noted from the presentation are as follows:

- Defects are created in a PMC material during processing as well as during service life.

- In a PMC material, defects are of many types and they are usually more in number.
- Several NDT techniques are available for defect detection in PMCs.
- NDT techniques for PMCs are primarily extensions, with necessary adaptations, of those originally developed for metals.
- NDT techniques have their own unique advantages and disadvantages; while ultrasonic testing and radiographic testing are the most common NDT techniques used in PMCs other methods, too, have their share of use.

EXERCISE PROBLEMS

12.1 What are the different types of defects encountered in PMCs? How are they different from those in metals?

12.2 What are the common nondestructive testing techniques employed in PMCs? Prepare a brief note giving the principles behind each of these techniques and their applicability in the detection of specific defects.

12.3 What are the two different techniques employed in ultrasonic testing? What are the different ways of data representation? Give a brief note.

12.4 Give a brief note on the basic principle of radiographic testing. What are the different techniques of radiographic testing that are used in composites?

12.5 Give a comparative note on the advantages and disadvantages of ultrasonic testing and radiographic testing.

REFERENCES AND SUGGESTED READING

1. P. K. Mallick, Nondestructive tests, *Composites Engineering Handbook* (P. K. Mallick, ed.), Marcel Dekker Inc., New York, 1997, pp. 1147–1181.
2. R. B. Heslehurst, *Defects and Damage in Composite Materials and Structures*, CRC Press, Boca Raton, FL, 2014.
3. T. S. Jones, Nondestructive evaluation methods for composites, *Handbook of Composites* (S. T. Peters, ed.), second edition, Chapman & Hall, London, 1998, pp. 838–856.
4. R. Prakash, Non-destructive testing of composites, *Composites*, 11(4), 1980, 217–224.
5. I. G. Scott and C. M. Scala, A review of non-destructive testing of composite materials, *NDT International*, 15(2), 1982, 75–86.
6. W. N. Reynolds, Nondestructive testing (NDT) of fiber-reinforced composite materials, *Materials & Design*, 5(6), 1984–1985, 256–270.
7. R. D. Adams and P. Cawley, A review of defect types and nondestructive testing techniques for composites and bonded joints, *NDT International*, 21(4), 1988, 208–222.
8. K. Senthil, A. Arockiarajan, R. Palaninathan, B. Santhosh, and K. M. Usha, Defects in composite structures: Its effects and prediction methods—A comprehensive review, *Composite Structures*, 106, 2013, 139–149.
9. M. E. Ibrahim, Nondestructive evaluation of thick-section composites and sandwich structures: A review, *Composites Part A: Applied Science and Manufacturing*, 64, 2014, 36–48.
10. M. R. Jolly, A. Prabhakar, B. Sturzu, K. Hollstein, R. Singh, S. Thomas, P. Foote, and A. Shaw, Review of non-destructive testing (NDT) techniques and their applicability to thick walled composites, *Procedia CIRP*, 38, 2015, 129–136.
11. S. Gholizadeh, A review of non-destructive testing methods of composite materials, *Procedia Structural Integrity*, 1, 2016, 50–57.
12. R. L. Crane, Nondestructive inspection of composites, *Comprehensive Composite Materials, Volume 5, Test Methods, Nondestructive Evaluation, and Smart Materials* (A. Kelly, editors-in-chief, L. Carlsson, R. L. Crane and K. Uchino, vol. eds.), Elsevier, Amsterdam, the Netherlands, 2000, pp. 259–320.
13. B. R. Tittmann, Ultrasonic inspection of composites, *Comprehensive Composite Materials, Volume 5, Test Methods, Nondestructive Evaluation, and Smart Materials* (A. Kelly, editors-in-chief, L. Carlsson, R. L. Crane and K. Uchino, vol. eds.), Elsevier, Amsterdam, the Netherlands, 2000, pp. 259–320.
14. E. G. Henneke, II, Ultrasonic nondestructive evaluation of advanced composites, *Non-Destructive Testing of Fiber Reinforced Plastics Composites, Volume 2* (J. Summerscales, ed.), Elsevier Applied Science, London, 1990, pp. 55–159.

15. ASM International Committee on Nondestructive Testing of Composites, Nondestructive testing, *ASM Handbook, Volume 21, Composites* (D. B. Miracle and S. L. Donaldson, vol. chairs), ASM International, Ohio, 2001, pp. 699–725.
16. R. L. Crane, D. Hagemair, and R. Fassbender, Radiographic inspection of composites, *Comprehensive Composite Materials, Volume 5, Test Methods, Nondestructive Evaluation, and Smart Materials* (A. Kelly, editors-in-chief, L. Carlsson, R. L. Crane and K. Uchino, vol. eds.), Elsevier, Oxford, 2000, pp. 325–344.
17. A. F. Blom and P. A. Gradin, Radiography, *Non-Destructive Testing of Fiber Reinforced Plastics Composites, Volume 1* (J. Summerscales, ed.), Elsevier Applied Science, London, 1987, pp. 1–23.
18. R. H. Bossi, K. D. Friddell, and A. R. Lowrey, Computed tomography, *Non-Destructive Testing of Fiber Reinforced Plastics Composites, Volume 2* (J. Summerscales, ed.), Elsevier Applied Science, London, 1990, pp. 201–252.
19. M. Arrington, Acoustic emission, *Non-Destructive Testing of Fiber Reinforced Plastics Composites, Volume 1* (J. Summerscales, ed.), Elsevier Applied Science, London, 1987, pp. 25–63.
20. M. Wevers and M. Surgeon, Acoustic emission and composites, *Comprehensive Composite Materials, Volume 5, Test Methods, Nondestructive Evaluation, and Smart Materials* (A. Kelly, editors-in-chief, L. Carlsson, R. L. Crane and K. Uchino, vol. eds.), Elsevier, Oxford, 2000, pp. 345–357.
21. W. Choi and H.-D. Yun, Non-destructive evaluation (NDE) of composites: Use of acoustic emission (AE) techniques, *Non-Destructive Evaluation (NDE) of Polymer Matrix Composites—Techniques and Applications* (V. M. Karbhari, ed.), Woodhead Publishing, Cambridge, 2013, pp. 557–573.
22. K. E. Puttick, Thermal NDT methods, *Non-Destructive Testing of Fiber Reinforced Plastics Composites, Volume 1* (J. Summerscales, ed.), Elsevier Applied Science, London, 1987, pp. 65–103.
23. R. Prakash, Eddy-current testing, *Non-Destructive Testing of Fiber Reinforced Plastics Composites, Volume 2* (J. Summerscales, ed.), Elsevier Applied Science, London, 1990, pp. 299–325.
24. D. Francis, Non-destructive evaluation (NDE) of composites: Introduction to shearography, *Non-Destructive Evaluation (NDE) of Polymer Matrix Composites—Techniques and Applications* (V. M. Karbhari, ed.), Woodhead Publishing, Cambridge, 2013, pp. 56–83.
25. Y. Y. Hung, L. X. Yang, and Y. H. Huang, Non-destructive evaluation (NDE) of composites: Digital shearography, *Non-Destructive Evaluation (NDE) of Polymer Matrix Composites—Techniques and Applications* (V. M. Karbhari, ed.), Woodhead Publishing, Cambridge, 2013, pp. 84–115.
26. F. Taillade, M. Quiertant, and K. Benzarti, Non-destructive evaluation (NDE) of composites: Using shearography to detect bond defects, *Non-Destructive Evaluation (NDE) of Polymer Matrix Composites—Techniques and Applications* (V. M. Karbhari, ed.), Woodhead Publishing, Cambridge, 2013, pp. 542–556.

13

Metal Matrix, Ceramic Matrix, and Carbon/Carbon Composites

13.1 CHAPTER ROAD MAP

Various aspects of PMCs have been covered in the chapters so far and we are close to moving on to the design of composite products. However, before that, we need to familiarize with some essential concepts related to the other three broad classes of composites—MMCs, CMCs, and C/C composites. These composites have their own characteristics and unique applications and play major roles in the overall field of composites technology. In fact, to a large extent, they complement each other and help the designer achieve an optimal solution. Thus, it is desirable that a composites engineer or scientist has introductory knowledge in these areas as well. With this in mind, this chapter is devoted to present brief discussions on these three classes of composites. Basic characteristics including advantages and disadvantages, constituent raw materials, viz. reinforcements and matrices, manufacturing methods, and applications are addressed.

This chapter can be taken up immediately after Chapter 1. However, familiarity with the topics on constituent materials and composites manufacturing discussed in Chapters 9 and 10 is desirable.

13.2 INTRODUCTION

Work on MMCs was initiated around the 1950s; however, significant development and growth took place much later—after the 1980s. Today, MMCs have their own place in the R&D and industry with numerous applications in aerospace, automotive, ground transportation, electronics, and other commercial sectors. CMCs and C/C composites are relatively new compared to the PMCs and MMCs and they are known for their exceptionally high-temperature capabilities.

Characteristic features, constituent reinforcing and matrix materials, manufacturing processes, and applications in respect of MMCs, CMCs, and C/C composites are addressed at a level deemed sufficient for a product development engineer in the field of PMCs. In-depth discussion is beyond the scope of this book. For further details, the interested reader can consult available articles and texts in the fields of MMCs, CMCs, and C/C composites. For instance, a thorough treatment of the characteristic behavior of MMCs, their constituents, viz. reinforcements and matrix, processing techniques, application, current industrial status, etc. is given in References 1–3. Further information on various general and specific areas related to MMCs can also be found in many articles, for instance, References 4–13. Carbon fibers are an important reinforcing material not only in PMCs but also in MMCs, CMCs, and C/C composites; extensive work has been done in this field and information regarding structure, properties, precursor materials and carbon fiber production, carbon fiber composites, and application can be found in several texts (see, for instance, References 14–16). Various aspects of CMCs, including general characteristics, reinforcing materials and matrix materials, fabrication process, and applications, are presented in References 17–19. CMCs are also discussed in a number of articles [19–22]. C/C composites are sometimes treated as a special case of CMCs where the matrix and reinforcement are both carbon. A number of text books and articles are available solely devoted to discuss characteristics, production of raw materials, preforming and fabrication of products, applications, and other related aspects of C/C composites [23–27].

13.3 METAL MATRIX COMPOSITES

13.3.1 Characteristics of MMCs

As noted in the introductory chapter, in an MMC material, a metal or an alloy is the continuous phase in which the reinforcement is embedded. The reinforcements can be either continuous or discontinuous. Based on the geometry of the reinforcements, four broad classes of MMCs can be identified, as follows:

- Particulate-reinforced MMCs
- Short fiber- and whisker-reinforced MMCs
- Continuous fiber-reinforced MMCs
- Monofilament-reinforced MMCs

Particulate reinforcements are either metallic or ceramic particles where no dimension of the reinforcements is more than about five times the other two dimensions. Particulate-reinforced MMCs are isotropic and they exhibit generally inferior mechanical properties as compared to the other types of MMCs. However, their processing is cheap and simple.

Short fibers and whiskers are discontinuous reinforcements with aspect ratio (length-to-diameter ratio) more than five. These MMCs are generally anisotropic and the extent of anisotropy increases with the nominal aspect ratio of the fibers. However, with specialized powder metallurgy routes and preform making, isotropic properties can also be achieved in these MMCs. Property characteristics of short fiber/whisker-reinforced MMCs are dominated by the matrix and they are pretty similar to or marginally better than the particulate-reinforced MMCs.

Continuous fibrous reinforcements are tows of Al_2O_3, SiC, carbon, boron, etc. (A tow is a bundle of several hundred or thousand fibers.) Fibers are embedded into the matrix in a certain direction and it results in highly anisotropic properties of the MMC. High fiber volume fractions are possible and composite properties are dominated to a large extent by the fibers. Strength and stiffness are very high in the fiber direction. Across the fibers, the matrix plays a dominant role and the transverse properties are similar to those of particulate-, short fiber-, or whisker-reinforced MMCs.

Monofilaments used are mainly boron, silicon carbide, etc. These are very similar to the continuous fibers, the primary difference being the diameter of the reinforcements. Monofilaments are larger in diameter and are available as individual filaments wound in spools.

Addition of reinforcements in a monolithic metal greatly improves its mechanical and other properties and MMCs have a number of advantages over monolithic metals as well as PMCs. There are certain disadvantages as well. Some of the key advantages and disadvantages are as follows:

Advantages:

1. *High specific strength and stiffness (w.r.t. monolithic metals)*: MMCs possess higher specific strength and stiffness compared to their monolithic counterparts. For example, a representative range of specific tensile strengths of monolithic

aluminum and its alloys is 22–259 MPa/g/cc, whereas the corresponding figures for SiC/Al MMC are of the order of 72–259 MPa/g/cc. As a result, in general, MMCs are structurally more efficient than monolithic metals.

2. *Low thermal expansion*: Average CTEs of MMCs are relatively lower than those of the monolithic metals and PMCs. As a result, issues related to thermal mismatch and thermal stresses and strains are less severe.
3. *High fatigue resistance*: MMCs exhibit higher fatigue resistance—higher fatigue strength and longer fatigue life—compared to unreinforced metals.
4. *High service temperature*: Common polymers such as epoxy, polyester, etc. degrade at high temperatures. Generally, they have low T_g and low service temperatures. As a result, PMCs cannot be exposed to high temperatures during their service. On the other hand, metals and MMCs have relatively higher service temperatures.
5. *Tailorable properties*: In an MMC, it is possible to significantly tailor the properties such as CTE, stiffness parameters, etc. by choosing the type and proportion of the reinforcements. It gives the designer a great deal of flexibility, which is not available with monolithic metals.
6. *High transverse strength and modulus*: Transverse strength and modulus of composite material are directly influenced by the matrix material. Metals typically possess higher tensile and compressive strengths and moduli than those of unreinforced common polymers. Thus, transverse strength and modulus of MMCs are higher than those of unidirectional PMCs.
7. *High shear strength and modulus*: In-plane shear strength and modulus are also matrix-dominated properties and metals typically possess higher shear strength and modulus than those of unreinforced common polymers. Thus, like transverse strength and modulus, shear strength and modulus of MMCs are higher than those of unidirectional PMCs.
8. *Moisture absorption*: Metals and MMCs do not absorb moisture and, as a result, issues like swelling due to moisture absorption, delamination, etc., which affect PMCs, are not present in MMCs.
9. *High electrical and thermal conductivities*: MMCs are highly conducting both electrically as well as thermally. Thus, they are suitable in applications where thermal and electrical conductivities are desired.
10. *Ease of joining*: Bolted and threaded joints can be easily incorporated in MMCs. Welded joints are also possible. On the other hand, the provision of mechanical joints is a major concern in PMCs.
11. *Resistance to most radiations*: MMCs are also generally resistant to most radiations like UV radiations, etc.; unlike PMCs, they do not degrade when exposed to such radiations.

Disadvantages:

1. *High densities*: The densities of MMCs are higher than those of polymers. For example, typical densities of carbon/epoxy, glass/epoxy, and Kevlar/epoxy composites are 1.5 g/cm^3, 1.8 g/cm^3, and 1.4 g/cm^3, respectively, whereas the typical density of an aluminum matrix composite is 2.8 g/cm^3.
2. *Low specific strength and stiffness (w.r.t. PMCs)*: The high densities of MMCs result in lower specific strength and modulus as compared to PMCs.
3. *High processing costs*: MMC processing methods are generally complex and more expensive than most PMC processing methods.
4. *Reduced ductility and fracture toughness (w.r.t. monolithic metals)*: Reinforcing a monolithic metal may reduce its ductility and fracture toughness. As a result, MMCs are likely to be more brittle than the monolithic metals.

13.3.2 Matrix Materials for MMCs

The two constituent materials in an MMC are the matrix and the reinforcements. Various metals and their alloys have been used as matrix in MMCs. In this respect, aluminum and its alloys have been the most common. However, other metals such as titanium, magnesium, copper, lead, and cobalt have also been used as matrix materials. The choice of a matrix material is dependent on the intended use and it is greatly influenced by the requirements of strength, temperature resistance, low density, and low cost. Also, in an MMC, ductility and fracture toughness are imparted primarily by the matrix and, accordingly, ductile matrix materials with high fracture toughness are generally chosen.

13.3.3 Reinforcing Materials for MMCs

The reinforcements can be either continuous or discontinuous, and they include continuous fibers, monofilaments, particulates, short fibers, and whiskers. Various metallic and ceramic materials are used as reinforcements in MMCs [1,5]. Some examples of reinforcing materials are as follows:

- *Particulates*: Metallic particles like tungsten and ceramic particles like silicon carbide, alumina, boron carbide, titanium boride, and titanium carbide
- *Short fibers and whiskers*: Alumina, silicon carbide, and silicon nitride
- *Continuous fibers*: Carbon fibers, boron fibers, alumina fibers, silicon carbide fibers, alumina-silicate fibers
- *Monofilaments*: Boron and silicon carbide

The basic objective of reinforcing a metallic material to form an MMC is to achieve a set of improved properties that are suitable in certain applications in a better way. These improved properties are often associated with light weight, high strength and stiffness, resistance to high temperatures, etc., and these are directly dependent on the choice of the reinforcing material. Some of the desirable characteristics of the reinforcing materials are as follows:

- Low density
- High mechanical properties
- Compatibility
- High temperature resistance
- Processability
- Cost

MMCs are often used in weight-sensitive applications, where high specific strength and stiffness are essential requirements. This is possible with low-density reinforcements of high strength and stiffness.

One of the most important factors in choosing the reinforcing material and matrix is the compatibility between the two. Both mechanical as well as chemical compatibility are important. There should be minimum thermal mismatch between the two phases. Also, there should not be undesirable chemical reaction at the interface between the reinforcements and the matrix that can lead to the formation of harmful intermetallic compounds and sites for possible crack initiation. Note that the interphase must be able to transfer loads efficiently between the matrix and reinforcements so that their capacities are exploited fully in the composite form.

13.3.4 Manufacturing Methods for MMCs

A good number of methods have been developed for the processing of MMCs [1–4,6]. All these methods involve essentially combining two distinct phases—a monolithic metallic matrix and a discontinuous reinforcement phase. Also, these methods, in a broad sense, have three sequential stages:

- Preprocessing
- Primary processing
- Secondary processing

Preprocessing is a stage that precedes the primary processing; in most cases, it involves certain suitable processes on the reinforcements. Primary processing is the actual stage of combining or forming the two separate phases to make the MMC. Secondary processing is the stage wherein the MMC is deformed, shaped, rolled, machined, hardened, or joined to form the final part.

MMC manufacturing methods can be broadly divided into four classes, each having its subclasses:

- Solid-state methods
  - Powder metallurgy methods
  - Consolidation diffusion bonding
- Liquid-state methods
  - Liquid metal infiltration
  - Stir casting
  - Spray casting
- Deposition methods
  - Electrodeposition
  - Spray deposition
  - Vapor deposition
- *In situ* methods
  - Reactive methods
  - Nonreactive methods

Table 13.1 schematically depicts the MMC manufacturing methods. Note that the table is only representative and not exhaustive. Also, dividing the MMC manufacturing methods into different categories with different stages is a matter of convenience of study. While distinct features and differences exist, the boundary lines may sometimes get blurred.

Solid-state methods: In the solid-state methods, both the reinforcement and matrix remain in the solid state and the two phases defuse and bond with each other at an elevated temperature to form the desired MMC [7]. The temperatures during solid-state methods, however, are relatively lower than the liquid-state methods and undesirable chemical reactions between the two phases are absent in these methods. Thus, solid-state methods are useful in situations where the matrix is reactive in the liquid state, for example, titanium matrix composites. (Molten titanium is very highly reactive and it can degrade nearly any possible reinforcing material.)

Solid-state methods can be further subdivided into two groups: powder metallurgy methods and consolidation diffusion bonding.

TABLE 13.1
MMC Manufacturing Methods

Manufacturing Methods	Typical Constituents	Typical Processes
Solid-State Methods		
Powder metallurgy methods	Reinforcements: Particles, short fibers and whiskers, for example, SiC, Al_2O_3, TiC, and B_4C Matrix: Powder, for example, Al, Fe, and Ni	Preprocessing: Blending, compaction Primary processing: Cold pressing with sintering, hot pressing, HIP Secondary processing: Extrusion, forging, rolling, machining, joining
Consolidation diffusion bonding	Reinforcements: Monofilaments, for example, SiC Matrix: Foil, for example, Ti	Preprocessing: Foil/monofilament stacking Primary processing: Diffusion bonding Secondary processing: Machining, joining
Liquid-State Methods		
Liquid metal infiltration	Reinforcements: Particles, short fibers and whiskers, for example, SiC and Al_2O_3 Matrix: Molten metal, for example, Al	Preprocessing: Preform making Primary processing: Infiltration Secondary processing: Extrusion, forging, rolling, machining, joining
Stir casting	Reinforcements: Particles, short fibers, and whiskers, for example, SiC and Al_2O_3 Matrix: Molten metal, for example, Al	Primary processing: Stir casting Secondary processing: Extrusion, forging, rolling, machining, joining
Spray casting	Reinforcements: Particles, short fibers, and whiskers, for example, SiC and Al_2O_3 Matrix: Molten metal droplets, for example, Al	Primary processing: Spray casting Secondary processing: Extrusion, forging, rolling, machining, joining
Deposition Methods		
Electrodeposition	Reinforcements: Particles, short fibers, continuous fibers, and monofilaments, for example, B Matrix: Electroplating solution of metals, for example, Ni	Primary processing: Deposition, winding of coated fibers/monofilaments, diffusion bonding Secondary processing: Machining, joining
Spray deposition	Reinforcements: Continuous fibers and monofilaments Matrix: Molten metal	Primary processing: Deposition, stacking of coated fibers/foils/monofilaments, diffusion bonding Secondary processing: Machining, joining
Vapor deposition	Reinforcements: continuous fibers and monofilaments Matrix: Molten metal	Primary processing: Deposition, winding of coated fibers, diffusion bonding Secondary processing: Machining, joining
In Situ Methods		
Reactive methods	Reinforcements: Particles, short fibers, and whiskers, for example, TiC Matrix: Powder, molten metal, for example, Al	Preprocessing: Blending Primary processing: Chemical reactions
Nonreactive methods	Reinforcements: Particles, whiskers, short fibers, and continuous fibers Matrix: Polyphase alloy	Primary processing: Unidirectional precipitation leading to the formation of reinforcements

Liquid-state methods: As the name suggests, in these methods, liquid metal is used as the matrix during processing. Three major groups of methods in this category are liquid metal infiltration process, stir casting, and spray casting.

Deposition methods: In these methods, the MMC part is made by the deposition of a coating material on the reinforcements followed by stacking or winding of the reinforcements and consolidation of the stacked or wound part. Reinforcements in the forms of particles, continuous fibers, monofilaments, and preforms can be used. A number of methods for metal coating of the reinforcements have been developed; notable among these methods are electrodeposition process, spray deposition, and vapor deposition.

In situ methods: In the *in situ* methods, the reinforcement phase is created *in situ* as a result of certain reactive or nonreactive process in the matrix phase. Depending on the nature of the process of reinforcement formation, these methods can be broadly placed into two groups—(i) reactive *in situ* methods and (ii) nonreactive *in situ* methods. These are one-step methods and difficulties associated with combining two different phases are greatly reduced.

Now, we shall briefly discuss some of the important MMC manufacturing methods.

13.3.4.1 Powder Metallurgy Methods

In the powder metallurgy methods, powders of the matrix material are blended with reinforcements [8]. The reinforcements are taken in the form of particles, short fibers, or whiskers and the blending process results in a homogeneous mix. The mix is then cold pressed followed by sintering. The cold-pressed green compact mass is a porous structure containing water vapor and volatile contaminants such as lubricants of mixing and blending additives. In the first stage of sintering, the water vapor and volatile contaminants are removed by a process of degassing. The second stage of the sintering process involves the consolidation of the green compact part at high temperature during which the particles diffuse to the powder matrix. As an alternative to the cold pressing and sintering route, the homogeneous mix of matrix and reinforcement is hot pressed. (Cold pressing is a process of consolidation at low temperature without simultaneous sintering. On the other hand, hot pressing is the process of powder consolidation at high temperature with simultaneous sintering.) In the secondary processing, the cold-pressed and sintered or hot-pressed MMC part is extruded, forged, or rolled to obtain the final finished product. Figure 13.1 schematically shows the powder metallurgy methods of MMC manufacture.

Short fibers and whiskers tend to get broken during the process of blending and consolidation and the final composite looks more like a particulate-reinforced composite. While aluminum alloy powders are commonly used as the matrix, different grades of steel and other metals such as nickel, titanium, etc. are also used. Common reinforcing materials are silicon carbide, aluminum oxide, etc.

The powder metallurgy methods are very popular in the field of MMCs. They are associated with several advantages as well as some disadvantages, which are as follows:

Advantages:

- Good mechanical properties
- Uniform fiber volume fractions can be achieved
- Useful for making isotropic MMCs
- No degradation of the reinforcements by chemical reactions with the matrix

Disadvantages:

- Oxide "skin" formation on Al particles may lead to reduced shear properties
- Useful primarily for making discontinuously reinforced MMCs only

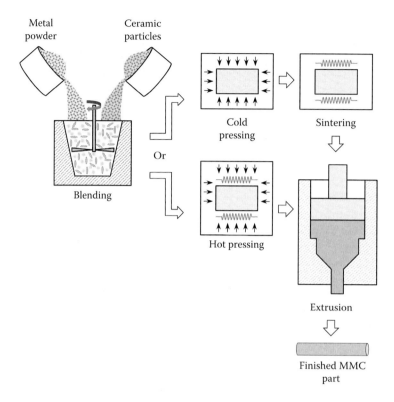

FIGURE 13.1 Schematic representation of the powder metallurgy methods. (With kind permission from Springer Science + Business Media: *Metal Matrix Composites*, second edition, 2013, N. Chawla and K. K. Chawla.)

- Fibers may get damaged during hot pressing
- The cost of production may be high when conventional powder metallurgy method is followed by subsequent extrusion and machining, etc.

13.3.4.2 Consolidation Diffusion Bonding

In this method, alternate layers of metallic foils and continuous reinforcing fibers are stacked in a certain predetermined order and the pack is then pressed at elevated temperature. During the process of hot pressing, the foils and the fibers defuse and a multilayered MMC part is formed. The process of consolidation diffusion bonding is schematically shown in Figure 13.2. A variant of the foil diffusion bonding technique is roll bonding, in which strips of two different metals are rolled and bonded to form a laminated composite material. Another variant of the foil diffusion technique is wire/fiber winding. In this process, continuous ceramic fibers and metallic wires are wound around a mandrel and pressed at elevated temperature.

The advantages associated with the consolidation diffusion bonding process are as follows:

Advantages:

- Possible to make MMCs using continuous fibers, monofilaments, and foils, resulting in high directional properties
- Controlled volume fraction can be achieved
- Minimum broken reinforcements
- Low porosity
- Negligible reaction between reinforcements and matrix

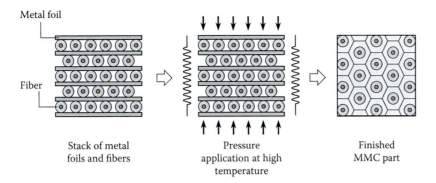

FIGURE 13.2 Schematic representation of the consolidation diffusion bonding methods. (With kind permission from Springer Science + Business Media: *Metal Matrix Composites*, second edition, 2013, N. Chawla and K. K. Chawla.)

Disadvantages:

- High processing temperature and pressure
- High cost of production
- Possibility of stress concentration in the MMC part due to improper fiber distribution

13.3.4.3 Liquid Metal Infiltration Process

In a liquid metal infiltration process, molten metal matrix is made to infiltrate into the reinforcement phase [9]. The reinforcement is a self-supporting body—either a preform of fibers or a porous body of particles or fibers. It is soaked into the molten metal, wherein the liquid fills the pores in the reinforcing body to make the desired MMC part. The process of infiltration of molten metal into the pores in the fiber body can be either spontaneous or forced.

Spontaneous infiltration, schematically represented in Figure 13.3, is a process in which no external force is used. However, capillary forces do not allow proper wetting of most reinforcements with molten metal and specific modification to the chemistry of the system is required.

Forced infiltration takes place under external pressure, which is generally applied either by a pressurized inert gas or by the piston of a hydraulic press. The schematic of this process by mechanical pressurization is shown in Figures 13.4. Another form of forced infiltration is vacuum-driven infiltration, in which the infiltration process

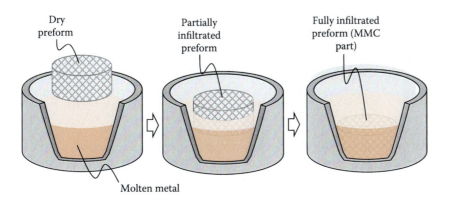

FIGURE 13.3 Schematic representation of the spontaneous infiltration process. (Adapted from V. J. Michaud, Liquid-state processing, *Fundamentals of Metal-Matrix Composites* (S. Suresh, A. Mortensen, and A. Needleman, eds.), Butterworth-Heinemann, Boston, MA, 1993, pp. 3–22.)

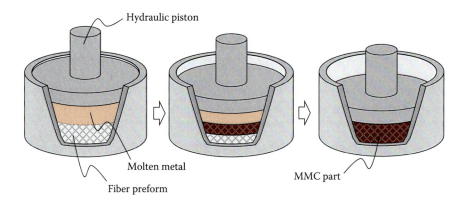

FIGURE 13.4 Schematic representation of the mechanically forced infiltration. (Adapted from V. J. Michaud, Liquid-state processing, *Fundamentals of Metal-Matrix Composites* (S. Suresh, A. Mortensen, and A. Needleman, eds.), Butterworth-Heinemann, Boston, MA, 1993, pp. 3–22.)

takes place under pressure created by vacuum around the preform. The application of pressure by a gas is limited to about 10 MPa. The mechanical means involves the application of much higher pressure at around 100 MPa and the pressure is maintained during solidification. The high pressure during infiltration under pressure by mechanical means results in pore-free MMC; however, the preform tends to get deformed or broken during the process.

The major advantages and disadvantages of the liquid metal infiltration methods are given below:

Advantages:

- Feasible to make complex and near net shape parts
- Rapid rate of production
- Low cost

Disadvantages:

- Reinforcements may degrade by chemical reaction with the molten metal
- Wettability of reinforcements molten metal can be difficult due to capillary forces

13.3.4.4 Stir Casting Method

In this method of MMC fabrication, reinforcing particles or short fibers are forcibly introduced into and mixed with a molten matrix metal by means of mechanical stirring (Figure 13.5). Stirring is followed by casting and solidification. Stir casting is a very simple and cost-effective method but it suffers from a drawback of relatively low volume fraction of the reinforcements. Further, the mix of the reinforcements in the matrix may remain nonhomogeneous coupled with the formation of clusters of particles/fibers. Rheocasting (a method of stirring the mix of reinforcements in matrix in a semisolid condition) is adopted to improve the distribution of the reinforcements in the metal matrix.

The principal advantages and disadvantages of stir casting methods can be listed as indicated below:

Advantages:

- Simple and cost-effective process
- Possible to remove undesirable impurities, oxides, and gases by vacuum application

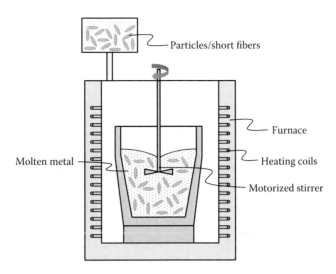

FIGURE 13.5 Schematic representation of the stir casting method. (Adapted from V. J. Michaud, Liquid-state processing, *Fundamentals of Metal-Matrix Composites* (S. Suresh, A. Mortensen, and A. Needleman, eds.), Butterworth-Heinemann, Boston, MA, 1993, pp. 3–22.)

Disadvantages:

- Low production rate
- Nonuniform fiber volume fraction
- Not suitable for continuous fibers and monofilaments
- Undesirable chemical reaction may take place between reinforcements and matrix

13.3.4.5 Spray Casting

The spray casting method is schematically shown in Figure 13.6. In this method, a liquid metal is atomized to form droplets and the liquid/semisolid droplets are collected on a substrate [10]. Discontinuous reinforcements in the forms of particulates, short fibers, and whiskers are introduced to the sprayed droplets and they are codeposited on the substrate to form the MMC part.

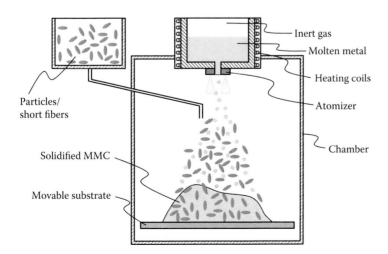

FIGURE 13.6 Schematic representation of the spray casting method. (Adapted from V. J. Michaud, Liquid-state processing, *Fundamentals of Metal-Matrix Composites* (S. Suresh, A. Mortensen, and A. Needleman, eds.), Butterworth-Heinemann, Boston, MA, 1993, pp. 3–22.)

Typically, a metal alloy is melted by induction heating and made to flow downward. Jets of an inert gas such nitrogen are directed onto the thin free-falling stream of liquid metal and forced through a ceramic nozzle that atomizes the liquid into droplets. Reinforcement particles such as SiC are sprayed on to the fine droplets and codeposition of particulate reinforcements and metallic matrix followed by rapid solidification takes place. An advantage of this method is its high rate of production. On the other hand, for uniform distribution of particulates in the MMC, utmost care and control of the environmental parameters in the atomization and particulate feeding have to be taken. As a result, sophisticated equipment may be needed, leading to a higher cost of production.

The important advantages and disadvantages of this method can be listed as follows:

Advantages:

- Rapid solidification and high rate of production
- Minimum chemical reaction between reinforcements and matrix

Disadvantages:

- Residual porosity cannot be eliminated during primary processing
- High cost of production

13.3.4.6 Deposition Methods

As noted earlier, three notable deposition methods are electrodeposition, spray deposition, and vapor deposition.

In the electrodeposition methods, an electroplating solution is utilized for depositing metallic matrix on the reinforcements. Figure 13.7 shows a schematic representation of the electrodeposition method involving winding of continuous fibers or monofilaments. A rotating mandrel is kept in a plating bath. The mandrel acts as the cathode and the electroplating solution as the anode. Continuous fibers are wound on the mandrel and the voltage difference across the cathode and anode results in simultaneous deposition of metal on the fibers. Several layers of fibers are wound and an MMC part with highly oriented reinforcements is obtained. A typical example of MMC by electrodeposition is B/Ni composites.

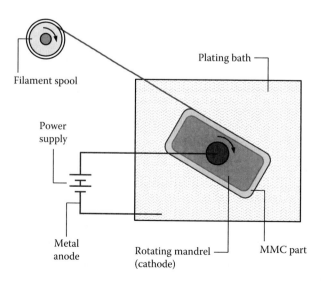

FIGURE 13.7 Schematic representation of the electrodeposition by winding.

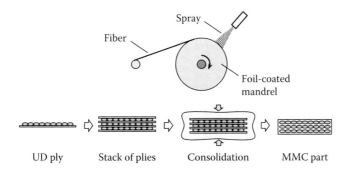

FIGURE 13.8 Schematic representation of spray deposition.

In spray deposition, typically fibers are wound on a drum covered with a foil and molten metal is sprayed onto the fibers. The wound fibers are cut open and removed from the drum to obtain a unidirectional ply. The plies are stacked and hot pressed. Figure 13.8 shows a typical spray deposition process.

In addition to the electrodeposition and spray deposition processes, vapor deposition methods are also employed in the manufacture of MMCs by the deposition of metal coatings on reinforcements. These methods include chemical vapor deposition (CVD) and physical vapor deposition (PVD).

Deposition methods are also associated with a number of advantages and disadvantages; the important ones are as follows:

Advantages:

- Possible to incorporate continuous fibers and monofilaments
- Controlled reinforcement orientation can be achieved
- Negligible reaction between reinforcements and matrix, and as a result, no formation of undesirable impurities and no reinforcement degradation

Disadvantages:

- Relatively complex process

13.3.4.7 *In Situ* Methods

As noted earlier, *in situ* methods are of two types—reactive and nonreactive.

In the reactive methods, multiple components are reacted in a controlled manner and the reinforcements and/or matrix are formed as a reaction output. The reinforcements that are typically particulates or whiskers get uniformly dispersed in the continuous metal and the resultant product is a particulate/whisker-reinforced MMC. A typical example of chemical reaction leading to the formation of MMC is TiC/Al composite from carbon, titanium, and aluminum, as follows:

$$C + Ti + Al \rightarrow TiC + Al$$

The reacting components can be reacted in different ways. For example, fine powders of ceramic and metallic materials can be blended and heated, making the metal powder to melt and react with the ceramic. Ceramic powders can also be externally added to a molten metal. Alternatively, a preform can be infiltrated with a molten metal; chemical reaction can take place either during infiltration or afterward. A schematic representation of the reactive *in situ* method by infiltration is shown in Figure 13.9.

In the nonreactive *in situ* methods (Figure 13.10), a polyphase alloy is heated and cooled to solidify in a controlled fashion that results in unidirectional precipitation

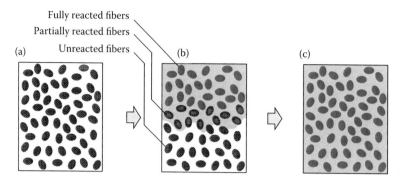

FIGURE 13.9 Schematic representation of reactive *in situ* method by infiltration of preform with molten metal. (a) Preform. (b) Partially infiltrated preform containing unreacted, partially reacted, and fully reacted fibers/particles. (c) Resultant MMC part. (Adapted from A. Evans, C. San Marchi, and A. Mortensen, *Metal Matrix Composites in Industry—An Introduction and a Survey*, Kluwer Academic Publishers, Boston, MA, 2003.)

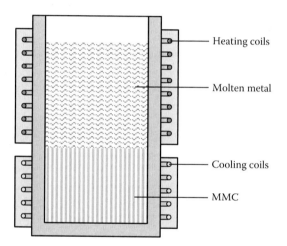

FIGURE 13.10 Schematic representation of the nonreactive *in situ* method. (With kind permission from Springer Science + Business Media: *Metal Matrix Composites*, second edition, 2013, N. Chawla and K. K. Chawla and adapted with permission from V. J. Michaud, Liquid-state processing, *Fundamentals of Metal-Matrix Composites* (S. Suresh, A. Mortensen, and A. Needleman, eds.), Butterworth-Heinemann, Boston, MA, 1993, pp. 3–22.)

and formation of the reinforcing phase in the continuous matrix. The reinforcements formed can be particles, short fibers, whiskers, or continuous fibers.

The *in situ* methods are not associated with any problems of wettability and the resulting MMCs possess good interface between the reinforcements and the matrix.

13.3.5 Applications of MMCs

MMCs have huge potential applications in a wide range of industrial sectors, of which, aerospace and automotive sectors are the two major sectors. In addition to these, other areas such as electronic packaging, recreational goods, etc. have significant applications of MMCs. Table 13.2 lists some of the important applications of MMCs in different fields [11–13]. Both continuously reinforced MMCs as well as discontinuously reinforced MMCs are in use. However, discontinuously reinforced aluminum (DRA) MMCs are the dominant ones. Discontinuously reinforced titanium matrix (DRTi) MMCs are among other similar examples. Common discontinuous reinforcements include particulate silicon carbide (SiC_p). Among the continuous reinforcements, carbon, alumina, boron, etc. have significant applications.

TABLE 13.2
Common Applications of MMCs

Sector	Applications	Remarks
Aerospace		
Aircraft	• Primary structures	
	– Fuel access door covers	Benefits: enhanced strength/stiffness, longer fatigue life, weight reduction Example: DRA, e.g., SiC_p/Al
	– Fuselage strut	Benefits: increased stiffness, cost reduction, enhanced damage tolerance Example: DRA, e.g., SiC_p/Al
	• Control components	
	– Ventral fins	Benefits: enhanced strength/stiffness, longer fatigue life, cost reduction Example: DRA, e.g., SiC_p/Al
	• Engine parts	
	– Nozzle actuator piston rod and link rod	Benefits: enhanced strength/stiffness and fatigue characteristics, weight saving Example: continuous fiber reinforced MMCs, e.g., SiC_{cf}/Ti
	– Fan-exit guide vane	Benefits: increased erosion resistance, increased resistance to ballistic damage, weight reduction, cost saving Example: DRA, e.g., SiC_p/Al, B_4C_p/Al
	• Subsystems	
	– Support racks	Benefits: enhanced isotropic stiffness and bearing strength Example: DRA, e.g., SiC_p/Al
	– Hydraulic manifolds	Benefits: high specific strength, better fatigue characteristics Example: DRA, e.g., SiC_p/Al and SiC_w/Al
Helicopter	• Rotor hub—Blade sleeves	Benefits: enhanced fatigue characteristics, weight reduction, cost saving Example: DRA, e.g., SiC_p/Al
Launch vehicles and missiles	• Wings and fins	Benefits: enhanced high-temperature capability, high specific strength/stiffness Example: DRA, e.g., SiC_p/Al
	• Covers for inertial navigation system	Benefits: CTE matching, weight saving, minimal machining, cost saving Example: DRA, e.g., SiC_p/Al
Automotive		
Car, bus, and truck	• Engine parts	
	– Piston	Benefits: better fatigue characteristics, increased wear resistance, better performance on account of low/matching CTE, low overall cost Example: DRA, e.g., SiC_p/Al, SiC_w/Al, B_4C_p/Al
	– Piston cylinder lining	Benefits: increased wear resistance, weight reduction, high thermal conductivity, longer life Example: continuous fiber–reinforced MMCs, e.g., C/Al, Al_2O_3/Al
	– Push rod and connecting rod	Benefits: high stiffness, high damping, weight reduction Example: DRA, e.g., SiC_p/Al, Al_2O_{3p}/Al
	– Intake and exhaust valves	Benefits: weight saving, fuel efficiency, durability Example: discontinuously reinforced titanium (DRTi) MMC with monoboride as reinforcements
	• Brake drum and brake rotor	Benefits: wear resistance, weight saving, high thermal conductivity Example: DRA like SiC_p/Al, Al_2O_{3p}/Al made by casting

(Continued)

TABLE 13.2 (*Continued*)
Common Applications of MMCs

Sector	Applications	Remarks
	• Drive shaft	Benefits: increased stiffness, weight reduction, corrosion resistance, cost saving
		Example: DRA, e.g., Al_2O_{3p}/Al
Others		
Electronic packaging systems	• Cores, substrate, carriers, housings	Benefits: controlled thermal expansion/CTE matching, cost reduction, weight saving
		Example: DRA, e.g., SiC_p/Al
Recreational and sporting goods	• Bicycle frames	Benefits: increased stiffness
		Example: DRA, e.g., Al_2O_{3p}/Al
	• Track shoe spike	Benefits: enhanced comfort
		Example: DRA, e.g., Al_2O_{3p}/Al and SiC_p/Al

13.4 CERAMIC MATRIX COMPOSITES

13.4.1 Characteristics of CMCs

By definition, CMCs are a class of composite materials, in which a monolithic ceramic material is reinforced with either continuous or discontinuous reinforcements; in other words, the matrix in these composites is a ceramic material. These are the newest class of composite materials that have gained acceptance after PMCs and MMCs. Ceramics, by themselves, are a class of nonmetallic, inorganic materials characterized by their high temperature resistance and low fracture toughness. Two broad classes of ceramics can be identified—first, traditional ceramics and second, advanced ceramics. Ceramics used in making artworks, pottery, bricks, tiles, etc. can be put in the first category. They are typically based on clay and glass (silica and feldspar). Advanced ceramics are the high-performance ceramics that are characterized by their exceptionally high temperature resistance, corrosion resistance, wear resistance, and hardness, but low fracture toughness. Common advanced ceramics include oxides and nonoxides (nitrides, carbides, and borides) of aluminum, silicon, titanium, and zirconium. Some intermetallic and pure elemental materials also fall in this category.

The primary drawback of a monolithic ceramic material is its low fracture toughness, and the primary objective of reinforcing it with suitable particles, whiskers, or fibers is to increase its fracture toughness. Thus, the inherent benefits associated with monolithic ceramics are also by and large present in CMCs, and in addition, CMCs are relatively tougher. The advantages and disadvantages associated with CMCs are listed below:

Advantages:

1. *High temperature resistance*: The service temperature of a material is the maximum temperature that it can be subjected to for an extended duration without degradation of its properties. It is an important parameter in any high-temperature application, and as class of materials, CMCs outclass both polymers and metals in this respect. For example, ceramics such as SiC, Si_3N_4, and Al_2O_3 have service temperatures around 1400–1500°C, whereas the maximum service temperatures for common metals and polymers are much lower, for example, it is in the order of 150–250°C for aluminum and 80–200°C for epoxy. In the case of composites, the service temperature is primarily a matrix-dominated property and CMCs typically possess very high service temperatures.
2. *Corrosion resistance*: Ceramics are known for their resistance to corrosion in a wide range of environments. Thus, CMCs are also highly resistant to corrosion.

Some ceramics such as SiC are practically inert to both acidic environment as well as basic environment. On the other hand, some ceramics may get affected by strong acids or strong bases.

3. *Hardness and wear resistance*: CMCs possess higher hardness and wear resistance even at high temperatures and they are suitable for making cutting tools. For example, the performance of alumina reinforced with silicon carbide whiskers composites is found to be three times better than conventional monolithic ceramic tools.
4. *Low density*: The densities of most ceramic matrices and their reinforcements fall typically within 2–4 g/cm^3. The resultant CMC densities are also relatively lower than that of metals and MMCs. Thus, CMCs are suitable for lightweight applications.
5. *High specific strength and stiffness*: CMCs possess high strength and stiffness, and due to their low densities, their specific strength and stiffness are typically higher than that of metals.
6. *Increased fracture toughness (w.r.t. monolithic ceramics)*: Monolithic ceramic materials are highly brittle with very low fracture toughness. The fracture toughness of these materials can be greatly increased by suitable incorporation of reinforcements. For example, the typical value of fracture toughness of monolithic silicon carbide (SiC) is about $3\,\mathrm{MPa}\sqrt{m}$, whereas it is of the order of $30\,\mathrm{MPa}\sqrt{m}$ for silicon carbide reinforced with silicon carbide fibers (SiC/SiC) composites. Several toughening mechanisms are possible—microcracking, crack bridging, crack deflection, crack branching, fiber pull-out, etc. Depending on the type of the reinforcements and matrix, one or more of these toughening mechanisms can come into play, which help avoid catastrophic failures in CMCs.

Disadvantages:

1. *High fabrication cost*: The processing of CMCs typically involves high temperatures. High processing temperatures lead to processing complexities and high cost.
2. *Low fracture toughness (w.r.t. MMCs and PMCs)*: CMCs, due to the inherent low fracture characteristics of the matrix materials, possess low fracture toughness compared to metals and PMCs.
3. *Thermal mismatch*: The matrix and the reinforcements in a CMC possess different thermal coefficients, and as a result, thermal stresses develop during processing.

13.4.2 Matrix Materials for CMCs

Various ceramics have been used as matrix in CMCs. They are either crystalline or noncrystalline. Various types of glasses are the examples of noncrystalline ceramics, whereas other ceramics are crystalline. Some common ceramics and their representative properties are listed in Table 13.3. The selection of the matrix material is a critical issue; in this regard, three broad aspects can be identified—(i) compatibility with the reinforcements, (ii) thermal stability, and (iii) processability. The matrix should be able to uniformly wet the reinforcements without causing any detrimental chemical reactions. Sometimes, even a minor reaction can result in gross degradation of the final performance of the CMC. For example, in whisker-reinforced CMCs, the reactive matrix can totally dissolve the reinforcements. Similarly, high thermal mismatch between the matrix and the reinforcements can lead to high thermal residual

TABLE 13.3
Representative Properties of Common Ceramic Matrix Materials

	Density (g/cm³)	Melting Point (°C)	CTE (10^{-6} m/m/°C)	Fracture Toughness (MPa$\sqrt{m}$)	Tensile Modulus (GPa)
Alumina (Al_2O_3)	3.9	2050	7–9	2–4	280–390
Silica (SiO_2)	2.2	1610	0.5	0.8	75
Zirconia (ZrO_2)	5.7	2760	7.9–13.5	2.8–8.5	205
Magnesium oxide (MgO)	3.6	2850	3.6	1.8	205–225
Silicon carbide (SiC)	3.2	1980	4.3–4.5	2.2–3.4	330–420
Silicon nitride (Si_3N_4)	3.1	1870	3.1	3.1–5.5	170–310
Glass	2.5	1500	3.5–8.9	0.5–2	60–70

stresses and low structural integrity. The melting points of ceramics are typically high, which implies high thermal stability. High melting points also imply greater processing complexities.

13.4.3 Reinforcing Materials for CMCs

Reinforcements used in CMCs are either continuous or discontinuous, and they include particulates, whiskers, short fibers, continuous fibers, and monofilaments. Some typical examples of reinforcing materials in CMCs are given in Table 13.4.

In a PMC, the primary function of the reinforcements is to provide strength and stiffness to the composite material. In an MMC too, strength and stiffness are core parameters in the set of properties that are targeted to be improved by reinforcing the monolithic metal. In a CMC, however, the characteristic function of the reinforcements is different; while strength and stiffness are important, the reinforcements are provided primarily to improve the fracture toughness. The compatibility of the reinforcements with the matrix is an essential requirement. Other desirable characteristics of the reinforcements are processability, high mechanical properties, high temperature resistance, low density, and low cost.

13.4.4 Manufacturing Methods for CMCs

Several methods have been developed for the processing of CMCs [17]. These methods can be classified in different ways. However, as noted in the section on MMCs, the basis of classification is largely a matter of convenience of discussion and understanding. Here, as in the case of MMCs, we shall use the same four major classes. Thus, the important CMC manufacturing methods can be grouped as follows:

TABLE 13.4
Common Reinforcing Materials in CMCs

Type of Reinforcements	Example
Particulates	Silicon carbide (SiC), alumina (Al_2O_3), titanium carbide (TiC), boron nitride (BN)
Whiskers	Alumina (Al_2O_3), silicon carbide (SiC), titanium boride (TiB_2)
Short fibers	Alumina (Al_2O_3), silicon carbide (SiC), alumina-silicate ($Al_2O_3 + SiO_2$), glass, carbon fibers
Continuous fibers	Carbon (C), boron (B), alumina (Al_2O_3), silicon carbide (SiC), alumina-silicate ($Al_2O_3 + SiO_2$)
Monofilaments	Boron (B), silicon carbide (SiC)

- Solid-state methods
  - Powder consolidation methods
  - Slurry infiltration methods
- Liquid-state methods
  - Liquid infiltration—melt infiltration and reactive liquid infiltration
  - Sol–gel technique
- Deposition methods
  - Chemical vapor infiltration (CVI) and chemical vapor deposition (CVD)
- *In situ* methods
  - Polymer infiltration and pyrolysis (PIP)
  - Reaction bonding processes

As in PMCs and MMCs, CMC manufacturing methods too involve essentially combining two distinct phases—a monolithic phase of ceramic matrix and discontinuous reinforcement phase. The matrix can be incorporated in the composite in different ways. The classification of CMC manufacturing methods given above is basically associated with the method of incorporation of the matrix.

In the solid-state methods, both the reinforcement and matrix remain in the solid state and the two phases defuse and bond with each other at elevated temperature to form the desired CMC. Typically, the matrix is incorporated in a powder form. Ceramic powders are either blended directly with the reinforcement phase or used to form a slurry. In the liquid-state methods, molten ceramic material is used to infiltrate a fiber preform. Deposition methods involve the deposition of the ceramic material on the fibers. Finally, in the *in situ* methods, the ceramic matrix is produced by either pyrolysis or other reaction phenomena *in situ*, that is, on the surface of the fibers.

13.4.4.1 Powder Consolidation Methods

Powder consolidation methods in CMCs are basically the same as the powder metallurgy methods in MMCs. Here, powders of the ceramic matrix material are blended with particulate, short fiber, or whisker reinforcements. The common particulate reinforcements used are SiC, TiC, ZrO_2, etc. Among the whisker reinforcements, SiC whiskers are one of the most common materials. Powders of Al_2O_3, ZrO_2, B_4C, etc. are commonly used as matrix materials. Blending of the reinforcement particles, whiskers, or short fibers with the matrix powders is of utmost importance and it can be done by employing one of the several mixing methods such as shear mixing, ultrasonic dispersion, milling, etc. The mix is then either cold pressed followed by sintering or hot pressed.

Cold pressing followed by sintering as well as hot pressing involve the application of high temperature and pressure. A notable disadvantage of these processes is the degradation of the reinforcements, especially by chemical reactions at the interface.

13.4.4.2 Slurry Infiltration

Slurry infiltration is a variant of the conventional powder consolidation by hot pressing. In this method, the matrix is used in the form of powder. A slurry is prepared by mixing the ceramic matrix powder in a liquid carrier (water or alcohol) and an organic binder. Sometimes, wetting agents are also added to the slurry. The process of CMC making in this method involves two stages—first, wetting of the reinforcements by the slurry, and second, matrix consolidation by hot pressing. In the first stage, the reinforcements, which can be either fiber tows or preform sheets, are made to pass through a slurry tank where they get impregnated in the slurry. The impregnated reinforcements are wound on a drum and the liquid carrier is allowed to evaporate. The impregnated fiber tows or the preform sheets are then cut and stacked in desired orientations. The organic binder is then burnt out. In the second stage, the stack of fiber tows of preform sheets

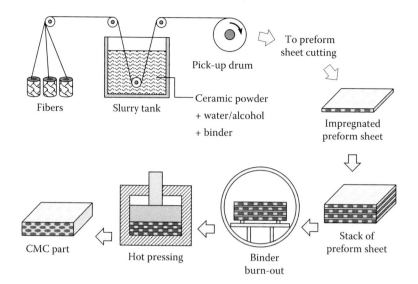

FIGURE 13.11 Schematic representation of slurry infiltration process. (Adapted with permission from K. K. Chawla, *Ceramics Matrix Composites*, second edition, Kluwer Academic Publishers, Boston, MA, 2002.)

is consolidated by hot pressing. Figure 13.11 shows the schematic representation of the slurry infiltration process.

Glass and glass–ceramic are the most commonly used matrix materials in the slurry infiltration method; relatively lower melting points of these matrices compared to other ceramic materials make them suitable for slurry infiltration. The major advantages and disadvantages of this method are

Advantages:

- Possible to incorporate continuous fibers
- Good mechanical properties
- Uniform fiber volume fractions
- Low porosity

Disadvantages:

- Not possible to use matrix materials with high melting points
- Fibers may get damaged during hot pressing

13.4.4.3 Liquid Infiltration

Liquid infiltration (Figure 13.12) in CMC processing is similar to resin transfer molding in PMCs and liquid metal infiltration in MMCs. Here, molten ceramic matrix is made to infiltrate into the preform kept in a closed chamber. Typically, the chamber is heated and the melt is forced under pressure by a piston into the preform at high temperature. Typical advantages and disadvantages of this process are

Advantages:

- Single-step process
    - Matrix can be formed in a single processing step of infiltration and repeated cycling is not needed.
- Varieties of preforms
    - Preform of any form, including fibers, whiskers, and particles, can be used.

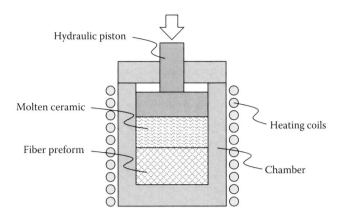

FIGURE 13.12 Schematic representation of liquid infiltration method for CMC manufacture. (Adapted with permission from K. K. Chawla, *Ceramics Matrix Composites*, second edition, Kluwer Academic Publishers, Boston, MA, 2002.)

Disadvantages:

- High melting points of ceramics
  - Complex process of manufacturing
  - Possibility of reaction between reinforcements and matrix
  - Possibility of reinforcement degradation
- Thermal mismatch between matrix and reinforcements
  - Differential shrinkage
  - Matrix cracking

Melt infiltration is done at high temperatures that can cause degradation of the reinforcements. Polymer precursors can be used to overcome this problem. Ceramic matrix can be obtained from such precursors at relatively low temperatures. However, it also leads to the generation of volatiles. The volatiles escape from the matrix, causing porosity and shrinkage. Reimpregnation, as in sol–gel process, is required for reducing porosity.

13.4.4.4 Sol–Gel Technique

In this process, a sol is made to infiltrate a fiber preform. The sol is a colloidal suspension of very fine ceramic particles. (The particles are so fine that no sedimentation takes place.) It is produced by chemically reacting a precursor material (e.g., a metal alkoxide), a solvent (e.g., an alcohol), a catalyst (e.g., an acid), and water. The preform soaked in the sol is allowed to dry, which leads to matrix shrinkage and formation of pores. To reduce the porosity, the infiltration–drying cycle is repeated several times till the desired density is achieved. Once the desired density is achieved, the preform is fired to obtain the final CMC. The sol–gel process is commonly used for making glass–ceramic matrix composites. It has its own advantages and disadvantages, as follows:

Advantages:

- Processing temperature is relatively low
  - Minimum damage of fiber preform
  - Low cost tooling
- Near net shape part can be made
  - Minimum machining
  - Complex part can be made
- Better control over fiber matrix composition

Disadvantages:

- High shrinkage
  - Matrix cracking can result
- Low yield
  - Repeated cycling needed
  - Long process

13.4.4.5 CVI and CVD

In this process, as shown in Figure 13.13, a fiber preform is kept in a heated reactor chamber. A carrier gas such as H_2 or He is made to pass through a container of gaseous reagents. The vapors of the reagents are carried by the carrier gas and pumped into the reactor chamber. They infiltrate the porous preform and react to form ceramic matrix vapor and other products. The ceramic matrix vapor gets deposited on the preform surfaces and the other products are diffused and carried away by the carrier gases. The process of deposition is continued till the pores get filled up reasonably uniformly and a solid compact CMC part is obtained. Various reagents have been used for forming various ceramic matrix vapors. For example, CH_3 and $SiCl_3$ can be used along with H_2 to form SiC.

Uniform deposition of the matrix vapors leading to uniform density throughout the preform is desirable for making a quality part. The rate of deposition is an important parameter in this respect. Too fast deposition may result in faster rate of matrix formation in the outer regions of the preform, where the pores get filled up quicker. As a result, the flow of gases into the interior regions gets interrupted. In such cases, several cycles of machining in the outer periphery followed by infiltration and deposition may be required. On the other hand, too slow deposition may take a long cycle time, leading to a high cost of fabrication.

Temperature and pressure affect the quality of deposition to a large extent. Depending on the temperature and pressure environments of the deposition process, the CVI process can have a number of variants, for example, (i) uniform temperature throughout the preform, (ii) temperature gradient across the preform thickness, (iii) uniform temperature together with pressure, (iv) temperature gradient together with pressure, etc. CVI under uniform temperature, called isothermal CVI, is typically a very slow process. On the other hand, forced infiltration under pressure and temperature gradient can greatly increase the rate of deposition.

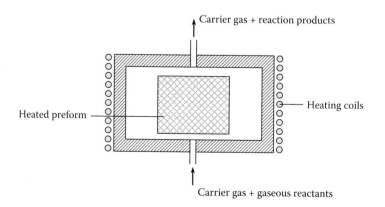

FIGURE 13.13 Schematic representation of the CVI process. (Adapted with permission from K. K. Chawla, *Ceramics Matrix Composites*, second edition, Kluwer Academic Publishers, Boston, MA, 2002.)

The CVI process has a number of advantages and disadvantages; some of the important ones are

Advantages:

- Relatively lower temperature and pressure
  - Minimum damage to the reinforcements
- Good mechanical properties
- Near net shape complex and large parts can be made

Disadvantages:

- Multiple cycles of infiltration processes are needed
  - Low yield
  - Slow and expensive
- Residual porosity cannot be removed completely

13.4.4.6 Polymer Infiltration and Pyrolysis

PIP is a relatively low-cost and low-temperature CMC processing technique, in which ceramic yielding polymeric precursors are used to infiltrate a fiber preform. (A preform can also be made using resin-impregnated fibers by common PMC manufacturing techniques.) The polymer is cured and then pyrolyzed. Pyrolysis of the polymeric precursor results in the formation of ceramic and some gaseous by-products. While the ceramic matrix remains in the preform and binds the fibers, the gaseous by-products escape through the pores. It leads to shrinkage and porosity, and several cycles of infiltration–pyrolysis are required till the desired density is achieved. The PIP process is schematically given in Figure 13.14.

The commonly used precursors include polymers containing various types of silane. (Silane is an inorganic compound containing one silicon atom and four hydrogen atoms, i.e., SiH_4.) These polymers are produced by dechlorination of chlorinated silane monomers that are easily available as by-products in the silicone industry. Typically, pyrolysis of the cured precursor polymer leads to the crystalline precipitation of ceramics such as SiC, Si_3N_4, and SiO_2 and evolution of gases such as SiO and CO.

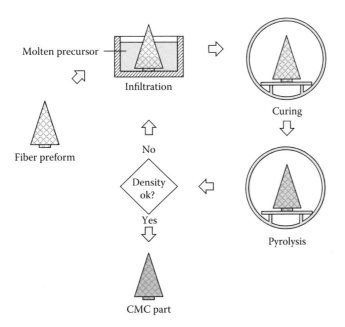

FIGURE 13.14 Schematic representation of polymer infiltration and pyrolysis process using preform.

The polymer infiltration phase is somewhat similar to PMC processing and suitable PMC manufacturing methods can be adopted for making the polymer-impregnated structure. Infiltration of the preform or the reinforcing structure takes place under capillary action at atmospheric pressure. The infiltration process can be aided by vacuum and positive pressure application.

Like other CMC manufacturing processes, the PIP process is also associated with some advantages as well as disadvantages; the important ones are as follows:

Advantages:

- Relatively low temperature process
  - Minimum fiber damage
  - Low process complexities
- Near net shape parts can be made

Disadvantages:

- Multiple infiltration–pyrolysis cycle
  - Long processing time
  - High processing cost
- Residual porosity

13.4.4.7 Reaction Bonding Processes

These processes have shown great promise, especially in the fabrication of silicon carbide-based composites. In these processes, molten silicon is made to infiltrate a carbon containing preform at high temperature. Infiltration takes place under either capillary action or pressure, and the chemical reaction between the carbon in the preform and liquid silicon results in the formation of silicon carbide. The primary advantages of these processes include minimum fiber damage and nearly nil matrix shrinkage. On the other hand, high residual porosity is a major disadvantage.

13.4.5 Applications of CMCs

CMCs have found applications in both aerospace and nonaerospace sectors [27]. In aerospace applications, performance is the driving force and high temperature resistance together with high specific strength/stiffness and enhanced damage tolerance are the essential requirements. There are many applications where a part is subjected to high temperatures. Most PMCs lose their strength and stiffness properties rapidly beyond about 100–150°C. High-performance metals and metallic alloys work up to about 800°C. Beyond this point, CMCs are the only solution. In this respect, C/C, SiC/Al_2O_3, SiC/SiC, and SiC/Si_3N_4 are some of the most notable CMCs.

In the nonaerospace applications, cost-effectiveness is also an important criterion. Resistance to corrosion and wear leading to long part life is a highly desirable characteristic.

Some of the common examples of the usage of CMCs are listed in Table 13.5 below.

13.5 CARBON/CARBON COMPOSITES

13.5.1 Characteristics of C/C Composites

The fibers in C/C composites can be either continuous or discontinuous. Depending on the number of directions in which the fibers are oriented, continuous fiber C/C composites can be either 1D or multidimensional. Multidimensional C/C composites can be of

TABLE 13.5
Applications of CMCs

Application	Remarks
a. Thermal protection system (TPS)	Space vehicles encounter very high temperatures (1500–1600°C) and thermal shocks during atmospheric reentry Example: TPS on nose cap, leading edges, wings, and flaps Material: C/C and C/SiC composites Benefits: reduced weight, reusability
b. Gas turbine components	Gas turbine components, including stationery as well as moving elements, are subjected to high temperatures (400–1200°C) and they are required to perform for long durations Example: combustion chamber liner, turbine blades, and turbine wheels Material: SiC/SiC and C/SiC composites Benefits: weight saving, increased turbine efficiency, fuel saving, increased life
c. Cutting tools	Cutting tools used in metal cutting operations like turning, milling, drilling, etc. are desired to be strong, wear resistant, tough, and thermally conductive Example: drill bits, inserts, etc. Material: Al_2O_3 matrix reinforced with particulates or whiskers of SiC, TiC, and ZrO_2 Benefits: better performance, increased fracture toughness, long life
d. Wear-resistant parts	Wear-resistant parts in automotive and other sectors demand good wear resistance and high-temperature capabilities Example: bearings, bushings, precision balls, liners, valves, pump housing wear rings, etc. Material: Al_2O_3 matrix reinforced with particulates of ZrO_2, SiC Benefits: long life, better performance
e. Braking system	Brake disks of aircrafts and racing cars are subjected to high wear and tear and exposed to high temperatures (~1500°C) and thermal shocks Example: brake disks and brake disk components Material: C/C and C/SiC composites Benefits: weight saving, greatly reduced wear and tear, resistant to corrosion due to road salts, effective braking in a wide range of environmental conditions
f. Rocket nozzle parts	Rocket nozzles are subjected to very high temperatures and highly erosive environment Example: nozzle throat, liners Material: C/C composites Benefits: weight saving, dimensional stability
g. Furnace components	Furnace components are subjected to very high temperatures and they tend to degrade and deform over time Example: burners, radiant tubes, flaps, ventilators, etc. Material: Si/SiC composites Benefits: long life, low maintenance, cost saving

two (2D), three (3D), or even higher dimensions (nD). These composites are made by using unidirectional carbon tows, tapes, and woven fabrics. On the other hand, discontinuous carbon fiber–reinforced composites do not have any preferred fiber orientation.

C/C composites are prone to oxidation and it is the matrix on which the rate of oxidation is dependent. The carbon matrix is formed typically from some precursor material. It can be amorphous or graphitic. Higher degree of graphitization makes it resistant to oxidation and thermally more conductive; however, it also makes the composite more brittle.

C/C composites have extremely high temperature resistance; they remain thermally stable in nonoxidizing environments at temperatures as high as 3000°C. They possess high strength and stiffness, and due to their low densities, their specific strength and stiffness characteristics are among the highest in composites. Carbon fibers have negative CTEs. As composite materials, they have low thermal coefficients, leading to generation of low residual thermal stresses. Wear resistance and better fracture toughness are among other advantages associated with C/C composites.

Processing of C/C composites typically involves high temperatures. High processing temperatures lead to processing complexities and high cost. Also, they are susceptible to oxidation at moderately high temperatures of 400–500°C. Thus, protective measures like coatings are needed in oxidizing environments.

The important advantages and disadvantages of C/C composites can be listed as follows:

Advantages:

- High temperature resistance
- High specific strength and stiffness
- Low CTE
- Wear resistance
- Higher fracture toughness (w.r.t. monolithic graphite)

Disadvantages:

- High fabrication cost
- Low oxidation resistance
- Low interlaminar properties
- Difficulty in jointing

Applications of C/C composites are found in a number of sectors including aerospace, defense, automotive, industrial, and biomedical. Typical requirements in these applications are structural stability at high temperature, thermal insulation, corrosion resistance, wear and tear resistance, and biocompatibility. The common applications are as follows:

- Aerospace and defense
  - Rocket nozzle—throat and liners
  - Reentry vehicle—thermal protection system
  - Aero-engine and turbine components
  - Brake disks and clutches
- Automotive
  - Engine pistons
- Industrial
  - Tools and dies
  - Refractory tiles
  - Heat pipes
- Biomedical application
  - Biocompatible implants

13.5.2 Manufacturing Methods for C/C Composites

The manufacturing methods for C/C composites can be broadly categorized into (i) chemical vapor infiltration (CVI) and (ii) pyrolysis of precursor materials. In either case, the manufacturing process involves, first, making of a fiber preform and, second, filling the gaps inside the preform with a carbon matrix. The preform can be made using unidirectional carbon tows, tapes, and rods or bidirectional woven/nonwoven fabrics. The fibers can be oriented in different directions to make 1D and multidirectional preforms (Figure 13.15).

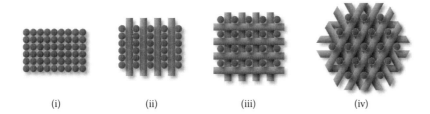

FIGURE 13.15 Schematic representation of fiber array in a preform. (i) 1D, (ii) 2D, (iii) 3D, and (iv) 4D.

In the CVI methods, the preform is infiltrated with a hydrocarbon gas such as methane, propane, benzene, etc. Carbon is deposited on the fiber surfaces of the heated preform by a process of thermal decomposition. Various variants of the CVI techniques are in vogue, of which, techniques involving isothermal conditions and temperature gradients are common. In the isothermal case, the gaseous hydrocarbon and the preform are maintained at a uniform temperature of around 1100°C. In this case, typically, a crust is formed on the outer periphery of the preform that prevents free flow of gas to the interior. As a result, repeated cycles of machining and infiltration are required till a desired density is achieved. On the other hand, in the temperature gradient techniques, temperature gradients are created across the preform thickness such that the interior of the preform is maintained at a high temperature of around 1100°C while the outer periphery is kept at a low temperature. In this case, crust formation does not take place and machining can be avoided.

In the pyrolysis methods, two types of precursor materials are used—pitch and thermosetting resins. Pitch is made from coal tar and petroleum; it can be either isotropic or mesophase. Molten pitch at about 300–400°C is infiltrated into a dry preform followed by carbonization at 1000°C, hot isostatic pressing at 80–100 MPa, and graphitization at 2700°C. In the case of resins, high carbon-yielding phenolic or epoxy resins are used for the impregnation of carbon fibers. Carbon prepreg laminates are vacuum bagged, cured under compaction pressure, and postcured. The laminate is then heated to around 800–1000°C in an inert atmosphere, where carbonization of the resin takes place. Volatiles are generated during carbonization, which escape from the laminate and cause porosity. Repeated cycles of infiltration and carbonization are required till the desired density is achieved.

Both the CVI and the pyrolysis methods have been successfully employed for making C/C composite products. In general, CVI methods are more suitable for thin parts, whereas the pyrolysis route is adopted for making thick parts.

13.6 SUMMARY

In this chapter, we have had a brief introduction to MMCs, CMCs, and C/C composites. These composite systems, as three major classes of composites other than PMCs, have their own characteristics in terms of their constituents, general mechanical and physical properties, manufacturing processes, and applications.

By definition, the matrix materials for these three classes of composites are metals, ceramics, and carbon. The reinforcements are either continuous or discontinuous and they can be of different physical forms, including particles, whiskers, short fibers, continuous fibers, and monofilaments. There are many types of manufacturing methods for MMCs, CMCs, and C/C composites. These methods revolve around combining the reinforcements with the matrix (or creating the reinforcements in the matrix) and typically they involve the application of high temperature and pressure. The constituents are incorporated in solid state, liquid state, or gaseous state, and depending on the physical state and other process characteristics, the manufacturing methods can be categorized into different groups. This categorization, however, is a matter of convenience and often the boundary lines between different groups tend to vanish.

Composites, as a whole, are used in both structural and thermal applications. As we had seen in Chapter 1, the structural applications of PMCs are primarily at room temperature. For high-temperature applications, to a large extent, the other three classes of composites complement each other. As a rough indicator, MMCs have a temperature resistance at a low-to-medium temperature range above room temperature. At medium-to-high temperature range CMCs and at high-to-very-high temperature range C/C composites are the solutions.

EXERCISE PROBLEMS

13.1 Write a brief note on the basic characteristics, constituent materials, advantages, and disadvantages of MMCs.
13.2 Write a short note on MMC manufacturing methods.
13.3 Write a short note on the application of MMCs.
13.4 What is the primary function of reinforcements in CMCs? Write a short note on the basic characteristics, advantages, and disadvantages of CMCs.
13.5 How are CMCs manufactured? Write a brief note on the principles of various CMC manufacturing methods.
13.6 Compare the basic characteristics and representative properties of PMCs, MMCs, CMCs, and C/C composites. Write a generic note on the selection of these composite materials in a design environment.

REFERENCES AND SUGGESTED READING

1. N. Chawla and K. K. Chawla, *Metal Matrix Composites*, second edition, Springer, New York, 2013.
2. K. U. Kainer (ed.), *Metal Matrix Composites—Custom-Made Materials for Automotive and Aerospace Engineering*, Wiley-VCH, Weinheim, 2003.
3. A. Evans, C. San Marchi, and A. Mortensen, *Metal Matrix Composites in Industry—An Introduction and a Survey*, Kluwer Academic Publishers, Boston, MA, 2003.
4. K. U. Kainer, Basics of metal matrix composites, *Metal Matrix Composites—Custom-Made Materials for Automotive and Aerospace Engineering* (K. U. Kainer, ed.), Wiley-VCH, Weinheim, 2003, pp. 1–54.
5. H. Dieringa and K. U. Kainer, Particles, fibers and short fibers for the reinforcement of metal materials, *Metal Matrix Composites—Custom-Made Materials for Automotive and Aerospace Engineering* (K. U. Kainer, ed.), Wiley-VCH, Weinheim, 2003, pp. 55–76.
6. V. I. Kostikov and V. S. Kilin, Metal matrix composites, *Handbook of Composites* (S. T. Peters, ed.), second edition, Chapman & Hall, London, 1998, pp. 291–306.
7. A. K. Ghosh, Solid-state processing, *Fundamentals of Metal-Matrix Composites* (S. Suresh, A. Mortensen, and A. Needleman, eds.), Butterworth-Heinemann, Boston, MA, 1993, pp. 23–41.
8. N. Hort and K. U. Kainer, Powder metallurgically manufactured metal matrix composites, *Metal Matrix Composites—Custom-Made Materials for Automotive and Aerospace Engineering* (K. U. Kainer, ed.), Wiley-VCH, Weinheim, 2003, pp. 243–276.
9. V. J. Michaud, Liquid-state processing, *Fundamentals of Metal-Matrix Composites* (S. Suresh, A. Mortensen, and A. Needleman, eds.), Butterworth-Heinemann, Boston, MA, 1993, pp. 3–22.
10. P. Krug and G. Sinha, Spray forming—An alternative manufacturing technique for MMC aluminum alloys, *Metal Matrix Composites—Custom-Made Materials for Automotive and Aerospace Engineering* (K. U. Kainer, ed.), Wiley-VCH, Weinheim, 2003, pp. 277–294.
11. M. J. Koczak, S. C. Khatri, J. E. Allison, and M. G. Bader, Metal-matrix composites for ground vehicle, aerospace and industrial applications, *Fundamentals of Metal-Matrix Composites* (S. Suresh, A. Mortensen, and A. Needleman, eds.), Butterworth-Heinemann, Boston, MA, 1993, pp. 297–326.
12. D. B. Miracle, Aeronautical applications of metal-matrix composites, *ASM Handbook, Vol. 21, Composites* (D. B. Miracle and S. L. Donaldson, vol. chairs), ASM International, Ohio, 2001, pp. 1043–1048.
13. W. H. Hunt, Jr. and D. B. Miracle, Automotive applications of metal-matrix composites, *ASM Handbook, Vol. 21, Composites* (D. B. Miracle and S. L. Donaldson, vol. chairs), ASM International, Ohio, 2001, pp. 1029–1032.

14. P. Morgan, *Carbon Fibers and Their Composites*, Taylor & Francis, Boca Raton, FL, 2005.
15. D. D. L. Chung, *Carbon Fiber Composites*, Butterworth-Heinemann, Boston, MA, 1994.
16. S.-J. Park, *Carbon Fibers*, Springer, Dordrecht, Heidelberg, New York, London, 2015.
17. K. K. Chawla, *Ceramics Matrix Composites*, second edition, Kluwer Academic Publishers, Boston, MA, 2003.
18. R. Warren (ed.), *Ceramic Matrix Composites*, Blackie & Son (Chapman & Hall), New York, 1992.
19. N. P. Bansal and J. Lamon (eds.), *Ceramic Matrix Composites: Materials, Modeling and Technology*, John Wiley & Sons, Hoboken, NJ, 2015.
20. M. F. Amateau, Ceramic composites, *Handbook of Composites* (S. T. Peters, ed.), second edition, Chapman & Hall, London, 1998, pp. 307–332.
21. D. W. Richerson, Ceramic matrix composites, *Composite Materials Handbook* (P. K. Mallick, ed.), Marcel Dekker, New York, 1997, pp. 983–1038.
22. J. R. Davis, Applications of ceramic matrix composites, *ASM Handbook, Vol. 21, Composites* (D. B. Miracle and S. L. Donaldson, vol. chairs), ASM International, Ohio, 2001, pp. 1101–1109.
23. G. Savage, *Carbon–Carbon Composites*, Springer, London, 1993.
24. E. Fitzer and L. M. Minocha, *Carbon Reinforcements and Carbon/Carbon Composites*, Springer, New York, 1998.
25. J. D. Buckley and D. D. Edie, *Carbon–Carbon Materials and Composites*, Noyes Publications, New Jersey, 1993.
26. J. D. Buckley, Carbon–carbon composites, *Handbook of Composites* (S. T. Peters, ed.), second edition, Chapman & Hall, London, 1998, pp. 333–351.
27. K. M. Kearns, Applications of carbon–carbon composites, *ASM Handbook, Vol. 21, Composites* (D. B. Miracle and S. L. Donaldson, vol. chairs), ASM International, Ohio, 2001, pp. 1067–1070.

14

Design of Composite Structures

14.1 CHAPTER ROAD MAP

A simplified schematic representation of a typical composite product development program is given in Figure 14.1. The different stages of product development are identified and it is clear that these stages are interrelated. Also, each of these stages can be associated with some topic in the broad field of composite materials technology. So far, we have had discussions on various topics on composites except design, which is the subject matter for discussion in this chapter. A point that can be noted is that design is a phase that comes fairly early in the overall product development program; but we have kept it for discussion in the last chapter. It is deliberate; it is a subject that demands reasonable level of insight into all other aspects of composites technology.

The concept of structural design needs some deliberation, and we shall begin our discussion with an attempt to define the term *design*. The process of structural design of a product has evolved over a long period and arguably there is no set procedure for a design problem; it is an art, yet certain set patterns and key features can be associated with it. Some of the fundamental features associated with composites design process are discussed. Laminate design and joint design are two crucial aspects in the overall subject of composite product design; these topics are addressed next. Finally, some representative composite products are presented as design examples.

14.2 INTRODUCTION

Design is a very common term in today's life. It is often used in association with some other terms, for example, graphic design, fashion design, content design, web design, structures design, and so on. Depending on the context, it takes different meanings in different fields. For example, in the field of graphic design, we are concerned with the creation of a graphic object by combining and arranging shapes, lines, pictures, and text in different styles. In a similar way, fashion design brings to our mind images of clothing with different styles and aesthetics. We can go on from one field to another and find the differences in meaning as well as similarities and establish three common features—requirements, resources, and constraints. A practical definition of the term *design* can be found by analyzing the significance of these three features.

First, a design is made to fulfil a certain requirement. Typically, the requirement is the creation of a product that does certain specific functions. Examples of such products are a printed advertisement in a magazine (graphic design), a designer dress (fashion design), a web page (web design), a multistoried building (building design), and so on. Needless to say that each of these products has its own specific function to perform.

Second, a design is made using certain resources. Broadly speaking, materials, technological know-how, and manpower constitute the overall resource base. These

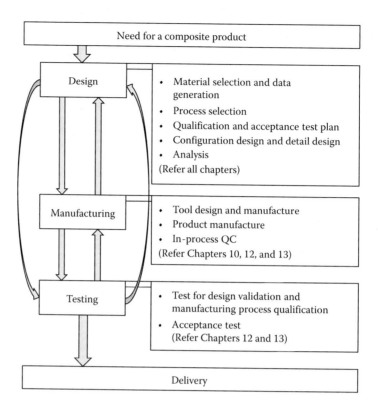

FIGURE 14.1 Schematic representation of a typical composite product development program.

resources are required for making the intended product as well as for carrying out the very act of designing.

Third, a design is made within certain constraints. As we shall see, design is a process of making an appropriate choice from among many possible alternatives. Often, this is restricted by considerations on functional, economic, aesthetic, and sociopolitical issues. Similar to the case of resources, the restrictions too apply to the product being designed as well as to the act of designing.

Obviously, given the broad nature of the process of designing, there is no universally acceptable definition of design. However, we can define design, especially from the point of view of convenience of discussion, *as a scheme that is used for making a product using certain resources within certain constraints to fulfil certain requirements.* Note that the requirements, resources, and the constraints vary widely from one sector to another.

Composites structural design is addressed in many standard texts, articles, and research papers. Discussions on the general features of the design of composites structures and process of design are addressed by several authors [1–5]. Laminate design and analysis are an essential part in any composite product design and many texts and articles are available on the subject [1–4,6–13]. Joints are involved in many composite products. They are typically the weak links in a composite structure and their behavior, failure modes, and design aspects are more complicated than those in metals [1,5,7,8,11–17]. Another critical aspect in design is stiffening of panels, which is required for higher stiffness in many structural applications. Composite stiffened structures are different from their metallic counterparts in respect of several issues related to the anisotropic nature of composites and their unique manufacturing methods; these issues associated with stiffened composite plates and shells are addressed by many [1,3,9,18,19]. Extensive work has been done in the areas of general design philosophy, design, analysis and optimization of laminates, design and analysis of joints, stiffened structures, fracture behavior, performance under fatigue loads and impact, and other related topics. While a detailed review of such works is not intended in this book, limited reference is made to indicate the direction of work that has been carried out.

14.3 BASIC FEATURES OF STRUCTURAL DESIGN

From the broad concept of design, now we move on to the design of structures or structural design. A structure is something that has a certain physical configuration, usually made up of a number of parts or elements such as beam, column, plate, shell, etc., such that it takes load and performs certain specific functions. The objective in structural design is to meet certain requirements to be achieved using available resources within specific constraints. In other words, here too, we have to deal with certain requirements, resources, and constraints. In this section, we dwell on these issues.

14.3.1 Requirements

A structure is designed to meet certain functional requirements. The functional requirements can vary widely from one type of structure to another—a bridge over a river is required to facilitate the movement of traffic across the river, a pressure vessel is required to contain a certain liquid or gas at high pressure, a space vehicle is required to deploy a satellite in space, and so on. While a structure has to meet its functional requirements, it is invariably subjected to certain loads during its life. The loads can be mechanical (e.g., force and bending moment), thermal (e.g., high temperature), environmental (e.g., chemical corrosion), etc.

We can identify the design requirements of a structure by studying the functions and the loads. For example, let us consider the case of a slab of a multistoried building. The primary functional requirement of the slab is to provide a firm, hard, and aesthetically acceptable surface for our use. There are other functional requirements: it should provide reasonable thermal insulation; should not allow seepage; should have provisions for hanging a ceiling fan; and so on. The slab, as we can see, is subjected to various loads during its life; these are the dead load (its own weight) and the live load (weight of human, furniture, etc.). Under these lateral loads, the slab undergoes bending and the primary structural function of the slab is to carry the loads without excessive bending during its entire life span such that the occupants feel safe and comfortable.

We can identify many design requirements that can be associated with various types of structures. Functional requirements are of widely varying nature, which depend on the application of the structure; on the other hand, the structural requirements can be broadly placed in two groups—strength and stiffness.

14.3.1.1 Strength

For any structure, the most common design requirement that comes to our mind is strength. The configuration of the structure has to be such that the stresses and strains under the design loads are under certain limiting values and material failure does not occur. The definition of material failure may vary from one material to another or from one application to another. For metallic materials, often, yielding of the material is considered as failure of the material, whereas, for composite materials, in most cases, fiber fracture represents a failure of the material. Thus, appropriate failure criterion (see Chapter 4) has to be employed so that the available margin before failure can be estimated.

14.3.1.2 Stiffness

There are many cases of structural applications, where design requirements are stiffness dependent. In the strength-based designs, stress and strain are the two parameters that are compared with certain specified limits, whereas, in the case of stiffness-based designs, the key parameters are deformation, vibration, and buckling load. For example, let us once again consider the slab of a multistoried building. If the slab is so designed such that the stresses and strains are within the specified limits but not the deflection that when we walk on it, it vibrates and gives a feeling of excessive deflection, the design is not acceptable! Similarly, for an axially loaded cylindrical shell under compression, buckling is likely to be the primary design requirement.

14.3.1.3 Other Design Requirements

In addition to the functional requirements and strength- and stiffness-based requirements, there are other design requirements that can be related to the intended life of the structure and certain specific structural characteristics such as energy absorption, thermal insulation, corrosion resistance, etc. Energy absorption is a design requirement in some applications where a structural part protects some other part by absorbing the incident energy. An example of such application is helmets where the impact energy is absorbed by the helmet material. Another example of energy absorption is rocket nozzle liners, which absorb heat and mechanical energy of the high-velocity gases and protects other structural nozzle parts by a process known as ablation. In many applications such as rocket motor casing, furnace walls, etc., thermal insulation is a critical design requirement. Similarly, in the chemical industry, offshore applications, naval applications, etc., corrosion resistance is of major concern.

A structure may be intended for either single use or multiple uses. A single-use structure may be intended for either immediate use or future use; when used in the future, it has to be stored after manufacture till the date of use. On the other hand, a structure intended for multiple uses may be put to regular maintenance. Clearly, a structure has a definite life span and it has to perform its specified functions satisfactorily over the entire life.

14.3.2 Resources

In structural design, we must consider the availability of the following resources—material, manufacturing technology, computing technology, and human resources.

14.3.2.1 Material

Materials are a crucial input in any structural design. Research in the field of materials science has resulted in the development of more efficient and high-performance materials with higher mechanical properties coupled with more useful physical characteristics. The availability of advanced materials has widened the horizon of the design process, leading to the realization of more efficient structures. The use of anisotropic high-performance composites such as CFRPs is an excellent example of this. Of course, the use of such materials has increased the complexity of the design and analysis process.

14.3.2.2 Manufacturing Technology

A design is not acceptable if a suitable manufacturing technology, including machinery and equipment, and tools and fixtures, is not available for translating it to a product. Innovations in the field of manufacturing have made it possible to design more efficient and reliable structures. Also, design must take into account the details of the manufacturing stages, including nondestructive testing, and in-process quality check steps have to be appropriately coupled with the actual manufacturing and assembly operations so that quality and reliability are ensured (see Reference 20 for a more detailed discussion on the concept of design for manufacturing).

14.3.2.3 Computing Technology

Design is a process of decision making aided by certain tools. Both analytical and numerical tools, as we had seen in the previous chapters, are available. Undoubtedly, in today's design world, finite element analysis is the most widely used tool, for which a number of commercially available software packages exist. There has been remarkable progress in the field of computing, which makes it possible to analyze complex structures quickly with added reliability. As a result, it is now possible to design more efficient and more reliable structures.

14.3.2.4 Human Resources

Arguably, the most important resource in the overall product development program is human resource. Technically competent manpower is essential for carrying out a design that considers all the aspects in the complete product development cycle. Process automation has reduced greatly the demand on human labor in certain manufacturing methods. Certain other methods remain labor dominated. The designer must take these aspects into account. After all, if a product cannot be realized for nonavailability of skilled work force economically, the design is not welcome and an alternate design that involves more automated process would be preferred.

A related technological progress is the advancement in the field of information technology. Information in the form of raw data as well as source of data is easily available today. It helps the designer greatly in making more informed decision in the design process.

14.3.3 Constraints

Structural design is carried out within certain constraints that arise, as indicated in the introduction, from functional, economic, aesthetic, and sociopolitical issues. The design engineer faces various constraints out of which some of the most common ones can be related to weight, cost, assembly requirements, and manufacturing feasibility. Of course, one can come across other relatively less common constraints. At this point, it is worth mentioning that sometimes there exists only a very fine distinction between design requirements and design constraints. Design constraints can be referred to as design requirements with conflicting nature.

14.3.3.1 Weight

In many cases, especially in aerospace applications, the functional utility of a product is reduced by higher weight. It is true in other sectors such as automotive as well. For example, the higher the structural weight of a space vehicle, the lower its payload weight, the higher the weight of an airplane, the higher its fuel cost, and so on. Thus, there are penalties associated with higher structural weight in many applications and in these cases there exist design constraints to keep the structural weight as low as possible. As mentioned earlier, design constraints can be viewed as design requirements with a conflicting nature. For example, in general, higher strength and stiffness requirements and lower weight requirements are contradictory to each other. To take a specific example, let us consider the case of a pressure vessel in a space vehicle, where, in order to increase the capacity to take higher internal pressure, the designer may be tempted to increase the shell thickness. Higher shell thickness is associated with weight penalty. Thus, higher internal pressure and lower weight requirements are contradictory to each other; a solution has to be found by efficiently utilizing high-performance materials and at times a compromise may be made by appropriately altering the specifications on pressure and weight.

14.3.3.2 Cost

In any application, an overall lower cost is welcome. There are some cases of national and strategic values, where, apparently, cost does not play a major role. In these cases, the sociopolitical cost of not having the product is generally very high and the cost of design and development is relatively lower. In most other applications, funds allocated for the development of a product are limited and cost is a major design constraint. Obviously, a design without economic value is of no use in these cases.

Cost has several elements—initial investment, operational cost, maintenance cost, etc. The initial investment includes plant and machinery, tools and fixtures, design and analysis, etc. These expenditures are primarily one-off in nature and they can be amortized over a period of time or over a certain number of products. Operational and maintenance costs are recurring in nature. The cost of raw materials, machine running cost, wages and salary, plant maintenance cost, etc. are some of the heads that need to be accounted for while doing costing of a product. The cost elements may vary grossly depending on the manufacturing methods. The design engineer must take all of these aspects so as to keep the overall cost within an acceptable limit.

14.3.3.3 Assembly Requirements

Quite often, a structure is built by assembling a number of components and subassemblies. Components and subassemblies are joined using adhesive and/or fasteners. Individual components must be designed under the constraint that the mating parts are compatible with each other at the interfaces. Assembly sequence is another consideration that imposes certain constraints on the design of the components.

14.3.3.4 Manufacturing Feasibility

Advanced manufacturing processes have greatly increased the horizon of possibilities in front of the design engineers and added immensely to the development of complex and highly efficient products. These processes, however, have their own limitations that put constraints. For successful component realization, it is important that these limitations are clearly understood and accounted for in the design process. For example, let us consider the composite manufacturing process of filament winding. No doubt, it is highly suitable for making axisymmetric components such as a cylindrical pipe, etc. and it allows automated efficient placement of impregnated fibers along the desired orientation. Helical winding, however, does not allow one to arbitrarily vary the fiber orientation along the axis of the component and the designer must keep this constraint in mind and choose fiber orientations that are windable. Another example is that of autoclave curing. It gives good consolidation and high-quality laminate. But it is not suitable for making a laminate of uniform thickness with good surface finish on both the sides. If good surface finish is required on both the sides of a uniformly thick laminate and autoclave curing is involved, then machining can be introduced and the resulting reduction in material properties due to fiber cutting may have to be accounted for in the design calculations. In other words, manufacturing feasibility and limitations should be considered during the design process and suitable corrective steps should be introduced as required.

14.4 DESIGN VERSUS ANALYSIS

Structural design and analysis are an essential phase in any product development program. A good understanding of the two terms and the differences between them should be clear to a structural engineer (Table 14.1).

First, structural design is a process in which we consider a number of possible configurations of the product and choose an acceptable configuration. It involves the alteration of shape and size of the structural elements and details of joints so that the design

TABLE 14.1
Design versus Analysis

Design	Analysis
• Structural configuration	
– Not known before design is started	– Known before analysis is started
– Chosen during the process of design (an output parameter)	– An input parameter
• Materials	
– Not known before design is started	– Known before analysis is started
– Selected during the process of design	– Material data used design are also used in analysis
– Material properties are either generated by testing or taken from available data sheets	
• Process	
– Not known before design is started	– Process details are generally not used directly in analysis
– Selected during the process of design	
– Process parameters are arrived at during process planning or process design	
• Specifications	
– Known before design is started	– Known input to the analysis process
– May be altered and finalized through an iterative process	
• General characteristics	
– Iterative process	– One-way process
– Indeterministic	– Deterministic
– Leader	– Follower
• Typical output	
– Structural configuration (shape and size of the product)	– Response of the structure to the specified loads (stress, strain, deflection, buckling factor, mode shape, and frequency)
– Joint details	
– Finalized specifications	– Estimated failure load
– Materials to be used in the manufacture of the product	– Estimated margin of safety
– Manufacturing process	

requirements are met in the best possible way. While the designer is expected to have an idea about the possible configurations of the product, the final configuration is not known at the start of the design process. In fact, to determine an acceptable configuration is the primary objective of design. On the contrary, the structural configuration is known beforehand and it is an input to the analysis process.

Second, a major design task is to select the materials to be used in the manufacture from among a number of candidate materials. Material properties are generally taken from available resources such as data sheet, handbook, etc. Alternatively, critical material data are also generated by experimental characterization. On the contrary, for analysis, these are input data.

Third, another major design task is to identify the manufacturing methods to be employed in the realization of the product. The detailed process parameters are determined as a part of process planning, which forms a part of the broad design exercise. Details of manufacturing methods and process parameters play a critical role in the design of composite structures. Analysis, on the other hand, is done often without paying any particular attention to the processing details.

Fourth, any structural design starts with the identification of the specifications in respect of functional and structural requirements. Typically, the broad specifications of the overall system are generated first followed by those of the subsystems and components. During the design exercise, the specifications are altered and frozen through a number of iterations. For analysis, the specifications in respect of loads, etc. are known input parameters.

Fifth, design is an iterative process of decision making; it is indeterministic in nature and it is the leader. On the other hand, analysis is a one-way process and it is deterministic in nature. It is the follower; it can be done using data generated by the design process after design has been completed.

Finally, a look at the type of the output data of design and analysis processes reveals that design gives us the structure in terms of its configuration, materials, and processes. On the other hand, analysis gives us the response behavior of the structure to the applied loads.

14.5 COMPOSITES STRUCTURAL DESIGN

The structural design of a composite product involves a number of steps that can be broadly identified as follows:

- Generation of specifications
- Materials selection
- Configuration design
- Analysis options
- Manufacturing process selection
- Testing and NDE options
- Design of part/laminate and joints

These steps are applicable to any structure in general but the details are greatly different in composite materials and structures. The exercise starts invariably with the generation of specifications. However, various other steps are not necessarily performed in sequence; they are all interrelated and the final design is made by a process of iteration. Figure 14.2 shows schematically the process of composites structures design.

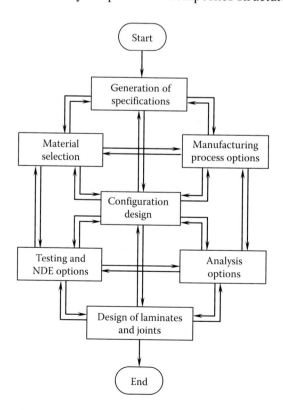

FIGURE 14.2 Composite structures design process.

14.5.1 Generation of Specifications

A composite product can be either an independent component or a part of a larger structure. Further, it can be either an all-composite component or an assembly of composite and metallic components. For an independent component, that is, a component that has an end use on its own, the design specifications are obtained directly from the consumers' needs. For example, an all-composite pleasure boat will be designed to specifications as per the taste, preference, and demands of consumers. On the other hand, a large number of composite products are used as parts/assemblies/subassemblies of other larger structures. In these cases, the specifications of the composite products are dictated by the specifications of the overall structure. For example, the rudder of an aircraft is a critical part that can be designed with composites, the specifications for which must however come from the overall specifications for the aircraft. Clearly, the design specifications of a product are derived from the basic need for it and the structural and environmental loads that it is likely to be subjected to during its life time.

The design specifications cover a wide range of issues related to the structure, which for the sake of convenience of discussion can be categorized as follows:

- Geometrical specifications
- Functional specifications
- Structural specifications
  - Structural loads
  - Environmental loads

Any structure has a certain shape and size, and it is invariably designed to certain specifications related to its geometry. In many cases, geometrical configurations follow a certain set of guidelines or patterns. For example, a pipeline is invariably circular in cross section, the wings of an aircraft are of airfoil cross section, and so on. However, the details in respect of diameter, length, thickness, etc., as applicable, must be specified. Note that these specifications depend on functional as well as structural requirements.

The functional specifications include all the tasks that the product being designed is required to perform. They are directly related to the functional requirements. Further, the life span over which the product is expected to serve, environmental conditions during its service life or perhaps the storage conditions, etc. may also be included as specifications.

While in service, a product is invariably subjected to certain loads that can be either structural or environmental. The structural loads are the forces, bending moments, torque, pressure, etc. that may act on the product. It is important to identify various possible load cases and carefully estimate the maximum possible levels of these loads under each load case during the entire life of the product. In addition to the structural loads, environmental loads such as thermal, chemical, etc. too need to be specified. Structural composite materials degrade at high temperatures. The rate of degradation is rather drastic above the glass transition temperature (T_g) and if the specified service temperature is above T_g, an appropriate thermal insulation is essential. Composites are also used as ablative liners and insulating liners, in which case thermal loads are required for designing the liner thickness. Polymers may become brittle at subzero temperatures and minimum service temperature is also an important design input. The exposure of a composite product to a corrosive environment such as chemical, saline, etc. may adversely affect its life; needless to say that such environmental loads must be specified and taken care of in the design calculations.

The generation and finalization of design specifications, very much like the overall design process, are also an iterative process. Certain specifications may result in nonoptimal design. Sometimes, certain specifications may put too tight restrictions on the manufacturing process to be implemented in an economical way. This is especially true in those cases, where a composite product is perhaps a part of a larger structure. In such a case, relaxing the specifications in one part and tightening in some other part may be necessary for an overall optimal solution. Clearly, this is possible by means of an iterative process.

14.5.2 Materials Selection

Materials selection is a critical task in any structural design exercise. It is especially true in composite structures due to the added complexity associated with the basic composite material system and its constituents. As we know, two constituents, viz. reinforcements and matrix, combine at the macroscopic level to form a composite material with unique properties; it implies that there are three broad areas to be considered in the selection of composite materials—selection of the composite material, selection of the fibers, and selection of the matrix.

14.5.2.1 Selection of the Composite Material

The selection of the basic composite material is the first step in the material selection process. Here, the designer chooses a material system (such as carbon/epoxy or glass/polyester, etc.) and its physical form (such as layered composite, flake composite, etc.) We can identify a number of factors that influence the selection of the basic composite material; these are listed in Table 14.2.

During the material selection process in the design of most structures, strength and stiffness are the two factors that possibly receive the maximum attention. However, for aerospace and other weight-sensitive applications, specific strength and specific stiffness properties are more important. Clearly, the density of a material is a critical parameter in such cases. Other material properties are also crucial, depending on the applications. Fatigue and fracture characteristics of materials are important in structures

TABLE 14.2
Factors Influencing Material Selection

Factors	Where Applicable
Strength and stiffness	Most structural applications
Fatigue strength	Structures under cyclic loading
Fracture toughness	Structures under impact loading
Density—specific strength and specific stiffness	Weight-sensitive applications as in aerospace
Glass transition temperature	Structures subjected to high temperatures
Thermal expansion	Structures under wide range of operating temperatures
Thermal insulation	As insulating layer to other structural parts under high temperatures
Thermal conductivity	Applications where accumulation of heat at a localized zone is not desirable
Electrical conductivity	Applications where accumulation of static charge at a localized zone is not desirable
Corrosion resistance	Structures subjected to corrosive environment as in chemical industry, marine applications, etc.
Wear resistance	Applications with moving parts
Resistance to fire	Applications with fire hazards, e.g., buildings, industrial plants
Manufacturing considerations	All applications
Cost	All applications

under cyclic loading and impact loading, respectively. The mechanical properties of most composite materials are adversely affected at elevated temperatures. Thus, glass transition temperature is a critical parameter if the service temperature of the structure is high. In certain applications, composite parts are used as an insulating layer to other structural metallic or composite parts. In these cases, thermal insulation is the design driver, and strength and stiffness characteristics are secondary requirements. Similarly, there are applications where the accumulation of heat or static charge is undesired; in such cases, thermal conductivity and electrical conductivity are the required material characteristics. Corrosion resistance is a major requirement in many applications such as chemical storage tanks, plants, marine applications, etc. Wear resistance is a desirable material property in applications with moving parts. On the other hand, fire resistance is a desirable property in applications with fire hazards. Manufacturing considerations in respect of materials and their physical form and characteristics must be given due respect while designing a composite structure. Finally, cost plays a crucial role in the selection of material in any structural application.

Clearly, while selecting materials, the material properties to be considered depend on the application of the product. It is also likely that in some applications, multiple requirements exist; in such cases, a practical approach is to prioritize the requirements and proceed from highest priority to the lower ones.

14.5.2.2 Selection of the Reinforcements

The selection of the reinforcing fibers is dictated by their functions. The primary function of the reinforcements is to impart strength and stiffness to the composite material in the fiber direction. The strength and stiffness properties of a unidirectional lamina are highly direction-oriented—being extremely high in the fiber direction and low across—and is essential to suitably orient the fibers so as to achieve the desired laminate properties. Next, the density of composite laminate and its specific strength and stiffness are directly dependent on the fiber density. Thus, it follows that fiber tensile strength and modulus and density are the most important parameters that influence fiber material selection. Among reinforcement materials, different varieties of carbon, glass, and aramid are the most commonly used. Carbon and aramid have low density and high tensile strength and modulus; they are used in most weight-sensitive and other high-performance applications, including aerospace, automobile, sports, etc. On the other hand, glass fibers have relatively higher density and low modulus; they are used in many ground systems and commercial applications.

In addition to strength and stiffness, there are other laminate characteristics that are dependent on the reinforcements. Thermal stability, thermal and electrical conductivity (or insulation), energy absorption, etc. are some of these characteristics. An excellent example of the use of composites for thermal stability is in satellites, where the negative CTE of carbon fiber is effectively utilized for compensating the positive CTE of resin matrix such that nearly zero thermal expansion or contraction is achieved across a wide range of temperature variations. Similarly, glass fibers are used in composites for imparting thermal insulation properties, carbon fibers for electrical conductivity and aramid fibers for impact resistance, and so on.

Another critical factor is the cost of reinforcing fibers. Carbon fibers are highly expensive and they are cost-effective mostly in high-end applications. This is also why glass fibers are almost universally used in any commercial application.

14.5.2.3 Selection of the Matrix

Reinforcements by themselves are worthless without the matrix; the matrix binds the reinforcements and gives shape to the composite material. Thus, in a composite material, although the matrix is the weak link, it is essential. It has its own specific functions and the factors involved in the selection of matrix can be directly linked to its functions. They are as follows:

- Fiber wettability and compatibility
- Manufacturing issues
- Environmental exposure
- High-temperature application
- Cost

The matrix must be compatible with the reinforcing fibers and capable of wetting them thoroughly. This is of critical importance as otherwise the basic load transfer mechanism is affected. Manufacturing issues play a major role in the selection of matrix material. Depending on the curing temperature and pressure requirements, the manufacturing process may become rather sophisticated and complex. For example, a room-temperature curing resin system needs hardly any additional setup for curing, whereas for certain other resin systems, we need an autoclave with high temperature, pressure, and vacuum application facilities and all the related consumables. Similarly, a resin system with high viscosity may not be acceptable for wet filament winding unless a suitable mechanism such as online heating arrangement is put in place. The matrix provides good protection to the reinforcing fibers against chemical attack and mechanical wear and tear. Clearly, environmental resistance and mechanical wear resistance are factors that influence matrix selection in such applications. Another major factor is the glass transition temperature of the matrix material for high-temperature applications. Finally, cost plays its own role in matrix selection. There are two aspects here. The first cost element is the basic cost of the resin system, which may vary greatly from one system to another. Second, the manufacturing facilities required, especially curing setup together with necessary consumables, can make a lot of difference in financial calculations and eventual selection of the matrix material.

14.5.3 Configuration Design

Configuration design is the process of choosing the shape and size of the structure, which involves the following:

- Selection of the overall shape and size of the structure
- Selection of various types of basic structural elements
- Selection of shape and size of each structural element
- Selection of proper joints

The basic structural elements frequently used in the design of a structure are rod, column, beam, plate, and shell. Solids of various shapes are also used sometimes. Similarly, other more complex structural elements that are essentially combination of the basic elements, for example, corrugated plates, can also be thought of. Further, in many structural applications with high stiffness requirements, plates and shells are used with stiffening members. Stiffened structures can be made either as bonded or fastened assembly or as an integral part. (We shall discuss stiffened structures in a little more detail in a later section.) Other special structural constructions peculiar to composites are honeycomb and sandwich constructions.

Configuration design is a very important step. Typically, an initial overall shape and size of the structure are chosen and various structural elements required are identified. The shape and size of each element are chosen through an iterative process that involves computation and selection, and the final structure is obtained as an assembly of various structural elements. Clearly, a proper scheme of joining the elements is essential and details of various joints have to be worked out.

14.5.4 Analysis Options

A design exercise is incomplete if we are not able to analyze it and make estimates of response of the structure to the applied loads. The response of a structure is measured in different ways, depending on the types of loads and the level of structural details. In general, structural analysis is done for the determination of the following:

- Static analysis—displacement, stress distribution, strain distribution
- Dynamic analysis—natural frequency, mode shape
- Stability analysis—buckling load
- Thermal analysis—temperature distribution

Structural analysis is done at different levels of details, under different types of loads, and with different objectives. At the overall assembly level, the objective of analysis is to find the overall response in terms of gross displacements, global buckling load, etc. and details in respect of local features such as cut-outs, joints, etc. are often ignored. On the other hand, at the individual component level or at the subassembly level, local details must be given due importance; local stress distribution around cut-outs or near some discontinuities, joint opening, and other local behavior of the structure are found out.

Finite element modeling is the most common tool for the analysis of any real-life structure. However, it may be noted that simple hand calculations, especially for basic structural elements, still play a critical role.

14.5.5 Manufacturing Process Selection

The realization of a composite product is largely process-dominant and the selection of an appropriate manufacturing process is a critical aspect that must be taken care of during design. It is especially true in composites due to the fact that during the development of a composite product, it is not only the product but also the material itself that gets made. Various processing options and manufacturing techniques are available in the field on PMCs (Chapter 10). By and large, the choice of the primary manufacturing process is obvious for some types of composite products. For example, for constant cross section unidirectional longitudinal members, pultrusion is almost universally used. Similarly, filament winding is an automatic choice for any pressure vessel, and so on. However, there still exist several factors that need to be considered for making a final decision in respect of manufacturing process selection. Some of these factors are

- Shape and size of the product
- Production requirement
- Tooling requirements
- Need and availability of skilled manpower
- Automation
- Reliability and repeatability
- Structural property requirement
- Cycle time
- Cost

The overall shape and size of the product are the first aspects that would come to our mind while making a selection of the manufacturing method. An associated factor is the complexity of the configuration. Certain processing techniques, such as compression molding, are suitable for small- to medium-sized parts, whereas certain others, for example, filament winding, can be used for making medium to large sizes. Similarly, certain techniques, such as press molding and pultrusion, are suitable for meeting the large-scale production requirement whereas certain others, for example, hand lay-up, can be used for the realization of a few components. Tooling requirement is a common feature for any composite processing technique. The nature of the tooling depends on the processing technique and the configuration of the part. It affects several aspects, including quality and cost, of component realization. Some of the composite processing techniques are heavily dependent on skilled manpower; obviously, these processes are not suitable if manpower is scarce. Automation is possible in some cases that results in reliability. Reliability and repeatability also vary depending on the processing techniques. For example, compression molded and pultruded parts are known for quality repeatability, whereas it is quite poor in wet hand laid-up parts. Another aspect of concern is structural property requirement. Filament winding, pultrusion, RTM, etc. can be employed to make highly directional composite parts. On the other hand, compression-molded randomly oriented components and open-molded CSM products are poor in structural properties. Finally, the cycle time of realization and overall cost play their own roles in the manufacturing process selection.

14.5.6 Testing and NDE Options

Testing and nondestructive evaluation options must be considered during the design of a composite product. In a product development program, tests on materials are conducted for both the generation of design data as well as in-process quality control purposes. Testing is also done at the product level for acceptance and qualification of the product. Similarly, NDE is done to detect the presence of flaws and deviations in the components and interfaces. The basic objective of carrying out testing and NDE is to infuse reliability and quality into the product. It is essential during product design itself to identify the needs and corresponding testing and NDE options at different levels of product development.

14.5.7 Design of Laminate and Joints

Arguably, the most important step in the design of a composite structure is the design of the laminates and joints. There are several aspects to be addressed and we address them below under two major headings—laminate design and joint design.

14.6 LAMINATE DESIGN

14.6.1 Scope of Laminate Design

A laminate is a layered composite structural element that is made by a number of laminae or plies. In a composite structure, a plate, a shell or panel, a beam, a stiffening rib, etc. are all some forms of laminates. Typically, the reinforcements in the laminae are oriented w.r.t. the coordinate system of the structural element and the laminae are stacked as per a certain ply sequence. The laminate structural response to applied loads can be greatly influenced by altering the ply sequence that involves primarily three variables, viz. ply material, ply thickness, and ply orientation. One more variable of critical importance is the fiber volume fraction, which depends mainly on the manufacturing process adopted. The shape of a laminate is obtained from the chosen configuration of the structure and various structural elements. On the other hand, laminate design basically involves the selection of ply sequence variables and an appropriate fiber volume fraction.

14.6.2 Laminate Design Concepts

Laminate design, being a part of an overall indeterministic design process, is a reasonably complex subject. It is particularly so in the case of composites due to their anisotropic nature and the presence of more numbers of elastic constants and strength parameters. The design process is made to look more complicated by the use of several terminologies that sound similar and are related to each other. They are basically related to loads, material properties, and safety factors. In this section, we dwell on these concepts and then proceed on to laminate design procedure in the next section.

14.6.2.1 Load Definitions

Various terms related to loads that are used by designers across various sectors can be grouped as follows:

- Applied loads—operating load, maximum operating load, limit load, and design limit load
- Failure loads—ultimate load
- Design loads—design ultimate load

Starting with the very first moment a product takes shape in the shop floor, it goes through various stages that include manufacturing, assembly, handling, storage, the specified use, etc. In each of these stages, it is subjected to certain loads. *Operating loads* are the general forces and moments that act on the product in its normal operating conditions. An obvious extension to this statement is that *maximum operating loads* are the maximum levels of the operating loads that can be expected during its entire life time. *Limit load* is another term that is used to refer to the maximum load that can be expected on the product during its entire life span. Loads expected on the product are often estimated by mathematical simulation, and from this point of view, limit load is also an estimated load to be used in design and often referred to as the *design limit load*.

Ultimate load is the maximum load that a structure can take before failure. Here, it is important to define the term *failure*. It may be noted that failure need not necessarily mean rupture of fibers or yield of metal or complete collapse of the structure. In simple words, it means inability of the structure to perform some or all of the designed tasks. Thus, the load at which a structure fails to perform its designed tasks satisfactorily is the ultimate load.

At this point, we need to make note of two aspects related to laminate failure. First, the ability of a structure to perform its tasks is judged basically from two angles—stress/strain and deflection. In the first case, stresses and strains are compared with the respective allowables, for which an appropriate failure criterion can be applied at the lamina level. In the second case, deflections are compared with specified maximum acceptable values. There are other parameters too that are required under specific loading conditions. For example, under compressive loading, stability is a primary concern and buckling load is compared with the applied load. In a dynamic loading environment, natural frequency is a critical parameter. The second point related to product failure is that a structure is made as an assembly of a number of structural elements and failure of any one element may or may not lead to failure of the whole structure. Further, in a composite laminate, failure process is progressive and ultimate load may correspond to first ply failure and last ply failure depending on failure perception.

Design ultimate load is the load for which design calculations are done such that the structure fails at this load level. If the design shows that the structure fails at a load above the design ultimate load, it is overdesigned, and if it fails at a lower load, it underdesigned. An efficient design is a case where failure takes place at the design load.

As mentioned earlier, various operating conditions may exist, of which a few may be critical that would perhaps envelop other redundant, benign, and less critical ones. It is necessary to identify various possible load cases and determine the critical load cases. Operating loads and design loads are identified or estimated and the corresponding design loads determined by applying appropriate factors. Design and analysis are done and the analysis results are compared with design allowables.

Another point that is worth mentioning is that the terminologies listed above are not exhaustive and other specific terms are also used in specific cases. For example, in the field of rocket motor casing design, maximum expected operating pressure (MEOP) is a very common term that is nothing but the maximum operating load we discussed above. Also, certain terms may be used to mean differently from what is stated here. It is important to go to the basics, understand the context in which a term is being used, and then concentrate on the design and analysis work.

14.6.2.2 Design Allowables

Design allowables refer to the material properties to be used in design and analysis. For composites, material properties are often generated by testing. Some of the key factors that should be kept in mind while arriving at the design allowables are as follows:

- Scatter in the sample test results
- Batch-to-batch variations of constituent properties
- Variations of material properties at the component level as compared to the sample level

Design allowables are basically statistical parameters derived from the database generated from sample test results. Certain amount of scatter is generally present in the test results; it is essential to test a sufficient number of samples and have a reasonable size of the database so as to ensure reliability of the design allowables. Typically, the data points follow a normal distribution from which mean and standard deviation are computed. A design allowable is derived as "mean $- 2 \times$ standard deviation" or "mean $- 3 \times$ standard deviation," which gives, respectively, above 95% and 99% probability that a test result would be above the derived design allowable.

The next source of uncertainty is from batch-to-batch variations in material properties. The reinforcing fibers and the matrix materials do vary in their properties from one batch to another. One way to counter this is to lower the computed design allowable by some appropriate factor. Alternatively, strict in-process quality control can be resorted to for screening of raw materials.

The third issue is especially true for composite materials. Composite materials, unlike metals, are built during the fabrication process. Composite shop floor practices may adversely affect properties of the materials that actually make a component. The ideal conditions of making a laminate for sample preparation are often absent during the fabrication of a composite product. Also, in certain cases of composite processing techniques, component-level ply construction is grossly different from what it is at the sample level. For example, in a filament-wound component, the ply construction in a hoop ply is the same as in a unidirectional test specimen, but the same is not true in a helical ply, especially near a cross-over junction. Similarly, fiber volume fraction at the component level is often different from (lower than) the sample-level fiber volume fraction. Thus, it is necessary that design allowables are suitably modified giving due considerations to the shop floor practices and processing technique.

14.6.2.3 Factor of Safety, Margin of Safety, Buckling Factor, and Knockdown Factor

Factors of safety are used in design to counter uncertainties associated with it. The uncertainties of different kinds can be grouped as

- Uncertainties related to material properties
- Uncertainties related to loads
- Uncertainties related to design and analysis assumptions

In the previous section, we had seen some of the factors that influence variations in material properties. Design allowables are obtained from test results or data sheet values by using suitable reduction factors. In some cases, the factor of safety associated with material properties is implicitly implied. For example, when using yield stress as the design strength for ductile materials, we indirectly use a reduction factor equal to the ratio of yield stress to ultimate stress. For uncertainties in load estimates, design and analysis assumptions and methodologies, the factor of safety is used is used in an explicit manner. Typically, it is defined as

$$\text{FoS} = \frac{\text{Design ultimate load}}{\text{Design limit load}} \quad (14.1)$$

For any major structure, the factor of safety (FoS) is a well-deliberated number that is arrived at by considering interests and views of various stakeholders. General guidelines exist to help decide on a suitable factor of safety for different loading conditions such as static load, cyclic load, impact, etc. In general, a high factor of safety reduces the risk of failure but at the cost of weight penalty and lower performance. A higher factor of safety is used when failure is complicated and catastrophic in nature, and cost of failure is high. On the other hand, when the cost of failure is low and failure is more predictable, and a lower factor of safety can be used. In-process quality control measures also play a role in this respect. Tighter material screening and process control help reduce shop floor uncertainties and a lower factor of safety can be used.

There are other terminologies as well that are used in similar contexts. One such term is *margin of safety*. The factor of safety is used in design calculations to arrive at various parameters such as thickness, ply orientation, etc. The factor of safety is always more than 1. The margin of safety, on the other hand, is used to indicate the margin available after the product has been designed. From the analysis of the designed product, we get the estimates of various design parameters and the available factor of safety is calculated as the ratio of the ultimate load-carrying capacity of the product to the design limit load. The margin of safety is then given by

$$\text{MoS} = \frac{\text{Ultimate load-carrying capacity}}{\text{Design limit load}} - 1 = \text{Available FoS} - 1 \quad (14.2)$$

Note that the margin of safety is always positive.

Similar to the concept of the factor of safety, there are two more factors that we frequently encounter—buckling factor and knockdown factor. Buckling is a stability problem and is associated with compression loading. Buckling factor is the ratio of the critical buckling load to the design limit load. Thus, buckling factor refers to the factor of safety in a stability problem. The factor of safety required in a buckling problem depends greatly on the geometrical configuration and nature of loading. Knockdown factor, on the other hand, is a reduction factor, obtained largely from experimental results, which is used for reducing the classical buckling load of a structure and the reduced theoretical buckling load is then compared with the design limit load.

14.6.3 Laminate Design Process

The laminate design process basically involves four steps:

- Initial laminate selection
- Laminate analysis and measurement
- Design criteria check
- Design refinement

The process is schematically shown in Figure 14.3. It is an important part of the overall composite structures design process and certain other aspects have to be addressed prior to laminate design. Thus, part configuration, materials data, manufacturing process details, and applied loads are known and these are essential design input. Note that we emphasize on the fact that these are *essential design input* that must be considered in totality. It is not uncommon to come across laminate design, which considers part configuration, material data, and loads but not the manufacturing process. A typical example is a part to be realized by filament winding containing helical layers. As we know, helical angles at different locations of the part would depend on

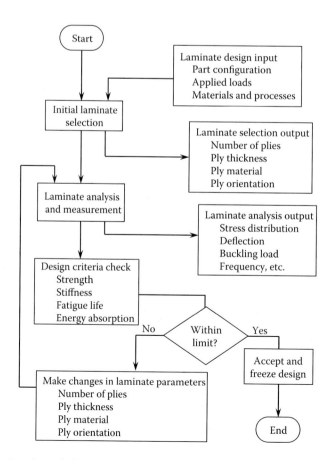

FIGURE 14.3 Laminate design process.

the geometry of the part. A laminate stacking sequence that is chosen based only on loads, material data, and part configuration but not the limitations of filament winding will most likely lead us to a situation where we have a good design on paper that cannot be translated to a product!

14.6.3.1 Laminate Selection

Laminate selection is a critical task but the starting point is somewhat arbitrary. We mentioned in one of the previous sections that laminate design involves the selection of primarily three variables—ply materials, ply thickness, and ply orientation. In addition, we have a fourth variable, viz. fiber volume fraction, which is typically governed by the processing technique. Material selection is done prior to coming to the stage of laminate design. However, issues such as forms of reinforcement—whether unidirectional or bidirectional—ply thickness, and orientation have to be addressed at this point.

Laminate performance is greatly influenced by the stacking sequence. As we had seen in Chapter 5, a number of special cases of laminate stacking sequence can be constructed. The significance of these stacking sequences in design is that some of the unnecessary coupling effects can be avoided as described in Table 14.3. In general, a symmetric laminate is the first choice as in this case unnecessary extension–bending coupling can be avoided and mathematical computational complexity is greatly reduced. However, strictly symmetric stacking sequence may be difficult in many real-life cases due to processing limitations, especially with ply drop-offs. On the other hand, in some specific cases, certain coupling effects can actually be used advantageously to meet design requirements.

The process of initial selection and subsequent refinement of stacking sequence can, however, be greatly simplified by considering the following two aspects:

- Practical aspects
- Invariant forms of laminate stiffness

First, infinite possibilities for the selection of stacking sequence exist theoretically, especially w.r.t. the angle of orientation of the plies. Thus, the initial choice of the laminate is to a large extent arbitrary, perhaps dictated by previous experience. While strength and stiffness characteristics of a laminate can be theoretically estimated,

TABLE 14.3
Special Cases of Laminates and Their Significance

Laminates	Remarks
Symmetric laminate	• No extension–bending coupling
	• No shear–twisting coupling
	• No extension–bending–shear coupling
	• No extension–twisting coupling
	• No shear–bending coupling
	• No bending–twisting coupling
Antisymmetric laminate	• No extension–shear coupling
	• No bending–twisting coupling
	• No twisting–curvature coupling
Balanced laminate	• No extension-shear coupling
Cross-ply laminate	• No extension–shear coupling
	• No extension–twisting coupling
	• No shear–bending coupling
	• No bending–twisting coupling
	• No twisting–curvature coupling

fatigue life is typically experimentally determined. The risk and cost involved in a totally new laminate are very high. Thus, stacking sequences, in general, and angles of orientation, in particular, are chosen from somewhat standard and familiar ranges of values. Also, changes made are often incremental.

Second, the invariant forms of laminate stiffness can be conveniently utilized in a design environment. We briefly discuss the concept in the following paragraphs.

14.6.3.1.1 Concept of Invariant Forms of Laminate Stiffness

Let us recall the equations for transformed reduced stiffnesses from Chapter 4:

$$\bar{Q}_{11} = Q_{11}\cos^4\theta + Q_{22}\sin^4\theta + 2(Q_{12}+2Q_{66})\sin^2\theta\cos^2\theta \tag{14.3}$$

$$\bar{Q}_{12} = (Q_{11}+Q_{22}-4Q_{66})\sin^2\theta\cos^2\theta + Q_{12}(\sin^4\theta+\cos^4\theta) \tag{14.4}$$

$$\bar{Q}_{16} = (Q_{11}-Q_{12}-2Q_{66})\sin\theta\cos^3\theta - (Q_{22}-Q_{12}-2Q_{66})\sin^3\theta\cos\theta \tag{14.5}$$

$$\bar{Q}_{22} = Q_{11}\sin^4\theta + Q_{22}\cos^4\theta + 2(Q_{12}+2Q_{66})\sin^2\theta\cos^2\theta \tag{14.6}$$

$$\bar{Q}_{26} = (Q_{11}-Q_{12}-2Q_{66})\sin^3\theta\cos\theta - (Q_{22}-Q_{12}-2Q_{66})\sin\theta\cos^3\theta \tag{14.7}$$

$$\bar{Q}_{66} = (Q_{11}-2Q_{12}+Q_{22}-2Q_{66})\sin^2\theta\cos^2\theta + Q_{66}(\sin^4\theta+\cos^4\theta) \tag{14.8}$$

These expressions show how the transformed reduced stiffness parameters of a globally orthotropic lamina vary depending on the orientation of the lamina. In a real-life design scenario, the engineer has to choose θ for each lamina in the laminate. Clearly, the above expressions are rather too complex to be of practical use. The problem can be overcome by employing the concept of invariant forms of laminate stiffness, which was originally introduced by Tsai and Pagano [21]. Now, by using various trigonometric identities, Equations 14.3 through 14.8 can be recast as

$$\bar{Q}_{11} = U_1 + U_2\cos 2\theta + U_3\cos 4\theta \tag{14.9}$$

$$\bar{Q}_{12} = U_4 - U_3\cos 4\theta \tag{14.10}$$

$$\bar{Q}_{16} = \frac{U_2}{2}\sin 2\theta + U_3\sin 4\theta \tag{14.11}$$

$$\bar{Q}_{22} = U_1 - U_2\cos 2\theta + U_3\cos 4\theta \tag{14.12}$$

$$\bar{Q}_{26} = \frac{U_2}{2}\sin 2\theta - U_3\sin 4\theta \tag{14.13}$$

$$\bar{Q}_{66} = U_5 - U_3\cos 4\theta \tag{14.14}$$

in which

$$U_1 = \frac{3Q_{11}+3Q_{22}+2Q_{12}+4Q_{66}}{8} \tag{14.15}$$

Design of Composite Structures

$$U_2 = \frac{Q_{11} - Q_{22}}{2} \tag{14.16}$$

$$U_3 = \frac{Q_{11} + Q_{22} - 2Q_{12} - 4Q_{66}}{8} \tag{14.17}$$

$$U_4 = \frac{Q_{11} + Q_{22} + 6Q_{12} - 4Q_{66}}{8} \tag{14.18}$$

$$U_5 = \frac{Q_{11} + Q_{22} - 2Q_{12} + 4Q_{66}}{8} \tag{14.19}$$

Note that the terms U_1, U_2, U_3, U_4, and U_5 are invariant of ply orientation θ and constants for a given material system. Note further that the transformed reduced stiffnesses $\bar{Q}_{ij}$ are not invariants of θ. However, by writing them in terms of the lamina stiffness invariants U_1, etc., it is easy to visualize the effect of θ on $\bar{Q}_{ij}$.

Next, the concept is extended from lamina to laminate, where the laminate stiffnesses A_{ij}, B_{ij}, and D_{ij} are obtained by expressing $\bar{Q}_{ij}$ in terms of the lamina stiffness invariants. At this point, we note the following:

- θ_k (angle of orientation for the kth lamina) is constant in each lamina, but it varies from lamina to lamina.
- $\bar{Q}_{ij}$ are constant in each lamina, but they vary from lamina to lamina.
- U_1, U_2, U_3, U_4, and U_5 are constant in all the plies if the material is the same.

Then, in the expressions for laminate stiffness matrices (see Chapter 5), we can write $\bar{Q}_{ij}$ as given by Equations 14.9 through 14.19. The terms U_1 to U_5 can be brought outside the thicknesswise summation and we can express the elements of [**A**], [**B**], and [**D**] matrices as shown below.

14.6.3.1.2 Extensional Stiffnesses

$$A_{11} = U_1 V_{1A} + U_2 V_{2A} + U_3 V_{3A} \tag{14.20}$$

$$A_{12} = U_4 V_{1A} - U_3 V_{3A} \tag{14.21}$$

$$A_{16} = \frac{U_2 V_{4A}}{2} + U_3 V_{5A} \tag{14.22}$$

$$A_{22} = U_1 V_{1A} - U_2 V_{2A} + U_3 V_{3A} \tag{14.23}$$

$$A_{26} = \frac{U_2 V_{4A}}{2} - U_3 V_{5A} \tag{14.24}$$

$$A_{66} = U_5 V_{1A} - U_3 V_{3A} \tag{14.25}$$

in which

$$V_{1A} = \sum_{k=1}^{n} (z_k - z_{k-1}) \tag{14.26}$$

$$V_{2A} = \sum_{k=1}^{n}(z_k - z_{k-1})\cos 2\theta_k \qquad (14.27)$$

$$V_{3A} = \sum_{k=1}^{n}(z_k - z_{k-1})\cos 4\theta_k \qquad (14.28)$$

$$V_{4A} = \sum_{k=1}^{n}(z_k - z_{k-1})\sin 2\theta_k \qquad (14.29)$$

$$V_{5A} = \sum_{k=1}^{n}(z_k - z_{k-1})\sin 4\theta_k \qquad (14.30)$$

14.6.3.1.3 Coupling Stiffnesses

$$B_{11} = U_1 V_{1B} + U_2 V_{2B} + U_3 V_{3B} \qquad (14.31)$$

$$B_{12} = U_4 V_{1B} - U_3 V_{3B} \qquad (14.32)$$

$$B_{16} = \frac{U_2 V_{4B}}{2} + U_3 V_{5B} \qquad (14.33)$$

$$B_{22} = U_1 V_{1B} - U_2 V_{2B} + U_3 V_{3B} \qquad (14.34)$$

$$B_{26} = \frac{U_2 V_{4B}}{2} - U_3 V_{5B} \qquad (14.35)$$

$$B_{66} = U_5 V_{1B} - U_3 V_{3B} \qquad (14.36)$$

in which

$$V_{1B} = \frac{1}{2}\sum_{k=1}^{n}\left(z_k^2 - z_{k-1}^2\right) \qquad (14.37)$$

$$V_{2B} = \frac{1}{2}\sum_{k=1}^{n}\left(z_k^2 - z_{k-1}^2\right)\cos 2\theta_k \qquad (14.38)$$

$$V_{3B} = \frac{1}{2}\sum_{k=1}^{n}\left(z_k^2 - z_{k-1}^2\right)\cos 4\theta_k \qquad (14.39)$$

$$V_{4B} = \frac{1}{2}\sum_{k=1}^{n}\left(z_k^2 - z_{k-1}^2\right)\sin 2\theta_k \qquad (14.40)$$

Design of Composite Structures

$$V_{5B} = \frac{1}{2}\sum_{k=1}^{n}\left(z_k^2 - z_{k-1}^2\right)\sin 4\theta_k \tag{14.41}$$

14.6.3.1.4 Bending Stiffnesses

$$D_{11} = U_1 V_{1D} + U_2 V_{2D} + U_3 V_{3D} \tag{14.42}$$

$$D_{12} = U_4 V_{1D} - U_3 V_{3D} \tag{14.43}$$

$$D_{16} = \frac{U_2 V_{4D}}{2} + U_3 V_{5D} \tag{14.44}$$

$$D_{22} = U_1 V_{1D} - U_2 V_{2D} + U_3 V_{3D} \tag{14.45}$$

$$D_{26} = \frac{U_2 V_{4D}}{2} - U_3 V_{5D} \tag{14.46}$$

$$D_{66} = U_5 V_{1D} - U_3 V_{3D} \tag{14.47}$$

in which

$$V_{1D} = \frac{1}{3}\sum_{k=1}^{n}\left(z_k^3 - z_{k-1}^3\right) \tag{14.48}$$

$$V_{2D} = \frac{1}{3}\sum_{k=1}^{n}\left(z_k^3 - z_{k-1}^3\right)\cos 2\theta_k \tag{14.49}$$

$$V_{3D} = \frac{1}{3}\sum_{k=1}^{n}\left(z_k^3 - z_{k-1}^3\right)\cos 4\theta_k \tag{14.50}$$

$$V_{4D} = \frac{1}{3}\sum_{k=1}^{n}\left(z_k^3 - z_{k-1}^3\right)\sin 2\theta_k \tag{14.51}$$

$$V_{5D} = \frac{1}{3}\sum_{k=1}^{n}\left(z_k^3 - z_{k-1}^3\right)\sin 4\theta_k \tag{14.52}$$

Note that the terms V_{iA}, V_{iB}, and V_{iD} are generally dependent on z_k and θ_k and they vary as indicated below:

$$
\begin{aligned}
&i = 1 \text{—constant} \\
&i = 2 \text{—}\cos 2\theta_k \\
&i = 3 \text{—}\cos 4\theta_k \\
&i = 4 \text{—}\sin 2\theta_k \\
&i = 5 \text{—}\sin 4\theta_k
\end{aligned}
$$

Thus, a much clearer picture of variation of laminate stiffnesses w.r.t. the angle of orientation is obtained, which is greatly helpful in selecting and modifying the stacking sequence of a laminate.

14.6.3.2 Laminate Analysis and Measurement

A real-life composite structure is generally a complex one made up of a number of parts and joints. Finite element modeling is the most common method for the analysis of such a structure at the assembly level. However, at the laminate level, classical laminate theory-based progressive failure analysis, netting model, and such other analysis tools are also sometimes used. All possible critical load cases are considered and laminate level loads and boundary conditions are applied.

The analysis process is deterministic, and various parameters such as displacement, stress/strain distribution, natural frequency, mode shape, buckling load and temperature distribution, fracture toughness, etc. are determined. Note that some of the parameters are highly influenced by joints and adjacent parts. Laminate analysis in respect of these parameters is often done as a comparative study. Also, the parameters sought to be determined are clearly dependent on the nature of the applied loads and laminate configuration. In many cases, the general state of stress or strain would tell us what we should seek to determine. For example, in a pressure vessel under internal pressure, the laminate is in a state of biaxial stress and membrane stresses and strains are the parameters we need to know. On the other hand, for a pressure vessel under external pressure, buckling becomes a critical parameter to be observed.

While most of the parameters we need can be found by analysis, certain parameters can be found by measurements. For example, fatigue life of a composite laminate is typically found by measurement. Similarly, ablative response, impact characteristics, and other specific features of a laminate are often experimentally determined.

14.6.3.3 Laminate Design Criteria

Laminate design criteria revolve around one or more of the following four aspects:

- Strength
- Stiffness
- Fatigue life
- Energy absorption

From a strength point of view, stress and strain outputs from laminate analysis are suitably used in appropriate failure criteria to verify the adequacy of the laminate against design allowables. In respect of stress and strains in a composite laminate, there are two aspects that need attention here—plywise variations and concentrations near cutouts and other discontinuities such as ply drop-offs. Stress/strain concentration near cutouts is well known in metallic structures and the same is valid in composites as well. Strain variation is gradual across various plies, but steep stepwise variation of stress is a characteristic feature of composite laminate. It is unfortunate but not uncommon to make mistakes of comparing surface stress/strain only or gross stress/strain with design allowables to determine the available factor of safety. Stiffness is reflected in the strain, deflection, and buckling load of the laminate. Strain, as we have seen above, are used in strength failure criteria. On the other hand, deflection is mostly specified as a limiting parameter based on functional requirements. Similarly, fatigue life and energy absorption may also be specified depending on the applications.

The application of design criteria verifies whether or not the chosen laminate is acceptable or not. If the strength, stiffness, fatigue life, and energy absorption characteristics are either too low or too high, modifications to the laminate design are made.

Modifications to the laminate, when required, are normally made in an incremental mode by adding or reducing plies. Another factor that needs to be considered is the weight, especially if weight budget is made at the laminate level itself. (Note that, with a view to controlling the weight of the overall composite product, weight budget can be made for the laminates and parts.)

14.7 JOINT DESIGN

14.7.1 Introduction

Joints are used for connecting structural elements of different kinds. They are generally considered as the weak links in any structure. Yet, most real-life structures do contain joints. Composites are known for their capabilities to be built as monolithic structures without joints. However, certain minimum joints cannot be avoided even in composite structures. Laminates in a composite structure can be designed to possess the highest strength and stiffness characteristics; however, if the joints that connect them are not adequate for efficient load transfer, the strength and stiffness of the laminates cannot be exploited properly and the overall structural performance reduces greatly. Here lies the importance of design of appropriate joints in a composite structure.

14.7.2 Types of Joints

Broadly, two types of joints are used in a composite structure—bonded joints and mechanically fastened joints. In addition to them, there are joints that are both bonded and mechanically fastened. Joints with special and product-specific features are also used in some cases. Thus, we can have a third category, and group different types of joints into three categories as follows:

- Bonded joints
- Mechanically fastened joints
- Special joints

Several subtypes, which can be either generic or product-specific in nature, can be found in each of the above. These subtypes are based on either configuration or process of realization of the joints. These are discussed in the following sections.

14.7.3 Bonded Joints

14.7.3.1 Introduction to Bonded Joints

As shown in Figure 14.4, bonded joints are designed with a thin layer of a suitable adhesive between two adherends. Various configurations of bonded joints are used [1,7,8,14,15]; some of the common ones are as follows (Figure 14.5):

- Single-lap and double-lap joints
- Single-stepped and double-stepped joints
- Single-bevel and double-bevel joints
- Single-scarf and double-scarf joints
- Single-butt strap and double-butt strap joints

Other configurations of bonded joints can also be made as variations of these joints listed above. In real-life structures, the joint configuration is dictated by the configurations of the parts being bonded. Parts of various configurations, including flat and

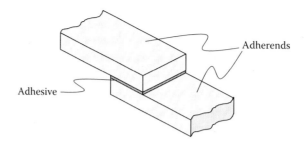

FIGURE 14.4 Typical bonded joint.

curved panels, cylinders, cones, and contoured parts, can be bonded as long as the mating surfaces have the same geometry.

In a composite joint, a composite laminate is bonded with another composite laminate, a metallic panel, or a rubber layer. Thus, based on the adherend materials, composite joints can be of the following types:

- Composite-to-composite joints
- Composite-to-metal joints
- Composite-to-rubber joints

The compatibility of the adhesive material with the adherend materials is important. Epoxy- and acrylic-based adhesives are most commonly used in bonded joints. During manufacturing, a suitable adhesive is applied on the adherends (i.e., the laminates that are adhesively joined) and cured. The adherends may be cured laminates or in the green condition; in the latter case, the adherends are cured along with the adhesives. From this point of view, the following two types of bonded joints are found:

- Joint with precured adherends
- Joint with cocured adherends

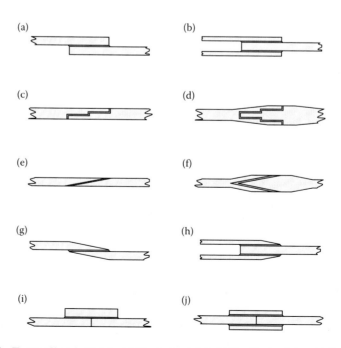

FIGURE 14.5 Types of bonded joints. (a) Single-lap joint. (b) Double-lap joint. (c) Single-stepped-lap joint. (d) Double-stepped-lap joint. (e) Single-bevel joint. (f) Double-bevel joint. (g) Single-scarf joint. (h) Double-scarf joint. (i) Single-butt-strap joint. (j) Double-butt-strap joint.

14.7.3.2 Failure Modes in Bonded Joints

A bonded joint is commonly designed to resist in-plane loads, where the primary load transfer mechanism is by shear. A distinct feature in all the "single joints" is that load transfer is eccentric, which results in bending of the joints. Thus, peel stresses are developed in the adhesive in these joints. The intensity of the peel stresses is the highest at the edges (Figure 14.6). On the other hand, load transfer is symmetric in all the "double joints"; in these joints, stresses in the adhesive are predominantly shear. A joint may also be subjected to out-of-plane loads, in which case, bending stresses develop in the adherends and peel stresses in the adhesive.

Depending on the nature of the applied loads, various failure modes are possible in a bonded composite joint, out of which the primary modes are [1,7,18]:

- Failure of the adherends
  - Adherend failure in tension
  - Adherend failure by interlaminar stress
  - Adherend failure by transverse stress
- Failure of the adhesive
  - Adhesive failure in shear
  - Adhesive failure in tension
- Failure of the adherend–adhesive interface
  - Interface failure in shear
  - Interface failure in tension

The failure modes broadly correspond to the adherends, adhesive, and the interface. The adherend failure can occur due to excessive longitudinal tensile stress, interlaminar stress, or transverse stress. While longitudinal tensile failure results from fiber fracture, interlaminar and transverse failures are due to matrix and fiber–matrix interface failure.

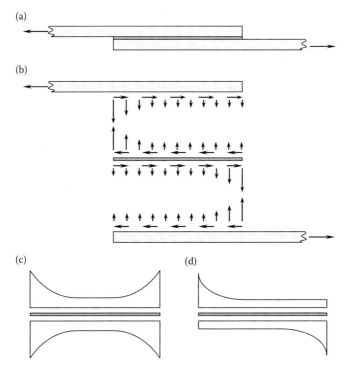

FIGURE 14.6 Schematic representation of load transfer mechanism in a bonded joint. (a) Single-lap joint in tension. (b) Free body diagrams. (c) Shear stress distribution in the adhesive. (d) Peel stress distributions.

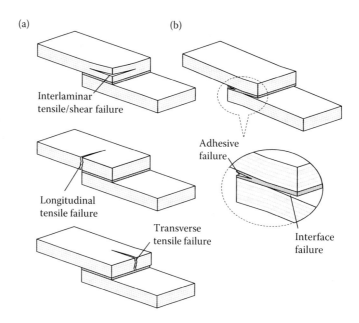

FIGURE 14.7 Failure modes in a bonded joint. (Adapted from F. L. Matthews and R. D. Rawlings, *Composite Materials: Engineering and Science*, CRC Press, Boca Raton, FL, 1999.)

Cohesive failure of the adhesive is a common failure mode in a bonded joint. This occurs when the adhesive mechanical properties are low. Failure of the adhesive can take place due to either excessive shear stress or out-of-plane tensile peel stress. Another mode of bonded joint failure is in the interface, which is common if surface preparation is not proper. Various failure modes are schematically depicted in Figure 14.7.

14.7.3.3 Advantages and Disadvantages of Bonded Joints

The advantages and disadvantages associated with bonded joints can be listed as follows [7,8,14]:

Advantages:

- Minimum stress concentration—The joint transfers load over a large surface area resulting in minimum stress concentration.
- Minimum weight penalty—Increase in weight on account of adhesive and other features associated with a bonded joint is generally low.
- No holes and cut-outs—No holes or cut-outs are required in a bonded joint; as a result, adherends are not weakened and stress concentration does not occur.
- Low cost—Adhesively bonded joints are economical and generally much easier to implement.

Disadvantages:

- No disassembly—Bonded joints are permanent in nature and disassembly is not possible.
- Crack propagation—The adhesive is isotropic in nature and, unlike the adherends, it does not contain any fibers that acts as crack-bridging mechanism. As result, an initial crack can easily propagate, causing catastrophic failure.
- Environmental effects—The adhesive can get adversely affected by environmental factors resulting in poor performance and loss of structural integrity of the joint.
- Surface preparation—Extensive surface preparation is needed so that proper bond between the adherends is created.

- Difficult quality control—Compared to mechanical joints, it is more difficult to confirm the joint quality of bonded joints by inspection.

14.7.3.4 General Design Considerations

The design of a composite bonded joint involves primarily three aspects:

- Configuration design of the joint
- Ply design of the laminates
- Selection of adhesive material

The basic question in any bonded joint is to determine the bond area required to effect efficient load transfer. The classical approach assumes a uniform adhesive shear stress and a linear relationship between joint strength and bond area. It is simple to follow and is the basis for some of the standard test procedures. However, this simplistic approach is not realistic. The stress distribution is highly nonlinear, especially near the ends, where the interlaminar shear stresses are very high. These stresses can be rather low and uniform at the interior portions in the bond area. Both analytical and numerical studies have been carried out for stress analysis and solutions have been proposed for different joint configurations (see, for instance, References 19 and 22–26). We shall not go into the details of joint design and rather address in a qualitative way some general design considerations that should be kept in mind while working out the details of a bonded joint.

Strength degradation: The shear strength of the adhesive in a bonded joint is affected by several factors such as adhesive layer thickness, surface preparation, etc. These factors cannot be controlled in a real-life structure as tightly as in a laboratory specimen. As a result, the strength obtained from laboratory specimens is often not achieved in bonded joints in a real-life structure. Thus, it is imperative that laboratory specimen test data are degraded by a suitable factor for use in the design of bonded joints. Typically, this factor is of the order of 0.6–0.8.

Complex failure modes: Composite joints are known for their complex failure modes that make failure prediction more difficult than the adherends. Generally, a composite product is designed to have a higher margin of safety at the joints than in the adherends. With this in mind, typically, adhesive material is selected so as to have about 50% higher shear strength than that of the adherends.

Ply sequence: The ply sequence of the adherends plays an important role in the performance of the joint. In general, it is preferable to use 0° plies (in the loading direction) next to the adhesive layer. On the other hand, using 90° plies (transverse to the loading direction) next to the adhesive layer is a poor practice as it can lead to delamination near the joint.

Configuration: The configuration of a joint is based on the basic configurations discussed in one of the previous sections. Single-lap joints are associated with eccentric loading and resultant bending, joint rotation, and high peel stresses. Thus, at structurally critical locations, double-lap joints are preferable.

14.7.4 Mechanical Joints

14.7.4.1 Introduction to Mechanical Joints

Mechanical joints are incorporated using basically three types of fasteners—bolts, screws, and rivets. A typical bolted joint configuration is schematically shown in Figure 14.8. Different types of bolts, screws, and rivets are in use. The dimensions

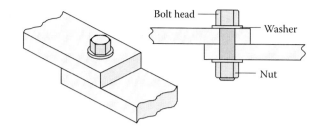

FIGURE 14.8 Typical bolted joint.

of various elements such as bolt heads, shank diameter, etc. are standardized, especially for metallic structures; we are not going into these details. It may, however, be noted that composites have their own specific characteristics that must be kept in mind while designing a mechanical joint. In general, shear properties of composites are not good. As a result, thread integrity is poor and it is preferable not to have any threads in composites. Also, high-performance fasteners, specially designed for composites applications, which are commercially available under various brand names, can be used; however, these fasteners are generally highly expensive.

Various generic configurations of mechanical joints are used [1,14]; these are given in Figure 14.9.

- Single-lap joint
- Double-lap joint
- Reinforced-edge joint
- Shimmed joint

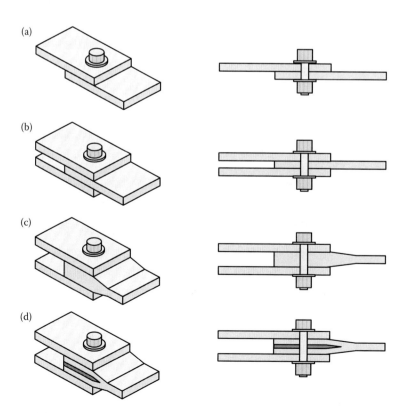

FIGURE 14.9 Types of mechanical joints. (a) Single-lap joint. (b) Double-lap joint. (c) Reinforced-edge joint. (d) Shimmed joint.

14.7.4.2 Failure Modes in Mechanical Joints

The laminates in a joint can be subjected to in-plane load, out-of-plane load, or a mixture of both. Depending on the nature of the laminate loads, the bolts may be in shear or tension. Often, the joint undergoes rotation due to eccentric loading, as in a single-lap joint, resulting in the generation of bending stress in the bolts.

Failure modes in composite mechanical joints are the same as those seen in metals; however, they are greatly influenced by ply sequence and characteristics peculiar to composites. A unidirectional laminate, depending on the direction of fibers, is likely to result in longitudinal tensile failure or shear failure. Tensile failure generally occurs due to high stress concentration near the bolt hole.

The primary failure modes associated with mechanical joints in composites are [1,7,14]:

- Laminate failure
  - Tension failure
  - Shear failure
  - Bearing failure
  - Shear-out or cleavage failure
  - Bolt pull-out failure
- Bolt failure
  - Shear failure
  - Bending failure

The tension failure of laminate occurs due to high tensile stress in the laminate along a plane through the bolt hole. The failure plane is normal to the loading direction. In a composite mechanical joint, the laminate net cross-sectional area gets reduced by the presence of the bolt holes. As a result, the net stress level increases. Also, holes and cut-outs are associated with stress concentration. Clearly, the joint strength in this mode (the maximum tensile load that the joint can take) is dependent on the reduced laminate cross-sectional area and the effective tensile strength of the laminate material.

Bearing failure is the failure of the laminate by bearing at the bolt hole. Bearing stress is dependent on bolt diameter and the laminate thickness, that is, the projected area of the hole.

The shear failure of the laminate is the failure by shear, which initiates from the hole edges, along two lines parallel to the loading direction. Shear-out or cleavage failure is a mixed-mode failure of the laminate by tension and shear.

Bolt pull-out is failure of the laminate by bearing and shear under the bolt head or nut that results in pulling out of the bolt and separation of the laminates. This mode of joint failure takes place when the joint is subjected to out-of-plane tensile loads due to bending, etc. Common laminate failure modes in a mechanical joint are schematically depicted in Figure 14.10.

In addition to the laminate failure modes, bolt failure too occurs in a mechanical joint. Most commonly, it takes place either in shear or in bending.

14.7.4.3 Advantages and Disadvantages of Mechanical Joints

Like the bonded joints, mechanically fastened joints also have their own advantages and disadvantages [7,8,14]; these are listed below:

Advantages:

- No surface preparation—Unlike bonded joints, mechanical joints do not need any surface preparation.

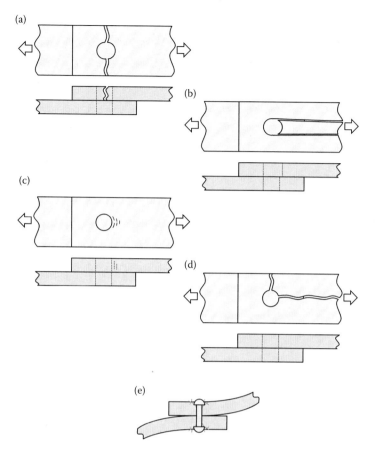

FIGURE 14.10 Typical laminate failure modes in mechanical joints. (a) Tension failure. (b) Shear failure. (c) Bearing failure. (d) Shear-out or cleavage failure. (e) Bolt pull-out failure.

- Disassembly—Mechanically joined (bolted) parts can be repeatedly assembled and disassembled.
- Simple inspection—It is comparatively simple to implement quality control inspection in a mechanically fastened joint.

Disadvantages:

- Stress concentration—Mechanical joints contain a number of holes that are invariably associated with stress concentration.
- Weight penalty—Increase in weight on account of fasteners, washers, etc. is generally high.
- High cost—Mechanical joints are relatively more expensive and difficult to implement.

14.7.4.4 General Design Considerations

Similar to the composite bonded joints, there are three primary aspects that need to be considered while designing a composite mechanical joint; these are

- Configuration design of the joint
- Ply design of the laminates
- Fasteners

The failure modes in bolted joints in composites are similar to those in metals, but they are more complex due to several factors peculiar to composites. Laminate tensile, shear, and bearing strengths depend on not only laminae properties but also

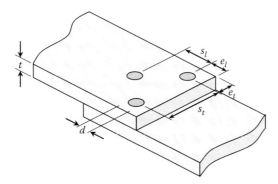

FIGURE 14.11 Geometrical parameters in a mechanical joint.

their stacking sequence. Composites do not plastically deform and defuse stresses, and thus stress concentration near a hole in a composite laminate is critical. Also, interlaminar stresses can develop near the free edges of a hole and adversely affect the structural integrity of a composite bolted joint. As in the case of bonded joints, both analytical and numerical studies have been carried out for stress analysis of bolted joints in composites [27–31]. The details of these works are not addressed here; instead, some general design considerations are discussed.

Configuration: Real-life joints generally involve improvisations of the basic configurations shown in Figure 14.9 and mentioned in Section 14.7.4.1. The performance of a mechanical joint is greatly influenced by its configuration. The geometrical parameters (Figure 14.11) that describe a mechanical joint are (i) laminate thickness (t), (ii) hole diameter (d), (iii) edge distances (e_l and e_t), and (iv) spacings (s_l and s_t). Failure modes depend greatly on these parameters.

Ply sequence: Failure modes in a composite mechanical joint are strongly influenced by the ply sequence of the laminates. A unidirectional lamina is very good in longitudinal strength but poor in transverse strength, in-plane shear strength, and bearing strength. Also, the laminate strength requirement around the holes in a mechanical joint is typically multidirectional. Consequently, unidirectional laminates are not preferred around a hole. Accordingly, unidirectional laminae are usually stacked in at different orientations and, often, bidirectional fabric reinforcements are introduced so as to increase the bearing strength.

Complex failure modes: Various possible failure modes in a mechanical joint are listed in Section 14.7.4.2. The anisotropic nature of composites makes it rather difficult to make an accurate prediction of the failure mode. In order to overcome this uncertainty, as in the case of bonded joints, mechanical joints are also typically designed with higher margins of safety than the laminates such that product failure takes place by laminate failure and not by joint failure.

14.7.5 Other Joints

In addition to the bonded and mechanical joints, there are other joints that are practiced in composite structures. For example, a joint may involve both bonding of the adherends as well as mechanical fastening. Also, in real structures, often composites are used together with metals. A joint, involving metals and composites, is generally associated with severe mismatch of stiffness. Stiffness mismatch may be present even between two composite adherends. In such cases, a layer of low modulus material such as rubber can be introduced between the two adherends for absorbing differential deformations of the adherends. Clearly, these joints have application-specific design requirements.

14.8 STIFFENED STRUCTURES

14.8.1 Introduction

Unstiffened thin shells typically do not possess sufficient strength and stiffness required in many applications. While the shell thickness can be increased to improve the mechanical characteristics, it is not generally an efficient solution in most weight-sensitive applications. In such cases, unstiffened shell structures (also referred to as monocoque structures) are not used and an efficient alternative is found in stiffened structures, in which the skin is stiffened by attaching stiffening members of various types.

A stiffener is basically a beam that is attached to the skin by methods such as bonding, stitching, riveting, *in situ* lay-up, etc. Both open sections as well as closed sections are used, of which commonly used cross sections include T-section, I-section, C-section, Z-section, box-section, hat-section, etc. The stiffeners are oriented in the longitudinal, axial, or meridional direction and circumferential, lateral, or transverse direction. Longitudinal stiffeners are typically known as longerons or stringers or spars, whereas the transverse stiffeners are called rings. In addition to these, in a relatively new concept of stiffened structures—referred to as grid-stiffened structures made by filament winding, stiffening members are also used at some angle to the longitudinal direction; in such a case, the stiffeners typically form a grid of triangles, diamonds, and hexagons [32,33].

In a stiffened structure, the skin can be stiffened either symmetrically or eccentrically. In many applications such as aircraft wing, aircraft and space vehicle fuselage, ship and submarine hull, etc., a clean exterior surface is required from aerodynamics or hydrodynamics points of view. In these cases, the shell is eccentrically stiffened on the inside. Also, when the stiffeners are provided on the same side of the skin, they typically cross each other at certain nodes. In conventional ring–stringer combination, usually, the rings are provided as continuous members, whereas the stringers are provided as continuous members between the rings only. On the other hand, in a grid-stiffened structure, the stiffeners are laid-up/wound in a continuous fashion, which eventually leads to thickness build-up at the nodes or cross-over locations.

14.8.2 Failure Modes in a Stiffened Structure

Stiffened structures are often used in applications that are subjected to bending, compression, and torsion loads. Under such loading environments, buckling is a common failure mode. Another failure mode peculiar to stiffened structure is skin–stiffener separation. In addition to these, the strength fracture of skin-stiffeners material may also occur. Under these three broad types of failure modes, various subtypes can be identified as follows [3]:

- Buckling
  - Global buckling
  - Stiffener column buckling
  - Local skin buckling
  - Local stiffener buckling or stiffener crippling
- Skin–stiffener separation
  - Skin–stiffener bond failure
  - Fastener joint failure
- Strength fracture
  - Skin material failure
  - Stiffener material failure

Global buckling is the overall buckling of the stiffened panel, in which both the skin and the stiffeners buckle simultaneously. In general, it leads to a catastrophic failure. Stiffener column buckling is also catastrophic in nature. Local skin buckling is buckling of the skin confined between stiffeners. In a similar way, local stiffener buckling involves localized buckling of flange or web. Local buckling of the skin or stiffener is normally considered as benign. Skin–stiffener separation results either due to interface bond failure or due to fastener joint failure by shear, tear, bearing and rivet pull-out, etc. The third major failure mode is due to overstressing of the skin or stiffener material and it is generally governed by first ply failure.

14.8.3 Design of Stiffeners

There are basically two elements in a stiffened structure—the skin and the stiffeners. Each of these elements has its own functions. Typically, the stiffeners provide bending stiffness. They also greatly enhance the buckling resistance of the structure. The skin, on the other hand, by its membrane action, carries tensile, compressive, and in-plane shear stresses. The proper design of a stiffened structure ensures efficient load sharing among the skin and stiffeners in such a way that failure takes place in a gradual manner.

Owing to the presence of the stiffeners, the number of design parameters is generally more in a stiffened structure. Some of the key parameters are listed below:

Skin:

- Ply sequence
- Thickness
- Manufacturing process

Stiffeners:

- Type of stiffeners
  - Orientation—axial, circumferential, angular, or a combination thereof
  - Continuous or discontinuous
- Cross-sectional shape
  - Open or closed
  - Cross section—C-section, box-section, etc.
  - Relative dimensions—height, width, and thickness
- Cross-sectional area
  - Cross-sectional area relative to skin thickness
  - Cross-sectional area relative to stiffener density
- Stiffener density—number of stiffeners of each type per unit length
- Manufacturing process
  - Prefabricated and bonded/stitched
  - *In situ* laid-up/wound

Failure modes, discussed in the previous section, depend on the choice of various design parameters indicated above. For example, if the stiffeners are highly stiff and widely spaced, skin local buckling is likely. It is also an important aspect in the design process to sequence the failure modes. For example, as mentioned before, local failures are generally considered as benign; accordingly, a stiffened panel is often so designed as to ensure local skin buckling before global buckling or stiffener crippling before stiffener column buckling.

On the other hand, from the above listing, it is clear that there are just too many parameters to be chosen. Simple approximate tools and other design thumb rules help in preliminary design and analysis, and invariably finite element analysis is carried out for final fine tuning and detailed performance estimate. Smeared stiffeners model is a simple approximate tool that is commonly used in preliminary design [32,34]. This approach is of great help in reducing the theoretically infinite numbers of design alternatives to limited numbers that can be handled in a practical design scenario. The basic methodology in smeared stiffeners modeling is to smear the stiffeners based on some criterion such as equivalent cross-sectional area, equivalent strain energy, etc. and thereby convert the stiffened structure into an equivalent monocoque structure, on which an appropriate analysis can be done. In general, a smeared stiffeners model is more reliable in structures with grids of dense stiffeners.

The total cross-sectional area of the stiffeners, spacing of stiffeners, skin thickness, and other details are generally chosen by adopting an iterative process so as to achieve the desired sequence of failure modes. Often, it is an act of maintaining balance. For example, if the stiffeners are too stiff and highly predominant as compared to the skin, almost the entire load is likely to be shared by the stiffeners only and the skin is going to be redundant. On the other hand, if the skin is too thick and the stiffeners are small and widely spaced, the entire load is likely to be shared by the skin alone and the stiffeners will only add to the weight!

The individual elements of a stiffener, viz. flanges, webs, crowns (flange-to-web meeting points), and trough angle (angle between web and normal to the panel), have their own functions. Typically, the flange and crown are designed for carrying the bending load and the web for shear. The trough angle is critical from the point of view of bending efficiency and twisting; too small a trough angle in a closed section such as hat-section may result in unexpected twisting.

14.9 OPTIMIZATION

Optimization is a frequently used term in connection with product design and development. Very often, it is used to imply a weight reduction process! However, it is much more than just a process of weight reduction. In fact, it is inherently associated with the very basic objective of any design. Design objective in any application is basically to make the *best* choice from among the many feasible design alternatives. The meaning of best design has to be understood clearly. In many aerospace applications, best design is the design with minimum weight. Other possible meanings of best design can be least cost, longest service life, maximum range, maximum speed, least fuel consumption, etc.

Optimization is the process of arriving at the best design choice. Thus, the first step in the process of optimization is to define clearly the design objective—the meaning of best design, which is referred to as the objective function or merit function. Optimization is a mathematical process; however, in practice, it is approached in two ways [1]:

- Mathematical methods
- Searching techniques

Several mathematical methods have been developed, for example, linear programming, nonlinear programming, integer programming, dynamic programming, Monte-Carlo method, etc. These methods are based on rational logic and they are applicable in the various fields of science and engineering. On the other hand, searching techniques are based on engineering judgment. In these methods, a certain feasible configuration is made and its response to the applied loads is estimated; a better configuration is searched based on the response of the previous configuration and the process is continued till a reasonable solution is obtained. These techniques are not based on rational logic and, in a strict sense they do not optimize the structure. However, the searching techniques are by far more common in practice than the mathematical methods.

14.10 DESIGN EXAMPLES

Today, composite structures of a wide variety are in use. The complexity of the configuration of these structures and applied loads vary greatly—from simple unidirectional structural element under uniaxial load, for example, a tension member, to complex 3D structures, for example, a cylindrical grid-stiffened shell under the combined action of axial force and bending moment. Quite clearly, it is not possible to discuss *all* of these. Our intention is to introduce the reader to the design of simple composite structural elements. Toward this, here, we consider the following representative categories of composite structural elements:

- One-dimensional structural elements under uniaxial loads
  - Tension member
  - Compression member
  - Torsion member
  - Beam
- Two-dimensional flat panels under in-plane loads
- Pressure vessel

14.10.1 Design of a Tension Member

A tension member is a simple slender rod, bar, or tube under either uniaxial static tension or uniaxial tension–tension fatigue. Such a structural element requires high strength and stiffness characteristics in its axial direction and is designed by providing reinforcements predominantly in the 0° direction. Often, based on manufacturing feasibility, reinforcements are provided at nonzero but small angle of orientation. Also, processing considerations may require reinforcements at 90° or other large angles, for example, hoop winding may be required for consolidation of 0° or near 0° helical windings during the manufacture of a tubular tension member by filament winding.

Matrix-dominated properties, namely, transverse and shear strength and stiffness characteristics are not of any particular importance in design calculations; however, the selection of a suitable resin is critical from a manufacturing point of view.

Broadly, two approaches can be adopted for the design of a tension member—micromechanics- and macromechanics-based approach.

14.10.1.1 Micromechanics-Based Approach

It is a greatly simplified approach, in which the effects of the matrix are ignored and design is carried out using longitudinal tensile strength and modulus of the reinforcing fibers. Further, all the fibers are considered to be aligned in the axial direction of the tension member.

Then, in the case of a strength-based design, the required area of cross section of the tension member can be obtained as

$$A = \frac{P}{\left(\sigma_{1c}^T\right)_{\text{ult}}} \approx \frac{P}{V_f \left(\sigma_{1f}^T\right)_{\text{ult}}} \tag{14.53}$$

in which
- A Area of cross section of the tension member (mm²)
- P Tensile force on the tension member (N)
- $(\sigma_{1c}^T)_{\text{ult}}$ Longitudinal tensile strength of unidirectional composite (N/mm²)
- $(\sigma_{1f}^T)_{\text{ult}}$ Longitudinal tensile strength of fiber (N/mm²)
- V_f Fiber volume fraction

(Note: Refer to Section 3.5.2.1 in Chapter 3 for the derivation of the composite longitudinal tensile strength from the tensile strength of fiber.)

In the case of a stiffness-based design, the required area of cross section of the tension member can be obtained as

$$A = \frac{PL}{E_{1c}\Delta} \approx \frac{PL}{V_f E_{1f}\Delta} \qquad (14.54)$$

in which

- A Area of cross section of the tension member (mm²)
- P Tensile force on the tension member (N)
- L Length of the tension member (mm)
- Δ Limiting axial deformation (mm)
- V_f Fiber volume fraction
- E_{1c} Longitudinal modulus of unidirectional composite (N/mm²)
- E_{1f} Longitudinal modulus of fiber (N/mm²)

(Note: Refer to Section 3.5.1.1 in Chapter 3 for the derivation of the composite longitudinal modulus from the modulus of the fiber.)

14.10.1.2 Macromechanics-Based Approach

In this approach, lamina strength and stiffness properties are used and different ply orientations can be incorporated. For simplicity, transverse tensile strength and shear strength of lamina are ignored.

Then, under strength-based consideration, the cross-sectional area can be obtained from the following static equilibrium equation:

$$P = \sum_{i=1}^{n} X_i^T A_i \cos^2 \theta_i \qquad (14.55)$$

in which

- P Tensile force on the tension member (N)
- A_i Area of cross section of the ith ply (mm²). The area of cross section of the tension member is obtained by summing up the A_i's.
- X_i^T Longitudinal tensile strength of the ith lamina (N/mm²)
- θ_i Orientation of ith lamina
- n Number of laminae

It is easy to see that Equation 14.55 can be suitably simplified when the tension member is designed with the laminae of the same material, same orientation, and equal thickness.

When the limiting axial elongation is specified, the lamina axial stiffness is required, which is obtained as

$$E_x = \left[\frac{\cos^4 \theta}{E_1} + \frac{\sin^4 \theta}{E_2} + \left(\frac{1}{G_{12}} - \frac{2\nu_{12}}{E_1} \right) \sin^2 \theta \cos^2 \theta \right]^{-1} \qquad (14.56)$$

(Note: Refer to Section 4.4 in Chapter 4 for the derivation of engineering constants of a generally orthotropic lamina.)

Design of Composite Structures

Each lamina undergoes axial deformation to the same extent and it can be shown that

$$\sum_{i=1}^{n} A_i E_{xi} = AE_x = \frac{PL}{\Delta} \qquad (14.57)$$

in which
- P Tensile force on the tension member (N)
- L Length of the tension member (mm)
- Δ Axial elongation of the tension member (mm)
- A_i Area of cross section of the ith ply (mm²)
- A Total area of cross section of the tension member (mm²)
- E_{xi} Axial modulus of the ith lamina (N/mm²)
- E_x Effective axial modulus of the tension member (N/mm²)
- n Number of laminae

The design of a tension member in itself is a simple affair; however, the joints that connect it to other structural elements are critical.

EXAMPLE 14.1

Design, adopting a micromechanics-based approach, a tension member of length 600 mm for carrying a static uniaxial tensile load of 200 kN. Consider the following as available material systems:

Carbon fiber reinforcement:

$$E_{1f} = 230 \text{ GPa}, \ \left(\sigma_{1f}^T\right)_{ult} = 4900 \text{ MPa}, \ \rho_f = 1.80 \text{ g/cm}^3$$

Kevlar fiber reinforcement:

$$E_{1f} = 125 \text{ GPa}, \ \left(\sigma_{1f}^T\right)_{ult} = 3600 \text{ MPa}, \ \rho_f = 1.45 \text{ g/cm}^3$$

Glass fiber reinforcement:

$$E_f = 75 \text{ GPa}, \ \left(\sigma_f^T\right)_{ult} = 3400 \text{ MPa}, \ \rho_f = 2.58 \text{ g/cm}^3$$

Cast epoxy resin:

$$E_m = 3.6 \text{ GPa}, \ \left(\sigma_m^T\right)_{ult} = 72 \text{ MPa}, \ \rho_m = 1.12 \text{ g/cm}^3$$

Take lower mass as a design criterion.

Solution

Let us first choose the manufacturing method for making the tension members. For the production of limited numbers, matched-die-molding can be conveniently employed for making laminates of suitable size from which the tension members of rectangular cross section can be obtained by parting appropriately.

For matched-die-molded unidirectional composites, a high fiber volume fraction of 0.6 can be achieved. Next, taking a factor of safety as 1.25, we can readily obtain the design tensile force as 250 kN. Then, for each of the three available fiber reinforcements, the total area of cross section of the tension member required is given by

Carbon fiber reinforcement:

$$\frac{250 \times 1000}{0.6 \times 4900} = 85.0 \text{ mm}^2$$

Kevlar fiber reinforcement:

$$\frac{250 \times 1000}{0.6 \times 3600} = 115.7 \text{ mm}^2$$

Glass fiber reinforcement:

$$\frac{250 \times 1000}{0.6 \times 3400} = 122.5 \text{ mm}^2$$

Let us take the ply thickness as 0.5 mm and consider the following cross sections:
Carbon fiber reinforcement:

$$10 \times 9.5 \text{ (area of c/s} = 95 \text{ mm}^2)$$

Kevlar fiber reinforcement:

$$11 \times 11 \text{ (area of c/s} = 121 \text{ mm}^2)$$

Glass fiber reinforcement:

$$12 \times 10.5 \text{ (area of c/s} = 126 \text{ mm}^2)$$

Note that the thickness of the tension member is a multiple of 0.5 mm. Now, we can estimate the density of each of the three material systems as follows:
Carbon/epoxy:

$$1.8 \times 0.6 + 1.12 \times 0.4 = 1.528 \text{ g/cm}^3$$

Kevlar/epoxy:

$$1.45 \times 0.6 + 1.12 \times 0.4 = 1.318 \text{ g/cm}^3$$

Glass/epoxy:

$$2.58 \times 0.6 + 1.12 \times 0.4 = 1.996 \text{ g/cm}^3$$

Then, we can compute the corresponding mass of the tension member as follows:

Carbon/epoxy:

$$\frac{10 \times 9.5 \times 600}{1000} \times 1.528 = 87.1 \, \text{g}$$

Kevlar/epoxy:

$$\frac{11 \times 11 \times 600}{1000} \times 1.318 = 95.7 \, \text{g}$$

Glass/epoxy:

$$\frac{12 \times 10.5 \times 600}{1000} \times 1.996 = 150.9 \, \text{g}$$

As we can see, carbon/epoxy gives us the lightest tension member. Thus, for weight-sensitive design, we choose carbon/epoxy as the material system.

Filament winding is a convenient technique for making unidirectional laminate. Here, we give a possible methodology with hypothetical process parameters. Toward this, a mandrel with flat surfaces can be used. Hoop winding is to be carried out for which first we need to estimate the required bandwidth and number of spools. Let us take the filament diameter and yield of the carbon fibers as 7 μm and 12 k, respectively. Then, the total cross-sectional area (A_f) of the filaments in one tow is given by

$$A_f = 12{,}000 \times \frac{\pi \times (7 \times 10^{-3})^2}{4} = 0.4618 \, \text{mm}^2$$

For fiber volume fraction $V_f = 0.6$, the total cross-sectional area (A_c) of the composite per tow is given by

$$A_c = \frac{0.4618}{0.6} = 0.7697 \, \text{mm}^2$$

Then, the bandwidth (B_w) for one tow for ply thickness of 0.5 mm is given by

$$B_w = \frac{0.7697}{0.5} = 1.54 \, \text{mm}$$

Note that for more number of tows, the corresponding bandwidth is obtained by multiplying the above figure by the number of tows; that is, bandwidth is given by $B_w = 1.54n$, n being the number of tows.

The number of spools is chosen primarily based on the size of the product being wound and the desired finish and accuracy. In general, the lower the number of spools, the better the finish and accuracy. Note further that for large number of spools, the bandwidth is high and actual hoop winding angle, which should be 90° in a strict sense, actually drifts away from 90°. On the other hand, too few spools may result in very high winding time and the choice of the number of spools turns out to be a compromise decision.

For small laminates, such as the one in this example, we can conveniently carry out hoop winding with two spools, that is, two tows. Let us then carry out

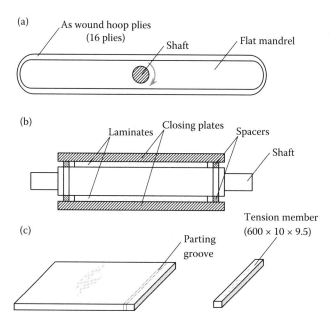

FIGURE 14.12 Schematic representation of the process of tension member fabrication in Example 14.1. (a) Hoop winding around a flat mandrel (end view). (b) Closure of the mandrel-cum-mold by plates (front view). (c) Laminate and final tension member.

hoop winding with two spools at a bandwidth of 3.1 mm. After the completion of 19 plies, the mandrel-cum-matched-die-mold is closed on the flat surfaces and curing is done at high temperature in an oven. Then, the tension members are obtained from the laminates by carrying out parting (see Figure 14.12).

Note: Refer to Chapter 10 for a more detailed discussion on filament winding parameters.

EXAMPLE 14.2

Consider the data given in Example 14.1. If the axial elongation is to be limited to 0.01 mm, design the tension member adopting a micromechanics-based approach.

Solution

Using Equation 14.54, the total area of cross section of the tension member required in each material system is given by

Carbon fiber reinforcement:

$$\frac{250 \times 600}{0.6 \times 230{,}000 \times 0.01} = 108.7 \text{ mm}^2$$

Kevlar fiber reinforcement:

$$\frac{250 \times 600}{0.6 \times 125{,}000 \times 0.01} = 200.0 \text{ mm}^2$$

Glass fiber reinforcement:

$$\frac{250 \times 600}{0.6 \times 75{,}000 \times 0.01} = 333.3 \text{ mm}^2$$

Design of Composite Structures

Let us take the ply thickness as 0.5 mm and consider the following cross sections:
Carbon fiber reinforcement:

$$11 \times 10.5 \text{ (area of c/s} = 115.5 \text{ mm}^2)$$

Kevlar fiber reinforcement:

$$14.5 \times 14 \text{ (area of c/s} = 203 \text{ mm}^2)$$

Glass fiber reinforcement:

$$19 \times 18 \text{ (area of c/s} = 342 \text{ mm}^2)$$

Then, using the densities estimated in Example 14.1, we can compute the corresponding mass of the tension member as follows:
Carbon/epoxy:

$$\frac{11 \times 10.5 \times 600}{1000} \times 1.528 = 105.9 \text{ g}$$

Kevlar/epoxy:

$$\frac{14.5 \times 14 \times 600}{1000} \times 1.318 = 160.5 \text{ g}$$

Glass/epoxy:

$$\frac{19 \times 18 \times 600}{1000} \times 1.996 = 409.6 \text{ g}$$

Axial elongation in each case is obtained as follows:
Carbon/epoxy:

$$\frac{250 \times 600}{0.6 \times 230,000 \times (11 \times 10.5)} = 9.41 \times 10^{-3} \text{ mm}$$

Kevlar/epoxy:

$$\frac{250 \times 600}{0.6 \times 125,000 \times (14.5 \times 14)} = 9.85 \times 10^{-3} \text{ mm}$$

Glass/epoxy:

$$\frac{250 \times 600}{0.6 \times 75,000 \times (19 \times 18)} = 9.75 \times 10^{-3} \text{ mm}$$

As we can see, all the three alternatives are acceptable from the point of view of axial elongation; however, carbon/epoxy gives us the minimum mass. Thus, we choose carbon/epoxy as the material system. The number of plies in this case is 21 and the processing method is similar to the one described in Example 14.1.

EXAMPLE 14.3

Design a tension member of length 800 mm and outer circular cross section to carry a tensile load of 750 kN. The material available is carbon/epoxy composite. Unidirectional carbon/epoxy composite properties at $V_f = 0.6$ are

$$E_1 = 130 \text{ GPa}, \quad X^T = 2400 \text{ MPa}, \text{ and } \rho_c = 1.53 \text{ g/cm}^3$$

Solution

Let us consider a tubular cross section and choose filament winding as the manufacturing process. We shall explore two ways to do filament winding of the tension member—axial winding and helical winding.

Axial Winding

In axial winding, the winding angle is zero. Unlike helical winding, it does not involve cross-overs and a fiber volume fraction of 0.6 can be expected. Then, the given unidirectional carbon/epoxy strength can be directly used, and taking a factor of safety of 1.25, the total area of cross section is readily obtained as

$$A_c = \frac{750 \times 1.25 \times 1000}{2400} = 390.6 \text{ mm}^2$$

Taking a nominal shell thickness as 4 mm, the mean diameter is obtained as

$$D_m = \frac{390.6}{\pi \times 4} = 31.1 \text{ mm}$$

We take the inner diameter as 30 mm and readily compute the required outer diameter as

$$D_o = \sqrt{30^2 + \frac{4 \times 390.6}{\pi}} = 37.4 \text{ mm}$$

which implies a shell thickness of 3.7 mm. The thickness provided, however, has to be an integer multiple of ply thickness. Accordingly, considering a ply thickness of 0.5 mm, we provide eight plies. From Example 14.1, we know that the required bandwidth for a ply thickness of 0.5 mm with two spools is 3.08 mm. However, it must also satisfy the following condition:

$$\frac{\pi \times D}{B_w} = n$$

in which n is an integer, which is nothing but the number of circuits per ply, and D and B_w are the inner diameter and bandwidth, respectively, for that ply. Note that diameter D increases by twice the ply thickness after the completion of each ply. Then, the parameters n and B_w for each ply can be tabulated as in Table 14.4.

In Table 14.4, D_i, n, B_w, t, and D_o are the inner diameter, number of circuits, bandwidth along the circumference, ply thickness, and outer diameter,

Design of Composite Structures 693

TABLE 14.4
Axial Winding Parameters (Example 14.3)

Ply	D_i	n	B_w	t	D_o
Ply-1 (hoop)	30.00	–	7.7	0.2	30.40
Ply-2 (axial)	30.40	31	3.08	0.50	31.40
Ply-3 (hoop)	31.40	–	7.7	0.2	31.80
Ply-4 (axial)	31.80	32	3.12	0.49	32.78
Ply-5 (hoop)	32.78	–	7.7	0.2	33.18
Ply-6 (axial)	33.18	33	3.16	0.49	34.16
Ply-7 (hoop)	34.16	–	7.7	0.2	34.56
Ply-8 (axial)	34.56	35	3.10	0.50	35.56
Ply-9 (hoop)	35.56	–	7.7	0.2	35.96
Ply-10 (axial)	35.96	37	3.05	0.50	36.96
Ply-11 (hoop)	36.96	–	7.7	0.2	37.36
Ply-12 (axial)	37.36	38	3.09	0.50	38.36
Ply-13 (hoop)	38.36	–	7.7	0.2	38.76
Ply-14 (axial)	38.76	40	3.04	0.51	39.78
Ply-15 (hoop)	39.78	–	7.7	0.2	40.18
Ply-16 (axial)	40.18	41	3.08	0.50	41.18
Ply-17 (hoop)	41.18	–	7.7	0.2	41.58

respectively, for the ply. Axial filament winding needs consolidation during winding and each of the above plies is consolidated by providing a hoop ply of 0.2 mm thickness. Thus, the total thickness turns out to be 5.8 mm, the cross section OD: 41.6 mm × ID: 30 mm, and the total mass 797 g.

Helical Winding

In helical winding, the winding angle is nonzero. It involves cross-overs and a lower fiber volume fraction is likely to result. V_f depends on a number of factors, and in general, $V_f = 0.5$ can be expected. The longitudinal modulus is linearly influenced by fiber volume fraction but not strength. However, for simplicity, we can consider a linear dependence of longitudinal tensile strength on fiber volume fraction.

A very important aspect in helical winding is the angle of winding. For a tension member, low angle helical winding is preferable. However, very small angle may put added requirement of consolidating hoop plies. Let us choose for the present design example 15° as the helical winding. Then, the total area of cross section is obtained as

$$A_c = \frac{750 \times 1.25 \times 1000}{(2400 \times 0.5 / 0.6) \times \cos^2 15°} = 502.4 \text{ mm}^2$$

Taking the inner diameter as 30 mm, the required outer diameter is readily computed as 39.2 mm. We choose to provide six helical plies of 0.8 mm thickness each at ±15°. Note that each helical ply is actually a compound ply of two subplies at +15° and −15° of thickness 0.4 mm.

We have seen in Example 14.1 that the bandwidth normal to the winding direction for a ply thickness of 0.5 mm for n spools is 1.54n. Accordingly, for two spools, the bandwidth normal to the meridian, that is, along the circumference, path is worked out as follows:

$$B_w = (1.54 \times 2) \times \left(\frac{0.5}{0.4}\right) \times \frac{1}{\cos 15°} = 4.0 \text{ mm}$$

TABLE 14.5
Helical Winding Parameters (Example 14.3)

Ply Description	D_i	n	B_w	t	D_o
Ply-1 (hoop)	30	–	7.7	0.2	30.4
Ply-2 (helical)	30.4	24	3.98	0.80	32.0
Ply-3 (helical)	32.0	25	4.02	0.80	33.6
Ply-4 (helical)	33.6	26	4.06	0.79	35.2
Ply-5 (helical)	35.2	28	3.95	0.81	36.8
Ply-6 (helical)	36.8	29	3.99	0.80	38.4
Ply-7 (helical)	38.4	30	4.02	0.80	40.0
Ply-8 (hoop)	40.0	–	7.7	0.20	40.4

which implies that the required number of circuits per ply is 23.6. We provide an integer number of circuits, say, 24, which in turn, implies a reduced bandwidth and increased ply thickness. The number of circuit for each ply is worked out and tabulated in Table 14.5.

In Table 14.5, D_i, n, B_w, t, and D_o are the inner diameter, number of circuits, bandwidth along the circumference, ply thickness, and outer diameter, respectively, for the ply. Note that we have provided two hoop plies.

The total thickness turns out to be 5.2 mm, the cross section OD: 40.4 mm × ID: 30 mm, and the total mass 672 g. (A reduced density of 1.46 is considered for a reduced fiber volume fraction.)

EXAMPLE 14.4

Determine the elongation of the tension member designed in Example 14.3. Consider the following data:

$$E_1 = 130\,\text{GPa},\ E_2 = 6.5\,\text{GPa},\ G_{12} = 5.0\,\text{GPa},\ \text{and}\ \nu_{12} = 0.25$$

If the axial elongation is to be limited to 0.02 mm, what should be the design modifications, if any?

Solution

Axial Winding

Using Equation 14.56, the axial moduli for the 0° and 90° plies are determined as follows:

$$E_{x,0°} = \left[\frac{1}{130} + 0 + 0\right]^{-1} = 130\,\text{GPa}$$

$$E_{x,90°} = \left[0 + \frac{1}{6.5} + 0\right]^{-1} = 6.5\,\text{GPa}$$

The areas of cross section of the 0° and 90° plies are calculated from the design data from Example 14.3, as follows:

$$A_{0°} = 448.7\,\text{mm}^2\ \text{and}\ A_{90°} = 202.3\,\text{mm}^2$$

Then, using Equation 14.57, the axial elongation is computed as follows:

$$\Delta = \frac{750 \times 1.25 \times 800}{448.7 \times 130{,}000 + 202.3 \times 6500} = 0.013 \text{ mm}$$

Helical Winding

Using Equation 14.56, the axial modulus for the 15° plies is determined as follows:

$$E_{x,15°} = \left[\frac{\cos^4 15°}{130} + \frac{\sin^4 15}{6.5} + \left(\frac{1}{5} - \frac{2 \times 0.25}{130}\right) \times \sin^2 15° \cos^2 15°\right]^{-1} = 50.9 \text{ GPa}$$

We have already found that the axial modulus of the 90° plies is 6.5 GPa. The areas of cross section of the 15° and 90° plies are calculated from the design data from Example 14.3, as follows:

$$A_{15°} = 530.8 \text{ mm}^2 \quad \text{and} \quad A_{90°} = 44.2 \text{ mm}^2$$

Then, using Equation 14.57, the axial elongation is computed as follows:

$$\Delta = \frac{750 \times 1.25 \times 800}{530.8 \times 50{,}900 + 44.2 \times 6500} = 0.028 \text{ mm}$$

The axial elongation can be reduced most conveniently by reducing the angle of winding. Let us consider a winding angle of 10°, which, it can be calculated to show, results in an axial modulus of 76.3 GPa. Then, without changing any other design parameters, it can be seen that the axial elongation comes down to 0.018 mm. Note, however, that with a reduced helical winding angle, it may become necessary to increase the consolidation hoop plies.

14.10.2 Design of a Compression Member

A compression member is typically a slender rod, bar, or tube under uniaxial compression. From the point of view of strength requirements, high compressive stresses must be avoided so as to prevent compression fracture. Similarly, axial stiffness is important when elastic shortening is to be limited within a specified value. Accordingly, Equations 14.55 and 14.57, in which compressive parameters have to be used in place of tensile, can be utilized for selecting the number, thickness, and orientation of plies. Thus,

$$P = \sum_{i=1}^{n} X_i^C A_i \cos^2 \theta_i \tag{14.58}$$

and

$$\sum_{i=1}^{n} A_i E_{xi} = AE_x = \frac{PL}{\Delta} \tag{14.59}$$

in which other parameters remaining the same,
  P Compressive force on the compression member (N)
  X_i^C Longitudinal compressive strength of the ith lamina (N/mm²)

Δ   Axial shortening of the compression member (mm)
E_{xi} Axial modulus of the ith lamina in compression (N/mm²)
E_x Effective axial modulus in compression (N/mm²)

The above relations give a simple means for making the design choice. However, most often, the design of such a slender compression member is driven by stability considerations, in which the primary objective is to avoid global buckling as well as local buckling. Ply sequences containing 0° plies combined with ±45° plies are common. Ply sequences containing only ±θ, in which θ is a small angle such as 15°, are also common.

The critical buckling load of an axially compression-loaded column with symmetric ply sequence of a rectangular cross section is given by (refer to Chapter 6 for derivations)

$$P_{cr} = k\pi^2 \left(\frac{E_{xx}^b I_{yy}}{l^2} \right) \quad (14.60)$$

in which

P_{cr} Critical buckling load (N)
E_{xx}^b Effective bending modulus (N/mm²)
I_{yy} Moment of inertia (mm⁴)
l Length (mm)

The coefficient k depends on boundary conditions of the compression member, as follows:

End Support Conditions	Value of k
Both ends simply supported:	1.0
Fixed-free column:	0.25
Fixed-fixed column:	4.0

Often, compression members are designed as slender tubular members. The design of a cylindrical shell under compression is controlled by its buckling behavior; the critical buckling load, in such a case, can be obtained using classical methods. However, the design of a slender tubular compression member is often simplified, in which its critical buckling load is given by Euler column buckling as follows:

$$P_{cr} = k\pi^2 \left(\frac{E_{xx} I_{yy}}{l^2} \right) \quad (14.61)$$

in which E_{xx} is the axial modulus of the composite and other parameters are as in Equation 14.60.

EXAMPLE 14.5

A platform weighting a total of 6400 kN is supported by four struts of height 4000 mm each. Assuming the total load is equally shared, design the compression members with circular cross section. Use the carbon/epoxy composite with the following material data:

$$E_1 = 130\,\text{GPa}, E_2 = 6.5\,\text{GPa}, G_{12} = 5.0\,\text{GPa}, \nu_{12} = 0.25, \text{and } X^C = 800\,\text{MPa}$$

Design of Composite Structures   697

Solution

Let us consider a tubular compression member composed primarily of ±15° plies. From a strength consideration, the required area of cross section is worked out as

$$A = \frac{1600 \times 1000}{800 \times \cos^2 15°} = 2144 \text{ mm}^2$$

Let us consider a hollow tube of thickness 11 mm and outer diameter 120 mm manufactured by filament winding. The tube is to be made by 10 helical plies of 1.0 mm each and two hoop plies of 0.5 mm each. The innermost and outermost plies are provided as hoop plies. The outermost hoop helps in consolidation and better finish. The innermost hoop makes it symmetric, and in certain cases, it helps prevent fiber pull-out. Each helical is actually a combination of two plies at +15° and −15°, each of 0.5 mm thickness.

Owing to the slenderness of the tube, its design is critical from a buckling point of view. Using Equation 14.56, the axial modulus for the 15° plies is determined as follows:

$$E_{xx,15°} = \left[\frac{\cos^4 15°}{130} + \frac{\sin^4 15}{6.5} + \left(\frac{1}{5} - \frac{2 \times 0.25}{130} \right) \times \sin^2 15° \times \cos^2 15° \right]^{-1} = 50.9 \text{ GPa}$$

Similarly,

$$E_{xx,90°} = 6.5 \text{ GPa}$$

The cross-sectional areas of the helical and hoop plies are 3424.3 and 342.4 mm², respectively. An effective axial modulus can be obtained by rule of mixture as

$$(E_{xx})^{eff} = 46.9 \text{ GPa}$$

The moment of inertia for the chosen area of cross section is obtained as

$$I_{yy} = \frac{\pi \times (120^4 - 98^4)}{4} = 90.42 \times 10^6 \text{ mm}^4$$

Then, considering pin ends, using Equation 14.60, the critical buckling load is estimated as follows:

$$P_{cr} = \frac{\pi^2 \times (46.9 \times 1000) \times (90.42 \times 10^6)}{4000^2 \times 1000} = 2616 \text{ kN}$$

which implies a buckling factor of 1.6.

Note: The design of a slender compression member is driven by stability. Buckling being an inherently catastrophic phenomenon, often a high buckling factor is desired. In our present case, the buckling factor can be increased by the following feasible means: (i) increase in the cross section, (ii) reduction in angle of winding, and (iii) use of fixed end connections.

14.10.3 Design of a Torsion Member

A torsion member is a slender structural element subjected to torsion. Automotive drive shaft is a typical example of a torsion member. While tubular structures are commonly used in these applications, sometimes, torsion members of other cross sections such as hollow square and rectangle are also used. Laminated composites with plies at $\pm 45°$ possess the maximum shear modulus and are most commonly used for the transmission of torques.

From strength and stiffness points of view, a torsion member is designed so as to keep the shear stresses and angle of twist within respective allowables. Thin tubular members can undergo torsional buckling and in such a case, stability becomes the design driver.

For a laminated composite tube, shear stress and angle of twist depend not only on the applied torque and geometry of the tube but also on the ply sequence. For balanced and symmetric ply sequence, the maximum shear stress and angle of twist are given by [35]

$$\tau_{xy} = \frac{T}{2\pi r^2 t} \quad (14.62)$$

and

$$\phi = \frac{T}{2\pi G_{xy} r^3 t} \quad (14.63)$$

in which

- τ_{xy} In-plane shear stress developed in the tube (N/mm²)
- ϕ Angle of twist per mm (mm^{-1})
- T Applied torque (N.m)
- r Mean radius of the tube (mm)
- t Wall thickness of the tube (mm)
- G_{xy} In-plane shear modulus (N/mm²)

On the other hand, the critical buckling torque is given by [36]

$$T_{cr} = \frac{122 k A_{11}^{3/8} D_{22}^{5/8} r^{5/4}}{5 l^{1/2}} \quad (14.64)$$

Note that x- and y-directions in the above expressions are along the axis of the tube and tangential to the circumference, respectively.

14.10.4 Design of a Beam

A beam is a slender structural element subjected to lateral loads. The various types of composite beams in common use are

- Solid cross sections
  - Rectangular (plies normal to the loading direction)
  - Rectangular (plies parallel to the loading direction)
- Thin-walled cross sections
  - Open ended, for example, T-section, I-section
  - Closed ended, for example, box-section
- Sandwich beam

Design of Composite Structures

TABLE 14.6
Isotropic Homogeneous Beam versus Composite Beam

Isotropic Homogeneous Beam (Rectangular Cross Section)	Composite Beam (Rectangular Cross Section, Plies Normal to Loads)
Stress distributions are continuous across the beam depth	Stress distributions are *not* continuous across the beam depth
Maximum normal stresses occur at the outermost faces	Maximum normal stresses can occur at any point in the beam cross section (not necessarily at the outermost faces)
Maximum shear stress occurs at the neutral plane	Maximum in-plane shear stress can occur at any point in the beam cross section (not necessarily at the midplane)
Deformation is in the plane of the applied loads only	Out-of-plane deformation can result due to various coupling effects depending on ply sequence

The response of a composite beam to applied loads differs grossly from that of an isotropic beam in respect of stress distribution and deformation patterns (Table 14.6). The differences must be kept in mind while designing a composite beam [37,38].

EXAMPLE 14.6

Design a cantilever beam of length 400 mm and width 25 mm to carry a tip lateral load of 5 N. Restrict the tip deflection within 1.0 mm. Use unidirectional carbon/epoxy prepreg with the following material data:

$$E_1 = 130 \text{ GPa}, E_2 = 6.5 \text{ GPa}, G_{12} = 5.0 \text{ GPa}, \nu_{12} = 0.25$$

Each prepreg ply is of 0.25 mm thickness.

Consider the following alternatives: (1) solid rectangular cross section (plies normal to the loading direction), (2) solid rectangular cross section (plies parallel to the loading direction), and (3) box-section.

Solution

Option-1

Solid rectangular cross section (plies normal to loading direction)

Let us consider a ply sequence consisting only 0° plies. For this ply sequence and the given material properties, the reduced stiffness matrix and transformed reduced stiffness matrix are given by

$$[Q] = [\bar{Q}] = \begin{bmatrix} 130{,}407.524 & 1630.094 & 0 \\ 1630.094 & 6520.376 & 0 \\ 0 & 0 & 5015.674 \end{bmatrix}$$

Let the height of the beam be denoted by h. Then, noting that the $\bar{Q}$ matrix is the same for all the plies, the bending stiffness matrix is given by

$$[D] = \frac{h^3}{12} \begin{bmatrix} 130{,}407.524 & 1630.094 & 0 \\ 1630.094 & 6520.376 & 0 \\ 0 & 0 & 5015.674 \end{bmatrix}$$

The bending compliance $[D^*]$ matrix can be obtained by using the equations described earlier (refer Equations 5.60 and 5.67, Chapter 5). The element in $[D^*]$ that we need for beam design is D_{11}^*. Note that for symmetric laminates, $[D^*]$ is simply the inverse of $[D]$. Then, D_{11}^* can be conveniently determined as follows (refer Equation 5.163, Chapter 5):

$$D_{11}^* = \frac{D_{22}D_{66} - D_{26}^2}{\text{Det}[D]}$$

With little arithmetic, we can show that

$$D_{11}^* = \frac{3}{32,500h^3}$$

and the effective bending modulus (refer Equation 6.15, Chapter 6)

$$E_{xx}^b = \frac{12}{h^3 D_{11}^*} = 130,000 \, \text{MPa}$$

The area moment of inertia about the bending axis is

$$I_{yy} = \frac{bh^3}{12}$$

Then, the tip deflection (δ) of the beam is given by (refer Equation 6.129, Chapter 6)

$$\delta = \frac{Pl^3}{3E_{xx}^b I_{yy}} = \frac{5 \times 400^3}{3 \times 130,000 \times (25 \times h^3/12)} = \frac{5120}{13h^3}$$

Now, we can readily find that $h \geq 7.3$ so that $\delta \leq 1.0$. Let us provide a beam thickness of 8.0 mm, that is, $h = 8.0$. Then,

$$[D] = \begin{bmatrix} 5,564,054.357 & 69,550.677 & 0 \\ 69,550.677 & 278,202.709 & 0 \\ 0 & 0 & 214,002.091 \end{bmatrix}$$

and the bending compliance matrix is

$$[D^*] = \begin{bmatrix} 180.288 & -45.072 & 0 \\ -45.072 & 3605.769 & 0 \\ 0 & 0 & 4672.852 \end{bmatrix} \times 10^{-9}$$

Axial bending stresses are given by (refer Equation 6.126, Chapter 6)

$$\sigma_{xx}^{(k)}(x,z) = \frac{P(l-x)z}{b}\left(\bar{Q}_{11}^{(k)}D_{11}^* + \bar{Q}_{12}^{(k)}D_{12}^* + \bar{Q}_{16}^{(k)}D_{16}^*\right)$$

Design of Composite Structures

Then, the maximum axial bending stresses are obtained as follows:

$$(\sigma_{xx})_{max} = \pm \frac{5 \times (400 - 0) \times 4}{25} \times (130,407.524 \times 180.288 - 1630.094 \times 45.072)$$
$$\times 10^{-9} = \pm 7.5 \text{ MPa}$$

The axial bending stresses are much within the respective tensile and compressive strengths.

The laminate for the beam is to be made by laying up 32 prepreg plies, all aligned in the same direction, in a matched-die-mold.

Note that we have verified the beam design only for longitudinal bending stresses. In a beam, interlaminar stresses may be high. The process of interlaminar stress determination is rather laborious as it involves sequential operations going through each ply in the laminate. In the present case, there are 32 plies and a manual method is not practical. For further details and demonstration, the reader may refer to Example 6.1.

Option-2

Solid rectangular cross section (plies parallel to loading direction)

With all 0° plies, the orientation of the plies w.r.t. the loading direction is insignificant and Option-2 is equivalent to Option-1. However, if plies at nonzero directions are used, plies parallel to the loading direction would result in higher bending stiffness.

Option-3: Box-Section

Let us consider a box-section with overall height of 20 mm, width of 25 mm, and flange/web thickness of 2 mm. Let us consider 0° plies for the flanges and the webs. The beam cross section is symmetric and the vertical distances from the centroid of the beam to the flange and web centroids are readily obtained as

$$z_{c1} = z_{c4} = 9 \text{ mm (flanges)}$$

and

$$z_{c2} = z_{c3} = 0 \text{ (webs)}$$

The extensional and bending stiffness matrices are determined as

$$[A] = \begin{bmatrix} 260,815.048 & 3260.188 & 0 \\ 3260.188 & 13,040.752 & 0 \\ 0 & 0 & 10,031.348 \end{bmatrix}$$

$$[D] = \begin{bmatrix} 86,938.349 & 1086.729 & 0 \\ 1086.729 & 4346.917 & 0 \\ 0 & 0 & 3343.783 \end{bmatrix}$$

The compliance matrices are

$$[A^*] = \begin{bmatrix} 3.846 & -0.962 & 0 \\ -0.962 & 76.923 & 0 \\ 0 & 0 & 99.687 \end{bmatrix} \times 10^{-6}$$

$$[D^*] = \begin{bmatrix} 11.538 & -2.885 & 0 \\ -2.885 & 230.769 & 0 \\ 0 & 0 & 299.062 \end{bmatrix} \times 10^{-6}$$

The effective flexural rigidity of the beam is given by (refer Equation 6.208, Chapter 6)

$$E_{xx}^{fl} I_{yy} = \left(\frac{25 \times 9^2}{3.846} + \frac{25}{11.538} + \frac{16^2}{3 \times 3.846} \times \frac{16}{2} + \frac{25}{11.538} \right) \times 10^6 = 708.355 \times 10^6$$

The tip deflection is then given by

$$\delta = \frac{Pl^3}{3E_{xx}^b I_{yy}} = \frac{5 \times 400^3}{3 \times 708.355 \times 10^6} = 0.15 \text{ mm}$$

It can be seen that the box-section offers a far more efficient solution in terms of lower tip deflection at a lower mass.

14.10.5 Design of a Flat Panel under In-Plane Loads

Flat panels under in-plane loads are designed for either strength or stiffness requirements. Under tensile loads, stiffness requirements arise out of possible functional restriction on axial and lateral deformations. On the other hand, buckling becomes a critical aspect under compression and in-plane shear.

In the design of a flat panel under biaxial loads, theoretically, there exist infinite possibilities of ply sequences. In practice, most designs are done using standard ply orientations such as ±45°, ±30°, ±60°, etc. in addition to 0° and 90° plies and the ply sequence options are greatly simplified. The plies are generally stacked to form a symmetric laminate so as to avoid unnecessary extension–bending coupling. However, the question of apportioning thicknesses to different plies w.r.t. the total thickness remains, and it is a critical one. In this respect, carpet plots help greatly, especially in the preliminary design. A carpet plot is a graphical representation of a dependent parameter w.r.t. certain independent variables. Carpet plots are available for various parameters for various material systems. Concepts of composites mechanics (refer Chapters 3 through 5) can be made use of and these plots can be generated. Figures 14.13 through 14.18 are six carpet

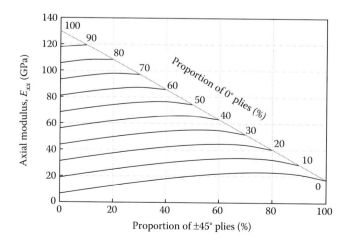

FIGURE 14.13 Carpet plot for axial modulus, E_{xx} (refer Example 14.7).

Design of Composite Structures 703

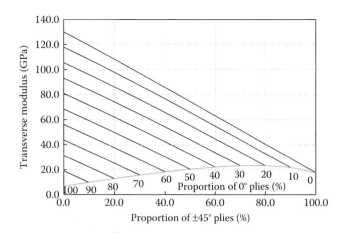

FIGURE 14.14 Carpet plot for transverse modulus, E_{yy} (refer Example 14.7).

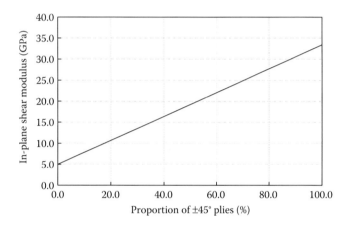

FIGURE 14.15 Carpet plot for in-plane shear modulus, G_{xy} (refer Example 14.7).

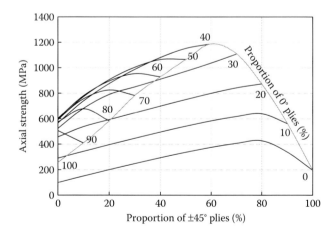

FIGURE 14.16 Carpet plot for axial tensile strength (refer Example 14.7).

plots generated for moduli and strengths of carbon/epoxy laminate. Material data are given in Example 14.7. It is easy to see that plots with other combinations of ply orientations can also be constructed. These plots show the dependence of the moduli on relative proportions of 0°, ±45°, and 90° plies. Note, however, that in-plane shear modulus does not depend on the proportion of 0° and 90° plies. Note, further, that laminate strengths depend not only on the proportions of the plies of different orientations but also on applied loads.

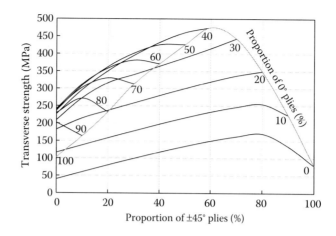

FIGURE 14.17 Carpet plot for transverse tensile strength (refer Example 14.7).

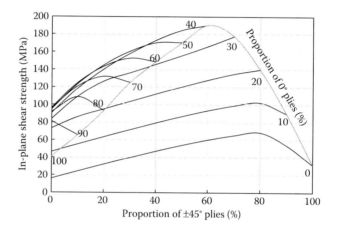

FIGURE 14.18 Carpet plot for in-plane shear strength (refer Example 14.7).

EXAMPLE 14.7

Design a rectangular flat panel of length 1200 mm and width 800 mm subjected to in-plane tensile force of 5000 and 3000 kN, respectively, in the longitudinal and lateral directions in addition to an in-plane shear force of 1200 kN acting on the longitudinal edge. The normal and shear deformations are to be restricted to 8 mm (axial), 10 mm (transverse), and 1°, respectively. Use unidirectional carbon/epoxy prepreg with the following material data:

$$E_1 = 130\,\text{GPa}, E_2 = 6.5\,\text{GPa}, G_{12} = 5.0\,\text{GPa}, \nu_{12} = 0.25, \text{ and } X^T = 2000\,\text{MPa}$$

Each prepreg ply is of 0.15 mm thickness.

Solution

Taking a factor of safety of 1.25, the force resultants acting on the panel are

$$N_{xx} = \frac{5000 \times 1000 \times 1.25}{800} = 7812.5\,\text{N/mm}$$

$$N_{yy} = \frac{3000 \times 1000 \times 1.25}{1200} = 3125\,\text{N/mm}$$

Design of Composite Structures

$$N_{xy} = \frac{1200 \times 1000 \times 1.25}{1200} = 1250 \, \text{N/mm}$$

For these force resultants and the given material properties, carpet plots for laminate moduli and strengths are generated and given in Figures 14.13 through 14.18. The laminates in these figures consist of symmetric combinations of $0°$, $\pm 45°$, and $90°$ plies at different proportions.

Now, let us choose the ply sequence based on stiffness and strength requirements. The maximum normal and in-plane shear strains that can be allowed are

$$(\varepsilon_{xx})_{max} = \frac{8}{1200} = 6667 \times 10^{-6}$$

$$(\varepsilon_{yy})_{max} = \frac{10}{800} = 12,500 \times 10^{-6}$$

$$(\gamma_{xy})_{max} = \frac{1 \times \pi}{180} = 17,453 \times 10^{-6}$$

Let the laminate thickness be t. Then, the overall normal and in-plane shear stresses are

$$(\sigma_{xx})_{tot} = \frac{7812.5}{t}$$

$$(\sigma_{yy})_{tot} = \frac{3125}{t}$$

$$(\tau_{xy})_{tot} = \frac{1250}{t}$$

The overall normal and in-plane shear moduli required are obtained as follows:

$$(E_{xx})_{tot} = \frac{7812.5/t}{6667 \times 10^{-6}} = \frac{1.172 \times 10^6}{t}$$

$$(E_{yy})_{tot} = \frac{3125/t}{12,500 \times 10^{-6}} = \frac{0.25 \times 10^6}{t}$$

$$(G_{xy})_{tot} = \frac{1250/t}{17,453 \times 10^{-6}} = \frac{0.072 \times 10^6}{t}$$

We can see that the required axial stiffness is 4.7 times the required transverse stiffness and the required transverse modulus is about 3.5 times the required in-plane shear modulus. Obviously, it demands a high proportion of $0°$ plies. Let us consider a laminate consisting of plies at $0°$, $\pm 45°$, and $90°$ at the ratio 80:10:10. From Figures 14.13 through 14.15, we see that 80% $0°$ plies together with 10% $\pm 45°$ and 10% $90°$ plies provide $(E_{xx})_{tot} = 108 \, \text{GPa}$, $(E_{yy})_{tot} = 22 \, \text{GPa}$, and $(G_{xy})_{tot} = 7.8 \, \text{GPa}$. The corresponding strengths are $X^T \approx 680 \, \text{MPa}$, $Y^T \approx 270 \, \text{MPa}$, and $S \approx 110 \, \text{MPa}$.

Now, based on stiffness requirements, the required laminate thickness is obtained as the maximum of the following:

$$\frac{1.172 \times 10^6}{108 \times 10^3} = 10.9, \quad \frac{0.25 \times 10^6}{22 \times 10^3} = 11.4, \quad \text{and} \quad \frac{0.072 \times 10^6}{7.8 \times 10^3} = 9.2$$

And, based on strength requirements, the required laminate thickness is obtained as the maximum of the following:

$$\frac{7812.5}{680} = 11.5, \quad \frac{3125}{270} = 11.6, \quad \text{and} \quad \frac{1250}{110} = 11.4$$

Let us then consider a laminate of total thickness 12 mm, in which individual thicknesses for different ply orientations are as follows: 0°→9.6 mm (64 plies), +45°→0.6 mm (4 plies), −45°→0.6 mm (4 plies), and 90°→1.2 mm (8 plies).

We have chosen the number of plies and their orientations. However, the ply sequence is yet to be chosen. Note that, for a given proportion of plies, ply sequence does not influence in-plane normal and shear deformations and lamina stresses, as long as symmetry is maintained. However, shear buckling characteristics can be improved by providing ±45° plies on the outer faces. Let us then choose the following ply sequence for our panel:

$$\left[45_2^\circ / -45_2^\circ / 0_{16}^\circ / 90_4^\circ / 0_{16}^\circ\right]_s$$

For this ply sequence and the given material properties, the reduced stiffness matrix and transformed reduced stiffness matrices are given by

$$[\boldsymbol{Q}] = [\overline{\boldsymbol{Q}}]^{(0^\circ)} = \begin{bmatrix} 130{,}407.524 & 1630.094 & 0 \\ 1630.094 & 6520.376 & 0 \\ 0 & 0 & 5015.674 \end{bmatrix}$$

$$[\overline{\boldsymbol{Q}}]^{(45^\circ)} = \begin{bmatrix} 40{,}062.696 & 30{,}031.348 & 30{,}971.787 \\ 30{,}031.348 & 40{,}062.696 & 30{,}971.787 \\ 30{,}971.787 & 30{,}971.787 & 33{,}416.928 \end{bmatrix}$$

$$[\overline{\boldsymbol{Q}}]^{(-45^\circ)} = \begin{bmatrix} 40{,}062.696 & 30{,}031.348 & -30{,}971.787 \\ 30{,}031.348 & 40{,}062.696 & -30{,}971.787 \\ -30{,}971.787 & -30{,}971.787 & 33{,}416.928 \end{bmatrix}$$

$$[\overline{\boldsymbol{Q}}]^{(90^\circ)} = \begin{bmatrix} 6520.376 & 1630.094 & 0 \\ 1630.094 & 130{,}407.524 & 0 \\ 0 & 0 & 5015.674 \end{bmatrix}$$

Then, for the chosen ply sequence, the laminate extensional stiffness and compliance matrices are obtained as

Design of Composite Structures

$$[A] = \begin{bmatrix} 1{,}307{,}811.917 & 53{,}642.633 & 0 \\ 53{,}642.633 & 267{,}159.662 & 0 \\ 0 & 0 & 94{,}269.593 \end{bmatrix}$$

$$[A^*] = \begin{bmatrix} 0.771 & -0.155 & 0 \\ -0.155 & 3.774 & 0 \\ 0 & 0 & 10.608 \end{bmatrix} \times 10^{-6}$$

Laminate middle surface strains are

$$\begin{Bmatrix} \varepsilon_{xx}^0 \\ \varepsilon_{yy}^0 \\ \gamma_{xy} \end{Bmatrix} = \begin{bmatrix} 0.771 & -0.155 & 0 \\ -0.155 & 3.774 & 0 \\ 0 & 0 & 10.608 \end{bmatrix} \times \begin{Bmatrix} 7812.5 \\ 3125 \\ 1250 \end{Bmatrix} \times 10^{-6} = \begin{Bmatrix} 5539 \\ 10{,}583 \\ 13{,}260 \end{Bmatrix} \times 10^{-6}$$

The laminate curvatures are zero, which means

$$\begin{Bmatrix} \varepsilon_{xx} \\ \varepsilon_{yy} \\ \gamma_{xy} \end{Bmatrix} = \begin{Bmatrix} 5539 \\ 10{,}583 \\ 13{,}260 \end{Bmatrix} \times 10^{-6}$$

at all locations in the laminate thickness
Deformations are

$$\text{Axial deformation} = 1200 \times 5539 \times 10^{-6} = 6.6 \, \text{mm}$$

$$\text{Transverse deformation} = 800 \times 10{,}583 \times 10^{-6} = 8.5 \, \text{mm}$$

$$\text{In-plane shear deformation} = \frac{180 \times 13{,}260 \times 10^{-6}}{\pi} = 0.76°$$

Global stresses in the plies are

$$\begin{Bmatrix} \sigma_{xx} \\ \sigma_{yy} \\ \tau_{xy} \end{Bmatrix}^{(0°)} = \begin{bmatrix} 130{,}407.524 & 1630.094 & 0 \\ 1630.094 & 6520.376 & 0 \\ 0 & 0 & 5015.674 \end{bmatrix} \begin{Bmatrix} 5539 \\ 10{,}583 \\ 13{,}260 \end{Bmatrix} \times 10^{-6} = \begin{Bmatrix} 739.6 \\ 78.0 \\ 66.5 \end{Bmatrix} \text{MPa}$$

$$\begin{Bmatrix} \sigma_{xx} \\ \sigma_{yy} \\ \tau_{xy} \end{Bmatrix}^{(45°)} = \begin{bmatrix} 40{,}062.696 & 30{,}031.348 & 30{,}971.787 \\ 30{,}031.348 & 40{,}062.696 & 30{,}971.787 \\ 30{,}971.787 & 30{,}971.787 & 33{,}416.928 \end{bmatrix} \begin{Bmatrix} 5539 \\ 10{,}583 \\ 13{,}260 \end{Bmatrix} \times 10^{-6}$$

$$= \begin{Bmatrix} 950.4 \\ 1001.0 \\ 942.4 \end{Bmatrix} \text{MPa}$$

$$\begin{Bmatrix} \sigma_{xx} \\ \sigma_{yy} \\ \tau_{xy} \end{Bmatrix}^{(-45°)} = \begin{bmatrix} 40,062.696 & 30,031.348 & -30,971.787 \\ 30,031.348 & 40,062.696 & -30,971.787 \\ -30,971.787 & -30,971.787 & 33,416.928 \end{bmatrix} \begin{Bmatrix} 5539 \\ 10,583 \\ 13,260 \end{Bmatrix} \times 10^{-6}$$

$$= \begin{Bmatrix} 129.0 \\ 179.6 \\ -56.2 \end{Bmatrix} \text{MPa}$$

$$\begin{Bmatrix} \sigma_{xx} \\ \sigma_{yy} \\ \tau_{xy} \end{Bmatrix}^{(90°)} = \begin{bmatrix} 6520.376 & 1630.094 & 0 \\ 1630.094 & 130,407.524 & 0 \\ 0 & 0 & 5015.674 \end{bmatrix} \begin{Bmatrix} 5539 \\ 10,583 \\ 13,260 \end{Bmatrix} \times 10^{-6} = \begin{Bmatrix} 53.4 \\ 1389.1 \\ 66.5 \end{Bmatrix} \text{MPa}$$

Then, the local stresses in the plies are obtained as follows:

$$\begin{Bmatrix} \sigma_{11} \\ \sigma_{22} \\ \tau_{12} \end{Bmatrix}^{(0°)} = \begin{bmatrix} 1 & 0 & 0 \\ 0 & 1 & 0 \\ 0 & 0 & 1 \end{bmatrix} \begin{Bmatrix} 739.6 \\ 78.0 \\ 66.5 \end{Bmatrix} = \begin{Bmatrix} 739.6 \\ 78.0 \\ 66.5 \end{Bmatrix} \text{MPa}$$

$$\begin{Bmatrix} \sigma_{11} \\ \sigma_{22} \\ \tau_{12} \end{Bmatrix}^{(45°)} = \begin{bmatrix} 0.5 & 0.5 & 1 \\ 0.5 & 0.5 & -1 \\ -0.5 & 0.5 & 0 \end{bmatrix} \begin{Bmatrix} 950.4 \\ 1001.0 \\ 942.4 \end{Bmatrix} = \begin{Bmatrix} 1918.1 \\ 33.3 \\ 25.3 \end{Bmatrix} \text{MPa}$$

$$\begin{Bmatrix} \sigma_{11} \\ \sigma_{22} \\ \tau_{12} \end{Bmatrix}^{(-45°)} = \begin{bmatrix} 0.5 & 0.5 & -1 \\ 0.5 & 0.5 & 1 \\ 0.5 & -0.5 & 0 \end{bmatrix} \begin{Bmatrix} 129.0 \\ 179.6 \\ -56.2 \end{Bmatrix} = \begin{Bmatrix} 210.5 \\ 98.1 \\ -25.3 \end{Bmatrix} \text{MPa}$$

$$\begin{Bmatrix} \sigma_{11} \\ \sigma_{22} \\ \tau_{12} \end{Bmatrix}^{(90°)} = \begin{bmatrix} 0 & 1 & 0 \\ 1 & 0 & 0 \\ 0 & 0 & -1 \end{bmatrix} \begin{Bmatrix} 53.4 \\ 1389.1 \\ 66.5 \end{Bmatrix} = \begin{Bmatrix} 1389.1 \\ 53.4 \\ -66.5 \end{Bmatrix} \text{MPa}$$

The deformations are within the limits and the stresses are within the material allowables. Also, the stresses in the individual plies are all tensile and thus buckling is unlikely.

14.10.6 Design of a Pressure Vessel under Internal Pressure

14.10.6.1 Introduction

Filament-wound composite pressure vessels are a major class of composite products that are routinely used in aerospace, defense, and commercial applications. Various types of pressure vessels are solid rocket motor casing, air bottle, hydrogen fuel tank, CNG tank, LPG tank, etc. As a matter of fact, solid rocket motor casings used for housing solid fuel in aerospace vehicles have been the primary factor responsible for the growth of filament winding from its nascent stage in the 1960s to an advanced highly automated technology today.

14.10.6.2 Advantages of Composite Pressure Vessels

There are several advantages of filament-wound composite pressure vessels over conventional isotropic materials. Given below is a list of some of these advantages:

- *Light weight*: The primary factor that makes composites the first choice for pressure vessels is their high strength-to-weight ratios. The performance of a pressure vessel is expressed in terms of a factor known as performance factor (η) defined as

$$\eta = \frac{pV}{W} \qquad (14.65)$$

  in which p is the internal pressure, V the enclosed volume, and W the weight. η is consistently higher for composite pressure vessels than metallic and it is a good measure for the comparison of the vessel design. Note that operating pressure, proof pressure, or design pressure can be used for p. Similarly, either pressure vessel weight or total weight including those of the pressure vessel and other associated parts can be used for W. However, the parameters p, V, and W must be used in a consistent manner.
- *Design flexibility*: In most cases, the implementation of certain modifications becomes necessary during the development period of a pressure vessel. These modifications include marginal changes in the ply sequence, inner contour, metallic end fittings, etc. The design as well as the manufacturing methodology of a composite pressure vessel is typically amenable to such changes.
- *Stress corrosion cracking*: The phenomenon of crack growth in a corrosive environment leading to unexpected failure under tensile stresses is called as stress corrosion cracking. It is typically common in metallic pressure vessels, but absent in composite pressure vessels.
- *Efficient utilization of material properties*: The axial stresses in the cylindrical portion of a pressure vessel are typically half of the circumferential stresses. In an isotropic material, material properties are equal in all directions and the shell thickness is provided to meet the circumferential stresses. It is easy to see, then, that in a pressure vessel designed with isotropic material, the material is underutilized as far as its strength in the axial direction is concerned. On the other hand, in the case of a composite pressure vessel, the plies can be oriented in such a way as to achieve an axial strength, which is about half of the circumferential strength. In this way, it is possible to exploit the material more efficiently.
- *Shorter cycle time*: The overall cycle time of development and production of a composite pressure vessel is much shorter than that of its metallic counterpart.
- *Low cost*: The tools and fixtures required for the development and production of composite pressure vessels are rather simple and very cheap when compared with metallic pressure vessels.

14.10.6.3 Configuration of a Pressure Vessel

The essential part of a pressure vessel is a cylindrical shell with two end domes. (In certain special cases, conical shell is also used.) Typically, two axial openings—one at each end dome—are provided. Each opening is reinforced with metallic fitting commonly known as polar boss. The metallic end fittings have primarily two functions to perform—first, they provide rigid support during winding for the helical windings to take reversals around them and second, they act as inlet/outlet and provide features for assembly of other components such as nozzle, inlet and outlet ports, etc.

The pressure vessel shell, depending on its functional requirements, is provided with other features such as supporting brackets, etc. Rocket motor casings are a major type of products within the broad class of pressure vessels. In a rocket motor casing, cylindrical or conical shells commonly referred to as skirts are provided. They are either *in situ* wound on or prefabricated and bonded to the pressure vessel shell.

The final performance of a pressure vessel is greatly influenced by its configuration. Some of the important features of the configuration design are as follows.

14.10.6.3.1 Pole Openings

The pole openings have significant bearing on the overall design of a pressure vessel. The implications are both structural as well as functional. Some of the key points are briefly discussed below:

First, the helical winding path is influenced by the relative pole opening diameters. The helical fibers take reversal around the polar boss. The imaginary circle, with its center on the axis, on which the helical path reversal point lies, is referred to as the composite opening. The polar boss design obviously dictates the composite pole opening diameter. Now, the significance of relative pole opening diameters is that geodesic winding is possible only if the two composite pole openings are equal. Typically, a rocket motor casing has unequal pole openings, as a result of which, the helical path is nongeodesic. Other parameters remaining constant, the extent of deviation from the geodesic path, and thus degree of winding difficulty depends on the ratio of the composite pole opening diameters. It is preferable to keep this ratio as small as possible, that is, closer to unity.

Second, the composite pole opening diameters have direct bearing on the helical winding angles. For optimal design, the lamina strengths should be fully exploited. Note that the helical plies provide strength and stiffness in both the meridional as well as circumferential directions. In the cylindrical portion of a pressure vessel, it is possible to orient the helical plies in such a way as to minimize the requirement of circumferential or hoop plies. As a matter of fact, for equal composite pole opening diameters, that is, for a geodesic path, the optimal helical winding angle in the cylinder is given by

$$\alpha_{cyl} = \tan^{-1}\sqrt{2} \approx 54.736° \tag{14.66}$$

for which the corresponding ratio of the geodesic composite pole opening radius to the cylinder radius is

$$\frac{r_0}{r} = \sin(\tan^{-1}\sqrt{2}) \approx 0.816 \tag{14.67}$$

In other words, for a composite pole opening radius, given by Equation 14.67, helical plies are sufficient to bear both the meridional as well as circumferential stresses. Note, however, that for such a pole opening size, the dome design and polar boss design may not be optimal.

Third, a pressure vessel is typically designed for a closed condition. The internal pressure, acting on the area enclosed by the composite pole opening, results in a concentrated force that acts on the edge of the composites at the pole. This edge loading results in bending stresses in the dome, which require additional reinforcements accompanied by weight penalty. The severity of the bending stresses is directly dependent on the pole opening diameter and thus it is preferable to have smaller pole openings.

Fourth, the larger the pole opening, the larger and heavier the polar boss. From this point of view, too, it is preferable to have small pole openings.

The pole opening diameters are often decided based on functional requirements. And like any other design process, it is also an iterative process and a final decision is a compromise decision that meets functional requirements at acceptable structural margins of safety.

14.10.6.4 End Domes

The meridional shapes of the end domes are crucial in the design of a pressure vessel. Various dome shapes in use include (i) simple geometrical shapes such as hemispherical, elliptical, ellipsoidal, and tori-spherical and (ii) analytically generated shapes such as isotensoid and constant deviation contour, etc. Geometrical shapes are easy to implement, but not optimal structurally [39]. They are most commonly used in smaller pressure vessels.

Isotensoid contour is an optimal contour; it is possible to carry out geodesic winding on it and the stresses are uniform. Recall, however, that geodesic winding is governed by the famous Clairut's principle $r \sin \alpha = C$, in which r and α are the local radius and angle of winding, respectively, and C is a constant ($= r_0$, the radius at the pole opening). It can be seen that the pole openings at the two ends must be equal for geodesic winding. In many pressure vessels, most notably in rocket motor casings, the pole openings are unequal and the winding path becomes nongeodesic (also called modified geodesic). Nongeodesic winding tends to be unstable and slips away from its designated path unless sufficient friction holds it back [40]. In other words, nongeodesic winding is feasible as long as the required friction coefficient is less than the available friction coefficient. The required friction coefficient is high for large deviation from geodesic path and vice versa, and it depends on the contour geometry. Constant deviation contour is a dome contour that is generated by an iterative analytical process such that the required friction coefficient is the least and uniform along the contour.

14.10.6.5 Metallic End Fittings

The metallic end fittings are designed to meet the following requirements:

- To facilitate smooth load transfer between the composite shell and metallic end fittings
- To provide features for joining other elements such as valve, nozzle, etc. to the pressure vessel
- To provide sufficient landing area for the helical fibers to take reversal during winding

The flange of the end fitting has to provide sufficient area so that the bearing stresses in the composite are within limit. In general, the flange width should be at least 1.5 times the width of a fiber band. A major failure mode in a pressure vessel is boss blow out during pressure testing. For avoiding such failure, as a thumb rule, the outer diameter of the end fitting is commonly extended to the inflection point or above it [41]. The flange thickness is typically the highest at the root and is gradually decreased to a minimum at the tip. Empirical relations are available for initial sizing of the flange thickness, which is fine tuned during finite element analysis.

Various configurations, available for giving provisions for attachment of other elements, can be broadly grouped into three types—threaded joint, axially bolted joint, and radially bolted joint.

14.10.6.6 Ply Design

Various types of reinforcing plies used in composite pressure vessel shell include the following:

- Continuous winding
  - Helical winding
  - Polar winding
  - Hoop winding
- Local lay-up
  - Fabric lay-up
  - Unidirectional sheet lay-up

Under internal pressure, the pressure vessel shell is in a state of predominantly membrane stress. Helical winding provides strength and stiffness in the meridional direction as well as circumferential direction. Polar winding also provides strength and stiffness in both the directions, and it can be used in place of helical winding. Helical winding, however, gives greater flexibility and is more common, especially in larger pressure vessels. Hoop winding is provided in the cylindrical region for providing additional circumferential strength and stiffness. In the dome regions, hoop winding is not feasible and local plies, often referred to as doilies, are laid-up for providing necessary additional reinforcements in the circumferential direction as well as meridional direction. Doily lay-up can be in the form of either bidirectional fabric lay-up or unidirectional sheet lay-up. Bidirectional fabric plies are useful in regions where meridional and circumferential stresses are of the same order as well as in regions where it is difficult to accurately estimate stresses. Unidirectional sheet plies, on the other hand, provide an efficient way of orienting the reinforcements.

Netting model

Netting model is a simple tool commonly used for determining the required thicknesses of meridional and circumferential reinforcements. It assumes that the laminate stiffness and strength are provided totally by the longitudinal stiffness and strength of the laminae; transverse and shear properties are ignored [2,42]. Then, considering the free-body diagram of a representative element under static equilibrium, it can be shown that

$$2\left(X_h^T t_h \cos^2 \alpha + X_m^T t_m\right)\cos \beta = pr \tag{14.68}$$

$$\left(X_h^T t_h \sin^2 \alpha + X_c^T t_c\right)\cos \beta = pr \tag{14.69}$$

in which

X_h^T, X_m^T, X_c^T Longitudinal tensile strengths of helical, meridional, and circumferential (hoop) plies

t_h, t_m, t_c Thicknesses of helical, meridional, and circumferential (hoop) plies

α Angle of winding

β Slope of the local tangent to the meridional contour

p Internal pressure

r Local radius

In a strength-based design, meridional strength and stiffness are provided totally by helical plies and no meridional reinforcements are generally provided, that is, $t_m = 0$.

Then, from the above Equations 14.68 and 14.69, helical ply thickness and hoop ply thickness can be determined. On the other hand, when additional meridional stiffness and strength are required, t_h and t_c are determined for known or trial value of t_m.

Another issue that needs mention here is the helical ply thickness variation w.r.t. local radius and angle of winding. Typically, helical thickness increases with decrease in local radius and increase in angle of winding; it is quite appreciable near the pole openings where fiber reversal takes place. An expression that gives an approximate estimate is as follows:

$$rt_h \cos\alpha = K \tag{14.70}$$

where K is a constant. During the design process, helical thickness is provided at a key cross-sectional location—typically in the cylinder with the minimum winding angle—and K is determined. Note, however, that it is only an approximate estimate; it gives an infinite thickness at the pole opening, which obviously is not correct. Other more theoretical, empirical, and semiempirical accurate models are available; these are generally complex, and often the designer depends on his experience and experimental data for helical ply thickness, especially near the pole openings.

EXAMPLE 14.8

Design the composite shell of a cylindrical pressure vessel with an inner diameter of 200 mm and a cylindrical length of 800 mm. Internal pressure is 20 MPa. The end domes are hemispherical in shape and reinforced by polar end fittings that have equal composite pole opening diameters of 40 mm (Figure 14.19).

Use unidirectional carbon/epoxy with the following material data:

$$E_1 = 130\,\text{GPa},\ E_2 = 6.5\,\text{GPa},\ G_{12} = 5.0\,\text{GPa},\ \nu_{12} = 0.25,\ X^T = 2000\,\text{MPa}$$

Solution

Let us provide meridional stiffness to the pressure vessel entirely by helical windings, that is, we design the shell using helical and hoop plies only. With equal

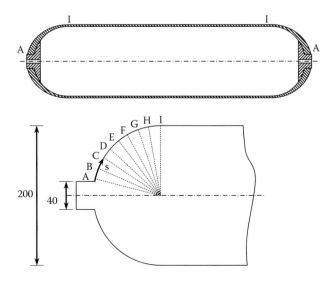

FIGURE 14.19 Geometrical details of a pressure vessel (refer Example 14.8).

TABLE 14.7

Required Thickness along the Meridional Contour (Example 14.8)

Location	s (mm)	r (mm)	β (°)	α (°)	Thickness Required t_h (mm)	t_c (mm)
A	4.6	24.5	75.821	54.736	3.00	0
B	21.2	40.1	66.344	29.897	1.33	1.67
C	37.7	54.7	56.866	21.463	1.16	1.85
D	54.2	67.7	47.388	17.182	1.10	1.90
E	70.8	78.9	37.911	14.684	1.07	1.93
F	87.3	87.9	28.433	13.146	1.06	1.95
G	103.9	94.6	18.955	12.208	1.05	1.95
H	120.4	98.6	9.478	11.699	1.04	1.96
I	136.9	100	0	11.537	1.04	1.96

composite pole openings, geodesic winding is feasible, in which the cylinder angle of winding is given by

$$\alpha_{cyl} = \sin^{-1}\left(\frac{40}{200}\right) = 11.537°$$

Then, taking a factor of safety of 2.0, using Equations 14.68 and 14.69, the required helical and hoop thicknesses at various points along the meridional contour are determined and tabulated in Table 14.7. Note that point A is the point of inflection, where the helical angle is given by $\alpha_i = \tan^{-1}\sqrt{2}$ and the required hoop thickness is zero.

Let us provide in the cylindrical portion, three helical plies of 0.4 mm thickness each and four hoop plies of 0.5 mm each as per the following ply sequence:

$$[90°/\pm\alpha/90°/\pm\alpha/90°/\pm\alpha/90°]$$

Helical plies cover the entire pressure vessel—from one pole opening to the other. The helical thickness provided increases from the equator toward the pole opening and the same is determined using Equation 14.70 and tabulated in Table 14.8.

TABLE 14.8

Thickness Provided along the Meridional Contour (Example 14.8)

Location	s (mm)	r (mm)	β (°)	α (°)	Thickness Provided t_h (mm)	t_c (mm)
A	4.6	24.5	75.821	54.736	8.31	2
B	21.2	40.1	66.344	29.897	3.38	2
C	37.7	54.7	56.866	21.463	2.31	2
D	54.2	67.7	47.388	17.182	1.82	2
E	70.8	78.9	37.911	14.684	1.54	2
F	87.3	87.9	28.433	13.146	1.37	2
G	103.9	94.6	18.955	12.208	1.27	2
H	120.4	98.6	9.478	11.699	1.22	2
I	136.9	100	0	11.537	1.20	2

Hoop winding, on the other hand, is not feasible in the domes and the need for circumferential reinforcement is met by local lay-ups (doilies). It is seen from the first table that the requirement of circumferential reinforcement gradually decreases as we move from the equator toward the pole and it vanishes at the inflection point and beyond. Note that the above calculations are based on the netting model, which does not consider interaction of the shell with polar bosses and other elements. Also, during doily lay-up, owing to its very geometrical shape of the doily developments, it is not feasible to orient the local reinforcement accurately everywhere. To accommodate such uncertainties, we provide 2-mm-thick circumferential reinforcements (i.e., same as at the cylindrical region) everywhere in the dome. The dome plies can be modified by finite element analysis.

14.11 SUMMARY

This chapter is devoted to the design of composite structures; design here is looked upon as a means for composite product development. Different stages of product development are identified; it is found that these stages are interrelated and the common thread is design. A review of the basic concepts of design, composites structural design, laminate design, joint design, stiffened structures, and several representative composite examples are presented. The key points can be enumerated as follows:

- The term *design* can be defined as a scheme that is used for making a product using certain *resources* within certain *constraints* to fulfil certain *requirements*.
- The requirements can be in respect of strength, stiffness, and other functional needs.
- The resources can be related to materials, manufacturing technologies, computational facilities, and human resources.
- The constraints during the design process are imposed typically from weight, cost, and assembly and manufacturing feasibility.
- It is important to differentiate between the terms *design* and *analysis*.
- Composite structural design involves many aspects; the important ones are generation of specifications, materials selection, configuration design, analysis options, manufacturing process selection, testing and NDE options, and design of laminate and joints.
- Laminate design is a central aspect in the composite structural design process. It basically involves the selection of ply sequence variables and an appropriate fiber volume fraction.
- Three key concepts of laminate design are load definitions, design allowables, and design factors.
- Laminate design process involves three critical steps—laminate selection, laminate analysis and measurements, and application of laminate design criteria.
- Joints are generally considered as the weak links in any structure; most real-life structures do contain joints. The final structural and functional performance of a structure often depends on the joints, and here lies the importance of the design of appropriate joints in a structure.
- Joints are broadly of two types—bonded joints and mechanical joints; they have their unique failure modes and are associated with their own advantages and disadvantages.
- Stiffened structures are typically efficient, but generally they are associated with more design complexities.

- Optimization is inherently associated with the very basic objective of any design. In brief, it is the process of arriving at the best design choice.
- Optimization is a mathematical process; however, in practice, it is approached in two ways—mathematical methods and searching techniques.
- A real-life structure can have many types of structural elements under different types of loads. Common examples include tension member, compression member, torsion member, beam, flat panel, etc.

EXERCISE PROBLEMS

14.1 Write a brief note defining the term *design*.

14.2 What are the general requirements, resources, and constraints in structural design? Consider the following broad classes of products and identify the requirements, resources, and constraints in each case: (i) aircraft wing, (ii) helicopter rotor blade, (iii) rocket motor case, (iv) sports car bumper, (v) storage tank, and (vi) pleasure boat hull.

14.3 Make a comparative note to explain the two terms—*design* and *analysis*.

14.4 Consider two products—helmets for two-wheeler riders and observation towers in an oil exploration firm engaged in remote area operations. Carry out the following:
   i. Generate specifications. Make suitable assumptions wherever necessary.
   ii. Suggest suitable materials and manufacturing processes. Justify your choice.
   iii. Carry out qualitative configuration design. Suggest analysis options.
   iv. Propose an acceptance plan.

14.5 Write a short note to explain the (i) various terms related to load—operating load, limit load, etc., (ii) various factors—factor of safety, margin of safety, etc., and (iii) design allowables.

14.6 Consider a laminate required to carry in-plane tensile load in one direction. Suggest a ply sequence if, for applying the tensile load, (i) metallic strips are bonded along the opposite edges and (ii) metallic strips are riveted along the opposite edges.

14.7 For a carbon/epoxy laminate, determine the invariant terms U_1, U_2, U_3, U_4, and U_5. Assume the following material properties: $E_1 = 150$ GPa, $E_2 = 8$ GPa, $G_{12} = 6$ GPa, and $\nu_{12} = 0.25$.

14.8 In the problem in Exercise 14.7 above, if the laminate is subjected to equal normal force resultants and the other force resultant and moment resultants are zero, determine the ply sequence of a three-ply symmetric laminate so as that normal deformation in the *x*-direction is half the normal deformation in the *y*-direction and in-plane shear deformation is zero.

14.9 What are the different types of bonded and bolted joints? Write a brief comparative note on the advantages and disadvantages of these joints.

14.10 Determine the average bearing stress in the hole of a single-lap pin joint. The following data are given: laminate ply sequence is $[0°/\pm45°/90°]_s$, each ply being 0.5 mm in thickness; hole diameter is 4 mm, and the applied load is 75 N. Is the joint safe, if the bearing strength for the above ply sequence is experimentally found as 12 MPa?

14.11 Write a brief note on failure modes in bonded and bolted joints.

14.12 What are the possible failure modes in a stiffened structure?

14.13 Design a minimum mass tension member of rectangular cross section to carry a uniaxial static tensile load of 45 kN. Adopt a micromechanics-based approach and give details of the fabrication method suggested. Also, suggest a possible joint with a metallic adopter at the ends. Consider the following materials:

Carbon fiber reinforcement : $E_{1f} = 240\,\text{GPa}, \left(\sigma_{1f}^T\right)_{ult} = 5400\,\text{MPa}$ and

$$\rho_f = 1.78\,\text{g/cm}^3$$

Kevlar fiber reinforcement : $E_{1f} = 120\,\text{GPa}, \left(\sigma_{1f}^T\right)_{ult} = 3500\,\text{MPa}$ and

$$\rho_f = 1.44\,\text{g/cm}^3$$

Glass fiber reinforcement : $E_f = 70\,\text{GPa}, \left(\sigma_f^T\right)_{ult} = 3200\,\text{MPa}$ and

$$\rho_f = 2.54\,\text{g/cm}^3$$

Cast epoxy resin : $E_m = 2.5\,\text{GPa}, \left(\sigma_m^T\right)_{ult} = 65\,\text{MPa}$ and

$$\rho_f = 1.15\,\text{g/cm}^3$$

14.14 Design the tension member of Exercise 14.13, if the axial elongation is limited to 0.02 mm.

14.15 Design a tension member of circular cross section to carry a load of 1000 kN. Use carbon/epoxy composite with material properties given in Exercise 14.13. Assume $V_f = 0.6$. Consider filament winding and pultrusion as the possible manufacturing method.

14.16 Design a compression member of length 6000 mm to carry an axial compressive load of 800 kN. Consider fixed-free end conditions and use carbon/epoxy composite with the following material data:

$$E_1 = 150\,\text{GPa}, E_2 = 8.0\,\text{GPa}, G_{12} = 6.0\,\text{GPa}, \nu_{12} = 0.25, \text{ and } X^C = 850\,\text{MPa}$$

14.17 Design a torsion member of length 800 mm to transmit a torque of 4.5 kN · m. Consider simply supported end conditions and use carbon/epoxy composite with the following material data:

$$E_1 = 150\,\text{GPa}, E_2 = 8.0\,\text{GPa}, G_{12} = 6.0\,\text{GPa}, \nu_{12} = 0.25, \text{ and } X^C = 850\,\text{MPa}$$

Also, the static shear strength of a laminate with ±45° plies is experimentally found to be 500 MPa.

14.18 Design a cantilever beam of length 600 mm and width 40 mm to carry a uniformly distributed load of 0.1 N/mm. Use unidirectional carbon/epoxy prepreg with the following material data:

$$E_1 = 140\,\text{GPa}, E_2 = 8\,\text{GPa}, G_{12} = 6\,\text{GPa}, \text{ and } \nu_{12} = 0.25$$

Assume the thickness of prepreg plies as 0.5 mm. Use box-section. The tip deflection of the beam is to be restricted to a maximum of 1.2 mm.

14.19 Consider a rectangular panel of size 1500 mm × 1200 mm subjected to in-plane tensile forces of 7500 kN and 5000 kN, respectively, in the longitudinal and lateral directions and an in-plane shear force of 1500 kN acting on the longitudinal edge. The laminate is to be designed with 0°, ±45°, and 90° plies. Construct the carpet plots and design the panel. The normal and shear deformations are to be restricted to 8 mm (axial), 10 mm (transverse), and 1°, respectively. The material properties are given as

$$E_1 = 140\,\text{GPa}, E_2 = 8\,\text{GPa}, G_{12} = 6\,\text{GPa}, \nu_{12} = 0.25, X^T = 1800\,\text{MPa}$$

14.20 Design a carbon/epoxy pipe of diameter 100 mm to contain an internal pressure of 100 MPa. Consider the material data given in Exercise 14.19.

REFERENCES AND SUGGESTED READING

1. R. M. Jones, *Mechanics of Composite Materials*, second edition, Taylor & Francis, New York, 1999.
2. D. V. Rosato, *Designing with Reinforced Composites: Technology—Performance—Economics*, Hanser, Munich, 1997.
3. C. Kassapoglou, *Design and Analysis of Composite Structures: With Applications to Aerospace Structures*, John Wiley & Sons, West Sussex, UK, 2010.
4. G. Eckold, *Design and Manufacture of Composite Structures*, Woodhead Publishing, Cambridge, 1994.
5. F. J. Schwan, Design of structure with composites, *Handbook of Composites* (S. T. Peters, ed.), second edition, Chapman & Hall, London, 1998, pp. 709–735.
6. Z. Gürdal, R. T. Haftka, and P. Hajela, *Design and Optimization of Laminated Composite Materials*, John Wiley & Sons, New York, 1999.
7. B. D. Agarwal, L. J. Broutman, and K. Chandrashekhara, *Analysis and Performance of Fiber Composites*, third edition, John Wiley & Sons, New York, 2006.
8. P. K. Mallick, *Fiber-Reinforced Composites—Materials, Manufacturing, and Design*, third edition, CRC Press, Boca Raton, FL, 2013.
9. E. J. Barbero, *Introduction to Composite Materials Design*, second edition, CRC Press, Boca Raton, FL, 2011.
10. J. M. Seng, Laminate design, *Handbook of Composites* (S. T. Peters, ed.), second edition, Chapman & Hall, London, 1998, pp. 686–708.
11. M. G. Bader, Material selection, preliminary design, and sizing for composite laminates, *Composite Materials Handbook* (P. K. Mallick, ed.), Marcel Dekker, New York, 1997, pp. 1207–1219.
12. P. K. Mallick, Design considerations for laminated composites, *Composite Materials Handbook* (P. K. Mallick, ed.), Marcel Dekker, New York, 1997, pp. 1221–1238.
13. S. Reeve, Introduction to engineering mechanics, analysis, and design, *ASM Handbook, Vol 21, Composites* (D. B. Miracle and S. L. Donaldson, vol. chairs), ASM International, Ohio, 2001, pp. 197–229.
14. F. L. Matthews and R. D. Rawlings, *Composite Materials: Engineering and Science*, CRC Press, Boca Raton, FL, 1999.
15. M. Mukhopadhyay, *Mechanics of Composite Materials and Structures*, Universities Press, Hyderabad, India, 2009.
16. A. Benator, J. Gillepspie Jr., and K. Kedward, Joining of composites, *Advanced Composites Manufacturing* (T. G. Gutowski, ed.), John Wiley & Sons, New York, 1997, pp. 487–512.
17. L. J. Hart-Smith, Bolted and bonded joints, *ASM Handbook, Vol 21, Composites* (D. B. Miracle and S. L. Donaldson, vol. chairs), ASM International, Ohio, 2001, pp. 271–289.
18. L. J. Hart-Smith, Design of adhesively bonded joints, *Joining Fiber-Reinforced Plastics* (F. L. Mathews, ed.), Elsevier Applied Science, London, 1987.
19. L. J. Hart-Smith, *Adhesive Bonded Double Lap Joints*, NASA Langley Contractor Report, NASA CR-112235, 1973.
20. S. K. Mazumdar, *Composites Manufacturing—Materials, Product and Process Engineering*, CRC Press, Boca Raton, 2002.
21. S. W. Tsai and N. J. Pagano, Invariant properties of composite materials, Composite Materials Workshop (S. W. Tsai, J. C. Halpin, and N. J. Pagano, eds.), St. Louis, Missouri, 13–21 July 1967, Technomic, Westport, Connecticut, 1968, pp. 233–253.
22. L. J. Hart-Smith, *Adhesive Bonded Single Lap Joints*, NASA Langley Contractor Report, NASA CR-112236, 1973.
23. L. J. Hart-Smith, *Adhesive Bonded Scarf and Stepped-Lap Joints*, NASA Langley Contractor Report, NASA CR-112237, 1973.
24. R. B. Heslehust and L. J. Hart-Smith, The science and art of structural adhesive bonding, *SAMPE Journal*, 8(2), 2002, 60–71.
25. C. C. Chamis and P. L. N. Murthy, Simplified procedures for designing adhesively bonded composite joints, *Journal of Reinforced Plastics and Composites*, 10(1), 1991, 29–41.
26. F. Mortensen and O. T. Thomsen, Analysis of adhesive bonded joints: A unified approach, *Composites Science and Technology*, 62(7–8), 2002, 1011–1031.
27. L. J. Hart-Smith, Design and empirical analysis of bolted or riveted joints, *Joining Fiber-Reinforced Plastics* (F. L. Mathews, ed.), Elsevier Applied Science, London, 1987.
28. C. C. Chamis, Simplified procedures for designing composite bolted joints, *Journal of Reinforced Plastics and Composites*, 9(6), 1990, 614–626.
29. A. Olmedo, C. Santiuste, and E. Barbero, An analytical model for predicting the stiffness and strength of pinned-joint composite laminates, *Composites Science and Technology*, 90(10), 2014, 67–73.

30. A. Olmedo, C. Santiuste, and E. Barbero, An analytical model for the secondary bending prediction in single-lap composite bolted joints, *Composite Structures*, 111, 2014, 354–361.
31. T. S. Ramamurthy, Recent studies on the behavior of interference fit pins in composite plates, *Composite Structures*, 13(2), 1989, 81–99.
32. V. V. Vasiliev, V. A. Barynin, and A. F. Rasin, Anisogrid lattice structures—Survey of development and application, *Composite Structures*, 54(2–3), 2001, 361–370.
33. V. V. Vasiliev and A. F. Razin, Anisogrid composite lattice structures for space-craft and aircraft application, *Composite Structures*, 76(1–2), 2006, 182–189.
34. R. M. Jones, Buckling of circular cylindrical shells with multiple orthotropic layers and eccentric stiffeners, *AIAA Journal*, 6(12), 1968, 2301–2305.
35. J. M. Whitney and J. C. Halpin, Analysis of laminated anisotropic tubes under combined loading, *Journal of Composite Materials*, 2(3), 1968, 360–367.
36. G. J. Simitses, Instability of orthotropic cylindrical shells under combined torsion and hydrostatic pressure, *AIAA Journal*, 5(9), 1967, 1463–1467.
37. L. P. Kollar and G. S. Springer, *Mechanics of Composite Structures*, Cambridge University Press, Cambridge, 2003.
38. M. E. Tuttle, *Structural Analysis of Polymeric Composite Materials*, Marcel Dekker, New York, 2004.
39. S. Koussios, Integral design for filament winding—Materials, *Winding Patterns, and Roving Dimensions for Optimal Pressure Vessels, Composite Filament Winding* (S. T. Peters, ed.), ASM International, Ohio, 2011, pp. 19–34.
40. A. P. Priestley, Programming techniques, *Computer-Aided Manufacturing, and Simulation Software, Composite Filament Winding* (S. T. Peters, ed.), ASM International, Ohio, 2011, pp. 35–47.
41. H. Reynolds, Pressure vessel design, *Fabrication, Analysis, and Testing, Composite Filament Winding* (S. T. Peters, ed.), ASM International, Ohio, 2011, pp. 115–148.
42. B. W. Tew, Preliminary design of tubular composite structures using netting theory and composite degradation factors, *Journal of Pressure Vessel Technology*, 117(4), 1995, 390–394.

Index

A

Ablation, 652
Acoustic emission (AE), 597, 608
    advantages and disadvantages, 611
    data acquisition, 609–610
    data analysis, 610–611
    felicity effect, 609
    Kaiser effect, 609
    sources, 608–609
    test setup, 609
Acoustic impedance of material, 599, 600
Acrylonitrile, 472
Active thermographic method, 612
    lock-in thermography, 612–613
    pulse thermography, 612
    vibrothermography, 613
Active thermography, 611
Additives, 463
Adhesive characterization, 586, 587
Advanced manufacturing processes, 654
AE, *see* Acoustic emission
Aerospace composite applications, 462
AFRP, *see* Aramid fiber-reinforced plastic
Alloy, 5
Almansi strain, 27–28
Alumina (Al_2O_3), 7, 479
Alumina–silica (Al_2O_3–SiO_2), 479
American Society for Mechanical Engineers (ASME), 541
American Society for Testing and Materials (ASTM), 541
    ASTM D2095, 587
    ASTM D2344, 571–572
    ASTM D3039, 557
    ASTM D3410, 561, 562
    ASTM D3479, 581
    ASTM D3528, 587–588
    ASTM D4255, 569–571
    ASTM D5034, 547
    ASTM D 5379, 567–569
    ASTM D5528, 577–578
    ASTM D5656, 587
    ASTM D5961, 585
    ASTM D6272, 573–575
    ASTM D638, 550, 556
    ASTM D6641, 561
    ASTM D695, 550–551, 561
    ASTM D790, 572–573
    ASTM E143, 551
Analysis options, 661
Analytical methods, 269, 331, 347
Angle of twist, 430, 698
Angle-ply laminate, 248
Angle-ply laminated plate
    antisymmetric, 372–374
    antisymmetric angle-ply laminated simply supported plate, 382–384, 392–395
    symmetric, 368–369
    symmetric angle-ply laminated simply supported plate, 377–379
Anisotropic high-performance composites, 652
Anisotropic materials, 9, 14, 61
    Hill's criterion for, 173
Antisymmetric angle-ply laminated plate, 372–374
Antisymmetric angle-ply laminated simply supported plate
    buckling, 382–384
    vibration, 392–395
Antisymmetric cross-ply laminated plate, 369–372
Antisymmetric cross-ply laminated simply supported plate
    buckling, 379–382
    vibration, 389–392
Antisymmetric laminate, 246–247
Applied loads, 663
Approximating function, 404–407, 424
    for general boundary conditions, 367–368
Aramid fiber-reinforced plastic (AFRP), 17
Aramid fibers, 476, 531; *see also* Carbon fibers; Glass fibers
    applications, 478
    dry-jet spinning process, 477
    forms, 477
    production, 476–477
    properties, 477–478
    staple fibers, 477
    types, 476
Aramid yarns, 477
Areal density of fabric, 544
AR-glass fibers, 468
ASME, *see* American Society for Mechanical Engineers
Assembly, 417, 424–425
    boundary conditions, 421
    of element matrices, 418
    expanded element stiffness matrix, 419–420
    requirements, 654
ASTM, *see* American Society for Testing and Materials
Attenuation, 600
    coefficient, 605
Autoclave, 495, 518
    cure cycle, 519
    curing, 654
    process, *see* Prepreg lay-up process
Automation and skilled manpower needs, 525–526
Auxiliary equation, 362
Average stress criterion, 583

B

Balanced fabrics, 480
Balanced laminate, 247
Bar element, 425–430
Barrier coat, 495
Barrier film, 519
Base epoxy resin, 459
    curatives, 460
    DGEBA, 459
    epoxide group, 460
    modifiers, 460
    properties, 460–462
Beam, 269
    theory, 433
Beam vibration
    fixed-fixed beam, 324–325
    fixed-free beam, 323–324
    governing equations, 320–322
    simply supported beam, 322–323
Bearing
    failure, 679
    strength, 585–586
Bending
    antisymmetric angle-ply laminated plate with all edges, 372–374
    antisymmetric cross-ply laminated plate with all edges, 369–372
    governing equations for, 334–339
    moments per unit length, 205
    solutions for bending of laminated plates, 355
    specially orthotropic plate with all edges, 355–360, 364–368
    specially orthotropic plate with two opposite edges, 360–363
    stiffnesses, 671–672
    strain energy, 353
    symmetric angle-ply laminated plate with all edges, 368–369
Biaxial extensometer, 557–558
Bidirectional lamina, 135, 144–145
Bisphenol A fumarate resins, 463
Bleeder, 519
BMC, *see* Bulk molding compounds
Body forces, 43, 410
Bolted joint, 585, 678
    bearing/by-pass strength, 586
    bearing strength, 585–586
    fastener pull-through strength, 586
    shear-out strength, 586
Bonded joints, 586, 673, 674; *see also* Mechanical joints
    adhesive characterization, 587
    advantages and disadvantages, 676–677
    characterization, 587–588
    composite joints, 674
    configurations, 673
    failure modes in, 675–676
    general design considerations, 677
    load transfer mechanism in, 675
Boron carbide (B_4C), 7
Boron fibers, 478–479
Boundary conditions in laminated plate, 344
    clamped boundary condition, 346
    simply supported boundary condition, 345–346
Boundary value problem, 346
Bounding techniques, 123
Box-section, 309–311
Braiding, 481
Branched polymer, 457
Breaking strength of fabric, 547
British Standards (BS), 541
BS, *see* British Standards
"B-stage", 483

Buckling, 269, 311–312, 340, 682; *see also* Column buckling
    antisymmetric angle-ply laminated simply supported plate, 382–384
    antisymmetric cross-ply laminated simply supported plate, 379–382
    differential equations, 340
    displacements, 342, 348, 381, 383–384
    equations for laminated plates, 339–342
    factor, 665–666
    governing equations for, 339–342
    mode, 340
    problems, 348, 349–350
    solutions for buckling of laminated plates, 374
    specially orthotropic simply supported plate, 374–377
    stress resultants, 342
    symmetric angle-ply laminated simply supported plate, 377–379
Bulk molding compounds (BMC), 482–483
Buoyancy method, 542
Bushing, 468
By-pass strength, 586

C

CADFILL®, 509
CADWIND®, 509
Camera, 606
Cantilever beam
    under point load, 291–292
    under uniformly distributed load, 293–294
Carbide, 530
Carbon, 645
    prepreg laminates, 645
Carbon/carbon composites (C/C composites), 5, 6, 619, 642
    characteristics, 642–644
    manufacturing methods, 644–645
Carbon/epoxy composites, 18
Carbon/epoxy prepregs, 495
Carbon fiber-reinforced plastic (CFRP), 17, 474, 476, 597, 652
Carbon fibers, 9, 470, 659; *see also* Aramid fibers; Glass fibers
    applications, 474–476
    forms of carbon fiber reinforcements, 473, 474
    from pitch, 472–473
    production, 471–473
    properties, 473–474, 475
    from Rayon, 473
    tensile modulus *vs.* tensile failure strain, 475
    tensile modulus *vs.* tensile strength, 475
    types, 470–471
Carbonization, 472, 473
Cartesian coordinate systems, 28, 202
    O-123, 88
    for stress/strain transformation, 46
Cathode ray tube (CRT), 600
Cauchy elastic material, 58
Cauchy's stress principle, 43–44
CC, *see* Compliance calibration
CCA approach, *see* Composite cylindrical assemblage approach
C/C composites, *see* Carbon/carbon composites
Center of gravity (c.g.), 516
Ceramic matrix composites (CMCs), 5, 6, 619, 634
    applications, 642, 643
    characteristics, 634–635
    CVD, 640–641
    CVI, 640–641
    liquid infiltration, 638–639
    manufacturing methods, 636
    matrix materials, 635–636
    polymer infiltration and pyrolysis, 641–642
    powder consolidation methods, 637
    reaction bonding processes, 642
    reinforcing materials, 636
    slurry infiltration, 637–638
    sol–gel technique, 639–640
Ceramics, 6, 634
    fibers, 479
CFRP, *see* Carbon fiber-reinforced plastic
c.g., *see* Center of gravity
C-glass fibers, 468
Chang's index, 528
Characteristic equation, 51, 319, 367, 368, 423
Chemical vapor deposition (CVD), 631, 637, 640–641
Chemical vapor infiltration (CVI), 637, 640–641, 644, 645
Chlorendic resins, 463
Choleski method, 422
Chopped strand mat (CSM), 18
Chopped strands, 480
Circumferential winding, *see* Hoop windings
Clairut's rule, 509–510
Clamped boundary condition, 346
Classical 3D elasticity formulations, 201
Classical Brookfield viscometers, 548
Classical laminated plate theory (CLPT), 200, 201
    constitutive relations in, 210–223
    kinematics, 202–205
    kinetics, 205–209
    Kirchhoff hypothesis, 201
Classical laminated shell theory (CLST), 200, 223
    constitutive relations in, 230–231
    geometry of middle surface, 224–226
    kinematics, 226–228
    kinetics, 228–230
Classical laminate theory (CLT), 197, 200
Closed mold processes, 491, 497; *see also* Continuous molding processes; Open mold processes
    compression molding process, 497–499
    RTM process, 499–501
    selection of manufacturing methods, 521
CLPT, *see* Classical laminated plate theory
CLST, *see* Classical laminated shell theory
CLT, *see* Classical laminate theory
CMCs, *see* Ceramic matrix composites
CMEs, *see* Coefficients of moisture expansion
CNC system, *see* Computer numerical control system
Cobalt naphthenate (CoNap), 463
Codazzi conditions, 226
Coefficient of thermal expansion (CTE), 9
Coefficients of moisture expansion (CMEs), 82
Coefficients of thermal expansion (CTEs), 82, 116–117
Cohesive failure of adhesive, 676
Cold pressing, 625
Column buckling, 311
    fixed-fixed column, 317–319
    fixed-free column, 316–317
    governing equations, 312–314
    simply supported column, 314–315

Common nylon, 476
Compatibility conditions, 42
Compatibility equations, 42
Complex failure modes, 677, 681
Compliance calibration (CC), 577
Compliance matrix, 140
Component level
    acceptance tests, 595
    full-scale component testing, 588
    subscale component testing, 588
    tests, 539, 540, 588
Composite cylindrical assemblage approach (CCA approach), 124
Composites, 3–4; *see also* Machining of composites; Testing of composites
    advantages, 10–13
    applications, 14–18
    C/C composites, 6
    characteristics, 4, 8–9
    classification, 5
    CMCs, 6
    comparison of common engineering materials mechanical properties, 12
    disadvantages, 13–14
    flake composites, 8
    functions of reinforcements and matrix, 8–9
    joints, 674
    laminated composites, 8
    material, 4–5
    MMCs, 5–6
    particulate composites, 7
    PMCs, 5, 15–17
    sandwich composites, 8
    short fiber composites, 7
    stiffened structures, 650
    technology, 489
    terminologies, 9–10
    3D composites, 8
    unidirectional, 8
Composite structures design, 649, 656
    basic features of structural design, 651–654
    composite product development program, 650
    configuration design, 660–661
    constraints, 653–654
    design examples, 685–715
    design *vs.* analysis, 654–656, 661
    generation of specifications, 657–658
    joints design, 662, 673–681
    laminate design, 662–673
    manufacturing process selection, 661–662
    materials selection, 658–660
    optimization, 684
    requirements, 651–652
    resources, 652–653
    stiffened structures, 682–684
    stiffening of panels, 650
    testing and NDE options, 662
Compression molding, 497, 498, 523
    advantages and disadvantages, 499
    processing steps, 497–498
    raw materials, 499
    tooling and capital equipment, 498
Compression testing, 560; *see also* Tension testing
    failure modes, 560–561
    test procedure and data reduction, 563
    test specimen and test fixture, 561–563
Compressive properties, 550–551
Compressive strength, 165, 563
Computed tomography, 607

Index

Computer numerical control system (CNC system), 509
Computing technology, 653
CoNap, *see* Cobalt naphthenate
Configuration design, 660–661
Consolidation, 490–491, 507
    diffusion bonding, 626–627
    pressure, 508
Constituent content, 552–554
Constituent materials, 4
Constitutive equations of lamina, 135
    generally orthotropic lamina, 144–152
    specially orthotropic lamina, 136–144
Constitutive modeling, 56
    elastic materials, 57–58
    generalized Hooke's Law, 58–69
    idealization of materials, 56–57
Constitutive relations, 24, 136–143
    analysis of laminate, 214
    in CLPT, 210, 230–231
    extensional, coupling, and bending compliance matrices, 213
    global strains, 218–219
    laminate stiffness matrices, 210–211, 212, 216–217
    stress transformation, 222–223
    transformations of strains, 220–222
    transformed reduced stiffness matrices, 215
Constraints, 653
    assembly requirements, 654
    cost, 654
    manufacturing feasibility, 654
    weight, 653
Contact molding processes, *see* Open mold processes
Continuous fibers, 5, 480
    C/C composites, 642–643
Continuous fibrous reinforcements, 620
Continuous molding processes, 491, 501;
    see also Closed mold processes; Open mold processes
    fiber placement, 504–506
    pultrusion, 501–503
    selection of manufacturing methods, 522
    tape winding, 503–504
Continuous molding processes, 523
Continuum, 23
Conventional metallic materials, 9
Conventional stiffened panel, 17
Coordinate system, 208–209
    O-123, 135
Coordinate transformation, 416–417
Corrosion resistance, 634–635, 659
Cost-effectiveness, 642
Cost, 654, 660
Coupling stiffnesses, 670–671
Coupon-level tests, 539, 595
Covalent bond, 457
Crack propagation, 577, 578, 580, 608, 676
    modes, 575
Cross-linked polymer, 457
Cross-ply laminate, 248
CRT, *see* Cathode ray tube
CSM, *see* Chopped strand mat
CTE, *see* Coefficient of thermal expansion
CTEs, *see* Coefficients of thermal expansion
Cubic symmetry, 68
Curatives, 460
Curing, 517; *see also* Filament winding
    of epoxy composites, 519–520
    of phenolic composites, 520
    tools and equipment, 517–518
    vacuum bagging, 518–519
Curing process, 460, 465
CVD, *see* Chemical vapor deposition
CVI, *see* Chemical vapor infiltration
Cylindrical coordinates, strain–displacement relations in, 40–41

D

Data acquisition, 609–610
Data analysis, 610–611
Data reduction, test procedure and, 557–559, 563
Data representation, 601
    A-scan, 601
    B-scan, 601–602
    C-scan, 602–603
DCB test, *see* Double-cantilever beam test
DCPD, *see* Dicyclopentadiene
Deflection of middle surface
    Levy method for bending, 360–363
    Navier method for bending, 355–358
    Ritz method for bending, 363–366
Deformation, 25
    gradient, 30–32
    of solid body, 28, 30
Delaminations, 575, 597
Density, 547–548
    of composites, 552
    of fiber, 542
    of material, 658
Deposition methods, 623, 625
    CMCs, 637
    MMCs, 630–631
Design, 649, 650, 653
    allowables, 664–665
    constraints, 653
    flexibility, 11
    limit load, 663
    loads, 663
    requirements, 651, 652
    ultimate load, 664
DETA, *see* Diethylene triamine
D-glass fibers, 468
DGEBA, *see* Diglycidyl ether of bisphenol-A
Dicyclopentadiene (DCPD), 463
Diethylene triamine (DETA), 460
Differential equation, 362
    buckling, 340
Differential scanning calorimetry (DSC), 549
    glass transition temperature by, 549–550
Diglycidyl ether of bisphenol-A (DGEBA), 459
Dimensional accuracy, 515
DIN, *see* German Institute for Standardization
Direct approach, 408
Discontinuous carbon fiber–reinforced composites, 643
Discontinuous fibers, 5
Discontinuously reinforced aluminum (DRA), 632
Discontinuously reinforced titanium matrix (DRTi), 632
Discretization, 403–404, 424
Displacement
    buckling, 342, 348, 381, 383–384
    global vector of nodal, 415
    gradient, 30–32
    nodal, 402, 412
    plane strain, 72–73
    plate middle surface, 358
    at point, 28–30
    under point load, 305
    torsional, 430
DMA, *see* Dynamic mechanical analysis
Dog-bone-shaped tensile test specimen, 556
Double-cantilever beam test (DCB test), 576–578
Double joints, 675
Dough molding compound (DMC), *see* Bulk molding compounds (BMC)
DRA, *see* Discontinuously reinforced aluminum
DRTi, *see* Discontinuously reinforced titanium matrix
Dry-jet spinning process, 476
Dry reinforcements, 502
Dry winding, 506–507
DSC, *see* Differential scanning calorimetry
Dust removal, 530–531
Dwell(s), 514
Dynamic analysis, 661
Dynamic mechanical analysis (DMA), 549
Dynamic thermography, 611

E

ECR-glass fibers, 468
ECT test, *see* Edge-cracked torsion test
Eddy current testing, 613
    advantages and disadvantages, 614
    based on phenomenon of electromagnetic induction, 613
    setup, 614
Edge-cracked torsion test (ECT test), 579
E-glass, 500
E-glass fibers, 13, 467, 470
    E-glass fiber-reinforced epoxy leaf springs, 18
    E-glass fiber-reinforced polyester composite laminates, 18
    E-glass fiber-reinforced SMCs, 18
Eigenvalue problems, 346, 348, 401, 423
Eigenvector, 354–355
Elastic constants, restrictions on, 143–144
Elastic deformation, 57
Elasticity-based models, 83, 123–124
Elastic materials, 57–58
Elastic moduli, 82, 84
    evaluation, 88
    Halpin–Tsai equations for, 125–128
    in-plane shear modulus, 95–99
    longitudinal modulus, 88–90
    major Poisson's ratio, 93–95
    transverse modulus, 90–93
Elastic stiffness matrix, 60, 63, 68
    symmetry of, 61
Electrodeposition methods, 630
Electromagnetic acoustic transducers (EMATs), 600
Electromagnetic spectrum, 604
Element-level properties, tests for, 582
    bolted joint, 585–586
    bonded joint, 586–588
    open-hole tests, 582–585
Element(s), 402–403
    characteristic matrices, 407–408
    characteristic matrices, 407–409
    coordinate system, 439, 443
    development, 425
    element-level tests, 582
    formulation, 424
    load vector, 415–416
    one-dimensional, 425–436
    stiffness matrix, 415, 418, 436, 439
    two-dimensional, 436–450

Elevated temperature curing, 461
EMATs, see Electromagnetic acoustic transducers
End-notched flexure test (ENF test), 576, 578–579
Energy
　absorption, 652
　of dilation, 173
　of distortion, 173
ENF test, see End-notched flexure test
Engineering constants of generally orthotropic lamina, 152
　example, 159–161
　in-plane shear stress, 157–158
　load cases, 153, 154
　modulus of elasticity in y-direction, 155–156
　normal stress in y-direction, 156
　of orthotropic material, 153
　Poisson's ratio in xy-plane, 154–155
　transformed stiffness matrix, 164
　variation of elastic modulus in x-direction, 161
　variation of Poisson's ratio with lamina angle, 162
　variation of shear coupling ratio, 163
Engineering shear strain, 27, 35, 40
Engineering strain, 26
Environmental loads, 657
Epoxy composites, curing of, 519–520
Epoxy resins, 459; see also Polyester resins
　applications, 462
　base epoxy resin, 459–462
　common categories of modifiers in, 460
　representative properties of, 461
Epoxy system, 461
Equilibrium equations, 53–54
　deformed configurations of differential plate element, 336
　differential plate element under transverse loading, 334
　force resultants, 335
　for laminated plate bending, 334
　moment equilibrium, 337–338
　moment resultants, 335
　static equilibrium equations, 339
Equilibrium problems, 401, 422
Essential design input, 666
Eularian description of motion, 29
Extended chain polyethylene fibers (Extended chain PE fibers), 479
Extensional stiffnesses, 669–670
Extension–shear coupling, 242
External concentrated forces, 410–411

F

Fabric(s), 480–481
　areal density, 544
　areal density of, 544
　balanced, 480
　breaking strength of, 547
　construction, 544
　nonwoven, 480
　porous, 519
　unidirectional, 481
　woven, 480, 481
Factor of safety (FoS), 665–666
Failure, 663
　loads, 663
Failure analysis of laminate
　FPF and LPF, 252
　progressive failure analysis, 252–260

Failure criteria
　design and analysis of composite structures, 179
　maximum strain criterion, 180–181, 183–184
　maximum strain failure criterion, 170–172
　maximum stress criterion, 182–183
　maximum stress failure criterion, 167–170, 180
　Tsai–Hill criterion, 181, 185
　Tsai–Hill failure criterion, 172–175
　Tsai–Wu criterion, 181–182, 183, 184, 185–187
　Tsai–Wu failure criterion, 175–179
Failure modes, 560–561
　in bonded joints, 675–676
　in mechanical joints, 679
　in stiffened structures, 682–683
Fastener pull-through strength, 586
Fatigue, 580
　testing, 580–581
Felicity
　effect, 609
　ratio, 609
FEM, see Finite element method
Female mold, 493
Fiber placement, 504
　advantages and disadvantages, 505–506
　processing steps, 504–505
　raw materials, 505
　tooling and capital equipment, 505
Fiber(s), 8, 620
　in C/C composites, 642–643
　density, 542
　fiber-reinforced composites, 13
　fiber-reinforced laminated composites, 8
　fracture, 651
　mass fraction, 86, 553
　microbuckling in extensional mode, 110
　microbuckling in shear mode, 110
　orientation in V-notch beam shear test specimen, 569
　packing, 85
　strengths of, 84
　volume fraction, 85, 553, 662
Filament(s), 480
　diameter, 543
　filament-wound phenolic resin-based pipes, 466
Filament winding, 506, 654, 661; see also Curing
　advantages and disadvantages, 517
　angle of winding, 511
　computational aspects, 509
　consolidation, 507
　geodesic windings, 509–510
　helical windings, 510–511
　hoop windings, 510–511
　impregnation, 506–507
　lay-up, 507
　nongeodesic windings, 509–510
　paths, 511
　polar windings, 510–511
　processing steps, 508–509
　programmed bandwidth and physical bandwidth, 512
　programming basics, 511–514
　programming for helical windings, 513–514
　programming for hoop winding, 512–513
　raw materials, 514
　solidification, 507–508

　tooling and capital equipment, 515–516
　winding programs, 506
Fill, see Weft
Fillers, 500, 502
Film radiography, 606
Finite element
　analysis, 653
　modeling, 661, 672
Finite element equations by variational approach, 409
　element load vector, 415–416
　external concentrated forces, 410–411
　matrices and vectors in element potential energy expression, 413
　matrices and vectors in structure potential energy expression, 414
　nodal displacements, 412
Finite element method (FEM), 400
　approximating function and shape function, 404–407
　assembly, 417–421
　basic concepts in, 402
　basic finite element procedure, 424–425
　coordinate transformation, 416–417
　derivation of element characteristic matrices, 408–409
　development of elements, 425
　discretization, 403–404
　element characteristic matrices and vectors, 407–408
　elements and nodes, 402–403
　one-dimensional elements, 425–436
　solution methods, 421–424
　two-dimensional elements, 436–450
　typical applications, 401
Finite strain
　physical meaning of finite strain tensor components, 38–40
　at point, 36
　tensor, 36–38
　theory, 32–33
Fire resistance, 659
First-order shear deformation theory (FSDT), 200
First law of thermodynamics, 54
First ply failure (FPF), 252
Fixed-fixed beam, 324–325
Fixed-fixed column, 317–319
Fixed-free beam, 323–324
Fixed-free column, 316–317
Fixed beam
　under point load, 286–289
　under uniformly distributed load, 289–291
Flake composites, 8
Flaws, 466
Flexural modulus, 574
Flexural testing, 572; see also Shear testing
　four-point flexure test, 573–575
　three-point flexure test, 572–573
Flow-cup viscometer, 549
Focal point on target, 605
Forced infiltration, 627
Forced vibration, 320
Force resultants, 205–209, 228–230, 333–334
FoS, see Factor of safety
Fourier series expansion, 361, 375
Fourier sine series, 348, 356–357
Four-point flexure test, 573–575
FPF, see First ply failure
Fracture modes, 575

Index

Fracture test methods, 579–580
Fracture toughness test, 575
    DCB test, 577–578
    ENF test, 578–579
    fracture modes, 575
    fracture test methods, 579–580
Free vibration, 320, 343
    Navier method for, 384–386, 389–395
    Ritz method for, 387–389
    Ritz solution for, 386–387
Free vibration problems, 348
FSDT, *see* First-order shear deformation theory
Full layerwise theories, 263
Full-scale component testing, 588
Functional requirements, 651
    design, 652
    stiffness, 651–652
    strength, 651
Functional requirements, 651, 652
Functional specifications, 657
Fundamental natural frequency, 386

G

Gamma rays, 604
    radiography, 606
Gauss condition, 226
Gaussian elimination method, 422
Gel coat, 492
General-purpose resins (GP resins), 463
General beam element, 434–436
Generalized Hooke's law, 21, 58
    anisotropic materials, 61
    cubic symmetry, 68
    fourth-order tensor, 58–59
    isotropic materials, 68–69
    monoclinic materials, 61–63
    orthotropic materials, 63–66
    symmetry of elastic stiffness matrix, 61
    symmetry of stress and strain tensors, 59–61
    transversely isotropic materials, 67–68
Generally orthotropic lamina, 136, 144, 148
    engineering constants, 152–164
    global strains, 151
    hygrothermal effects in, 189–193
    under in-plane loading, 149, 150
    local stresses, 152
    stress transformation, 145
    transformation matrix, 146
    transformed reduced stiffness matrix, 147, 148
General purpose fibers (GP fibers), 467, 470
Generation
    of output, 425
    of specifications, 657–658
Geodesic windings, 509–510
Geometrical specifications, 657
Geometric boundary conditions, 314, 344
Geometry of middle surface, 224–226
German Institute for Standardization (DIN), 541
Glass fiber-reinforced plastics (GFRP), 4, 17, 493
Glass fibers; *see also* Aramid fibers; Carbon fibers
    applications, 469, 470
    chemical composition of types of, 467
    forms, 469
    GP fibers, 467
    production, 468–469
    properties, 469, 470
    specialty glass fibers, 468
    types, 467–468
Glass/polyester composites, 18
Glass transition, 549
Glass transition temperature (T_g), 461, 549, 659
    by differential scanning calorimetry, 549–550
Glass/vinyl ester composites, 18
Global buckling, 683
Global coordinate system, 135
Global stiffness matrix, 415, 418
Global stresses, 206
Global vector
    of nodal displacements, 415
    of nodal loads, 415
Glycol, 463
Governing equations, 23–24
    in solid mechanics, 25
GP fibers, *see* General purpose fibers
GP resins, *see* General-purpose resins
Green elastic materials, 58
Green's finite strain tensor, 38
Green strain, 27
Grid-stiffened panel, 17
Grid-stiffened structures, 682

H

Halpin–Tsai equations, 124–125
    for elastic moduli, 125
    fiber cross section and fiber packing in, 126
    in-plane shear modulus, 127–128
    longitudinal modulus, 125–126
    major Poisson's ratio, 126
    transverse modulus, 126
Hand lay-up, 662
Hardeners, *see* Curatives
Heat conduction problems, 402
Heat deflection temperature (HDT), 463
Helical windings, 510–511, 654
    programming for, 513–514
HET acid, *see* Hexachlorocyclopentadiene acid
Hevea brasiliensis (*H. brasiliensis*), 458
Hexachlorocyclopentadiene acid (HET acid), 463
High-voltage electrical pulses, 600
Higher order shear deformations theories, 263
High performance at low weight, 476
H-method, 406
Homogeneous material, 9–10
Homogeneous solution, 362
Honeycomb sandwich, 17, 18
Hooke's law for orthotropic materials, 63
Hoop windings, 510–511
    programming for, 512–513
Human resources, 653
Hygro-thermo-mechanical properties, 135
Hygrothermal constitutive relations, 231–236
Hygrothermal effects, 187
    coefficients of moisture expansion of laminate, 236–240
    coefficients of thermal expansion of laminate, 236–240
    in generally orthotropic lamina, 189–193
    hygrothermal constitutive relations, 231–236
    in laminate, 231
    in specially orthotropic lamina, 188–189
Hygrothermal properties, 135
Hygrothermal strains, 231, 232
Hyperelastic materials, 58
Hypothesis, 202

I

Idealization of materials, 56–57
Illinois Institute of Technology Research Institute (IITRI), 563
Impregnation, 490
Incisions, 458
Indian Standards Institute (ISI), 541
Indicial notation, 30
Infinitesimal cuboid in equilibrium, 53
Infinitesimal shear strain, 35
Infinitesimal strain
    at point, 33–36
    theory, 32–33
Infrared camera, 611
Infrared thermography, 611
    advantages and disadvantages, 613
    in field of NDT, 611
    types of active thermographic methods, 612–613
Initial value problems, 346, 401
Injection molding compounds, 483
In-plane shear
    failure, 253
    force per unit length, 205
    modulus, 95–99, 127–128
    strength, 114–116, 165
    stress, 157–158
    tests, 565
In-plane stresses, 276–277
    Levy method for bending, 363
    Navier method for bending, 358–359
    Ritz method for bending, 366–367
In-plane uniaxial compressive loads
    antisymmetric angle-ply laminated simply supported plate, 382–384
    antisymmetric cross-ply laminated simply supported plate, 379–382
    specially orthotropic simply supported plate, 374–377
    symmetric angle-ply laminated simply supported plate, 377–379
In situ methods, 623, 625
    CMCs, 637
    MMCs, 631–632
Integration identities, 355
Interlaminar normal stress, 260, 359
Interlaminar shear
    stresses, 359
    tests, 565
Interlaminar stresses, 260–261, 277–279
    Levy method for bending, 363
    Navier method for bending, 359–360
    Ritz method for bending, 366–367
International Organisation for Standardization (ISO), 541
Interpolation function, 406
Invariant forms of laminate stiffness, 668–669
I-section, 307–309
ISI, *see* Indian Standards Institute
ISO, *see* International Organisation for Standardization
Isophthalic polyester resins, 463
Isotropic beam, 299
Isotropic fibers, 84
Isotropic materials, 9, 68–69; *see also* Orthotropic materials
    plane strain problem in, 73–74
    plane stress problem in, 71
Iterative methods, 423–424

J

Jacobi methods, 423–424
Joints design, 662, 673; *see also* Laminate design
  bonded joints, 673–677
  mechanical joints, 677–681
  other joints, 681
  types, 673

K

Kaiser effect, 609
Kinematics, 24
  boundary condition, 344
  of CLPT, 202–205
  of CLST, 226–228
  compatibility conditions, 42
  deformation gradient and displacement gradient, 30–32
  displacement at point, 28–30
  finite strain at point, 36–40
  infinitesimal strain and finite strain theories, 32–33
  infinitesimal strain at point, 33–36
  normal strain and shear strain, 25–26
  rigid body, 24–25
  strain–displacement relations in cylindrical coordinates, 40–41
  strain measurement types, 26–28
  transformation of strain tensor, 41–42
Kinetics, 24, 43
  Cauchy's stress principle and stress vector, 43–44
  of CLPT, 205–209
  of CLST, 228–230
  equilibrium equations, 53–54
  forces on body, 43
  principal stresses, 50–53
  state of stress at point and stress tensor, 44–46
  stress tensor–stress vector relationship, 49–50
  transformation of stress tensor, 46–49
Kirchhoff force, 345
Kirchhoff hypothesis, 201, 261
Knockdown factor, 665–666

L

Lagrangian description of motion, 29
Lagrangian finite strain tensor, 38
Lame′ parameters, 224
Lamina, 10, 82, 134, 198, 539
  constitutive equations, 135–152
  elasticity-based models, 123–124
  engineering constants of generally orthotropic lamina, 152–164
  hygrothermal effects, 187–193
  macromechanics, 135
  mechanics of materials-based models, 88–123
  micromechanics, 84–88
  micromechanics models, 83–84
  principal nomenclature, 79–82, 133–134
  semiempirical models, 124–128
  standard test methods for mechanical parameters, 555
  strength, 164–187
  tests for mechanical properties, 555
Laminae, nonmechanical tests on, 552
  constituent content, 552–554
  density of composites, 552
  standard test methods for nonmechanical parameters of composites, 552
  void content, 554–555
Lamina/laminate properties, tests for, 551
  compression testing, 560–565
  fatigue testing, 580–581
  flexural testing, 572–575
  fracture toughness test, 575–580
  shear testing, 565–572
  tension testing, 556–560
  tests for, 581–582
Laminate, 10
  analysis and measurement, 672
  coordinate system O-xyz, 135
Laminated beam bending-solid rectangular cross section, 272, 294
  assumptions and restrictions, 272–273
  cantilever beam under point load, 291–292
  cantilever beam under uniformly distributed load, 293–294
  carbon/epoxy, 296–299
  effective longitudinal stress, 296
  fixed beam under point load, 286–289
  fixed beam under uniformly distributed load, 289–291
  governing equations, 273–276
  in-plane stresses, 276–277
  interlaminar stresses, 277–279
  simply supported beam under point load, 279–284
  simply supported beam under uniformly distributed load, 285–286
  for symmetric laminate, 295
Laminated beam bending-thin-walled cross section, 299
  box-section, 309–311
  I-section, 307–309
  T-section, 299–307
Laminated composites, 8, 269
  cross-sectional configurations, 270–271
  cylindrical bending of laminated plate, 271
  1D laminated structural element, 270, 271
  plates, 332
  principal nomenclature, 269–270
  rectangular general plate element with, 447–450
  solid rectangular cross sections, 271
Laminate design, 662, 663; *see also* Joints design
  buckling factor, 665–666
  criteria, 672–673
  design allowables, 664–665
  factor of safety, 665–666
  knockdown factor, 665–666
  laminate analysis and measurement, 672
  laminate selection, 667–672
  load definitions, 663–664
  margin of safety, 665–666
  process, 666–673
  scope, 662–663
Laminated plates, analytical solutions for
  boundary conditions in laminated plate, 344–346
  governing equations for bending, buckling, and vibration of laminated plates, 334–344
  laminated composite plates, 332
  rectangular laminated plate under general loading, 333–334
  solution methods, 346–355
  solutions for bending of laminated plates, 355–374
  solutions for buckling of laminated plates, 374–384
  solutions for vibration of laminated plates, 384–395
Laminate, macromechanics of
  angle-ply, 248
  antisymmetric, 246–247
  balanced, 247
  cases of, 240
  classical laminated shell theory, 223–231
  classification of laminate analysis theories, 200–201
  CLPT, 201–223
  codes, 198–200
  cross-ply, 248
  failure analysis, 252–260
  hygrothermal effects in, 231–240
  interlaminar stress, 260–261
  laminate analysis, 260
  layerwise theories, 263–264
  principal nomenclature, 197–198
  quasi-isotropic, 248–252
  shear deformation theories, 261–263
  significance of stiffness matrix terms, 240–242
  single-ply, 242–244
  symmetric, 244–246
Laminate selection, 667
  bending stiffnesses, 671–672
  coupling stiffnesses, 670–671
  extensional stiffnesses, 669–670
  invariant forms of laminate stiffness, 668–669
  special cases of laminates and significance, 667
Laminate stiffness
  invariant forms of, 668–669
  matrices, 210–211, 212, 216–217
Large deformations, 32–33
Large deformation theory, *see* Finite strain theory
Large strain theory, *see* Finite strain theory
Last ply failure (LPF), 252
Lay-up, 490, 504, 507, 524
Layered composites, 6
Layerwise theories, 201, 263–264
Levy method for bending, 349–350
  deflection of middle surface, 360–363
  in-plane stresses, 363
  interlaminar stresses, 363
Limit load, 663
Linear polymer, 457
Linear voltage differential transformer (LVDT), 572
Liquid-state methods, 623, 625, 637
Liquid displacement method, 547
Liquid infiltration, 638–639
Liquid metal infiltration process, 627–628
Load definitions, 663–664
Local coordinate system, 135
Local fiber buckling, 108
Local skin buckling, 683
Local stresses and strains, 253
Lock-in thermography, 612–613
Loft template technique, 493
Logarithmic strain, *see* True strain
Longerons, *see* Longitudinal stiffeners
Longitudinal compressive failure, 252

Index

Longitudinal compressive strength, 107–110, 165
Longitudinal modulus, 88–90, 125–126
Longitudinal stiffeners, 682
Longitudinal tensile failure, 252–253
Longitudinal tensile strength, 100–107, 165
Low-molecular-weight resins, 465
Low-viscosity resin system, 514
LPF, *see* Last ply failure
LVDT, *see* Linear voltage differential transformer

M

Machining of composites, 529; *see also* Composites; Testing of composites
　critical aspects of, 530–531
　delamination of laminate during drilling, 531
　edge machining, 529
　holes and other features, 530
　for part making, 530
　requirements of, 529
　surface machining, 529–530
Macromechanics, 10
　macromechanics-based approach, 686–687
Macromechanics of laminate
　angle-ply, 248
　antisymmetric, 246–247
　balanced, 247
　cases of, 240
　classical laminated shell theory, 223–231
　classification of laminate analysis theories, 200–201
　CLPT, 201–223
　codes, 198–200
　cross-ply, 248
　failure analysis, 252–260
　hygrothermal effects in, 231–240
　interlaminar stress, 260–261
　laminate analysis, 260
　layerwise theories, 263–264
　principal nomenclature, 197–198
　quasi-isotropic, 248–252
　shear deformation theories, 261–263
　significance of stiffness matrix terms, 240–242
　single-ply, 242–244
　symmetric, 244–246
Magnetic flux, 613
Major Poisson's ratio, 93–95, 126
Male mold, 493
Mandrel, 508
　extraction, 515
Manufacturing
　feasibility, 654
　issues, 660
　process selection, 661–662
　techniques, 12
　technology, 652
Margin of safety (MoS), 665–666
Mass absorption, 605
Mass per unit area, 544
Matched-die-mold process, *see* Compression molding
Material(s), 3, 652
　axis strengths, 166
　compliance matrix, 61
　composite, 658–659
　coordinate system, 135
　description of motion, 29
　factors influencing, 658
　failure, 651
　matrix, 659–660
　point, 23
　reinforcements, 659
　selection, 658
　types, 4
Mathematical methods, 684
Mathematical models, 23, 56
Matrix, 4
　burning method, 552, 553
　characteristics and functions of, 8–9
　crazing, 596–597
　density, 547–548
　glass transition temperature, 549–550
　mass fraction, 553
　matrix-dominated properties, 685
　mechanical tests on matrix, 550–551
　nonmechanical tests on, 547
　nonmechanical tests on matrix, 547–550
　selection, 659–660
　strengths of, 84
　tests on, 547
　viscosity, 548–549
　volume fraction, 553
　volume fraction, 85
Mats, 480–481
Maximum expected operating pressure (MEOP), 664
Maximum operating loads, 663
Maximum strain
　criterion, 180–181
　failure criterion, 165, 170–172
Maximum stress
　criterion, 182–183
　failure criterion, 165, 167–170, 180
MBT, *see* Modified beam theory
MCC, *see* Modified compliance calibration
Mechanical joints, 677; *see also* Bonded joints
　advantages and disadvantages, 679–680
　bolted joint, 678
　failure modes in, 679, 680
　general design considerations, 680–681
　geometrical parameters in, 681
　types, 678
Mechanical strains, 231–232
Mechanical tests on matrix, 550; *see also* Nonmechanical tests on matrix
　compressive properties, 550–551
　shear properties, 551
　tensile properties, 550
Mechanical tests on reinforcement, 544; *see also* Nonmechanical tests on reinforcement
　breaking strength of fabric, 547
　standard test methods for mechanical parameters of reinforcements, 544
　tensile properties by single-filament tensile testing, 544–546
　tensile properties by tow tensile testing, 546–547
Mechanics, 23
Mechanics of materials-based models, 83, 88
　elastic moduli evaluation, 88–99
　evaluation of moisture coefficients, 119–123
　evaluation of strengths, 99–116
　evaluation of thermal coefficients, 116–119
MEKP, *see* Methyl ethyl ketone peroxide
Melt infiltration, 639
Membrane
　prebuckled configuration, 340
　strain energy, 353
MEOP, *see* Maximum expected operating pressure
Merit function, *see* Objective function
Mesh generation, 403, 424
Mesophase, 473
Meta-aramids, 476
Metal, 5
　alloy, 630
　metallic particles, 7
Metal matrix composites (MMCs), 5–6, 619, 620
　applications, 632–634
　characteristics, 620–621
　consolidation diffusion bonding, 626–627
　deposition methods, 630–631
　liquid metal infiltration process, 627–628
　manufacturing methods for, 623
　matrix materials for, 622
　powder metallurgy methods, 625–626
　reinforcing materials for, 622
　in situ methods, 631–632
　spray casting, 629–630
　stir casting method, 628–629
Methyl ethyl ketone peroxide (MEKP), 463
Micromechanics, 10, 82, 84
　assumptions and restrictions, 84
　elastic moduli and strengths of fibers and matrix, 84
　mass fractions, 86–87
　micromechanics-based approach, 685–686
　RVE, 87–88
　variables, 84
　volume fractions, 84–86
MIL, *see* United States Defence Standard
Minimum potential energy principle, 123, 340, 350–352, 366, 379, 389, 409, 414
Mixed-mode bending (MMB), 579–580
MMB, *see* Mixed-mode bending
MMCs, *see* Metal matrix composites
Modified beam theory (MBT), 577
Modified compliance calibration (MCC), 577
Modifiers, 460
Moisture absorption, 621
Moisture coefficients, evaluation of, 119–123
Moisture content, 543
Moisture expansion coefficients of laminate, 236–240
Molding compounds, 482
　BMC, 482–483
　injection, 483
　SMC, 482, 483
Molten titanium, 623
Moment equilibrium, 337–338
Moment resultants, 205–209, 228–230, 333
Monoclinic materials, 61–63
Monocoque structures, *see* Unstiffened shell structures
Monofilaments, 620
Monolithic metals, 620–621
MoS, *see* Margin of safety
MPD-I, *see* Poly-*m*-phenylene isophthalamide
Multidirectional lamina, 135

N

Natural boundary conditions, 314, 345
Natural fibers, 479–480
Natural frequencies, 395
Natural frequency, 663
Natural rubber, 459
Natural strain, *see* True strain

Navier method, 347–349
　for bending, 369–374
　for buckling, 374–376, 379–382, 382–384
　deflection of middle surface, 355–358
　for free vibration, 384–386, 389–395
　in-plane stresses, 358–359
　interlaminar stresses, 359–360
NDE, see Nondestructive evaluation/examination
NDT, see Nondestructive testing
Negative CTE, 477
Negative work done, 364–365
Netting models, 83
Newton's second law of motion, 53
Nodal displacements, 402, 412
　vector, 440, 444
Nodes, 402–403
NOL ring test, 559–560
Nondestructive evaluation/examination (NDE), 595, 662
Nondestructive testing (NDT), 595
　acoustic emission, 608–611
　defects in PMCs, 596–597
　eddy current testing, 613–614
　infrared thermography, 611–613
　radiographic testing, 604–608
　shearography, 615
　techniques, 597–598
　ultrasonic testing, 598–604
Nongeodesic windings, 509–510
Nonhomogeneous material, 10
Nonmechanical tests on laminae, 552
　constituent content, 552–554
　density of composites, 552
　standard test methods for nonmechanical parameters of composites, 552
　void content, 554–555
Nonmechanical tests on matrix, 547; see also Mechanical tests on matrix
　density, 547–548
　glass transition temperature, 549–550
　viscosity, 548–549
Nonmechanical tests on reinforcement, 541; see also Mechanical tests on reinforcement
　areal density of fabric, 544
　density of fiber, 542
　fabric construction, 544
　filament diameter, 543
　moisture content, 543
　standard test methods, 542
　Tex, 543–544
Nonmetallic flakes, 8
Nonmetallic particles, 7
Nonporous film, 519
Nonreactive in situ methods, 631–632
Nonuniform wetting, 528
Nonwoven fabrics, 480
Nonzero extension, 242
Normal forces per unit length, 205
Normal section, 224
Normal strain, 25–26
Novolac phenolic resins, 465
Numerical integration methods, 424
Numerical methods, 347

O

Objective function, 684
One-dimension (1D)
　approach, 26–28
　strain in bar, 26–27
One-dimensional elements (1D elements), 402–404
　bar element, 425–430
　general beam element, 434–436
　planar beam element, 431–434
　torsion element, 430–431
Open mold processes, 491, 521; see also Closed mold processes; Continuous molding processes
　prepreg lay-up, 494–495
　rosette lay-up, 496–497
　selection of manufacturing methods, 521
　spray-up, 495–496
　wet lay-up, 492–494
Operating loads, 663
Optimization, 684
Orthophthalic polyester resins, 463
Ortho resins, 463
Orthotropic lamina, 139
　strength of, 165–166
Orthotropic materials, 9, 63–66; see also Isotropic materials
　plane strain problem in, 73
　plane stress problem in, 70–71
Orthotropic plate
　specially orthotropic plate with all edges, 355–360, 364–368
　specially orthotropic plate with two opposite edges, 360–363
　specially orthotropic simply supported plate, 374–376
　specially orthotropic simply supported plate, 384–386
Oscillatory motion, 343
Ostwald's U-tube viscometer, 549

P

PAN, see Polyacrylonitrile
Para-aramids fibers, 476, 478
Partial layerwise theories, 263
Partial ply degradation, 253
Particle, 7, 23
Particulate composites, 7
Particulate reinforcements, 620
Particulate silicon carbide (SiC_p), 632
Passive thermography, 611
Pay-out-eye, 507, 516
PCD, see Polycrystalline diamond
Performs, 481–482
Periodic motion, 319
Phased composites, 6
Phenolic resins, 465–466, 520
　phenolic resin-based engineering plastics, 466
Phenolics, 465
　curing of phenolic composites, 520
Phenomenological approach, 23
Physical approach, 23
"Physical bandwidth", 512
Physical forms of reinforcements, 480
　common 2D weave styles, 481
　continuous and short fibers, 480
　fabrics and mats, 480–481
　molding compounds, 482–483
　performs, 481–482
　prepregs, 483–484
Physical problem, 401–402
Physical vapor deposition (PVD), 631
Piezoelectric material, 600
PIP, see Polymer infiltration and pyrolysis
Pitch, 472–473, 645
Planar beam element, 431–434
Plane elasticity, 21
Plane elasticity problems, 69
　plane strain, 72–74
　plane stress, 69–71
Plane strain, 72
　displacement, 72–73
　problem in isotropic materials, 73–74
　problem in orthotropic materials, 73
Plane stress problem, 69
　in isotropic materials, 71
　in orthotropic materials, 70–71
　thin plate, 69–70
Plate-bending
　equilibrium equations, 335
　problem, 347–348
　Ritz method for, 350–351
　theory, 441
Plate buckling, Ritz method for, 352–355
Plate middle surface displacement, 358
Plate theory, 438
Plural laminae, 198
Ply, 198
　degradation, 253
　sequence, 677, 681
PMC manufacturing methods, 489; see also Polymer matrix composites (PMCs)
　automation and skilled manpower needs, 525–526
　closed mold processes, 497–501, 521
　composites manufacturing methods, 490, 491
　composites technology, 489
　configuration of product, 522
　consolidation, 490–491
　continuous molding processes, 501–506, 522
　cost, 526–527
　curing, 517–520
　cycle time, 526
　filament winding, 506–517
　impregnation, 490
　lay-up, 490
　machining of composites, 529–531
　manufacturing process selection, 520
　open mold processes, 491–497, 521
　process modeling, 527–529
　process parameters, 527
　production requirement, 524–525
　reliability and repeatability, 524
　size of product, 522–523
　solidification, 491
　structural property requirement, 523
　surface finish, 523–524
　tooling requirements, 525
PMCs, see Polymer matrix composites
P-method, 406
Point load
　cantilever beam under, 291–292
　displacement under, 305
　fixed beam under, 286–289
　simply supported beam under, 279–284
Point stress criterion, 583
Poisson's effect, 83, 90
Poisson's ratio, 143, 559
Polar windings, 510–511
Polyacrylonitrile (PAN), 470
　PAN-based carbon fibers, 471–472
Polycrystalline diamond (PCD), 530

Index

Polyester, 13
Polyester oligomer, 462
    additives, 463
    properties, 463–464
    solvent, 463
Polyester resins, 462; *see also* Epoxy resins
    applications, 464, 465
    polyester oligomer, 462–464
    properties of unreinforced, 464
Polyisoprene, 458
Polymer infiltration and pyrolysis (PIP), 637, 641–642
Polymerization methods, 472
Polymer matrix composites (PMCs), 5, 457, 619
    applications, 15–17
    common reinforcements for, 467–480
    common thermosets for, 459–466
    defects in, 596–597
    NDT techniques for, 598
    physical forms of reinforcements, 480–484
    polymers, 457–459
    reinforcements, 466
    representative properties of fibers and conventional structural materials, 466
Polymer(s), 457
    comparison of thermoplastics with thermosets, 458
    infiltration phase, 642
    rubber, 458–459
    thermoplastics, 458
    thermosets, 458
Poly-*m*-phenylene isophthalamide (MPD-I), 476
Polynomial coefficients, 440, 444
Polynomial function, 404
Poly-*p*-phenylene terephthalamide (PPD-T), 476
Poly vinyl alcohol (PVA), 481, 492
Porosity, 596
Porous fabric, 519
Potential energy, 409–410
    functional, 386
Pot life, 461
Powder consolidation methods, 637
Powder metallurgy methods, 625–626
PPD, *see* *p*-phenylene diamine
PPD-T, *see* Poly-*p*-phenylene terephthalamide
p-phenylene diamine (PPD), 476
Preimpregnated roving prepregs, 483
Prepreg lay-up, 524, 526
    process, 494–495
    vacuum bagging of, 518
Prepregs, 483–484
Preprocessing, 623
Primary magnetic field, 613–614
Primary processing, 623
Principal directions, 225
Principal normal sections, 225
Principal radii of curvature, 225
Principal stresses, 166
Principle of conservation of energy, *see* First law of thermodynamics
Probe, 613
Process automation, 653
Processing techniques, 5
Process modeling, 527–529
"Programmed bandwidth", 512
Progressive failure analysis, 252
    force and moment resultants, 255
    local stresses at different internal pressures, 258
    maximum strain failure criterion, 257
    ply degradation, 253
    strength ratios at different internal pressures, 258
    transformed reduced stiffness matrix, 259
    transverse and shear stiffnesses, 260
Propagation problems, *see* Transient problems
Propylene oxide, 463
PT, *see* Pulse thermography
Pulse-echo technique, 601
Pulse generator, 600
Pulse thermography (PT), 612
Pultrusion, 501, 523, 525
    advantages and disadvantages, 502–503
    processing steps, 501–502
    raw materials, 502
    tooling and capital equipment, 502
PVA, *see* Poly vinyl alcohol
PVD, *see* Physical vapor deposition
Pyrolysis
    methods, 645
    of precursor materials, 644

Q

Quality, 596
Quasi-isotropic laminate, 248–252

R

Radiation, 13
Radiographic testing, 604; *see also* Ultrasonic testing
    advantages and disadvantages, 607
    characteristic features of X-rays and gamma rays, 604
    computed tomography, 607
    gamma ray radiography, 606
    real-time radiography, 606
    setup, 605
    x-ray radiography, 605–606
Rail shear test, 569–571
Randomly oriented lamina, 135
Raw materials, 14
Rayleigh–Ritz method, 424
Rayon, carbon fiber from, 473
Reaction bonding processes, 642
Reaction injection molding (RIM), 501
Reactive *in situ* methods, 631
Real-time radiography, 606
Rectangular bending plate element, 439–443
Rectangular general plate element, 443–447
    with laminated composites, 447–450
Rectangular laminated plate under general loading, 333–334
Rectangular membrane element, 436–439
Reinforced reaction injection molding (RRIM), 501
Reinforcements, 4, 466, 627
    characteristics and functions, 8–9
    mechanical tests on reinforcement, 544–547
    nonmechanical tests on reinforcement, 541–544
    selection, 659
    tests on, 541
Reinforcing factors, 125
Reinforcing material, 4
Representative volume element (RVE), 83, 87–88, 89, 124
Resin
    bath, 502
    system, 660
    transfer molding, 481
Resin transfer molding (RTM), 18, 499, 500
    advantages and disadvantages, 500–501
    processing steps, 499–500
    raw materials, 500
    tooling and capital equipment, 500
Resole phenolic resins, 465
Resources, 652
    computing technology, 653
    human resources, 653
    manufacturing technology, 652
    materials, 652
Restriction, 202
Reuter's matrix, 190
Rigid body, 24–25
RIM, *see* Reaction injection molding
Rings, *see* Transverse stiffeners
Ritz approximation function, 376, 379
Ritz method, 350
    for bending, 364–369
    for buckling, 376–377, 377–379
    for free vibration, 387–389
    for plate bending, 350–351
    for plate buckling, 352–355
    useful integration identities, 355
Ritz solution for free vibration, 386–387
R-method, 406
RMS techniques, *see* Root-mean-square techniques
Rocket nozzle liners, 652
Root-mean-square techniques (RMS techniques), 610
Rosette lay-up, 496–497, 526
Roving, 480
RRIM, *see* Reinforced reaction injection molding
RTM, *see* Resin transfer molding
Rubber, 458–459
Rudder of aircraft, 657
"Rule of mixtures", 89
RVE, *see* Representative volume element

S

SAE, *see* Society of Automotive Engineers
Sandwich composites, 8
Saturated acid, 463
SCRIMP, *see* Seemann composite resin infusion molding process
Searching techniques, 684
Secondary processing, 623
Seemann composite resin infusion molding process (SCRIMP), 501
Semiempirical models, 83, 124
    Halpin–Tsai equations, 124–125, 125–128
S-glass fibers, 468
Shape functions, 404–406, 430
Shear coupling, 153
    ratios, 153
Shear deformation theories, 200, 261–263
Shear failure
    of laminate, 679
    mode, 110
Shear force, 287
Shearography, 597, 615
Shear-out strength, 586
Shear properties, 551

Shear strain, 25–27, 35, 153, 566
    components in finite strain tensor, 40
    engineering, 27, 35, 40
    excessive in-plane, 181, 184
    in-plane, 171
    infinitesimal, 35
    tensorial, 27
    true, 28
    ultimate in-plane, 116, 171
Shear strength, 165
Shear stress, 43, 153
Shear testing, 565; see also Flexural testing
    rail shear test, 569–571
    short beam shear test, 571–572
    uniaxial tension test of ±45° laminate, 565–567
    V-notch beam shear test, 567–569
Sheet molding compounds (SMCs), 18, 466, 482, 483
Short beam shear test, 571–572
Short fibers, 480, 620, 625
    composites, 7
Short high-voltage electrical pulses, 600
Sign convention for stresses and strengths, 165
Silane, 641
Silicon carbide (SiC), 7, 479, 635
Silicon nitride (Si_3N_4), 7
Simple harmonic motion, 319
Simply supported beam
    beam vibration, 322–323
    under point load, 279–284
    under uniformly distributed load, 285–286
Simply supported boundary condition, 345–346
Simply supported column, 314–315
Single-filament tensile testing, 545
    tensile properties by, 544–546
Single generally orthotropic ply, 244
Single isotropic ply, 242–243
"Single joints", 675
Single-ply laminate, 242
    single generally orthotropic ply, 244
    single isotropic ply, 242–243
    single specially orthotropic ply, 243–244
Single specially orthotropic ply, 243–244
Sink–float method, 542
Slurry infiltration, 637–638
Small deformation, 32–33
    theory, see Infinitesimal strain theory
Small strain theory, see Infinitesimal strain theory
SMCs, see Sheet molding compounds
Smeared stiffeners model, 684
Society of Automotive Engineers (SAE), 541
Sol–gel technique, 639–640
Solid circular cross sections, 271
Solidification, 491, 507–508
Solid mechanics, 23
    constitutive modeling, 56–69
    fundamental principles and governing equations, 23–24
    kinematics, 24–42
    kinetics, 43–54
    plane elasticity problems, 69–74
    principal nomenclature, 21–23
    spatial point, material point, and configuration, 23
    thermodynamics, 54–56
Solid rectangular cross sections, 271
Solid-state methods, 623, 637
Solution, 425
Solution methods, 346, 421–424

Levy method, 349–350
Navier method, 347–349
Ritz method, 350–355
Solvent, 463
Spars, see Longitudinal stiffeners
Spatial description of motion, 29
Spatial point, 23
Specially orthotropic lamina, 136
    constitutive relation, 136–143
    hygrothermal effects in, 188–189
    restrictions on elastic constants, 143–144
Specially orthotropic plate
    with all edges, 355–360, 364–368
    with two opposite edges, 360–363
Specially orthotropic simply supported plate
    buckling, 374–377
    vibration, 384–386, 386–387
Specialty glass fibers, 468
Specimen in open-hole compressive strength test, 584
Spinning, 472
Spontaneous infiltration, 627
Spray casting, 629–630
Spray deposition, 631
Spray-up, 495–496, 524
Spun fibers, 473
SRIM, see Structural reaction injection molding
Stability, 663
    analysis, 661
Stabilization, 472
Stacking sequence of laminate, 240
State of stress at point, 44–46
Static analysis, 661
Static boundary condition, 345
Static equilibrium equations, 339, 686
Static force equilibrium, 335–336
Steady-state problems, see Equilibrium problems
Stiffened structures, 660, 682
    design of stiffeners, 683–684
    failure modes in, 682–683
Stiffener column buckling, 683
Stiffening of panels, 650
Stiffness, 100
    matrix, 141
    properties, 135
    significance of stiffness matrix terms, 240–242
Stir casting method, 628–629
Stitching, 481
Strain, 25, 651–652
    Almansi strain, 27–28
    engineering strain, 26
    green strain, 27
    measurement types, 26
    true strain, 26–27
Strain–displacement relations, 202–205, 226–228, 430, 433, 438, 441, 446, 448
    in cylindrical coordinates, 40–41
Strain energy, 173, 353, 364–365, 409
    density function, 56
    expression, 449–450
    release rate, 575, 578
Strain tensor
    symmetry of, 59–61
    transformation, 41–42
Strand, 480
Strength(s), 164, 651
    degradation, 677
    evaluation of, 99
    failure criteria, 167–187
    of fibers and matrix, 84

    in-plane shear strength, 114–116
    longitudinal compressive strength, 107–110
    longitudinal tensile strength, 100–107
    of orthotropic lamina, 165–166
    parameters, 82
    strength-based designs, 651–652
    strength parameters, 99–100
    transverse compressive strength, 113–114
    transverse tensile strength, 110–113
Stress, 25, 43, 651–652
    analysis problems, 402
    concentration factor, 582
    principal, 50–53
    transformation, 145, 165–166
    waves, see Ultrasonic waves
Stress–strain curve, 563
Stress–strain relations, 136
Stress tensor, 44–46
    stress tensor–stress vector relationship, 49–50
    symmetry of, 59–61
    transformation of, 46–49
Stress vector, 43–44
    stress tensor–stress vector relationship, 49–50
Stringers, see Longitudinal stiffeners
Structural analysis, 661
Structural design, 653, 654
Structural loads, 657
Structural property requirement, 523
Structural reaction injection molding (SRIM), 18, 501
Structural requirements, 651
Structural specifications, 657
St. Venant's compatibility equations, see Compatibility equations
S-2 glass fibers, 468
Subscale component testing, 588
Surface displacements, middle, 343–344
Surface finish, 523–524
Surface forces, 43, 410
Symmetric angle-ply laminated plate, 387–389
    bending, 368–369
    vibration, 387–389
Symmetric angle-ply laminated simply supported plate, 377–379
Symmetric laminate, 244–246
Symmetry
    of elastic stiffness matrix, 61
    of stress and strain tensors, 59–61
Synthetic rubbers, 459

T

Tape winding, 503–504
TCl, see Terephthaloyl chloride
Tensile modulus, 545
Tensile properties, 550
    by single-filament tensile testing, 544–546
    by tow tensile testing, 546–547
Tensile strength, 165, 545, 558
Tension member design, 685
    examples, 687–715
    macromechanics-based approach, 686–687
    micromechanics-based approach, 685–686
Tension testing, 556; see also Compression testing
    NOL ring test, 559–560
    test procedure and data reduction, 557–559
    test specimen and specimen preparation, 556–557

Index

TEPA, see Tetraethylene pentamine
Terephthalic polyester resins, 463
Terephthaloyl chloride (TCl), 476
Testing, 662
Testing of composites, 537; see also Composites; Machining of composites
  building block approach, 539
  objectives, 538–539
  in polymer matrix composites, 540
  standards, 540–541
  tests at component level, 588
  tests for element-level properties, 582–588
  tests for lamina/laminate properties, 551–582
  tests on matrix, 547–551
  tests on reinforcement, 541–547
Test specimen
  and specimen preparation, 556–557
  and test fixture, 561–563
Tetraethylene pentamine (TEPA), 460
Tex, 543–544
Thermal analysis, 661
Thermal coefficients, evaluation of, 116–119
Thermal expansion, 515
  coefficients of laminate, 236–240
Thermal insulation, 652, 659
Thermal stresses, 187–188
Thermodynamics, 24, 54–56
Thermographic NDT methods, 611
Thermomechanical analysis (TMA), 549
Thermoplastics, 458
Thermosets, 458
Thermosetting resins, 645
Third-order shear deformation theory (TSDT), 200
Three-dimension (3D)
  composites, 8
  elasticity theories, 201
  elements, 402–404
  linear elasticity problem, 69
Three-point flexure test, 572–573
Through-transmission technique, 600
Titanium carbide (TiC), 7
Titanium diboride (TiB_2), 7
TMA, see Thermomechanical analysis
Tooling requirement, 662
Torsional displacement, 430
Torsion element, 430–431
Total ply degradation, 253
Towpregs, see Preimpregnated roving prepregs
Tow tensile testing, tensile properties by, 546–547
Transducer, 600
Transformation
  matrix, 61, 146, 416–417
  methods, 423–424
  of strain tensor, 41–42
  of stress tensor, 46–49
  transformed reduced stiffness matrix, 147, 259
Transient problems, 401, 424
Transverse
  compressive failure, 253
  compressive strength, 113–114, 165
  deflections, 370, 380, 390
  isotropic materials, 67–68
  loading, 371
  modulus, 90–93, 126
  stiffeners, 682
  tensile failure, 110, 253
  tensile strength, 110–113, 165
Triethylene triamine (TETA), 460
True strain, 26–27
Tsai–Hill critaerion, 181, 185
Tsai–Hill failure criterion, 165, 172–175
Tsai–Wu criterion, 181–182, 185–187
  comparison of failure load by different failure criteria of lamina, 184
  failure criterion, 165, 175–179
  failure load as per maximum stress failure criterion of lamina, 183
TSDT, see Third-order shear deformation theory
T-section, 299–307
Tube current, 605
Tube voltage, 605
Twisting moment per unit length, 205
Two-dimensional elements (2D elements), 402–404
  rectangular bending plate element, 439–443
  rectangular general plate element, 443–447
  rectangular general plate element with laminated composites, 447–450
  rectangular membrane element, 436–439
Two-dimensional structure (2D structure), 8

U

Ultimate laminate failure (ULF), 252
Ultimate load, 663
Ultra-high modulus carbon fibers, 470
Ultrasonic testing; see also Radiographic testing
  advantages and disadvantages, 603–604
  data representation, 601–603
  equipment, 599–600
  pulse-echo technique, 601
  through-transmission technique, 600
  ultrasonic waves, 598–599
Ultrasonic waves, 598–599, 609
Ultraviolet (UV), 5–6
  radiation, 13
Uniaxial tension test, 565–567
Unidirectional composites, 8
Unidirectional fabrics, 481
Unidirectional lamina, 87–88, 135, 144
Uniformly distributed load
  cantilever beam under, 293–294
  fixed beam under, 289–291
  simply supported beam under, 285–286
United States Defence Standard (MIL), 541
Unsaturated polyester resins, see Polyester resins
Unstiffened shell structures, 682
UV, see Ultraviolet

V

Vacuum-assisted resin transfer molding (VARTM), 501
Vacuum bagging, 518–519
Vacuum bagging process, see Prepreg lay-up process
Variational approach, 408
  finite element equations by, 408–416
VARTM, see Vacuum-assisted resin transfer molding
Vectors, 407–408
  of strain components, 60
Vibration, 319–320, 343; see also Beam vibration
  antisymmetric angle-ply laminated simply supported plate, 392–395
  antisymmetric cross-ply laminated simply supported plate, 389–392
  equations for laminated plates, 343–344
  problem, 348–349
  solutions of laminated plates, 384
  specially orthotropic simply supported plate, 384–386, 386–387
  symmetric angle-ply laminated plate, 387–389
Vibrothermography, 612, 613
Vinyl ester resins, 464–465
Viscosity, 548–549
V-notch beam shear test, 567–569
Void content, 554–555
Voids volume fraction, 85, 553
Volume fractions, 84–86
von Mises yield criterion, 173

W

Warp, 480
Waveform of AE signals, 610
Wear resistance, 659
Weft, 480
Weight, 653
Weighted residual methods, 409
Wet
  rerolled rovings, 506–507
  spinning process, 476
  winding, 506–507
  wound systems, 508
Wet lay-up process, 492, 526
  advantages and disadvantages, 493–494
  basic raw materials, 493
  using composite mold and plaster pattern, 493
  processing steps, 492
  tooling and capital equipment, 493
Wetting, see Impregnation
Whiskers, 479, 620, 625
Woven fabrics, 480, 481

X

X-rays, 604
  radiography, 605–606

Y

Yarns, 480

Z

Zirconite ($Zr-Al_2O_3$), 479